NEUTRON PHYSICS

BY

K. H. BECKURTS and K. WIRTZ

KERNFORSCHUNGSZENTRUM KARLSRUHE

Translated by: L. Dresner, *Oak Ridge National Laboratory*

WITH 293 FIGURES

SPRINGER-VERLAG · NEW YORK INC.

1964

Completely revised Edition of "Elementare Neutronenphysik"
by K. WIRTZ *and* K. H. BECKURTS, *published in 1958 in German*

Library of Congress Catalog Card Number 64-25646
Printed in Germany

Printed by Universitätsdruckerei H. Stürtz AG, Würzburg

Titel Nr. 1235

From the Preface to the German Edition

This book is based upon a series of lectures I have occasionally given at the University of Göttingen since 1951. They were meant to introduce the students of experimental physics to the work in a neutron physics laboratory dealing with the problem of measuring neutron flux, diffusion length, Fermi age, effective neutron temperature, absorption cross sections and similar problems. Moreover, these lectures were intended to prepare the students for a subsequent lecture covering the physics of nuclear reactors. The original character of this series of lectures has been retained in the book. It is intended for use by students as well as anyone desiring to work on neutron physics measurements. The first half mainly covers the theory of neutron fields, i.e. essentially diffusion and slowing down theory. The second half is largely concerned with measurements in neutron fields. The appendix contains information and data which, in our experience, are frequently required in a neutron laboratory. The field of nuclear physics proper is briefly touched upon in the first two chapters, but only to the extent necessary for the understanding of the following chapters. The multitude of applications of neutron radiation has not been covered.

The conclusion of this manuscript coincided with the end of my long period of activity with the Max-Planck-Institut für Physik at Göttingen. To Professor HEISENBERG I owe thanks for his advice and suggestions for many of the subjects treated here. Thanks are also due to many of my young colleagues of the Göttingen group whose work has been included in this book. K. H. BECKURTS carefully revised my original manuscripts; some of the chapters were rewritten or written anew by him. I would like to thank M. KÜCHLE for his critical perusal of the manuscript.

Karlsruhe, January 1958 K. WIRTZ

Preface

Due to the rapid expansion of neutron physics over the past few years, the size of this book had to be increased considerably. Nearly all chapters had to be rewritten and six additional chapters were compiled. The subject matter has now been subdivided into four parts: Part I briefly deals with those fundamentals and experimental methods of nuclear physics most important for neutron physics. Part II relates to the theory of neutron fields, while special methods of measurement in neutron fields are treated in Parts III and IV. An Appendix contains data on cross sections, resonance integrals, etc. The remarks contained in the first edition about radiation protection when handling neutron sources were omitted, because excellent presentations of the subject have been published in the meantime.

The basic character of the German edition has been in essence retained. In particular, we have only slightly altered the title of the book. It is known that neutron physics today primarily means the problems of neutron interaction with nuclei or molecules, whereas in this book mainly the production, theory and measurement of neutron fields in moderating substances is treated.

This new edition was written almost exclusively by K. H. BECKURTS. Thanks are due to W. PÖNITZ for compiling the tables and data, supervising the preparation of the illustrations, and checking the manuscripts. Professor H. GOLDSTEIN (Columbia University) was kind enough to check part of the proofs and to give many useful hints. Valuable help in the revision of some chapters was rendered by W. GLÄSER, J. KALLFELZ, E. KIEFHABER, M. KÜCHLE, and W. REICHARDT. Special thanks are due to L. DRESNER (Oak Ridge National Laboratory) for the tedious work of translating the manuscript which was originally written in German.

Karlsruhe, May 1964 The Authors

Contents

Part I: Production and Nuclear Interaction of Neutrons

Part I

Production and Nuclear Interaction of Neutrons

1. Physical Properties of the Neutron

1.1. Historical

In the year 1930, BOTHE and BECKER discovered artificial nuclear γ-radiation. They bombarded Li, Be, B, and other light elements with α-particles from a polonium preparation and with a needle counter observed γ-rays, whose energies they determined by absorption in lead as roughly equal to those of natural γ-rays. The γ-emission from the bombardment of Li was ascribed to nuclear excitation through inelastic scattering of the α-particles; that of boron was explained by the fact that the residual C^{13} nucleus from the $B^{10}(\alpha, p)C^{13}$ reaction can be left in an excited state. Now, as then, both interpretations are considered correct. In the case of beryllium, which exhibited the most penetrating γ-radiation, BOTHE and BECKER hypothesized that nuclear excitation through capture of the α-particle occurred.

The reaction $Be + \alpha$ was subsequently investigated in detail by JOLIOT and CURIE. Using an ionization chamber, they found in addition to the γ-radiation already observed by BOTHE a penetrating radiation which could liberate protons with kinetic energies around 5 Mev from hydrogenous substances. Because of the large mass and energy of the particles liberated (presumably by the Compton effect), the hypothesis that this radiation was γ-radiation led to incredibly high γ-ray energies.

From these and some additional observations, CHADWICK drew the conclusion in 1932 that besides the γ-radiation already found by BOTHE, a hitherto unknown neutral particle with a mass approximately equal to that of the proton is produced. He called the new particle the neutron (symbol "n") and described the production of neutrons from beryllium irradiated with α-particles by the equation:

$$Be^9 + \alpha \rightarrow C^{12} + n,$$

frequently written as follows:

$$Be^9(\alpha, n)C^{12}.$$

Since free neutrons had not previously been observed in nature, it was tempting to suppose that they are not stable as free particles. On the other hand, the discovery of the neutron gave us the key to resolving many questions of nuclear structure. In particular, HEISENBERG concluded in 1932 that atomic nuclei contain only protons and neutrons as building blocks. In this way, the old picture, in which nuclei contained protons and electrons and which led to great theoretical difficulties, could be discarded. The neutron was assumed to be a stable particle in the nucleus, and this assumption has stood the test of time until today.

1.2. Properties of Free Neutrons

CHADWICK himself had concluded from the kinetic energy of the recoil protons liberated by neutrons, that the mass of the neutron is approximately equal to that of the proton. An exact determination of the neutron mass can be made, as CHADWICK and GOLDHABER first showed, from the binding energy of the deuteron: the mass m_d of the deuteron is equal to the sum $m_n + m_p$ of the neutron and proton masses diminished by the mass defect corresponding to the binding energy E_d. Thus

$$m_n = m_d - m_p + E_d/c^2 .$$

E_d can be determined either from the energy of the γ-ray produced by capture of a neutron by a proton (according to CHUPP *et al.*, $E_d = 2.225 \pm 0.003$ Mev) or from the threshold energy for the photo-disintegration of the deuteron (according to NOYES *et al.*, $E_d = 2.227 \pm 0.003$ Mev). From E_d and the known values of m_d and m_p, m_n can easily be calculated. Table 1.2.1 contains the masses and some other properties of the neutron and several other light particles.

Table 1.2.1. *Properties of Some Light Particles*

Particle	Mass in Grams	Mass in Atomic Weight Units[1] u	Charge	Spin	Magnetic Moment
Neutron . .	1.674663×10^{-24}	1.008665	$< 10^{-18}\, e$	$\frac{1}{2}$	-1.913148 NM[2]
Proton. . .	1.672357×10^{-24}	1.007276	e	$\frac{1}{2}$	2.79276 NM
H-Atom . .	1.673268×10^{-24}	1.007825	—	—	—
Deuteron .	3.343057×10^{-24}	2.013554	e	1	0.857407 NM
Electron . .	9.1081×10^{-28}	5.4859×10^{-4}	$-e$	$\frac{1}{2}$	1.0011596 M

[1] $1\,\text{u} = \frac{1}{12}$ of the mass of the C^{12}-atom $= 1.660277 \times 10^{-24}$ g. According to the equivalence $E = mc^2$ of mass and energy, 1 u corresponds to an energy of 931.441 Mev.
[2] $1\,\text{NM} = e\hbar/2m_p c = 0.505038 \times 10^{-23}$ erg/gauss. $1\,\text{M} = e\hbar/2m_e c = 0.927249 \times 10^{-20}$ erg/gauss.

It happens that the neutron is 0.840×10^{-3} u heavier than the hydrogen atom. For this reason, it is possible for a neutron to undergo a β-decay into a proton and an electron with a maximum energy of 0.840×0.9314 Mev $= 782$ kev. This β-decay of the free neutron was first observed by SNELL in 1948. The maximum energy determined by ROBSON is 782 ± 13 kev, in good agreement with the value calculated from the mass difference. According to SPIVAK, the half-life of the free neutron against this decay is 11.7 ± 0.3 min. Thus, compared with the processes with which we shall be concerned in the following, the neutron is long-lived and from the standpoint of these processes should be considered as a stable particle.

The spin of the neutron is $\frac{1}{2}$. Since the deuteron has spin 1 and no orbital angular momentum and since the proton has spin $\frac{1}{2}$, it follows that the neutron must have spin $\frac{1}{2}$ or $\frac{3}{2}$. One can exclude the value $\frac{3}{2}$ with certainty on the basis of various experiments, among them the energy dependence of the cross section for neutron-proton scattering.

That the neutron has a magnetic moment was first proven in 1940 by ALVAREZ and BLOCH. The more exactly measured value in the table was determined by COHEN, CORNGOLD, and RAMSEY by a refined Rabi-method. On the basis of this magnetic moment, the neutron can interact magnetically with the atoms in ferromagnetic substances[1]. This so-called "magnetic scattering" of neutrons, which is of importance for many solid-state investigations and for the production of polarized neutrons, will not be treated in this book.

Various experiments have been performed to determine the electric charge of the neutron. SHAPIRO and ESTULIN tried to deflect neutrons with strong electric fields and found no effect within the limits of accuracy of their apparatus (10^{-12} e). FERMI and MARSHALL obtained an upper limit to the charge of 10^{-18} e from neutron scattering experiments on xenon.

1.3. Fundamentals of Nuclear Reactions with Neutrons

If neutrons collide with atomic nuclei, they can produce a variety of nuclear reactions, depending on their energy. Generally, a distinction is made between scattering processes, in which a neutron makes an elastic or inelastic collision with a nucleus, and absorption processes, in which the neutron disappears into the nucleus and various secondary radiations appear. For the quantitative characterization of neutron reactions, the interaction cross section is used.

1.3.1. Cross Section, Mean Free Path

Let a collimated neutron current J — J is the number of neutrons which cross a 1-cm² surface perpendicular to the beam direction per second — fall on a material which contains N identical atomic nuclei per cm³. The number of "events" (scattering and absorption) per sec and cm³ is

$$\psi = J N \sigma . \tag{1.3.1}$$

Here it has been assumed that the current penetrates the sample practically without attenuation. The proportionality constant σ is called the interaction cross section. σ has the dimension cm², as follows from the other quantities in Eq. (1.3.1): $J\,(\mathrm{cm^{-2}\,sec^{-1}})$, $N\,(\mathrm{cm^{-3}})$, $\psi\,(\mathrm{cm^{-3}\,sec^{-1}})$. As a measure for cross sections the unit barn is used:

$$1\ \text{barn} = 10^{-24}\ \text{cm}^2 . \tag{1.3.2}$$

Scattering and absorption cross sections are usually distinguished from one another (σ_s, σ_a, respectively); each of these cross sections is composed of partial cross sections such as elastic and inelastic scattering, radiative capture, fission, etc. (cf. Sec. 1.3.2). The sum of all partial cross sections is called the *total* cross section; thus

$$\sigma_t = \sigma_s + \sigma_a . \tag{1.3.3}$$

Frequently, use is made of the quantity

$$\sigma N = \Sigma , \tag{1.3.4}$$

[1] The interaction with the magnetic moments of atomic nuclei is negligibly small.

called the *macroscopic* cross section. It specifies the cross section per cm³. It can also be considered as the probability that the neutron is scattered or absorbed in a path 1 cm long. The physical significance of Σ will become still clearer later.

The number of atoms N in one cm³ of a substance is obtained from

$$N = \frac{\text{density}}{\text{atomic weight}} \times \text{Avogadro's number}. \tag{1.3.5}$$

For graphite of density 1.6 g/cm³ one finds, for example,

$$N = \frac{1.6}{12} \times 6.025 \times 10^{23}\ \text{atoms/cm}^3 = 8.02 \times 10^{22}\ \text{atoms/cm}^3. \tag{1.3.6}$$

If the scattering or absorbing substance contains molecules with atoms of different cross sections, in place of Eq. (1.3.1) the following equation holds:

$$\psi = J N_{\text{mol}} \sum_i n_i \sigma_i \tag{1.3.7}$$

where

$$N_{\text{mol}} = \frac{\text{density}}{\text{molecular weight}} \times \text{Avogadro's number} \tag{1.3.8}$$

is the number of molecules in a cm³ and n_i is the number of atoms of type i with cross section σ_i in the molecule. The cross sections are thus additive. We will, however, see that there are exceptions to this rule.

We now advance a consideration that will lead us to the concept of the *mean free path*. Let a neutron encounter a substance. The probability that an atom in the substance is hit is proportional to the path Δx in the substance and equals $\Sigma \Delta x$. Accordingly, the probability that the neutron traverses the path Δx without a collision is $1 - \Sigma \Delta x$. The probability of traversing the distance $n \cdot \Delta x = x$ without a collision is

$$(1 - \Sigma \Delta x)^n = (1 - \Sigma \Delta x)^{\frac{x}{\Delta x}} = (1 - \Sigma \Delta x)^{\frac{\Sigma x}{\Sigma \Delta x}}. \tag{1.3.9}$$

When $\Delta x \to 0$ with x remaining constant (so that $n \to \infty$), this probability becomes

$$\lim_{\Delta x \to 0} (1 - \Sigma \Delta x)^{\frac{\Sigma x}{\Sigma \Delta x}} = e^{-\Sigma x}. \tag{1.3.10}$$

Thus a fraction $e^{-\Sigma x}$ of the incident neutrons traverse the distance x without collision, or in other words, $e^{-\Sigma x}$ is the probability that a neutron traverses the distance x without a collision. The probability that the neutron collides with an atom in the differential distance dx after traversing the distance x is $\Sigma\, dx\, e^{-\Sigma x}$. The probability that it collides after any arbitrary path length is

$$\int_0^\infty e^{-\Sigma x} \Sigma\, dx = 1. \tag{1.3.11}$$

We do not, however, want to know that a collision surely happens in an infinite interval, but rather how long on the average the interval is after which the neutron collides. To find this average path, one multiplies all paths with the frequency of their occurrence, sums, and divides by their total number, i.e., by

Eq. (1.3.11). One then obtains the following result for the "average" path length for a collision:

$$\lambda = \frac{\int_0^\infty x\, e^{-\Sigma x} \Sigma\, dx}{\int_0^\infty e^{-\Sigma x} \Sigma\, dx} = \int_0^\infty x\, e^{-\Sigma x} \Sigma\, dx = \frac{1}{\Sigma}. \qquad (1.3.12)$$

Thus we obtain the important result that the mean free path equals the reciprocal of the macroscopic cross section:

$$\lambda = \frac{1}{\Sigma} = \frac{1}{N\sigma}. \qquad (1.3.13)$$

We distinguish between the mean free path for scattering,

$$\lambda_s = \frac{1}{\Sigma_s} = \frac{1}{N\sigma_s}, \qquad (1.3.14)$$

and for absorption,

$$\lambda_a = \frac{1}{\Sigma_a} = \frac{1}{N\sigma_a}. \qquad (1.3.15)$$

Thus

$$\frac{1}{\lambda} = \Sigma_t = \frac{1}{\lambda_s} + \frac{1}{\lambda_a}. \qquad (1.3.16)$$

If a neutron moves with a constant velocity v, the *mean time between two collisions* is

$$\tau = \frac{\lambda}{v}. \qquad (1.3.17)$$

The number of such collisions per second is

$$\frac{1}{\tau} = \frac{v}{\lambda} = v\Sigma. \qquad (1.3.18\,\text{a})$$

If there are n neutrons of velocity v per cm^3 of a substance, the number ψ of events per sec and cm^3 is

$$\psi = \frac{n}{\tau} = nv\Sigma. \qquad (1.3.18\,\text{b})$$

1.3.2. Classification of Neutron Reactions

For the purposes of the following discussion we divide all neutrons in three groups according to their kinetic energy[1]

$$E = \tfrac{1}{2} m v^2 \qquad (1.3.19)$$

namely, slow neutrons ($E < 1000$ ev), intermediate energy neutrons (1 kev $< E <$ 500 kev), and fast neutrons (0.5 Mev $< E <$ 20 Mev). Neutrons with higher energies do not appear in nuclear reactors. Furthermore, we distinguish light nuclei ($A < 25$), intermediate nuclei ($25 < A < 80$), and heavy nuclei ($A > 80$).

The following two types of reactions can occur:

Compound Nucleus Reactions. The incident neutron unites with the target nucleus forming a relatively long-lived ($\gtrsim 10^{-17}$ sec) compound nucleus, whose

[1] If E is measured in ev and v in m/sec, then $E = 5.21 \times 10^{-9}\, v^2$ and $v = 1.38 \times 10^4 \sqrt{E}$.

excitation energy equals the sum of kinetic energy[1] and the binding energy (7—10 Mev for intermediate nuclei and 6—7 Mev for heavy nuclei) of the captured neutron. The decay of the compound nucleus can occur in various ways:

A neutron with the same energy as the originally captured neutron can be emitted. This process is called *compound elastic scattering* or sometimes *resonance scattering*; the latter name should be used only in the region where the cross section shows a resonance behaviour. In the neutron-nucleus center-of-mass system, the energy of the neutron does not change. The process is thus an elastic collision.

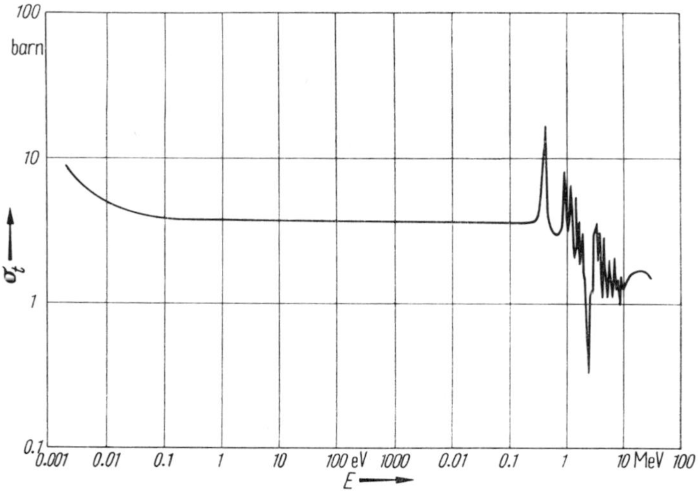

Fig. 1.3.1. Total cross section of Oxygen as a function of neutron energy

The excitation energy of the compound nucleus can be given up by emission of one or several γ-rays. This phenomenon is called *radiative capture* or the (n, γ) process. The resulting nucleus is frequently unstable against β-decay.

For sufficiently high excitation energies, the compound nucleus can emit charged particles or two neutrons [(n, α), (n, p), (n, np), $(n, 2n)$ reactions]. A neutron with a kinetic energy smaller than that of the incident neutron can also be emitted; in this case, the residual nucleus remains in an excited state which subsequently decays by γ-ray emission *(inelastic scattering)*. Finally, *fission* can occur in the heaviest nuclei.

Direct Reactions. Nuclear reactions can proceed directly, i.e. without the formation of a compound nucleus. For neutrons in the energy range considered here, by far the most important direct reaction is *direct elastic scattering*, i.e. elastic scattering without formation of a compound nucleus. Direct elastic scattering is frequently identified with *potential scattering*, i.e. the deflection of the incident neutron by the (real) nuclear potential which represents the average of all interactions with other nucleons.

This identification is not correct if the incident neutron can be absorbed into compound states. This absorption can be visualized as being caused by an imaginary potential. Thus the nuclear potential becomes complex; direct elastic scattering must be considered as potential

[1] To be precise, the kinetic energy in the neutron-nucleus center-of-mass system; for heavy nuclei the difference is unimportant.

scattering by a complex potential, not just by its real part. Detailed discussions of this problem can be found in the papers on the optical model mentioned in Sec. 1.3.5.

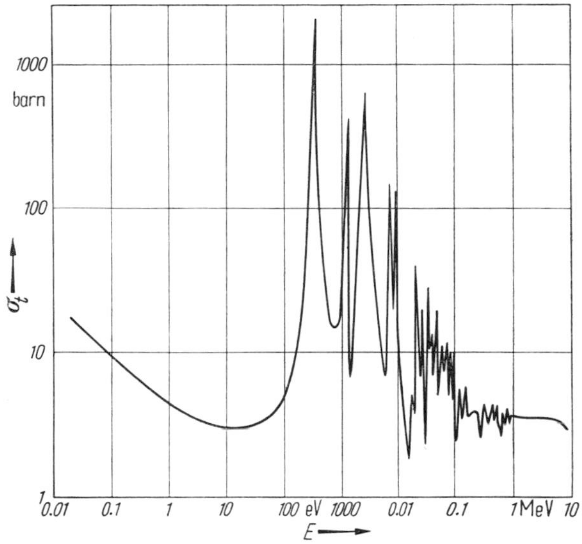

Fig. 1.3.2. Total cross section of Manganese as a function of neutron energy

While direct elastic scattering can always occur, compound nucleus formation is a resonance reaction, i.e., a compound nucleus can only be formed if the sum of the binding and kinetic energies of the incident neutron corresponds to an excited state of the compound nucleus. The cross sections of atomic nuclei are, for this reason, composed of a slowly varying cross section for direct elastic scattering and a cross section for compound nucleus reactions, which has sharp maxima at the resonance energies. If the resonances lie close together, the cross section for compound nucleus formation may exhibit a continuous variation with energy.

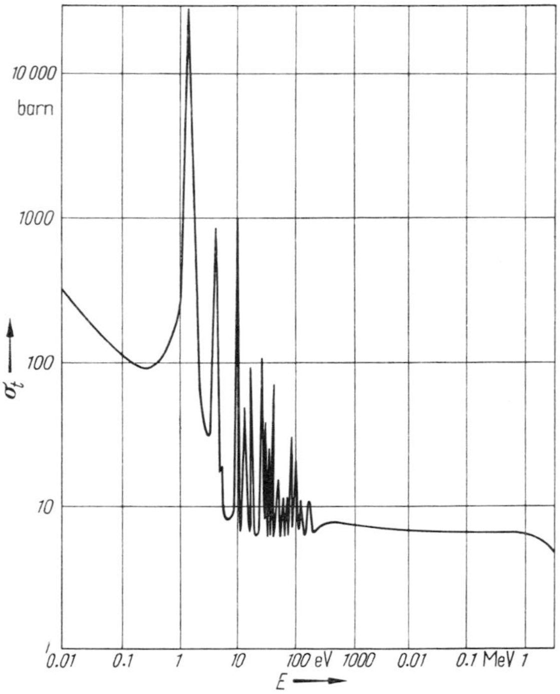

Fig. 1.3.3. Total cross section of Indium as a function of neutron energy

On the basis of these simple characteristics, some important conclusions concerning the energy dependence of the total cross section may be drawn. Figs. 1.3.1, 1.3.2, and 1.3.3 show the total cross sections of oxygen, manganese, and indium as functions of energy. As a light nucleus, oxygen has only few, widely spaced excited states; for this reason, its cross section shows only single, widely spaced resonances above 0.5 Mev. The predominant reaction in the low and intermediate energy range is potential scattering. The cross section for potential scattering is about four times the geometric cross section (it is always a few barns). Nearly all light nuclei show a similar behavior[1]. Manganese, an intermediate nucleus, exhibits resonances which are well separated by an average interval of several kev all the way up to high energies. The energy variation of the cross section near the resonances can be described by the *Breit-Wigner formula* (see Sec. 1.3.4). In the case of indium, in contrast, the resonances are very close together. At neutron energies as low as several kev the resonances already overlap to a considerable extent, and the variation of the cross section becomes continuous.

1.3.3. Properties of the Compound States

Corresponding to its finite lifetime τ, there exists a width (energy uncertainty) for each compound state

$$\Gamma = \frac{\hbar}{\tau} . \qquad (1.3.20)$$

$\Gamma/\hbar = 1/\tau$ is the probability per second that the compound state decays. It is composed additively of the probabilities for the various "decay channels" such as γ-emission, re-emission of a neutron, emission of charged particles, etc. Thus one writes Γ, the *total width*, as the sum of all *partial widths*:

$$\Gamma = \Gamma_\gamma + \Gamma_n + \Gamma_\alpha + \cdots . \qquad (1.3.21)$$

Γ_γ/Γ is then the probability that the decay of the compound nucleus takes place through the emission of γ-radiation, Γ_n/Γ the probability for decay through re-emission of a neutron, etc. Γ_γ is called the *radiation width*, Γ_n the *neutron width*, etc. In general, the Γ_i vary from resonance to resonance and among the individual nuclei. However, some general regularities do exist.

The radiation width Γ_γ of a given nucleus changes only slightly from resonance to resonance. From measurements, Γ_γ has been found to be around 0.2 ev for intermediate nuclei; for heavy nuclei, Γ_γ decreases to some 30 mev (millielectron-volts).

The neutron width exhibits a very complicated behavior. For a given nucleus, it varies quite strongly from resonance to resonance; however, on the average it shows a clear increase with neutron energy. For light nuclei, Γ_n is of the order of kev, for intermediate nuclei of the order of ev, and for heavy nuclei of the order of mev. Most slow- and intermediate-neutron resonances are so-called s-wave resonances, i.e., the captured neutron transmits no orbital angular momentum to the compound nucleus. In this case, a reduced neutron width can be introduced for each level through the relation

$$\Gamma_n = \Gamma_{n0} \sqrt{E} \quad (E \text{ in ev}) . \qquad (1.3.22)$$

The Γ_{n0} derived from the various resonances of one nucleus are distributed according to a certain statistical law (the so-called Porter-Thomas distribution) around a mean value $\bar{\Gamma}_{n0}$ that is characteristic of the particular nucleus. Reduced widths can also be defined for p-, d-, etc.- wave resonances, but the corresponding expressions are more complicated. For heavy nuclei and low neutron energies, Γ_γ is $\gg \Gamma_n$, i.e., resonance capture far outweighs resonance scattering. With increasing

[1] Important exceptions are B^{10}, Li^6, and He^3, for which strongly exothermic (n, α) and (n, p) reactions can occur. In this regard, see Sec. 1.4.1.

energy, the fraction of scattering increases. For light nuclei Γ_n is $\gg \Gamma_\gamma$, i.e., the resonances are predominantly scattering resonances.

Γ_n and Γ_γ are particularly distinguished among the partial widths, since resonance scattering and radiative capture are possible for *all* neutron-induced compound nucleus reactions. (n, p) and (n, α) reactions in most cases and the $(n, 2n)$ reaction and inelastic scattering in all cases first occur above a particular threshold energy of the incident neutron. The partial widths Γ_α, Γ_p, etc. are zero below the threshold and above it exhibit a behavior similar to that of the neutron width, i.e., they show statistical fluctuations from level to level and increase on the average. Because of the Coulomb force, the partial widths for the emission of charged particles decrease strongly with increasing mass number. As a rule, at high neutron energies the partial widths for inelastic scattering and $(n, 2n)$ reactions are the predominant ones, since with increasing energy an ever greater number of excited states of the residual nucleus becomes available for inclastic excitation.

Some of the heaviest nuclei (particularly important are U^{233}, U^{235}, Pu^{239}, and Pu^{241}) are fissionable by slow neutrons. As a result their fission widths Γ_f are different from zero. The Γ_f show statistical fluctuations from level to level. For U^{235}, Γ_f is about 50 mev ($\Gamma_\gamma = 36$ mev and $\overline{\Gamma}_{n0} = 0.01$ mev); for Pu^{239}, Γ_f is about 130 mev ($\Gamma_\gamma = 39$ mev and $\overline{\Gamma}_{n0} = 0.38$ mev).

1.3.4. The Breit-Wigner Formula

In the neighborhood of a single, isolated resonance, the variation of the (n, γ) cross section can be described with the help of the Breit-Wigner formula. If we restrict ourselves to s-wave resonances of a nucleus with zero spin[1], the following formula applies in the neighborhood of a resonance of energy E_R:

$$\sigma_{n,\gamma}(E) = \pi \lambdabar^2 \frac{\Gamma_n \Gamma_\gamma}{(E - E_R)^2 + (\Gamma/2)^2} \,. \tag{1.3.23}$$

(For a derivation, see, for example, Blatt and Weisskopf). Here

$$\lambdabar = \frac{\lambda}{2\pi} = \frac{\hbar}{mv} \tag{1.3.24}$$

is the reduced de Broglie wavelength of the neutron[2]. The cross section has a maximum at $E = E_R$, where it equals

$$\sigma^0_{n,\gamma} = 4\pi \lambdabar_R^2 \cdot \frac{\Gamma_n \Gamma_\gamma}{\Gamma^2} \,. \tag{1.3.25}$$

At $E = E_R \pm \frac{\Gamma}{2}$, $\sigma_{n,\gamma}$ has approximately half its maximum value; Γ thus has the significance of the half-value width of the resonance.

As a typical example, Fig. 1.3.4 shows the cross section of rhodium for very slow neutrons. A resonance at 1.23 ev can clearly be recognized. The curve has

---

[1] The general case of non-zero spin is discussed in Sec. 4.1.2.

[2] The following relations hold between the de Broglie wavelength λ (cm) and the neutron energy E (ev):

$$\lambda = \frac{2.86 \times 10^{-9}}{\sqrt{E}}\,; \quad E = \frac{8.17 \times 10^{-18}}{\lambda^2} \,.$$

been calculated with the help of the Breit-Wigner formula. At very low energy the cross section shows an increase which is proportional to $1/\sqrt{E}$. This famous $1/v$-law for the capture cross section can be explained by the Breit-Wigner formula:

As $E \to 0$, according to Eq. (1.3.23) (when $E_R \gg \Gamma$)

$$\sigma_{n,\gamma}(E) \approx \pi \lambda\!\!\!^{-2} \frac{\Gamma_n \Gamma_\gamma}{E_R^2}. \tag{1.3.26 a}$$

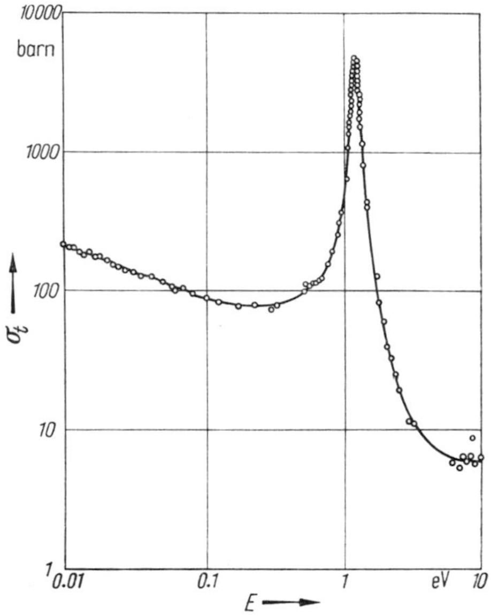

Fig. 1.3.4. Total cross section of Rhodium as a function of neutron energy. o Measurements; ——— Breit-Wigner fit

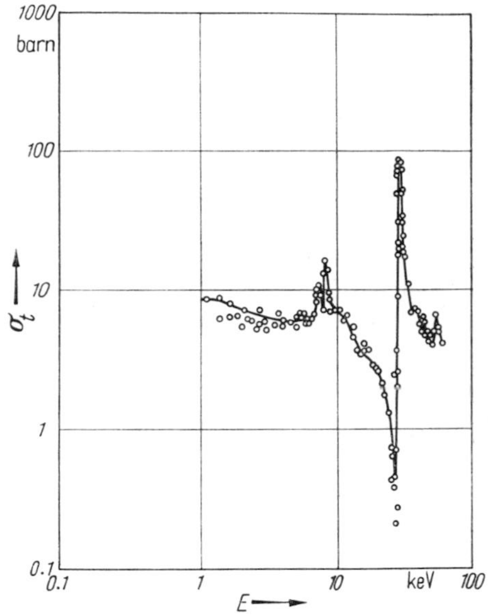

Fig. 1.3.5. Total cross section of Iron as a function of neutron energy. The dip just below the 29 kev resonance is due to interference of resonance and potential scattering

From this equation it follows by use of Eq. (1.3.22) that

$$\left.\begin{aligned}\sigma_{n,\gamma}(E) &= \pi \lambda\!\!\!^{-2} \frac{\Gamma_{n0}\Gamma_\gamma}{E_R^2} \sqrt{E} \\ &= \pi \lambda\!\!\!^{-2}_R \frac{\Gamma_{n0}\Gamma_\gamma}{E_R} \frac{1}{\sqrt{E}}.\end{aligned}\right\} \tag{1.3.26 b}$$

The $1/v$-law has the following physical significance: Neutrons with $E \approx 0$ can be captured with the formation of a compound state. Since the "resonance condition" is not exactly fulfilled, the "transmission factor" $\frac{\Gamma_n \Gamma_\gamma}{(E-E_R)^2 + \left(\frac{\Gamma}{2}\right)^2}$ occurring in Eq. (1.3.23) is very small. However, since the "contact cross section" $\pi \lambda\!\!\!^{-2}$ tends to very large values as $E \to 0$, the capture cross section resulting from the product of both factors becomes large. One can also intuitively explain the $1/v$-law by saying that the probability of capture of a neutron is proportional to the time it spends in the neighborhood of the nucleus. The $1/v$-law for the slow-neutron capture cross section holds for nearly all nuclei[1]. In many cases it is

[1] The only exceptions to this rule are the nuclei in which a resonance occurs in the immediate neighborhood of $E = 0$.

possible to calculate the $1/v$-part of the cross section from the parameters of the observed resonances. Frequently, however, one must take into account so-called negative-energy resonances, i.e., excited states of the compound nucleus whose energy is less than the binding energy of the last neutron.

The Breit-Wigner formula for other reactions is the same as Eq. (1.3.23) except that Γ_γ is replaced by another appropriate partial width. The simple Breit-Wigner formula is not very accurate, however, for describing the fission cross section. The multi-level formula, which takes into account the mutual influence of several resonances, must then be used (cf. RAINWATER).

Because of the interference of potential and resonance scattering, the Breit-Wigner formula for the description of elastic scattering has a somewhat different form. If we write the cross section for potential scattering (which dominates at large distances from the resonances) in the form

$$\sigma_{\text{pot}} = 4\pi a^2 \quad (a = \text{effective nuclear radius for potential scattering}) \tag{1.3.27}$$

then, in the neighborhood of a low energy resonance,

$$\sigma_s(E) = \pi \bar{\lambda}^2 \left| 2\frac{a}{\bar{\lambda}} + \frac{\Gamma_n}{(E - E_R) + (i\Gamma/2)} \right|^2. \tag{1.3.28}$$

Eq. (1.3.28) says that in scattering, the amplitudes not the intensities of the scattered waves must be added. A detailed discussion can be found in FELD among others. Expansion of Eq. (1.3.28) yields

$$\sigma_s(E) = \pi \bar{\lambda}^2 \frac{\Gamma_n^2}{(E - E_R)^2 + (\Gamma/2)^2} + 4\pi a^2 + 4\pi \bar{\lambda} \Gamma_n a \frac{E - E_R}{(E - E_R)^2 + (\Gamma/2)^2}. \tag{1.3.29}$$

The first term describes pure resonance scattering and is the same as Eq. (1.3.23) if one replaces Γ_γ by Γ_n. The second term describes pure potential scattering; for $|E - E_R| \gg \Gamma$, σ_s goes over into $4\pi a^2 = \sigma_{\text{pot}}$. The third term describes the interference between potential and resonance scattering. For $E < E_R$, it is negative, i.e., the interference is destructive. Potential and resonance scattering can even virtually cancel one another out; Fig. 1.3.5 shows this for the case of the 29 kev resonance in iron.

1.3.5. Cross Sections in the Continuum Region

The individual resonance structure of the cross section goes over into a continuum at high energies for intermediate nuclei and at intermediate energies for heavy nuclei. In many cases, this is because the resolving power of the neutron spectrometers used for cross section measurements is not sufficient at high energies to separate the individual resonances.

Fig. 1.3.6 shows semi-schematically the average cross section as a function of energy and mass number for many intermediate and heavy nuclei. A continuous decrease of the cross section with increasing energy and an increase with increasing mass number is clearly recognizable[1]. Furthermore, a certain giant resonance structure is recognizable in the cross section, i.e., certain resonance-like maxima (as a function of E at fixed A and as a function of A at fixed E) are evident. The *optical model* developed by FESHBACH, PORTER, and WEISSKOPF gives an explanation of this behavior; however, we shall not go more deeply into this matter.

[1] $\sigma_t/\pi R^2$ is plotted, with R taken from Eq. (1.3.31).

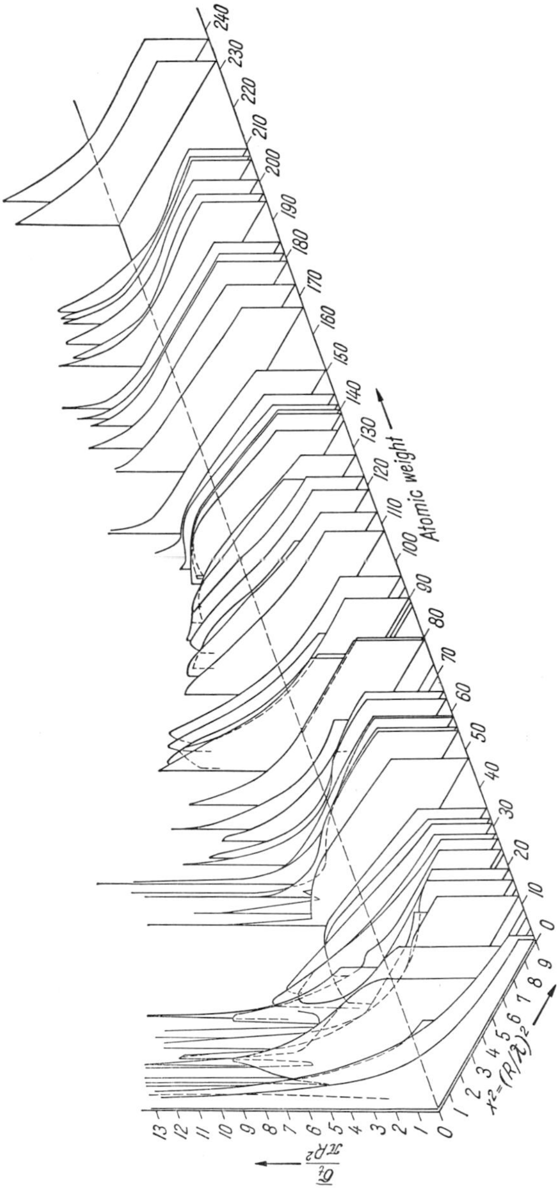

Fig. 1.3.6. Total cross sections of the elements as a function of neutron energy and atomic number (FESHBACH, PORTER and WEISSKOPF)

At very high energies the total cross section approaches twice the geometric cross section. Then

$$\sigma_t(E) \approx 2\pi(R + \lambda\!\!\!^{-})^2 \tag{1.3.30}$$

applies, where R is the "geometric" nuclear radius calculated from the relation[1]

$$R = R_0 \cdot A^{\frac{1}{3}}, \; R_0 \approx 1.4 \times 10^{-13}\,\text{cm}. \tag{1.3.31}$$

[1] A detailed discussion of the validity of Eq. (1.3.30) is given by D. W. MILLER in MARION and FOWLER, Vol. II, p. 1027 ff.

The total cross section given by Eq. (1.3.30) is composed of equally large cross sections σ_s for direct elastic scattering and σ_c for compound nucleus formation. Compound elastic scattering is negligible for high energies. $\sigma_c = \pi(R+\lambda)^2$ signifies that the nucleus is "black" to very fast neutrons, i.e., it absorbs all incident neutrons. The scattering cross section $\sigma_s = \pi(R+\lambda)^2$ describes the so-called *shadow scattering*, which is strongly anisotropic and directed mainly forwards.

Fig. 1.3.7 again schematically summarizes the various types of reactions and cross sections discussed in this section.

	Slow neutrons $E<1000$ ev	Intermediate energy neutrons 1 kev$<E<500$ kev	Fast neutrons 0.5 Mev$<E<20$ Mev
Light nuclei $A<25$	Potential scattering	Separated Resonance	resonances, $\bar{D}\approx 0.1\ldots 1$ Mev scattering; $n,p\text{-}n$, α- and $n,2n$-reactions
Intermediate nuclei $25<A<80$	Separated Resonance Potential scattering	resonances; $\bar{D}\approx 1\text{—}100$ kev scattering, radioactive capture	overlapping continuum resonances $n,p\text{-}n$, α- and $n,2n$-reactions; inelastic scattering
Heavy nuclei $A>80$	Separated resonances, $\bar{D}\approx 1\text{—}100$ ev Radiative capture	overlapping resonances	continuum Inelastic scattering $n,2n$-reactions

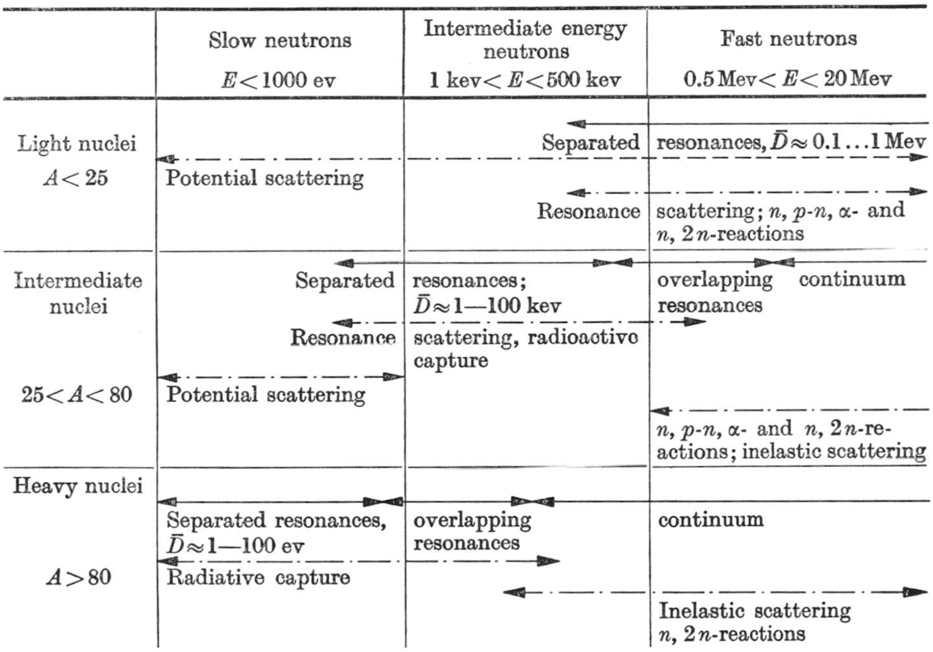

Fig. 1.3.7. Systematics of neutron reactions

1.4. Experimental Cross Sections

In this section, we shall discuss some cross sections which are particularly important for the following chapters. Detailed graphical and tabular representations of cross sections can be found in "Neutron Cross Sections" by D. J. HUGHES and R. B. SCHWARTZ (BNL-325, 1958) and "Angular Distributions" by M. D. GOLDBERG, V. M. MAY and J. R. STEHN (BNL-400, Vol. I and II, 1962). A series of special tables of cross sections for reactor calculations have been set up by various authors, among them PARKER, HOWERTON, and SCHMIDT.

1.4.1. Cross Sections of B^{10}, Li^6, and He^3

The cross sections of B^{10}, Li^6, and He^3 exhibit a behavior different from that of the other light nuclei. All three substances have significance for neutron detection.

Natural boron consists of the isotopes B^{11}, with an abundance of 80.2%, and B^{10}, with an abundance of 19.8%. Neutron capture in B^{10} leads to an exothermic (n, α) reaction:

$$B^{10}+n \begin{cases} \to Li^7+\alpha+2.79\text{ Mev } (6.1\%) \\ \to Li^{7*}+\alpha+2.31\text{ Mev } (93.9\%) \\ \qquad \searrow Li^7+\gamma+0.478\text{ Mev.} \end{cases}$$

The branching ratio depends somewhat on the energy; the percentages specified above apply for slow neutrons. The cross section for this reaction is 3840 ± 11 barns at $v_0=2200$ m/sec[1]. It follows a $1/v$-law very accurately in the energy range from 0 to 10^4 ev. The scattering cross section of B^{10} for slow neutrons is 4.0 barn.

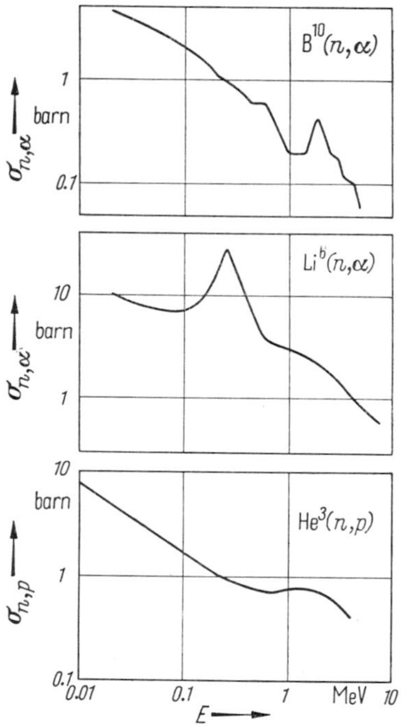

Fig. 1.4.1. The cross sections of some neutron-induced reactions on light nuclei as a function of energy

Li^6, which has an abundance of 7.52% in the natural element, undergoes the following reaction with neutrons:

$$Li^6+n\to H^3+\alpha+4.786\text{ Mev}.$$

The cross section for this reaction is 936 ± 6 barn at 2200 m/sec. It follows the $1/v$-law in the range $0<E<1$ kev and exhibits a broad resonance at 250 kev (see Fig. 1.4.1). The scattering cross section for slow neutrons is 1.4 barn.

Natural helium contains He^3 with an abundance of $1.3\cdot10^{-4}$%. He^3 undergoes the following reaction

$$He^3+n\to H^3+p+0.764\text{ Mev}.$$

The cross section for this reaction is 5327 ± 10 barn at $v_0=2200$ m/sec; its behavior at higher energies is plotted in Fig. 1.4.1. The scattering cross section for slow neutrons is 0.8 barn.

1.4.2. Cross Sections of Some Moderator Substances

Fig. 1.4.2 shows the total cross sections of *hydrogen* and *deuterium* as functions of energy. The total cross sections are practically equal to the respective elastic scattering cross sections in both these nuclei. The capture cross sections at $v_0=2200$ m/sec are 0.328 barn for hydrogen and 0.46 millibarn (mbarn) for deuterium and both cross sections follow the $1/v$-law[2].

---

[1] $v=2200$ m/sec is the most probable velocity in a Maxwellian velocity distribution at 20.4 °C (see Sec. 5.3). It is conventional to specify slow-neutron cross sections at this velocity, which corresponds to an energy of 0.0253 ev.

[2] A direct proof of the validity of the $1/v$-law for the capture cross section of light nuclei has not as yet been offered. However, since no resonances appear at low energies, this assumption seems good.

At neutron energies above 3.339 Mev the deuteron can be split by collision with a neutron:

$$n + H^2 \rightarrow H^1 + 2n.$$

The contribution of the cross section for this reaction to the total cross section is very small, however (it increases from zero at the threshold to about 100 mbarn at 10 Mev). Figs. 1.4.3 and 1.4.4 show the angular distribution of elastically scattered neutrons in the center-of-mass system at various energies. For hydrogen, the scattering in the center-of-mass system is isotropic at all energies of interest. In contrast, deuterium exhibits a considerable anisotropy for neutron energies above about 0.2 Mev.

We shall come back to the energy dependence of the scattering cross section at low energies again in Sec. 1.4.4.

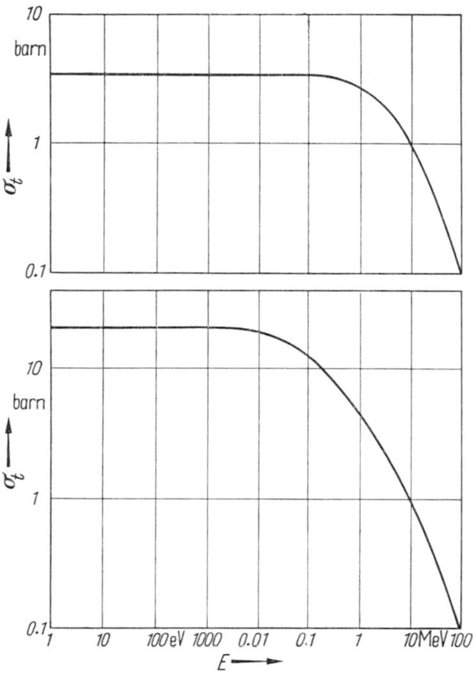

Fig. 1.4.2. Total cross section of Deuterium (above) and Hydrogen (below) as a function of neutron energy

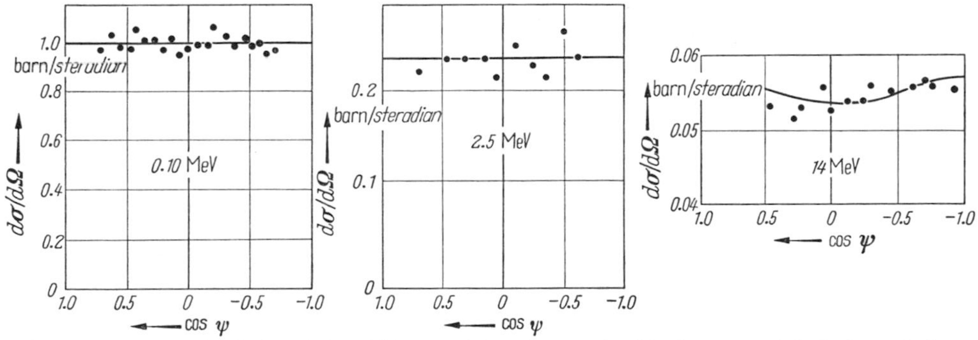

Fig. 1.4.3. Angular distribution (in the C. M. system) of neutrons scattered by Hydrogen. $d\sigma/d\Omega$ is the scattering cross section per unit solid angle, i.e. $1/2\pi$ times $\sigma_{n,n}(\cos\psi)$

Fig. 1.4.5 shows the total cross sections of *beryllium* and *carbon*. The capture cross sections are again very small. For $v_0 = 2200$ m/sec, $\sigma_a = 10$ mbarn in Be and 3.8 mbarn in C[1]; the $1/v$-law is valid. The cross section of beryllium has scattering resonances at 0.62, 0.81, and 2.73 Mev. Above a neutron energy of 1.85 Mev, an $(n, 2n)$ process is possible:

$$Be^9 + n \rightarrow Be^8 + 2n.$$

The cross section of carbon exhibits some resonances above 2 Mev. At neutron energies above 4.43 Mev, inelastic neutron scattering can occur. The angular

[1] This cross section can be greatly increased by impurities in the case of graphite (cf. Chapter 17).

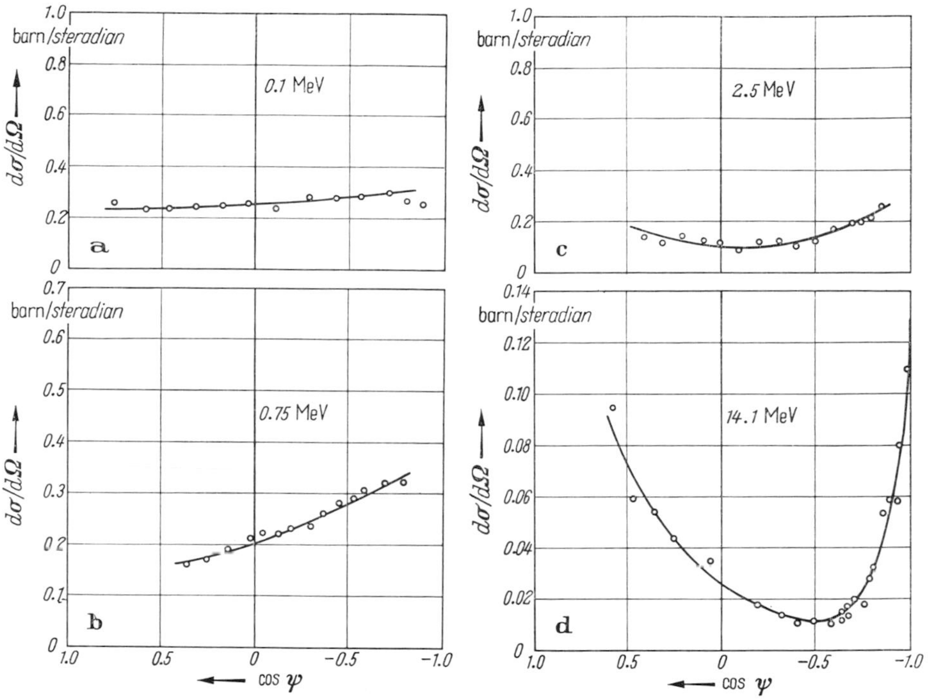

Fig. 1.4.4 a—d. Angular distribution (in the C. M. system) of neutrons scattered by Deuterium

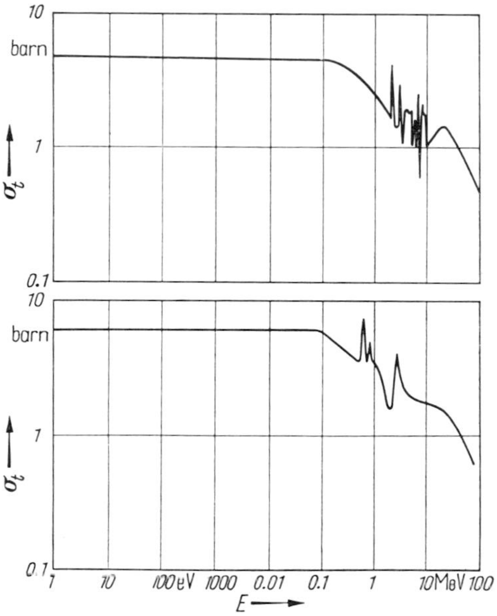

Fig. 1.4.5. Total cross section of Carbon (above) and Beryllium (below) as a function of neutron energy

distribution of the elastically scattered neutrons is isotropic in the center-of-mass system for energies up to 0.5 Mev in beryllium and up to 1.4 Mev in carbon. The behavior of the scattering cross section at low energy is discussed in Sec. 1.4.3.

The total cross section of oxygen has already been shown in Fig. 1.3.1. The capture cross section at $v_0 = 2200$ m/sec is < 0.2 mbarn. Above an energy of 0.3 Mev, scattering resonances appear. For neutron energies above 3.350 Mev, the reaction

$$O^{16} + n \rightarrow C^{13} + He^4$$

contributes significantly to the cross section (see Fig. 1.4.6 in this connection). The angular distribution of the elastically scattered neutrons becomes anisotropic in the center-of-mass system for energies above 0.3 Mev.

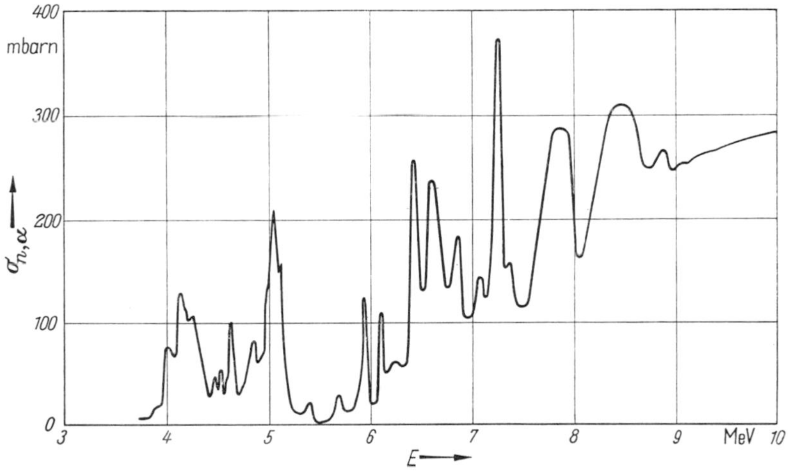

Fig. 1.4.6. The cross section for the $O^{16}(n, \alpha)\,C^{13}$ reaction as a function of neutron energy

1.4.3. The Scattering Cross Section of Crystalline Media at Low Energies

Fig. 1.4.7 shows the total cross section of crystalline beryllium at low energies. While the constant cross section for potential scattering applies above 0.025 ev, at lower energies oscillations appear; in particular, the cross section falls off sharply at 0.0052 ev. At lower energies the remaining cross section increases proportionally to $1/v$; it is obviously dependent on the temperature of the sample. These effects are connected with the crystalline structure of beryllium: the scattering of slow neutrons on beryllium takes place *coherently*, i.e., the scattered neutron waves from neighboring beryllium nuclei can interfere with one another. Since for neutron energies of about 0.01 to 0.1 ev the de Broglie wavelength of the neutrons is around 10^{-8} cm, that is, of the order of magnitude of the interatomic distances, the phenomenon of *diffraction*, well known in X-ray physics, appears.

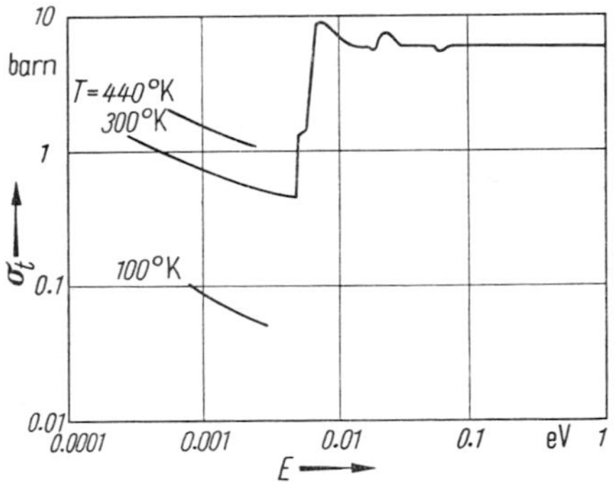

Fig. 1.4.7. The cross section of Beryllium at low neutron energies and various temperatures

The subject of *neutron optics* is based on these interference effects; however, neutron optics will not be treated in this book[1]. As is well known, diffraction in a crystal can be described formally as reflection from the lattice planes. If d is the distance between two lattice planes, then reflection only occurs when the relation

$$\left.\begin{aligned} n\lambda &= 2d\sin\Theta \\ n &= 1, 2, 3, 4, \ldots \end{aligned}\right\} \quad (1.4.1)$$

connects the glancing angle Θ, the lattice spacing d, and the wavelength λ. If neutrons of wavelength λ strike a polycrystal, then because of the arbitrary orientation of the individual microcrystals, lattice planes and glancing angles for which Eq. (1.4.1) is fulfilled always exist. If, however,

$$\lambda > 2d_{\max}, \quad (1.4.2)$$

where $d_{\max}$ is the largest lattice spacing occurring in the crystal (in Be $d_{\max}=1.98$ Å), Eq. (1.4.1) can no longer be satisfied. Then coherent scattering cannot take place. This consideration explains the cut-off shown in Fig. 1.4.7. This phenomenon also occurs in graphite (at $E=0.00183$ ev) and nearly all other crystalline media.

Neutron scattering is not coherent in all cases. For example, in an element that is composed of several isotopes with different scattering properties, the coherent scattering is different from the total scattering. For nuclei with spins different from zero, the scattering amplitude depends on the orientation of the neutron spin with respect to the nuclear spin. If it is very different for the two spin orientations, the part of the scattering capable of interference can be much smaller than the total scattering. When coherent effects play a role, one must characterize a scattering process not only by specifying the total scattering cross section, but also by specifying the coherent scattering cross section σ_{coh}, which characterizes the intensity of the coherently scattered radiation. While in many cases σ_s and σ_{coh} nearly agree (beryllium: $\sigma_s=7.54$ barn, $\sigma_{\mathrm{coh}}=7.53$ barn; carbon: $\sigma_s=5.53$ barn, $\sigma_{\mathrm{coh}}=5.50$ barn), strong deviations occur in *hydrogen* ($\sigma_s=81.5$ barn, $\sigma_{\mathrm{coh}}=1.79$ barn) and *vanadium* ($\sigma_s=5.13$ barn, $\sigma_{\mathrm{coh}}=0.032$ barn). Both these nuclei scatter practically purely incoherently[2].

Below the cut-off energy, the cross section is composed of the partial cross sections for absorption, incoherent scattering, and *thermal inelastic scattering*. In graphite and beryllium, the first two effects are negligible; below the cut-off energy, the cross section describes processes in which the neutron picks up energy from the lattice vibrations. For $T\to 0$, this cross section vanishes, as can be seen clearly in Fig. 1.4.7. We shall return to the problem of the inelastic scattering of slow neutrons again in Sec. 10.1.

1.4.4. The Scattering Cross Section of Free and Bound Nuclei

Fig. 1.4.8 shows by means of the total cross section of H_2O a further phenomenon important at low energies: Above 1 ev and up to 10 kev this cross section is

[1] See Bacon, G. E.: Neutron Diffraction 2nd edition; Oxford: Clarendon Press 1962. — Hughes, D. J.: Neutron Optics. New York: Interscience Publishers 1954.

[2] In this connection see the book of D. J. Hughes cited above.

constant and equal to the sum of the scattering cross sections of two free protons and an oxygen nucleus. Around 1 ev an increase of the scattering cross section begins which eventually approaches a value four times the cross section above 1 ev[1]. This increase is connected with the chemical binding of the protons:

Let us consider the collision of a neutron and a free (unbound) nucleus of mass M. This case is relevant when the neutron energy is large compared to the energy of the chemical binding of the nucleus. Let us denote the cross section for this collision with σ_{sf}. Let us now think of the nucleus rigidly bound to a molecule of mass M_{mol}. As the quantum mechanical treatment of this scattering problem shows, the scattering cross section of the bound atom varies directly with the square of the reduced mass ratio:

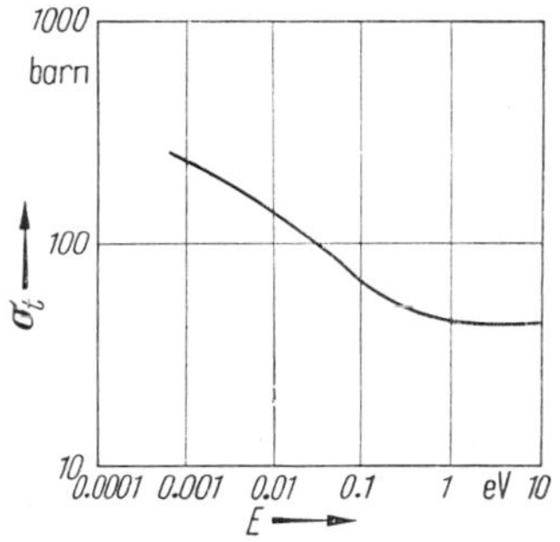

Fig. 1.4.8. The cross section of H_2O at low neutron energies

$$\sigma' = \sigma_{sf}\left(\frac{\mu'}{\mu}\right)^2 \tag{1.4.3}$$

where

$$\mu' = \frac{m_n M_{\text{mol}}}{m_n + M_{\text{mol}}}, \tag{1.4.4}$$

$$\mu = \frac{m_n M}{m_n + M}. \tag{1.4.5}$$

Because $M_{\text{mol}} > M$, μ' is always $> \mu$. Of particular interest is the case of completely rigid binding ($M_{\text{mol}} \to \infty$), which, for example, can be realized in solids. Then

$$\sigma' = \sigma_{sb} = \left(\frac{m_n}{\mu}\right)^2 \sigma_{sf} \tag{1.4.6}$$

or in terms of the mass number $A \approx M/m_n$

$$\sigma_{sb} = \left(\frac{A+1}{A}\right)^2 \sigma_{sf}. \tag{1.4.7}$$

The cross section defined by Eq. (1.4.7) is called the scattering cross section for bound atoms.

It is now easy to explain the cross section of water shown in Fig. 1.4.8: For $E > 1$ ev the protons are effectively free and the scattering cross section is equal to the sum of the atomic scattering cross sections σ_{sf}. With decreasing energy the chemical binding makes itself felt and the "effective mass" of the proton increases and with it the scattering cross section according to Eq. (1.4.3). At very low energies the protons can be considered as rigidly bound to the oxygen atoms; then $M_{\text{mol}}/M = 18$ ($\gg 1$). Thus according to Eq. (1.4.7) one obtains an increase of the cross section by a factor of 4. The same effect exists in all hydrogenous compounds. D_2O shows a similar but less marked behavior.

1.4.5. Slow Neutron Cross Sections of the Elements

Table 1.4.1 gives some slow-neutron absorption and scattering cross sections. The values of the absorption cross section apply to $v_0 = 2200$ m/sec; in some cases

[1] Because of the Doppler effect the actually observed increase is somewhat steeper: see Sec. 10.1.

a g-factor is given after the cross section. This signifies that the particular cross section deviates from the $1/v$-law. The use of the g-factor is explained in Sec. 5.3.

Table 1.4.1. *Scattering and Absorption Cross Sections of the Elements for Slow (2200 m/sec) Neutrons*

Element	σ_a (barn)	$\bar{\sigma}_s$ (barn)	Element	σ_a (barn)	$\bar{\sigma}_s$ (barn)
${}_{1}$H	0.328±0.002	38±4 (gas)	${}_{47}$Ag	64.5±0.6	6±1
${}_{2}$He	0	0.8±0.2		g=1.004	
${}_{3}$Li	70.4±0.4	1.4±0.3	${}_{48}$Cd	2537±9	7±1
${}_{4}$Be	0.010±0.001	7±1		g=1.338	
${}_{5}$B	758±4	4±1	${}_{49}$In	194±2	2.2±0.5
${}_{6}$C	3.73±0.07	4.8±0.2		g=1.020	
	$\cdot 10^{-3}$		${}_{50}$Sn	0.625±0.015	4±1
${}_{7}$N	1.88±0.05	10±1	${}_{51}$Sb	5.7±1.0	4.3±0.5
${}_{8}$O	$<2\cdot 10^{-4}$	4.2±0.3	${}_{52}$Te	4.7±0.1	5±1
${}_{9}$F	$<1\cdot 10^{-2}$	3.9±0.2	${}_{53}$I	6.22±0.25	3.6±0.5
${}_{10}$Ne	0.032±0.009	2.4±0.3	${}_{54}$Xe	74±1	4.3±0.4
${}_{11}$Na	0.531±0.008	4.0±0.5	${}_{55}$Cs	28±1	7±1
${}_{12}$Mg	0.063±0.003	3.6±0.4	${}_{56}$Ba	1.2±0.1	8±1
${}_{13}$Al	0.241±0.003	1.4±0.1	${}_{57}$La	8.9±0.2	9.3±0.7
${}_{14}$Si	0.16±0.02	1.7±0.3	${}_{58}$Ce	0.73±0.08	2.8±0.5
${}_{15}$P	0.20±0.02	5±1	${}_{59}$Pr	11.3±0.2	4.0±0.4
${}_{16}$S	0.52±0.02	1.1±0.2	${}_{60}$Nd	49.9±2.2	16±3
${}_{17}$Cl	33.8±1.1	16±3	${}_{62}$Sm	5828±30	
${}_{18}$A	0.66±0.04	1.5±0.5		g=1.638	
${}_{19}$K	2.07±0.07	1.5±0.3	${}_{63}$Eu	4406±30	8±1
${}_{20}$Ca	0.44±0.02	3.2±0.3		g=0.999	
${}_{21}$Sc	24±1	24±2	${}_{64}$Gd	46617±100	
${}_{22}$Ti	5.8±0.4	4±1		g=0.888	
${}_{23}$V	5.00±0.01	5±1	${}_{65}$Tb	46±3	
${}_{24}$Cr	3.1±0.2	3.0±0.5	${}_{66}$Dy	940±20	100±20
${}_{25}$Mn	13.2±0.1	2.3±0.3	${}_{67}$Ho	65±3	
${}_{26}$Fe	2.62±0.06	11±1	${}_{68}$Er	173±17	
${}_{27}$Co	37.1±1.0	7±1	${}_{69}$Tm	127±4	7±3
${}_{28}$Ni	4.6±0.1	17.5±1	${}_{70}$Yb	37±4	12±5
${}_{29}$Cu	3.81±0.03	7.2±0.7	${}_{71}$Lu	112±5	
${}_{30}$Zn	1.10±0.02	3.6±0.4	${}_{72}$Hf	101.4±0.5	8±2
${}_{31}$Ga	2.80±0.13	4±1		g=1.020	
${}_{32}$Ge	2.45±0.20	3±1	${}_{73}$Ta	21.0±0.7	5±1
${}_{33}$As	4.3±0.2	6±1	${}_{74}$W	19.2±1.0	5±1
${}_{34}$Se	11.7±0.1	11±2	${}_{75}$Re	86±4	14±4
${}_{35}$Br	6.82±0.06	6±1	${}_{76}$Os	15.3±0.7	15.3±1.5
${}_{36}$Kr	31±2	7.2±0.7	${}_{77}$Ir	440±20	
${}_{37}$Rb	0.73±0.07	5.5±0.5	${}_{78}$Pt	8.8±0.4	10±1
${}_{38}$Sr	1.21±0.06	10±1	${}_{79}$Au	98.6±0.3	9.3±1.0
${}_{39}$Y	1.31±0.08			g=1.005	
${}_{40}$Zr	0.185±0.004	8±1	${}_{80}$Hg	374±5	20±5
${}_{41}$Nb	1.16±0.02	5±1		g=0.998	
${}_{42}$Mo	2.70±0.04	7±1	${}_{81}$Tl	3.4±0.5	14±2
${}_{43}$Tc	22±3	5±1	${}_{82}$Pb	0.170±0.002	11±1
${}_{44}$Ru	2.56±0.12	6±1	${}_{83}$Bi	0.034±0.002	9±1
${}_{45}$Rh	149±4	5±1	${}_{90}$Th	7.56±0.11	12.6±0.2
${}_{46}$Pd	8.0±1.5	3.6±0.6	${}_{92}$U	7.68±0.07	8.3±0.2
				g=0.99	

The scattering cross sections are averages over a room-temperature Maxwell velocity distribution. The values were mainly taken from the tabulation of Hughes and Schwartz, although newer values were used in some places.

Chapter 1: References

General

AMALDI, E.: The Production and Slowing Down of Neutrons. In: Handbuch der Physik, Bd. XXXVIII/2, p. 1. Berlin-Göttingen-Heidelberg: Springer 1959.

BLATT, J. M., and V. F. WEISSKOPF: Theoretical Nuclear Physics. New York: John Wiley & Sons 1952.

EVANS, R. D.: The Atomic Nucleus. New York-Toronto-London: McGraw-Hill 1955.

FELD, B. T.: The Neutron. In: Experimental Nuclear Physics, vol. II, p. 209. New York: John Wiley & Sons 1953.

HUGHES, D. J.: Pile Neutron Research. Cambridge: Addison-Wesley Publishing Company, Inc. 1953.

— Neutron Cross Sections. London-New York-Paris: Pergamon Press 1957.

MARION, J. B., and J. L. FOWLER (ed.): Fast Neutron Physics; Part II: Experiments and Theory; New York: Interscience Publishers 1963.

WEINBERG, A. M., and E. P. WIGNER: The Physical Theory of Neutron Chain Reactors. Chicago: Chicago University Press 1958.

WLASSOW, N. A.: Neutronen. Köln: Hoffmann 1959.

Special

BOTHE, W., u. H. BECKER: Z. Physik **66**, 289 (1930). CHADWICK, J.: Proc. Roy. Soc. (London), Ser. A **136**, 692 (1932). CURIE, J., and F. JOLIOT: J. Physique Radium **4**, 21 (1933). HEISENBERG, W.: Z. Physik **77**, 1 (1932).	Discovery of the Neutron.
CHADWICK, J., and M. GOLDHABER: Proc. Roy. Soc. (London), Ser. A **151**, 479 (1935). CHUPP, E. L., R. W. JEWELL, and W. JOHN: Phys. Rev. **121**, 234 (1961). NOYES, J. C., *et al.*: Phys. Rev. **95**, 396 (1954).	Binding Energy of the Deuteron.
ROBSON, J. M.: Phys. Rev. **83**, 349 (1951). SNELL, A. H.: Phys. Rev. **78**, 310 (1950). SPIVAK, P. E., *et al.*: Nucl. Phys. **10**, 395 (1958).	β-Decay of the Neutron.
ALVAREZ, L. W., and F. BLOCH: Phys. Rev. **57**, 111 (1940). COHEN, V. W., N. R. CORNGOLD, and N. F. RAMSEY: Phys. Rev. **104**, 283 (1956).	Magnetic Moment of the Neutron.
FERMI, E., and L. MARSHALL: Phys. Rev. **72**, 1139 (1947). SHAPIRO, I. S., and I. V. ESTULIN: Soviet Physics JETP **3**, 626 (1956).	Charge of the Neutron.
BURCHAM, W. E.: Nuclear Reactions, Levels, and Spectra of Light Nuclei. KINSEY, B. B.: Nuclear Reactions, Levels, and Spectra of Heavy Nuclei. RAINWATER, J.: Resonance Processes by Neutrons. In: Handbuch der Physik, Bd. XL. Berlin-Göttingen-Heidelberg: Springer 1957.	Discussion of Nuclear Reactions Involving Neutrons.
FESHBACH, H.: In: Nuclear Spectroscopy, part B, p. 1033. New York and London: Academic Press 1960. — C. E. PORTER, and V. F. WEISSKOPF: Phys. Rev. **96**, 448 (1954). SÜSSMANN, G., u. F. MEDINA: Fortschr. d. Physik **4**, 297 (1956).	Optical Model.
GOLDBERG, M. D., V. M. MAY and J. R. STEHN: Angular Distributions in Neutron-Induced Reactions, Vol. I and II. BNL-400, Second Edition (1962). HOWERTON, R. J.: Tabulated Differential Neutron Cross Sections, part I: UCRL-5226 (1956); part II: UCRL-5351 (1958); part III: UCRL-5573 (1961). HUGHES, D. J., and B. CARTER: Angular Distributions. BNL-400 (1958). —, and R. B. SCHWARTZ: Neutron Cross Sections. BNL-325 (1958) and Supplement 1960. PARKER, K.: AWRE 0-27/60 (1960); 0-71/60 (1960); 0-79/63 (1963) and 0-82/63 (1963). SCHMIDT, J. J.: KFK-120 (1963).	Tables and Curves of Cross Sections.

2. Neutron Sources

2.1. Neutron Production by Nuclear Reactions

Because of their short lifetime, free neutrons do not occur in nature and must be artificially produced. A simple method consists of separating them from nuclei in which they are particularly loosely bound. There is a variety of reactions which lead to neutron production. In such reactions compound nuclei excited[1] with the sum of the binding energy and the kinetic energy (in the center-of-mass system) of the projectiles first are formed by bombardment of target nuclei with α-particles, protons, deuterons, or γ-rays. If the excitation energy is larger than the binding energy of the "last neutron" in the compound nucleus, then a neutron is very likely to be emitted. The remaining excitation energy is distributed as kinetic energy between the neutron and the residual nucleus. The residual nucleus can remain excited and later return to the ground state by γ-emission.

Table 2.1.1. *The Binding Energy of the Last Neutron in Light Nuclei**

Nucleus	Binding Energy (Mev)	Nucleus	Binding Energy (Mev)
H^2	2.225	C^{12}	18.720
H^3	6.258	C^{13}	4.937
He^3	7.719	C^{14}	8.176
He^4	20.577	N^{13}	20.326
He^5	–0.956	N^{14}	10.553
Li^6	5.663	N^{15}	10.834
Li^7	7.253	N^{16}	2.500
Li^8	2.033	O^{15}	13.222
Be^8	18.896	O^{16}	15.669
Be^9	1.665	O^{17}	4.142
Be^{10}	6.814	O^{18}	8.047
B^9	18.575	F^{18}	9.141
B^{10}	8.440	F^{19}	10.442
B^{11}	11.456	F^{20}	6.599
C^{11}	13.092		

* The binding energies were calculated using the tables of EVERLING *et al.*, Nucl. Phys. **18**, 529 (1960).

In order to survey the possibilities for neutron production, let us consider the binding energy of the last neutron, i.e., the separation energy of a neutron, for various light nuclei. The values given in Table 2.1.1 are partly derived from mass differences, partly from the energetics of nuclear reactions.

It turns out that the separation energies of nuclei that can be thought of as composed of α-particles (He^4, Be^8, C^{12}, O^{16}) are particularly large; these nuclei are very stable (with the exception of Be^8, which is unstable against decay into two α-particles). Conversely, a neutron added to such a nucleus is particularly loosely bound. Beyond oxygen this behavior is less marked; for intermediate nuclei the separation energy averages 7—10 Mev and for heavy nuclei around 6—7 Mev. For neutron production, the light nuclei play the predominant role, since, because of the strong Coulomb barrier, reactions of charged particles with heavy nuclei first become likely at very high energies.

2.1.1. The Various Types of Reactions

(α, n) *Reactions*

$$ {}_zX^A + {}_2He^4 \to {}_{z+2}X^{A+3} + n + Q\,. $$

The Q-value can be >0 (exothermic reaction) or <0 (endothermic reaction).

[1] Many neutron-producing reactions proceed directly, i.e., without the formation of a compound nucleus. One important example of such a reaction is deuteron stripping. The results of our following energetic considerations also apply in this case.

Examples:

$$Be^9 + He^4 \rightarrow C^{12} + n + 5.704 \text{ Mev}$$

$$B^{11} + He^4 \rightarrow N^{14} + n + 0.158 \text{ Mev}$$

$$Li^7 + He^4 \rightarrow B^{10} + n - 2.790 \text{ Mev}.$$

The excitation energy of the nucleus resulting from α-particle capture is about 10 Mev; for this reason, (α, n) reactions, as can be seen by comparison with Table 2.1.1, are sometimes exothermic, sometimes endothermic.

(d, n) *Reactions*

$$_zX^A + {}_1H^2 \rightarrow {}_{z+1}X^{A+1} + n + Q.$$

Examples:

$$H^3 + H^2 \rightarrow He^4 + n + 17.588 \text{ Mev}$$

$$Li^7 + H^2 \rightarrow Be^8 + n + 15.028 \text{ Mev}$$

$$C^{12} + H^2 \rightarrow N^{13} + n - 0.282 \text{ Mev}.$$

Because of the small binding energy of the deuteron, a very highly excited compound nucleus is always formed by its capture; consequently, nearly all (d, n) reactions are exothermic.

(p, n) *Reactions*

$$_zX^A + {}_1H^1 \rightarrow {}_{z+1}X^A + n + Q.$$

Examples:

$$Li^7 + H^1 \rightarrow Be^7 + n - 1.646 \text{ Mev}$$

$$H^3 + H^1 \rightarrow He^3 + n - 0.764 \text{ Mev}.$$

In a (p, n) reaction, there is produced from a nucleus $_zX^A$ the same nucleus $_{z+1}X^A$ which would arise from β-decay of $_zX^A$. Let us assume that such a β-decay is possible and that the maximum β-energy is E_β. For the Q-value of the (p, n) reaction

$$Q = E_\beta - Q_n \tag{2.1.1}$$

evidently applies, where $Q_n = 0.782$ Mev is the Q-value of the β-decay of the neutron. Tritium, mentioned above, is a β-emitter with a maximum β-energy of 18 kev; thus $Q = 18 \text{ kev} - 782 \text{ kev} = -764$ kev. All (p, n) reactions on stable nuclei are thus endothermic; the threshold energy is at least 0.782 Mev.

(γ, n) *Reactions (Nuclear Photoeffect)*

$$_zX^A + \gamma \rightarrow {}_zX^{A-1} + n + Q.$$

Examples:

$$Be^9 + \gamma \rightarrow Be^8 + n - 1.666 \text{ Mev}$$

$$H^2 + \gamma \rightarrow H^1 + n - 2.225 \text{ Mev}.$$

2.1.2. Energetic Considerations

With the help of the energy and momentum theorems of classical mechanics, several important relations can be derived. It will first be shown that as a consequence of momentum conservation the threshold energy E_s of an endothermic reaction is larger than the Q-value: Let a projectile with mass m_G and velocity v ($E=\frac{1}{2}m_G v^2$) strike a stationary target nucleus. Then

$$\frac{m_G}{2}v^2=\frac{m_z}{2}v_z^2+Q \tag{2.1.2}$$

must hold in order that the reaction can occur (m_z = mass of the compound nucleus, v_z = velocity of the compound nucleus). According to the theorem of momentum conservation

$$m_G v = m_z v_z \tag{2.1.3a}$$

so that

$$v_z^2=\left(\frac{m_G}{m_z}\right)^2 v^2. \tag{2.1.3b}$$

Thus, according to Eq. (2.1.2),

$$\left.\begin{aligned}\frac{m_G}{2}v^2-\frac{m_z}{2}\left(\frac{m_G}{m_z}\right)^2 v^2&=Q\\ \frac{m_G}{2}v^2\left(1-\frac{m_G}{m_z}\right)&=Q\\ E_s&=\frac{Q}{1-\dfrac{m_G}{m_z}}.\end{aligned}\right\} \tag{2.1.4}$$

The threshold energy E_s for the (p, n) reaction on Li^7, for example, is 1.881 Mev.

Let us now consider a bit further the question of how the energy released in an exothermic reaction is distributed between the neutron and the residual nucleus. For this purpose we first consider a stationary compound nucleus which decays into a neutron of mass m_n and a residual nucleus of mass m_r. In this case,

$$\frac{m_n}{2}v_n^2+\frac{m_r}{2}v_r^2=Q \tag{2.1.5}$$

and

$$m_n v_n = m_r v_r. \tag{2.1.6}$$

Thus

$$\left.\begin{aligned}\frac{m_n}{2}v_n^2\left(1+\frac{m_n}{m_r}\right)&=Q\\ E_n=\frac{Q}{1+\dfrac{m_n}{m_r}}&=\frac{m_r}{m_z}Q.\end{aligned}\right\} \tag{2.1.7}$$

The energy of the neutron in the reaction $H^3(d, n)He^3$ (with zero deuteron energy) accordingly amounts to $\frac{4}{5}\times 17.588$ Mev ≈ 14.1 Mev.

If the primary particle has a non-negligible kinetic energy E — this is the usual case — the neutron energy is dependent on the angle of emission with respect to the direction of the primary particle. Let us calculate the neutron energy in the simple limiting cases of emission at 0° (forward direction) and 180°. The energy conservation theorem states:

$$\frac{m_G}{2}v^2+Q=E+Q=\frac{m_n}{2}v_n^2+\frac{m_r}{2}v_r^2. \tag{2.1.8}$$

The momentum conservation theorem states:

$$m_G v=\pm(m_n v_n - m_r v_r). \tag{2.1.9}$$

Here the plus sign holds for emission at 0° and the minus sign for emission at 180°. Thus it follows:

$$E_n = \frac{m_n}{2} v_n^2 = \frac{Q}{1+\frac{m_n}{m_r}} + E\frac{m_G m_n}{(m_r+m_n)^2}\left\{2+\frac{m_r(m_n+m_r)}{m_G m_n}\left[1-\frac{m_G}{m_r}\right]\right.$$
$$\left. \pm 2\sqrt{1+\frac{m_r(m_n+m_r)}{m_G m_n}\left[\frac{Q}{E}+\left(1-\frac{m_G}{m_r}\right)\right]}\right\}. \qquad (2.1.10)$$

As we shall see in Sec. 2.5.1, Eq. (2.1.10) is a special case of the general equation, Eq. (2.5.1), which holds for arbitrary angles of emission as well as for exothermic and endothermic reactions.

Let us now apply Eq. (2.1.10) to the reaction $H^3(d, n)He^4$: For a deuteron energy $E=0$, $E_n=14.1$ Mev in all directions. It follows from Eq. (2.1.10), however, that at $E=0.500$ Mev

$E_n=15.79$ Mev in the forward direction

$E_n=12.77$ Mev in the backward direction.

For the (α, n) reaction on beryllium with α-particles from radium C′ (7.680 Mev) one finds the neutron energies

$E_n=13.1$ Mev in the forward direction

$E_n=\ 7.7$ Mev in the backward direction.

The energies in all other directions always lie between these two extreme values.

In this derivation we have not calculated relativistically. Since for a neutron of 10 Mev, $v_n/c=0.14$, the errors caused by this neglect are small. Precision calculations, particularly at high energies, must be carried out relativistically. Special considerations regarding the (γ, n) reaction follow in Sec. 2.3.1.

2.1.3. Calculation of the Strength of Neutron Sources

The determination of the yield of neutron-producing nuclear reactions is carried out with the help of the cross sections introduced in Sec. 1.3.1. If a current J ($cm^{-2}\,sec^{-1}$) of protons, deuterons, or α-particles traverses a target substance which contains N identical atoms per cm^3, the number dQ of neutrons produced per cm^2 and sec in a layer of thickness dx is

$$dQ = J\sigma N\,dx. \qquad (2.1.11)$$

Here σ (barn) is the cross section for the neutron-producing nuclear reaction. In integrating Eq. (2.1.11) to determine the total yield, one must take into account, on one hand, that the cross sections of the neutron-producing nuclear reactions vary rapidly with the energy of the initiating particles, and, on the other hand, that the initiating particles are slowed down very rapidly by Coulomb interaction with the electrons in the target substance (ranges often only a few microns). The slowing down of charged particles in matter is characterized by the energy-dependent slowing-down power dE/dx (ev cm^{-1}). The following equation holds for the range R of a particle that strikes a target substance with the energy E_0:

$$R = \int_0^{E_0} \frac{dE}{dE/dx}. \qquad (2.1.12)$$

According to Eqs. (2.1.11) and (2.1.12), the total neutron source strength per cm² of target surface of a *thick* target due to bombardment by charged particles of energy E_0 is:

$$Q = J N \int_0^{E_0} \frac{\sigma(E)}{dE/dx}\, dE. \tag{2.1.13}$$

If one introduces the *yield* η $(=Q/J)$, i.e., the number of neutrons produced per primary particle, then one finally obtains

$$\eta = N \int_0^{E_0} \frac{\sigma(E)}{dE/dx}\, dE. \tag{2.1.14}$$

Thus in order to calculate the yield of a neutron source, not only the cross section of the neutron-producing reaction, but also the slowing-down power of the target substance, must be known[1]. If σ depends only weakly on the energy, it can be taken before the integral to give

$$\eta = N\sigma \int_0^{E_0} \frac{dE}{dE/dx} = N\sigma R = \frac{R}{\lambda} \tag{2.1.15}$$

where λ is the mean free path of the primary particle for a nuclear collision. Eq. (2.1.15) can also be used for yield estimates of strongly energy-dependent reactions, if a suitable mean value of σ is used.

For the production of monoenergetic neutrons, one generally uses *thin* targets, i.e., targets in which the energy loss of the primaries is small. If ΔE is the "thickness" of such a target, then

$$\eta = N \int_{E_0-\Delta E}^{E_0} \frac{\sigma(E)}{dE/dx}\, dE \approx N \frac{\sigma(E_0)\,\Delta E}{(dE/dx)_{E_0}} \tag{2.1.16}$$

holds for the yield due to bombardment with primaries of energy E_0.

2.2. Radioactive (α, n) Sources

2.2.1. The (Ra—Be) Source

A strong, isotropic neutron source which approximates a point source and which is nearly constant in time can be produced with the $Be^9(\alpha, n)C^{12}$ reaction if one uses the strong α-activity of natural radium. This neutron source is frequently found in laboratories working with weak neutron intensities and often serves as a standard neutron source. Some disadvantages of the (Ra—Be) source are the strong γ-radiation accompanying the neutrons and the inhomogeneity of the neutron energy. Fig. 2.2.1 shows the decay series of radium. In one gram of pure ${}_{88}Ra^{226}$, 3.7×10^{10} radium atoms decay per sec (1 curie). Each time an energetic α-particle is emitted. Within several months, pure Ra^{226} is in

[1] Frequently, instead of the slowing-down power, the "atomic slowing-down cross section" $\varepsilon = \frac{dE/dx}{N}$ (ev cm²) is specified; then, simply, $\eta = \int_0^{E_0} \frac{\sigma(E)}{\varepsilon}\, dE$. For numerical values of ε see Whaling.

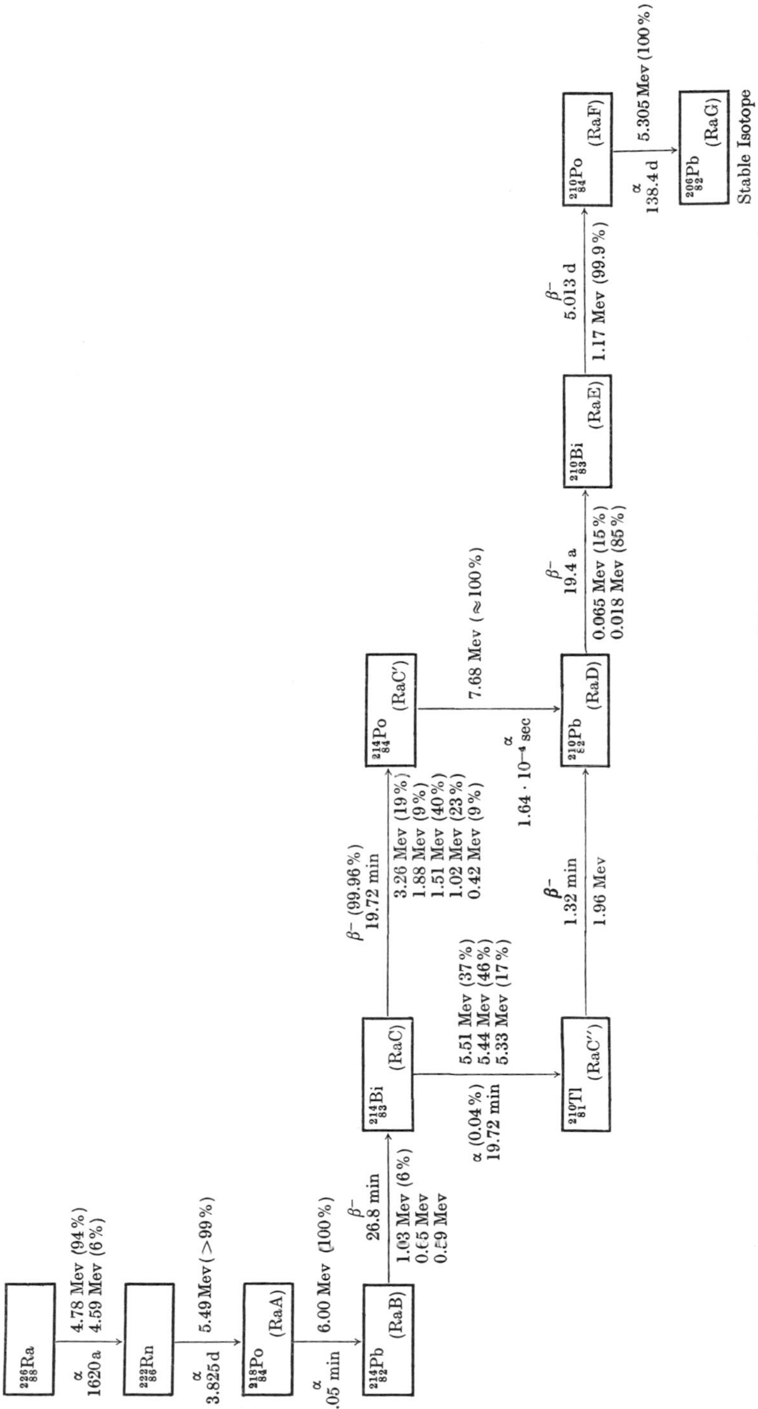

Fig. 2.2.1. The decay series of Radium

equilibrium with all its daughter substances up to but not including radium-D. Up to this time, a vanishingly small quantity of radium-E is formed. After a month, the α-radiation and therefore also the neutron intensity of a (Ra—Be) source becomes practically constant. Since among the daughter substances there are three other α-emitters, the α-activity increases to 4 curies altogether. The α-rays of the three daughter products are even more energetic than that of radium itself[1].

Table 2.2.1. *γ-Rays from Radium and Its Daughter Products*

Energy (Mev)	Yield (γ-Rays per α-decay in Ra^{226})	Energy of the Neutron from the $Be^9(\gamma, n)Be^8$-Reaction (kev)
1.690	0.0224	21
1.761	0.143	84
1.820	0.024	137
2.090	0.022	377
2.200	0.059	475
2.420	0.025	670

Among the daughter products are also found β-emitters, which leave behind highly excited nuclei that emit γ-rays. Table 2.2.1 gives the intensities and energies of the more energetic of these γ-rays according to LATYSHEV.

For $E_\gamma > 1.666$ Mev, the γ-rays can produce neutrons through the $Be^9(\gamma, n)Be^8$ reaction; this reaction contributes several % to the source strength of a (Ra—Be) preparation. The third column of Table 2.2.1 gives the energies of the individual photoneutron groups.

Fig. 2.2.2 shows a cross section through a typical (Ra—Be) source. The actual source body consists of a mixture of radium bromide and beryllium powder which is compressed under high pressure. It is carefully soldered into a shell of brass or nickel, which is usually enclosed in a second shell for safety reasons (radon). Radium and beryllium are mixed as a rule in a weight ratio of 1:5; sources with as much as 5 grams of radium have been manufactured. Great care is a prerequisite for the manufacture of a (Ra—Be) source whose strength is to be quite constant in time.

Fig. 2.2.2. Cross section through a typical (Ra—Be) source

The yield of a (Ra—Be) source can be estimated with the help of Eq. (2.1.15). Since the weight ratio of radium to beryllium is 1:5, the atomic ratio is 1:125, and it can be assumed that the path of an α-particle traverses beryllium only. The range of a 5-Mev α-particle in Be of density 1.75 g/cm³ (the average density of the compressed bodies) is about 30 microns. Fig. 2.2.3 shows the cross section for the (α, n) reaction in beryllium as a function of the α-energy according to HALPERN. From this curve, we obtain a mean value for all four α-radiating

[1] The increase in the neutron intensity of a freshly made neutron source can be described for the first few weeks approximately by

$$Q = Q_0 \frac{1 + 5(1 - e^{-t/3.8})}{6}$$

where t is the time in days elapsed since the closure of the source. The neutron intensity increases to about six times the initial value, although the α-activity only increases about four times. The reason for this is that the α-decay of the daughter products is more energetic than that of the radium and the cross section of the (α, n) reaction increases sharply with energy.

components of the source of 300 mbarn. For Be of density 1.75 g/cm³ there then results $\lambda = 30$ cm and thus

$$\left.\begin{aligned}\eta &= \frac{R}{\lambda}\\ &= \frac{3\times10^{-3}}{30} = 10^{-4}.\end{aligned}\right\} \quad (2.2.1)$$

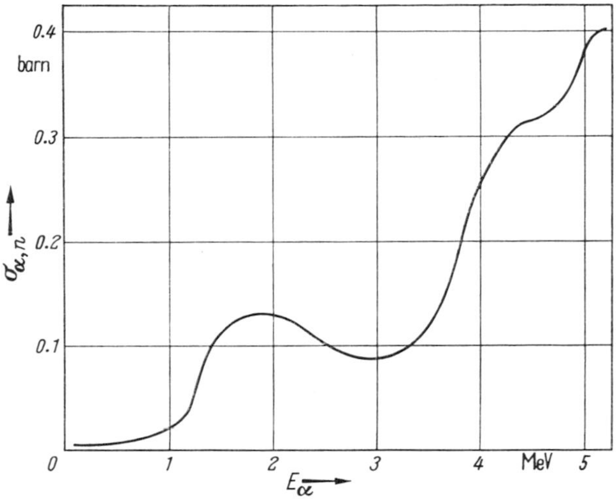

Fig. 2.2.3. The cross section of the $Be^9(\alpha, n)C^{12}$ reaction as a function of α-particle energy

1 gram of radium in equilibrium with its daughter substances excluding RaD has an α-activity of 4 curies and therefore produces

$$4\times3.7\times10^{10}\times\\ \times10^{-4}\text{ neutrons/sec}\\ = 1.5\times10^{7}\text{ neutrons/sec}.$$

The actually observed source strengths of (Ra—Be) sources ($1.2-1.7\times10^7$ n/sec/g Ra^{226}) are of this order of magnitude. The methodology of absolute source strength measurements is discussed in Sec. 14.3.

Fig. 2.2.4 shows the spectrum of a (Ra—Be) source according to the measurements of various authors as well as a calculated energy distribution due to Hess. The energy distribution varies continuously up to a maximum energy of 13.1 Mev, corresponding to the largest α-energy (RaC′) of 7.68 Mev. The most probable energy is nearly 4 Mev; the average energy is about 5 Mev.

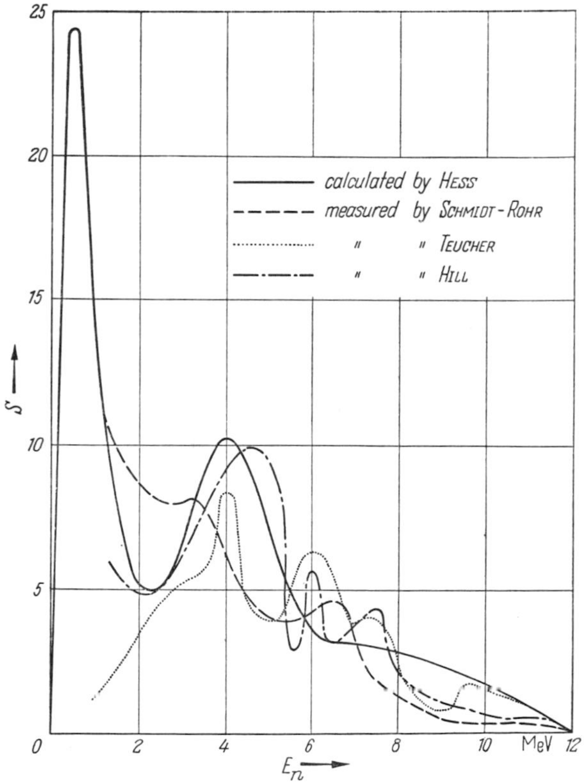

Fig. 2.2.4. The neutron energy spectrum of a (Ra—Be) source (from various authors)

The continuous variation of the spectrum can be explained on four grounds:

1. As a result of the continuous slowing down of the α-particles in beryllium, (α, n) reactions occur at all energies between 0 and 7.68 Mev.

2. To each α-energy belong various neutron energies because of the different possible directions of emission.

3. The C^{12}-nucleus can be left in an excited state (the first two excited states of C^{12} lie at 4.43 and 7.65 Mev). This explains the large number of neutrons with energies smaller than 5.4 Mev, the kinetic energy of the neutrons produced by zero energy α-particles when the C^{12}-nucleus is left in the ground state.

4. Low energy neutrons can be produced by the photoeffect in beryllium and by the reactions

$$Be^9 + He^4 \rightarrow He^4 + Be^8 + n - 1.665 \text{ Mev}$$

$$Be^9 + He^4 \rightarrow 3\,He^4 + n - 1.571 \text{ Mev}.$$

Furthermore, it can easily be seen that the various experimental results deviate sharply from one another and also do not fit the spectrum calculated by Hess. In the energy range under 1 Mev, where according to Hess a clear maximum should occur, measurements could hitherto not be carried out.

2.2.2. Other Neutron Sources of the (α, n) Type

Po^{210} (RaF) is a well-known α-emitter, which has a 138.5-day half-life and an α-energy of 5.305 Mev. RaF can either be separated as a daughter product of radium, or it can be produced by the irradiation of Bi^{209} with neutrons in a nuclear reactor:

$$Bi^{209}(n, \gamma)Bi^{210} \xrightarrow[5d]{\beta^-} Po^{210}.$$

Po^{210} has the great advantage that it does not emit β- or γ-rays. Its short lifetime is disturbing. About 2.5×10^6 neutrons/sec per curie of Po^{210} can be obtained from a (Po—Be) source. Fig. 2.2.5 shows the energy spectrum of (Po—Be) neutrons.

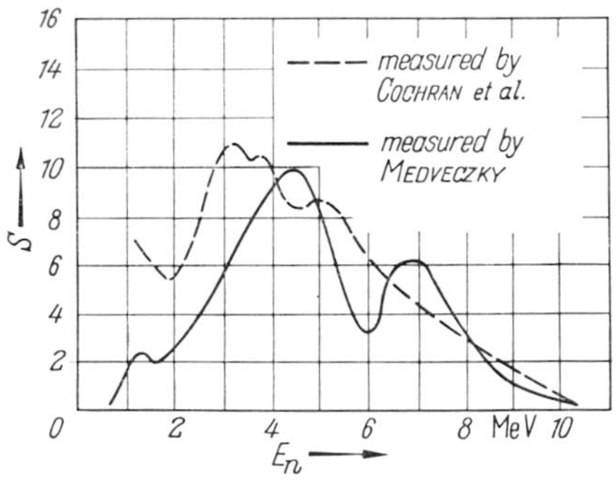

Fig. 2.2.5. The neutron energy spectrum of a (Po—Be) source

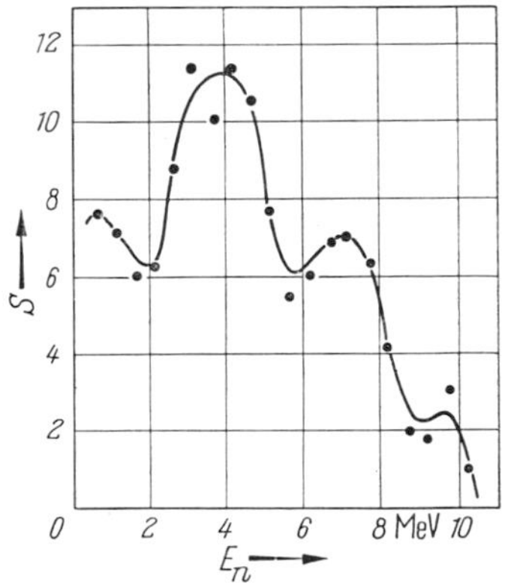

Fig. 2.2.6. The neutron energy spectrum of a (Pu—Be) source

The average neutron energy is about 4 Mev; the maximum energy is somewhat less than in the case of the (Ra—Be) source because of the smaller α-energy. In a (Po—Be) source, numerous neutrons are also produced in reactions which leave the C^{12}-nucleus in an excited state. Since an excited state in C^{12} decays in less than 10^{-15} sec with emission of one or two γ-rays, in many cases a γ-ray is emitted nearly simultaneously with a neutron; since no other γ-radiation is present in Po^{210}, it is easy to demonstrate neutron-γ coincidences. This possibility has occasionally been used for time-of-flight measurements.

Recently, plutonium-239, which has become available in large quantities because of its technological significance, has been used for the manufacture of neutron sources. Pu^{239} decays with a half-life of 24360 years; the α-energies are 5.15, 5.13, and 5.10 Mev. Plutonium can be alloyed with beryllium to make $PuBe_{13}$. Such a (Pu—Be) neutron source emits 8.5×10^4 n/sec/g of plutonium, according to RUNNALLS and BOUCHER. Fig. 2.2.6 shows the spectrum of (Pu—Be) neutrons observed by STEWART. Among the advantages of (Pu—Be) sources are these: (i) because of the existence of a plutonium-beryllium alloy, the sources are highly reproducible; (ii) they emit no penetrating γ-radiation; and (iii) they have a very long lifetime. Disturbing for many applications is the small specific neutron emission and the fact that in a neutron field the source strength can change as a result of fission in the Pu^{239}.

Table 2.2.2. (α, n) *Reactions on Light Nuclei*

Target	Q-Value (Mev)	Yield: Neutrons per 10^6 α-Particles from Po^{210} (Thick Targets)[1]
Li^6	—3.977	0
Li^7	—2.790	2.6
Be^9	5.704	80
B^{10}	1.061	13
B^{11}	0.158	26
C^{13}	2.215	10
O^{17}	0.589	
O^{18}	—0.700	29
F^{19}	—1.949	12
Na^{23}	—2.971	1.5
Mg^{24}	—7.192	0
Mg^{25}	2.655	6.1
Mg^{26}	0.036	
Al^{27}	—2.652	0.74

[1] Data of ROBERTS.

Table 2.2.2 contains useful information concerning a series of further (α, n) reactions which can be used for radioactive neutron sources.

2.3. Radioactive (γ, n) Sources

In contrast to the radioactive (α, n) sources, which emit a continuous spectrum of neutrons, nearly monoenergetic neutrons can be produced in photoneutron sources by use of monochromatic γ-rays. Since the γ-energy of radioactive substances rarely exceeds 3 Mev, only the (γ, n) reactions in beryllium ($Q=-1.665$ Mev) and deuterium ($Q=-2.225$ Mev) come into consideration; various natural and artificial radioactive isotopes serve as γ-emitters. Disadvantages of photoneutron sources are their small yield and the usually short half-life of the γ-emitters. Because of their strong, penetrating γ-radiation, work with large photoneutron sources requires special protective measures.

2.3.1. Energetic Consideration of the (γ, n) Reaction

First, it will be shown that also in a (γ, n) reaction the threshold energy E_s is greater than the Q-value. Let a γ-ray of energy E hit a stationary target nucleus. Then

$$E=E_s=\frac{m_t}{2}v_t^2+Q \tag{2.3.1}$$

must hold (energy conservation theorem), if the reaction is to occur at all. According to the momentum conservation theorem

$$\frac{E_s}{c}=m_t v_t \tag{2.3.2}$$

(c = the velocity of light). It then follows that

$$E_s=|Q|+\frac{E_s^2}{2m_t c^2}\approx|Q|\left[1+\frac{|Q|}{1862\text{ Mev}\cdot A}\right]. \tag{2.3.3}$$

Here A is the mass number of the target nucleus; the numerical factor arises from the conversion of mass to energy. Thus, E_s is always greater than Q, but the difference is negligibly small. For deuterium, for example, $E_s = 1.0006 \times |Q|$.

Let us consider, furthermore, what energies the neutrons emitted in the forward and backward directions in a (γ, n) reaction have. If we denote the mass and velocity of the neutron by m_n and v_n and the mass and velocity of the residual nucleus by m_r and v_r, then according to the energy conservation theorem

$$E_n = E - |Q| - \frac{m_r}{2} v_r^2 \tag{2.3.4}$$

and according to the momentum conservation theorem

$$\frac{E}{c} = \pm (m_n v_n - m_r v_r). \tag{2.3.5}$$

Here the plus sign holds when the neutron is emitted at 0° to the direction of the incident γ-ray and the minus sign when the neutron is emitted at an angle of 180°. If terms of the order of $E/\big((m_n + m_r)c^2\big)$ are neglected, the calculation yields

$$E_n = \frac{A-1}{A} \left[E - |Q|\right] \pm E \cdot \sqrt{\frac{2(A-1)(E-|Q|)}{931 \text{ Mev} \cdot A^3}}. \tag{2.3.6}$$

Since all emission angles occur in the photoneutron sources used in practice, Eq. (2.3.6) asserts that even with the use of monochromatic γ-rays the neutron spectrum is not strictly monochromatic. According to Eq. (2.3.6) the energy spread is

$$\left.\begin{aligned} \Delta E_n &= 2E \times \\ &\times \sqrt{\frac{2(A-1)(E-|Q|)}{931 \text{ Mev} \cdot A^3}}. \end{aligned}\right\} \tag{2.3.7}$$

By using the γ-radiation of Na^{24} ($E = 2.757$ Mev) one has, for example, $E_n = 265 \pm 33$ kev for deuterium and $E_n = 969 \pm 5.1$ kev for beryllium.

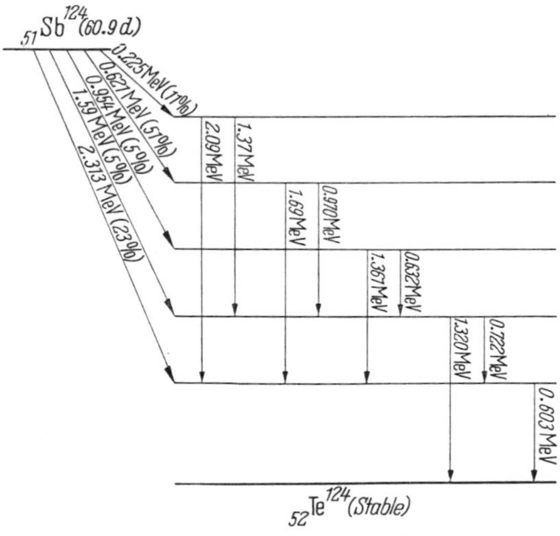

Fig. 2.3.1. Decay scheme of Sb^{124}

2.3.2. The (Sb—Be) Source

Fig. 2.3.1 shows the decay scheme of antimony-124, which is produced by neutron irradiation of antimony-123. 48% of all β-decays lead to an excited state of Te^{124}, which decays by emitting a 1.692-Mev γ-ray. Natural antimony consists of 42.75% Sb^{123} and 57.25% Sb^{121}. The cross section of Sb^{123} for the formation of the 60.9-day activity is 2.5 barn at $v_0 = 2200$ m/sec[1].

[1] By irradiation to saturation in a thermal neutron flux of 10^{13} cm^{-2} sec^{-1}, one obtains an activity of the 1.692-Mev γ-ray of about 2 curies/g of natural antimony.

Fig. 2.3.2 shows a cross section through a typical (Sb—Be) neutron source. The inner antimony cylinder can be taken out of the beryllium mantle. Thus it is possible to turn the source on and off at will. From such a source, up to 10^7 neutrons per sec and per curie of the 1.692-Mev activity of antimony can be obtained. It is the most widely used photoneutron source. The short 60.9-day half-life is disturbing; however, it is always possible to "charge" the source up again by irradiation in a nuclear reactor. From Eq. (2.3.6) it follows that with $E = 1.692$ Mev the neutron energy E_n is 26 ± 1.5 kev. Experimentally, $E_n = 24 \pm 2.2$ kev was found (SCHMITT).

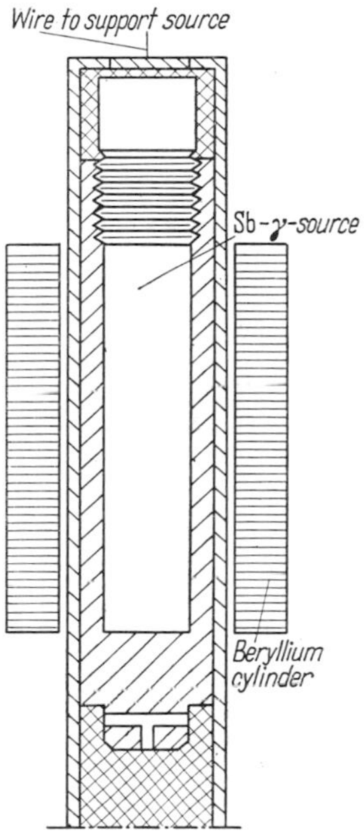

Fig. 2.3.2. (Sb—Be) source (WATTENBERG)

2.3.3. Other Photoneutron Sources

Table 2.3.1 contains useful data on other photoneutron sources due to HANSON[1]. The values given in the column marked Y for yield refer to a standard arrangement in which 1 gram of D_2O or beryllium is located at a distance of 1 cm from 1 curie of the particular γ-radiator. The calculation of the yield of a practical source is carried out with the equation

$$Q = 4\pi \varrho t Y \qquad (2.3.8)$$

where ϱ is the density and t the effective thickness of the surrounding D_2O or Be mantle. Among the natural γ-radiators are mesothorium (Ra^{228}, half-life 6.7 y,

Table 2.3.1. *Photoneutron Sources*

γ-Emitter	$T_{\frac{1}{2}}$	γ-Energy (Mev)	γ-Rays per Decay	Target	$\bar{E}_n$ in kev from Eq. (2.3.6)	Measured Values of E_n in kev	Yield $Y \times 10^{-4}$
Na^{24}	15.0 h	2.757	1.00	Be	969	830	24—29
		2.757	1.00	D_2O	265	220	12; 14
Al^{28}	2.27 min	1.782	1.00	Be	103		
Cl^{38}	37.29 min	2.15	0.47	Be	430		
Mn^{56}	2.576 h	1.77	0.30	Be	93	150	
		2.06	0.20	Be	350	300	2.9
		2.88	0.01	Be	1076		
		2.88	0.01	D_2O	350	220	0.31
Ga^{72}	14.1 h	1.87	0.08	Be	181		5.9
		2.21	0.33	Be	484		3.7
		2.51	0.26	Be	750		
		2.51	0.26	D_2O	140	130	4.6; 6.9
As^{76}	26.7 h	1.77	0.2	Be	93		
		2.06	0.1	Be	350		
Y^{88}	104 d	1.853	0.995	Be	166	160	10
		2.76	0.005	Be	972		13
		2.76	0.005	D_2O	265		0.3
In^{116}	54 min	2.090	0.25	Be	377	300	0.82
La^{140}	40.2 h	2.51	0.04	Be	747	620	0.23; 0.34
		2.51	0.04	D_2O	140	130	0.68; 0.97
Pr^{144}	17.3 min	2.185	0.02	Be	462		0.08

[1] See MARION and FOWLER, p. 32.

γ-energies: 1.8, 2.2, and 2.6 Mev) and the daughter products of radium; the contribution of the (γ, n) reaction to the neutron yield of a (Ra—Be) source has already been explained in Sec. 2.2.1. A pure $Ra(\gamma,n)Be$ neutron source is a useful standard source, since it is very stable and reproducible and has the long half-life of radium.

2.4. Fission Neutron Sources

Some of the heaviest atomic nuclei decay by spontaneous fission. Since in fission several neutrons are always produced, these heavy nuclei can be used as neutron sources. Table 2.4.1 contains useful data concerning some of the transuranics. With the exception of californium-252, all of the nuclei are strong α-emitters; the neutron spectra of such sources show contributions from neutrons which are produced by (α, n) reactions in impurities. Furthermore, the neutron spectrum of such sources resembles the fission neutron spectrum of U^{235} (see Fig. 2.6.1). In practice, californium-252 and plutonium-240 sources are particularly widely used. The source strength of these sources also increases when they are exposed to neutron irradiation, owing to additional neutron-induced fission.

Table 2.4.1. *Spontaneous Fission Neutron Sources*

Nucleus	Half-Life	α-Particles per Fission	Neutrons per Fission	Neutrons per mg per sec[1]
Pu^{236}	2.85 y	1.3×10^{9}	1.9	26
Pu^{238}	89.4 y	5.5×10^{8}	2.0	2.2
Pu^{240}	6600 y	1.9×10^{7}	2.1	1.1
Pu^{242}	3.79×10^{5} y	1.9×10^{5}	2.3	1.7
Cm^{242}	162.5 d	1.6×10^{7}	2.3	1.7×10^{4}
Cm^{244}	18.4 y	7.6×10^{5}	2.6	9×10^{3}
Cf^{252}	2.6 y	—	3.5	2.7×10^{9}

[1] Without (α, n) neutrons, whose yield depends on the purity of the source.

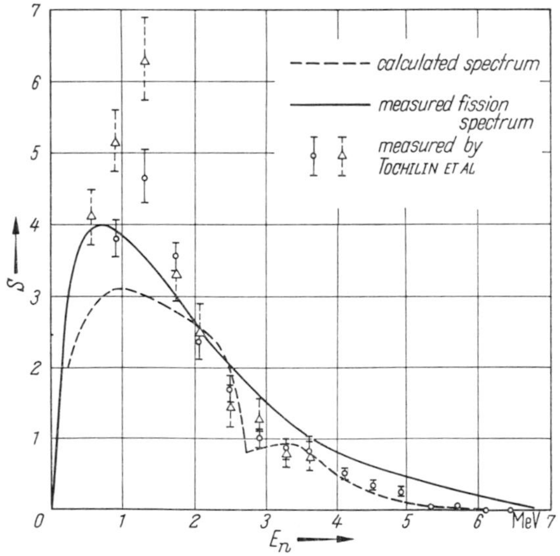

Fig. 2.4.1. The neutron energy spectrum of a mock fission source (from MARION and FOWLER, loc. cit.)

It is possible to manufacture radioactive (α, n) sources whose neutron spectrum resembles the fission neutron spectrum. The (α, n) reaction in F^{19} yields neutrons with an average energy of about 2 Mev, which is close to the average energy of fission neutrons. According to TOCHILIN and ALVES, a *mock fission source* can be prepared from a mixture of polonium, fluorine, and boron, with small additional quantities of lithium and beryllium. Fig. 2.4.1 shows the neutron spectrum of such a source.

2.5. Neutron Production with Artificially Accelerated Particles

As neutron sources, accelerators have advantages over radioactive preparations. The achievable intensity lies many orders of magnitude above that of the radioactive sources. Monoenergetic neutrons of practically any desired energy can be produced with good resolution. Finally, the source can be pulsed by interrupting the beam, so that time-of-flight measurements become possible. According to which of these three considerations stands in the foreground of interest, different kinds of accelerator facilities can be set up:

(a) Monoenergetic neutrons can be produced in (p, n) and (d, n) reactions with accelerated protons and deuterons in the *van de Graaff generator*. Occasionally, cyclotrons with variable and particularly homogeneous energy have been used for this purpose.

(b) The (d, n) reaction is already quite productive at low deuteron energy (<1 Mev) in deuterium, tritium, lithium, and beryllium. In combination with *simple low-energy accelerators* which produce relatively high currents, these reactions result in strong continuous or pulsed neutron sources.

(c) Extremely strong pulsed neutron sources can be obtained if the bremsstrahlung from *electron linear accelerators* is used to produce (γ, n) reactions. Such sources are mainly used for *time-of-flight spectrometers*. Occasionally, (p, n) and (d, n) neutrons from cyclotrons and synchrocyclotrons are used for this purpose.

2.5.1. Monoenergetic Neutrons from (p, n) Reactions

Since for the method of production now to be described, the dependence of the neutrons produced in a nuclear reaction on the direction of emission plays a great role, Eq. (2.1.10) will first be generalized to arbitrary angles ϑ of emission and to endothermic reactions. If we denote the energy and mass of the emitted neutron by E_n and m_n, the energy and mass of the projectile nucleus by E and m_G, and the masses of the target and residual nuclei by m_t and m_r ($m_r \approx m_t + m_G - m_n$), then

$$E_n = E \frac{m_G m_n}{(m_n + m_r)^2} \left\{ 2\cos^2\vartheta + \frac{m_r(m_r + m_n)}{m_G m_n} \left[\frac{Q}{E} + \left(1 - \frac{m_G}{m_r}\right)\right] \pm 2\cos\vartheta \sqrt{\cos^2\vartheta + \frac{m_r(m_r+m_n)}{m_G m_n}\left[\frac{Q}{E} + \left(1 - \frac{m_G}{m_r}\right)\right]} \right\}. \tag{2.5.1}$$

For endothermic reactions, $Q < 0$ and the reaction begins at the threshold energy

$$E_s = |Q| \frac{m_G + m_t}{m_t}. \tag{2.5.2}$$

[cf. Eq. (2.1.4)]. At this primary energy, neutrons of zero energy in the center-of-mass system are produced; in the laboratory system they move in the forward direction with the velocity of the center of mass. Their energy is

$$E_{ns} = E_s \frac{m_G m_n}{(m_n + m_r)^2}. \tag{2.5.3}$$

With increasing primary energy the neutrons are emitted in a cone around the forward direction whose apex angle is given by

$$\cos\vartheta_0 = \left| \sqrt{\frac{m_r(m_n + m_r)}{m_G m_n}\left[\frac{|Q|}{E} - \left(1 - \frac{m_G}{m_r}\right)\right]} \right|. \tag{2.5.4}$$

Inside this cone two neutron energies, corresponding to the plus and minus sign in Eq. (2.5.1), belong to each direction ϑ. With further increase of the primary energy, the cone widens; the energy of one of the neutron groups increases, that of the other decreases[1]. The energy of the second group reaches the value zero at a primary energy

$$E'_s = \frac{m_r}{m_t - m_n} |Q| . \tag{2.5.5}$$

The apex angle is $\vartheta_0 = 90°$ at this primary energy; this means that neutrons are emitted in the entire forward hemisphere. Further increase in the primary energy causes neutrons to be emitted in all directions. Then only the plus sign holds in Eq. (2.5.1), i.e., the relation between primary energy, angle, and neutron energy becomes unique. According to Eq. (2.5.5), E'_s represents the threshold energy for the production of monoenergetic neutrons. It is usual to represent the somewhat involved content of Eq. (2.5.1) with the help of a simple nomogram (Figs. 2.5.1 and 2.5.2).

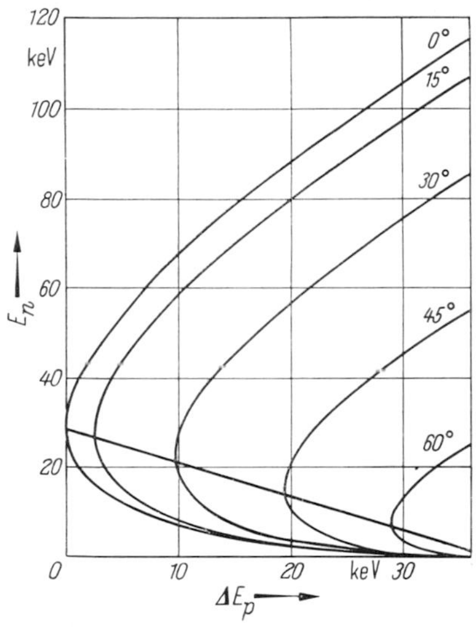

Fig. 2.5.1. The relation between the proton energy and the neutron energy and emission angle close to the $Li^7(p, n)Be^7$ threshold

The most frequently used (p, n) reaction for the production of monoenergetic neutrons is

$$Li^7 + p \rightarrow Be^7 + n - 1.646 \text{ Mev}.$$

The threshold energy E_s for this reaction is 1.881 Mev. At this proton energy, neutrons are emitted in the forward direction with $E_{ns} = 30$ kev. Fig. 2.5.1 shows the relation between proton energy ($\Delta E_p = E - E_s$), angle of emission, and neutron energy in the primary energy range $E_s < E < E'_s$ according

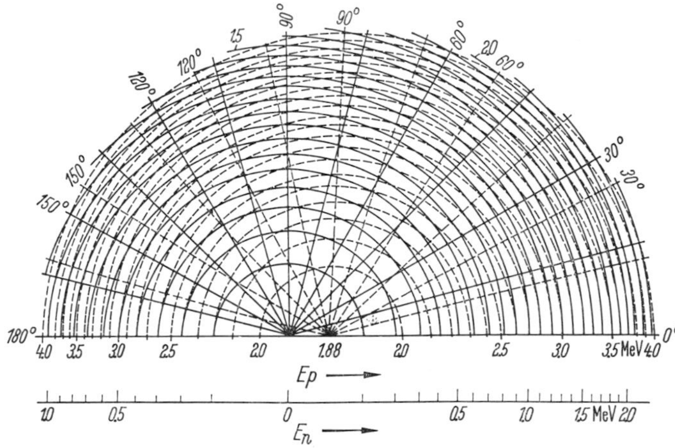

Fig. 2.5.2. McKibben nomogram of the $Li^7(p, n)Be^7$ reaction

[1] The existence of the second neutron group is connected with the fact that in the neighborhood of the threshold the neutrons emitted in the backward direction in the center-of-mass system also appear in the forward direction in the laboratory system. One can see this easily, if one derives Eq. (2.5.1) with the help of the transformation from the center-of-mass system to the laboratory system. In this connection, see, for example, AMALDI, p. 90ff.

to HANNA. The threshold energy E_s' for the emission of monoenergetic neutrons is 1.920 Mev; at this proton energy, the neutrons emitted in the forward direction have an energy of 120 kev. If the proton energy is further increased, the neutron energy increases corresponding to Eq. (2.5.1); at a proton energy of 2.378 Mev (a neutron energy in the forward direction of 652 kev) a reaction

$$\mathrm{Li}^7 + p \rightarrow \mathrm{Be}^{7*} + n - 2.076\ \mathrm{Mev}$$
$$\searrow$$
$$\mathrm{Be}^7 + \gamma + 430\ \mathrm{kev}$$

becomes possible, i.e., the Be^7 nucleus can be left in an excited state. At higher energy, a low energy neutron group corresponding to this reaction appears whose intensity gradually increases to 10% of that of the main group (at $E = 5$ Mev). Therefore, one can produce monoenergetic neutrons in the forward direction in the energy range 120—650 kev by means of the $\mathrm{Li}^7(p, n)\mathrm{Be}^7$ reaction. Monoenergetic neutrons of lower energies appear in other directions[1]; for a given primary energy $> E_s'$, E_n decreases continuously with increasing ϑ according to Eq. (2.5.1). The neutron yield also falls off sharply with angle, so that measurements in the backward direction pose considerable difficulties. Detailed tables of the dependence of the neutron energy on ϑ and E have been given by GIBBONS and NEWSON[2]. Fig. 2.5.2 shows the connection between the proton and neutron energies in the form of a nomogram (MCKIBBEN).

The energy homogeneity of the (p, n) neutrons emitted at a given angle is limited by the energy inhomogeneity of the primaries. With a well-stabilized van de Graaff generator, ion beams can be produced whose energy fluctuates by less than 1 kev; by the use of special energy filters, this value can be decreased to 250 ev. In order not to introduce any additional energy inhomogeneity due to slowing down of the primaries, thin targets in which the energy loss of the primaries is small are used. Such targets can be made by careful evaporation of lithium metal or lithium fluoride onto tantalum or platinum backing. They must be carefully cooled under proton bombardment. A good vacuum ($\approx 10^{-6}$ mm Hg) in the target chamber of the van de Graaff generator is a prerequisite for the use of such targets. The condensation of pumping oils, which leads to the formation of carbon deposits, is particularly disturbing. This holds for all targets, particularly for the tritium-zirconium targets to be discussed later.

The calculation of the yield of the $\mathrm{Li}^7(p, n)\mathrm{Be}^7$ reaction can be carried out according to the formulae of Sec. 2.1.3. Data on the energy-dependent cross section and the slowing-down power of Li-metal can be found in MARION and FOWLER. Fig. 2.5.3 shows the neutron yield in the forward direction (per steradian and μcoulomb of protons) for a 40-kev-thick, lithium-metal target as a function of the proton energy according to HANSON, TASCHEK, and WILLIAMS.

A much used (p, n) reaction is

$$\mathrm{H}^3 + p \rightarrow \mathrm{He}^3 + n - 0.764\ \mathrm{Mev}.$$

The threshold energy E_s is 1.019 Mev; the energy of the neutrons emitted in the forward direction at the threshold is 63.7 kev. Above a proton energy E_s' of

[1] With the exception of the 30-kev group, which appears exactly at the threshold.

[2] See MARION and FOWLER, p. 136ff.

1.148 Mev (neutron energy: 286.5 kev in the forward direction), the relation between angle and neutron energy becomes single-valued. He^3 has no excited states; for this reason the $H^3(p, n)He^3$ reaction makes possible the production of neutrons over a wide energy range (up to about 4 Mev). At proton energies above 5 Mev, secondary processes (like $H^3+p \rightarrow H^2+H^2$) come into play. Detailed tables of reaction kinematics can be found in FOWLER and BROLLEY. As targets one frequently uses the so-called gas targets, that is, chambers filled with tritium gas and sealed with a thin window facing the accelerator. Extremely thin aluminum, nickel, or molybdenum foils (1—10 mg/cm^2) are used as window material; they can support tritium pressures up to 1 atmosphere with currents up to several μamp. The energy loss of the primaries in such a window is actually quite considerable ($\approx$100 kev); however, the additional energy inhomogeneity it causes is small. The use of tritium-zirconium or tritium-titanium targets, which we shall describe in Sec. 2.5.2, is quite usual.

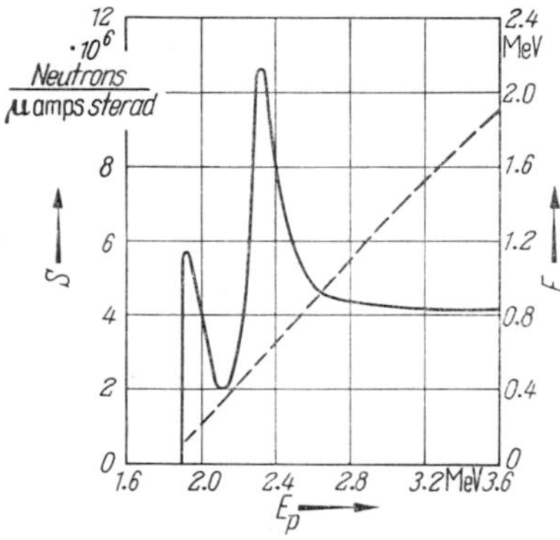

Fig. 2.5.3. The neutron yield of a 40 kev-thick Lithium metal target in the forward direction as a function of proton energy. The dotted line shows the corresponding neutron energy (right scale)

Occasionally the reactions

$$Sc^{45}+p \rightarrow Ti^{45}+n-2.840 \text{ Mev}$$

$$V^{51}+p \rightarrow Cr^{51}+n-1.536 \text{ Mev}$$

are used for neutron production. They offer the possibility of producing relatively low-energy neutrons in the forward direction [for scandium $E_n(0°)=5.49$ kev at $E_s'=2.904$ Mev; for vanadium $E_n(0°)=2.3$ kev at $E_s'=1.567$ Mev]; the neutron yield however is very small.

2.5.2. (*d*, *n*) Neutron Sources

The reactions

$$H^2+H^2 \rightarrow He^3+n+3.265 \text{ Mev}$$

and

$$H^2+H^3 \rightarrow He^4+n+17.588 \text{ Mev}$$

frequently serve for the production of monoenergetic neutrons of high energy. The calculation of the dependence of the neutron energy on the primary energy is done with Eq. (2.5.1); since $Q>0$, only the plus sign can stand before the radical and the relation between E, E_n, and ϑ is always unique. Fig. 2.5.4 shows the connection between E, E_n, and ϑ for both reactions; the $H^2(d, n)He^3$ reaction is suitable for neutron production in the range 2—10 Mev and the $H^3(d, n)He^4$ reaction from 12 to over 20 Mev. Exact tables can be found in FOWLER and BROLLEY. Gas targets and very frequently tritium- (or deuterium-) titanium (or zirconium) targets are used as thin targets. Such targets are manufactured as follows: A thin layer of zirconium or titanium is evaporated onto a base of copper, silver, or tungsten. Then, after careful out-gassing, the layer is loaded with gas by slow cooling in a tritium or deuterium atmosphere. By this process as many as 1.5 tritium or deuterium atoms can be adsorbed per zirconium or titanium atom. Such targets are very productive and with good cooling can stand high ion currents.

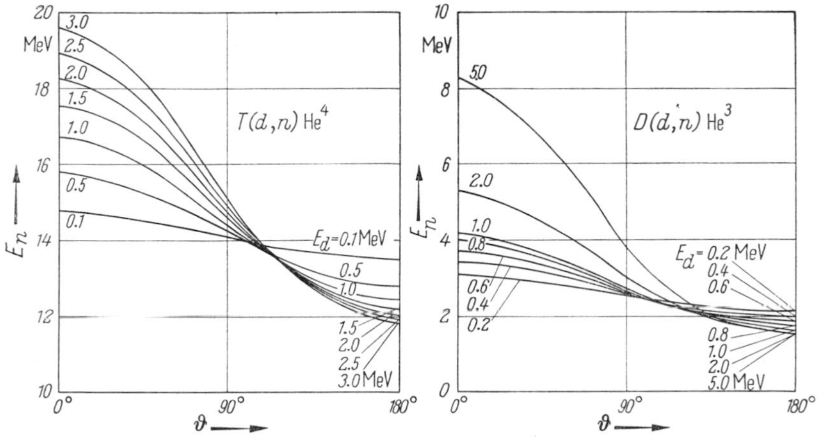

Fig. 2.5.4. The relation between deuteron energy and the neutron energy and emission angle for the reactions $H^3(d, n)He^4$ and $H^2(d, n)He^3$ (from AMALDI, loc. cit.)

Thin deuterium targets can also be manufactured in the form of heavy ice targets; to do this D_2O vapor is condensed on a base cooled with liquid air. Fig. 2.5.5 shows the cross sections for the (d, n) reaction on H^2 and H^3 as well as the slowing-down power of a tritium-titanium target as functions of the deuteron energy; these data allow the calculation of the yields of thin targets with Eq. (2.1.16)[1].

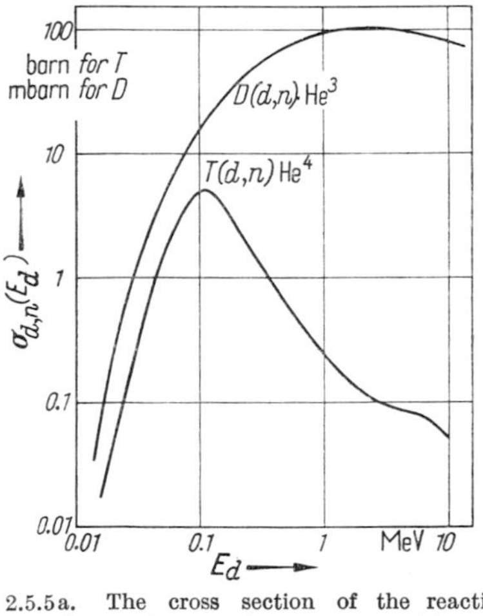

Fig. 2.5.5a. The cross section of the reactions $H^2(d, n)He^3$ and $H^3(d, n)He^4$ as a function of deuteron energy

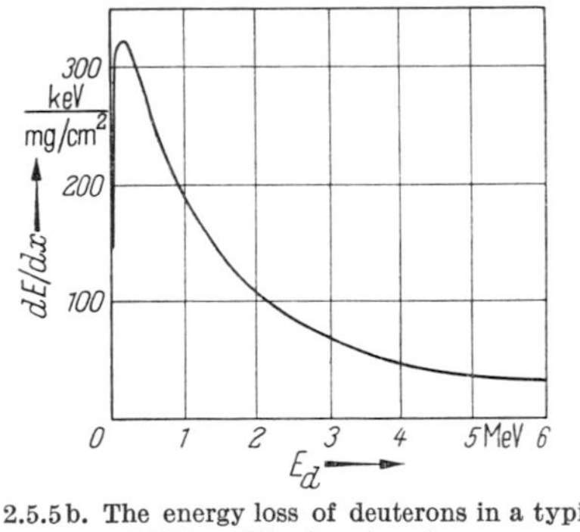

Fig. 2.5.5b. The energy loss of deuterons in a typical Tritium-Titanium target

When the energy homogeneity of the neutrons is not important, thick targets, whose thickness is larger than the range of the primary particle, are used. For the $H^3(d, n)He^4$ reaction thick tritium-zirconium or tritium-titanium targets are used, and for the $H^2(d, n)He^3$ reaction thick heavy ice, deuterium-zirconium, or

[1] If it is desired to calculate the angular dependence of the neutron intensity, the angular dependence of the cross sections is necessary. In this regard, see FOWLER and BROLLEY.

deuterium-titanium targets are used. Frequently, self-targets are used, i.e., metal foils are loaded with deuterium by the deuteron beam of the accelerator[1].

Fig. 2.5.6 shows the yield of various (d, n) reactions on thick targets as a function of the deuteron energy. The $H^2(d, n)He^3$ reaction and particularly the $H^3(d, n)He^4$ reaction are very productive even for very small deuteron energies. Thus about 10^{11} neutrons per sec can be obtained by bombardment of a fresh tritium-titanium target with a 0.5-mamp current of 300-kev deuterons[2]. For this reason, small neutron generators with voltages between 100 and 300 kev and high deuteron currents are frequently used as strong continuous or pulsed sources of $H^3(d, n)He^4$ neutrons.

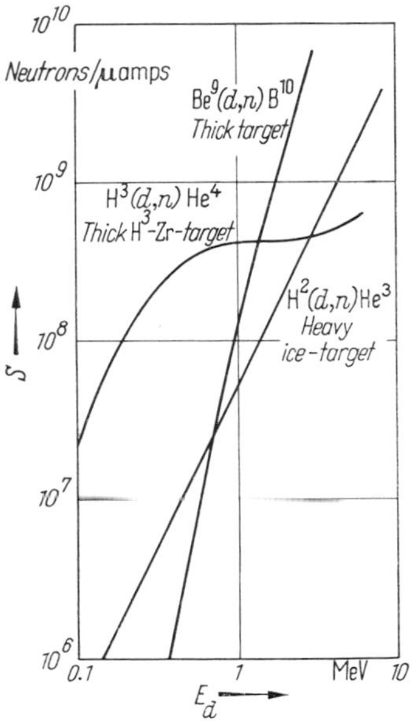

Fig. 2.5.6. The thick target neutron yield from various (d, n) reactions (from the Bulletin "H" of the High Voltage Engineering Corporation; Burlington)

At higher deuteron energies the reaction

$$Be^9 + H^2 \rightarrow B^{10} + n + 4.362 \text{ Mev}$$

is frequently used. The neutron spectrum from this reaction is broad, since the B^{10} nucleus can remain in an excited state. Beryllium metal is used as the target. These targets are very tough, since they can be well cooled. Occasionally, the (d, n) reaction on Li^7 is used:

$$Li^7 + H^2 \rightarrow Be^8 + n + 15.028 \text{ Mev}$$

$$Li^7 + H^2 \rightarrow 2\,He^4 + n + 15.122 \text{ Mev}$$

$$Li^7 + H^2 \rightarrow He^5 + He^4 + 14.165 \text{ Mev}$$

$$\searrow$$

$$He^4 + n + 0.958 \text{ Mev.}$$

It likewise does not give monoenergetic neutrons. The yield is good even for low deuteron energies; however, the manufacture of durable lithium targets causes difficulty.

2.5.3. Utilization of the Bremsstrahlung from Electron Accelerators

With modern travelling-wave linear accelerators, very large electron currents can be accelerated to energies $E = 50-100$ Mev. If these electrons strike a target, they give rise to intense bremsstrahlung whose spectrum is continuous up to an energy $h\nu = E$. Since the energy of a large part of the bremsstrahlung quanta exceeds the binding energy of neutrons in the target substance, neutrons are produced by (γ, n) processes. Fig. 2.5.7 shows the yield of a thick target as a function of the electron energy. It increases very sharply with the electron energy; the yield is largest for uranium, for which not only the (γ, n) reaction but also photofission contributes to the neutron emission. For this reason, thick

[1] According to FIEBIGER, about $\frac{1}{2}$ of the source strength of a heavy ice target is obtained with a deuteron current of 50 μamp on gold.

[2] A well-cooled tritium target can withstand a heat dissipation of 100 watts/cm². Its lifetime is about 1 mamp-hour; after this quantity of current, the yield has fallen to some $\frac{1}{10}$ of its initial value.

uranium targets are mainly used in practice; by bombardment of a thick uranium target with an average current of 40-Mev electrons of 1 mamp, a yield of about 10^{14} neutrons per second is obtained. The construction of targets for such high heat ratings (40 kw in the above example) presents special difficulties (see WIBLIN and POOLE). The neutron energy spectrum of a uranium target bombarded with energetic electrons resembles the spectrum of fission neutrons; the average neutron energy is about 2 Mev.

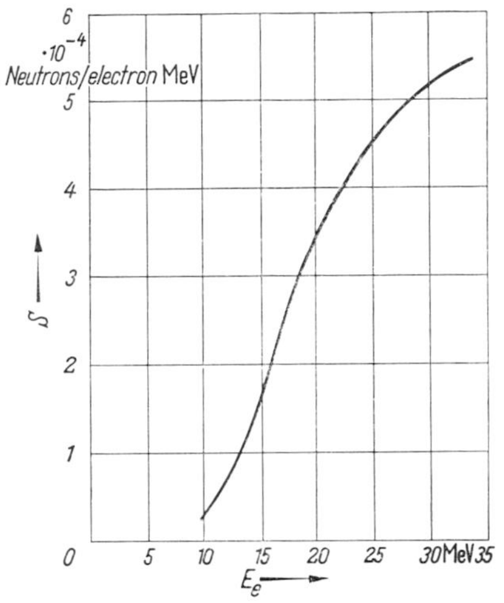

Fig. 2.5.7. The neutron yield from a thick Uranium target upon electron bombardment (from BARBER and GEORGE)

A special feature of the electron linear accelerator is the fact that the electron acceleration, and thus the neutron production, does not occur continuously, but rather in pulses. A typical modern linear accelerator yields up to 500 electron pulses per second with a length which can be adjusted between 0.01 and 5 μsec; the electron current during the pulse can be as much as 1 amp. Occasionally, betatrons have also been used as pulsed neutron sources; the attainable intensity, however, is very much smaller than that of linear accelerators.

2.5.4. Production of Quasi-Monoenergetic Neutrons with the Time-of-Flight Method

In Secs. 2.5.1 and 2.5.2, it was shown how particular monoenergetic groups of neutrons could be produced over a very wide range of energies up to more than 20 Mev with the help of van de Graaff generators and several (p, n) and (d, n) reactions. If one ignores the difficult backward-angle method with the $Li^7(p, n)Be^7$ reaction and the low-yield (p, n) reactions on vanadium and scandium, the lower limit of this range of energies is at 120 kev. Once can, however, span the energy range 0.01—120 kev with accelerator neutrons, if one can succeed in sorting the neutrons out of a continuous energy distribution on the basis of their time-of-flight.

Let a neutron source emit short neutron pulses with a broad energy distribution. Let the neutron detector be at a distance l from the source. Since neutrons of energy E require a time

$$t = \frac{l}{v} = \frac{l}{\sqrt{2E/m}} \tag{2.5.6}$$

to traverse the flight path l, a unique relation exists between the neutron energy and the arrival time of the neutrons at the detector, providing only that the length of the neutron pulse is small compared to the time of flight. In this way, energy measurements on neutrons from continuous sources are possible. Since in place of the detector one can use another measuring instrument and detect the nuclear reactions caused by the neutrons, one can investigate neutron reactions in the same way as with purely monoenergetic sources[1].

[1] Naturally, one cannot use an activation method.

In practice, the detector is connected to a "multi-channel analyzer", which separately registers the number Z_i of events occurring in the equal time intervals $t_{i+1}-t_i$; thus the entire neutron spectrum can be obtained in one measurement. If the flight time is given in μsec, the flight path in meters, and the neutron energy in ev, then the important relation

$$\frac{t}{l}=\frac{72.3}{\sqrt{E}} \tag{2.5.7}$$

holds. The accuracy with which energy measurements are possible with the time-of-flight method is connected with the uncertainties Δt in the flight time and Δl in the flight path. Δt is composed of the pulse width of the neutron source as well as the time resolution of the detector and the associated electronics. As a rule, the uncertainty in the flight path length can be neglected in comparison with the uncertainty in the flight time; in this case the relation

$$\Delta E=\left|\frac{\partial E}{\partial t}\right|\Delta t=2E\frac{\Delta t}{t} \tag{2.5.8}$$

holds. With the help of Eq. (2.5.7) one finds ($\Delta t/l$ in μsec/m; E in ev)

$$\Delta E=0.028\frac{\Delta t}{l}E^{\frac{3}{2}}. \tag{2.5.9}$$

We now briefly discuss the most important applications of the time-of-flight method (cf. also Sec. 2.6.3).

5—120 kev: We have seen in Sec. 2.5.1 that in this energy range the $Li^7(p, n)Be^7$ reaction has a good yield in the forward direction, but, because of the existence of the second neutron group, does not give monoenergetic neutrons. By means of the time-of-flight method it is easily possible to separate the two neutron groups from one another. However, if the necessary equipment for such time-of-flight measurements is available, it is easier to make neutrons with a continuous spectrum by use of thick lithium targets[1] and to sort them in energy according to the time-of-flight method. Then one can span the entire energy range in one measurement. This technique was developed by Good and co-workers, who used the proton beam of a van de Graaff generator pulsed by electrostatic deflection as a proton source. For a pulse length of 5 nsec and a flight path of 5 meters ("resolving power" $\Delta t/l=1$ nsec/m), $\Delta E=28$ ev at $E=10$ kev according to Eq. (2.5.9); thus, $\Delta E/E=2.8\times10^{-3}$. A detailed description of the relevant experimental methods can be found in Neiler and Good[2].

0.001—10 kev: If the uranium target of a linear electron accelerator such as has been described in Sec. 2.5.3 is surrounded with a moderator (plastic, plexiglass, etc.), the emitted neutrons can be slowed down to low energy. Fig. 2.5.8 shows the facility for neutron spectroscopy at the linear accelerator in Harwell[3]. Long, evacuated neutron flight tubes, at whose ends neutron counters

[1] For example, if a thick lithium target is bombarded with 1.910-Mev protons, neutrons with energies between 1 and 105 kev are obtained in the forward direction, as shown in Fig. 2.5.1.

[2] See Marion and Fowler, p. 509ff.

[3] In order to increase the source strength, the target of the Harwell linear accelerator is surrounded by a booster zone of enriched uranium.

or detectors for special neutron reactions are found, go out in all directions. Because of the high neutron intensity, flight paths up to a length of 200 m can be used. The time uncertainty Δt is essentially determined by the fluctuations

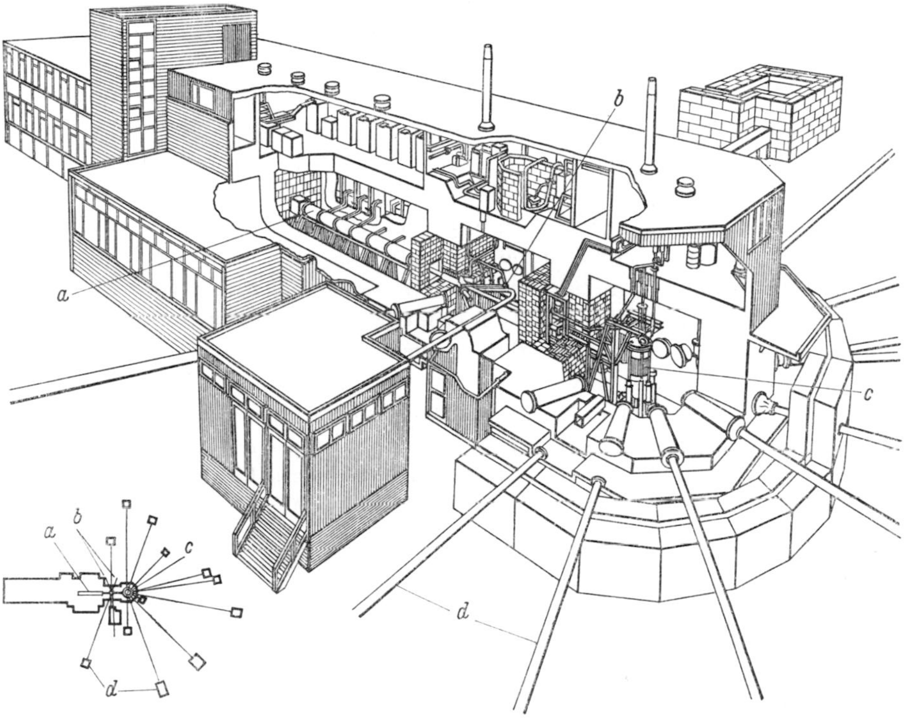

Fig. 2.5.8. The Harwell linear accelerator facility. a Accelerating waveguides; b auxiliary target; c main target with booster; d neutron flight channels (reproduced with the kind permission of E. R. RAE, HARWELL)

in the slowing-down time of the neutrons to energy E; according to Sec. 9.1.1, the relation

$$\Delta t = \frac{\sqrt{3}}{\Sigma_s \cdot v} = \frac{1.8}{\sqrt{E}} \qquad (\Delta t \text{ in } \mu\text{sec}, E \text{ in eV}) \tag{2.5.10}$$

holds approximately for water and plastic. For $E=100$ ev, one obtains $\Delta t = 0.18\,\mu$sec; by neglecting further uncertainties and assuming a flight path of 100 meters, one finds $\Delta E = 0.05$ ev, that is, $\Delta E/E = 5\times 10^{-4}$.

Clearly, very good resolution can be achieved with this method. For this reason, time-of-flight spectrometers based on linear accelerators are being operated in several laboratories. The Nevis synchrocyclotron of Columbia University, which can produce neutrons through (p, n) reactions with energetic protons, also is used as a time-of-flight spectrometer.

Time-of-flight measurements in the Mev range have occasionally been carried out using the natural phase bunching in cyclotrons. A typical cyclotron (here the 42″ cyclotron of the University of Michigan; see GRISMORE and PARKINSON) yields 10^7 4-nsec pulses of 25-Mev deuterons per sec. By bombardment of a thick beryllium target, neutrons with a broad energy spectrum from 0—25 Mev are produced; and with time-of-flight measurements with about a 5-meter flight path, "monoenergetic" neutrons can be produced. In order to avoid overlap between individual measurement cycles, the frequency of the neutron pulses must be drastically reduced, which requires special measures. Since, according to Secs. 2.5.1 and 2.5.2, truly monoenergetic neutrons can be produced in this energy range[1], this method is only seldom used; for higher neutron energies, however, it is very promising (BOWEN).

2.6. Principles of Neutron Production in Research Reactors

Let us consider a typical research reactor with a thermal power of 10 Mw. Since about 200 Mev of energy are liberated in one fission, about 3×10^{17} fissions occur per sec. About 2.5 neutrons are produced per fission; thus about 7.5×10^{17} neutrons are produced per sec in the reactor. To be sure, only a vanishingly small fraction of this extremely high source strength is available for neutron experiments.

It is not the purpose of this section to introduce the reader to the nature of the chain reaction or to the innumerable physical and technological problems of the various types of reactors. Rather, it is the purpose of this section to describe the most important types of radiation in typical research reactors, the various irradiation facilities, as well as some of the auxiliary equipment of neutron spectroscopy. There is a series of good introductions to reactor technology in which the individual reactor types are described in more detail.

2.6.1. The Various Types of Radiation

The neutron radiation in the interior of a reactor is characterized by the neutron flux Φ (see Sec. 5.1). Φ is the number of neutrons which pass in all directions through a sphere of cross section $\pi R^2 = 1$ cm^2, per sec. In first approximation, all flight directions are equally likely, i.e., the neutron velocity distribution is isotropic. The neutrons in a reactor have a broad energy distribution, reaching from 0.001 ev to over 10 Mev. It is characterized by the spectrum $\Phi(E)$. $\Phi(E)\,dE$ is the flux of neutrons with energies between E and $E + dE$. Within the spectrum, the following three regions are distinguished:

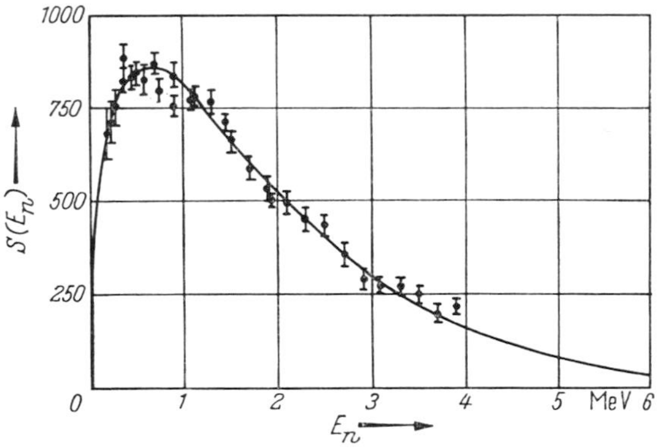

Fig. 2.6.1. The energy spectrum of neutrons produced in thermal neutron fission of U^{235}

a) $E > 0.5$ Mev — the region of fast neutrons. Here $\Phi_f = \int_{0.5\,\text{Mev}}^{\infty} \Phi(E)\,dE$. In this energy range, $\Phi(E)$ approximately follows the energy distribution of the neutrons produced in fission (cf. Fig. 2.6.1). The fission spectrum can be approximated by

$$N(E) \approx e^{-E} \sinh \sqrt{2E}, \quad (E \text{ in Mev}). \tag{2.6.1}$$

[1] It should be noted that with one of the usual 5.5-Mev van de Graaff generators the (d, n) reaction on deuterium yields a maximum of 8.7-Mev neutrons and the (d, n) reaction on tritium a minimum of 11.6-Mev neutrons. The intervening gap can only be bridged with a variable energy cyclotron or a tandem van de Graaff.

b) $0.2\ \text{ev} < E < 0.5\ \text{Mev}$ — the region of epithermal or resonance neutrons.

$$\Phi(E)\,dE = \frac{\Phi_{\text{epi}}}{E}\,dE. \tag{2.6.2}$$

In this energy range, the spectrum is predominantly determined by the neutrons being slowed down by elastic collision with the nuclei of the moderator substance. Φ_{epi} only depends weakly on the energy and can be taken constant as a first approximation.

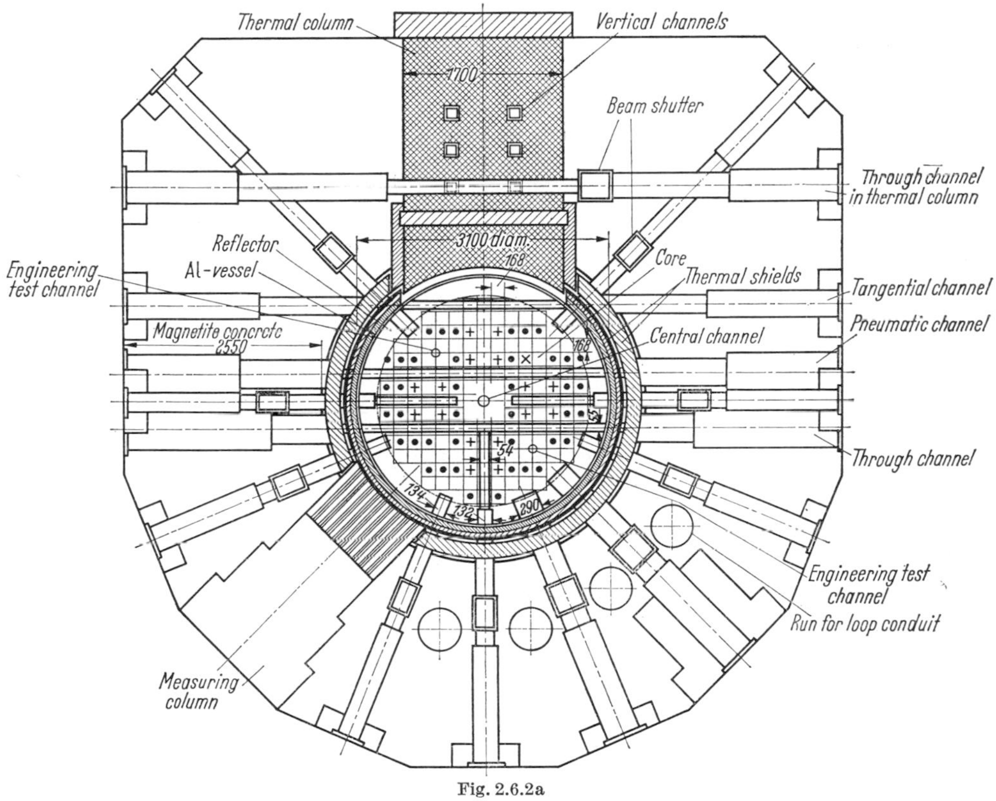

Fig. 2.6.2a

Fig. 2.6.2a and b. Horizontal (a) and vertical (b) cross section of the Karlsruhe FR 2 Research Reactor (lengths are in mm)

c) $E < 0.2\ \text{ev}$ — the region of thermal neutrons.

$$\Phi(E)\,dE = \Phi_{\text{th}}\,\frac{E}{kT}\,e^{-E/kT}\,\frac{dE}{kT}. \tag{2.6.3}$$

Thermal neutrons are nearly in thermodynamic equilibrium with the thermal motion of the moderator atoms; their energy distribution can frequently be approximated by a Maxwell distribution with a neutron temperature somewhat higher than the moderator temperature. The following equation holds for the total flux of thermal neutrons:

$$\Phi_{\text{th}}\int_0^{0.2\,\text{ev}}\frac{E}{kT}\,e^{-E/kT}\,\frac{dE}{kT} \approx \Phi_{\text{th}}\int_0^{\infty}\frac{E}{kT}\,e^{-E/kT}\,\frac{dE}{kT} = \Phi_{\text{th}}. \tag{2.6.4}$$

This division is relatively rough; under special circumstances the spectra in all three energy ranges show considerable deviation from these simple behaviors[1]. For the purpose of orientation, however, this three-fold division is sufficiently accurate.

The intensity of each of these radiation components depends on the point inside the reactor being considered, the type of reactor, and the power at

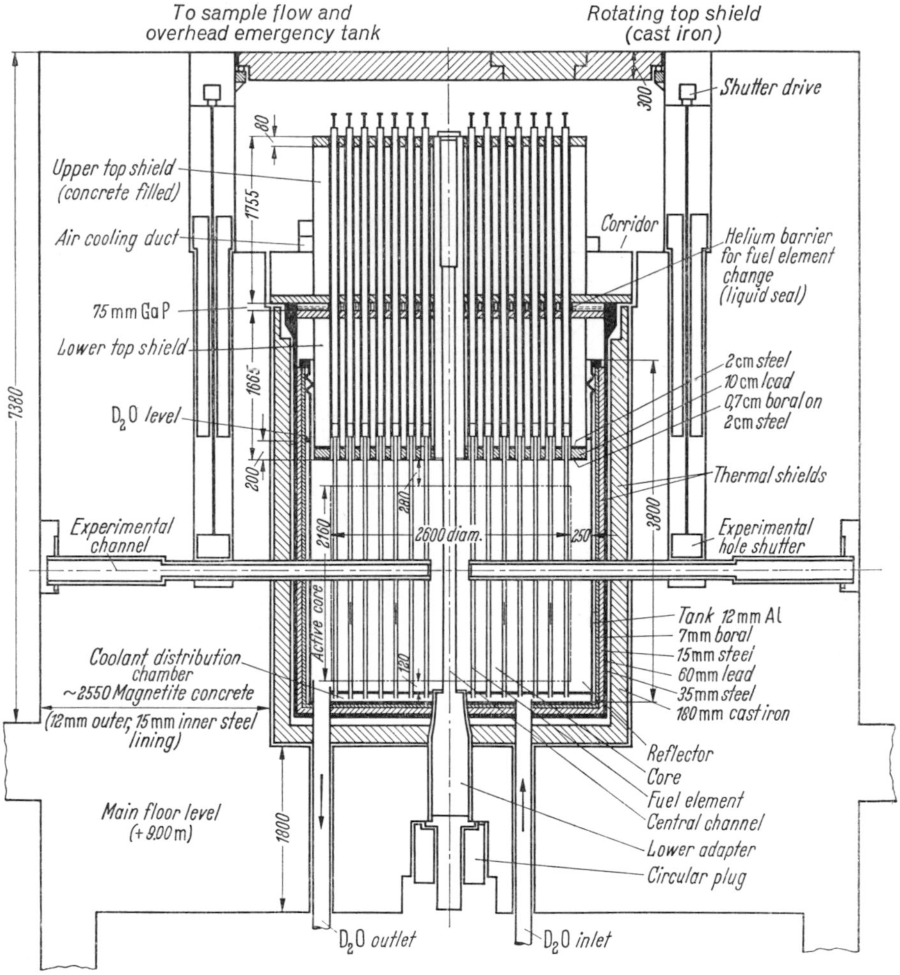

Fig. 2.6.2 b

which the reactor is operated. With respect to the spatial distribution of the radiation in the reactor, we shall content ourselves with the remark that the intensities of the various radiation components have maxima in the middle of the active zone and fall off fairly quickly as one approaches the surface. We give more details about the radiation field in the center of the reactor only for two representative types:

[1] We shall discuss the low-energy end of the spectrum in more detail in Sec. 10.2. Information on fast and epithermal neutron spectra in reactors can be found in Rowlands, in Håkansson, and in Wright, among others.

(a) Heavy-water-moderated research reactors fuelled with natural uranium. Fig. 2.6.2 shows, by way of example, the Karlsruhe research reactor FR-2. Other reactors of this type are the Canadian research reactors NRX and NRU, the French reactor EL-2, and the Swiss research reactor Diorit.

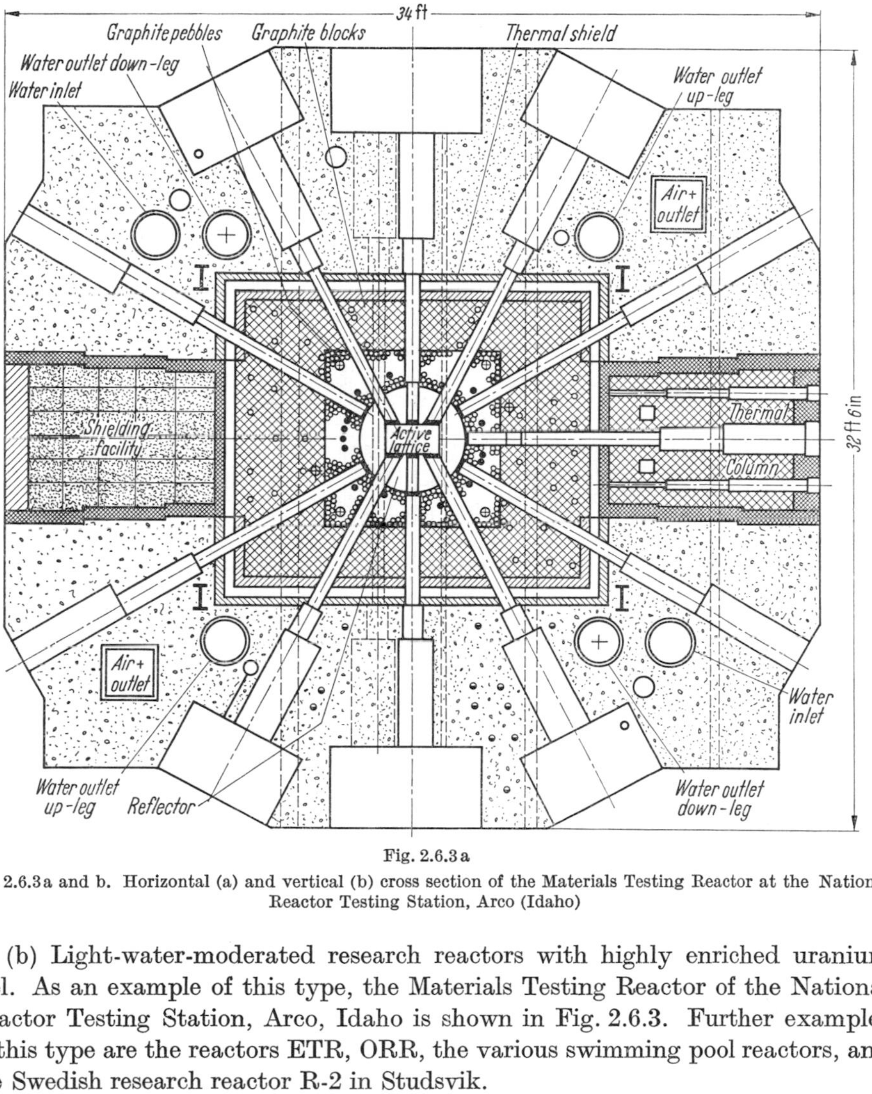

Fig. 2.6.3a

Fig. 2.6.3a and b. Horizontal (a) and vertical (b) cross section of the Materials Testing Reactor at the National Reactor Testing Station, Arco (Idaho)

(b) Light-water-moderated research reactors with highly enriched uranium fuel. As an example of this type, the Materials Testing Reactor of the National Reactor Testing Station, Arco, Idaho is shown in Fig. 2.6.3. Further examples of this type are the reactors ETR, ORR, the various swimming pool reactors, and the Swedish research reactor R-2 in Studsvik.

Table 2.6.1. *Radiation Levels in Typical Research Reactors*

Type	Φ_f (cm^{-2}sec^{-1})	Φ_{epi} (cm^{-2}sec^{-1})	Φ_{th} (cm^{-2}sec^{-1})	Φ_γ (Mev cm^{-2}sec^{-1})
a	2×10^{12}	10^{12}	3×10^{13}	10^{12}
b	10^{14}	10^{13}	10^{14}	5×10^{13}

Table 2.6.1 shows the values of the individual radiation components in the middle of the active zone, i.e., at the flux maxima, for both of these reactor types. These values are referred to a reactor power of 10 Mw; at 1 Mw one must multiply with the factor 0.1, etc. These values must only be considered rough guide

values, since even for a given basic reactor type, the ratio of flux to power depends on the size of the reactor and the composition of the fuel. However, these values are sufficiently accurate for the purposes of orientation. It happens that considerably higher values occur throughout with reactors of type b than with type a. This

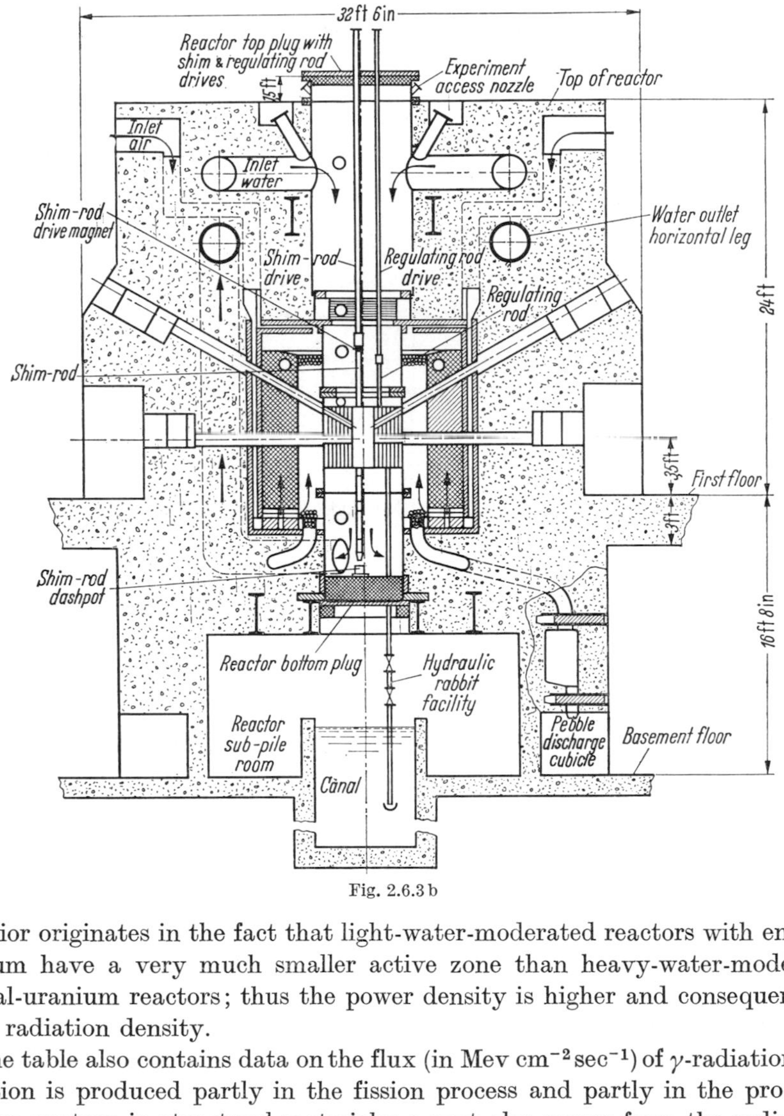

Fig. 2.6.3 b

behavior originates in the fact that light-water-moderated reactors with enriched uranium have a very much smaller active zone than heavy-water-moderated, natural-uranium reactors; thus the power density is higher and consequently so is the radiation density.

The table also contains data on the flux (in Mev $cm^{-2} sec^{-1}$) of γ-radiation. This radiation is produced partly in the fission process and partly in the process of neutron capture in structural materials; a part also comes from the radioactive daughter products of the primary fission fragments.

2.6.2. Experimental Facilities

We have to distinguish between devices for the irradiation of substances in the reactor and devices with which radiation for experiments is taken out of the reactor. The latter are particularly of interest in neutron physics. For the extraction of radiation, beam holes are used. Beam holes are usually cylindrical aluminum pipes which either reach through the shield in the horizontal direction to the reflector or to the core or pass entirely through the reactor (cf. Figs. 2.6.2 and 2.6.3). The number of neutrons that can be obtained with such a canal can be easily estimated as follows: If Φ is the flux on the front surface F of the beam hole, then $\Phi F/4$ is the number of neutrons entering the front surface of the beam tube per second[1]. If l is the distance between the front surface of the tube and the outer edge of the shield (tube exit), then the neutron current density J ($cm^{-2}\,sec^{-1}$) at the tube exit is

$$J = \frac{\Phi F}{4\pi l^2}. \tag{2.6.5}$$

When $l = 300$ cm, $\Phi = 10^{13}$ $cm^{-2}\,sec^{-1}$, and $F = 50$ cm^2, $J = 5 \times 10^8$ $cm^{-2}\,sec^{-1}$.

In the extraction of radiation with a beam tube, the fact that one simultaneously gets all the components of the reactor radiation, i.e., thermal, epithermal, and fast neutrons, as well as γ-rays, represents a fundamental difficulty. Various methods have been developed to suppress individual components; thus one can diminish the intensity of the thermal group with cadmium or boron filters, that of the epithermal and fast groups with moderator layers, and that of the γ-radiation with lead or bismuth filters. A particularly effective measure for the diminution of γ-radiation consists in the use of a fully penetrating beam tube into which a small scattering H_2O, D_2O or graphite sample is placed; this sample mainly scatters neutrons, and the γ-intensity at the tube exit remains small. The production, shielding, and optimization of neutron radiation from research reactors altogether represent a difficult problem, into which we cannot go further here[2].

Another important experimental facility associated with research reactors is the thermal column (see Fig. 2.6.2). A thermal column is a prism of highly purified graphite (cross section about 2×2 m^2) stretching from the surface of the core to the edge of the outer shield of the reactor. The neutrons leaking out of the core are strongly moderated but only weakly absorbed in this graphite column, so that a very pure thermal neutron field is created. With the help of beam tubes penetrating the thermal column, pure thermal neutron beams can be produced. With use of the thermal column, there arises the possibility of producing in pure form (but only with a considerable loss in intensity) at least one of the components of the reactor radiation.

Occasionally, particularly in shielding experiments, neutron converters are used. In such experiments, one allows a beam of slow neutrons from the reactor to fall on a plate of enriched uranium. One thereby obtains a strong source of fission neutrons; contamination by slow primary neutrons can easily be eliminated by cadmium, or even better, by boron filters.

2.6.3. Auxiliary Apparatus for the Production of Monoenergetic Neutrons

For the measurement of cross sections with slow neutrons there has been developed a series of auxiliary devices that permit the separation of monochromatic neutrons from the continuous reactor spectrum. The most important devices

---

[1] Cf. Sec. 5.1.1.

[2] Cf. P. A. Egelstaff (ed.): Tailored Neutron Beams, J. Nucl. Energy Parts A & B **17**, No. 4/5 (1963).

are the crystal spectrometer, the crystal filter, the mechanical monochromator, and the chopper.

Crystal spectrometer, energy range 0.01—10 ev (Fig. 2.6.4). The crystal spectrometer is based on the principle of Bragg reflection of neutrons [cf. Eq. (1.4.1)]. The reflection takes place from a single crystal whose surface has been cut parallel to the desired lattice planes; frequently, however, the crystal is also used in transmission. The incident and reflected beams are carefully collimated (angular divergence

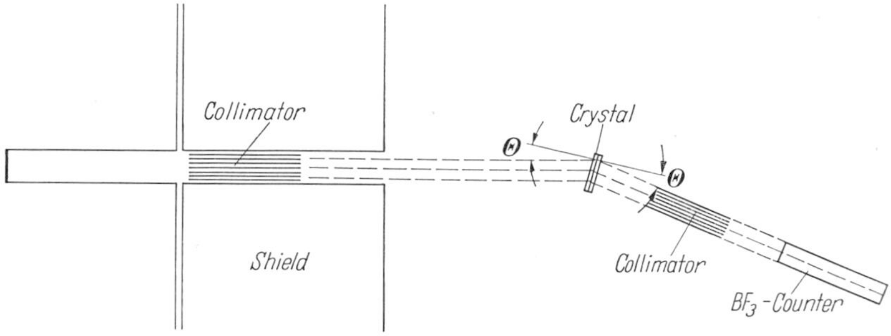

Fig. 2.6.4. Crystal spectrometer. In this case, the crystal is used in transmission

$\Delta\Theta$ about 0.1° or less) with the help of a Soller collimator in order to make the energy inhomogeneity ΔE of the neutrons as small as possible. The first order of reflection is used, since the reflected intensity falls off like $1/n^2$. Frequently used lattice planes are: Be 1231, $d=0.732$ Å; NaCl 240, $d=1.26$ Å; Cu 111, $d=2.08$ Å; LiF 111, $d=2.32$ Å. For $d=2.32$ Å and $E=1$ ev, $\Theta=3.5°$; at 100 ev, Θ would equal 0.35°. Since the energy inhomogeneity ΔE varies according to

$$\frac{\Delta E}{E}=2\cdot\cot\Theta\cdot\Delta\Theta\approx 2\frac{\Delta\Theta}{\Theta} \tag{2.6.6}$$

one would obtain $\Delta E/E=0.6$ at 100 ev with $\Delta\Theta=0.1°$. The behavior of Be is somewhat more favorable because of the considerably smaller lattice spacing; however, in no case can one obtain good resolution above 10 ev with a crystal spectrometer. Work at higher energies is also more difficult for reasons of intensity, since the reactor spectrum $\Phi(E)$ incident on the spectrometer and the crystal reflectivity both vary like $1/E$; thus the reflected intensity varies like $1/E^2$. At very low energies, i.e., below the maximum of the Maxwell distribution, higher orders of reflection appear: Eq. (1.4.1) shows that for given d and Θ in addition to the main reflection ($n=1$), neutrons with 4, 9, etc. times the main energy appear in the reflected beam. In this case, special measures for suppression of the higher reflections are necessary (e.g., prior use of a rough monochromator).

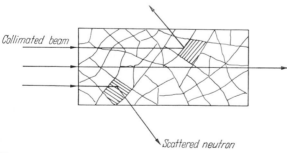

Fig. 2.6.5. Crystal filter. The neutrons above the cut-off energy always find lattice planes oriented in such a way that a Bragg reflection can occur

Crystal Filter (Fig. 2.6.5). The cut-off phenomenon in crystalline scatterers discussed in Sec. 1.4.3 can be used to construct a somewhat coarse but very simple monochromator. A neutron beam from a reactor is allowed to filter through a polycrystalline beryllium column (cross section about $5 \times 5\ \text{cm}^2$) about 25 cm long. Neutrons with energies above the cut-off energy are then scattered out of

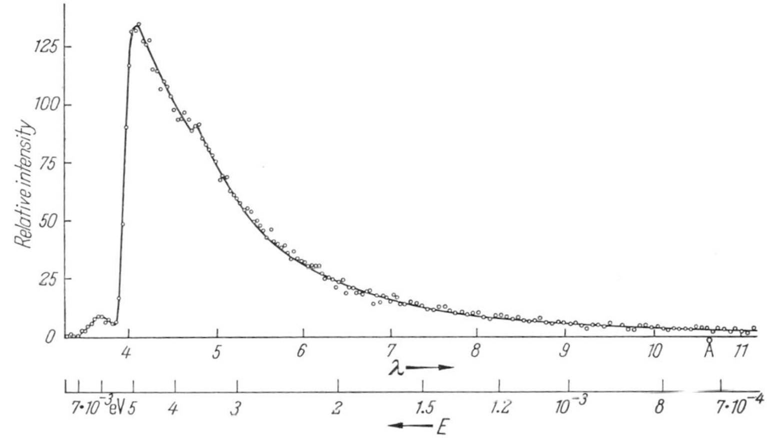

Fig. 2.6.6. The intensity distribution of thermal neutrons after filtering through a polycrystalline Beryllium column [LARSSON et al.]. The figure shows a time-of-flight distribution recorded with a thin $\frac{1}{v}$ detector; in order to convert it to a flux $\Phi(E)$ all values must be multiplied with $\lambda^2 \sim \frac{1}{E}$

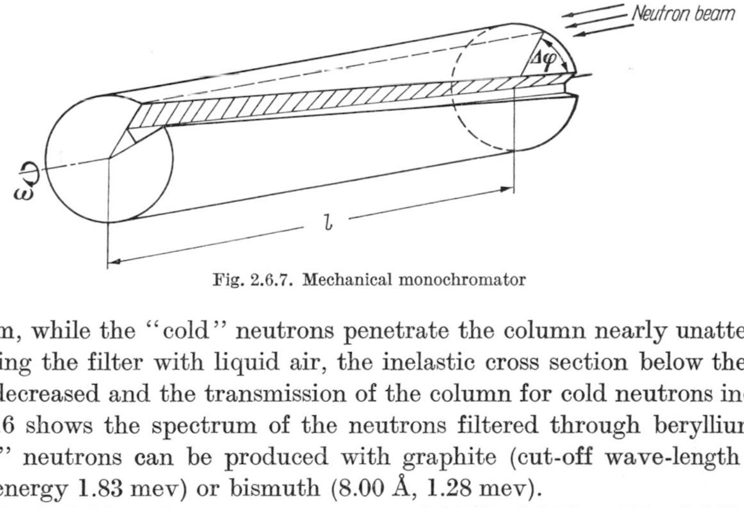

Fig. 2.6.7. Mechanical monochromator

the beam, while the "cold" neutrons penetrate the column nearly unattenuated. By cooling the filter with liquid air, the inelastic cross section below the cut-off can be decreased and the transmission of the column for cold neutrons increased. Fig. 2.6.6 shows the spectrum of the neutrons filtered through beryllium. Still "colder" neutrons can be produced with graphite (cut-off wave-length 6.04 Å, cut-off energy 1.83 mev) or bismuth (8.00 Å, 1.28 mev).

Mechanical Monochromator, energy range 0.0005—0.005 ev (Fig. 2.6.7). A slow neutron can obviously only penetrate the rotor shown in Fig. 2.6.8 when the relation

$$\frac{\Delta\varphi}{\omega} = \frac{l}{v} \tag{2.6.7}$$

is fulfilled. Here ω is the angular velocity of the rotor and v is the neutron velocity. For $l = 50$ cm, $\Delta\varphi = 30°$, and $\omega = 105\ \text{sec}^{-1}$ (1000 rpm), it follows that $v = 100$ m/sec, i.e., monoenergetic neutrons with an energy of 0.52 mev are produced. The energy homogeneity is very bad in comparison with that attainable with a crystal spectrometer; however, the transmission for neutrons with the "right" energy is very high. Cadmium is usually used as the absorbing material; for this reason such a monochromator is rather transparent to neutrons with $E > 0.25$ ev. Occasionally, mechanical monochromators are combined with beryllium filters; pure mev neutron beams can then be produced. The intensity of such a source can be increased considerably if a very cold moderator (a vessel filled with liquid hydrogen) is introduced into the reactor, thereby shifting the intensity maximum of the neutron energy distribution into the mev range.

Choppers. A chopper does not produce pure monoenergetic neutrons, but rather it chops a continuous neutron beam into short pulses. This makes possible time-of-flight measurements on the neutrons, as we already saw in Sec. 2.5.4.

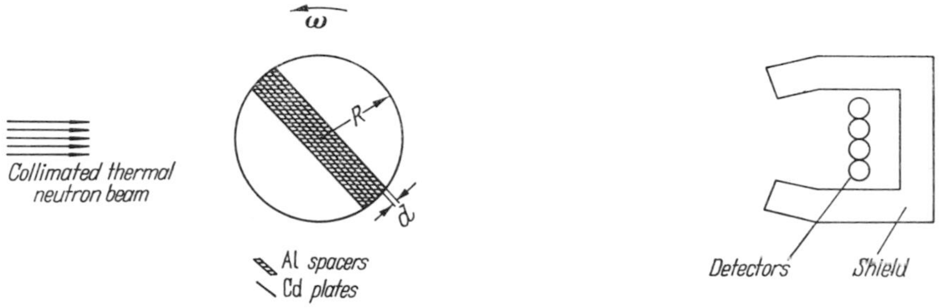

Fig. 2.6.8. Slow chopper

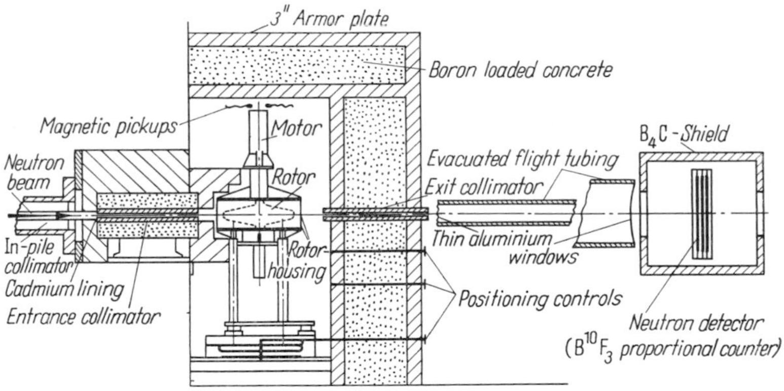

Fig. 2.6.9. Fast chopper at the Oak Ridge National Laboratory

We distinguish slow and fast choppers. In *slow* choppers, cadmium is used as the shielding material (Fig. 2.6.8); they are suitable for spectroscopy below 0.25 ev. The equation

$$\Delta t = \frac{d}{R\omega} \tag{2.6.8}$$

obviously holds for the pulse length produced by the rotor shown in Fig. 2.6.8. With $d = 1$ mm, $R = 5$ cm, and $\omega = 1050\ \text{sec}^{-1}$, we obtain $\Delta t = 20\ \mu\text{sec}$; with a flight path of 5m, the resolution is $\Delta t/l = 4\ \mu\text{sec/m}$, i.e., $\Delta E = 3$ mev for $E = 0.1$ ev. In *fast* choppers, the rotor consists of heavy materials (iron, nickel) in order that it be opaque to neutrons to as high an energy as possible. If one uses these materials to the limit of their strengths, with heavy rotors and very high rates of revolution, pulse lengths down to 1 μsec can be achieved. Fig. 2.6.9 shows schematically the chopper facility at the Oak Ridge Research Reactor. With a pulse length of 0.7 μsec and a maximum flight path of 180 m, the best attainable resolution is about 4×10^{-3} μsec/m, corresponding to $\Delta E = 3.5$ ev at 1 kev. In the past, such chopper facilities made great contributions to the study of neutron resonances. Today, they are greatly surpassed in attainable resolution and intensity by the time-of-flight spectrometers associated with linear accelerators.

Chapter 2: References

General

AMALDI, E.: The Production and Slowing Down of Neutrons. In: Handbuch der Physik, Bd. XXXVIII/2, p. 1, especially § 31—60. Berlin-Göttingen-Heidelberg: Springer 1959.

FELD, B. T.: The Neutron. In: Experimental Nuclear Physics, vol. II, p. 209, New York: John Wiley & Sons 1953, especially chap. 3: Sources and Detectors; Neutron Spectroscopy.

MARION, J. B., and J. L. FOWLER: Fast Neutron Physics, part I: Techniques. New York: Interscience Publishers 1960, especially Sec. I: Neutron Sources.

WLASSOW, N. A.: Neutronen. Köln: Hoffmann 1959, especially chap. II: Neutronenquellen.

Special[1]

WHALING, W.: Handbuch der Physik Bd. XXXIV, p. 193. Berlin-Göttingen-Heidelberg: Springer 1958 (Slowing-Down Power of Target Substances).

LATYSHEV, G. D.: Rev. Mod. Phys. **19**, 132 (1947) (γ-Rays from Radium and Its Daughter Products).

HALPERN, I.: Phys. Rev. **76**, 248 (1949).
RISSER, J. R., J. E. PRICE, and C. M. CLASS: Phys. Rev. **105**, 1288 (1957).
} Cross Section of the $Be^9(\alpha, n)$ Reaction.

HILL, L. D.: AECD 1945 (1947).
TEUCHER, M.: Z. Physik **126**, 410 (1949).
SCHMIDT-ROHR, U.: Z. Naturforsch. **8a**, 470 (1953).
} Measurement of the Energy Distribution of (Ra—Be) Neutrons.

HESS, W. N.: Ann. Phys. **6**, 115 (1959) (Calculation of the Spectra of Various (α, n) Sources).

ANDERSON, M. E., and W. H. BOND: Nucl. Phys. **43**, 330 (1963).
BREEN, R. J., M. R. HERTZ, and D. U. WRIGHT: Mound Laboratory Report MLM 1054 (1956).
COCHRAN, R. G., and K. M. HENRY: Rev. Sci. Instr. **26**, 757 (1955).
MEDVECZKY, L. L.: Acta Phys. Acad. Sci. Hung. **6**, 261 (1956).
WHITMORE, B. G., and W. B. BAKER: Phys. Rev. **78**, 799 (1950).
} Measurement of the Energy Distribution of (Po—Be) and (Pu—Be) Neutrons.

RUNNALLS, O. I. C., and R. R. BOUCHER: Can. J. Phys. **34**, 949 (1949).
STEWART, LEONA: Phys. Rev. **98**, 740 (1955).
} (Pu—Be) Neutron Sources.

ROBERTS, J. H.: Los Alamos Report MDDC 731 (1947).
WATTENBERG, A.: Phys. Rev. **71**, 497 (1947).
} Yields of the (α, n) Reactions on Various Light Nuclei.

RUSSEL, B., *et al.*: Phys. Rev. **73**, 545 (1948).
SANDMEIER, H. A.: Kerntechnik **3**, 167; **3**, 215 (1961).
SCHMITT, H. W.: Nucl. Phys. **20**, 220 (1960).
WATTENBERG, A.: Nucl. Sci. Series **6** (1949).
} Photoneutron Sources.

CASWELL, R. S., *et al.*: NBS Handbook No. 72, p. 12 (1960) (Spontaneous Fission as a Neutron Source).

[1] In this and many subsequent chapters, frequent reference is made to papers given at the 1955 or 1958 United Nations Conferences on the Peaceful Uses of Atomic Energy at Geneva. They are quoted like the following examples: Geneva 1955 P/574 Vol. 4 p. 74 (which means that this is paper number 574 given at the 1955 conference, published in Vol. 4 p. 74 of the respective proceedings), or Geneva 1958 P/59 Vol. 14 p. 266 (which means paper number 59 given at the 1958 conference, published in Vol. 14 p. 266 of the respective proceedings).

MARTIN, D. S.: AECD 3077 (1946).
TOCHILIN, E., and R. V. ALVES: USNRDL-TR 201 (1958); cf. also Health Physics 1, 332 (1958).
— Mock Fission Sources.

FOWLER, J. L., and J. E. BROLLEY: Rev. Mod. Phys. **28**, 103 (1956).
GIBBONS, J. H.: Phys. Rev. **102**, 1574 (1956).
HANNA, R. C.: Phil. Mag. VII, **46**, 383 (1955).
HANSON, A. O., R. F. TASCHEK, and J. H. WILLIAMS: Rev. Mod. Phys. **21**, 635 (1949).
McKIBBEN, J. L.: Phys. Rev. **70**, 101 (1946).
— Monoenergetic Neutrons from (p, n) and (d, n) Reactions.

AMALDI, E., L. R. HAFSTAD, and M. A. TUVE: Phys. Rev. **51**, 896 (1937).
Bulletin "H" of the High Voltage Engineering Corporation, Burlington, Massachusetts.
— Yields of (d, n) Reactions with Thick Targets.

FIEBIGER, K.: Z. Naturforsch. **11**a, 607 (1956) (Self-Targets for the $H^2(d, n)He^3$ Reaction).

BLOMSÖ, E., and G. v. DARDEL: Appl. Sci. Res. B **3**, 35 (1952).
EYRICH, W.: Nukleonik **4**, 167 (1962).
PECK, R. A., and H. P. EUBANK: Rev. Sci. Instr. **26**, 444 (1955).
— Description of Small Neutron Generators.

BALDWIN, G. C., E. R. GAERTTNER, and M. L. YEATER: Phys. Rev. **104**, 1652 (1956).
BARBER, W. C., and W. D. GEORGE: Phys. Rev. **116**, 1551 (1959).
— Neutron Yields of Various Targets when Bombarded by Energetic Electrons.

SPAEPEN, J. (ed.): Neutron Time of Flight Methods; Proceedings of the 1961 EANDC Symposium, Brussels: Euratom 1961 (Time-of-Flight Method).
GOOD, W. M., J. H. NEILER, and J. H. GIBBONS: Phys. Rev. **109**, 926 (1958) (Time-of-Flight Measurements with $Li^7(p, n) Be^7$ Neutrons).
POOLE, M. J., and E. R. WIBLIN: Geneva 1958 P/59 Vol. 14 p. 266 (Harwell Linac Neutron Spectrometer).
YEATER, M. L., E. R. GAERTTNER, and G. C. BALDWIN: Rev. Sci. Instr. **28**, 514 (1957) (Use of a Betatron as a Neutron Spectrometer).

HAVENS, W. W.: Geneva 1955 P/574 Vol. 4 p. 74.
RAINWATER, J., *et al.*: Rev. Sci. Instr. **31**, 481 (1960).
— The Columbia Synchrocyclotron as a Neutron Spectrometer.

BOWEN, P. H., *et al.*: Nucl. Phys. **22**, 640 (1961).
GRISMORE, R., and W. C. PARKINSON: Rev. Sci. Instr. **28**, 245 (1957).
— Time-of-Flight Measurements in the Mev Range with Cyclotrons.

HUGHES, D. J.: Pile Neutron Research. Cambridge: Addison-Wesley 1953.
— Reactor Techniques, in: Handbuch der Physik, Bd. XLIV/2, p. 330ff. Berlin-Göttingen-Heidelberg: Springer 1959.
— Reactors as Neutron Sources.

HÅKANSSON, R.: Nukleonik **4**, 233 (1962).
ROBINSON, M. T., O. S. OEN, and D. K. HOLMES: Nucl. Sci. Eng. **10**, 61 (1961).
ROWLANDS, G.: J. Nucl. Energy A **11**, 150 (1959/60).
WRIGHT, S. B.: AERE-R 4080 (1962).
— Calculation of Fast and Epithermal Neutron Spectra in Reactors.

BORST, L. B., and V. L. SAILOR: Rev. Sci. Instr. **24**, 141 (1953) (Crystal Spectrometers).
EGELSTAFF, P. A., and R. S. PEASE: J. Sci. Instr. **31**, 207 (1954) (Crystal Filters).

HOLT, N.: Rev. Sci. Instr. **28**, 1 (1957).
LOWDE, R. D.: J. Nucl. Energy **A**, **11**, 69 (1959/60).
— Mechanical Monochromators.

WEBB, F. J.: Nucl. Sci. Eng. **9**, 120 (1961).
— J. Nucl. Energy A & B **17**, 187 (1963).
— Low-Temperature Moderators.

EGELSTAFF, P. A.: J. Nucl. Energy **1**, 57 (1954/55).
LARSSON, K. E., *et al.*: J. Nucl. Energy **6**, 222 (1958).
— Arkiv Fysik **16**, 199 (1959).
— Slow Choppers.

BLOCK, R. C., G. G. SLAUGHTER, and J. A. HARVEY: Nucl. Sci. Eng. 8, 112 (1960).
SEIDL, F. G., *et al.*: Phys. Rev. **95**, 476 (1954).
ZIMMERMANN, R. L., *et al.*: Nucl. Instr. Meth. **13**, 1 (1961).
— Fast Choppers.

3. Neutron Detectors

Since the neutron is an electrically neutral particle, it does not interact with the usual detectors employed for ionizing radiation; it is therefore detected indirectly using nuclear reactions that produce charged secondary particles. There are two different kinds of such reactions. In the first kind, the charged nuclear particles are produced instantaneously, as, for example, in the $B^{10}(n, \alpha)Li^7$ or $He^3(n, p)H^3$ reactions or in neutron-proton scattering. The *promptly* indicating neutron detectors, which are treated in this chapter, depend on these reactions. The second kind of nuclear reaction forms the basis of the *radioactive indicators* (probes), which are particularly important for measurements in neutron fields. These indicators are based on reactions that lead to the production of radioactivity (β^+-, β^--, or γ-radiation) which can be measured after removal of the indicator from the neutron field. They are discussed in detail in Chapters 11, 12, and 13.

3.1. Neutron Detection with the (n, α) Reactions in B^{10} and Li^6 [1]

3.1.1. The Boron Counter

The boron counter is the simplest and most widely used device employed for the detection of neutrons. It generally has the form of a proportional counter filled with boron triflouride (BF_3) gas which has been highly enriched (up to 96%) in boron-10. Boron trimethyl ($B(CH_3)_3$) has also been tested as a filling gas. Less usual are the so-called boron-coated counters, i.e., counters which have a boron-containing coating on their walls and which are filled with an ordinary counting gas.

Fig. 3.1.1 shows the pulse height spectrum of a BF_3 counter irradiated with thermal neutrons. The maxima corresponding to the two branches of the $B^{10}(n, \alpha)Li^7$ reaction are easily recognizable. Some of the α-particles or Li^7-nuclei produced near the walls or in the ends of the counter can leave the sensitive volume of the counter or strike the walls before they give up all of their energy as ionization energy to the gas. As a consequence of these end and wall effects, the principal maximum shows a tail of smaller pulses. For very small pulse heights, the spectrum again increases; these small pulses are due to γ-rays which produce electrons in the chamber walls. It is possible to distinguish neutrons and γ-rays easily by use of a suitably biased integral discriminator[2]. If the gas contains impurities, the spectrum can be smeared out considerably by electron capture, making the separation of neutrons and γ-rays more difficult. For this reason,

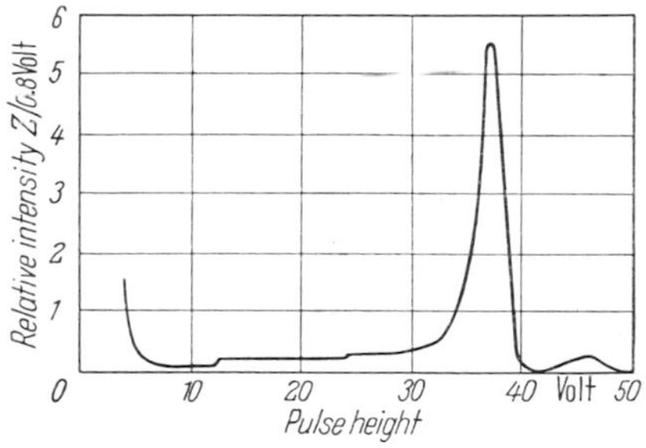

Fig. 3.1.1. A somewhat idealized pulse height spectrum from a BF_3-counter bombarded with thermal neutrons

[1] The Q-values and cross sections for both reactions were discussed in Sec. 1.4.1.

[2] The right setting of the discriminator can be found from the counter characteristic. The characteristic must show a plateau whose slope in a good counter is only about 1—2% per 100 volts of counter voltage.

great care must be taken to provide gas of high purity in the manufacture of boron counters[1].

If p is the gas pressure in a boron counter (in atmospheres of $B^{10}F_3$) and l is its length (in cm), then the detection efficiency for a neutron of energy E (in ev) incident in the longitudinal direction is

$$\varepsilon = 1 - \exp\left(-1.7 \times 10^{-2} \frac{pl}{\sqrt{E}}\right). \tag{3.1.1}$$

When $p=1$ atm and $l=20$ cm, $\varepsilon \approx 0.90$ for thermal neutrons ($E=0.0253$ ev) but is only 0.03 for 100-ev neutrons. Clearly, the ordinary boron counter is suitable only for the detection of very slow neutrons.

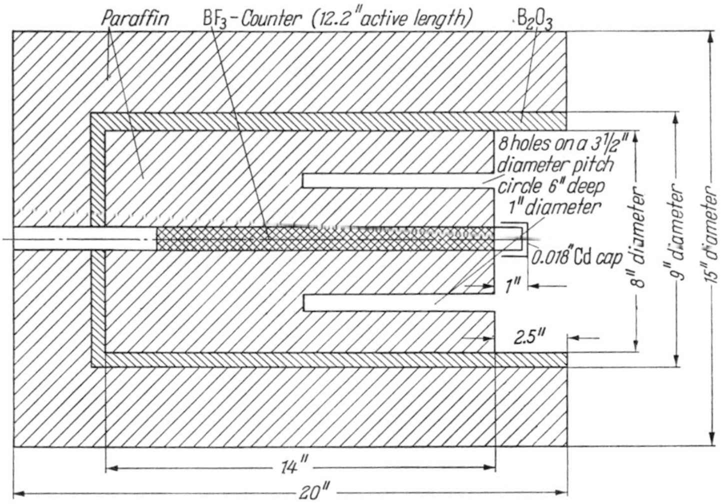

Fig. 3.1.2. The long counter developed by McTaggart

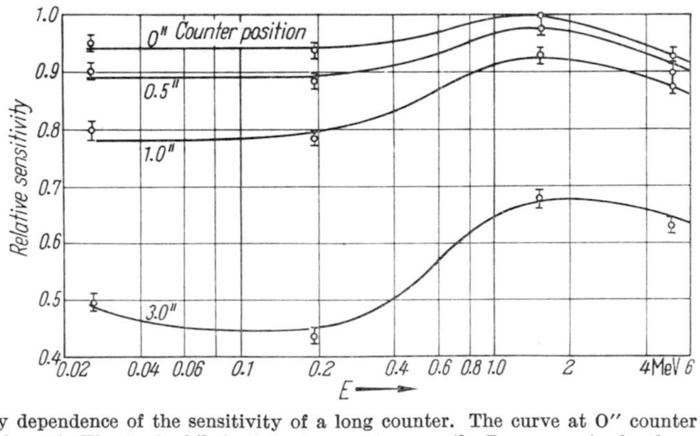

Fig. 3.1.3. Energy dependence of the sensitivity of a long counter. The curve at O" counter position refers to the assembly as shown in Fig. 3.1.2 while in the subsequent curves the Boron counter has been shifted to the left

In order to detect faster neutrons with a boron counter, the neutrons must first be moderated. According to Hanson and McKibben, the best way to do this is with a so-called *long counter*. A long counter is a standard arrangement of

[1] Today, good BF_3 counters are commercially available.

a boron counter in a paraffin moderator which permits the detection of fast neutrons with an efficiency nearly independent of the energy over a wide range of energies. Fig. 3.1.2 shows a long counter developed by McTaggart and Fig. 3.1.3 shows the energy dependence of its efficiency, which fluctuates by only 3% between 25 kev and 5 Mev. The construction of such a counter can be simplified: for example, the holes in the front surface can be omitted. When this is done, however, the efficiency falls off at energies below 1 Mev. Without the B_2O_3 layer and the outside paraffin coat, the counter becomes sensitive to neutrons incident from the side. Above 5 Mev, the sensitivity of the long counter gradually falls off. However, as Graves and Davis have shown, a flat variation of the efficiency in the range 1—14 Mev can be achieved by pulling the boron counter back from the front surface of the moderator.

3.1.2. Boron Scintillators

The basic disadvantages of the BF_3 counter are its low efficiency at energies above 1 ev and its relatively poor resolution time (about 1 μsec; in time-of-flight measurements with slow neutrons, there is a further time uncertainty $\Delta t = l/v$, since one does not know where in the counter the neutron is captured). These disadvantages can largely be circumvented by use of scintillation counters; however, experimental costs become higher and discrimination against γ-rays becomes more difficult.

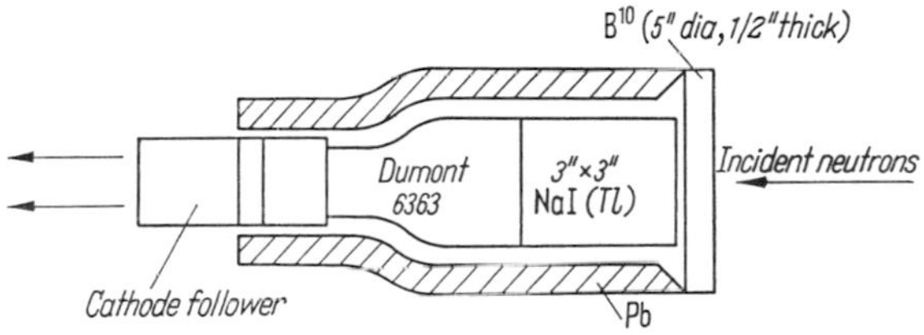

Fig. 3.1.4. Boron slab counter (Marion and Fowler, loc. cit.)

In the detector shown in Fig. 3.1.4, the 478-kev γ-rays produced by neutron capture in a boron-10 plate are detected by a NaI crystal. By use of the so-called "fast-slow" coincidence technique, resolution times of the order of nsec can be achieved; for this reason, the detector is particularly suitable for time-of-flight measurements. A boron-10 plate about 1 cm thick is black to all incident neutrons with energies up to about 10 kev. The detector efficiency for the 478-kev γ-rays is about 10%. Discrimination against background γ-radiation is difficult.

Simple, relatively γ-insensitive neutron scintillators can be made by melting $B_2^{10}O_3$ and zinc sulphide together. Such counters can be manufactured 1 or 2 mm thick. Their efficiency is about equal to that of a BF_3 counter. They are particularly suitable for time-of-flight measurements with slow neutrons; the resolution time is about 0.25 μsec. Because of the thinness of the counters, there is no additional time uncertainty.

Neutron detectors which are sensitive over a wide range of energies can be made by dissolving boron compounds in liquid scintillators. Bollinger and Thomas describe a detector which uses a mixture of toluol and methyl borate (50% each) as the solvent and 4 g/l of 2—5 diphenyloxazol as the scintillator. In such a solution, fast neutrons are slowed down by collision with the protons before they are captured by the boron. For a 2-cm thick detector, the efficiency is about 36% at 2 kev; the resolution time, which is largely determined by the slowing-down process, is about 0.3 μsec. Considerable difficulty is caused by the fact that in such a scintillator the light yield of heavy charged particles is about fifty times smaller than that of electrons; the pulse height due to the 2.30 Mev released in the $B^{10}(n, \alpha)Li^7$ reaction is about the same as that of 40-kev electrons. For this reason, suppression of multiplier noise as well as discrimination against γ-rays is difficult.

3.1.3. Detection Techniques with Lithium-6

Lithium-6 is mainly used in the form of europium-activated LiI single crystals, which are commercially available and which have a scintillation behavior similar to that of NaI(Tl) single crystals. Capture of thermal neutrons produces in the pulse height spectrum a sharp line ($\Delta E/E \approx 6\%$) corresponding to the Q-value of 4.78 Mev. (Because of the smaller light yield of the heavy charged particles compared to electrons, the line appears in the same place as that due to 4.1-Mev electrons.) By use of a window or an integral discriminator, γ-rays with energies less than 4.1 Mev can be separated out. The resolution time is about 0.3 μsec; the efficiency for slow neutrons is high (the efficiency of a 1-cm thick crystal is still about 60% at 10 ev). LiI single crystals can also be used for fast neutron spectroscopy (in the energy range 5—15 Mev). If E is the energy of a neutron which initiates an (n, α) reaction, then $E + 4.78$ Mev is the total energy deposited in the crystal. Thus, the pulse height spectrum can be translated into the energy spectrum of the incident neutrons if the cross section for the $Li^6(n, \alpha)H^3$ reaction is known. The relation between the pulse height and the neutron energy is particularly affected by the fact that the light yield depends on the distribution of the reaction energy between the triton and the α-particle; in order to achieve good resolution, the crystal must be cooled to the temperature of liquid nitrogen (Murray).

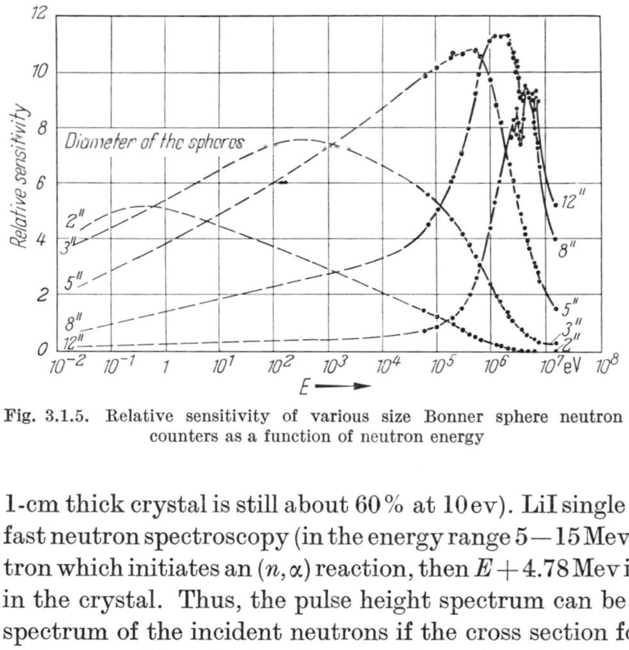

Fig. 3.1.5. Relative sensitivity of various size Bonner sphere neutron counters as a function of neutron energy

For the detection of fast neutrons with Li^6I, Bramblett, Ewing, and Bonner have suggested the use of a set of "standard" spheres of some moderating substance. If neutrons with an only approximately known energy spectrum are to be detected, a small Li^6I single crystal is placed successively in the middle of polyethylene spheres with diameters of 2, 3, 5, 8, and 12 inches, and the counting rate is recorded. Fig. 3.1.5 shows the experimentally determined efficiencies of these spheres as functions of the energy. With increasing diameter, the maximum sensitivity obviously shifts to higher energies. If the results of measurements with the various spheres in an unknown spectrum are combined, conclusions about the shape of this spectrum are possible.

Recently, it has proved possible to manufacture so-called glass scintillators by melting Li_2^6O, Al_2O_3, Ce_2O_3, and SiO_2 together. These counters are fast (resolution times ≈ 5 nsec can be achieved) and have high sensitivities: the efficiency of a 1.5-inch scintillator for 1-kev neutrons is about 25%. The pulse height spectrum exhibits a maximum at an equivalent electron energy of about 1.6 Mev with about 25% resolution; by use of an integral discriminator, the counter can be made completely insensitive to γ-rays with energies under 1.3 Mev (FIRK, SLAUGHTER, and GINTHER). Glass scintillators can also be manufactured using B_2O_3; conversely, the mixture with ZnS described in Sec. 3.1.2 can also be used with lithium (STEDMANN).

3.2. The Detection of Fast Neutrons and the Measurement of Their Energy Using Recoil Protons

If a neutron of energy E strikes a stationary proton, the proton receives a kinetic energy[1]

$$E_p = E\cos^2\vartheta. \tag{3.2.1}$$

Here ϑ is the scattering angle of the proton (in the laboratory system) measured from the direction of incidence of the neutron. E_p can take all values between 0 (glancing collision, $\vartheta = 90°$) and E (head-on collision, $\vartheta = 0°$). Because the scattering is isotropic in the center-of-mass system all proton energies between 0 and E are equally probable. In other words, when a hydrogenous material is bombarded with neutrons of energy E, the spectrum of recoil protons is given by:

$$\left.\begin{aligned} f(E_p)\,dE_p &= \frac{dE_p}{E} \qquad & E_p &\leq E \\ &= 0 & E_p &> E\,. \end{aligned}\right\} \tag{3.2.2}$$

If a neutron current with an inhomogeneous energy distribution $J(E)$ is incident on a hydrogenous substance, the recoil proton spectrum is given by

$$f(E_p) = \text{const.} \int_{E_p}^{E_{\max}} J(E)\,\sigma_H(E)\,\frac{dE}{E}\,. \tag{3.2.3}$$

Thus

$$J(E) \sim \frac{E}{\sigma_H(E)} \left|\frac{df}{dE_p}\right|_{E_p=E}\,. \tag{3.2.4}$$

The above relations are important for the utilization of recoil protons in neutron spectroscopy; if, for a given scattering, E_p and ϑ are simultaneously measured, the neutron energy can be determined. Several neutron spectrometers, which will be discussed in Sec. 3.2.2, depend on this principle. If the measurements are limited only to the energies of the recoil protons, the neutron spectrum can be determined by differentiation of the observed recoil proton energy distribution; in practice, this method provides the necessary accuracy only in the rarest cases.

[1] For the kinematics of neutron-proton collisions, see Sec. 7.1.

3.2.1. Simple Recoil Proton Counters

Fig. 3.2.1 shows a typical proportional counter used for the detection of fast neutrons. H_2 or methane at pressures up to 2 atm can be used for the gas filling. The range of the recoil protons in these gases is relatively large (in hydrogen at 1 atm, $R = 2.5$ cm at 0.5 Mev); end and wall effects play an important role in this counter, and the pulse height spectrum deviates sharply from the simple rectangular form given by Eq. (3.2.2) (particularly for bombardment with high-energy neutrons). For this reason, the determination of unknown neutron spectra is hardly possible. On the other hand, with a known spectrum, this counter is

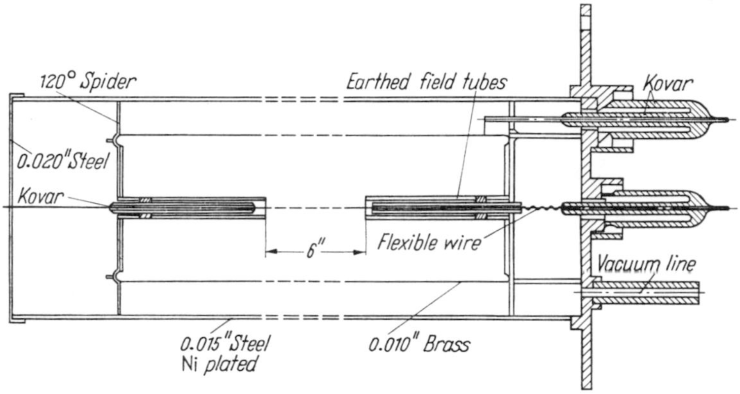

Fig. 3.2.1. Proton recoil proportional counter (MARION and FOWLER, loc. cit.)

well adapted to the measurement of fast neutron fluxes in the range 30 kev to 3 Mev. It can be transformed into a threshold detector by use of an electronic pulse height discriminator; then only such pulses are registered which correspond to recoil proton energies $> E_0$. According to Eq. (3.2.2) the efficiency is given by

$$\left.\begin{aligned} &\varepsilon = 0 && E < E_0 \\ &\varepsilon \sim \frac{E - E_0}{E}\,\sigma_H(E) && E > E_0 \end{aligned}\right\} \qquad (3.2.5)$$

if end and wall effects are neglected. The absolute value of the efficiency is about 1% for a 1-Mev neutron incident in the longitudinal direction when the counter is filled with CH_4 at a pressure of 1.5 atm.

A very much higher efficiency and very much smaller wall and end effects can be realized if recoil proton scintillation counters are used for detection. Organic crystals (anthracene, stilbene), liquid scintillators, and plastic scintillators are suitable for this purpose. For example, the detection efficiency for a 100-kev neutron in a 1-cm-thick stilbene crystal is 60% and sinks to about 4% for 10-Mev neutrons. Neutrons with energies under 30 kev cannot be detected, since the pulses they produce are lost in the "noise" of the photomultiplier. These counters are well adapted to time-of-flight measurements with fast neutrons; the best achievable resolution time is around 1 nsec. The pulse height spectrum produced by irradiation with monoenergetic neutrons does not have the simple rectangular form given by Eq. (3.2.2), since for protons the relation between light yield and energy is not linear. For this reason, it is not possible to determine the energy distribution of the incident neutrons from the pulse height spectrum using Eq. (3.2.4); however, corrections can be made for the nonlinearity of the crystal (Broek and Anderson). A serious difficulty in working with recoil proton scintillation counters is that they are about as sensitive to γ-radiation as they are to neutrons. Lately, it has become possible to reduce the γ-sensitivity of liquid scintillators and organic crystals by several orders of magnitude using the pulse-shape discrimination method (Brooks).

According to Hornyak, a simple, γ-insensitive recoil proton scintillator can be made by melting zinc sulphide together with lucite or perspex. Such *Hornyak-buttons* do not have a sharp pulse spectrum and consequently are best for survey purposes.

The sensitivity of most recoil proton detectors is only slightly dependent on the direction of the incident neutrons, and the detectors must be well shielded if laterally incident neutrons are not to be counted. It is, however, possible to manufacture scintillation counters with a strongly direction-dependent sensitivity. For this purpose, one uses the fact that the recoil protons arising from the scattering of fast neutrons are strongly collimated in the forward direction in the laboratory system. With a plastic scintillator composed of individual rodlets whose thickness is less than the range of the recoil protons, Stetson and Berko have achieved, with suitable discriminator settings, sensitivities for longitudinally incident neutrons up to 1000 times greater than for laterally incident neutrons. Strong direction sensitivity can also be achieved with spark counters (Krüger).

3.2.2. Neutron Spectroscopy with Recoil Protons

The simultaneous observation of the energy and angle of emission of a recoil proton is possible in a nuclear emulsion. In working with nuclear emulsions, both

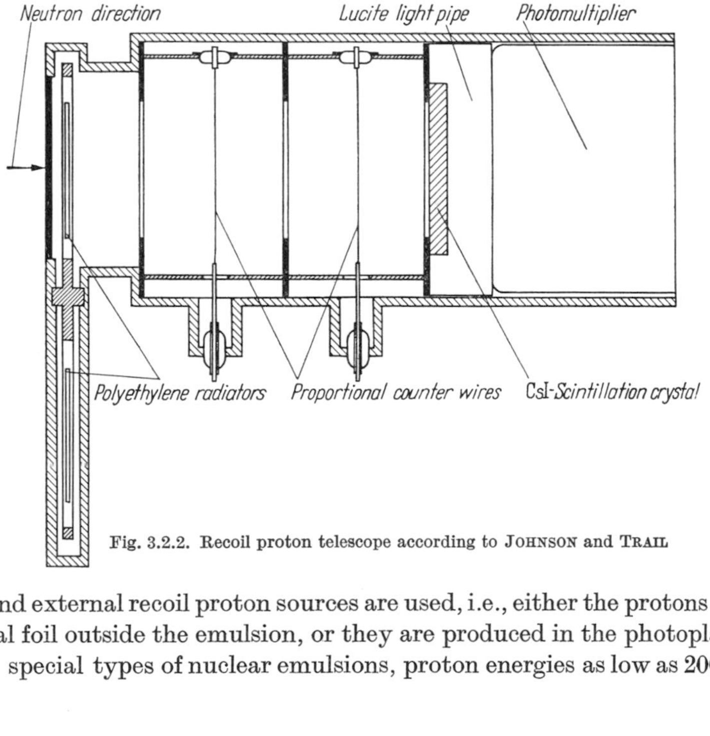

Fig. 3.2.2. Recoil proton telescope according to Johnson and Trail

internal and external recoil proton sources are used, i.e., either the protons originate in a special foil outside the emulsion, or they are produced in the photoplate itself. By use of special types of nuclear emulsions, proton energies as low as 200 kev can be measured. For this reason, the photoplate method is very frequently used to investigate neutron spectra in the energy range 0.2—20 Mev. For the determination of neutron energies in the range 50—500 kev, recoil proton tracks in cloud chambers are occasionally used.

There are various types of directly indicating neutron spectrometers. In the device of JOHNSON and TRAIL, shown in Fig. 3.2.2, recoil protons are ejected from a thin plastic foil. The protons emitted in the forward direction traverse two proportional counters and finally stop in a NaI(Tl) crystal. An event is registered if a triple coincidence between both proportional counters and the NaI crystal occurs. The proton energy can be determined from the pulse height of the scintillation counter; small corrections must be applied for the energy loss in the proportional counters. With this instrument, an energy resolution of about 5% of the neutron energy can be achieved in the energy range 1—20 Mev; the efficiency is small ($\approx 10^{-5}$) since the plastic radiating foil must be extremely thin. According to SCHMIDT-ROHR and also PERLOW, energy measurements can be performed in the energy range 50 kev to 1 Mev by collimation of the recoil protons produced in the gas volume of a proportional counter. Another instrument for making energy measurements, which resembles the well-known two-crystal Compton spectrometer for γ-rays, is illustrated in Fig. 3.2.3. The energy distribution of the recoil protons produced in the first scintillator is measured; however, an event is registered only if the scattered neutron enters a second crystal oriented at an angle ψ with respect to the direction of incidence of the neutron (i.e. if a delayed coincidence occurs). It follows that

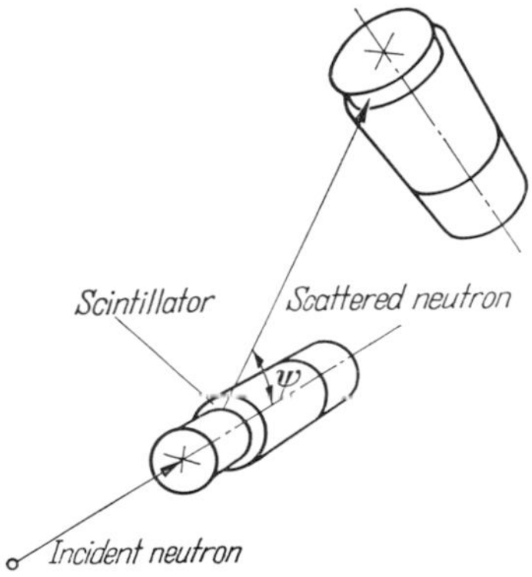

Fig. 3.2.3. Recoil proton coincidence spectrometer (MARION and FOWLER, loc. cit.)

$$E = E_p / \sin^2 \psi \qquad (3.2.6)$$

and after correction for the nonlinearity of the light yield, the neutron spectrum can be determined. This method commands but a modest sensitivity and resolution.

3.3. Other Methods of Detection

3.3.1. The Fission Counter

In the fission of a heavy nucleus, the fission fragments have a total kinetic energy of the order of 160 Mev. It is therefore possible to make extremely γ-insensitive fission chambers for neutron detection. Such detectors are important for neutron measurements in reactors (where there is a high γ-background), for measurements of the fission cross sections (Sec. 4.4.1), and as threshold detectors. Fig. 3.3.1 shows the fission cross section of some heavy nuclei as functions of the neutron energy; clearly discernible thresholds can be recognized at 0.26 Mev in U^{234}, 0.32 Mev in Np^{237}, 0.70 Mev in U^{236}, 1.3 Mev in U^{238}, and 1.5 Mev in Th^{232}. Fission chambers clad with these isotopes are frequently used with fast neutrons as threshold detectors, or more generally as spectral indicators[1]. The thermally fissionable isotopes U^{233}, U^{235}, and Pu^{239} also serve occasionally as spectral indicators in thermal neutron fields[2].

Fission chambers are generally filled with argon, and are operated in the ionization chamber range. The fissionable material is placed as a coating on the electrodes. In order to obtain a good pulse height spectrum, not more than 1 mg/cm² may be deposited, since otherwise the self-absorption of the fission products becomes noticeable. Because of this small surface density of uranium, it is difficult to make fission chambers with high detection efficiencies. In individual special cases, multiplate chambers with electrode surfaces of several thousand cm² have been built. When plutonium is used, the total quantity of fissionable material is limited to 10—20 mg because of the strong α-activity. As a rule, chambers for cross section measurements are built as simple parallel-plate chambers whose structure is particularly easy to examine, since they are to be used for as accurate an absolute counting as is possible. Electrodes in the form of coaxial cylinders are frequently used for flux measurements. A very detailed discussion of fission chambers can be found in an article by LAMPHERE[3].

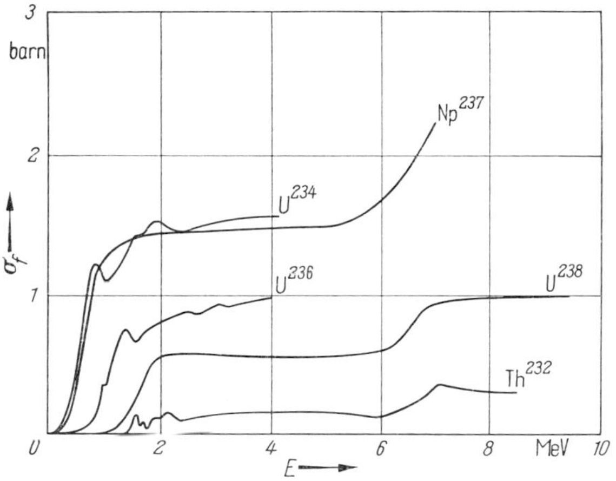

Fig. 3.3.1. The fission cross section of some heavy nuclei as a function of neutron energy

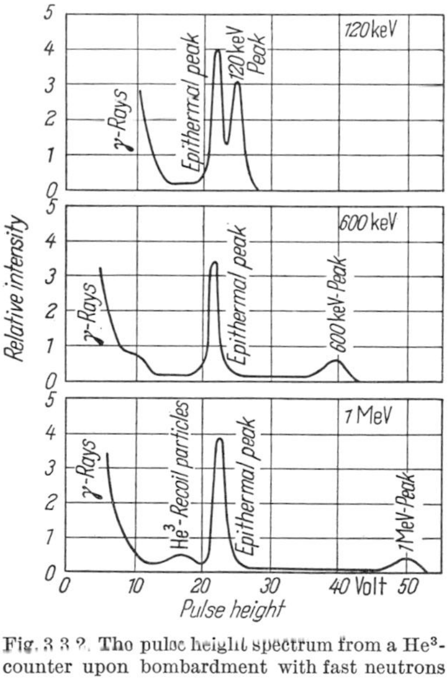

Fig. 3.3.2. The pulse height spectrum from a He^3-counter upon bombardment with fast neutrons

3.3.2. The He^3 Detector

Upon irradiation with neutrons of energy E, a proportional counter filled with helium-3 yields pulses whose height is proportional to $E+Q$. Q (=764 kev) is the reaction energy of the He^3 (n, p) H^3 reaction, which has been discussed in Sec. 1.4.1. This reaction is frequently used to investigate neutron spectra in the 200 kev to 2 Mev energy range. A cylindrical counter described by BATCHELOR, AVES, and

[1] Cf. Chapter 13.
[2] Cf. Sec. 15.2.
[3] MARION and FOWLER, p. 449.

SKYRME has a diameter of 2—1/8 inches, a sensitive length of 4 inches, and is filled with 27 cm of He^3 and 164 cm of krypton (to reduce wall and end effects). Fig. 3.3.2 shows the pulse height spectra measured with this counter under bombardment with 0.12-, 0.6-, and 1.0-Mev neutrons. At higher neutron energies, the pulses arising from the He^3-recoil nuclei become noticeably perturbing: in an elastic collision between a neutron of energy E_n and a stationary helium nucleus the latter receives an energy between 0 and $(\frac{3}{4})E_n$. For this reason, a very broad spectrum which reaches energies in the Mev range cannot easily be investigated with the He^3 spectrometer[1].

Chapter 3: References

General

ALLEN, W. D.: Neutron Detection. London: George Newnes Ltd. 1960.

BARSCHALL, H. H.: Detection of Neutrons. In: Handbuch der Physik, Bd. XLV, p. 437ff. Berlin-Göttingen-Heidelberg: Springer 1958.

MARION, J. B., and J. A. FOWLER (ed.): Fast Neutron Physics, part I: Techniques. New York: Interscience Publishers 1960, especially Sec. II: Recoil Detection Methods, and Sec. III: Detection by Neutron-Induced Methods.

ROSSI, B., and H. STAUB: Ionization Chambers and Counters. New York-Toronto-London: McGraw-Hill 1949.

Special[2]

TONGIORI, C. V., S. HAYAKAWA, and M. WIDGOFF: Rev. Sci. Instr. **22**, 899 (1951).

FERGUSON, G. A., and F. E. JABLONSKI: Rev. Sci. Instr. **28**, 893 (1957).

FOWLER, I. L., and P. R. TUNNICLIFFE: Rev. Sci. Instr. **21**, 734 (1950).

HAUSER, U. H.: Z. Naturforsch. **7a**, 781 (1952).

} BF_3 and $B(CH_3)_3$ Counters.

ALLEN, W. D., and A. T. G. FERGUSON: Proc. Phys. Soc. (London) A, **70** 639 (1957).

GRAVES, E. R., and R. W. DAVIS: Phys. Rev. **97**, 1205 (1955).

HANSON, A. O., and M. L. MCKIBBEN: Phys. Rev. **72**, 673 (1947).

MCTAGGART, M. H.: AWRE-Report NR/Al/59 (1958).

TOCHILIN, E., and R. V. ALVES: Nucl. Instr. Meth. **8**, 225 (1960).

} Long Counter.

BOLLINGER, L. M., and G. E. THOMAS: Rev. Sci. Instr. **28**, 489 (1957).

KOONTZ, P. G., G. R. KEEPIN, and J. E. ASHLEY: Rev. Sci. Instr. **26**, 352 (1955).

MUEHLHAUSE, C. O., and G. E. THOMAS: Phys. Rev. **85**, 926 (1952).

RAE, E. E., and E. M. BOWEY: Proc. Phys. Soc. (London) A **66**, 1073 (1953).

SUN, K. H., P. R. MALMBERG, and F. A. PECJAK: Nucleonics **14** No. 7, 46 (1956).

} Boron Scintillators.

BRAMBLETT, R. L., R. I. EWING, and T. W. BONNER: Nucl. Instr. Meth. **9**, 1 (1960).

FIRK, F. W. K., G. G. SLAUGHTER, and R. J. GINTHER: Nucl. Instr. Meth. **13**, 313 (1961).

MURRAY, R. B.: Nucl. Instr. Meth. **2**, 237 (1958).

STEDMANN, R.: Rev. Sci. Instr. **31**, 1156 (1960).

} Li^6 Neutron Detectors.

---

[1] However, as FERGUSON (EANDC-UK-17, 1962) has shown, recoil nuclei can be eliminated from the spectrum, if coincidence measurements between the two charged particles arising from the $He^3(n, p)H^3$-reaction are made. SAYRES (EANDC-33, 1963) has used pulse shape discrimination to extend the range of the He^3 detector to 8 Mev.

[2] Cf. footnote on page 53.

BROOKS, F. D.: In: Neutron Time-of-Flight Methods, p. 389. Brussels: Euratom 1961 (Neutron Detectors for Time-of-Flight Measurements).

ALLEN, W. D., and A. T. G. FERGUSON: Proc. Phys. Soc. (London) A **70**, 639 (1952).
SKYRME, T. H. R., P. R. TUNNICLIFFE, and A. G. WARD: Rev. Sci. Instr. **23**, 204 (1952).
Recoil Protons in Proportional Counters.

BROEK, H. W., and C. E. ANDERSON: Rev. Sci. Instr. **31**, 1063 (1960).
CRANBERG, L., R. K. BEAUCHAMP, and J. S. LEVIN: Rev. Sci. Instr. **28**, 89 (1957).
GETTNER, M., and W. SELOVE: Rev. Sci. Instr. **31**, 450 (1960).
HARDY, J. E.: Rev. Sci. Instr. **29**, 705 (1958).
MCCRARY, J. H., H. L. TAYLOR and T. W. BONNER: Phys. Rev. **94**, 808 (1954).
Recoil Protons in Scintillation Counters.

BROOKS, F. D.: Nucl. Instr. Meth. **4**, 151 (1959).
OWEN, R. B.: Nucleonics **17** No. 9, 92 (1959).
n-γ Separation by Pulse-Shape Discrimination.

HANDLOSER, J., and W. E. HIGINBOTHAM: Rev. Sci. Instr. **25**, 98 (1954).
HORNYAK, W. F.: Rev. Sci. Instr. **23**, 264 (1952).
Hornyak Button.

KRÜGER, G.: Nukleonik **1**, 237 (1959).
STETSON, R. F., and S. BERKO: Nucl. Instr. Meth. **6**, 94 (1960).
Direction Sensitivity of Recoil Proton Detectors.

DECKERS, H.: Nucl. Instr. Meth. **13**, 224 (1961).
NERESON, N., and F. REINES: Rev. Sci. Instr. **21**, 534 (1950).
ROSEN, L.: Nucleonics **11** No. 7, 32 (1953); Geneva 1955 P/582 Vol. 4 p. 97.
Observation of Recoil Protons in Photographic Plates.

EGGLER, C., *et al.*: Nucl. Sci. Eng. **1**, 391 (1956) (Observation of Recoil Protons in a Cloud Chamber).

BAME, S. J., *et al.*: Rev. Sci. Instr. **28**, 997 (1957).
BENENSON, R. E., and M. B. SHURMAN: Rev. Sci. Instr. **29**, 1 (1958).
CHAGNON, P. R., G. E. OWEN, and L. MADANSKY: Rev. Sci. Instr. **26**, 1156 (1955).
COCHRAN, R. G., and K. M. HENRY: Rev. Sci. Instr. **26**, 757 (1955).
GILES, R.: Rev. Sci. Instr. **24**, 986 (1953).
NERESON, M., and S. DARDEN: Phys. Rev. **89**, 775 (1953).
PERLOW, G. J.: Rev. Sci. Instr. **27**, 460 (1956).
SCHMIDT-ROHR, U.: Z. Naturforsch. **8a**, 470 (1953).
Recoil Proton Spectrometers.

BAER, W., and R. T. BAYARD: Rev. Sci. Instr. **24**, 138 (1953).
HICKMAN, G. D.: Nucl. Instr. Meth. **13**, 190 (1961).
LAMPHERE, R. W.: Phys. Rev. **91**, 655 (1953).
ROHR, R. C., E. R. ROHRER, and R. L. MACKLIN: Rev. Sci. Instr. **23**, 595 (1952).
Construction of Fission Chambers.

BATCHELOR, R., R. AVES, and T. H. R. SKYRME: Rev. Sci. Instr. **26**, 1037 (1955).
BATCHELOR, R., and H. R. MCHYDER: J. Nucl. Energy **3**, 7 (1956).
Helium-3 Detector.

4. Measurement of Cross Sections

In this chapter, a short survey will be given of the most important *differential* methods for determining cross sections. By differential methods is meant methods in which a substance is bombarded by a neutron beam and the cross section determined from the attenuation or scattering of the neutrons or from the intensity of the secondary radiation. As a rule, cross sections are measured as functions of the energy of monochromatic neutrons; occasionally, however, average values in a well-defined spectrum (Maxwellian spectrum of thermal neutrons, $1/E$-spectrum, fission spectrum) are measured. In later chapters, various *integral* methods for determining cross sections which are based on the interaction of a substance with a neutron field will be discussed in detail.

4.1. Transmission Experiments

4.1.1. Determination of the Total Cross Section

Fig. 4.1.1 schematically shows the experimental arrangement for a simple transmission experiment. If Z^0 is the counting rate at the detector with no sample, then after insertion of a sample of thickness d the counting rate is

$$Z^+ = Z^0 \cdot e^{-N \sigma_t d}. \tag{4.1.1}$$

If $T = Z^+/Z^0$ is defined as the transmission of the sample, the total cross section σ_t can be expressed as

$$\sigma_t = \frac{1}{Nd} \ln \frac{1}{T}. \tag{4.1.2}$$

In order that Eq. (4.1.1) apply rigorously, scattered neutrons must not reach the detector, i.e., the experiments must be carried out with "good geometry". Good geometry is best achieved by locating the detector as far from the source as possible and by making the detector and the sample small. In transmission experiments, the background of neutrons scattered from the walls of the laboratory must be carefully controlled; otherwise, the counting rate Z^+ can be badly in error, especially in the case of strong attenuation by the sample. The background is measured by placing a "shadow cone", which removes all the neutrons of the primary beam, between the source and the detector. In all other respects, the determination of σ_t is very easy and can be carried out rather accurately. The special merit of the transmission method is that it yields forthwith an absolute value of the cross section. In Chapter 1, we already familiarized ourselves with some typical results of this kind of $\sigma_t(E)$-measurement.

Fig. 4.1.1. Principle of a transmission experiment

As a rule, only transmission measurements with monoenergetic neutrons are meaningful. If the neutron beam has a broad energy spectrum $J(E)$, a simple interpretation of the experiment is no longer possible, since spectral changes during transmission through the sample affect the sensitivity of the detector[1]. Only if the sample is thin $\left(N d \sigma_t(E) \ll 1\right)$ and the detector efficiency independent of the energy do the following equations hold:

$$Z^+ = Z^0 e^{-N \bar{\sigma}_t d} \tag{4.1.3}$$

where

$$\bar{\sigma}_t = \frac{\int \sigma_t(E) J(E)\, dE}{\int J(E)\, dE} \tag{4.1.4}$$

i.e., only then is an effective cross section which is an average over the incident spectrum measured.

4.1.2. Determination of Resonance Parameters

Of particular interest are transmission measurements in the resonance region, where the cross sections change rapidly with energy. Fig. 4.1.2 shows a transmission curve covering a wide energy range that is typical of the curves obtained

[1] However, the process can be reversed, i.e., with a known cross section, information about the spectrum can be obtained from transmission measurements (cf. Sec. 15.2).

with a chopper. The cross section $\sigma_t(E)$ can be calculated from the energy-dependent transmission with the help of Eq. (4.1.2). However, with this procedure, the true form of the cross section is not obtained, but rather a form which in places is considerably modified by the Doppler effect (Sec. 7.4.3) and the limited resolving power of the spectrometer. Basically, the interesting thing is not the determination of the energy variation of the total cross section $\sigma_t(E)$, but rather the determination of the Breit-Wigner parameters of the resonances. For this reason, one tries to determine these quantities as directly as possible from the measured transmission curves, and in this way to eliminate perturbations caused

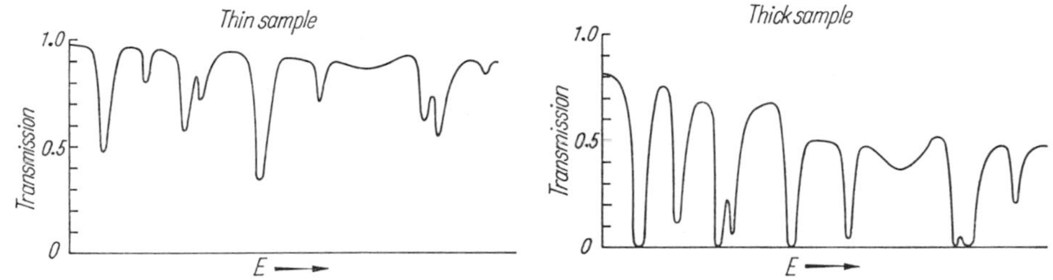

Fig. 4.1.2. Typical transmission curves in the resonance region

by the Doppler effect and the limited instrumental resolution. There are various methods for determining the resonance parameters directly; here we briefly discuss only the simple area method.

For the sake of simplicity, let us consider a non-fissionable nucleus (only Γ_γ and $\Gamma_n \neq 0$), and let us neglect potential scattering and thus also interference effects. In the neighborhood of a resonance, the following equation holds for the total cross section (cf. Sec. 1.3.4):

$$\sigma_t(E) = \pi \lambda\!\!\!^{-2} \frac{\Gamma_n \Gamma}{(E-E_R)^2 + (\Gamma/2)^2} \,. \tag{4.1.5a}$$

By introducing the abbreviation

$$\sigma_0 = \sigma_t(E_R) = 4\pi \lambda\!\!\!^{-2}_R \frac{\Gamma_n}{\Gamma} \tag{4.1.5b}$$

we can write Eq. (4.1.5a) as

$$\sigma_t(E) = \frac{\sigma_0}{1 + \left(\frac{E-E_R}{\Gamma/2}\right)^2} \,. \tag{4.1.6}$$

The transmission of a sample which contains $Nd = n$ atoms/cm² can then be written

$$T(E) = \exp\left[-\frac{n\sigma_0}{1 + \left(\frac{E-E_R}{\Gamma/2}\right)^2}\right] \tag{4.1.7}$$

The area under the transmission curve is

$$A = \int\limits_{E_R-\varepsilon}^{E_R+\varepsilon} \left\{1 - \exp\left[-\frac{n\sigma_0}{1 + \left(\frac{E-E_R}{\Gamma/2}\right)^2}\right]\right\} dE \,. \tag{4.1.8}$$

Here ε must be chosen large enough so that nearly the entire resonance is included in the integration. Then we can set $\varepsilon=\infty$ in the integration without introducing any significant error; the integration can now be carried out and gives

$$A(n\sigma_0, \Gamma)=\frac{\pi}{2}\, n\sigma_0 \Gamma e^{-n\sigma_0/2}\left[I_0(n\sigma_0/2)+I_1(n\sigma_0/2)\right] \tag{4.1.9a}$$

where I_0 and I_1 are modified Bessel functions of the first kind. In the limiting cases of extremely thin and extremely thick foils, Eq. (4.1.9a) simplifies to

$$\text{thin foils, } n\sigma_0 \ll 1: \quad A=\frac{\pi}{2}\, n\sigma_0 \Gamma. \tag{4.1.9b}$$

This result can be obtained directly from Eq. (4.1.8), since for $n\sigma_0 \ll 1$ the integrand can be replaced by the exponent.

$$\text{thick foils, } n\sigma_0 \gg 1: \quad A=\Gamma\sqrt{\pi n\sigma_0}. \tag{4.1.9c}$$

This result — at least within a numerical factor — can also be directly obtained from Eq. (4.1.8). When $n\sigma_0 \gg 1$, the exponent is large and the integrand is equal to unity if $(2(E-E_R)/\Gamma)^2 \ll n\sigma_0$; on the other hand, if $(2(E-E_R)/\Gamma)^2 \gg n\sigma_0$ the exponent and therefore the integrand vanish. The integral is thus approximately equal to the "band width" $\Delta E=\Gamma\sqrt{n\sigma_0}$.

Thus from transmission measurements on thin and thick foils and graphical integration of the transmission curves, two different combinations of the parameters σ_0 and Γ can be obtained. From these combinations, σ_0 and Γ can individually be found. Using Eq. (4.1.5b), Γ_n can then be found, and finally Γ_γ can be calculated as the difference of Γ and Γ_n. An advantage of this method is the fact that although the resolution function of the spectrometer changes the form of the transmission curve (broadens and flattens it), the area A under the transmission curve does not change. This can easily be shown by direct calculation, but becomes physically evident when one realizes that A is proportional to the total number of neutrons absorbed and scattered by the resonance, which naturally is not affected by the resolving power. However, A can change if the resonance is appreciably broadened by the Doppler effect (Sec. 7.4.3). In this case, the relations between A, σ_0, and Γ are summarized in diagrams published by PILCHER *et al.* However, Eq. (4.1.9b) and Eq. (4.1.9c) still hold in the limiting cases $n\sigma_0 \ll 1$ and $n\sigma_0 \gg 1$. Thus the area method always permits a simple determination of the resonance parameters from transmission measurements on thick and thin samples.

A serious complication arises from the fact that the form of the Breit-Wigner formula hitherto used only holds for reactions in which the target nucleus has zero spin and the neutron has zero orbital angular momentum. In the general case, a statistical weight factor

$$g=\frac{2J+1}{2(2I+1)} \tag{4.1.10}$$

appears in the Breit-Wigner formula; here, I is the spin of the target nucleus and J is the spin of the compound nucleus. For s-wave resonances ($l=0$), $J=I+\frac{1}{2}$ or $I-\frac{1}{2}$. As a rule, the spin J of the compound nucleus is not known; therefore, the g-factor appears as an additional unknown. Eq. (4.1.6) still holds for the total cross section, but now σ_0 is given by

$$\sigma_0=4\pi\lambda\!\!\!^{-2}_R\, g\,\frac{\Gamma_n}{\Gamma}. \tag{4.1.11}$$

From analysis of the transmission curves σ_0 and Γ can be found as above, but now only the combination $g\Gamma_n$ can be obtained from them. If it is desired to find Γ_n and thus to determine Γ_γ, g must be known. For example, if $I=\frac{1}{2}$, then according to Eq. (4.1.10), the possible values of g are $\frac{1}{4}$ or $\frac{3}{4}$; thus, a considerable uncertainty exists. In some cases, the appropriate J- and g-values can be guessed, as for example, when $\Gamma_n \gg \Gamma_\gamma$ or when the assumption of the smaller of the two possible g-values leads to a value of $\Gamma_n > \Gamma$. In most cases, however, Γ_n, Γ_γ, or J must be determined directly by additional measurements. For this purpose, resonance scattering or resonance capture may be studied, or J determined through bombardment of polarized nuclei by polarized neutrons. However, all these experiments are very difficult to carry out; thus the g-factor introduces a complicating feature into all resonance cross section measurements.

4.1.3. The Determination of the Nonelastic Cross Section

Fig. 4.1.3 shows an arrangement frequently used to measure the nonelastic cross section

$$\left.\begin{aligned}\sigma_{ne} &= \sigma_t - \sigma_{n,n}\\ &= \sigma_{n,n'} + \sigma_{n,\gamma} + \sigma_{n,f} + \cdots\end{aligned}\right\} \quad (4.1.12)$$

of heavy nuclei for fast neutrons. The substance to be investigated surrounds the source, which is assumed to emit isotropically, in the form of a concentric spherical shell of thickness d. At a distance l large compared to the sphere radius R is found the detector, which is operated as a threshold detector and does not respond to any neutrons which have suffered an inelastic collision. If Q is the source strength, then in the absence of any spherical shell, the current J at the detector is given by

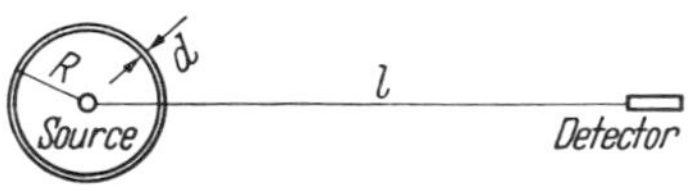

Fig. 4.1.3. Principle of a spherical shell transmission experiment

$$J = \frac{Q}{4\pi l^2}\,. \quad (4.1.13)$$

When the spherical shell is in place the current is given by

$$J' = \frac{Q}{4\pi l^2} \cdot e^{-Nd\sigma_t} + J_s \quad (4.1.14)$$

where J_s is the current of neutrons elastically scattered in the shell. The inelastically scattered neutrons are not recorded and therefore can be neglected. For thin shells, $N\sigma_t d \ll 1$, and because $l \gg R$, it clearly follows that

$$J_s = \frac{QNd\sigma_{n,n}}{4\pi l^2}\,. \quad (4.1.15)$$

Thus

$$J' = \frac{Q}{4\pi l^2}(1 - Nd\sigma_t) + \frac{Q}{4\pi l^2} Nd\sigma_{n,n} \quad (4.1.16)$$

or

$$J' = \frac{Q}{4\pi l^2} e^{-Nd\sigma_{ne}}\,. \quad (4.1.17)$$

Thus a transmission experiment with a thin spherical shell yields the nonelastic cross section directly. As Bethe, Beyster, and Carter have shown, under special circumstances, Eq. (4.1.17) holds even when the condition $Nd\sigma \ll 1$

is not fulfilled. It can furthermore be shown that the neutron paths in this experiment can be reversed, i.e., that Eq. (4.1.17) also holds when the spherical shell surrounds the detector. Use is made of this fact in practice, since most (p, n) or (d, n) sources do not emit isotropically, whereas isotropically sensitive detectors are easy to make.

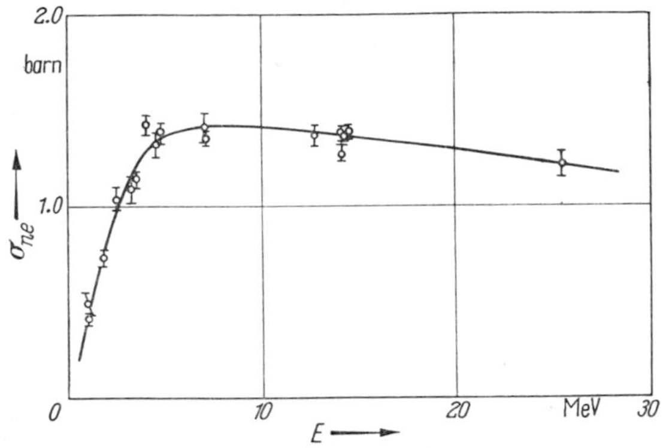

Fig. 4.1.4. The non-elastic cross section of Iron as a function of neutron energy

Fig. 4.1.4 shows the non-elastic cross section of iron, which exhibits a behavior typical of most nuclei, determined by means of a spherical shell experiment. The cross section rises steeply until an energy of several Mev; then it is fairly constant over a wide energy range and is approximately equal to the geometric cross section. The total cross section is about twice as big (cf. the discussion at the end of Sec. 1.3). At high energies, in many cases, the nonelastic cross section is equal to the cross section for inelastic scattering, since competing reactions can be neglected. The spherical shell method thus is a simple means for the determination of the cross section for inelastic scattering.

A further important application of this method of measurement is the following: In non-fissionable nuclei, absorption is the only process competing with neutron scattering below the energy at which inelastic scattering begins. Therefore, a measurement of the spherical shell transmission directly yields the absorption cross section, which in most cases is equal to the cross section for radiative capture. This procedure is rather inaccurate if σ_a is not larger than σ_s.

4.2. Scattering Experiments

4.2.1. Elastic Scattering and Angular Distributions

Fig. 4.2.1 shows a typical arrangement for the study of elastic neutron scattering: The neutrons from the source strike a scattering sample whose dimensions must be small compared to the scattering mean free path in order that the results not be perturbed by multiple scattering. Oriented at a variable angle ϑ_0 is a detector, which is carefully shielded against the direct neutron beam as well as the background of neutrons scattered in the room. If $\Delta\Omega$ is the solid angle subtended by the detector, then with a sample volume V and a detector efficiency ε the counting rate is

$$Z = J\,V N \frac{1}{2\pi}\,\sigma_{n,n}(\cos\vartheta_0)\,\varepsilon\,\Delta\Omega\,. \tag{4.2.1}$$

$\sigma_{n,n}(\cos\vartheta_0)$ is the "differential scattering cross section"; the total scattering cross section is obtained from it by integration

$$\sigma_{n,n} = \int_0^\pi \sigma_{n,n}(\cos\vartheta_0)\,\sin\vartheta_0\,d\vartheta_0\,. \tag{4.2.2}$$

According to Eqs. (4.2.1) and (4.2.2), the absolute values of J and ε are involved in an absolute measurement of $\sigma_{n,n}$. One can, however, easily make relative measurements with respect to a standard whose scattering cross section is known from other measurements (e.g., for many light nuclei, $\sigma_{n,n}=\sigma_t$ since $\sigma_{ne} \ll \sigma_{n,n}$). In a frequently employed variant of this method, the scatterer has the form of a circular ring, upon whose axis lie both the source and detector. In this case, the direct radiation is shielded by a shadow cone, which lies between the source and the detector. In scattering experiments, there frequently arises the necessity of eliminating the effect of the inelastically scattered neutrons on the experimental results. For this purpose, one of the threshold detectors discussed in Chapter 3 can be used; frequently, the time-of-flight method, whose application to scattering measurements is discussed in Sec. 4.2.2, is used.

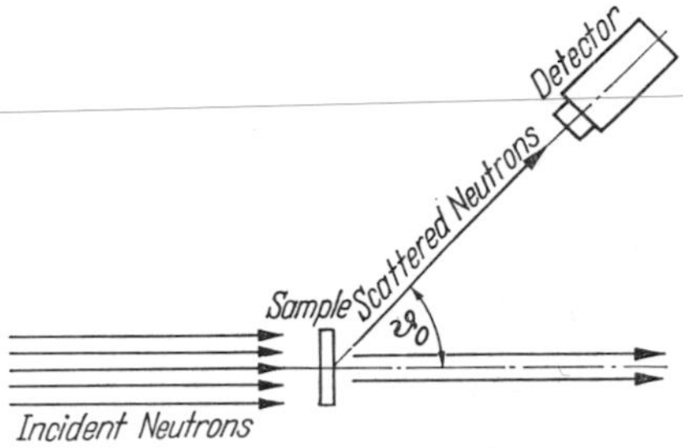

Fig. 4.2.1. Principle of an experiment to determine the angular distribution in neutron scattering

If the scattering substance is suitable for use as the filling gas of a proportional counter (hydrogen or the noble gases), a fundamentally different method for investigating the angular distribution in neutron scattering becomes possible. On the basis of the laws of elastic collision, there is a direct relation between the energy distribution of the recoil nuclei and the angular distribution of the neutrons (Sec. 7.1). If the pulse spectrum of the recoil nuclei in a proportional counter is measured with a pulse height analyzer, then after correction for the wall and end effects, the angular distribution can be obtained.

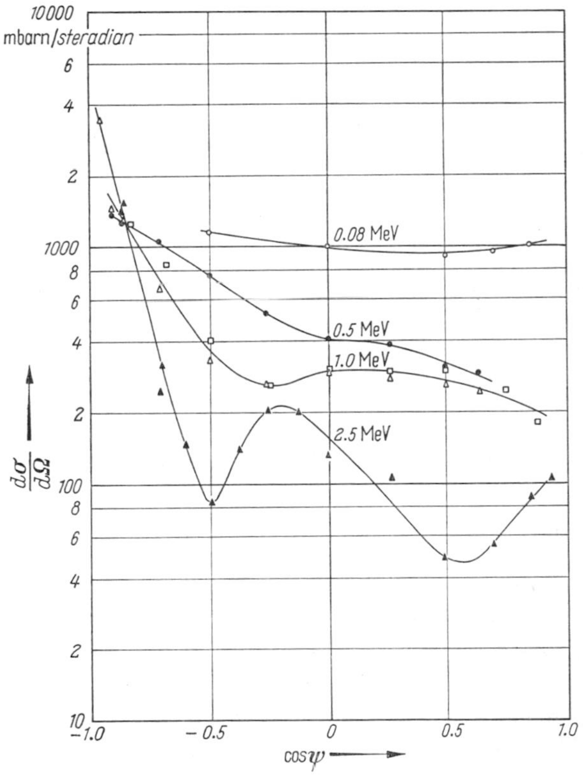

Fig. 4.2.2. Angular distribution (in the C. M. system) of neutrons scattered elastically by Iron

The angular distributions of neutrons elastically scattered from some light nuclei were discussed in Sec. 1.4.2. Fig. 4.2.2 shows the angular distribution of the neutrons scattered elastically from iron, which is typical of the heavier nuclei. Although for low energies the scattering is nearly isotropic in the center-of-mass system, with increasing energy a strong anisotropy becomes evident: the forward direction is ever more preferred. In many cases, the experimentally observed angular distributions of elastically scattered neutrons can be described by the optical model, which will not be further treated here.

4.2.2. Inelastic Scattering

An inelastic scattering process is characterized by specification of a differential scattering cross section $\sigma_{n,n'}(E', E, \cos\vartheta_0)$; here E' is the energy of the neutron before and E the energy after the collision. ϑ_0 is the scattering angle. A measurement of this cross section can be made with the apparatus shown in Fig. 4.2.1. In addition, however, the energy E after the collision must be measured. In principle, the neutron spectrometers discussed in Chapter 3 can be used for these measurements, but in most cases their resolution and sensitivity are inadequate. In most cases, one uses the time-of-flight method, whose principle can be explained by means of Fig. 4.2.3. With the help of one of the (p, n) or (d, n) reactions discussed in Chapter 2, a van de Graaff generator produces monochromatic neutrons of energy E'. By use of an electrostatic deflection mechanism, the proton or deuteron beam is pulsed in such a way that very short pulses (a few nsec long) are produced with high frequency (10^5—10^6 per sec). These neutrons are scattered by the sample, and the time-of-flight spectrum of the neutrons scattered at an angle ϑ_0 is recorded with the help of a "fast" detector and a multichannel time analyzer. The energy spectrum of the scattered neutrons can be calculated from the time-of-flight spectrum with the formulas given in Sec. 2.5.5; for this purpose, the energy dependence of the detector efficiency must of course be known. The resolution and the neutron intensity necessary for good measurement can generally be achieved only with difficulty. Among the important perturbations are time-dependent background effects which are traceable to room-scattered neutrons. If one succeeds in completely determining the differential scattering cross section $\sigma_{n,n'}(E', E, \cos\vartheta_0)$ in this way, one can obtain the inelastic scattering cross section $\sigma_{n,n'}(E', E)$ by integration over all angles and the total inelastic scattering cross section $\sigma_{n,n'}(E')$, which can also be measured by the spherical shell method, by further integration over all energies.

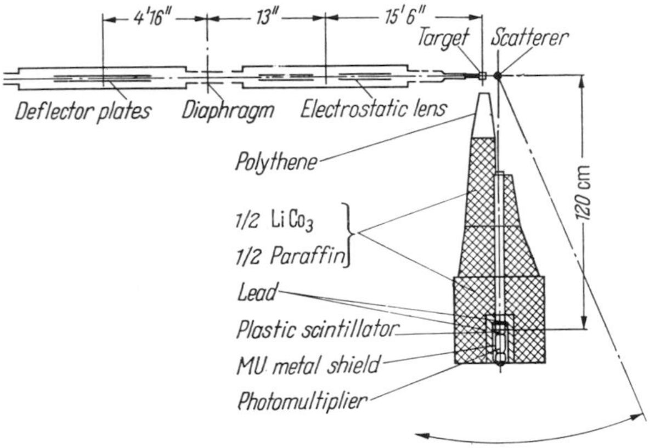

Fig. 4.2.3. Time-of-flight assembly used by CRANBERG for inelastic scattering measurements

If only a few levels of the target nucleus contribute to the inelastic scattering, the inelastic excitation cross section $\sigma_{n,n'}(E', E_J)$ can be measured as a function of the neutron energy E' by measuring the intensity of the γ-radiation accompanying the deexcitation of the state with a NaI(Tl) crystal[1]. Here, as in the time-of-flight method, the determination of the absolute value of the inelastic scattering cross section is particularly difficult.

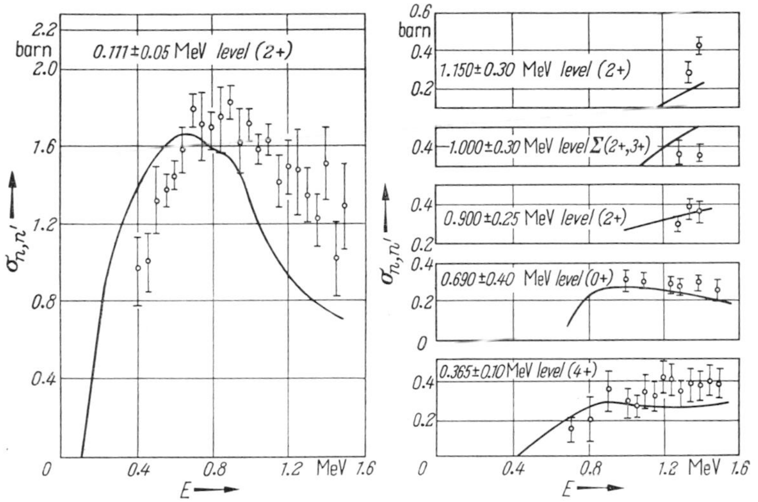

Fig. 4.2.4. The inelastic excitation cross section for the first levels of W^{184}. ϕ Experimental values; —— theoretical fit [from BNL 818 (T-317)]

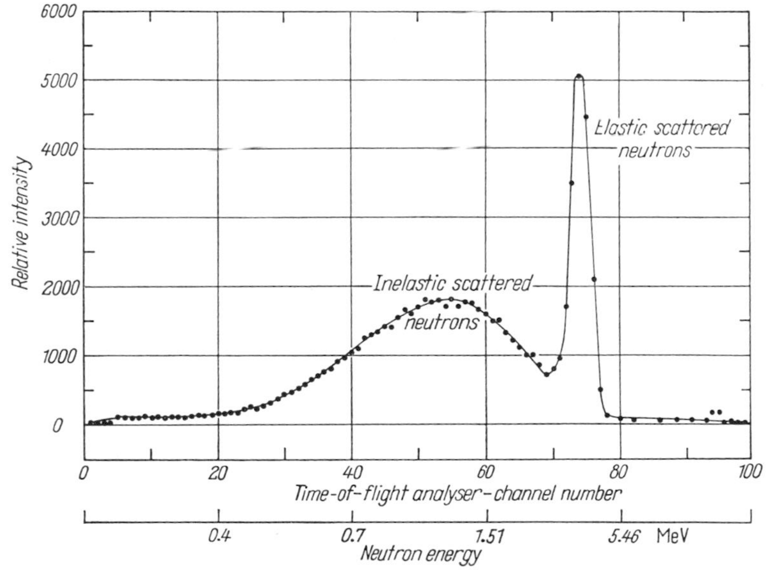

Fig. 4.2.5. The energy spectrum of 5-Mev neutrons scattered inelastically through 90° by Tantalum (Buccino et al., unpublished)

Figs. 4.2.4 and 4.2.5 show some results of inelastic scattering experiments. For low energies, it is possible to separate the contributions of individual levels of the target nucleus from one another, and thus to determine the inelastic excitation

[1] Here E_J is the energy of the target nucleus excited by the inelastic collision; the energy of the scattered neutron is given simply by $E = E' - E_J$ if the recoil energy of the nucleus is neglected.

cross sections $\sigma_{n,n'}(E', E_J)$. At high incident energies, so many levels contribute to the inelastic cross section that a separation into individual excitation cross sections is no longer meaningful. Experience shows that the process of emission of the neutron from the compound nucleus can then be treated approximately as a nuclear evaporation process; thus

$$\sigma_{n,n'}(E', E) = \sigma_{n,n'}(E') \frac{E}{E_0^2} e^{-E/E_0}. \tag{4.2.3}$$

The "nuclear temperature" E_0 can be determined from the energy distribution of the inelastically scattered neutrons; E_0 depends on E' and has an order of magnitude of 1 Mev. The angular distribution of the inelastically scattered neutrons can be taken in first approximation to be isotropic.

4.3. Experiments for Determining Reaction Cross Sections

Here we consider all reactions except fission: thus we consider (n, α), (n, p), (n, γ), $(n, 2n)$, etc. reactions. There is no single method for determining the cross sections of all these reactions; the various procedures which are usual can be classified as follows:

(a) If $\sigma_{n,x} \approx \sigma_t$, the cross section can be determined from a transmission experiment. This is the case for many (n, γ) reactions with slow neutrons and, e.g., for the slow neutron (n, α) reaction on B^{10}. If the scattering cross section cannot be neglected compared with the reaction cross section, spherical shell transmission experiments can be undertaken.

(b) The cross sections for reactions which lead to the absorption of a neutron can often be determined using the integral methods to be discussed later. These integral methods can be applied even when the reaction cross section is very small compared to the scattering cross section.

(c) If the reaction leads to a stable residual nucleus, the yield and therefore the cross section can be determined by mass spectrometric analysis. This method is applicable in only a few cases and only after extremely strong irradiation.

(d) In many cases, the reaction leads to a radioactive residual nucleus whose activity after the neutron irradiation can be determined by means of its electron, positron, or γ-radiation. If a thin sheet ($N d \sigma_{\text{act}} \ll 1$) of the substance being studied is irradiated for a time t_1 by a neutron current J ($\text{cm}^{-2}\,\text{sec}^{-1}$), its activity A (the number of decays per sec and cm^2 of surface) at a time t_2 after the end of the irradiation[1] is given by

$$A = J N \sigma_{\text{ac}}\, d (1 - e^{-\lambda t_1}) e^{-\lambda t_2}. \tag{4.3.1}$$

Here $\lambda = 0.693/T_{\frac{1}{2}}$ is the radioactive decay constant of the residual nucleus. If J and A can be absolutely determined[2], σ_{act} can be calculated easily. This activation method is the most frequently used procedure for measuring reaction cross sections. Nearly all high-energy (n, p), (n, α), and $(n, 2n)$ cross sections have been determined in this way. We will acquaint ourselves with some examples in Chapter 13.

---

[1] Cf. Sec. 11.1.

[2] Cf. Chapter 14.

(e) If it is possible to incorporate the target substance in a detector (e.g., as a gas, such as He^3, or as a component of a scintillation crystal, such as LiI or CsI, etc.), the (n, p) and (n, α) reactions can be studied by direct observation of the pulse height spectra from the detectors. (n, p) and (n, α) cross sections have occasionally been determined by direct detection of the protons and α-particles ejected from extremely thin foils.

(f) The cross section for radiative capture can be determined by direct observation of the γ-radiation arising from the neutron bombardment of a target substance. Since the transition from the excited compound state to the ground state usually does not take place directly but rather passes through various intermediate states, the emitted γ-spectrum is complex. It can change sharply with changing energy of the captured neutron, since different intermediate states come into play. The number of γ-rays emitted is therefore not a meaningful measure of the capture cross section. On the other hand, if all the γ-rays arising from one capture process are summed up in a large scintillation counter (a large tank of a liquid scintillator) and only those pulses whose height corresponds to the excitation energy of the compound nucleus (i.e., the sum of the kinetic energy and the neutron binding energy in the compound nucleus) are accepted, then a counting rate proportional to the capture cross section is obtained. With this technique, the (n, γ) cross sections of nuclides which are not activated by radiative capture can be measured. Furthermore, the technique can be applied in connection with the time-of-flight method in the energy range from 10 ev to 100 kev, where truly monoenergetic neutron sources for activation experiments are not available.

With the exception of the transmission method and some of the integral methods, none of the methods mentioned above gives an absolute value for the cross section; in order to obtain an absolute value, the incident neutron flux as well as the reaction rate must be determined absolutely. Both measurements are difficult, although the flux measurement is the more difficult of the two, and for this reason, the published values for reaction cross sections contain many inaccuracies. In many cases, measurements are limited to determination of the cross sections relative to some standard cross section.

4.4. Experiments on Fissionable Substances

Although multiplying media are not treated in this book, the most important methods for the determination of the parameters of fissionable substances will be discussed here briefly, because they are among the important techniques of measurement in neutron physics. Among the properties of a fissionable substance, the most interesting are the fission cross section $\sigma_{n,f}(E)$, the capture cross section $\sigma_{n,\gamma}(E)$, and the average number $\bar{\nu}$ of secondary neutrons emitted in fission. The capture-to-fission ratio

$$\alpha(E) = \frac{\sigma_{n,\gamma}(E)}{\sigma_{n,f}(E)} \tag{4.4.1}$$

and the neutron yield per neutron absorbed in the fissionable material

$$\eta(E) = \frac{\bar{\nu}(E)\,\sigma_{n,f}(E)}{\sigma_{n,f}(E) + \sigma_{n,\gamma}(E)} = \frac{\bar{\nu}(E)}{1 + \alpha(E)} \tag{4.4.2}$$

are also frequently used. In principle, there are independent methods of measuring all five of these parameters, but they cannot be applied to all nuclei in all energy ranges. However, since only three of these parameters are independent, three suitable measurements are enough.

4.4.1. Measurement of Fission Cross Sections

The principle of a fission cross section measurement is simple: the fissionable substance is placed in one of the fission chambers described in Sec. 3.3.1 and bombarded with neutrons. If the fission rate can be determined absolutely and if the neutron current and the total amount of fissionable material present are known, $\sigma_{n,f}$ can be easily calculated. In practice, the absolute determination of

Table 4.4.1. *The Fission Cross Section of U^{235} for Slow Neutrons ($v_0 = 2200$ m/sec)*

Author	Result	Method
DERUYTTER (1961)	587 ± 6 barn	Absolute counting of the fissions in a 4π-fission chamber. Flux measurement by activation of Au^{197}
FRIESEN, SEPPI, and LEONARD[1] (1956)	557 barn	Absolute counting of the fissions in a fission chamber. Flux measurement by activation of Au^{197}
RAFFLE (1959)	590 ± 12 barn	(same as above)
BIGHAM (1958)	568 ± 7 barn	(same as above)
SAPLAKOGLU (1958)	605 ± 6 barn	Fission counting by means of the secondary neutrons. Absolute measurement of the flux by total absorption in a B^{10} plate

[1] Quoted by G. C. HANNA in AECL-873 (1960).

the fission rate, the determination of the quantity of uranium or plutonium present, and the measurement of the flux prove difficult. Although, as a rule, the energy variation of $\sigma_{n,f}$ can be accurately determined, our knowledge of the absolute values of fission cross sections is still unsatisfactory in many ways. Table 4.4.1 illustrates this by means of some new results for $\sigma_{n,f}(E = 0.0253\text{ ev})$ of U^{235}.

In order to determine the fission cross section in the resonance region, the Breit-Wigner parameters Γ_n, Γ, g, and the fission width Γ_f must be known. Γ_f follows immediately from combining a transmission and a fission rate measurement. Transmission measurements with thick and thin foils give σ_0 and Γ, according to Sec. 4.1.2. If one determines the variation of $\sigma_{n,f}(E)$ over a resonance and calculates the area under the curve, one then obtains

$$A_f = \int n\sigma_{n,f}(E)\,dE = \frac{\pi}{2}\,n\sigma_0\Gamma_f. \tag{4.4.3}$$

By combining this area with the transmission data, one can immediately calculate Γ_f. Frequently, instead of the fission cross section, the quantity η is determined in the resonance region; at the maximum of a resonance

$$\frac{\eta}{\bar{\nu}} = \frac{\Gamma_f}{\Gamma_f + \Gamma_\gamma} \tag{4.4.4}$$

and Γ_f can again be obtained by combination with the transmission data. The description of the resonances in fissionable substances by the Breit-Wigner single-level formula is not very accurate, however, and lately people have tried to fit the measured cross sections with the multilevel formula.

4.4.2. The Determination of $\bar{\nu}$, η, and α

In order to measure $\bar{\nu}$, the average number of secondary neutrons emitted in fission, the substance being investigated is placed in a fission chamber which is then bombarded by neutrons. If all fissions are registered and if Z is the total fission rate, then the number of secondary neutrons emitted is given by

$$Q = Z\bar{\nu}. \tag{4.4.5}$$

Q is determined by one of the standard methods for measuring source strengths discussed in Chapter 14. The majority of $\bar{\nu}$-measurements hitherto carried out have been measurements relative to $\bar{\nu}$ for the thermal fission of U^{235}.

η is measured in accordance with its definition as the ratio of the number of neutrons emitted by a fissionable sample to the number of neutrons absorbed. At low neutron energies, such a measurement is simple: since $\sigma_{n,f}+\sigma_{n,\gamma} \gg \sigma_s$, all incident neutrons are absorbed in a sufficiently thick sample, and one needs only to measure the intensity of the incident neutron beam and the number of secondary neutrons. At higher energies, on the other hand, a thin sample is used in order to keep the scattering small, and the number of neutrons absorbed is determined from the transmission. The number of secondary neutrons is determined as in $\bar{\nu}$-measurements by a standard method to be discussed later.

The capture-to-fission ratio can be determined by placing a fissionable sample in the middle of a large tank of liquid scintillator and bombarding it with mono-energetic neutrons from a pulsed source. Each time a neutron is captured, the absorption of the capture γ-rays produces in the scintillator a pulse whose height is proportional to the total excitation energy of the compound nucleus. However, if a fission occurs, there first occurs a pulse due to the prompt fission γ-radiation, followed, after a delay of 10 to 100 μsec, by another pulse due to the capture of the thermalized fission neutrons. With suitable electronic circuitry, capture and fission events can be distinguished and α measured directly. Another method is based on the mass spectrometric analysis of a sample which has been irradiated in a reactor for a long time; in this method, only an average value of α over the often only poorly known reactor spectrum is obtained. Finally, the ratio $\sigma_{n,f}/\sigma_t$ can be determined if the total fission rate and the transmission of a sample are simultaneously measured. For very slow neutrons $\sigma_s \ll \sigma_t$, and $1+\alpha$ can be obtained after only a small correction.

$\sigma_{n,f}$, $\sigma_{n,\gamma}$, α, $\bar{\nu}$, and η at 2200 m/sec for U^{233}, U^{235}, and Pu^{239} are tabulated in Table 4.1.2. Since only three of these five parameters are independent, the various experimental values must be adjusted to satisfy the relations 4.4.1 and 4.4.2 in order to obtain a "consistent" set of parameters. This procedure is not free from arbitrariness, as we shall show with the following example concerning U^{235} [1]:

[1] It should be noted that other authors, e.g., SHER and FELDERBAUM, employed considerably more involved averaging procedures than those used here.

η: The most trustworthy measured value by far is that determined by MACKLIN *et al.* by the manganese bath method (Chapter 13); they obtained 2.077 ± 0.010.

$1+\alpha$: From $\sigma_{n,f}/\sigma_t$ SAFFORD and MELKONIAN obtained $1+\alpha = 1.160 \pm 0.010$[1]. The value $\bar{\nu} = 2.41 \pm 0.02$ can be derived from these two data using Eq. (4.4.2). This value is close to the directly determined values (KENWARD *et al.*: 2.420 ± 0.037; MOAT *et al.*: 2.39 ± 0.05; COLVIN and SOWERBY: 2.43 ± 0.037).

Table 4.4.2. *Parameters of Fissionable Substances at $v_0 = 2200$ m/sec*

Nucleus	$\sigma_{n,f}$ (barn)	$\sigma_{n,\gamma}$ (barn)	$1+\alpha$	η	$\bar{\nu}$	Source
U^{233}	524 ± 4	49 ± 2	1.0935 ± 0.0038	2.291 ± 0.009	2.50 ± 0.012	EVANS and FLUHARTY
U^{235}	586 ± 7	94 ± 5	1.160 ± 0.010	2.077 ± 0.010	2.410 ± 0.020	See the text
Pu^{239}	748.2 ± 4.9	281.9 ± 6	1.377 ± 0.011	2.093 ± 0.014	2.882 ± 0.016	SHER and FELDERBAUM

From a transmission experiment, BLOCK, SLAUGHTER, and HARVEY found a value of 680 ± 6 barn for the absorption cross section $\sigma_a = \sigma_{n,f} + \sigma_{n,\gamma}$; using the best value of α given above, one finds according to Eq. (4.4.1) a fission cross section of 586 ± 7 barn and a capture cross section of 94 ± 5 barn. This indirectly determined value of the fission cross section agrees very well with the directly determined values of RAFFLE and of DERUYTTER.

Chapter 4: References

General

HUGHES, D. J.: Neutron Cross Sections. London-New York-Paris: Pergamon Press 1957.

MARION, J. B., and J. L. FOWLER (ed.): Fast Neutron Physics, part II: Experiments and Theory. New York: Interscience Publishers 1963.

SPAEPEN, J. (ed.): Neutron Time-of-Flight Methods. Proceedings of an EANDC Symposium, Brussels: Euratom 1961. Especially Session II: Examples of the Use of Time-of-Flight Methods.

Special[2,3]

Transmission Experiments with Fast Neutrons:

NERESON, N., and S. DARDEN: Phys. Rev. **89**, 775 (1953); **94**, 1678 (1954) and LA-1655 (1954).

OKAZAKI, A., *et al.*: Phys. Rev. **93**, 461 (1954).

WALT, M., *et al.*: Phys. Rev. **89**, 1271 (1953).

Transmission Experiments in the Resonance Range:

BJERRUM-MØLLER, H., F. J. SHORE, and V. L. SAILOR: Nucl. Sci. Eng. **8**, 183 (1960).

CARPENTER, R. T., and L. M. BOLLINGER: Nucl. Phys. **21**, 66 (1960).

NEWSON, H., *et al.*: Ann. Phys. **8**, 211 (1959); **14**, 346, 365 (1961).

ROSEN, J. L., *et al.*: Phys. Rev. **118**, 687 (1960).

Methods for Determining Resonance Parameters:

LANE, R. O., N. F. MOREHOUSE, and D. L. PHILLIPS: Nucl. Instr. Meth. **9**, 87 (1960).

LYNN, J. E.: Nucl. Phys. **7**, 599 (1957).

—, and E. R. RAE: J. Nucl. Energy **1**, 418 (1957).

PILCHER, V. E., J. A. HARVEY, and D. J. HUGHES: Phys. Rev. **103**, 1342 (1956).

---

[1] Actually, a value of 1.171 was obtained at $E = 0.00291$ ev; this value was then corrected to 2200 m/sec, using the known energy dependence of the cross sections.

[2] This field has an exceptionally copious literature, and the works cited here represent only a limited selection made principally on methodological grounds.

[3] Cf. footnote on page 53.

Fox, J. D., *et al.*: Phys. Rev. **110**, 1472 (1958).
London, H. H., and E. R. Rae: Phys. Rev. **107**, 1333 (1957).
Rae, E. R., *et al.*: Nucl. Phys. **5**, 89 (1958).
Wood, R. E.: Phys. Rev. **104**, 1425 (1956).

Determination of the g-Factor by Scattering and Capture Measurements.

Bernstein, S., *et al.*: Phys. Rev. **94**, 1243 (1954).
Postma, H., *et al.*: Phys. Rev. **126**, 979 (1962).
Sailor, V. L., *et al.*: Bull. Am. Phys. Soc. **6**, 275 (1961).
Stolovy, A.: Phys. Rev. **118**, 211 (1960).

Determination of the Spin of the Capture State Through Polarization Experiments.

Bethe, H. A., J. R. Beyster, and R. E. Carter: J. Nucl. Energy **3**, 207, 273 (1956); **4**, 3, 147 (1957).
Beyster, J. R., *et al.*: Phys. Rev. **98**, 1216 (1955).
Graves, E. R., and R. W. Davis: Phys. Rev. **97**, 1205 (1955).
MacGregor, M. H., R. Booth, and W. P. Ball: Phys. Rev. **130**, 1471 (1963); **108**, 726 (1957).

Measurement of the Non-Elastic Cross Section by Spherical Shell Transmission.

Macklin, R. L., H. W. Schmitt, and J. H. Gibbons: ORNL-2022 (1956); Phys. Rev. **102**, 797 (1956).
Schmitt, H. W., and C. W. Cook: Nucl. Phys. **20**, 202 (1960).

Measurement of Absorption Cross Sections by Spherical Shell Transmission.

Beyster, J. R., M. Walt, and E. W. Salmi: Phys. Rev. **104**, 1319 (1956).
Cohn, H. O., and J. L. Fowler: Phys. Rev. **114**, 194 (1959).
Coon, J. H., *et al.*: Phys. Rev. **111**, 250 (1958).
Gilboy, W. B., and J. H. Towle: Nucl. Phys. **42**, 86 (1963).
Langsdorf, A., R. O. Lane, and J. E. Monahan: Phys. Rev. **107**, 1077 (1957).
Seagrave, J. D., L. Cranberg, and J. E. Simmons: Phys. Rev. **119**, 1981 (1960).

Investigations of Elastic Scattering and Its Angular Distribution.

Cranberg, L.: LA-2177 (1959).
—, and J. S. Levin: Phys. Rev. **103**, 343 (1956); **109**, 2063 (1958).
Reitmann, D., C. A. Engelbrecht, and A. M. Smith: Phys. Rev. (in press).
Smith, A. B.: Z. Physik **175**, 242 (1963).

Measurement of Inelastic Scattering by the Time-of-Flight Method.

Day, R. B.: Phys. Rev. **102**, 767 (1956).
—, and M. Walt: Phys. Rev. **117**, 1330 (1960).
Freeman, Joan: Progr. Nucl. Phys. **5**, 37 (1956).
van Loef, J. J., and D. A. Lind: Phys. Rev. **101**, 103 (1956).

Measurement of Inelastic Scattering by Measurement of the Associated γ-Radiation.

Hürlimann, T., and P. Huber: Helv. Phys. Acta **28**, 33 (1955).
Kern, B. D., W. E. Thompson, and J. M. Ferguson: Nucl. Phys. **10**, 226 (1959).
Rapaport, J., and J. J. van Loef: Phys. Rev. **114**, 565 (1959).
Terrell, J., and D. M. Holm: Phys. Rev. **109**, 2031 (1958).

Investigation of (n, p) Reactions by Activation.

Grundl, J. A., R. L. Henkel, and B. L. Perkins: Phys. Rev. **109**, 425 (1958).
Schmitt, H. W., and J. Halperin: Phys. Rev. **121**, 827 (1961).

Investigation of (n, α) Reactions by Activation.

Bayhurst, B. P., and R. J. Prestwood: LA-2493 (1962); Phys. Rev. **121**, 1438 (1961).
Ferguson, J. M., and W. E. Thompson: Phys. Rev. **118**, 228 (1960).

Investigation of $(n, 2n)$ Reactions by Activation.

Bame, S. J., and R. L. Cubitt: Phys. Rev. **113**, 256 (1959).
Cox, S. A.: Phys. Rev. **122**, 1280 (1961).
Hanna, R. C., and B. Rose: J. Nucl. Energy A **8**, 197 (1959).
Macklin, R. L., N. H. Lazar, and W. S. Lyon: Phys. Rev. **107**, 504 (1957).

Investigation of (n, γ) Reactions by Activation.

Block, R. C., *et al.*: In: Neutron Time-of-Flight Methods, p. 203. Brussels: Euratom 1961.
Diven, B. C., J. Terrel, and A. Hemmendinger: Phys. Rev. **120**, 556 (1960).
Gibbons, J. H., *et al.*: Phys. Rev. **122**, 182 (1961).
Macklin, R. L., J. H. Gibbons, and T. Inada: Nucl. Phys. **43**, 353 (1963).

Direct Measurement of the (n, γ) Cross Section.

Allen, W. D., and R. L. Henkel: Progr. Nucl. Energy Series I Vol. 2 p. 1 (1958). Lamphere, R. W.: Phys. Rev. **104**, 1654 (1956). —, and R. E. Greene: Phys. Rev. **100**, 763 (1955). Schmitt, H. W., and R. B. Murray: Phys. Rev. **116**, 1575 (1959).	Measurement of the Fast-Neutron Fission Cross Section.
Bowman, C. D., G. F. Auchampaugh, and S. C. Fultz: Phys. Rev. **130**, 1482 (1963). Egelstaff, P. A., and D. J. Hughes: Progr. Nucl. Energy Series I Vol. 1 p. 55 (1956). Havens, W. W., *et al.*: Phys. Rev. **116**, 1538 (1959). Shore, F. J., and V. L. Sailor: Phys. Rev. **112**, 191 (1958).	Fission Measurements in the Resonance Region.
Moore, M. S., and C. W. Reich: Phys. Rev. **118**, 718 (1960). Vogt, E.: Phys. Rev. **118**, 724 (1960).	Multilevel Fit of the Fission Cross Section in the Resonance Region.
Bigham, C. B., *et al.*: Geneva 1958 P/204 Vol. 16 p. 125. Deruytter, A. J.: J. Nucl. Energy A & B **15**, 165 (1961). Raffle, J. F.: J. Nucl. Energy A **10**, 8 (1959). Saplakoglu, A.: Geneva 1958 P/1599 Vol. 16 p. 103.	Measurement of the Fission Cross Section of U^{235} for Slow Neutrons.
Colvin, D. W., and M. G. Sowerby: EANDC-(UK) 3 (1960). Diven, B. C., *et al.*: Phys. Rev. **101**, 1012 (1956). Kalashnikova, V. I., *et al.*: In: Proc. Conf. Acad. Sci. USSR, "Peaceful Uses of Atomic Energy", Phys. and Math. Sci., New York: Consultants Bureau 1955, p. 123. Kenward, C. J., R. Richmond, and J. E. Sanders: AERE E/R 2212 (1958). Moat, A., D. S. Mather, and M. H. McTaggart: J. Nucl. Energy A & B **15**, 102 (1961).	Measurement of $\bar{\nu}$.
Gaerttner, E. R., *et al.*: Nucl. Sci. Eng. **3**, 758 (1958). Macklin, R. L., *et al.*: Nucl. Sci. Eng. **8**, 210 (1960). Palevsky, H., *et al.*: J. Nucl. Energy **3**, 177 (1956).	Measurement of η.
Cocking, S. J.: J. Nucl. Energy **6**, 285 (1957/58). Diven, B. C., J. Terrell, and A. Hemmendinger: Phys. Rev. **109**, 144 (1958). Kanne, W. R., H. B. Stewart, and F. A. White: Geneva 1955 P/595 Vol. 4 p 315. Safford, G. J., and E. Melkonian: Phys. Rev. **113**, 1285 (1959).	Measurement of α.
Block, R. C., G. G. Slaughter, and J. A. Harvey: Nucl. Sci. Eng. **8**, 112 (1960). Safford, G. J., W. W. Havens, and B. M. Rustad: Nucl. Sci. Eng. **6**, 433 (1959).	Slow-Neutron Absorption Cross Section of U^{235}.
Evans, J. E., and R. G. Fluharty: Nucl. Sci. Eng. **8**, 66 (1960). Leonard, B. R.: ANL-6122, 227 (1959); HW-69342 (1961). Safford, G. J., and W. W. Havens: Nucleonics **17** No. 11, 134 (1959). Sher, R., and J. Felderbaum: Least Square Analysis of the 2200 m/sec Parameters of U^{233}, U^{235} and Pu^{239}. BNL-722 (T-256-92-94-1) (1962).	Determination of "Consistent" Sets of Parameters for U^{233}, U^{235}, and Pu^{239}.

Part II

The Theory of Neutron Fields

5. Neutron Fields

Let us consider sources in a scattering medium which emit neutrons with an energy E or with an energy distribution $S(E)$. These neutrons exchange kinetic energy with the atoms of the scattering substance through elastic and frequently also inelastic collisions. If the neutrons are emitted with energies higher than the kinetic energy of the thermal motion of the scattering atoms, they lose energy in successive collisions until they are in equilibrium with the thermal motion of their surroundings. Their energies then assume, or at least approximate, a thermal distribution. Neutrons with such an energy distribution are called "thermal neutrons".

There is no scattering medium which does not absorb neutrons, particularly slow neutrons. In applied neutron physics, great importance is attached to substances which (i) scatter neutrons strongly, i.e., have large scattering cross sections Σ_s, (ii) moderate neutrons strongly, i.e., have low atomic weight, and (iii) absorb neutrons weakly, i.e., have small absorption cross sections ($\sigma_a \ll \sigma_s$). Such substances are good *moderators*; among the best are H_2O, D_2O, beryllium, beryllium oxide, graphite, and various hydrogenous organic compounds. By using uranium and plutonium, media can be produced in which production of neutrons occurs in place of absorption. However, multiplying media, which form the basis for the construction of nuclear reactors, will not be treated in this book.

The neutron intensity (which will be rigorously defined below) in a scattering and absorbing medium depends on the strength of the source, and, according to the behavior of the source, can be stationary or non-stationary; it can, furthermore, give rise to diffusion currents. As a rule, the general treatment of the diffusion process leads to difficult mathematical problems. We will therefore study some simple approximations. The totality of neutrons in a scattering medium, characterized by their distribution in space, time, and energy, is called the neutron diffusion field or, more simply, the neutron field.

5.1. Characterization of the Field

5.1.1. Neutron Flux, Neutron Density, and Neutron Current

Let us consider a volume element $dV = dx\,dy\,dz$ of the scattering medium with the position vector $\boldsymbol{r}$. Let $n(\boldsymbol{r}, \boldsymbol{\Omega}, E)\,dV\,d\Omega\,dE$ be the number of neutrons in this volume element whose flight direction, characterized by the unit vector $\boldsymbol{\Omega}$, lies in the differential solid angle $d\Omega$ around $\boldsymbol{\Omega}$, and whose kinetic energy lies between E and $E+dE$. $n(\boldsymbol{r}, \boldsymbol{\Omega}, E)$ (cm^{-3} steradian^{-1} ev^{-1}) is thus the number per cm^3, i.e., the density, of neutrons with energies in a unit interval at E and flight directions in a unit solid angle around $\mathbf{\Omega}$. Specifying this *differential density*, which can also depend on time, is sufficient to describe a neutron field. The total number of neutrons with a given flight direction is obtained by integration over all energies:

$$n(\mathbf{r}, \mathbf{\Omega})\, dV\, d\Omega = \int_0^\infty n(\mathbf{r}, \mathbf{\Omega}, E)\, dV\, d\Omega\, dE. \tag{5.1.1}$$

$n(\mathbf{r}, \mathbf{\Omega})$ is called the *vector density*. It replaces $n(\mathbf{r}, \mathbf{\Omega}, E)$ when we are dealing with a neutron field of constant energy. If we now integrate over all flight directions as well, we obtain the total number of neutrons in the volume element dV at the point $\mathbf{r}$:

$$\left.\begin{aligned} n(\mathbf{r})\, dV &= \int_{4\pi} n(\mathbf{r}, \mathbf{\Omega})\, dV\, d\Omega \\ &= \int_{4\pi}\int_0^\infty n(\mathbf{r}, \mathbf{\Omega}, E)\, dV\, d\Omega\, dE. \end{aligned}\right\} \tag{5.1.2}$$

$n(\mathbf{r})$ is the *density* of neutrons at the point $\mathbf{r}$.

Next, we introduce the *differential neutron flux*, defined by

$$F(\mathbf{r}, \mathbf{\Omega}, E)\, d\Omega\, dE = n(\mathbf{r}, \mathbf{\Omega}, E)\, v\, d\Omega\, dE \tag{5.1.3}$$

where $v = \sqrt{2E/m}$ is the neutron velocity. It is the number of neutrons at the point $\mathbf{r}$ with energies between E and $E + dE$ and flight directions in the differential solid angle $d\Omega$ around $\mathbf{\Omega}$ which penetrate a surface of area 1 cm^2 perpendicular to the direction $\mathbf{\Omega}$ in one second. By integration of the differential flux $F(\mathbf{r}, \mathbf{\Omega}, E)$ (cm^{-2} c^{-}se^{1} steradian^{-1} ev^{-1}) over energy, we obtain the *vector flux* $F(\mathbf{r}, \mathbf{\Omega})$, which is the number of neutrons which penetrate a 1-cm^2 surface perpendicular to the direction $\mathbf{\Omega}$ through the differential solid angle $d\Omega$ per second. Finally,

$$\Phi(\mathbf{r}) = \int_{4\pi} F(\mathbf{r}, \mathbf{\Omega})\, d\Omega = n(\mathbf{r})\, \bar{v} \tag{5.1.4}$$

is called the *flux* of neutrons[1]. In practice, the flux Φ (cm^{-2} sec^{-1}) is the datum most frequently used for describing neutron fields, and for this reason it is important to make its physical significance clear.

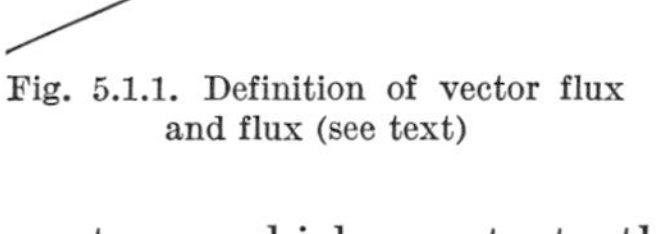

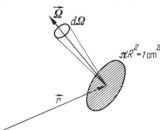

Fig. 5.1.1. Definition of vector flux and flux (see text)

Let us consider a circular disc with a surface area $\pi R^2 = 1$ cm^2 whose center is rigidly fixed to the point $\mathbf{r}$ (Fig. 5.1.1). $F(\mathbf{r}, \mathbf{\Omega})\, d\Omega$ is the number of neutrons which penetrate the disc through the solid angle element $d\Omega$ around the direction of its normal $\mathbf{\Omega}$ per sec. In this picture, the integration by which the flux is formed can be represented by rotation of the disc in all directions keeping the center fixed. In such a rotation, the disc describes a sphere of cross section $\pi R^2 = 1$ cm^2 and of surface $4\pi R^2 = 4$ cm^2. Thus the flux Φ is the number of neutrons which penetrate this sphere per second from all sides. It follows from this fact that in an isotropic neutron field, i.e., a field in which all flight directions are equally represented, $\Phi/2$ neutrons penetrate a 1-cm^2 surface per second. For,

[1] In a monoenergetic neutron field, $\bar{v}$ is simply the velocity; but in a polyenergetic field, $\bar{v}$ is the average of the velocity over the energy spectrum of the density.

in the isotropic field equally many neutrons pass through each surface element of the sphere introduced above. Since altogether $2\,\Phi$ neutrons pass through 4 cm² per second (each neutron passes once from the outside to the inside and once from the inside to the outside), $\Phi/2$ neutrons go through 1 cm² per second. We will see later that this conclusion also holds in weakly anisotropic neutron fields.

In most cases, F can be represented as a function of only one angle ϑ that is measured from a "distribution axis" with respect to which the field is considered to be rotationally symmetric. In such a case, we can always expand the quantity $F(\boldsymbol{r}, \boldsymbol{\Omega})$ in Legendre polynomials, as a rule with considerable advantage insofar as the mathematical treatment of the problem is concerned:

$$F(\boldsymbol{r}, \boldsymbol{\Omega}) = \frac{1}{4\pi} \sum_{l=0}^{\infty} (2l+1) F_l(\boldsymbol{r})\, P_l(\cos\vartheta). \tag{5.1.5}$$

The first four Legendre polynomials are:

$$P_0 = 1 \qquad P_1 = \cos\vartheta$$

$$P_2 = \tfrac{1}{2}(3\cos^2\vartheta - 1) \qquad P_3 = \tfrac{1}{2}(5\cos^3\vartheta - 3\cos^2\vartheta).$$

The equation

$$F_l(\boldsymbol{r}) = 2\pi \int_0^{\pi} F(\boldsymbol{r}, \boldsymbol{\Omega}) P_l(\cos\vartheta) \sin\vartheta\, d\vartheta \tag{5.1.6}$$

holds for the quantities $F_l(\boldsymbol{r})$. In particular,

$$F_0(\boldsymbol{r}) = 2\pi \int_0^{\pi} F(\boldsymbol{r}, \boldsymbol{\Omega}) \sin\vartheta\, d\vartheta = \int_{4\pi} F(\boldsymbol{r}, \boldsymbol{\Omega})\, d\Omega = \Phi(\boldsymbol{r}). \tag{5.1.7}$$

The second expansion coefficient also has a physical significance. In order to see this, let us introduce an important new quantity, the *current density* J in the direction of the distribution axis. The magnitude of this vector is the net number of neutrons which penetrate a 1-cm² surface perpendicular to the distribution axis per second. Thus

$$J(\boldsymbol{r}) = \int_{4\pi} F(\boldsymbol{r}, \boldsymbol{\Omega}) \cos\vartheta\, d\Omega = 2\pi \int_0^{\pi} F(\boldsymbol{r}, \boldsymbol{\Omega}) \cos\vartheta \sin\vartheta\, d\vartheta. \tag{5.1.8}$$

By comparison with Eq. (5.1.6) it follows that $J(\boldsymbol{r}) = F_1(\boldsymbol{r})$. Thus the first two terms of the expansion (5.1.5) can be written:

$$F(\boldsymbol{r}, \boldsymbol{\Omega}) = \frac{1}{4\pi} \Phi(\boldsymbol{r}) + \frac{3}{4\pi} J(\boldsymbol{r}) \cos\vartheta \tag{5.1.9}$$

and it will turn out that in many cases the vector flux $F(\boldsymbol{r}, \boldsymbol{\Omega})$ can be approximated by these two terms with adequate accuracy.

5.1.2. The General Transport Equation

The space and time behavior of a neutron field can be described by setting up a "neutron balance". The number of neutrons $n(\boldsymbol{r}, \boldsymbol{\Omega}, E)\, dV\, d\Omega\, dE$ in a volume element dV with flight directions in the solid angle $d\Omega$ and energies in the interval dE can change for the following reasons:

1. Leakage out of dV:

$$\operatorname{div}\big(\boldsymbol{\Omega} F(\boldsymbol{r}, \boldsymbol{\Omega}, E)\big)\, dV\, d\Omega\, dE = \boldsymbol{\Omega}\cdot\operatorname{grad} F(\boldsymbol{r}, \boldsymbol{\Omega}, E)\, dV\, d\Omega\, dE:$$

2. Loss due to absorption and scattering into other directions[1].

$$\Sigma_t F(\boldsymbol{r}, \boldsymbol{\Omega}, E)\, dV\, d\Omega\, dE; \quad \Sigma_t = \Sigma_a + \Sigma_s \qquad \text{[cf. Eq. (1.3.18b)]}.$$

3. Gain due to in-scattering of neutrons from other directions and energy intervals:

$$\int_{4\pi}\int_0^\infty \Sigma_s(\boldsymbol{\Omega}'\to\boldsymbol{\Omega}, E'\to E) F(\boldsymbol{r}, \boldsymbol{\Omega}', E')\, d\Omega'\, dE'\, dV\, d\Omega\, dE.$$

Here $\Sigma_s(\boldsymbol{\Omega}'\to\boldsymbol{\Omega}, E'\to E)\, d\Omega\, dE$ is the cross section for a process in which a neutron with the direction $\boldsymbol{\Omega}'$ and the energy E' is scattered into the element of solid angle $d\Omega$ around $\boldsymbol{\Omega}$ and the energy interval dE at E. Since we are only considering isotropic substances in which the total scattering cross section does not depend on the flight direction of the neutron, it follows that

$$\int_{4\pi}\int_0^\infty \Sigma_s(\boldsymbol{\Omega}'\to\boldsymbol{\Omega}, E'\to E)\, d\Omega\, dE = \Sigma_s(E'). \tag{5.1.10}$$

If the neutron scattering is isotropic in the laboratory system, then obviously

$$\Sigma_s(\boldsymbol{\Omega}'\to\boldsymbol{\Omega}, E'\to E) = \frac{1}{4\pi}\,\Sigma_s(E'\to E) \tag{5.1.11}$$

where $\Sigma_s(E'\to E)\, dE$ is the cross section for scattering processes which transfer neutrons of energy E' into the energy range $(E, E+dE)$.

4. Production of neutrons by sources in dV [source density $S(\boldsymbol{r}, \boldsymbol{\Omega}, E)$]:

$$S(\boldsymbol{r}, \boldsymbol{\Omega}, E)\, dV\, d\Omega\, dE.$$

The sum of all these contributions gives the time rate of change of the differential density:

$$\left.\begin{aligned} \frac{\partial n(\boldsymbol{r}, \boldsymbol{\Omega}, E)}{\partial t} &= \frac{1}{v}\frac{\partial F(\boldsymbol{r}, \boldsymbol{\Omega}, E)}{\partial t} \\ &= -\boldsymbol{\Omega}\cdot\operatorname{grad} F(\boldsymbol{r}, \boldsymbol{\Omega}, E) - \Sigma_t(E) F(\boldsymbol{r}, \boldsymbol{\Omega}, E) \\ &\quad + \int_{4\pi}\int_0^\infty \Sigma_s(\boldsymbol{\Omega}'\to\boldsymbol{\Omega}, E'\to E) F(\boldsymbol{r}, \boldsymbol{\Omega}', E')\, d\Omega'\, dE' \\ &\quad + S(\boldsymbol{r}, \boldsymbol{\Omega}, E). \end{aligned}\right\} \tag{5.1.12}$$

This integro-differential equation in seven variables (three position, two direction, one energy, and one time coordinate) is called the transport or Boltzmann equation. Together with appropriate boundary conditions, it determines the vector flux arising from a given source distribution. The most important boundary conditions occur at

(a) the interface G between two scattering media A and B. By reason of continuity, the following equation must clearly hold for all $\boldsymbol{r}_G$, $\boldsymbol{\Omega}$, and E:

$$F_A(\boldsymbol{r}_G, \boldsymbol{\Omega}, E) = F_B(\boldsymbol{r}_G, \boldsymbol{\Omega}, E) \tag{5.1.13}$$

[1] Even for the highest achievable neutron densities, neutron-neutron collisions do not occur; thus we need only consider nuclear collisions.

(b) the interface between a scattering medium and a vacuum or a totally absorbing medium. Since no neutrons can return to the medium through this interface,

$$F(\boldsymbol{r}_G, \boldsymbol{\Omega}, E) = 0 \tag{5.1.14}$$

must obviously hold for all inwardly directed $\boldsymbol{\Omega}$.

Eq. (5.1.12) can be simplified in some important special cases. To these belongs the time- and energy-independent case. For this case, the transport equation takes the form

$$\boldsymbol{\Omega} \cdot \operatorname{grad} F(\boldsymbol{r}, \boldsymbol{\Omega}) + \Sigma_t F(\boldsymbol{r}, \boldsymbol{\Omega}) = \int_{4\pi} \Sigma_s(\boldsymbol{\Omega}' \to \boldsymbol{\Omega}) F(\boldsymbol{r}, \boldsymbol{\Omega}')\, d\Omega' + S(\boldsymbol{r}, \boldsymbol{\Omega}) \tag{5.1.15}$$

and describes the diffusion of monoenergetic neutrons in a medium which contains stationary sources and in which a collision causes no change in energy. Eq. (5.1.15) will be treated further in several subsequent sections of this chapter and again in Chapter 6. Also important is the time- and space-independent case. Here the term $\boldsymbol{\Omega} \cdot \operatorname{grad} F$ in Eq. (5.1.12) drops out, and it becomes possible to integrate over all angles. One then obtains

$$(\Sigma_a + \Sigma_s)\, \Phi(E) = \int_0^\infty \Sigma_s(E' \to E)\, \Phi(E')\, dE' + S(E). \tag{5.1.16}$$

Here $\Phi(E)$ is the neutron flux per unit energy at E. Eq. (5.1.16) describes the moderation of neutrons in an infinite medium with homogeneously distributed sources; it will be considered further in Chapters 7 and 10. In Chapter 8, we will study the case of stationary but space- and energy-dependent neutron fields, i.e., the case of neutron diffusion with energy exchange with the atoms of the scattering medium. The transport equation for this case is

$$\left.\begin{aligned} &\boldsymbol{\Omega} \cdot \operatorname{grad} F(\boldsymbol{r}, \boldsymbol{\Omega}, E) + \Sigma_t F(\boldsymbol{r}, \boldsymbol{\Omega}, E) \\ &\quad = \int_{4\pi}\int_0^\infty \Sigma_s(\boldsymbol{\Omega}' \to \boldsymbol{\Omega}, E' \to E) F(\boldsymbol{r}, \boldsymbol{\Omega}', E')\, d\Omega'\, dE' + S(\boldsymbol{r}, \boldsymbol{\Omega}, E). \end{aligned}\right\} \tag{5.1.17}$$

Finally, in Chapter 9, we will study time-dependent neutron fields, and in particular time- and space- as well as time- and energy-dependent fields; for description of these fields, the terms $\frac{1}{v}\frac{\partial F}{\partial t}$ and $\frac{1}{v}\frac{\partial \Phi}{\partial t}$ must appear on the left-hand sides of Eqs. (5.1.15) and (5.1.16), respectively.

5.1.3. Integral Form of the Transport Equation

For many applications, the transport equation can be written with advantage in an integral form[1]. Let us consider a circular disc with a surface area of 1 cm^2 (Fig. 5.1.2), and ask for the number of neutrons crossing this surface per second with directions in the element of solid angle $d\Omega$ around the normal to the disc. The contribution of the volume element $R^2\, dR\, d\Omega$ is obviously equal to

$[\int F(\boldsymbol{r} - R\boldsymbol{\Omega}, \boldsymbol{\Omega}') \Sigma_s(\boldsymbol{\Omega}' \to \boldsymbol{\Omega})\, d\Omega' + S(\boldsymbol{r} - R\boldsymbol{\Omega}, \boldsymbol{\Omega})]\, R^2\, dR\, d\Omega$ times the

[1] We limit ourselves here to the energy- and time-independent case; however, it is possible easily to transform the general Eq. (5.1.12) to an integral form.

probability that the neutrons reach the surface of the disc. The latter is equal to the solid angle multiplied by the probability $e^{-\Sigma_t R}$ that no collision occurs along the path R. If we divide by the element of solid angle $d\Omega$, it then follows that

$$F(\boldsymbol{r},\boldsymbol{\Omega})=\int_0^\infty\Big[\int_{4\pi}\Sigma_s(\boldsymbol{\Omega}'\to\boldsymbol{\Omega})F(\boldsymbol{r}-R\boldsymbol{\Omega},\boldsymbol{\Omega}')\,d\Omega'+S(\boldsymbol{r}-R\boldsymbol{\Omega},\boldsymbol{\Omega})\Big]e^{-\Sigma_t R}\,dR\,. \quad (5.1.18)$$

The integration extends to infinity if the scattering medium is infinite; in a finite medium, one integrates to the value $R_{\max}$ at which the straight line $\boldsymbol{r}-R\boldsymbol{\Omega}$ reaches the surface. Eq. (5.1.18) can also be derived directly from the transport Eq. (5.1.15).

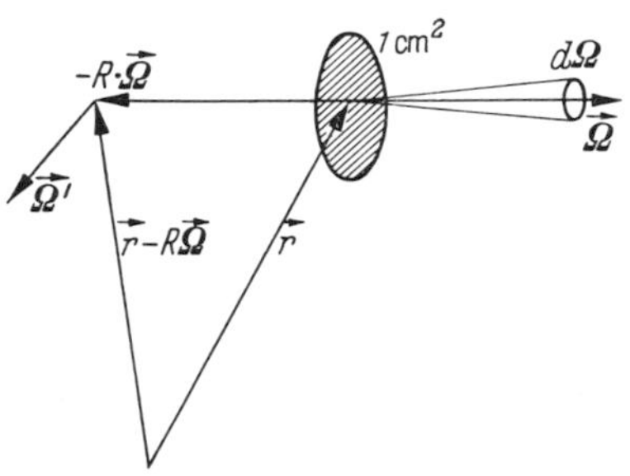

Fig. 5.1.2. Derivation of the integral form of the transport equation (see text)

From now on we will assume that the spatially distributed sources radiate isotropically, i.e., that

$$S(\boldsymbol{r},\boldsymbol{\Omega})=\frac{1}{4\pi}S(\boldsymbol{r})\,. \quad (5.1.19)$$

This assumption applies to most cases in practice, so that Eq. (5.1.19) implies no severe loss of generality. If we assume for the moment furthermore that the neutron scattering is isotropic in the laboratory system, i.e., that $\Sigma_s(\boldsymbol{\Omega}'\to\boldsymbol{\Omega})=\frac{1}{4\pi}\Sigma_s$, the integral transport equation assumes a particularly simple form. Now we can integrate over $\boldsymbol{\Omega}'$ on the right-hand side and obtain

$$F(\boldsymbol{r},\boldsymbol{\Omega})=\frac{1}{4\pi}\int_0^\infty[\Sigma_s\Phi(\boldsymbol{r}-R\boldsymbol{\Omega})+S(\boldsymbol{r}-R\boldsymbol{\Omega})]e^{-\Sigma_t R}\,dR\,. \quad (5.1.20)$$

If we replace $\boldsymbol{r}-R\boldsymbol{\Omega}$ on the right-hand side by a new position coordinate $\boldsymbol{r}'$ and furthermore replace $dR\,d\Omega=R^2\,dR\,d\Omega/R^2$ by $dV'/(\boldsymbol{r}-\boldsymbol{r}')^2$, we can integrate over $\boldsymbol{\Omega}$ and obtain

$$\Phi(\boldsymbol{r})=\int_{\text{all }\boldsymbol{r}'}[\Sigma_s\Phi(\boldsymbol{r}')+S(\boldsymbol{r}')]\frac{e^{-\Sigma_t|\boldsymbol{r}-\boldsymbol{r}'|}}{4\pi(\boldsymbol{r}-\boldsymbol{r}')^2}\,dV'\,. \quad (5.1.21)$$

In contrast to the transport Eq. (5.1.15), this equation, which can also easily be derived directly, no longer contains the direction coordinate $\boldsymbol{\Omega}$; for this reason it is easier to solve in some practical cases than Eq. (5.1.15).

5.2. Treatment of the Stationary, Energy-Independent Transport Equation. Elementary Diffusion Theory

In this section, we treat the time- and energy-independent Eq. (5.1.15), which describes the diffusion of monoenergetic neutrons in a scattering medium in which a collision causes no energy loss, or whose cross sections Σ_t and $\Sigma_s(E'\to E)$ do not depend on E and E'. On first sight, this situation may not appear to have much significance, but we will see later that it can be used to describe the diffusion of thermal neutrons. Also we will study methods of solution applicable to this simple case which can be generalized to the energy-dependent case.

5.2.1. Development in Spherical Harmonics; Fick's Law

It will be our aim to derive from Eq. (5.1.15) a simple diffusion equation of the FICK's law type, i.e., one which represents a linear relation between the current density introduced in Eq. (5.1.8) and the gradient of the neutron flux. Again we assume that the neutron field is rotationally symmetric around a distribution axis; furthermore, we assume plane symmetry and let the distribution axis be the x-axis. Eq. (5.1.15) then becomes

$$\cos\vartheta \frac{\partial F(x,\mathbf{\Omega})}{\partial x} + \Sigma_t F(x,\mathbf{\Omega}) = \int \Sigma_s(\mathbf{\Omega}' \to \mathbf{\Omega}) F(x,\mathbf{\Omega}')\, d\Omega' + \frac{1}{4\pi} S(x). \quad (5.2.1)$$

Next we note that in an isotropic substance the scattering cross section $\Sigma_s(\mathbf{\Omega}'\to\mathbf{\Omega})$ depends only on the angle ϑ_0 between the directions $\mathbf{\Omega}$ and $\mathbf{\Omega}'$. Therefore, in analogy to Eq. (5.1.5), we can develop $\Sigma_s(\mathbf{\Omega}'\to\mathbf{\Omega})$ in a series of Legendre polynomials of $\cos\vartheta_0 = \mathbf{\Omega}\cdot\mathbf{\Omega}'$:

$$\Sigma_s(\mathbf{\Omega}'\to\mathbf{\Omega}) = \frac{1}{4\pi} \sum_{l=0}^{\infty} (2l+1) \Sigma_{sl} P_l(\cos\vartheta_0), \quad (5.2.2)$$

$$\Sigma_{sl} = 2\pi \int_0^{\pi} \Sigma_s(\mathbf{\Omega}'\to\mathbf{\Omega}) P_l(\cos\vartheta_0) \sin\vartheta_0\, d\vartheta_0. \quad (5.2.3)$$

The first expansion coefficient is given by

$$\Sigma_{s0} = 2\pi \int_0^{\pi} \Sigma_s(\mathbf{\Omega}'\to\mathbf{\Omega}) \sin\vartheta_0\, d\vartheta_0 = \int_{4\pi} \Sigma_s(\mathbf{\Omega}'\to\mathbf{\Omega})\, d\Omega_0 = \Sigma_s. \quad (5.2.4)$$

Σ_s is the total scattering cross section. Furthermore,

$$\left.\begin{aligned} \Sigma_{s1} &= 2\pi \int_0^{\pi} \Sigma_s(\mathbf{\Omega}'\to\mathbf{\Omega}) \cos\vartheta_0 \sin\vartheta_0\, d\vartheta_0 \\ &= \int_{4\pi} \Sigma_s(\mathbf{\Omega}'\to\mathbf{\Omega}) \cos\vartheta_0\, d\Omega_0 = \Sigma_s \overline{\cos\vartheta_0}. \end{aligned}\right\} \quad (5.2.5)$$

Eq. (5.2.5) defines the important *average cosine of the scattering angle* in the laboratory system.

We can now substitute the expansion (5.2.2) into Eq. (5.2.1). However, we must express the Legendre polynomials $P_l(\cos\vartheta_0)$ in terms of the angles ϑ', φ', and ϑ, φ (cf. Fig. 5.2.1). According to the addition theorem for the Legendre polynomials

$$\left.\begin{aligned} P_l(\cos\vartheta_0) &= P_l(\cos\vartheta) P_l(\cos\vartheta') \\ &\quad + 2 \sum_{m=1}^{l} \frac{(l-m)!}{(l+m)!} P_l^m(\cos\vartheta) P_l^m(\cos\vartheta') \cos m(\varphi'-\varphi). \end{aligned}\right\} \quad (5.2.6)$$

Here the P_l^m are associated Legendre polynomials. If we insert Eq. (5.2.2) and Eq. (5.2.6) into Eq. (5.2.1) and carry out the integration over φ' on the right-hand side, the terms containing the P_l^m drop out, and we obtain

$$\left.\begin{aligned} &\cos\vartheta \frac{\partial F(x,\mathbf{\Omega})}{\partial x} + \Sigma_t F(x,\mathbf{\Omega}) \\ &= \frac{1}{2} \sum_{l=0}^{\infty} (2l+1) \Sigma_{sl} P_l(\cos\vartheta) \int_0^{\pi} F(x,\mathbf{\Omega}') P_l(\cos\vartheta') \sin\vartheta'\, d\vartheta' + \frac{1}{4\pi} S(x). \end{aligned}\right\} \quad (5.2.7)$$

Next we expand the vector flux F in Legendre polynomials exactly as in Sec. 5.1.1 and break the expansion off after the second term, i.e., we set

$$F(x, \mathbf{\Omega}) = \frac{1}{4\pi}\,\Phi(x) + \frac{3}{4\pi}\,J(x)\cos\vartheta. \tag{5.1.9}$$

Truncating the series after the second term leads to a relation which is called the *elementary diffusion equation*. We will study the realm of its applicability in detail later. If we substitute Eq. (5.1.9) into Eq. (5.2.7) and integrate once over $d\Omega$ and once over $\cos\vartheta\, d\Omega$, we obtain after some simple rewriting

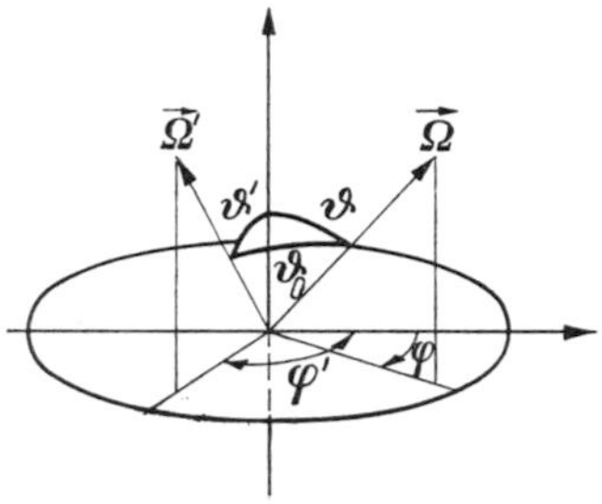

Fig. 5.2.1. Relation between the angles ϑ, ϑ', ϑ_0, φ and φ'

$$\frac{dJ}{dx} + \Sigma_a \Phi = S(x) \tag{5.2.8}$$

and

$$\frac{1}{3}\,\frac{d\Phi}{dx} + (\Sigma_t - \Sigma_{s1})J = 0. \tag{5.2.9}$$

Both of these important relations have simple physical significance: Eq. (5.2.8) is a continuity condition which says that in one cm³ of a stationary neutron field the losses through leakage and absorption are equal to the number of neutrons provided by the source (S). This relation is independent of the special approximation 5.1.9; this is physically obvious and can be easily proven formally if one notes that if additional terms in the expression for F are kept they drop out in the integration over $\mathbf{\Omega}$. From Eq. (5.2.9), one obtains

$$J = -\frac{1}{3(\Sigma_s(1-\overline{\cos\vartheta_0}) + \Sigma_a)}\,\frac{d\Phi}{dx} = -D\,\frac{d\Phi}{dx}. \tag{5.2.10}$$

This is Fick's famous law of diffusion, viz., that the current density is proportional to the flux gradient. The constant of proportionality D (cm) is called the *diffusion coefficient*; if the *transport cross section*

$$\Sigma_{tr} = \Sigma_s(1-\overline{\cos\vartheta_0}) \tag{5.2.11}$$

is introduced, D can be expressed as

$$D = \frac{1}{3(\Sigma_{tr} + \Sigma_a)}. \tag{5.2.12}$$

In contrast to Eq. (5.2.8), Eq. (5.2.10) does represent an approximation which comes about from truncating the expansion of F after the second term. If one keeps the higher terms in the Legendre polynomial expansion of F, there then results, as can be shown by an easy calculation, the exact expression

$$J = -\frac{1}{3(\Sigma_{tr} + \Sigma_a)}\,\frac{d}{dx}\left[\Phi(x)\left(1 + 2\,\frac{F_2(x)}{\Phi(x)}\right)\right]. \tag{5.2.13}$$

If we combine Eqs. (5.2.8) and (5.2.10), we obtain the *elementary diffusion equation*

$$D\,\frac{d^2\Phi}{dx^2} - \Sigma_a \Phi(x) + S(x) = 0. \tag{5.2.14}$$

If we now introduce the *diffusion length*

$$L=\sqrt{D/\Sigma_a}=\sqrt{\frac{1}{3\Sigma_a(\Sigma_{tr}+\Sigma_a)}} \tag{5.2.15}$$

we obtain

$$\frac{d^2\Phi}{dx^2}-\frac{1}{L^2}\Phi+\frac{S(x)}{D}=0 \tag{5.2.16a}$$

or in the general three-dimensional case

$$\nabla^2\Phi-\frac{1}{L^2}\Phi+\frac{S(\boldsymbol{r})}{D}=0. \tag{5.2.16b}$$

We now have a simple differential equation which permits the calculation of the flux arising from a specified source distribution. Before we consider any applications of Eq. (5.2.16), however, we will discuss the question of the validity of this equation and the question of boundary conditions.

5.2.2. Asymptotic Solution of the Transport Equation

There is one important case in which an exact solution to the transport equation can easily be obtained, namely, the case of the so-called asymptotic distribution in an infinite medium. The asymptotic neutron distribution is the neutron distribution at very great distances from the sources; only neutrons which have already had many collisions contribute to the asymptotic flux. We will derive this asymptotic solution for the special case of plane symmetry and will be able to draw important conclusions from it concerning the validity of elementary diffusion theory.

In the interest of simplicity we assume that the neutron scattering is isotropic in the laboratory, i.e., that all the Σ_{sl} except $\Sigma_{s0}=\Sigma_s$ are zero. In this case, the one-dimensional, source-free transport equation is

$$\cos\vartheta\frac{\partial F}{\partial x}+\Sigma_t F=\frac{1}{2}\Sigma_s\int_0^\pi F(x,\boldsymbol{\Omega}')\sin\vartheta'\,d\vartheta'. \tag{5.2.17}$$

Next we assume a solution of the form

$$F(x,\boldsymbol{\Omega})=e^{-\varkappa x}f(\cos\vartheta) \tag{5.2.18}$$

with $\varkappa$ as yet undetermined. Substitution into Eq. (5.2.17) yields

$$f(\cos\vartheta)\{\Sigma_t-\varkappa\cos\vartheta\}=\tfrac{1}{2}\Sigma_s\int_0^\pi f(\cos\vartheta')\sin\vartheta'\,d\vartheta' \tag{5.2.19a}$$

i.e.,

$$f(\cos\vartheta)=\frac{1}{2}\Sigma_s\frac{\int_0^\pi f(\cos\vartheta')\sin\vartheta'\,d\vartheta'}{\Sigma_t-\varkappa\cos\vartheta}. \tag{5.2.19b}$$

Integration over $\sin\vartheta\,d\vartheta$ gives a relation between $\varkappa$, Σ_s, and Σ_t:

$$\int_0^\pi f(\cos\vartheta)\sin\vartheta\,d\vartheta=\frac{1}{2}\Sigma_s\int_0^\pi f(\cos\vartheta')\sin\vartheta'\,d\vartheta'\cdot\int_0^\pi\frac{\sin\vartheta\,d\vartheta}{\Sigma_t-\varkappa\cos\vartheta}. \tag{5.2.20}$$

Setting $\cos\vartheta=\mu$ and $\sin\vartheta\,d\vartheta=-d\mu$ we obtain

$$1=\frac{1}{2}\Sigma_s\int_{-1}^{+1}\frac{d\mu}{\Sigma_t-\varkappa\mu}$$

i.e.,

$$1=\frac{\Sigma_s}{2\varkappa}\ln\frac{1+\varkappa/\Sigma_t}{1-\varkappa/\Sigma_t}. \tag{5.2.21}$$

Eq. (5.2.21) permits the determination of $\varkappa$ from Σ_s and Σ_t. Fig. 5.2.2 shows $\varkappa/\Sigma_t$ as a function of Σ_s/Σ_t. For $\Sigma_s/\Sigma_t\approx 1$, i.e., $\Sigma_a\ll\Sigma_s$ (weak absorption), $\varkappa\ll\Sigma_t$ and the logarithm can be expanded. One then obtains

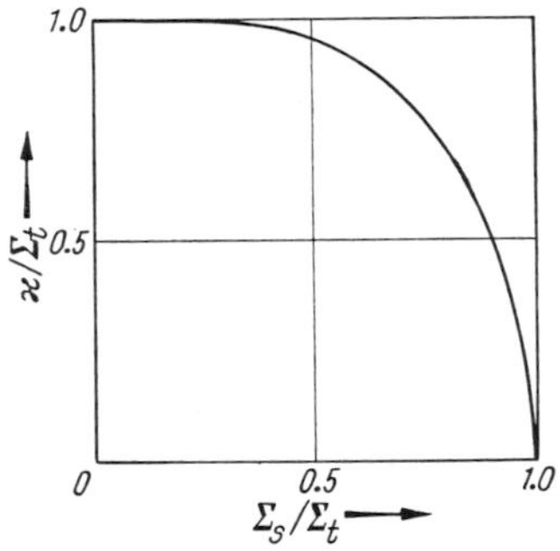

Fig. 5.2.2. The dependence of $\varkappa/\Sigma_t$ on Σ_s/Σ_t

$$\varkappa^2=3\Sigma_a\Sigma_t\left(1-\frac{4}{5}\frac{\Sigma_a}{\Sigma_t}\right) \tag{5.2.22a}$$

which becomes

$$\varkappa^2=3\Sigma_a\Sigma_t \tag{5.2.22b}$$

in the limit $\Sigma_a/\Sigma_s\to 0$. Since for each root $\varkappa$ of Eq. (5.2.21) there is also a root $-\varkappa$, the general asymptotic solution of the transport equation is

$$\left.\begin{aligned}F(x,\boldsymbol{\Omega})=&\frac{A\Sigma_s}{4\pi}e^{-\varkappa x}\frac{1}{\Sigma_t-\varkappa\cos\vartheta}+\\&+\frac{B\Sigma_s}{4\pi}e^{\varkappa x}\frac{1}{\Sigma_t+\varkappa\cos\vartheta}\end{aligned}\right\} \tag{5.2.23a}$$

where A and B are constants. The corresponding expression for the flux is

$$\Phi(x)=2\pi\int_0^\pi F(x,\boldsymbol{\Omega})\sin\vartheta\,d\vartheta=A\,e^{-\varkappa x}+B\,e^{\varkappa x}. \tag{5.2.23b}$$

Now we can investigate the validity of FICK's law for this asymptotic solution. We begin by calculating the current density from Eq. (5.2.23a) with the help of Eq. (5.1.8):

$$J=\frac{\Sigma_a}{\varkappa}A\,e^{-\varkappa x}-\frac{\Sigma_a}{\varkappa}B\,e^{\varkappa x}. \tag{5.2.24a}$$

From Eq. (5.2.23b) it then follows that

$$J=-\frac{\Sigma_a}{\varkappa^2}\frac{d\Phi}{dx}. \tag{5.2.24b}$$

Thus there is always a relation of the FICK's law type between the current density and the flux gradient in the asymptotic range[1]. In particular, it follows in the limiting case of vanishing absorption that

$$J=-\frac{1}{3(\Sigma_s+\Sigma_a)}\frac{d\Phi}{dx} \tag{5.2.24c}$$

which agrees with Eq. (5.2.10) (for isotropic scattering $\overline{\cos\vartheta_0}=0$). Thus elementary diffusion theory is valid asymptotically in weakly absorbing media. This conclusion could have been obtained directly from Eq. (5.2.23b): obviously the flux given by this equation is a solution of the differential equation

$$\frac{d^2\Phi}{dx^2}-\varkappa^2\Phi=0 \tag{5.2.25}$$

[1] This means that the ratio $F_2(x)/\Phi(x)$ appearing in Eq. (5.2.13) does not depend on x.

(or $\nabla^2 \Phi - \varkappa^2 \Phi = 0$ in the three-dimensional case), which in view of Eq. (5.2.22b) agrees with the elementary diffusion Eq. (5.2.16a) in the limiting case $\Sigma_a \ll \Sigma_s$. Thus for weak absorption, $\varkappa$ is equal to the reciprocal diffusion length.

One may ask why the case of vanishingly small absorption is particularly denoted as "elementary diffusion theory". Obviously, Eqs. (5.2.24b) and (5.2.25) hold for the asymptotic distribution for arbitrary absorption, and one could formulate a diffusion theory valid for arbitrarily large absorption in which $1/\varkappa$ would appear in place of the diffusion length and $\varkappa^2/\Sigma_a$ in place of the diffusion constant. However, as we shall see in Sec. 5.2.3, the asymptotic solution of the transport equation in strongly absorbing media has no practical significance, since then even at large distances from the source, non-asymptotic parts of the solution arising from the source predominate.

The results above can also be obtained for the case of weakly anisotropic scattering, if the scattering cross section is represented by the first two terms in its expansion in Legendre polynomials. The angular distribution $f(\cos\vartheta)$ and the transcendental equation for the determination of $\varkappa$ then become somewhat more complicated; in the case $\Sigma_a/\Sigma_s \ll 1$, one obtains the equation

$$\varkappa^2 = 3\Sigma_a \Sigma_t (1 - \overline{\cos\vartheta_0}) \left[1 - \frac{4}{5}\frac{\Sigma_a}{\Sigma_t} + \frac{\Sigma_a \overline{\cos\vartheta_0}}{\Sigma_t(1 - \overline{\cos\vartheta_0})}\right] \tag{5.2.26}$$

for $\varkappa^2$. As $\Sigma_a/\Sigma_t \to 0$, $1/\varkappa$ again approaches the diffusion length defined in Eq. (5.2.15); Eqs. (5.2.24) and (5.2.25) remain the same. In particular, one comes to the same conclusions regarding the validity of elementary diffusion theory as in the case of isotropic scattering.

5.2.3. The Neutron Flux in the Vicinity of a Point Source

We next consider an isotropic point source in an infinite medium and investigate how the neutron flux behaves at small distances from the source. For isotropic scattering in the laboratory system, the integral transport equation takes the form

$$\Phi(\boldsymbol{r}) = \int [\Sigma_s \Phi(\boldsymbol{r}') + Q\,\delta(\boldsymbol{r}')] \frac{e^{-\Sigma_t|\boldsymbol{r}-\boldsymbol{r}'|}}{4\pi(\boldsymbol{r}-\boldsymbol{r}')^2}\, dV'. \tag{5.2.27}$$

The integration over the δ-function can immediately be carried out and yields

$$\Phi(r) = Q\frac{e^{-\Sigma_t r}}{4\pi r^2} + \int \Sigma_s \Phi(r') \frac{e^{-\Sigma_t|\boldsymbol{r}-\boldsymbol{r}'|}}{4\pi(\boldsymbol{r}-\boldsymbol{r}')^2}\, dV'. \tag{5.2.28}$$

The first term represents the contribution to the flux of the neutrons which arrive at the point $\boldsymbol{r}$ without having made a collision. At large distances from the source, the integration of the second term will again lead to an asymptotic solution which because of the spherical symmetry has the form[1]

$$\Phi_{\mathrm{As}}(r) \sim \frac{e^{-\varkappa r}}{r}$$

with $\varkappa$ taken from Eq. (5.2.21). In order to obtain a solution for all r, Eq. (5.2.28) must be solved; this can easily be done by a Fourier transformation method.

---

[1] Cf. Sec. 6.1.1.

We will not discuss this method of solution here (cf. e.g., CASE, PLACZEK, and DE HOFFMANN), but only summarize the results . They are

$$\Phi(r) = \Phi_{\text{transient}}(r) + \Phi_{\text{As}}(r) \tag{5.2.29}$$

with

$$\Phi_{\text{As}}(r) = \frac{Q\beta\varkappa^2}{4\pi\Sigma_a}\frac{e^{-\varkappa r}}{r} \tag{5.2.30a}$$

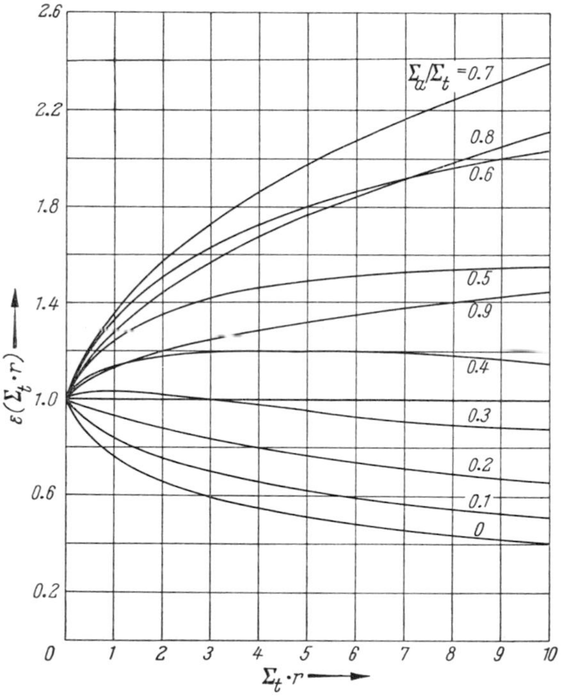

Fig. 5.2.3. The dependence of $\varepsilon(\Sigma_t \cdot r)$ on $\Sigma_t \cdot r$ at various absorptions

where

$$\beta = 2\frac{\Sigma_a}{\Sigma_s}\frac{\Sigma_t^2 - \varkappa^2}{\varkappa^2 - \Sigma_a\Sigma_t} \tag{5.2.30b}$$

and

$$\left.\begin{aligned}&\Phi_{\text{transient}}(r) = \\ &\frac{Q}{4\pi r^2}e^{-\Sigma_t r}\varepsilon(\Sigma_t r).\end{aligned}\right\} \tag{5.2.31}$$

Here ε is a slowly varying function which is of the order of magnitude of unity; it is plotted in Fig. 5.2.3 for several values of Σ_a/Σ_t.

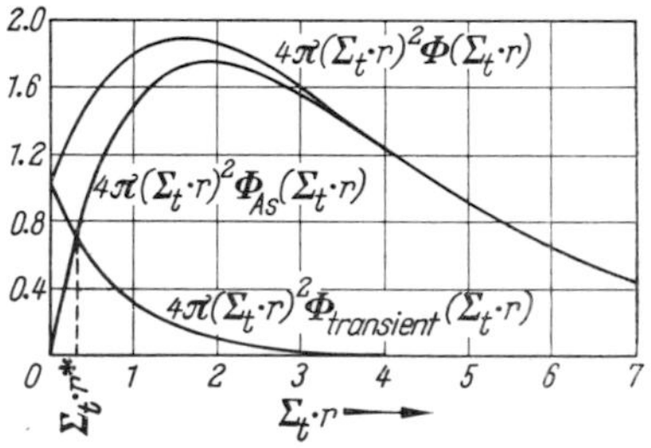

Fig. 5.2.4. Comparison of the asymptotic and the transient solutions for $\Sigma_a/\Sigma_t = 0.1$

Plotted in Fig. 5.2.4 is the total flux as well as both summands on the right-hand side of Eq. (5.2.29) for $\Sigma_a/\Sigma_t = 0.1$. It can easily be seen that $\Phi_{\text{transient}}(r)$ contributes noticeably to the flux only in the immediate neighborhood of the source; for large distances the asymptotic solution dominates. For the distance r^* at which both parts are equally large, the complete calculation gives

$$r^* = \frac{0.031}{\Sigma_t}e^{2\Sigma_t/\Sigma_s}. \tag{5.2.32}$$

According to this relation, for $\Sigma_a \ll \Sigma_s$, $r^* = 0.22/\Sigma_t$. On the other hand, the situation is totally different for strong absorption. If, for example, $\Sigma_a = 0.9\,\Sigma_t$, i.e., $\Sigma_t/\Sigma_s = 10$, $r^* = (1.48 \times 10^7)/\Sigma_t$, i.e., the asymptotic state is never reached in any practical sense.

The quantity β introduced in Eq. (5.2.30b) has a special significance. To see what it is, let us ask for the number Z of neutrons which are absorbed per second from the asymptotic distribution. It is

$$\left.\begin{aligned} Z &= \int \Sigma_a \Phi_{\mathrm{As}}\, dV = \int_0^\infty \Sigma_a \Phi_{\mathrm{As}}(r)\, 4\pi r^2\, dr \\ &= Q\beta\varkappa^2 \int_0^\infty r e^{-\varkappa r}\, dr = Q\beta. \end{aligned}\right\} \tag{5.2.33}$$

Since the total number of neutrons absorbed per sec must be equal to the source strength Q, β gives the fraction of the neutrons from the source which enter the asymptotic distribution. If $\Sigma_a/\Sigma_s \ll 1$, then as can easily be verified,

$$\beta = 1 - \frac{4}{5}\,\frac{\Sigma_a}{\Sigma_t} \tag{5.2.34}$$

i.e. as $\Sigma_a/\Sigma_s \to 0$ *all* the neutrons enter the asymptotic distribution.

One can also apply all these considerations to media with weakly anisotropic scattering and obtain qualitatively the same results. In this way, important conclusions can be drawn about the validity of elementary diffusion theory: In an infinite, weakly absorbing medium, elementary diffusion theory applies beyond a distance of 1 to 2 mean free paths from an isotropic point source. It gives not only the variation of the flux, but also its absolute value within an error of the order of magnitude of Σ_a/Σ_t. This statement also holds for arbitrarily shaped sources such as surfaces, etc. which can be thought of as composed of point sources. If we now notice that localized strong absorbers, as well as surfaces through which neutrons can escape from the medium, can be represented by suitably distributed negative point sources, we then see that in any weakly absorbing medium there exists an asymptotic neutron field which can be described by the elementary diffusion equation beyond a distance of 1 to 2 mean free paths from localized sources or boundary surfaces.

Thus we have answered the question of the validity of the elementary diffusion equation except in the case of spatially distributed sources, which appear in Eq. (5.2.16a) as the source density $S(x)$. That the conclusions stated above regarding the applicability of elementary diffusion theory also hold when the medium contains homogeneously distributed sources can be shown in the following way: Let us assume that we have found an exact solution $F(\boldsymbol{r}, \boldsymbol{\Omega})$ to the transport equation in a medium in which there is a homogeneous source distribution $S/4\pi$. Then $F^*(\boldsymbol{r}, \boldsymbol{\Omega}) = F(\boldsymbol{r}, \boldsymbol{\Omega}) - S/4\pi\Sigma_a$ is a solution of the source-free transport equation. Eq. (5.2.25) holds for the flux $\Phi^* = \int F^*(\boldsymbol{r}, \boldsymbol{\Omega})\, d\Omega$ in the asymptotic range; therefore

$$\nabla^2\Phi - \varkappa^2\Phi = -\frac{\varkappa^2}{\Sigma_a}\, S \tag{5.2.35}$$

which becomes Eq. (5.2.16b) in the limit $\Sigma_a/\Sigma_t \to 0$, holds for $\Phi = \Phi^* + \dfrac{S}{\Sigma_a}$. This proof, which has been given here for a spatially constant source density, can also be generalized to slowly varying source densities.

5.2.4. Neutron Flux at the Surface of a Scattering Medium

Now that we know in what range elementary diffusion theory is valid, we must investigate the boundary conditions which apply at the interface between a scattering medium and a vacuum or "black" absorber. Let us consider a medium which extends in the x-direction from $x=0$ to $x=\infty$ and from $-\infty$ to $+\infty$ in the y- and z-directions. Let there be a plane source at $x=x_0 \gg \frac{1}{\Sigma_t}$. Let the half-space $x<0$ be empty; this means that a neutron which leaves the medium through the plane $x=0$ cannot return. If we take $\Sigma_a=0$ and assume isotropic scattering, the transport equation becomes

$$\cos\vartheta \frac{\partial F}{\partial x} + \Sigma_s F = \frac{1}{2} \Sigma_s \int_0^\pi F \sin\vartheta' \, d\vartheta' \tag{5.2.36a}$$

and must be solved with the boundary condition

$$F(\boldsymbol{\Omega}, x=0)=0 \quad \text{for} \quad 0<\vartheta<\pi/2. \tag{5.2.36b}$$

We can easily specify the general form of the asymptotic solution; since $\Sigma_a=0$ and $\varkappa^2=0$, for $x \gg 1/\Sigma_s$ but $\ll x_0$ we have

$$\frac{\partial^2 \Phi_{\text{As}}}{\partial x^2}=0 \tag{5.2.37}$$

i.e.,

$$\Phi_{\text{As}}=x+d$$

where d is arbitrary[1].

If the flux is to be known for all x, Eq. (5.2.36a) must be solved with the boundary condition (5.2.36b). The detailed treatment of this mathematically demanding problem of transport theory, called the Milne problem, can be found, for example, in the book of DAVISON; there the following results are given:

$$\Phi(x)=x+d+\Phi_{\text{transient}}(x) \tag{5.2.38a}$$

with

$$d=0.7104/\Sigma_s \tag{5.2.38b}$$

and

$$\Phi_{\text{transient}}=-\frac{0.2436}{\Sigma_s}\{E_2(\Sigma_s x)-0.9211\,E_3(\Sigma_s x)\}. \tag{5.2.38c}$$ [2]

Fig. 5.2.5 shows the variation of the flux in the neighborhood of the surface. The asymptotic state is clearly reached after about one mean free path. The quantity d is called the *extrapolated endpoint*, since the extrapolation of Φ_{As} vanishes at $x=-d$. Frequently, d is also called the *extrapolation length*, but this is incorrect. The extrapolation length is defined for an arbitrarily shaped surface as

$$\lambda=\frac{\Phi_{\text{As}}(\boldsymbol{r}_G)}{(\partial\Phi_{\text{As}}/\partial n)_{\boldsymbol{r}_G}} \tag{5.2.39}$$

where the differentiation is in the normal direction. Here, and in many practical cases, d and λ are identical; however, we will also encounter exceptions.

The Milne problem can also be treated for an absorbing medium. For the asymptotic part of the solution

$$\Phi_{\text{As}}(x)=\sinh\varkappa(x+d) \tag{5.2.40}$$

---

[1] The general solution is actually $a(x+d)$, but we choose the source strength so that $a=1$.

[2] The exact form of the transient solution is more complicated; Eq. (5.2.38c) is an approximation due to LE CAINE which never deviates from the exact solution by more than 0.3%. For the E_n functions, cf. appendix III.

holds. The extrapolated endpoint d is plotted as a function of Σ_a/Σ_t in Fig. 5.2.6. For $\Sigma_a \ll \Sigma_t$, d is again $0.7104/\Sigma_s$. It is important to note that in an absorbing medium the extrapolation length λ and the extrapolated endpoint d are not the same if d is not $\ll 1/\varkappa$.

In the case of weakly anisotropic scattering, we obtain the same results, save that the scattering cross section is everywhere replaced by the transport cross section $\Sigma_{tr} = \Sigma_s(1 - \overline{\cos\vartheta_0})$. In particular, the extrapolated endpoint in a weakly absorbing medium is given by

$$d = \frac{0.7104}{\Sigma_s(1-\overline{\cos\vartheta_0})} = \frac{0.7104}{\Sigma_{tr}}. \tag{5.2.41}$$

Thus in the application of elementary diffusion theory the boundary condition at the surface of a vacuum or a black absorber is that the neutron flux vanish on the extrapolated boundary; if $\boldsymbol{n}$ is the outward normal, then

$$\Phi(\boldsymbol{r}_G + d\cdot\boldsymbol{n}) = 0. \tag{5.2.42}$$

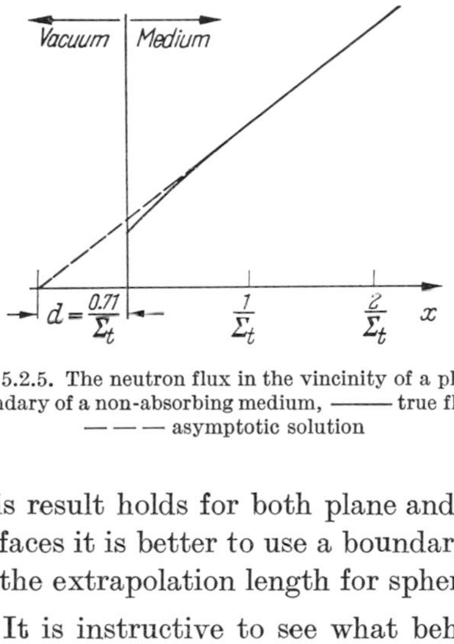

Fig. 5.2.5. The neutron flux in the vicinity of a plane boundary of a non-absorbing medium, ——— true flux; — — — asymptotic solution

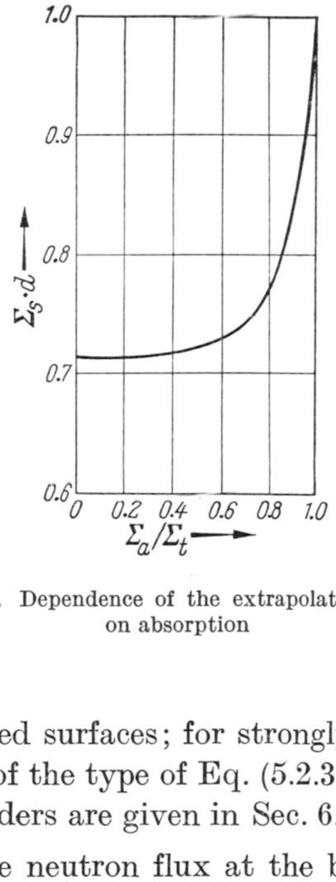

Fig. 5.2.6. Dependence of the extrapolated endpoint on absorption

This result holds for both plane and mildly curved surfaces; for strongly curved surfaces it is better to use a boundary condition of the type of Eq. (5.2.39). Data on the extrapolation length for spheres and cylinders are given in Sec. 6.3.2.

It is instructive to see what behavior for the neutron flux at the boundary follows from a naive application of elementary diffusion theory. We start from the defining equation for the neutron current density, Eq. (5.1.8), which we resolve according to

$$J = J^+ - J^- \tag{5.2.43a}$$

into a component J^+ in the direction of the distribution axis and a component J^- in the opposite direction. Here

$$J^+ = 2\pi \int_0^{\pi/2} F(x, \boldsymbol{\Omega}) \cos\vartheta \sin\vartheta \, d\vartheta, \tag{5.2.43b}$$

$$J^- = -2\pi \int_{\pi/2}^{\pi} F(x, \boldsymbol{\Omega}) \cos\vartheta \sin\vartheta \, d\vartheta. \tag{5.2.43c}$$

If according to the approximation of elementary diffusion theory we now set[1]

$$F(x, \mathbf{\Omega}) = \frac{1}{4\pi}\,\Phi(x) - \frac{1}{4\pi\Sigma_{tr}}\frac{d\Phi}{dx}\cos\vartheta$$

then

$$\left.\begin{aligned} J^+ &= \frac{\Phi(x)}{4} - \frac{1}{6\Sigma_{tr}}\frac{d\Phi}{dx} \\ J^- &= \frac{\Phi(x)}{4} + \frac{1}{6\Sigma_{tr}}\frac{d\Phi}{dx}\,. \end{aligned}\right\} \qquad (5.2.44)^2$$

These relations are exact if the conditions of applicability of elementary diffusion theory are fulfilled, i.e., in our case, if we are several mean free paths from the boundary. If one assumes for the sake of argument that elementary diffusion theory holds right up to the surface of the medium, then one must take

$$J^+(x) = 0 \quad \text{at} \quad x = 0 \qquad (5.2.45\text{a})$$

as the boundary condition. With the help of Eq. (5.2.44) it follows that

$$\left(\frac{\Phi(x)}{d\Phi/dx}\right)_{x=0} = \frac{2}{3\Sigma_{tr}} = \frac{0.667}{\Sigma_{tr}}\,. \qquad (5.2.45\text{b})$$

Thus we obtain $0.667/\Sigma_{tr}$ for the extrapolation length, which is identical with the extrapolated endpoint in this case, instead of the exact value of $0.7104/\Sigma_{tr}$.

5.2.5. Improved Approximations for the Solution of the Transport Equation

Exact solution of the transport equation is only possible in a few cases (e.g., for simple geometry) and even then sometimes only at the cost of considerable mathematical labor. We have seen in the last section that in weakly absorbing media the neutron field can be described by elementary diffusion theory beyond a certain distance from sources and boundaries; this theory is exact in the limiting case $\Sigma_a/\Sigma_t \to 0$ and represents an approximation which becomes progressively worse the larger Σ_a/Σ_t is. In many practical cases, $\Sigma_a/\Sigma_t \ll 1$, and for this reason we will frequently use diffusion theory (in this connection cf. Chapter 6). In those cases in which the conditions for the application of diffusion theory are not fulfilled, we must look for better approximation methods. The two most important methods are the following.

P_N-Approximation (Method of Spherical Harmonics). In this approximation, the vector flux $F(\mathbf{r}, \mathbf{\Omega})$ is expanded in Legendre polynomials as in Sec. 5.2.1. However, the expansion is not broken off after the second term but rather after the *N-th* term, $F_N(x)$. Clearly the approximation is better, the higher N is. By multiplication of the transport equation with each of the $P_l(\cos\vartheta)$ and integration, the following $N+1$ equations are obtained:

$$\left.\begin{aligned} &\frac{dF_1}{dx} + \Sigma_t F_0 = \Sigma_{s0} F_0 + S_0(x) \\ \frac{l+1}{2l+1}\frac{dF_{l+1}}{dx} + \frac{l}{2l+1}&\frac{dF_{l-1}}{dx} + \Sigma_t F_l = \Sigma_{sl} F_l, \qquad l = 1, 2, \ldots, N-1 \\ \frac{N}{2N+1}&\frac{dF_{N-1}}{dx} + \Sigma_t F_N = \Sigma_{sN} F_N\,. \end{aligned}\right\} \qquad (5.2.46)$$

[1] Since Σ_a must be $\ll \Sigma_t$ for the application of diffusion theory, we can neglect Σ_a in comparison with Σ_{tr} in the expression for the diffusion coefficient.

[2] $J^+ + J^- = \Phi/2$ is the number of neutrons which penetrate one cm^2 per second from both sides; cf. Sec. 5.1.1.

This is a system of linear differential equations for the $F_l(x)$ which must be solved in combination with appropriate boundary conditions; the formulation of the boundary conditions is a characteristic difficulty of the P_N-method which we will not consider here.

S_N-Approximation (Method of Discrete Ordinates; WICK's *Method).* In this method the transport equation is again reduced to a system of linear differential equations. However, this reduction is not accomplished by expanding F in a series of spherical harmonics, but rather by replacing the integration over the angle ϑ in the transport equation by a summation over a discrete number of directions. In order to understand this method, let us consider in the interest of simplicity an isotropically scattering, source-free medium with plane symmetry, introduce $\mu = \cos\vartheta$ as the direction variable, and replace the vector flux $F(x, \mathbf{\Omega})$ by $F(x, \mu) = 2\pi F(x, \mathbf{\Omega})$. The transport equation then becomes

$$\mu \frac{\partial F}{\partial x} + \Sigma_t F = \frac{1}{2} \Sigma_s \int_{-1}^{+1} F(x, \mu')\, d\mu'. \tag{5.2.47}$$

If instead of the continuous direction variable μ, discrete directions μ_i, $i = 1, \ldots, N$, are introduced and $F(x, \mu_i)$ denoted by $F_i(x)$, the integral equation can be replaced by the system of differential equations

$$\mu_i \frac{\partial F_i}{dx} + \Sigma_t F_i = \frac{1}{2} \Sigma_s \sum_{K=1}^{N} g_K F_K(x) \tag{5.2.48}$$

where the g_K are suitably chosen weights. In WICK's method, the directions μ_K and the weight factors g_K are chosen so that the best possible approximation to the integral is achieved via the Gauss integration method. In CARLSON's S_N-method, the interval $-1 < \mu < 1$ is usually divided into equal parts and the vector flux approximated by a linear function of μ in each part. Eq. (5.2.48) is then integrated with specified boundary conditions for the $F_K(x)$; with a large number N of directions, this is only possible with the help of an electronic computer.

Details of this and other methods can be found in the literature, particularly in the book of DAVISON and in the work of CARTER and ROWLANDS.

5.3. Thermal Neutrons

If one introduces neutrons into an infinite, non-absorbing medium, they will remain there an infinitely long time. After a time, a true equilibrium will necessarily be established between the neutrons and the thermal motion of the scattering atoms: the neutron energies will assume a Maxwell distribution with the temperature of the scattering medium. In practice, such an ideal case does not occur; all media capture neutrons and are finite. Since all practical sources emit neutrons with energies higher than thermal energies, a certain time elapses before a neutron is "thermalized". Thus there is a certain probability that the neutron is lost by absorption or leakage before its complete thermalization; that means, however, that true equilibrium is never reached. This would be so even if the source could emit neutrons with an equilibrium distribution, because all cross sections are energy dependent. However, in a good moderator that is sufficiently large and only weakly absorbs, the neutron energy distribution at the end of the slowing-down process *approximates* an equilibrium distribution.

Since it is conceptually useful, in this section we shall study some properties of an idealized thermal neutron field in which the neutrons have a true equilibrium distribution. The present results can be looked upon as a first approximation; in Chapter 10, we shall study the properties of thermal neutrons in more detail.

5.3.1. Maxwell Distribution of Energies

According to the Maxwell distribution law, the number of neutrons dn per cm^3 with energies between E and $E+dE$ is

$$dn = n(E)\,dE = \frac{2\pi n}{(\pi kT)^{\frac{3}{2}}}\, e^{-E/kT}\sqrt{E}\, dE \tag{5.3.1}$$

or

$$\frac{n(E)\,dE}{n} = \frac{2}{\sqrt{\pi}}\, e^{-E/kT}\sqrt{\frac{E}{kT}}\,\frac{dE}{kT} \tag{5.3.2}$$

where $n = \int_0^\infty n(E)\,dE$ is the (total) density. It follows that the average energy is

$$\bar{E} = \frac{\int_0^\infty E\, n(E)\,dE}{\int_0^\infty n(E)\,dE} = \int_0^\infty \frac{2}{\sqrt{\pi}}\, e^{-E/kT}\left(\frac{E}{kT}\right)^{\frac{3}{2}} dE = \frac{3}{2}\,kT. \tag{5.3.3}$$

Using $E = mv^2/2$ and $dE = mv\,dv$, we find that the distribution of velocities is

$$\frac{n(v)\,dv}{n} = \frac{4}{\sqrt{\pi}}\, e^{-\frac{mv^2}{2kT}}\,\frac{mv^2}{2kT}\sqrt{\frac{m}{2kT}}\,dv. \tag{5.3.4}$$

The most probable velocity is

$$v_T = \sqrt{\frac{2kT}{m}}. \tag{5.3.5}$$

Furthermore,

$$\frac{m}{2}\,v_T^2 = kT. \tag{5.3.6}$$

Frequently we write $kT = E_T$ [1], so that

$$\frac{m}{2}\,v_T^2 = E_T. \tag{5.3.7}$$

When $T = 293.6$ °K (20.4 °C), $E_T = kT = 0.0253$ ev, and $v_T = 2200$ m/sec $= v_0$.

The average velocity is

$$\bar{v} = \frac{\int_0^\infty v\, n(v)\,dv}{\int_0^\infty n(v)\,dv} = \frac{4}{\sqrt{\pi}}\int_0^\infty e^{-(v/v_T)^2}\left(\frac{v}{v_T}\right)^3 dv = \frac{2}{\sqrt{\pi}}\,v_T. \tag{5.3.8}$$

Therefore, $\bar{v}^2 = \frac{4}{\pi}\,v_T^2$, while $\overline{v^2} = \frac{3}{2}\,v_T^2$.

---

[1] E_T is not the most probable energy; rather from Eq. (5.3.2) it follows that the most probable energy is $kT/2 = E_T/2$.

The distribution of the neutron flux is given by

$$\frac{\Phi(E)\,dE}{\Phi}=e^{-E/E_T}\frac{E}{E_T}\frac{dE}{E_T} \tag{5.3.9}$$

or

$$\frac{\Phi(v)\,dv}{\Phi}=2e^{-(v/v_T)^2}\left(\frac{v}{v_T}\right)^3\frac{dv}{v_T}. \tag{5.3.10}$$

Here

$$\Phi=\int_0^\infty n(v)\,v\,dv=n\bar{v}=\frac{2}{\sqrt{\pi}}\,n v_T \tag{5.3.11}$$

is the total flux. Fig. 5.3.1 shows $n(E)/n$ and $\Phi(E)/\Phi$ as functions of E/E_T. It is easy to see that compared to the density, the flux is always shifted to higher energies.

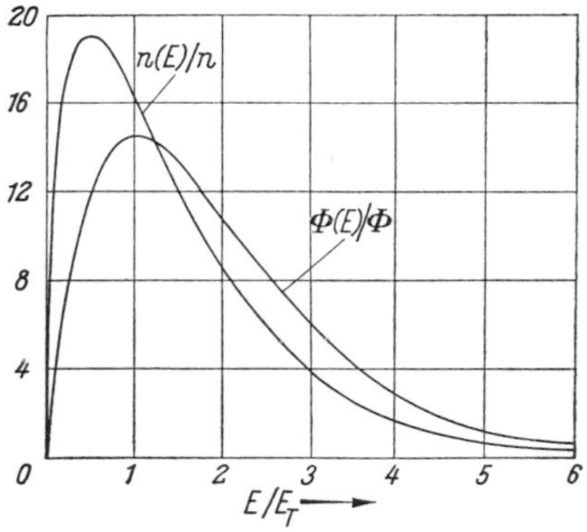

Fig. 5.3.1. The Maxwell distributions of neutron flux and density

It is worthwhile to derive some general formulas for averaging expressions of the type $(E/E_T)^l$ over the density and flux distributions. Averaging over the density gives

$$\overline{\left(\frac{E}{E_T}\right)^l}=\int_0^\infty\frac{n(E)}{n}\left(\frac{E}{E_T}\right)^l dE=\frac{2}{\sqrt{\pi}}\int_0^\infty x^{l+\frac{1}{2}}e^{-x}\,dx=\frac{\Gamma(l+\frac{3}{2})}{\Gamma(\frac{3}{2})} \tag{5.3.12}$$

while averaging over the flux gives

$$\overline{\left(\frac{E}{E_T}\right)^l}=\int_0^\infty\frac{\Phi(E)}{\Phi}\left(\frac{E}{E_T}\right)^l dE=\int_0^\infty x^{l+1}e^{-x}\,dx=\Gamma(l+2). \tag{5.3.13}$$

Here $\Gamma(\alpha)=\int_0^\infty t^{\alpha-1}e^{-t}dt$ is the gamma function[1]. Some special values of the gamma function are $\Gamma(\frac{1}{2})=\sqrt{\pi}$; $\Gamma(1)=1$; $\Gamma(\frac{3}{2})=\sqrt{\pi}/2$; $\Gamma(2)=1$; $\Gamma(\frac{5}{2})=3\sqrt{\pi}/4$.

5.3.2. Reaction Rates in Thermal Neutron Fields

If a substance with the cross section $\Sigma(E)$ is placed in a thermal neutron field, the number of reactions per cm³ and per sec is given by

$$\psi=\int_0^\infty\Sigma(E)\,\Phi(E)\,dE. \tag{5.3.14}$$

This reaction rate can be written as the product of the total neutron flux Φ and an average cross section $\overline{\Sigma}$; the latter is obtained from an average over the energy distribution of the flux:

$$\psi=\overline{\Sigma}\Phi;\quad \overline{\Sigma}=\int_0^\infty\Sigma(E)\left(\frac{E}{E_T}\right)e^{-E/E_T}\frac{dE}{E_T}. \tag{5.3.15}$$

In the important special case of a $1/v$-absorber, $\Sigma_a(E)=\Sigma_a(E_T)\sqrt{\frac{E_T}{E}}$, and

[1] The gamma function is tabulated in Jahnke-Emde-Lösch: Tables of Higher Functions. Stuttgart: Teubner 1960.

it follows from Eq. (5.3.13) that

$$\bar{\Sigma}_a = \frac{\sqrt{\pi}}{2}\,\Sigma_a(E_T). \tag{5.3.16}$$

The average cross section of a $1/v$-absorber is thus smaller by a factor of $\sqrt{\pi}/2 = 0.886$ than the cross section at the energy E_T; this has to be taken into account when the cross sections given in tables are applied to thermal neutrons. It is easy to see from Eq. (5.3.8) that this average cross section is equal to the cross section at the average neutron velocity $\bar{v} = \frac{2}{\sqrt{\pi}} v_T$; thus

$$\psi = \bar{\Sigma}_a \Phi = \Sigma_a(\bar{v})\, n \cdot \bar{v}. \tag{5.3.17}$$

From this it follows that for a given neutron density n, the reaction rate of a $1/v$-absorber does not depend on the energy $E_T = kT$ [1].

Occasionally, one reads works in which the neutron flux is not given, as it is here, by $\Phi = n\bar{v}$ but rather by nv; the reaction rate of a $1/v$-absorber is then obtained by using the cross section corresponding to this velocity. Thus v can be arbitrarily chosen; in the so-called Westcott convention (cf. Chap. 12), for example, v is always $v_0 = 2200$ m/sec, i.e., the most probable velocity at a temperature of 293.6 °K, even when the actual temperature T is different from this value. The reaction rate of a $1/v$-absorber is then $n v_0 \Sigma_a(v_0)$, so that the cross section at this velocity, which is what is specified in most tables, can be used directly. If the expression $\Phi = n\bar{v}$, which is usual here, is used for the flux, then for any temperature it follows that

$$\bar{\Sigma}_a = \frac{\sqrt{\pi}}{2}\sqrt{\frac{293.6}{T}}\,\Sigma_a(v_0). \tag{5.3.18}$$

If a cross section does not follow the $1/v$-law, we frequently use the g-factor, defined as follows: Let $\psi = \Sigma_a \Phi$, with

$$\bar{\Sigma}_a = g(T)\frac{\sqrt{\pi}}{2}\sqrt{\frac{293.6}{T}}\,\Sigma_a(v_0). \tag{5.3.19}$$

From this equation, it follows that

$$g(T) = \frac{2}{\sqrt{\pi}}\sqrt{\frac{T}{293.6}}\,\frac{1}{\Sigma_a(v_0)}\int_0^\infty \Sigma_a(E)\left(\frac{E}{kT}\right)e^{-E/kT}\frac{dE}{kT}. \tag{5.3.20}$$

For a $1/v$-absorber, $g(T) = 1$, as can easily be verified. If one uses the density instead of the flux, the reaction rate is given by

$$\psi = n v_0 \Sigma_a(v_0)\, g(T). \tag{5.3.21}$$

The g-factor can be obtained as a function of temperature, if Eq. (5.3.20) is integrated graphically or numerically over the experimental variation of the cross section at low energies. If the cross section can be represented by the Breit-Wigner formula in this energy range, the integration can also be carried out analytically. Westcott gives detailed tables of $g(T)$ for many elements.

---

[1] Even in the case of an arbitrary spectrum $n(E)$, the reaction rate of a $1/v$-absorber does not depend on the spectrum; for it is always the case that

$$\psi = \int n(E)\, v\frac{K}{v}\, dE = K\int n(E)\, dE = Kn.$$

5.3.3. The Elementary Diffusion Equation for Thermal Neutrons

If the energy distribution of the neutrons in a medium does not depend on position, it is always possible to describe the diffusion process by an energy-independent one-group equation in which the cross sections are suitable averages over the spectral distribution. In particular, we can derive an elementary diffusion equation for thermal neutrons. For this purpose, we begin with the energy-, space-, and angle-dependent transport Eq. (5.1.17), and assume for simplicity plane symmetry, isotropic sources, and isotropic scattering in the laboratory system:

$$\left.\begin{aligned} &\cos\vartheta \frac{\partial F(x,\mathbf{\Omega},E)}{\partial x} + \Sigma(E) F(x,\mathbf{\Omega},E) \\ &\quad = \frac{1}{2}\int_0^\pi\int_0^\infty \Sigma_s(E'\to E) F(x,\mathbf{\Omega}',E') \sin\vartheta'\, d\vartheta'\, dE' + \frac{1}{2} S(x,E). \end{aligned}\right\} \quad (5.3.22)$$

The further treatment of Eq. (5.3.22) will be exactly like that of the energy-independent equation in Sec. 5.2.1. Again we set

$$F(x,\mathbf{\Omega},E) = \frac{1}{4\pi}\Phi(x,E) + \frac{3}{4\pi} J(x,E)\cos\vartheta. \quad (5.3.23)$$

If we substitute this in Eq. (5.3.22) and integrate once over $d\Omega$ and once over $\cos\vartheta\, d\Omega$ we obtain

$$\frac{\partial J(x,E)}{\partial x} + \Sigma_t \Phi(x,E) = S(x,E) + \int_0^\infty \Sigma_s(E'\to E)\,\Phi(x,E')\,dE' \quad (5.3.24\text{a})$$

and

$$\frac{1}{3}\frac{\partial \Phi(x,E)}{\partial x} + \Sigma_t(E) J(x,E) = 0. \quad (5.3.24\text{b})$$

By combination of both these equations, there arises an energy-dependent diffusion equation

$$\left.\begin{aligned} &-\frac{1}{3\Sigma_t(E)}\nabla^2\Phi(x,E) + \Sigma_t(E)\,\Phi(x,E) \\ &\quad = S(x,E) + \int_0^\infty \Sigma_s(E'\to E)\,\Phi(x,E')\,dE'. \end{aligned}\right\} \quad (5.3.25)$$

This equation is still free from special assumptions concerning the spectrum. In subsequent chapters, we will frequently use it in various forms to calculate the spectrum in space-dependent slowing-down and thermalization problems.

We now make the assumption that in a weakly absorbing medium the flux $\Phi(x,E)$ can always be written in the form

$$\Phi(x,E) = \Phi(x)\frac{E}{(kT)^2} e^{-E/kT}. \quad (5.3.26)$$

If we substitute this in Eq. (5.3.25), we find that an integration over all energies[1] yields

$$-\bar{D}\nabla^2\Phi(x) + \bar{\Sigma}_a\Phi(x) = S(x). \quad (5.3.27)$$

[1] Here use has been made of the fact that $\int_0^\infty \Sigma_s(E'\to E)\,dE = \Sigma_s(E')$ and thus cancels the term containing the scattering cross section.

Here $S(x)=\int_0^\infty S(x, E)\,dE$,

$$\overline{D}=\int_0^\infty \left(\frac{E}{kT}\right)\frac{1}{3\Sigma_t(E)}\,e^{-E/kT}\,\frac{dE}{kT} \tag{5.3.28}$$

and

$$\overline{\Sigma}_a=\int_0^\infty \left(\frac{E}{kT}\right)\Sigma_a(E)\,e^{-E/kT}\,\frac{dE}{kT}\,. \tag{5.3.29}$$

Thus the effective diffusion parameters D and Σ_a for thermal neutrons are found by averaging the energy-dependendent parameters over the Maxwell energy distribution.

In the case of anisotropic scattering in which the average cosine of the scattering angle depends only weakly on the energy, we have

$$\left.\begin{aligned} D&=\int_0^\infty \left(\frac{E}{kT}\right)\frac{1}{3(\Sigma_{tr}(E)+\Sigma_a(E))}\,e^{-E/kT}\,\frac{dE}{kT}\\ &\approx\int_0^\infty \left(\frac{E}{kT}\right)\frac{1}{3\Sigma_{tr}(E)}\,e^{-E/kT}\,\frac{dE}{kT}=\frac{\lambda_{tr}}{3}\end{aligned}\right\} \tag{5.3.30}$$

where λ_{tr} is the thermal neutron transport mean free path. The diffusion length is then given by

$$L^2=\frac{\overline{D}}{\overline{\Sigma}_a}=\frac{\lambda_{tr}\,\lambda_a}{3} \tag{5.3.31}$$

(with $\lambda_a=1/\overline{\Sigma}_a$=the mean free path for absorption).

Frequently, the diffusion equation is written not in terms of Φ but in terms of $n=\Phi/\overline{v}$; in terms of n it is

$$\overline{D}\overline{v}\nabla^2 n-\overline{\Sigma}_a\overline{v}n=-S \tag{5.3.32}$$

or with the *diffusion constant* $D_0=\overline{D}\overline{v}=\frac{\lambda_{tr}}{3}\overline{v}$ (cm²/sec) and the average lifetime against absorption $l_0=1/\overline{v}\,\overline{\Sigma}_a$

$$D_0\nabla^2 n-n/l_0=-S\,. \tag{5.3.33}$$

Chapter 5: References

General

AMALDI, E.: Loc. cit., especially §. 112—123.

CASE, K. M., F. DE HOFFMANN, and G. PLACZEK: Introduction to the Theory of Neutron Diffusion, vol. I. LOS ALAMOS, N. M.: Los Alamos Scientific Laboratory 1953.

DAVISON, B.: Neutron Transport Theory. Oxford: Clarendon Press 1957.

SALMON, J., *et al.*: Théorie Cinétique de Neutrons Rapides. Paris: Presses Universitaires de France 1961.

WEINBERG, A. M., and E. P. WIGNER: The Physical Theory of Neutron Chain Reactors. Chicago: Chicago University Press 1958, especially Chap. IX, Transport Theory and Diffusion of Monoenergetic Neutrons.

Special

BOTHE, W.: Z. Physik **118**, 401 (1942); **119**, 493 (1942) (Transport Equations).
LE CAINE, J.: Phys. Rev. **72**, 564 (1947).
MARK, C.: Phys. Rev. **72**, 558 (1947).
MARSHAK, R. E.: Phys. Rev. **72**, 47 (1947).
PLACZEK, G., and W. SEIDEL: Phys. Rev. **72**, 550 (1947).
} Treatment of the Milne Problem.
WEINBERG, A. M.: AECD-3405 (1949) (Transport Theory and the Diffusion Equation).
KOFINK, W.: ORNL-2334, 2358 (1957) (The P_N-Method).
CARLSON, B. G.: LA-1891 (1955) (The S_N-Method).
WICK, G. C.: Z. Physik **121**, 702 (1943) (WICK's Method).
CARTER, C., and G. ROWLANDS: J. Nucl. Energy A & B **15**, 1 (1961) (Various Approximate Methods for the Treatment of the Stationary, Energy-Independent Transport Equation).
WESTCOTT, C. H.: AECL-1101 (1960) (Cross Sections for Thermal Neutrons).

6. Applications of Elementary Diffusion Theory

In this chapter, we shall extend the elementary diffusion theory that was developed in Secs. 5.2. and 5.3 and apply it to some important special problems. We shall mainly be interested in the diffusion of thermal neutrons in weakly absorbing "good" moderators, such as H_2O, D_2O, graphite, beryllium, beryllium oxide, and various hydrogenous materials (paraffin, plexiglass, etc.). In these media, the conditions for the application of diffusion theory are well fulfilled; as long as we stay one or two mean free paths away from boundaries and sources, the results of diffusion theory will be quite accurate. We specifically assume that the neutron sources emit neutrons with a thermal energy distribution.

6.1. Extension of the General Theory

6.1.1. Elementary Derivation of Fick's Law

According to GLASSTONE and EDLUND, FICK's law and elementary diffusion theory can be derived without knowledge of the general transport equation. Let us consider an isotropically scattering, weakly absorbing ($\Sigma_a \ll \Sigma_s$) medium and calculate the current of neutrons which penetrates the element of surface dF per second (Fig. 6.1.1). In the first place,

$$J^- \, dF = \frac{dF}{4\pi} \Sigma_s \int_0^\infty \int_0^{2\pi} \int_0^{\pi/2} \Phi(\mathbf{r}) e^{-\Sigma_s r} dr \, d\varphi \cos\vartheta \sin\vartheta \, d\vartheta \tag{6.1.1}$$

is the number of neutrons streaming from $+z$ to $-z$ in a second; furthermore,

$$J^+ \, dF = \frac{dF}{4\pi} \Sigma_s \int_0^\infty \int_0^{2\pi} \int_{\pi/2}^{\pi} \Phi(\mathbf{r}) e^{-\Sigma_s r} dr \, d\varphi \, |\cos\vartheta| \sin\vartheta \, d\vartheta \tag{6.1.2}$$

is the number of neutrons streaming from $-z$ to $+z$ per second. If we develop Φ in a Taylor series around the origin,

$$\Phi(x, y, z) = \Phi(0) + x\left(\frac{\partial \Phi}{\partial x}\right)_{\mathbf{r}=0} + y\left(\frac{\partial \Phi}{\partial y}\right)_{\mathbf{r}=0} + z\left(\frac{\partial \Phi}{\partial z}\right)_{\mathbf{r}=0} \tag{6.1.3}$$

and substitute into Eqs. (6.1.1) and (6.1.2), an integration yields[1]

$$\left.\begin{aligned} J^- &= \frac{1}{4}\,\Phi(0) + \frac{1}{6\Sigma_s}\left(\frac{\partial \Phi}{\partial z}\right)_{\boldsymbol{r}=0} \\ J^+ &= \frac{1}{4}\,\Phi(0) - \frac{1}{6\Sigma_s}\left(\frac{\partial \Phi}{\partial z}\right)_{\boldsymbol{r}=0}. \end{aligned}\right\} \tag{6.1.4}$$

FICK's law follows immediately from these equations:

$$J = J^+ - J^- = -\frac{1}{3\Sigma_s}\left(\frac{\partial \Phi}{\partial z}\right). \tag{6.1.5}$$

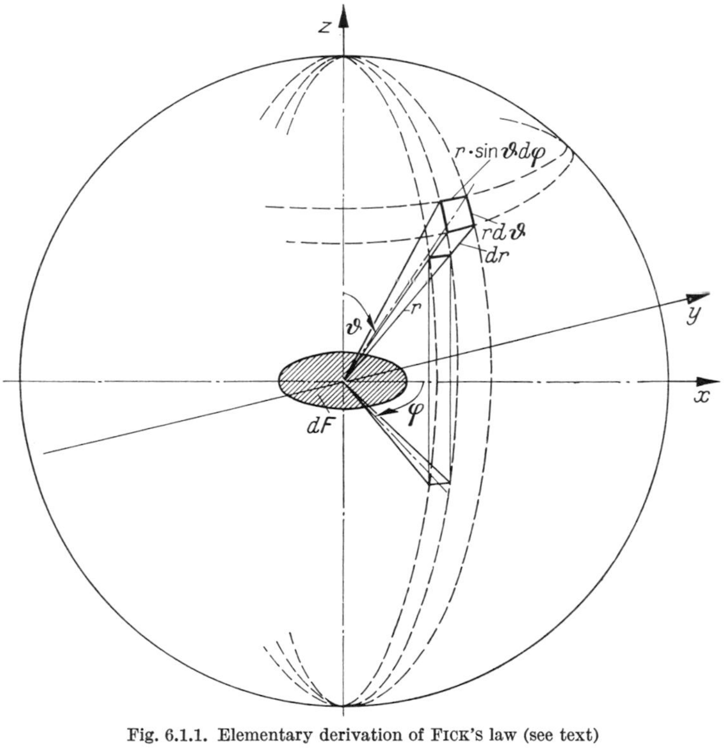

Fig. 6.1.1. Elementary derivation of FICK's law (see text)

If we set $D = 1/3\Sigma_s$, FICK's law can be written in the familiar form

$$\boldsymbol{J} = -D \cdot \operatorname{grad} \Phi. \tag{6.1.6}$$

For Eq. (6.1.5) to hold, the linear approximation (6.1.3) must be a good representation of the flux over an interval of several mean free paths, i.e.,

$$\frac{1}{\Sigma_s}\,\frac{\partial \Phi}{\partial x} \ll \Phi \tag{6.1.7}$$

must hold. This is certainly not the case near sources or boundaries; it is also not

[1] In general, $\Phi(\boldsymbol{r}) = \Phi(0) + \boldsymbol{r}\cdot\operatorname{grad}\Phi = \Phi(0) + r\cos\vartheta\left(\frac{\partial \Phi}{\partial z}\right)_{\boldsymbol{r}=0} + r\sin\vartheta\sin\varphi\left(\frac{\partial \Phi}{\partial y}\right)_{\boldsymbol{r}=0} + r\sin\vartheta\cos\varphi\left(\frac{\partial \Phi}{\partial x}\right)_{\boldsymbol{r}=0}$, but the terms containing $\sin\varphi$ and $\cos\varphi$ drop out in the integration over φ.

the case in a strongly absorbing medium. If we now write a neutron balance for one cm^3, viz.,

$$\underbrace{\operatorname{div} \boldsymbol{J}}_{\text{loss or gain through diffusion}} + \underbrace{\Sigma_a \Phi}_{\text{loss through absorption}} = \underbrace{S(\boldsymbol{r})}_{\text{gain from sources}} \tag{6.1.8}$$

we immediately obtain the elementary diffusion equation

$$-D\nabla^2\Phi + \Sigma_a \Phi = S(\boldsymbol{r}). \tag{6.1.9}$$

6.1.2. Diffusion Parameters for Thermal Neutrons

Table 6.1.1 contains the diffusion parameters of various moderators for thermal neutrons ($T = 293.6$ °K). These values have been obtained from various experimental data which occasionally differ from one another rather strongly and therefore cannot be considered very accurate. The absorption cross sections of graphite and beryllium can be greatly increased by hardly noticeable impurities; the values given here are for pure moderators. If we are dealing with a neutron field with a temperature other than 293.6 °K, we can easily correct the absorption cross section, but we must newly determine the diffusion constant. In the discussion of methods of measurement for L and $\bar{D}$, we will learn of values measured at other temperatures.

Table 6.1.1. *Diffusion Parameters for Thermal Neutrons* ($T_0 = 293.6$ °K)

Substance	Density (g/cm^3)	$\bar{D}$ (cm)	$\Sigma_a(\bar{v})$ (cm^{-1})	L (cm)	λ_{tr} (cm)	$D_0 = \bar{D}\bar{v}$ (cm^2/sec)	$l_0 = 1/\bar{v}\,\Sigma_a(\bar{v})$ (sec)
H_2O	1.00	0.144	1.89×10^{-2}	2.755	0.431	0.357×10^5	2.13×10^{-4}
D_2O (pure)	1.10	0.810	0.31×10^{-4}	161	2.43	2.01×10^5	0.130
D_2O (99.8 mole %)	1.10	0.802	0.70×10^{-4}	107	2.41	1.99×10^5	5.76×10^{-2}
Graphite	1.60	0.858	0.311×10^{-3}	52.5	2.574	2.13×10^5	1.29×10^{-2}
Beryllium	1.85	0.495	1.10×10^{-3}	21.2	1.48	1.23×10^5	3.65×10^{-3}
Beryllium Oxide .	2.96	0.471	5.30×10^{-4}	29.8	1.41	1.17×10^5	7.60×10^{-3}
Paraffin	0.87	0.109	2.26×10^{-2}	2.19	0.327	2.70×10^4	1.78×10^{-4}
Plexiglass	1.18	0.136	1.73×10^{-2}	2.80	0.411	3.4×10^4	2.32×10^{-4}
Dowtherm A[1] . .	1.06	0.198	1.15×10^{-2}	4.15	0.594	4.92×10^4	3.48×10^{-4}
Zirconium Hydride[2]	3.48	0.233	1.52×10^{-2}	3.93	0.699	5.79×10^4	2.66×10^{-4}

[1] Diphenyl oxide and diphenyl mixed in a ratio of 1:0.36. The melting point is 12.3 °C.
[2] $ZrH_{1.7}$.

6.1.3. The Neutron Flux Near a Point Source: The Significance of the Diffusion Length

The diffusion length introduced by the relation $L = \sqrt{D/\Sigma_a}$ has a simple physical significance. Let us consider the neutron flux near a point source of strength Q in an infinite medium. Because of the spherical symmetry of the system, the diffusion equation takes the form

$$\frac{d^2\Phi}{dr^2} + \frac{2}{r}\frac{d\Phi}{dr} - \frac{\Phi}{L^2} = 0 \tag{6.1.10}$$

or

$$\frac{1}{r}\frac{d^2}{dr^2}(r\Phi) - \frac{1}{L^2}\Phi = 0.$$

Its general solution is

$$\Phi(r) = \frac{A}{r} e^{-r/L} + \frac{B}{r} e^{r/L}. \qquad (6.1.11)$$

Since the neutron flux cannot become infinite as $r \to \infty$, $B = 0$. The constant A can be determined from the requirement that in the stationary state all the neutrons emitted from the source must be absorbed in the medium:

$$\left.\begin{aligned} Q &= \int \Sigma_a \Phi(r)\, dV = \Sigma_a \int_0^\infty \frac{A}{r} e^{-r/L} 4\pi r^2 dr \\ &= 4\pi \Sigma_a A \int_0^\infty e^{-r/L}\, r\, dr = 4\pi\, \Sigma_a A L^2. \end{aligned}\right\} \qquad (6.1.12)$$

Thus finally,

$$\Phi(r) = \frac{Q}{4\pi \Sigma_a L^2} \frac{e^{-r/L}}{r} = \frac{Q}{4\pi D} \frac{e^{-r/L}}{r}. \qquad (6.1.13)$$

Next, we seek the average distance $\bar{r}$ from the source at which a neutron is absorbed. The probability that a neutron is absorbed between r and $r + dr$ is $\Sigma_a \Phi(r)\, 4\pi\, r^2\, dr/Q$. Thus

$$\left.\begin{aligned} \bar{r} &= \frac{1}{Q} \int_0^\infty r\, \Sigma_a \Phi(r)\, 4\pi\, r^2\, dr \\ &= \frac{1}{L^2} \int_0^\infty r^2 e^{-r/L}\, dr = 2L. \end{aligned}\right\} \qquad (6.1.14)$$

The mean squared distance from the source at which a neutron is captured is

$$\overline{r^2} = \frac{1}{L^2} \int_0^\infty r^3\, e^{-r/L}\, dr = 6L^2. \qquad (6.1.15)$$

Thus the diffusion length specifies within a numerical factor the average distance between the place where a neutron is produced and the place where it is absorbed. This average distance must not be confused with the average distance actually traversed by the neutron; the latter is equal to the mean free path for absorption, and because of the neutron's zig-zag motion is very much greater than the mean separation between the site of production and the site of absorption.

6.2. Solution of the Diffusion Equation in Simple Cases

6.2.1. Simple Symmetrical Sources in Infinite Media

Point Source: See Sec. 6.1.3.

Plane Source: Let the source, which lies in the (x, y)-plane, emit Q neutrons per cm^2 per second. Then for $z \neq 0$,

$$\frac{d^2\Phi}{dz^2} - \frac{1}{L^2}\Phi = 0.$$

This equation has the solution

$$\begin{aligned} \Phi(z) &= A\, e^{-z/L} \quad \text{for} \quad z > 0 \\ &= A\, e^{z/L} \quad \text{for} \quad z < 0. \end{aligned}$$

The constant A again follows from a neutron balance:

$$Q = A\left\{\int_{-\infty}^{0} \Sigma_a e^{z/L}\,dz + \int_{0}^{\infty} \Sigma_a e^{-z/L}\,dz\right\} = 2\Sigma_a A L .$$

Thus

$$\Phi(z) = \frac{Q}{2\Sigma_a L}\, e^{-|z|/L} . \tag{6.2.1}$$

Line Source: Let the source, which lies along the z-axis of a cylindrical coordinate system, emit Q neutrons per cm per second. Then for $r \neq 0$,

$$\frac{d^2\Phi}{dr^2} + \frac{1}{r}\frac{d\Phi}{dr} - \frac{1}{L^2}\Phi = 0 .$$

This equation can be solved in terms of the modified Bessel functions of zeroth order, $I_0(r/L)$ and $K_0(r/L)$ [1]. These functions are plotted in Fig. 6.2.1. Since $I_0(r/L)$ diverges as $r/L \to \infty$, we find

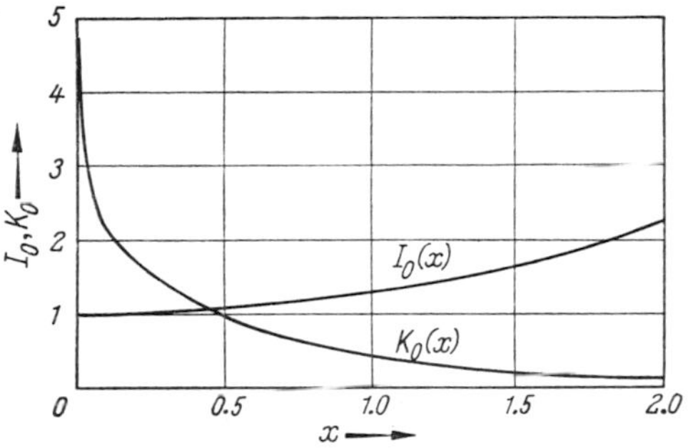

Fig. 6.2.1. The modified Bessel functions of zeroth order

$$\Phi(r) = A\,K_0(r/L)$$

$$Q = \int_0^\infty \Sigma_a \Phi(r)\, 2\pi r\, dr = 2\pi\, A\, \Sigma_a L^2 \int_0^\infty \frac{r}{L} K_0\left(\frac{r}{L}\right)\frac{dr}{L}$$
$$= 2\pi\, A\, \Sigma_a L^2 .$$

Thus

$$\Phi(r) = \frac{Q}{2\pi \Sigma_a L^2} K_0(r/L) . \tag{6.2.2}$$

Spherical Shell Source: Let the spherical shell have the radius r' and emit Q neutrons per second; thus the source density per cm² of surface is $Q/4\pi r'^2$. We take

$$\Phi_>(r) = A\,\frac{e^{-r/L}}{r} \qquad (r > r')$$

$$\Phi_<(r) = B\,\frac{\sinh(r/L)}{r} \qquad (r < r')$$

for the solution of the spherically symmetric diffusion Eq. (6.1.10). At $r = r'$ both solutions must be equal, i.e.,

$$A\,e^{-r'/L} = B \sinh(r'/L) .$$

Furthermore, we can express the source strength in terms of the current density at $r = r'$:

$$(|J_>| + |J_<|)_{r=r'} = \frac{Q}{4\pi r'^2} = D\left(\frac{d\Phi_<}{dr} - \frac{d\Phi_>}{dr}\right)_{r=r'} .$$

[1] See, for example, WATSON: Theory of Bessel Functions. Cambridge: Cambridge Univ. Press 1952.

It follows from these two relations that

$$A=\frac{QL}{4\pi Dr'}\sinh(r'/L) \qquad B=\frac{QL}{4\pi Dr'}e^{-r'/L}.$$

Thus

$$\Phi(r)=\frac{QL}{8\pi Drr'}\left\{e^{-|r-r'|/L}-e^{-|r+r'|/L}\right\} \tag{6.2.3}$$

for all r.

Cylindrical Shell Source: Let the source of radius r' emit Q neutrons per second and cm of length; the surface density of the source is thus $Q/2\pi r'$. Let the axis of the cylinder define the z-axis of a system of cylindrical coordinates. As a solution of the cylindrically symmetric diffusion equation we take

$$\Phi_>(r)=A\,K_0(r/L) \qquad (r>r')$$

$$\Phi_<(r)=B\,I_0(r/L) \qquad (r<r').$$

At $r=r'$ the two solutions must be equal, i.e.,

$$A\,K_0(r'/L)=B\,I_0(r'/L).$$

Again we can relate the source strength to the current density at the source:

$$(|J_>|+|J_<|)_{r=r'}=\frac{Q}{2\pi r'}=D\left(\frac{d\Phi_<}{dr}-\frac{d\Phi_>}{dr}\right)_{r=r'}.$$

From these relations it follows that

$$A=\frac{Q}{2\pi D}I_0(r'/L) \qquad B=\frac{Q}{2\pi D}K_0(r'/L)$$

and thus

$$\left.\begin{aligned}\Phi(r)&=\frac{Q}{2\pi D}I_0(r'/L)\,K_0(r/L) \qquad (r>r')\\ \Phi(r)&=\frac{Q}{2\pi D}K_0(r'/L)\,I_0(r/L) \qquad (r<r').\end{aligned}\right\} \tag{6.2.4}$$

6.2.2. General Source Distributions: The Diffusion Kernel

In practice, thermal neutron sources do not have the forms assumed in the above examples; most often, spatially distributed neutron sources are present (slowed-down neutrons). However, as will now be shown, we can always replace a source distribution by a system of sources of the kind discussed above.

In an infinite medium the flux at the point $\boldsymbol{r}$ due to a unit point source (strength: 1 neutron/sec) at $\boldsymbol{r}'$ is given by

$$\Phi(\boldsymbol{r})=\frac{1}{4\pi D}\frac{e^{-|\boldsymbol{r}-\boldsymbol{r}'|/L}}{|\boldsymbol{r}-\boldsymbol{r}'|}.$$

When a spatically distributed source is present, we can consider the sources in the volume element dV' to be a point source of strength $S(\boldsymbol{r}')\,dV'$; their contribution to the flux at $\boldsymbol{r}$ is

$$d\Phi(\boldsymbol{r})=\frac{S(\boldsymbol{r}')\,dV'}{4\pi D}\frac{e^{-|\boldsymbol{r}-\boldsymbol{r}'|/L}}{|\boldsymbol{r}-\boldsymbol{r}'|}.$$

Integration over all $\boldsymbol{r}'$ gives the flux at $\boldsymbol{r}$:

$$\Phi(\boldsymbol{r})=\int S(\boldsymbol{r}')\frac{e^{-|\boldsymbol{r}-\boldsymbol{r}'|/L}}{4\pi D|\boldsymbol{r}-\boldsymbol{r}'|}\,dV'. \tag{6.2.5}$$

The expression

$$G_{pt}(\boldsymbol{r}, \boldsymbol{r}') = \frac{e^{-|\boldsymbol{r}-\boldsymbol{r}'|/L}}{4\pi D|\boldsymbol{r}-\boldsymbol{r}'|} \tag{6.2.6}$$

is called the point diffusion kernel. Eq. (6.2.5) permits the calculation of the neutron flux of an arbitrary source distribution $S(\boldsymbol{r}')$.

Frequently the source distribution has some symmetry; then in place of the point diffusion kernel (6.2.6) another diffusion kernel adapted to this symmetry may appear. Under some circumstances this replacement can considerably simplify the integration over the kernel. If the source density depends only on z' (plane symmetry), then the source distribution can be replaced by plane sources of strength $S(z')\,dz'$ and the flux expressed as

$$\Phi(z) = \int\limits_{-\infty}^{+\infty} S(z') \frac{e^{-|z-z'|/L}}{2\Sigma_a L} dz'. \tag{6.2.7}$$

The diffusion kernel in this case is

$$G_{pl}(z, z') = \frac{e^{-|z-z'|/L}}{2\Sigma_a L}. \tag{6.2.8}$$

A spherically symmetric source distribution $S(r')$ can be replaced by a system of spherical shells of strength $4\pi r'^2\, dr'\, S(r')$; the flux is then given by

$$\Phi(r) = \int\limits_{0}^{\infty} S(r') \frac{L}{8\pi D r r'} \left\{e^{-|r-r'|/L} - e^{-|r+r'|/L}\right\} 4\pi\, r'^2\, dr'. \tag{6.2.9}$$

In this case, the diffusion kernel is

$$G_{ss}(r, r') = \frac{L}{8\pi D r r'} \left\{e^{-|r-r'|/L} - e^{-|r+r'|/L}\right\}. \tag{6.2.10}$$

An axially symmetric source distribution can be replaced by a system of cylindrical shell sources of strength $2\pi r'\, dr'\, S(r')$; then

$$\Phi(r) = \int\limits_{0}^{\infty} S(r')\, G_{As}(r, r')\, 2\pi\, r'\, dr' \tag{6.2.11}$$

with

$$G_{As}(r, r') = \frac{1}{2\pi D} \begin{cases} K_0(r/L)\, I_0(r'/L) & r > r' \\ I_0(r/L)\, K_0(r'/L) & r < r'. \end{cases} \tag{6.2.12}$$

Finally, the source strength may depend only on the coordinates (r', φ') of a point in a plane. The source distribution can then be replaced by a system of line sources of strength $S(r', \varphi')\, r'\, dr'\, d\varphi'$ and the flux expressed as

$$\Phi(r, \varphi) = \int\limits_{0}^{\infty} \int\limits_{0}^{2\pi} S(r', \varphi')\, G_L(r, \varphi;\, r', \varphi')\, r'\, dr'\, d\varphi' \tag{6.2.13}$$

with

$$G_L(r, \varphi;\, r', \varphi') = \frac{1}{2\pi D} K_0(\varrho/L). \tag{6.2.14}$$

Here

$$\varrho = \sqrt{r^2 + r'^2 - 2 r r' \cos(\varphi - \varphi')}$$

is the distance between the source point (r', φ') and the point (r, φ) being considered.

6.2.3. Diffusion Problems with Simple Boundary Conditions

The five infinite-medium diffusion problems considered in Sec. 6.2.1 will now be solved in finite media which have the same symmetry as the source. The boundary condition will be that the neutron flux shall vanish on the extrapolated boundary of the medium, i.e., that $\Phi(\boldsymbol{r}_G + \boldsymbol{n} \cdot d)$ shall equal zero. Here $d = 0.71\,\lambda_{tr}$[1]. To solve the diffusion equation we use the following fact: The neutron flux in a finite medium bounded by the surface $\boldsymbol{r}_G$ arising from prescribed sources in the medium is equal to the neutron flux in an infinite medium with the same inner sources and suitable additional negative sources outside the surface $\boldsymbol{r}_G$. We therefore take for the flux the sum of the solution obtained in Sec. 6.2.1 and an as yet undetermined multiple of the flux distribution of a fictitious surface source on the outer surface. The coefficient t of the latter component of the flux will be determined from the boundary conditions.

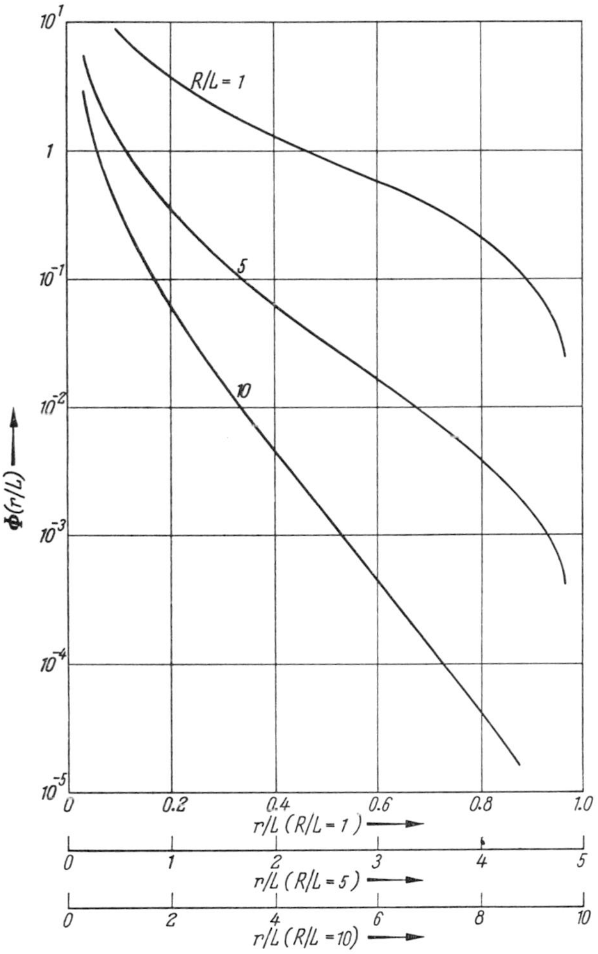

Fig. 6.2.2. The neutron flux around a point source at the center of a sphere

Point Source at the Center of a Sphere: Let R be the effective radius of the sphere, i.e., the actual radius increased by the distance d. Then

$$\Phi(r) = \frac{Q}{4\pi \Sigma_a L^2} \frac{e^{-r/L}}{r} + t \frac{\sinh(r/L)}{r}.$$

From $\Phi(R) = 0$ it follows that

$$t = -\frac{Q}{4\pi \Sigma_a L^2} \frac{e^{-R/L}}{\sinh(R/L)}$$

and thus

$$\Phi(r) = \frac{Q}{4\pi \Sigma_a L^2 \sinh(R/L)} \frac{\sinh\left(\frac{R-r}{L}\right)}{r}. \tag{6.2.15}$$

[1] This relation was introduced in Sec. 5.2.4 for monoenergetic neutrons; the expression follows from simple averaging over the thermal energy distribution of the flux. It represents only a first approximation, which is rather bad in media with a strongly energy-dependent scattering cross section, like H_2O. We will return to the question of the extrapolated endpoint for thermal neutrons in Chapter 10.

Fig. 6.2.2 shows this solution for several values of R/L; it lets us see the effect of the surface on the flux distribution. It is instructive to calculate the fraction B of the neutrons absorbed inside the sphere. B is given by

$$B=\frac{\int_0^R \Sigma_a \Phi(r) 4\pi r^2\, dr}{Q}=1-\frac{R/L}{\sinh(R/L)}. \qquad (6.2.16)\,[1]$$

For $R=L$, 15% of all the neutrons are absorbed in the sphere, while for $R=10\,L$, 99.9% are absorbed.

Plane Source at the Center of a Slab: Let the slab be infinite in the x- and y-directions and let it have an effective thickness equal to $2a$. Let the source lie in the plane $z=0$. We take

$$\Phi(z)=\frac{Q}{2\Sigma_a L}\, e^{-|z|/L}+t\cosh(z/L).$$

It follows from the boundary conditions $\Phi(a)=\Phi(-a)=0$ that

$$t=\frac{-\frac{Q}{2\Sigma_a L}\, e^{-a/L}}{\cosh(a/L)}$$

so that

$$\Phi(z)=\frac{Q}{2\Sigma_a\cdot L\cosh(a/L)}\sinh\left(\frac{a-|z|}{L}\right). \qquad (6.2.17)$$

Line Source on the Axis of an Infinitely Long Cylinder of Effective Radius R:

$$\Phi(r)=\frac{Q}{2\pi D}\left[K_0(r/L)-\frac{K_0(R/L)}{I_0(R/L)}\, I_0(r/L)\right]. \qquad (6.2.18)$$

Spherical Shell Source of Radius r′ Concentric with a Sphere of Radius R:

$$\left.\begin{aligned}\Phi(r)&=\frac{QL}{4\pi D r'}\,\frac{\sinh\left(\frac{R-r'}{L}\right)}{\sinh(R/L)}\,\frac{\sinh(r/L)}{r} && (r<r')\\ \Phi(r)&=\frac{QL}{4\pi D r'}\,\frac{\sinh(r'/L)}{\sinh(R/L)}\,\frac{\sinh\left(\frac{R-r}{L}\right)}{r} && (r>r').\end{aligned}\right\} \qquad (6.2.19)$$

Infinitely Long Cylindrical Shell Source of Radius r′ Concentric with an Infinite Cylinder of Radius R:

$$\left.\begin{aligned}\Phi(r)&=\frac{Q}{2\pi D}\,\frac{I_0(r/L)}{I_0(R/L)}\left\{I_0(R/L)K_0(r'/L)-K_0(R/L)I_0(r'/L)\right\} && (r<r')\\ \Phi(r)&=\frac{Q}{2\pi D}\,\frac{I_0(r'/L)}{I_0(R/L)}\left\{I_0(R/L)K_0(r/L)-K_0(R/L)I_0(r/L)\right\} && (r>r').\end{aligned}\right\} \qquad (6.2.20)$$

As before, we can derive diffusion kernels for the treatment of general source distributions from these solutions.

---

[1] Strictly speaking, the integral should only extend to $R-d$; however, if R is not too small, i.e., if R is somewhat $>L$, the error is small.

6.2.4. The Neutron Distribution in a Prism "Pile"

In many experiments, media in the form of prisms or cylinders — piles, so-called — are used. Next, let us consider a prism infinitely long in the z-direction which contains a point source located at the point x', y', z'. The relevant diffusion equation is

$$\nabla^2 \Phi - \frac{1}{L^2}\Phi = -\frac{Q}{D}\delta(x-x')\delta(y-y')\delta(z-z')$$

and must be solved under the boundary conditions:

$$\Phi(x, y, z) = 0 \quad \text{for} \quad x=0, \quad x=a, \quad y=0, \quad y=b.$$

In addition, the flux must vanish in the limits $z \to \pm\infty$. Let us develop Φ in a Fourier series:

$$\Phi(x, y, z) = \sum_{l,m=1,2,3\cdots} \Phi_{l,m}(z) \sin\frac{l\pi x}{a} \sin\frac{m\pi y}{b}. \tag{6.2.21}$$

This expression always fulfills the boundary conditions in the x- and y-directions. Now let us expand the δ-functions:

$$\delta(x-x')\delta(y-y') = \sum_{l,m} \Delta_{l,m} \sin\frac{l\pi x}{a} \sin\frac{m\pi y}{b}$$

with

$$\left.\begin{aligned}\Delta_{l,m} &= \frac{4}{ab}\int_0^a\int_0^b \delta(x-x')\delta(y-y') \sin\frac{l\pi x}{a} \sin\frac{m\pi y}{b}\,dx\,dy\\ &= \frac{4}{ab}\sin\frac{l\pi x'}{a}\sin\frac{m\pi y'}{b}.\end{aligned}\right\} \tag{6.2.22}$$

If we substitute Eqs. (6.2.21) and (6.2.22) into the diffusion equation, carry out the indicated differentiations with respect to x and y, multiply by $\sin\frac{l\pi x}{a}\sin\frac{m\pi y}{b}$, and integrate over the cross section of the pile, we obtain the differential equation

$$\frac{d^2\Phi_{l,m}}{dz^2} - \frac{1}{L_{l,m}^2}\Phi_{l,m} = -\frac{4}{ab}\frac{Q}{D}\sin\frac{l\pi x'}{a}\sin\frac{m\pi y'}{b}\,\delta(z-z') \tag{6.2.23a}$$

with

$$\frac{1}{L_{l,m}^2} = \frac{1}{L^2} + \pi^2\left[\frac{l^2}{a^2} + \frac{m^2}{b^2}\right] \tag{6.2.23b}$$

for the Fourier components $\Phi_{l,m}(z)$.

Formally, this equation is identical with that for the flux $\Phi_{l,m}$ that arises from a plane source of strength $\frac{4Q}{ab}\sin\frac{l\pi x'}{a}\sin\frac{m\pi y'}{b}$ located at $z=z'$ in an infinite medium whose diffusion length is $L_{l,m}$; its solution is thus

$$\Phi_{l,m}(z) = \frac{2L_{l,m}Q}{Dab}\sin\frac{l\pi x'}{a}\sin\frac{m\pi y'}{b}\,e^{-|z-z'|/L_{l,m}}.$$

Finally then

$$\Phi(x, y, z) = \frac{2Q}{Dab}\sum_{l,m} L_{l,m}\sin\frac{l\pi x'}{a}\sin\frac{m\pi y'}{b}\,e^{-|z-z'|/L_{l,m}}\sin\frac{l\pi x}{a}\sin\frac{m\pi y}{b}. \tag{6.2.24}$$

According to Eq. (6.2.23b) the relaxation lengths $L_{l,m}$ decrease with increasing l, m; thus for large distances from the source the flux is given by

$$\Phi(x, y, z) \sim \sin\frac{\pi x}{a}\sin\frac{\pi y}{b}\, e^{-|z-z'|/L_{11}}.$$

In other words, the flux falls off exponentially in the z-direction, just as in the case of an infinite medium. However, instead of the diffusion length, there appears a relaxation length $L_{11}=L/\sqrt{1+\pi^2 L^2(1/a^2+1/b^2)}$, which accounts for the lateral leakage of neutrons and which is smaller the smaller the cross sectional dimensions of the pile.

Eq. (6.2.24) with $Q=1$ gives the diffusion kernel for the calculation of the neutron flux due to an arbitrary source distribution in an infinite pile.

In order to calculate the neutron flux in a finite pile (Fig. 6.2.3), we once again start with a Fourier series. Eq. (6.2.23) must then be solved with the boundary condition $\Phi_{l,m}(z)=0$ for $z=0$ and $z=c$. As we can easily check with the help of the method developed in Sec. 6.2.3, the result is

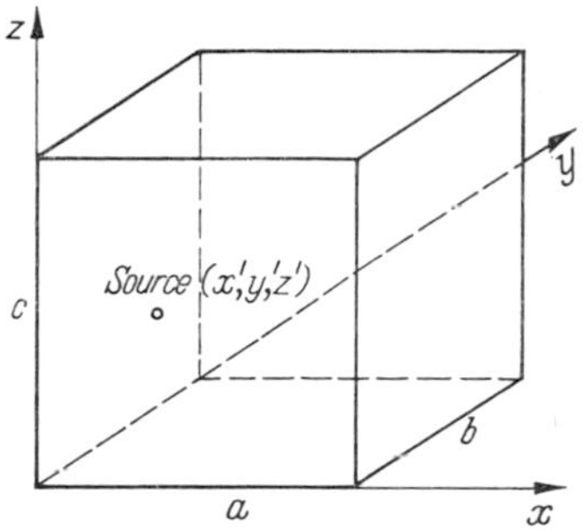

Fig. 6.2.3. A rectangular pile

$$\left.\begin{aligned}\Phi_{l,m}(z) &= \frac{4L_{l,m}Q}{Dab}\sin\frac{l\pi x'}{a}\sin\frac{m\pi y'}{b}\frac{\sinh(z'/L_{l,m})}{\sinh(c/L_{l,m})}\sinh\left(\frac{c-z}{L_{l,m}}\right) \quad (z>z')\\ \Phi_{l,m}(z) &= \frac{4L_{l,m}Q}{Dab}\sin\frac{l\pi x'}{a}\sin\frac{m\pi y'}{b}\frac{\sinh\left(\frac{c-z'}{L_{l,m}}\right)}{\sinh(c/L_{l,m})}\sinh(z/L_{l,m}) \quad (z<z').\end{aligned}\right\} \quad (6.2.25)$$

If the source is sufficiently far from the upper surface, then far from the source the flux is given by

$$\Phi(x, y, z) \sim \sin\frac{\pi x}{a}\sin\frac{\pi y}{b}\sinh\left(\frac{c-z}{L_{11}}\right).$$

When $\frac{c-z}{L_{11}} \geqq 2$, the sinh can be replaced for all practical purposes by $\frac{1}{2}e^{(c-z)/L_{11}}$; thus the effect of the boundary on the flux distribution is no longer palpable at a distance of several relaxation lengths, and the flux falls off exponentially just as it does in an infinitely long pile.

6.3. Some Special Diffusion Problems

6.3.1. Media in Contact: The Albedo

Fig. 6.3.1 shows a simple arrangement of two different media. At $z=0$ there is a plane source which is surrounded on both sides by slabs of thickness l_I. Adjacent to these slabs are slabs of another substance of thickness $l_{II}-l_I$. We take

$$\Phi_I(z) = \frac{Q}{2\Sigma_{a_I}L_I}e^{-z/L_I} + t_I\cosh(z/L_I), \quad (6.3.1\,\mathrm{a})$$

$$\Phi_{II}(z) = t_{II}\sinh\left(\frac{l_{II}-|z|}{L_{II}}\right) \quad (6.3.1\,\mathrm{b})$$

for the neutron flux. This form of the flux already fulfills the boundary condition that the flux vanish at $z = l_{II}$. To determine the constants t_I and t_{II} we apply the condition that the flux and current be continuous at the interface between the media:

$$\left.\begin{aligned} \Phi_I(l_I) &= \Phi_{II}(l_I) \\ -D_I\left(\frac{d\Phi_I}{dz}\right)_{l_I} &= -D_{II}\left(\frac{d\Phi_{II}}{dz}\right)_{l_I}. \end{aligned}\right\} \tag{6.3.2}$$

This leads to

$$\left.\begin{aligned} t_I &= \frac{\frac{Q}{2}\, e^{-l_I/L_I}\left[\sinh\left(\frac{l_{II}-l_I}{L_{II}}\right) - \frac{D_{II}}{\Sigma_{a_I} L_I L_{II}}\cosh\left(\frac{l_{II}-l_I}{L_{II}}\right)\right]}{\frac{D_{II}}{L_{II}}\cosh\left(\frac{l_I}{L_I}\right)\cosh\left(\frac{l_{II}-l_I}{L_{II}}\right) + \frac{D_I}{L_I}\sinh\left(\frac{l_I}{L_I}\right)\sinh\left(\frac{l_{II}-l_I}{L_{II}}\right)} \\ t_{II} &= \frac{\frac{Q}{2}}{\frac{D_{II}}{L_{II}}\cosh\left(\frac{l_I}{L_I}\right)\cosh\left(\frac{l_{II}-l_I}{L_{II}}\right) + \frac{D_I}{L_I}\sinh\left(\frac{l_I}{L_I}\right)\sinh\left(\frac{l_{II}-l_I}{L_{II}}\right)} \end{aligned}\right\} \tag{6.3.3}$$

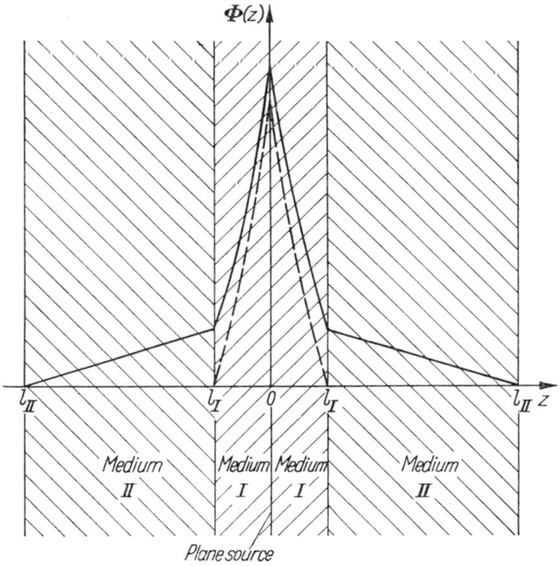

Fig. 6.3.1. The neutron flux around a plane source in a reflected (———) and unreflected (— — —) slab

Fig. 6.3.1 shows the variation of the neutron flux in a typical case. For comparison, the flux for the case in which no reflector surrounds medium I is also plotted. We can see that the neutron flux is increased everywhere in medium I by the reflection of neutrons at the interface. The effect of medium II can very easily be described by a reflection coefficient, the so-called *albedo*, at the I—II interface. The albedo β is the ratio of the current J^- flowing from II to I to the current J^+ flowing from I to II. If J^- and J^+ are now expressed in terms of the properties of medium II, then (cf. Eq. [5.2.44])

$$\beta = \left\{\frac{\frac{\Phi}{4} + \frac{D_{II}}{2}\frac{d\Phi}{dz}}{\frac{\Phi}{4} - \frac{D_{II}}{2}\frac{d\Phi}{dz}}\right\}_{z=l_I} = \frac{1 + 2D_{II}\left(\frac{d\Phi/dz}{\Phi}\right)_{z=l_I}}{1 - 2D_{II}\left(\frac{d\Phi/dz}{\Phi}\right)_{z=l_I}}. \tag{6.3.4}$$

If we denote the reflector thickness $l_{II} - l_I$ by b, it follows from Eq. (6.3.1b) that $\left(\frac{d\Phi/dz}{\Phi}\right)_{z=l_I} = -\frac{1}{L_{II}}\coth\left(\frac{b}{L_{II}}\right)$ and thus that

$$\beta = \frac{1 - \frac{2D_{II}}{L_{II}}\coth\left(\frac{b}{L_{II}}\right)}{1 + \frac{2D_{II}}{L_{II}}\coth\left(\frac{b}{L_{II}}\right)}. \tag{6.3.5}$$

For a thick reflector, $b \gg L$, $\coth\left(\frac{b}{L_{II}}\right) \approx 1$ and

$$\beta = \frac{1 - 2D_{II}/L_{II}}{1 + 2D_{II}/L_{II}} \approx 1 - 4D_{II}/L_{II}. \tag{6.3.6}$$

If the albedo of a reflector is known, we need not solve the diffusion equation in the reflector in order to calculate the flux in medium I; instead we replace the reflector by the condition that the ratio of the inward and outward currents at the interface be equal to β. Then Eq. (6.3.5) holds in slab geometry; however, Eq. (6.3.6) for thick reflectors also holds in spherical and cylindrical geometry. Table 6.3.1 shows that quite thin layers of heavy water and graphite reflect well, while H_2O because of its comparatively large absorption is not as good a reflector of thermal neutrons.

Table 6.3.1. *The Albedo of Various Scatterers for Thermal Neutrons in Plane Geometry*

Material	Reflector Thickness			
	$b=\infty$	$b=60$ cm	$b=40$ cm	$b=20$ cm
H_2O	0.811	0.811	0.811	0.811
D_2O . . .	0.981	0.947	0.922	0.853
Beryllium .	0.911	0.910	0.907	0.881
Graphite . .	0.936	0.923	0.903	0.834

6.3.2. A Strong Absorber in a Neutron Field[1]

Let the neutron flux in an infinite medium be constant and equal to Φ_0. If we now introduce a piece of a strongly absorbing substance like cadmium into the neutron field, it will act as a negative source and depress the neutron flux near it. If it is a spherical absorber of radius R, the neutron field at a distance r from its center is given by

$$\Phi(r) = \Phi_0 - \frac{A}{r}\, e^{-r/L}. \tag{6.3.7}$$

The constant A can be determined from the boundary condition at the surface of the absorber. According to Eq. (5.2.39)

$$\frac{\Phi(R)}{(d\Phi/dr)_{r=R}} = \lambda. \tag{6.3.8}$$

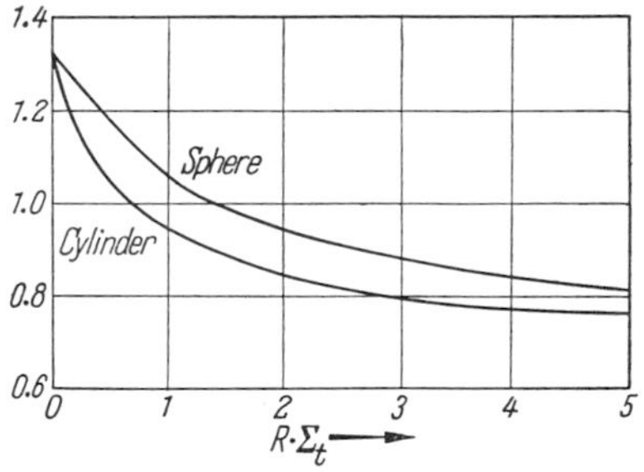

Fig. 6.3.2. The extrapolation length for spheres and cylinders (Davison and Kushneriuk)

The extrapolation length λ depends on the radius of curvature of the absorber; Fig. 6.3.2 shows λ as a function of the radius of curvature for spheres and cylinders (Davison). For $R \to \infty$, λ approaches $0.7104/\Sigma_t$; the limiting value for $R=0$ is $4/3\,\Sigma_t$. These curves are exact for monoenergetic neutrons in a non-absorbing, isotropically scattering medium; they can also be used as a first approximation for thermal neutrons in a weakly absorbing medium with weakly anisotropic scattering if $1/\Sigma_s$ is replaced by λ_{tr}. It then follows from Eqs. (6.3.7) and (6.3.8) that for a spherical absorber

$$A = -\frac{\Phi_0 R e^{R/L}}{\lambda\left[\frac{1}{R} + \frac{1}{L}\right] + 1} \tag{6.3.9}$$

[1] See also Secs. 11.2 and 11.3.

and thus

$$\Phi(r)=\Phi_0\left[1-\frac{1}{\lambda\left(\frac{1}{R}+\frac{1}{L}\right)+1}\,\frac{R}{r}\,e^{(R-r)/L}\right]. \tag{6.3.10}$$

The number Z of neutrons captured by the absorber per second can be obtained from the current at $r=R$:

$$Z=4\pi R^2 D\left(\frac{d\Phi}{dr}\right)_{r=R}=\pi R^2\Phi_0\,\frac{4D\left(\frac{1}{R}+\frac{1}{L}\right)}{\lambda\left(\frac{1}{R}+\frac{1}{L}\right)+1}. \tag{6.3.11}$$

For $R\ll\lambda$ (and thus $R\ll L$), $Z=\pi R^2\Phi_0$, since then $\lambda=\frac{4}{3}\lambda_{tr}=4D$; Z is then equal to the number of neutrons incident on a sphere of radius R in the unperturbed flux. For larger R, Z is always less than $\pi R^2\Phi_0$, since a flux depression occurs near the sphere. In particular, in the limiting case $R\gg L$, Z becomes $\pi R^2\cdot\Phi_0 4D/L$.

We can easily make a similar analysis for an infinitely long cylindrical absorber. For absorbers which do not have these simple symmetrical forms, the calculation of the neutron flux is a challenging mathematical problem.

Chapter 6: References

General

FERMI, E.: (Ed. J. G. BECKERLEY): Neutron Physics. AECD-2664 (1951), especially chap. VII: The Distribution of Slow Neutrons in a Medium.

GLASSTONE, S., and M. C. EDLUND: The Elements of Nuclear Reactor Theory. New York: D. Van Nostrand 1952, especially chap. V: The Diffusion of Neutrons.

Special

Solution of the Diffusion Equation:

PLACZEK, G.: In: The Science and Engineering of Nuclear Power, vol. II, p. 77. Cambridge: Addison-Wesley 1949.

WALLACE, P. R.: Nucleonics 4, No. 2, 30 (1949); 4, No. 3, 48 (1949).

—, and J. LECAINE: AECL-336 (1943).

Extrapolation Length for Spheres and Cylinders:

Reactor Physics Constants. ANL-5800 (1959), p. 167.

DAVISON, B., and S. A. KUSHNERIUK: MT-214 (1946).

7. Slowing Down

As a rule, neutrons produced in nuclear reactions have energies far above the thermal energy range. When such fast neutrons collide with the atoms of a scattering medium, loss of energy occurs simultaneously with the diffusion process. The collisions can be either elastic or inelastic. As long as the neutron energy is greater than about 1 ev, the struck atoms can be considered free and at rest before the collision. This is no longer the case at lower energies, where the chemical binding of the atoms of the scatterer and their thermal motion affect the slowing-down process. It is the aim of slowing-down theory to determine the space and energy distribution of the neutron flux arising from a given distribution of sources. In what follows, we shall treat three distinct aspects of this problem: To begin with, in this chapter we shall consider slowing down in the space-independent case, i.e., in an infinite medium with uniformly distributed sources. We shall take into account only elastic collisions with free atoms that are initially at rest, neglecting both the inelastic scattering of fast neutrons and the effects associated with the chemical binding and thermal motion of the atoms of the scatterer. In Chapter 8, we shall treat the space-dependent case under the same limitations, i.e., we shall study the simultaneous diffusion and slowing down of neutrons in a finite medium containing an arbitrary distribution of sources. Finally, in Chapter 10 we shall study the thermalization of neutrons, that is, the moderation of neutrons in the range of very low energies, where the chemical binding and thermal motion of the atoms must be taken into account. In those light elements that can be used as moderators, inelastic scattering plays no role at energies less than several Mev (cf. Sec. 1.4.2), and we need not take it into account in studying the slowing-down process[1].

The transport equation for the slowing down of neutrons in an infinite medium with uniformly distributed sources was introduced in Sec. 5.1.2. It is

$$(\Sigma_a+\Sigma_s)\,\Phi(E)=\int_0^\infty \Sigma_s(E'\to E)\,\Phi(E')\,dE'+S(E) \qquad (5.1.16)$$

and represents a simple balance equation for the neutrons in the unit energy interval around E: On the left stands the number of neutrons lost per cm^3 and second by absorption and scattering, while on the right stands the number of neutrons gained from the sources or by in-scattering from other energies. Before we treat this equation any further, we shall next calculate the cross section $\Sigma_s(E'\to E)$ with the help of the laws of elastic collisions[2].

7.1. Elastic Scattering and Moderation

7.1.1. Collision Dynamics

Let us consider an elastic collision between a neutron of mass 1 and energy E_1 and a free atomic nucleus of mass A that is initially at rest. Our main goal is to calculate the probability that after the collision the neutron has an energy between E_2 and E_2+dE_2. In addition, the angular distribution of the neutrons after the collision is of interest. For the considerations that follow, it is necessary to consider the collision both with respect to the laboratory frame of reference (L-system) and the center-of-mass frame of reference (C-system). The dynamics of the collision are shown in Fig. 7.1.1. Let us introduce the following notation:

v_1 = the velocity of the neutron before the collision
v_2 = the velocity of the neutron after the collision
v_m = the velocity of the center of mass
ϑ_0 = the scattering angle
} in the L-system,

v_1-v_m = the velocity of the neutron before the collision
v_a = the velocity of the neutron after the collision
ψ = the scattering angle.
} in the C-system.

[1] The situation is altogether different in fast reactors, which contain no moderator and in which inelastic scattering by heavy nuclei seriously affects the spectrum.

[2] Eq. (5.1.16) also holds for the spatially averaged flux in any homogeneous medium with arbitrarily distributed sources; however, the medium must be so large that no neutrons escape through its surface.

As is evident from Fig. 7.1.2,

$$v_2^2 = v_a^2 + v_m^2 + 2 v_a v_m \cos \psi \tag{7.1.1}$$

and

$$v_2 \cos \vartheta_0 = v_a \cos \psi + v_m. \tag{7.1.2}$$

It follows from the theorem of momentum conservation that the velocity of the center of mass is

$$v_m = \frac{v_1}{A+1}. \tag{7.1.3}$$

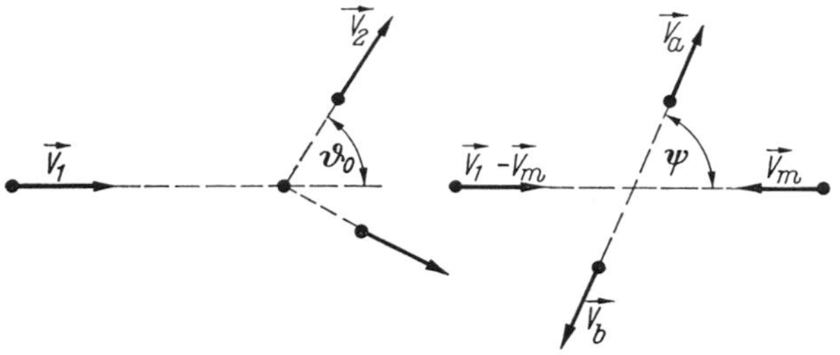

Fig. 7.1.1. Kinematics in the *L*- (left) and *C*- (right) system

The velocity of the neutron in the *C*-system before the collision is

$$v_1 - v_m = \frac{A}{A+1} v_1. \tag{7.1.4}$$

In the *C*-system, the nucleus has the velocity v_m of the center of mass in the *L*-system. Furthermore, in the *C*-system, the momenta of the neutron and the nucleus are equal and oppositely directed. After the collision, they must still be equal and oppositely directed. Since the kinetic energy in the *C*-system is also conserved during an elastic collision, the post-collision velocities of the neutron and nucleus must be the same as the pre-collision velocities. Thus

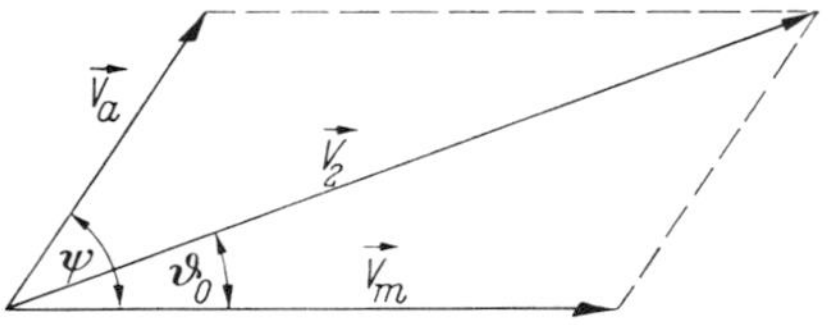

Fig. 7.1.2. The relation between $\vec{v}_a$, $\vec{v}_m$ and $\vec{v}_2$ (see text)

$$v_a = \frac{A}{A+1} v_1. \tag{7.1.5}$$

It now follows immediately from Eq. (7.1.1) that the energy $E_2 = v_2^2/2$ of the neutron after the collision in the *L*-system is

$$E_2 = E_1 \frac{A^2 + 2A \cos \psi + 1}{(A+1)^2}. \tag{7.1.6}$$

Furthermore, the scattering angle in the *L*-system is given by

$$\cos \vartheta_0 = \frac{A \cos \psi + 1}{\sqrt{A^2 + 2A \cos \psi + 1}}. \tag{7.1.7}$$

Next we introduce the auxiliary quantity

$$\alpha = \left(\frac{A-1}{A+1}\right)^2. \tag{7.1.8}$$

Then Eq. (7.1.6) becomes

$$\frac{E_2}{E_1} = \frac{1}{2} [(1+\alpha) + (1-\alpha) \cos \psi]. \tag{7.1.9}$$

The maximum energy loss occurs in a head-on collision, i.e., a collision for which $\psi = 180°$ and $\cos \psi = -1$; E_2 then equals αE_1. The energy cannot fall below αE_1 as the result of a single elastic collision. For hydrogen $\alpha = 0$, and $(E_2)_{\min} = 0$. For large *A*

$$\alpha \approx 1 - \frac{4}{A+2}. \tag{7.1.10}$$

7.1.2. Energy and Angular Distribution

Let $g(E_1 \to E_2)\, dE_2$ be the probability that a neutron of energy E_1 before the collision acquires an energy between E_2 and $E_2 + dE_2$ after the collision. Since E_2 is connected uniquely with the scattering angle ψ by Eq. (7.1.9),

$$g(E_1 \to E_2)\, dE_2 = p(\cos\psi)\, d\cos\psi \tag{7.1.11}$$

where $p(\cos\psi)\, d\cos\psi$ is the probability that the cosine of the scattering angle in the center-of-mass system lies between $\cos\psi$ and $\cos\psi + d\cos\psi$. As we already have seen in Sec. 1.4.2, for most good moderators the scattering in the C-system is isotropic below 1 Mev; in such a case, $p(\cos\psi) = \text{const.} = \frac{1}{2}$. Then, however,

$$g(E_1 \to E_2) = \frac{1}{2}\, \frac{d\cos\psi}{dE_2} = \frac{1}{(1-\alpha)E_1} \tag{7.1.12}$$

for $\alpha E_1 \leqq E_2 \leqq E_1$. For $E_2 > E_1$ and for $E_2 < \alpha E_1$, $g(E_1 \to E_2) = 0$. The probability $g(E_1 \to E_2)$ given by Eq. (7.1.12) will be used later for the solution of the transport equation, Eq. (5.1.16), where

$$\Sigma_s(E')\, g(E' \to E) \tag{7.1.13}$$

will be substituted for $\Sigma_s(E' \to E)$.

Just as the energy distribution after the collision can be calculated, so can the angular distribution of the neutrons after the collision in the laboratory system be calculated. For most transport problems, only the mean cosine of the scattering angle is of interest; for isotropic scattering in the C-system,

$$\overline{\cos\vartheta_0} = \int_{-1}^{+1} \frac{A\cos\psi + 1}{\sqrt{A^2 + 2A\cos\psi + 1}} \cdot \frac{1}{2}\, d\cos\psi = \frac{2}{3A}\,. \tag{7.1.14}$$

The forward direction is more strongly preferred the lighter the struck nucleus is. For heavy nuclei, $2/3A \approx 0$ and the scattering in the laboratory system is nearly isotropic.

When the scattering in the C-system is not isotropic, the expressions for $g(E_1 \to E_2)$ and $p(\cos\psi)$ become more complicated. For weakly anisotropic scattering,

$$p(\cos\psi) = \tfrac{1}{2}\left(1 + 3\,\overline{\cos\psi}\,\cos\psi\right) \tag{7.1.15}$$

and one finds in the same way as before that

$$\left.\begin{aligned} g(E_1 \to E_2) &= \frac{1}{(1-\alpha)E_1}\left[1 + 3\,\overline{\cos\psi} - \frac{6}{1-\alpha}\,\overline{\cos\psi}\,\frac{E_1 - E_2}{E_1}\right] \\ &\qquad (\text{for } \alpha E_1 \leqq E_2 \leqq E_1) \end{aligned}\right\} \tag{7.1.16a}$$

and

$$\overline{\cos\vartheta_0} = \frac{2}{3A} + \overline{\cos\psi}\left(1 - \frac{3}{5A^2}\right). \tag{7.1.16b}$$

7.1.3. The Average Logarithmic Energy Decrement

Knowledge of $g(E_1 \to E_2)$ now allows us to calculate the average energy of a neutron after a collision. It is given by

$$\overline{E_2} = \int_{\alpha E_1}^{E_1} E_2\, g(E_1 \to E_2)\, dE_2 = \int_{\alpha E_1}^{E_1} \frac{E_2\, dE_2}{(1-\alpha)E_1} = \frac{E_1}{2}(1+\alpha). \tag{7.1.17a}$$

The average energy loss per collision is then

$$\overline{\Delta E} = \overline{E_1 - E_2} = E_1 \frac{1-\alpha}{2}. \tag{7.1.17b}$$

The average *logarithmic* energy loss ξ is given by

$$\left.\begin{aligned} \xi &= \ln E_1 - \overline{\ln E_2} = \overline{\ln\left(\frac{E_1}{E_2}\right)} = \int_{\alpha E_1}^{E_1} \ln\left(\frac{E_1}{E_2}\right) g(E_1 \to E_2)\, dE_2 \\ &= \int_{\alpha E_1}^{E_1} \ln\left(\frac{E_1}{E_2}\right) \frac{dE_2}{(1-\alpha)E_1} = 1 + \frac{\alpha}{1-\alpha} \ln \alpha. \end{aligned}\right\} \tag{7.1.18}$$

The average logarithmic energy decrement is thus an energy-independent constant. This suggests the introduction of a logarithmic energy scale for the slowing-down process (cf. Sec. 7.2.1). ξ is equal to 1 for hydrogen, and for large A can be quite well approximated by

$$\xi \approx \frac{2}{A + \frac{2}{3}}. \tag{7.1.19}$$

(Cf. also Table 7.1.1.)

Knowledge of the quantity ξ makes it possible for us to estimate the average number n of collisions necessary to moderate a neutron with an initial energy E_Q to the energy E. Clearly,

$$n\xi = \ln\left(\frac{E_Q}{E}\right); \qquad n = \frac{\ln(E_Q/E)}{\xi}. \tag{7.1.20}$$

Table 7.1.1 contains α, ξ, and n (for moderation from 2 Mev to 0.0253 ev) for several nuclei. The values of n obtained from Eq. (7.1.20) are only approximate, since thermalization effects at low energies and non-isotropic scattering in the C-system at high energies have not been taken into account. What is more, Eq. (7.1.20) is only an approximation for n; in an exact calculation of the number of collisions, the energy distribution of the neutrons during the slowing-down process must be taken into account (DE MARKUS, KÜSTERS).

Table 7.1.1. *Slowing-Down Parameters of Various Substances*

Parameter	Substance							
	H	D	He	Li	Be	C	O	U
A	1	2	4	7	9	12	16	238
α	0	0.111	0.360	0.562	0.640	0.716	0.778	0.983
ξ	1.000	0.725	0.425	0.268	0.209	0.158	0.120	0.00838
n [2 Mev→0.0253 ev] according to Eq. (7.1.20) . .	18	25	43	67	86	114	150	2172

It follows from Eq. (7.1.16a) that in the case of weakly anisotropic scattering in the C-system

$$\xi = \xi_{\text{isotropic}} - 3\,\overline{\cos\psi}\left[\frac{1}{1-\alpha} - \frac{1}{2} + \frac{\alpha}{(1-\alpha)^2}\ln\alpha\right]. \tag{7.1.21}$$

We can estimate the effect of anisotropic scattering on the slowing-down process with Eq. (7.1.21). If the differential scattering cross section of deuterium at 0.75 Mev that is given in Fig. 1.4.4 is approximated by Eq. (7.1.15), a value of

0.067 is obtained for $\overline{\cos\psi}$. It then follows that $\xi=0.662$ compared to $\xi_{\text{isotropic}}=0.725$. The difference is small, and from now on we will not take anisotropic scattering into account. However, for very accurate calculations, the anisotropy of scattering at high energies must be included.

We now introduce two important parameters of the slowing-down process: the *slowing-down power* and the *moderating ratio*. A moderator slows neutrons down better the larger ξ is. In addition, Σ_s must be large in order that neutrons collide as often as possible. Therefore, the quantity $\xi\Sigma_s$ is called the slowing-down power. Table 7.1.2 gives $\xi\Sigma_s$ for several moderators. The values apply to the "plateau region" that stretches from about 1 ev to several kev; at higher energies $\xi\Sigma_s$ decreases (with Σ_s). At lower energies, the quantity ξ loses its meaning. In addition, a good moderator should capture only weakly, i.e., Σ_a should be small. Therefore, a better measure of the moderating properties is the quantity $\xi\Sigma_s/\Sigma_a$, the so-called moderating ratio. The moderating ratio is usually evaluated with the average capture cross section for thermal neutrons at room temperature.

Table 7.1.2. *Slowing-Down Power and Moderating Ratio*

Moderator	Density	Slowing-Down Power	Moderating Ratio
H_2O	1.00	1.35 cm^{-1}	71
D_2O (pure)	1.10	0.176 cm^{-1}	5670
D_2O (99.8 mole %)	1.10	0.178 cm^{-1}	2540
Graphite	1.60	0.060 cm^{-1}	192
Beryllium	1.85	0.158 cm^{-1}	143

7.2. Slowing Down in Hydrogen ($A=1$)

In this section, we shall study the transport Eq. (5.1.16) in hydrogen; the scattering cross section, which is given by Eq. (7.1.13), takes a particularly simple form for $A=1$. It will develop that in this case we can easily obtain closed solutions for $\Phi(E)$, while in the general case $A\neq 1$, we can only obtain approximate solutions. First, however, we shall introduce some additional concepts that are customary in the treatment of slowing-down problems.

7.2.1. Collision Density — Slowing-Down Density — Lethargy

The *collision density* $\psi(E)$ is defined as the number $(\Sigma_a+\Sigma_s)\,\Phi(E)$ of neutrons which undergo collision (scattering or absorption) per cm^3 and second. Using Eq. (7.1.13), Eq. (5.1.16) can be written in terms of $\psi(E)$ as follows:

$$\psi(E)=\int_E^{E/\alpha}\frac{\Sigma_s}{\Sigma_a+\Sigma_s}\,\psi(E')\,g(E'\to E)\,dE'+S(E)\,. \qquad (7.2.1)^1$$

The *slowing-down density* $q(E)$ is defined as the number of neutrons slowing down *past* the energy E per cm^3 and second. The probability that a neutron of initial energy $E'>E$ has an energy $E''<E$ after the collision is

$$G(E',E)=\int_{\alpha E'}^{E} g(E'\to E'')\,dE''\,. \qquad (7.2.2)$$

[1] In hydrogen, the upper limit of integration must be the highest source energy E_Q. The integral must also extend to E_Q in heavier nuclei in case $E>\alpha E_Q$.

The slowing-down density is then

$$q(E)=\int_{E}^{E/\alpha}\frac{\Sigma_s}{\Sigma_a+\Sigma_s}\,\psi(E')\,G(E',E)\,dE'. \tag{7.2.3}$$

If follows from Eq. (7.1.12) that the probability $G(E', E)$ is given by

$$G(E',E)=\int_{\alpha E'}^{E}\frac{dE''}{(1-\alpha)E'}=\frac{E-\alpha E'}{(1-\alpha)E'}. \tag{7.2.4}$$

$G(E', E)$ is zero for $E'<E$ and for $E'\geqq E/\alpha$. It follows from Eqs. (7.2.1) to (7.2.4) that

$$\frac{dq}{dE}=\frac{\Sigma_a}{\Sigma_a+\Sigma_s}\,\psi(E)-S(E). \tag{7.2.5}$$

Frequently, the slowing-down process is described with the help of a logarithmic energy scale. The use of such a scale is suggested by the constancy of the average logarithmic energy loss per collision. We use as a variable the so-called *lethargy*, defined by

$$u=\ln\left(\frac{E_Q}{E}\right). \tag{7.2.6}$$

Here E_Q is an arbitrary reference energy; in most practical applications, E_Q is taken to be the highest energy appearing in the source spectrum. Then $u=0$ at the beginning of the slowing-down process and during moderation increases continuously. Eqs. (7.2.1) to (7.2.5) can be written in terms of the lethargy as follows:

$$\psi(u)=\int_{u-\ln(1/\alpha)}^{u}\frac{\Sigma_s}{\Sigma_a+\Sigma_s}\,\psi(u')\,g(u'\to u)\,du'+S(u), \tag{7.2.7}$$[1]

$$q(u)=\int_{u-\ln(1/\alpha)}^{u}\frac{\Sigma_s}{\Sigma_a+\Sigma_s}\,\psi(u')\,G(u',u)\,du', \tag{7.2.8}$$

$$G(u',u)=\int_{u}^{u'+\ln(1/\alpha)}g(u'\to u'')\,du'', \tag{7.2.9}$$

$$\frac{dq}{du}=-\frac{\Sigma_a}{\Sigma_a+\Sigma_s}\,\psi(u)+S(u). \tag{7.2.10}$$

Here

$$\begin{aligned}g(u'\to u)&=g(E'\to E)\left|\frac{dE}{du}\right|=\frac{e^{-(u-u')}}{1-\alpha}\quad(\text{for } u-\ln(1/\alpha)\leqq u'\leqq u)\\&=0\quad\text{otherwise,}\end{aligned} \tag{7.2.11}$$

$$\begin{aligned}G(u',u)&=\int_{u}^{u'+\ln(1/\alpha)}\frac{e^{-(u-u'')}}{1-\alpha}\,du''=\frac{e^{-(u-u')}-\alpha}{1-\alpha}\quad(\text{for } u-\ln(1/\alpha)\leqq u'\leqq u)\\&=0\quad\text{otherwise.}\end{aligned} \tag{7.2.12}$$

---

[1] In hydrogen, the lower limit of integration must be $u=0$. In heavier moderators, the integration must also extend to $u=0$ if $u<\ln(1/\alpha)$.

It should be noted that

$$\psi(u)=\psi(E)\left|\frac{dE}{du}\right|=E\psi(E). \tag{7.2.13}$$

Naturally, $q(u)=q(E)$.

7.2.2. Calculation of the Slowing-Down Density and the Energy Spectrum

Because $\alpha=0$ in hydrogen, the equation for the collision density can be written

$$\psi(E)=\int_E^{E_Q}\frac{\Sigma_s}{\Sigma_a+\Sigma_s}\,\frac{\psi(E')}{E'}\,dE'+S(E). \tag{7.2.14}$$

A differential equation results from differentiation with respect to E:

$$\frac{d\psi(E)}{dE}=-\frac{\Sigma_s}{\Sigma_a+\Sigma_s}\,\frac{\psi(E)}{E}+\frac{dS(E)}{dE}. \tag{7.2.15}$$

This equation can be integrated directly, but it proves convenient first to rewrite it in terms of the slowing-down density and then to integrate it. It follows from Eqs. (7.2.3), (7.2.4), and (7.2.14) that

$$q(E)=E\left[\psi(E)-S(E)\right] \tag{7.2.16}$$

so that

$$\frac{dq}{dE}=\frac{\Sigma_a}{\Sigma_a+\Sigma_s}\,\frac{q(E)}{E}-\frac{\Sigma_s}{\Sigma_s+\Sigma_a}\,S(E). \tag{7.2.17}$$

Let us next consider the case of no absorption. From Eq. (7.2.17) we have

$$q(E)=\int_E^{E_Q} S(E')\,dE'. \tag{7.2.18}$$

Thus when there is no absorption, the slowing-down density at energy E is equal to the number of neutrons produced per cm^3 and second at energies above E. If the source density is monoenergetic, $S(E)=S\cdot\delta(E-E_Q)$, then for $E\leqq E_Q$,

$$q(E)=S.$$

This result clearly holds for all non-absorbing moderators; it is physically obvious and follows immediately from the general Eq. (7.2.5).

According to Eq. (7.2.16), the collision density in hydrogen is given by

$$\psi(E)=\frac{q(E)}{E}+S(E)$$

so that when $S(E)=S\cdot\delta(E-E_Q)$

$$\left.\begin{aligned}\psi(E)&=\frac{S}{E}+S\cdot\delta(E-E_Q)\\ \Phi(E)&=\frac{S}{\Sigma_s E}+\frac{S}{\Sigma_s}\,\delta(E-E_Q).\end{aligned}\right\} \tag{7.2.19}$$

Thus at all energies smaller than the source energy $\Phi(E)=S/\Sigma_s E$. Since the scattering cross section of hydrogen is constant in the energy range from 1 ev to 10^4 ev, the flux of moderated neutrons follows a $1/E$-law there.

Next we turn to the general case of slowing down with absorption. We shall be interested in either pure hydrogen or in a mixture of hydrogen and a heavy absorber whose nuclei do not contribute to the moderation. When the source density is monoenergetic, i.e., when $S(E)=S\cdot\delta(E-E_Q)$,

$$\left.\begin{aligned} dq &= \frac{\Sigma_a}{\Sigma_a+\Sigma_s}\,\frac{q(E)}{E}\,dE \\ \frac{d\ln q}{dE} &= \frac{\Sigma_a}{\Sigma_a+\Sigma_s}\,\frac{1}{E} \\ \ln q(E)-\ln q(E_Q) &= \int\limits_{E_Q}^{E} \frac{\Sigma_a}{\Sigma_s+\Sigma_a}\,\frac{dE'}{E'} \end{aligned}\right\} \tag{7.2.20}$$

if $E<E_Q$. Now $q(E_Q)=\left(\frac{\Sigma_s}{\Sigma_a+\Sigma_s}\right)_{E_Q}\cdot S$, so that

$$q(E)=\left(\frac{\Sigma_s}{\Sigma_a+\Sigma_s}\right)_{E_Q}\cdot S\cdot e^{-\int\limits_{E}^{E_Q}\frac{\Sigma_a}{\Sigma_a+\Sigma_s}\frac{dE'}{E'}}. \tag{7.2.21}$$

The expression

$$p(E)=\left(\frac{\Sigma_s}{\Sigma_a+\Sigma_s}\right)_{E_Q}\cdot e^{-\int\limits_{E}^{E_Q}\frac{\Sigma_a}{\Sigma_a+\Sigma_s}\frac{dE'}{E'}} \tag{7.2.22}$$

is called the *resonance escape probability*. $p(E)$ gives the probability that a neutron produced with energy E_Q will not be absorbed during moderation to the energy E. The factor $\left(\frac{\Sigma_s}{\Sigma_a+\Sigma_s}\right)_{E_Q}$ is the probability that a source neutron will not be absorbed on its first collision. Frequently, $\Sigma_a\ll\Sigma_s$ at the source energy, so that one simply has

$$p(E)=e^{-\int\limits_{E}^{E_Q}\frac{\Sigma_a}{\Sigma_a+\Sigma_s}\frac{dE'}{E'}}. \tag{7.2.23}$$

For a general source distribution the slowing-down density is

$$q(E)=\int\limits_{E}^{\infty}\frac{\Sigma_s(E')}{\Sigma_a(E')+\Sigma_s(E')}\,S(E')\,e^{-\int\limits_{E}^{E'}\frac{\Sigma_a}{\Sigma_a+\Sigma_s}\frac{dE''}{E''}}\,dE'. \tag{7.2.24}$$

We can then define the quantity

$$p(E)=\frac{q(E)}{\int\limits_{E}^{\infty}S(E')\,dE'} \tag{7.2.25}$$

as the resonance escape probability.

According to Eq. (7.2.16), the flux is given by

$$\Phi(E)=\frac{q(E)}{(\Sigma_a+\Sigma_s)E}+\frac{S(E)}{\Sigma_a+\Sigma_s}. \tag{7.2.26}$$

In a strong absorber, the flux will depart from the pure $1/E$-law because of the decrease of q with decreasing E. Furthermore, there will be a flux depression near each of the various resonances.

We shall now calculate the p-factor using Eq. (7.2.23) for two instructive special cases. First we consider a hypothetical absorber whose absorption cross section Σ_a is infinite in the interval $E_1 \leqq E' \leqq E_2$ and is zero elsewhere. Then for $E < E_1$ and $E_Q > E_2$,

$$p = e^{-\int_{E_1}^{E_2} \frac{dE'}{E'}} = \frac{E_1}{E_2}. \tag{7.2.27}$$

Thus, although the absorption cross section is infinitely large, neutrons are slowed down to energies $< E_1$; and indeed more are slowed down the narrower the "dangerous zone" is. This situation arises out of the fact that neutrons are moderated from energies $> E_2$ to energies $< E_1$ in *one* collision; thus they jump over the dangerous zone. These considerations are important for understanding the p-factor in an absorber whose cross section has very high resonances.

If the ratio of the scattering cross section to the absorption cross section is independent of energy and equal to β, then

$$p(E) = e^{-\int_{E}^{E_Q} \frac{1}{1+\beta} \frac{dE'}{E'}} = \left(\frac{E}{E_Q}\right)^{\frac{1}{1+\beta}}. \tag{7.2.28}$$

The neutron flux is then given by

$$\Phi(E) = \frac{S}{\Sigma_a} \left(\frac{E}{E_Q}\right)^{\frac{1}{1+\beta}} \frac{1}{(1+\beta)E}. \tag{7.2.29}$$

$\Phi(E)$ is thus proportional to $E^{-\beta/(1+\beta)}$, i.e., it increases more slowly than $1/E$ with decreasing energy.

7.3. Slowing Down in Heavy Media ($A \neq 1$)

7.3.1. Non-Absorbing Media

Let us next assume that the energy of the neutron source is again E_Q. Then if $S = 1$,

$$\psi(E) = \int_{E}^{E/\alpha} \psi(E') \frac{dE'}{(1-\alpha)E'} + \delta(E - E_Q). \tag{7.3.1}$$

Reduction to a simple differential equation is not possible here, since the energy also occurs in the upper limit of integration. However, by using a procedure developed by Placzek, one can solve the integral equation step-wise, i.e., first in the interval $\alpha E_Q < E < E_Q$, then in the interval $\alpha^2 E_Q < E < \alpha E_Q$, etc. The resulting expressions are complicated, and we content ourselves with showing the Placzek solution $\psi_0(u)$ for $A = 2$, 4, and 12 in Fig. 7.3.1[1]. In the interval $0 < u < 3 \ln \frac{1}{\alpha}$, i.e., $\alpha^3 E_Q < E < E_Q$, the collision density exhibits oscillations ("Placzek wiggles") around the asymptotic value $1/\xi$. These oscillations are connected with the fact that a neutron can lose at most an energy $(1-\alpha)E$ in any

[1] The Placzek solution describes only those neutrons which have already made a collision. In order to obtain the complete solution, it is necessary to add the contribution $\delta(u)$ of the uncollided source neutrons.

one collision. A discontinuity occurs in the collision density at $E=\alpha E_Q$, since source neutrons can at most reach the energy αE_Q on their first collision.

It can easily be shown that $\psi_0(E)=1/\xi E$ asymptotically. The slowing-down density $q(E)$, which in a non-absorbing medium is equal to the source density $S=1$ when $E \leqq E_Q$, is given by

$$\left.\begin{aligned} q &= \int_E^{E/\alpha} \psi(E') \frac{E-\alpha E'}{(1-\alpha)E'} \, dE' \\ &= \frac{1}{\xi} \int_E^{E/\alpha} \frac{E-\alpha E'}{(1-\alpha)E'} \frac{dE'}{E'} = \frac{1}{\xi}\left[1+\frac{\alpha}{1-\alpha} \ln \alpha\right] = 1 . \end{aligned}\right\} \tag{7.3.2}$$

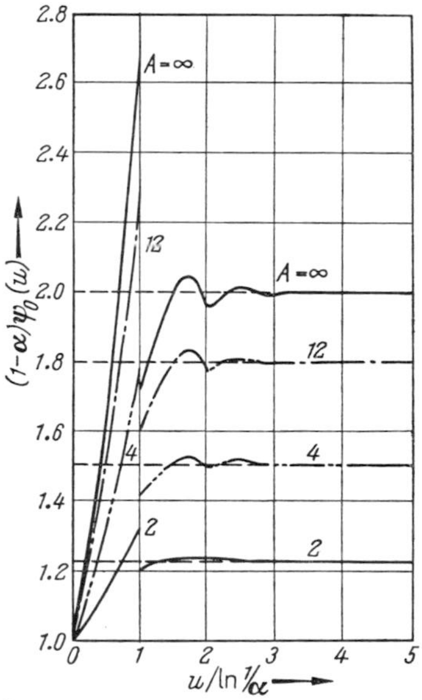

Fig. 7.3.1. The Placzek solution for the collision density for various atomic numbers of the moderator (from Weinberg and Wigner loc. cit.)

By simple substitution in Eq. (7.3.1) it can easily be shown that $\psi(E) \sim 1/\xi E$ is a solution.

We have now derived the important result that in the "asymptotic" range in a non-absorbing medium the slowing-down density $q(E)=S$ and the flux $\Phi(E)$ are connected by the relation

$$\Phi(E)=\frac{q}{\xi \Sigma_s E} . \tag{7.3.3}$$

Since the scattering cross sections of most moderators are constant between 1 ev and several kev, the flux has a $1/E$-spectrum in this "epithermal" region. Frequently, one writes

$$\Phi(E)\, dE=\Phi_{\text{epi}} \frac{dE}{E} ; \qquad \Phi_{\text{epi}}=\frac{q}{\xi \Sigma_s} . \tag{7.3.4}$$

For a given slowing-down density, Φ_{epi} is larger the smaller the slowing-down power of the moderator is. All these results obviously also hold for hydrogen, as we can easily see by comparison with Eq. (7.2.19).

If instead of a monoenergetic source at $E=E_Q$ there is a source distribution, then

$$\psi(E)=S(E)+\int_E^{E_{\max}} S(E')\psi_{E'}(E)\, dE' . \tag{7.3.5}$$

Here $\psi_{E'}(E)$ is the collision density that arises from a source of unit strength at E', in other words the Placzek solution discussed above. If $\psi_{E'}(E)$ is replaced by its asymptotic value $1/\xi E$, Eq. (7.3.5) becomes

$$\psi(E)=S(E)+\frac{1}{\xi E} \int_E^{E_{\max}} S(E')\, dE' . \tag{7.3.6a}$$

For energies which are smaller than the lowest energy in the source spectrum

$$\psi(E)=\frac{S}{\xi E} . \tag{7.3.6b}$$

Fig. 7.3.2 shows the flux $\Phi(u) = E\,\Phi(E)$ calculated by ROWLANDS for a fission neutron source in carbon; the strong flux variations in the neighborhood of 2 and 3 Mev are due to resonances in the carbon cross section. $\Phi(u)$ becomes constant for $E < 0.1$ Mev, and only below this energy does a $1/E$-spectrum appear. ROWLANDS showed in this case that neglect of the "Placzek wiggles", i.e., approximation of Eq. (7.3.5) by Eq. (7.3.6a), leads to a maximum error of about 10% near 10 Mev; as the energy decreases the approximation becomes progressively better, and for $E < 20$ kev is virtually exact.

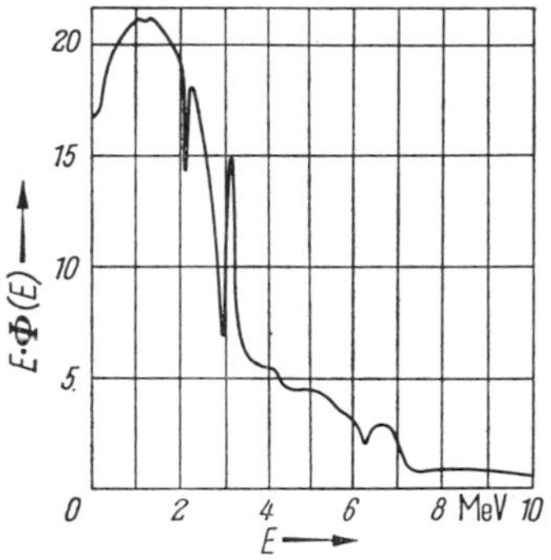

Fig. 7.3.2. $E\Phi(E)$ for a fission source in Carbon (ROWLANDS)

Next we shall study slowing down in a non-absorbing medium which consists of a *mixture* of several kinds of nuclei. The collision density is then defined as $\psi(E) = \Phi(E) \sum_{i=1}^{N} \Sigma_{si} = \Phi(E)\,\Sigma_s$, and in the asymptotic range

$$\psi(E) = \sum_{i=1}^{N} \int_{E}^{E/\alpha_i} \frac{\Sigma_{si}}{\Sigma_s} \psi(E') \frac{dE'}{(1-\alpha_i)E'}, \qquad (7.3.7\text{a})$$

$$q = \sum_{i=1}^{N} \int_{E}^{E/\alpha_i} \frac{\Sigma_{si}}{\Sigma_s} \psi(E') \frac{E-\alpha_i E'}{(1-\alpha_i)E'}\, dE'. \qquad (7.3.7\text{b})$$

If we assume that all the scattering cross sections vary similarly with energy, i.e., that the ratios Σ_{si}/Σ_s are independent of the energy, then $\psi(E) = \frac{\text{const.}}{E}$ is a solution of Eq. (7.3.7a). The constant can be determined from Eq. (7.3.7b):

$$q = \text{const.} \sum_{i=1}^{N} \frac{\Sigma_{si}}{\Sigma_s} \xi_i. \qquad (7.3.8\text{a})$$

If we introduce the average logarithmic energy decrement of the mixture of nuclei

$$\bar{\xi} = \frac{\sum_{i=1}^{N} \Sigma_{si} \xi_i}{\Sigma_s} \qquad (7.3.8\text{b})$$

it follows that

$$\psi(E) = \frac{q}{\bar{\xi} E}; \qquad \Phi(E) = \frac{q}{\bar{\xi} \Sigma_s E}. \qquad (7.3.9)$$

Thus in a mixture the slowing-down powers of the individual components combine additively.

7.3.2. Absorbing Media: WIGNER's Approximation

Lethargy is the most convenient variable for the treatment of slowing down in absorbing media; if there is a monoenergetic source at $u = 0$, then for $u > 0$

$$\psi(u) = \int_{u-\ln(1/\alpha)}^{u} \frac{\Sigma_s}{\Sigma_a + \Sigma_s} \psi(u') \frac{e^{-(u-u')}}{1-\alpha}\, du', \qquad (7.3.9\text{a})$$

$$q(u) = \int_{u-\ln(1/\alpha)}^{u} \frac{\Sigma_s}{\Sigma_a + \Sigma_s} \psi(u') \frac{e^{-(u-u')} - \alpha}{1-\alpha}\, du', \qquad (7.3.9\text{b})$$

$$\frac{dq}{du} = -\frac{\Sigma_a}{\Sigma_a + \Sigma_s} \psi(u). \qquad (7.3.9\text{c})$$

Recently, BEDNARZ has given an analytic solution of these equations for arbitrary Σ_s and Σ_a. His method of solution is mathematically difficult and leads to very complicated expressions. We restrict ourselves here to the discussion of several important approximate methods.

Let us imagine a monoenergetic source of unit strength at $u=0$ in a moderating medium. The slowing-down density $q(u)$ at u is then equal to the resonance escape probability $p(u)$ at u. The number of neutrons absorbed per cm³ and sec during moderation to lethargy u is $1-p(u)$. Now we take

$$\psi(u)=\frac{1}{\xi}-\frac{1-p(u)}{\xi}=\frac{p(u)}{\xi} \tag{7.3.10}$$

for the collision density. This is to be understood in the following sense: If the absorption were zero, $1/\xi$ would be the asymptotic collision density resulting from a source of unit strength. However, during moderation to lethargy u, $1-p(u)$ neutrons are absorbed. This absorption can be formally represented by a negative source of strength $1-p(u)$. The asymptotic collision density of this source is $\frac{1-p(u)}{\xi}$ and must be subtracted from the collision density $1/\xi$ which prevails in the absence of absorption.

Using Eq. (7.3.10), Eq. (7.3.9c) can immediately be integrated:

$$\left.\begin{aligned} \frac{dp}{du} &= -\frac{\Sigma_a}{\xi(\Sigma_a+\Sigma_s)}\,p(u),\\ p(u) &= e^{-\int\limits_0^u \frac{\Sigma_a}{\xi(\Sigma_a+\Sigma_s)}\,du'}. \end{aligned}\right\} \tag{7.3.11}$$

In terms of the energy these equations become

$$p(E)=e^{-\int\limits_E^{E_Q} \frac{\Sigma_a}{\xi(\Sigma_a+\Sigma_s)}\frac{dE'}{E'}}, \tag{7.3.12a}$$

$$\Phi(E)=\frac{q(E)}{\xi(\Sigma_a+\Sigma_s)E}\,. \tag{7.3.12b}$$

The decisive approximation in Eqs. (7.3.10—12) is the assumption everywhere of the asymptotic value $1/\xi$ for the collision density, i.e., the neglect of the "Placzek wiggles" in the first few collision intervals . We can, in fact, assume that the source is to be found at very high energies and that the absorption first appears at much lower energies, so that the non-asymptotic oscillations arising from the source play no role. However, the collision density arising from the "negative sources" also exhibits non-asymptotic oscillations, and we must write more exactly

$$\psi(u)=\frac{1}{\xi}+\int\limits_0^u \frac{dp}{du'}\,\psi_{u'}(u)\,du'. \tag{7.3.13}$$

Here $\psi_{u'}(u)$ is the collision density at u in a non-absorbing medium due to a unit source at u'. If $\psi_{u'}(u)$ is replaced by its asymptotic value $1/\xi$, Eq. (7.3.10) again results. We will return to Eq. (7.3.13), which was first formulated by WEINBERG and WIGNER and independently by CORNGOLD, in the next section.

Eqs. (7.3.12a) and (7.3.12b), which are consequences of Eq. (7.3.10), are therefore not exact; they are frequently called the WIGNER approximation. Only for hydrogen are they exact, as can be seen by comparison with Eqs. (7.2.22) and (7.2.26). As we have previously seen, in hydrogen there are no non-asymptotic oscillations in the collision density near the source. There are two more important cases in which Eqs. (7.3.12a) and (7.3.12b) are quite accurate. One is the case of very weak absorption, $\Sigma_a \ll \Sigma_s$. In this case, the deviation of the collision density from the collision density in a non-absorbing medium is so small that the non-asymptotic oscillations can be neglected. One often writes then

$$p(E) = e^{-\int\limits_E^{E_Q} \frac{\Sigma_a}{\xi \Sigma_s} \frac{dE'}{E'}} \tag{7.3.14a}$$

and

$$\Phi(E) = \frac{q(E)}{\xi \Sigma_s E}. \tag{7.3.14b}$$

Occasionally, these equations are also referred to as the FERMI approximation.

The other, by far more important case is that of absorption by sharp, well-separated resonances. If the width Δu of the resonance is small compared to the collision interval $\ln(1/\alpha)$ and if the distance between resonances is several collision intervals, Eqs. (7.3.12a) and (7.3.12b) represent an excellent approximation. In this case, the collision density inside the resonance region is

$$\psi_R(u) = \int\limits_{u-\ln(1/\alpha)}^{u} \frac{\Sigma_s}{\Sigma_a + \Sigma_s} \psi(u') \frac{e^{-(u-u')}}{1-\alpha} du' \approx \psi(u) \tag{7.3.15}$$

i.e., it is nearly equal to the constant collision density that would be present if the resonance were not there. This comes about because the contribution of collisions in the resonance region to the integral can be neglected $(\Delta u \ll \ln(1/\alpha))$. However, when the difference between the "perturbed" collision density and the "unperturbed" collision density is negligibly small, the effects of the non-asymptotic oscillations are also negligibly small, i.e., Eqs. (7.3.10—12) are very accurate.

Eq. (7.3.11) can be directly derived on the basis of these considerations: Let the slowing-down density for lethargies less than that of the first resonance be unity. Then $\psi_R(u) = \psi(u) = 1/\xi$ and $\Phi_R(u) = \frac{1}{\xi(\Sigma_a + \Sigma_s)}$. The probability that the neutron is absorbed in the first resonance is $\int\limits_{\Delta u} \Sigma_a \Phi(u')\, du' = \frac{1}{\xi} \int\limits_{\Delta u} \frac{\Sigma_a}{\Sigma_a + \Sigma_s} du'$. The probability that it escapes this fate is $1 - \frac{1}{\xi} \int\limits_{\Delta u} \frac{\Sigma_a}{\Sigma_a + \Sigma_s} du'$, or because $\Delta u \ll \xi$, $e^{-\frac{1}{\xi}\int\limits_{\Delta u} \frac{\Sigma_a}{\Sigma_a + \Sigma_s} du'}$. The probability that during moderation the neutron escapes capture in several successive resonances is

$$\left.\begin{aligned} p(u) &= e^{-\frac{1}{\xi}\int\limits_{\Delta u_1} \frac{\Sigma_a}{\Sigma_a + \Sigma_s} du'} \cdot e^{-\frac{1}{\xi}\int\limits_{\Delta u_2} \frac{\Sigma_a}{\Sigma_a + \Sigma_s} du'} \cdots \\ &= e^{-\frac{1}{\xi}\int\limits_0^u \frac{\Sigma_a}{\Sigma_a + \Sigma_s} du'}. \end{aligned}\right\} \tag{7.3.16}$$

The conditions required for the application of these formulas are thus rather well fulfilled in many practical cases; Eqs. (7.3.12a) and (7.3.12b) are widely used for the calculation of slowing down in homogeneous media. Eq. (7.3.12a) is also the starting point in the definition of the *resonance integral*, which will be further discussed in Sec. 7.4.

One can easily construct a case in which the Wigner approximation leads to a false result. For this purpose, let us assume that Σ_a is infinite in an energy interval $E_1 \leqq E' \leqq E_2$ and zero elsewhere. It then follows from Eq. (7.3.12a) that for $E < E_1$

$$p(E) = \left(\frac{E_1}{E_2}\right)^{1/\xi}.$$

Let us now consider the special case $E_1 = \alpha E_2$; then it follows that $p = \alpha^{1/\xi}$ ($\approx e^{-2}$ for large A). In fact, however, all the neutrons are absorbed between E_1 and E_2, since no neutron can jump over this interval. Thus p must be zero. Of course, the case just considered does not occur in practice. We shall nevertheless continue to be interested in better approximations to the resonance escape probability in cases in which the absorption cross section varies slowly with energy and cannot be neglected compared to the scattering cross section.

7.3.3. The Goertzel-Greuling Approximation

An integro-differential equation for the resonance escape probability can be obtained from Eqs. (7.3.9c) and (7.3.13):

$$\frac{dp}{du} = -\frac{\Sigma_a}{\Sigma_a + \Sigma_s}\left[\frac{p(u)}{\xi} + \int_0^u \frac{dp}{du'}\left\{\psi_{u'}(u) - \frac{1}{\xi}\right\} du'\right]. \tag{7.3.17}$$

In order to obtain this result the function $\psi_{u'}(u)$ has been decomposed into an asymptotic part and a non-asymptotic part as follows: $\psi_{u'}(u) = \frac{1}{\xi} + \left\{\psi_{u'}(u) - \frac{1}{\xi}\right\}$. Neglect of the non-asymptotic part leads immediately to the Wigner approximation of Sec. 7.3.2. Eq. (7.3.17) is the starting point of several higher-order approximation procedures for calculating the resonance escape probability. It has been solved by Weinberg and Wigner and by Corngold in various high-order approximations. We describe next a simple approximate solution due to Dresner.

As we have seen in Sec. 7.3.1, the non-asymptotic part of $\psi_{u'}(u)$ is appreciably different from zero only in the interval $u' < u < u' + 3\ln(1/\alpha)$. Therefore, we set by way of approximation

$$\left.\begin{aligned} \int_0^u \frac{dp}{du'}\left\{\psi_{u'}(u) - \frac{1}{\xi}\right\} du' &\approx \frac{dp}{du}\int_0^u \left\{\psi_{u'}(u) - \frac{1}{\xi}\right\} du' \\ &\approx \frac{dp}{du}\int_0^\infty \left\{\psi_0(u') - \frac{1}{\xi}\right\} du'. \end{aligned}\right\} \tag{7.3.18}$$

This approximation will be good if dp/du changes but little in one collision interval; thus it is bad when the absorption cross section has sharp resonances. The integral on the right-hand side has been evaluated by DRESNER[1], who gives as its value

$$\int_0^\infty \left\{\psi_0(u) - \frac{1}{\xi}\right\} du = \frac{\gamma}{\xi} - 1 \tag{7.3.19a}$$

with

$$\gamma = 1 - \frac{\alpha \ln^2(1/\alpha)}{2(1-\alpha)\xi}. \tag{7.3.19b}$$

($\gamma=1$ for $A=1$, 0.584 for $A=2$, and 0.138 for $A=12$.)

Using Eqs. (7.3.19a) and (7.3.18), Eq. (7.3.17) becomes

$$\frac{dp}{du} = -\frac{\Sigma_a}{\gamma \Sigma_a + \xi \Sigma_s} p(u), \tag{7.3.20a}$$

$$p(u) = e^{-\int_0^u \frac{\Sigma_a}{\gamma \Sigma_a + \xi \Sigma_s} du'} \tag{7.3.20b}$$

or finally

$$p(E) = e^{-\int_E^{E_Q} \frac{\Sigma_a}{\xi \Sigma_s + \gamma \Sigma_a} \frac{dE'}{E'}}. \tag{7.3.20c}$$

Eq. (7.3.20c) was first derived by another means by GOERTZEL and GREULING and for this reason is often called the Goertzel-Greuling approximation to the resonance escape probability. For a slowly varying capture cross section, it represents a considerably better approximation than the Wigner approximation.

An important case of slowly varying capture is the case $\Sigma_a(E) \sim 1/\sqrt{E}$. If the scattering cross section is constant, the integration can be explicitly carried out; if $E_Q = \infty$, the following expression results for the resonance escape probability:

$$p(E) = \left(1 + \frac{\gamma}{\xi} \frac{\Sigma_a(E)}{\Sigma_s}\right)^{-2/\gamma}. \tag{7.3.21}$$

7.4. Resonance Integrals

7.4.1. Definition of the Resonance Integral

Let us imagine an infinite medium containing a homogeneous distribution of high-energy sources whose strength is one neutron per cm³ and sec. Let an absorber substance also be homogeneously distributed with a density of N atoms per cm³. Let the absorber have an isolated resonance whose width is small compared to the collision interval. According to Sec. 7.3.2, the number of neutrons absorbed per cm³ and sec is

$$\psi_a = \frac{1}{\xi} \int_0^\infty \frac{\Sigma_a(u)}{\Sigma_a(u) + \Sigma_s(u)} du. \tag{7.4.1}$$

Let us now write ψ_a as

$$\psi_a = N I_{\text{eff}} / \xi \Sigma_p. \tag{7.4.2}$$

$1/\xi \Sigma_p$ is the neutron flux (per unit lethargy) which would be present if there were no resonance. Here

$$\Sigma_p = \Sigma_{sm} + N \sigma_{\text{pot}} \tag{7.4.3}$$

[1] Op. cit., p. 23.

where Σ_{sm} is the scattering cross section of the moderator and σ_{pot} is the microscopic cross section for potential scattering of the absorber.

I_{eff} (barn) is called the *effective resonance integral.* According to Eqs. (7.4.1) and (7.4.2) it is given by

$$I_{\text{eff}}=\int_0^\infty \frac{\Sigma_p \sigma_a(u)}{\Sigma_s(u)+\Sigma_a(u)}\,du=\int_0^\infty \frac{\sigma_a(u)\,du}{1+\dfrac{\sigma_s^{\text{Res}}(u)+\sigma_a(u)}{\sigma_p}}=\int \frac{\sigma_a(E)}{1+\dfrac{\sigma_s^{\text{Res}}(E)+\sigma_a(E)}{\sigma_p}}\frac{dE}{E}. \tag{7.4.4}$$

In the second step we have introduced the following notation

$$\sigma_p=\frac{\Sigma_p}{N}=\frac{\Sigma_{sm}}{N}+\sigma_{\text{pot}}. \tag{7.4.5a}$$

σ_p gives the scattering cross section (with the exception of the resonance scattering) per absorber atom. In addition,

$$\sigma_{\text{s}}^{\text{Res}}(u)=\frac{\Sigma_s(u)-\Sigma_p}{N} \tag{7.4.5b}$$

is the microscopic resonance scattering cross section, i.e., the scattering cross section of the absorber atom minus σ_{pot}.

The defining Eq. (7.4.4) holds for a single resonance. The integration need only extend over the resonance; but it can be extended from 0 to ∞ without introducing any error, since the absorption cross section vanishes everywhere outside the resonance. Usually, the effective resonance integral for a real absorber that has many narrow resonances is defined in the same way, provided only that distances between the resonances are sufficiently large. In this general case, ψ_a is no longer equal to $N I_{\text{eff}}/\xi\Sigma_p$ but rather is given by

$$\psi_a=1-e^{-N I_{\text{eff}}/\xi\Sigma_p}. \tag{7.4.6}$$

Only when the exponent is small does Eq. (7.4.2) apply. The resonance integrals defined here for homogeneous mixtures are frequently called *homogeneous* resonance integrals. *Heterogeneous* resonance integrals, which we will study in another connection in Chapter 12, are of greater practical importance. The following, somewhat formal considerations serve largely as preparation for the discussion in Chapter 12. Our goal here will be the calculation of the effective resonance integral and the resonance escape probability from the Breit-Wigner parameters of the resonances. Before we proceed, however, we must introduce a few additional concepts.

As Eq. (7.4.4) shows, for a given resonance absorber, I_{eff} is largest when $\sigma_p\to\infty$, i.e., when the absorber is present in the moderator in "infinite dilution"[1]. Then

$$I_{\text{eff}}=I_\infty=\int\sigma_a(u)\,du=\int\sigma_a(E)\frac{dE}{E}. \tag{7.4.7}$$

Usually, I_∞ is called the "resonance integral at infinite dilution" or simply the "resonance integral". I_∞ can easily be determined experimentally (cf. Chapter 12). I_{eff} decreases with increasing concentration of the absorber, i.e., with decreasing σ_p. This comes about because the flux at the resonance, which is proportional to $1/(\Sigma_s+\Sigma_a)$, has a depression which is deeper the smaller Σ_p is compared to Σ_a (self-shielding); at infinite dilution, there is no flux depression because $\Sigma_a\ll\Sigma_p$.

[1] Because $N\to 0$, ψ_a also approaches zero.

7.4.2. Calculation of the Resonance Integral with Help of the Breit-Wigner Formula

For the purposes of simplicity, the interference between potential and resonance scattering will be neglected. We write, as in Sec. 4.1.2

$$\sigma_a(E)+\sigma_s^{\mathrm{Res}}(E)=\frac{\sigma_0}{1+\left(\frac{E-E_R}{\Gamma/2}\right)^2} \tag{7.4.8a}$$

with

$$\sigma_0=4\pi\lambda\!\!\!^{-2}_R g\frac{\Gamma_n}{\Gamma}, \tag{7.4.8b}$$

$$\sigma_a(E)=\frac{\Gamma_\gamma}{\Gamma}\,\frac{\sigma_0}{1+\left(\frac{E-E_R}{\Gamma/2}\right)^2}. \tag{7.4.8c}$$

Then

$$I_{\mathrm{eff}}=\int\frac{\frac{\Gamma_\gamma}{\Gamma}\,\frac{\sigma_0}{1+\left(\frac{E-E_R}{\Gamma/2}\right)^2}}{1+\frac{\sigma_0/\sigma_p}{1+\left(\frac{E-E_R}{\Gamma/2}\right)^2}}\,\frac{dE}{E}. \tag{7.4.9}$$

Next we introduce $x=\frac{E-E_R}{\Gamma/2}$ as a new variable of integration and use the abbreviation $t=\sigma_p/\sigma_0$; if we notice that the integration can be extended from $-\infty$ to $+\infty$ and $1/E$ replaced by $1/E_R$ because of the sharpness of the resonance, we obtain

$$I_{\mathrm{eff}}=\frac{\sigma_p\Gamma_\gamma}{2E_R}\int\limits_{-\infty}^{+\infty}\frac{dx}{1+t(1+x^2)}. \tag{7.4.10}$$

The integral is transformed by the substitution $y^2=t x^2/(t+1)$ into

$$\frac{1}{\sqrt{t(1+t)}}\int\limits_{-\infty}^{+\infty}\frac{dy}{1+y^2}=\frac{\pi}{\sqrt{t(1+t)}}$$

so that I_{eff} becomes

$$I_{\mathrm{eff}}=\frac{\pi\sigma_p\Gamma_\gamma/2E_R}{\sqrt{t(1+t)}} \tag{7.4.11}$$

or because $t=\sigma_p/\sigma_0$

$$I_{\mathrm{eff}}=\frac{\pi\sigma_0\Gamma_\gamma/2E_R}{\sqrt{1+\sigma_0/\sigma_p}}=\frac{I_\infty}{\sqrt{1+\sigma_0/\sigma_p}}. \tag{7.4.12}$$

Here

$$I_\infty=\pi\sigma_0\Gamma_\gamma/2E_R=2\pi^2\lambda\!\!\!^{-2}_R g\frac{\Gamma_n\Gamma_\gamma}{\Gamma E_R} \tag{7.4.13}$$

is the limiting value at "infinite dilution". Eq. (7.4.12) shows the effect of self-shielding quite clearly. For very high absorber concentrations, $\sigma_p/\sigma_0\ll 1$ and $I_{\mathrm{eff}}\approx\sqrt{\sigma_p/\sigma_0}\,I_\infty$.

In the case of very high absorber concentrations, the formulas developed here are no longer immediately applicable. We assumed that the resonances were narrow, i.e., that their widths were small compared to a collision interval. As a rule, this requirement is fulfilled if the width being considered is the natural width of the resonance; in other words, the condition $\Gamma \ll (1-\alpha) E_R$ is fulfilled if the moderation is by light nuclei. However, the relevant width is not the natural width of the resonance but rather the width of that energy interval in which the resonance cross section $\frac{\sigma_0}{1+\left(\frac{E-E_R}{\Gamma/2}\right)^2}$ is larger than the constant scattering cross section σ_p. This width is of the order of magnitude of $\Gamma_{\text{eff}} \approx \Gamma \sqrt{\frac{\sigma_0}{\sigma_p}}$, and we must require that Γ_{eff} be $\ll (1-\alpha) E_R$, a condition that is not always fulfilled when the absorber concentration is high. This problem has been studied in detail by DRESNER among others. We shall restrict ourselves in what follows to the case in which Γ_{eff} is always $\ll (1-\alpha) E_R$.

Unfortunately, Eq. (7.4.11) contains still another serious oversimplification: The Breit-Wigner formula in the form hitherto discussed describes the cross sections for resonance reactions of neutrons with nuclei which are initially at rest. In order to use the Breit-Wigner formula in practice we must modify it to take into account the *Doppler effect,* which comes about because of the thermal motion of the absorber nuclei.

7.4.3. The Doppler Broadening of Resonance Lines

Let us bombard a group of atomic nuclei with neutrons of velocity v. If the struck nuclei are initially at rest, the reaction rate is proportional to $v\sigma(v)$. We assume, however, that the nuclei have a Maxwell distribution of velocities

$$P(V)\,dV = \left(\frac{M}{2\pi k T_0}\right)^{\frac{3}{2}} e^{-\frac{MV^2}{2kT_0}} 4\pi V^2\,dV. \tag{7.4.14}$$

The relative velocity is then

$$v_{\text{Rel}} = |\boldsymbol{V} - \boldsymbol{v}| \tag{7.4.15}$$

and the reaction rate is then proportional to

$$\int v_{\text{Rel}}\,\sigma(v_{\text{Rel}})\,P(v_{\text{Rel}})\,dv_{\text{Rel}} = \int_0^\infty |\boldsymbol{V}-\boldsymbol{v}|\,\sigma(|\boldsymbol{V}-\boldsymbol{v}|)\,P(V)\,dV \equiv v\sigma_{\text{eff}}(v). \tag{7.4.16}$$

In the last step we have introduced an *effective cross section* that must be used to describe the interaction of neutrons with nuclei in thermal motion. We can immediately see from Eq. (7.4.16) that when $\sigma(v) \sim 1/v$, $\sigma(v) = \sigma_{\text{eff}}(v)$, a fact which is intuitively clear. On the other hand, if σ is constant, $\sigma_{\text{eff}}(v)$ is by no means always equal to σ, as will be shown in Sec. 10.1. Here we are interested in the case of resonance reactions, in which $\sigma(v)$ is described by the Breit-Wigner formula. As one might guess, in this case the natural line is broadened, since neutrons incident off resonance can still "find" nuclei for which the resonance condition is fulfilled. A measure of the smearing out of the energy $E = mv^2/2$ of the incident neutron is given by the quantity

$$\frac{m}{2}\,\overline{[(V+v)^2 - (V-v)^2]} = 2mv\overline{V} = \frac{4}{\sqrt{\pi}} \sqrt{\frac{4EkT_0}{A}}. \tag{7.4.17}$$

Here A is the mass number of the struck nucleus. Thus there appears in place of the natural line width Γ an effective line width of the order of magnitude of $\sqrt{\frac{4E_R kT_0}{A}}$ (as long as the latter is not small compared to Γ, i.e., as long as the Doppler effect is not altogether negligible).

The calculation of $\sigma_{\text{eff}}(E)$ with the help of Eqs. (7.4.14) and (7.4.16) and the Breit-Wigner formula is straightforward but leads to very complicated expressions. For the important special case in which the resonance energy E_R is $\gg kT_0$ and the interference terms are neglected, the cross sections in the neighborhood of a resonance are given by

$$\left.\begin{aligned}\sigma_{a\,\text{eff}}(E) &= \sigma_0 \frac{\Gamma_\gamma}{\Gamma}\,\psi(\Theta, x)\\ \sigma_{s\,\text{eff}}^{\text{Res}}(E) &= \sigma_0 \frac{\Gamma_n}{\Gamma}\,\psi(\Theta, x).\end{aligned}\right\} \tag{7.4.18}$$

Here x is again $\frac{E-E_R}{\Gamma/2}$; furthermore

$$\Theta = \frac{\Gamma}{\Delta}; \qquad \Delta = \sqrt{\frac{4E_R kT_0}{A}} \tag{7.4.19}$$

and

$$\psi(\Theta, x) = \frac{\Theta}{2\sqrt{\pi}} \int_{-\infty}^{+\infty} \frac{e^{-\frac{1}{4}\Theta^2 (x-y)^2}}{1+y^2}\, dy. \tag{7.4.20}$$

The function $\psi(\Theta, x)$ now gives the line shape and appears in place of the natural line shape $(1+x^2)^{-1}$. When $T_0 \to 0$, i.e., when $\Delta \to 0$ and $\Theta \to \infty$, $\psi(\Theta, x)$ approaches the natural line shape ("weak Doppler broadening"). When $\Theta \to 0$, i.e., when $\Gamma \ll \Delta$, then for x not too large

$$\psi(\Theta, x) = \frac{1}{2}\sqrt{\pi}\,\Theta e^{-\frac{1}{4}\Theta^2 x^2} = \frac{1}{2}\sqrt{\pi}\,\frac{\Gamma}{\Delta}\, e^{-\left(\frac{E-E_R}{\Delta}\right)^2}. \tag{7.4.21}$$

In this case of "strong" Doppler broadening, the line shape is a Gaussian whose width is the *Doppler width* Δ that was introduced above on heuristic grounds.

When $x^2 \gg \frac{6}{\Theta^2}$, $\psi(\Theta, x)$ is always asymptotic to the natural line shape, i.e.,

$$\psi(\Theta, x) \sim (1+x^2)^{-1} + \text{higher terms in } (1+x^2)^{-1}. \tag{7.4.22}$$

In other words, sufficiently far from the line center the natural line shape always prevails. Dresner has given a detailed discussion of the properties of the ψ-function. The function has been tabulated by Rose *et al.* A detailed derivation of Eq. (7.4.20) can be found, for example, in Solbrig; also see Adler and Naliboff. From now on we shall drop the index "eff" on the cross section.

So far, our considerations have been based on the assumption that the absorber atoms have a Maxwell distribution of velocities. This would be the case if the absorption occurred in a perfect gas. In practice, the absorber is either a liquid or a solid, and the validity of our formula is in some doubt. However, Lamb has shown that if a solid body can be described either by the Debye or the Einstein model and if the temperature is not too low, the Doppler effect can be calculated with Eq. (7.4.14), save that in place of the actual thermodynamic temperature a slightly higher effective temperature appears.

7.4.4. Calculation of the Resonance Integral Taking the Doppler Effect into Account

Substitution of Eq. (7.4.18) into (7.4.4) yields

$$I_{\text{eff}} = \frac{\sigma_p \Gamma_\gamma}{2E_R} \int_{-\infty}^{+\infty} \frac{\psi(\Theta, x)}{\psi(\Theta, x) + t}\, dx. \tag{7.4.23}$$

It is not possible to evaluate this integral in closed form. DRESNER, as well as ADLER *et al.*, has calculated it numerically and tabulated it for various values of Θ and t; some of these results are shown in Fig. 7.4.1. For very high dilution and for very high concentration of the absorber nuclei, the curves approach the curve for $\Theta = \infty$, i.e., for $T = 0\,°\text{K}$; in between the Doppler effect always increases the resonance integral, and in fact increases it more the smaller Θ is, i.e., the larger the temperature is. This can be interpreted in the following way: At very high dilution, $t \gg 1$, and $\psi(\Theta, x)$ can be neglected compared to t in the denominator of Eq. (7.4.23). Then[1]

$$\left.\begin{aligned} I_{\text{eff}} &= \frac{\sigma_0 \Gamma_\gamma}{2E_R} \int_{-\infty}^{+\infty} \psi(\Theta, x)\, dx \\ &= \frac{\pi \sigma_0 \Gamma_\gamma}{2E_R} = I_\infty. \end{aligned}\right\} \tag{7.4.24}$$

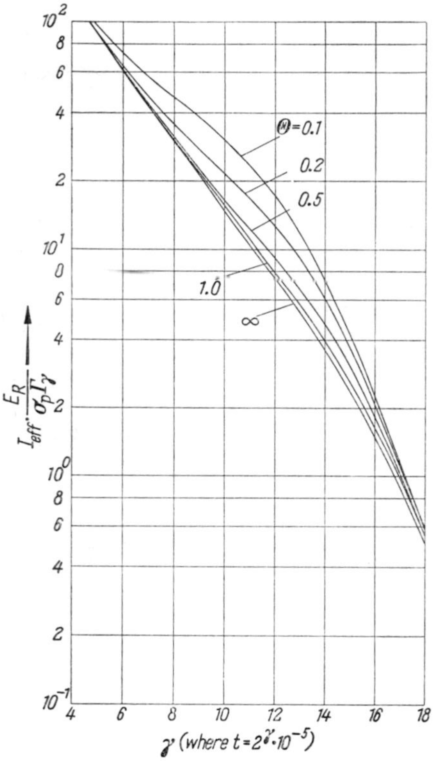

Fig. 7.4.1. The effective resonance integral as a function of dilution at various values of the Doppler parameter Θ

Eq. (7.4.24) says that at infinite dilution the Doppler effect does not affect the integrated resonance absorption. In this case, the energy dependence of the flux near the resonance is small, and the resonance integral is simply an integral over the cross section, which naturally is not influenced by the Doppler effect. At higher concentrations of the absorber nuclei, the flux shows a depression near the resonance. Owing to the Doppler effect, this depression is broader and flatter than in the case $T_0 = 0$; as a result, the self-shielding is smaller than in the case $T_0 = 0$ and the resonance integral is larger.

---

[1] $$\int_{-\infty}^{+\infty} \psi(\Theta, x)\, dx = \frac{\Theta}{2\sqrt{\pi}} \int_{-\infty}^{+\infty} dx \int_{-\infty}^{+\infty} dy\, \frac{e^{-\frac{1}{4}\Theta^2 (x-y)^2}}{1+y^2}$$

$$= \frac{\Theta}{2\sqrt{\pi}} \int_{-\infty}^{+\infty} \frac{dy}{1+y^2} \underbrace{\int_{-\infty}^{+\infty} du\, e^{-\frac{1}{4}\Theta^2 u^2}}_{\frac{2}{\Theta}\sqrt{\pi}} = \int_{-\infty}^{+\infty} \frac{dy}{1+y^2} = \pi.$$

In the case of very high concentration there is an extremely large flux depression, so that the central part of the resonance does not contribute at all to the value of the resonance integral. In the wings, which then alone contribute to the resonance integral, the resonance line has the natural line shape [see Eq. (7.4.22)]; for this reason, the resonance integral is not affected by the Doppler effect.

Chapter 7: References

General

DRESNER, L.: Resonance Absorption in Nuclear Reactors. Oxford-London-New York-Paris: Pergamon Press 1960.

WEINBERG, A. M., and E. P. WIGNER: The Physical Theory of Neutron Chain Reactors. Chicago: Chicago University Press 1958, especially Chap. X: Energy Spectrum During Moderation.

Special[1]

KÜSTERS, H.: Nukleonik 5, 33 (1963).
MARCUS, W. C. DE: Nucl. Sci. Eng. 5, 336 (1959).
} The Average Number of Collisions in Moderation by Elastic Collision.

PLACZEK, G.: Phys. Rev. 69, 423 (1946) (Slowing Down in Non-Absorbing Media).

ROWLANDS, G.: J. Nucl. Energy A 11, 150 (1960); A 13, 14 (1960) (Slowing Down of Fission Neutrons).

BEDNARZ, R.: Nucl. Sci. Eng. 10, 219 (1961) (Exact Solution of the Slowing-Down Equation).

CORNGOLD, N.: Proc. Phys. Soc. (London) A 70, 793 (1957).
GOERTZEL, G., and E. GREULING: Nucl. Sci. Eng. 7, 69 (1960).
KEANE, A.: Nucl. Sci. Eng. 10, 117 (1961).
WEINBERG, A. M., and E. P. WIGNER: BNL-433, 125 (1956).
} Calculation of the Resonance Escape Probability for $A \neq 1$.

WIGNER, E. P., *et al.*: J. Appl. Phys. 26, 260 (1955) (Resonance Integrals and Resonance Absorption).

ADLER, F. T., and Y. D. NALIBOFF: J. Nucl. Energy A & B 14, 209 (1961).
GOERTZEL, G.: Geneva 1955 P/613; Vol. 15 p. 472
ROSE, M., *et al.*: BNL-257 (1953); WAPD-SR 506 (1954).
SOLBRIG, A. W.: Nucl. Sci. Eng. 10, 167 (1961).
— Amer. J. Phys. 29, 257 (1961).
} Doppler Broadening of Resonance Lines.

ADLER, F.T., L.W. NORDHEIM, and G.W. HINMAN: Geneva 1958 P/1988; Vol. 16 p. 155.
DRESNER, L.: Nucl. Sci. Eng. 1, 68 (1956).
} Calculation of the Resonance Integral for Doppler-Broadened Resonances.

8. The Spatial Distribution of Moderated Neutrons

In this chapter we shall calculate the space and energy distribution of neutrons during the slowing-down process. Our goal will be to obtain the flux $\Phi(\boldsymbol{r}, E)$ or the slowing-down density $q(\boldsymbol{r}, E)$ arising from given sources. It will turn out that this general problem is considerably more difficult than the two special cases of it previously discussed, viz., diffusion without moderation and moderation without diffusion. We can easily see the reasons for the difficulty if we consider, for example, the neutron field due to a point source of fast neutrons in an infinite medium. At large distances from the source, the neutron flux is predominantly due to neutrons that have made no collisions or at most a few small-angle collisions that produce only a small energy loss. Their distribution of directions is strongly anisotropic, and a description of the diffusion process by FICK's law is no longer

[1] Cf. footnote on p. 53.

possible. For this reason, the treatment of "deep penetration" problems is particularly difficult. They are mainly of interest in shielding calculations, and we will consider them no further here[1] (cf. HOLTE as well as VERDE and WICK). At smaller distances — this limitation will be made more precise later — the distribution of neutron directions (with the exception of the energy range immediately below the source energy) is only weakly anisotropic, and with certain limitations FICK's law is valid. The flux can therefore be described approximately by an energy-dependent diffusion equation. Even this equation is not soluble in general, and we are forced to make further approximations. In "heavy" moderators (Be, graphite), a very simple approximation called age theory is possible, and we shall study it in detail in Sec. 8.2. For H_2O und D_2O, as well as other proton- and deuteron-containing substances, age theory gives very inaccurate results; and in Sec. 8.3 we shall become acquainted with the rudiments of some better approximations. Finally, in Sec. 8.4, we shall calculate the distribution of thermal neutrons due to specified fast neutron sources in several simple cases. However, before we proceed with our study of the various theories, we shall introduce an important empirical quantity, the so-called mean squared slowing-down distance.

8.1. The Mean Squared Slowing-Down Distance

8.1.1. Definition of $\overline{r_E^2}$

Let us consider an isotropic point source of fast neutrons in an infinite medium. Let us assume that we know the slowing-down density $q(r, E)$, which in this case is spherically symmetric and only depends on the distance r from the source. We can then characterize the slowing-down process by the quantity $\overline{r_E^2}$, the average of the square of the source distance at which the neutrons *pass* the energy[2] E. It is given by

$$\overline{r_E^2} = \frac{\int_0^\infty r^2 q(r, E)\, 4\pi r^2\, dr}{\int_0^\infty q(r, E)\, 4\pi r^2\, dr}. \qquad (8.1.1)$$

Frequently, we set

$$\overline{r_E^2} = 6 L_s^2 = 6\tau_E \qquad (8.1.2)$$

where L_s is called the *slowing-down length* and τ_E is called the *age* of the neutrons. The reason for the latter nomenclature will become clear later. A somewhat different definition of the mean squared slowing-down distance is based on the flux $\Phi(r, E)$ rather than on the slowing-down density $q(r, E)$:

$$\overline{r_E^{*2}} = \frac{\int_0^\infty r^2 \Phi(r, E)\, 4\pi r^2\, dr}{\int_0^\infty \Phi(r, E)\, 4\pi r^2\, dr}. \qquad (8.1.3)$$

$\overline{r_E^{*2}}$ is the mean square of the distance from the source at which the neutrons *possess* the energy E. It is somewhat greater than $\overline{r_E^2}$, but in most cases the difference is unimportant [cf. Eq. (8.1.15)].

---

[1] Some empirical data on deep penetration can be found in Chapter 16.

[2] i.e., are slowed down past the energy E.

The significance of these quantities is that they are a direct measure of the spreading out of the neutrons during moderation. They can easily and rather accurately be determined experimentally (Chapter 16) and can be compared directly with the predictions of the various approximate theoretical calculations.

8.1.2. Elementary Calculation of $\overline{r_E^2}$

Let us consider an infinite, non-absorbing medium in which neutrons slow down only by elastic collisions. Fig. 8.1.1 shows a typical path of a neutron during the

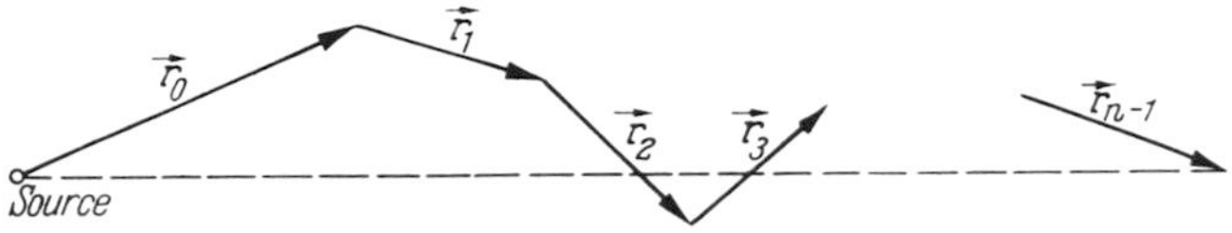

Fig. 8.1.1. A typical neutron path during the slowing-down process

moderation from the source energy E_Q to the energy E. Let r_ν be the free path of a neutron after its ν-th collision. Then

$$\overline{r_E^2} = \overline{\left(\sum_{\nu=0}^{n-1} \boldsymbol{r}_\nu\right)^2} = \overline{\sum_{\nu=0}^{n-1} r_\nu^2 + 2 \sum_{\nu=0}^{n-2} \sum_{\mu=\nu+1}^{n-1} \boldsymbol{r}_\nu \cdot \boldsymbol{r}_\mu}\,. \tag{8.1.4}$$

The summation only extends to $n-1$ because on the n-th collision the energy E is passed. If we denote the angle between $\boldsymbol{r}_\nu$ and $\boldsymbol{r}_\mu$ by $\vartheta_{\nu,\mu}$, then

$$\overline{r_E^2} = \overline{\sum_{\nu=0}^{n-1} r_\nu^2} + 2\,\overline{\sum_{\nu=0}^{n-2} r_\nu \sum_{\mu=\nu+1}^{n-1} r_\mu \cos\vartheta_{\nu,\mu}}\,. \tag{8.1.5}$$

The average is to be carried out

1. over the azimuthal scattering angles $\varphi_{\nu,\nu+1}$ of the individual scattering processes;
2. over the path lengths r_ν between the collisions;
3. over the polar scattering angles $\vartheta_{\nu,\nu+1}$; this is equivalent to an average over all E_ν, since the energy loss in a collision uniquely determines the scattering angle;
4. over all numbers of collisions that can lead from the source energy E_Q to the energy E.

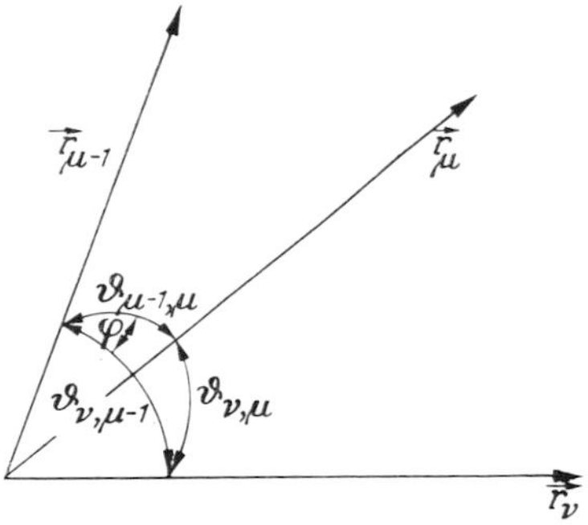

Fig. 8.1.2. The relation between $\vartheta_{\nu,\mu-1}$, $\vartheta_{\mu-1,\mu}$ and $\vartheta_{\nu,\mu}$

We begin with the average over the azimuthal scattering angles. The considerations connected with this average will make it possible to express $\cos\vartheta_{\nu,\mu}$ in terms of $\cos\vartheta_{\nu,\nu+1}$, $\cos\vartheta_{\nu+1,\nu+2}$, etc., i.e., in terms of the scattering angles for the individual collisions. For this purpose, we use a recursion procedure which can be made clear by means of Fig. 8.1.2. Let us suppose we already know $\cos\vartheta_{\nu,\mu-1}$ and let us calculate $\cos\vartheta_{\nu,\mu}$. If φ is the azimuthal scattering angle of the μ-th collision, then according to the cosine theorem of spherical trigonometry

$$\cos\vartheta_{\nu,\mu} = \cos\vartheta_{\nu,\mu-1}\cos\vartheta_{\mu-1,\mu} + \sin\vartheta_{\nu,\mu-1}\sin\vartheta_{\mu-1,\mu}\cos\varphi\,. \tag{8.1.6}$$

We must average this expression over all φ. Since all azimuthal angles are equally likely, the term containing $\cos\varphi$ drops out and we obtain

$$\cos\vartheta_{\nu,\mu}=\cos\vartheta_{\nu,\mu-1}\cos\vartheta_{\mu-1,\mu}. \tag{8.1.7a}$$

Thus, as we can easily see,

$$\cos\vartheta_{\nu,\mu}=\cos\vartheta_{\nu,\nu+1}\cos\vartheta_{\nu+1,\nu+2}\ldots\cos\vartheta_{\mu-1,\mu} \tag{8.1.7b}$$

where for purposes of simplicity we have dropped the overbar that signifies the averaging process.

The mean distance between two collisions is given by

$$\overline{r_\nu}=\frac{\int\limits_0^\infty r e^{-\Sigma_s(E_\nu)r}\,dr}{\int\limits_0^\infty e^{-\Sigma_s(E_\nu)r}\,dr}=\frac{1}{\Sigma_s(E_\nu)}=\lambda_s(E_\nu) \tag{8.1.8a}$$

while the mean squared distance is given by

$$\overline{r_\nu^2}=\frac{\int\limits_0^\infty r^2 e^{-\Sigma_s(E_\nu)r}\,dr}{\int\limits_0^\infty e^{-\Sigma_s(E_\nu)r}\,dr}=\frac{2}{\Sigma_s^2(E_\nu)}=2\lambda_s^2(E_\nu). \tag{8.1.8b}$$

Here E_ν is the energy of the neutron after the ν-*th* collision. Thus

$$\overline{r_E^2}=2\left[\overline{\sum_{\nu=0}^{n-1}\lambda_s^2(E_\nu)+\sum_{\nu=0}^{n-2}\lambda_s(E_\nu)\sum_{\mu=\nu+1}^{n-1}\lambda_s(E_\mu)\cos\vartheta_{\nu,\nu+1}\ldots\cos\vartheta_{\mu-1,\mu}}\right]. \tag{8.1.9}$$

Since the scattering angle and the energy loss are uniquely connected by the laws of elastic collisions, we could now express $\cos\vartheta_{\nu,\nu+1}$ in terms of E_ν and $E_{\nu+1}$. We would then still have to carry out the averages over the distribution of the E_ν and over the distribution of the collision numbers. Except in the case of pure hydrogen, this procedure is very difficult to carry out. Therefore, from now on let us consider a simple approximate procedure which is valid in heavy moderators ($A\gg 1$). In heavy moderators, the slowing down is due to many collisions each of which produces only a very small energy loss; the energy distribution after the ν-*th* collision is then rather sharply concentrated around the average value $\overline{E}_\nu$. If we now average over the scattering angles and notice that $\overline{\cos\vartheta_{\nu,\nu+1}}=\overline{\cos\vartheta_{\nu+1,\nu+2}}$ etc. $=2/3A$, then we obtain

$$\overline{r_E^2}=2\left[\sum_{\nu=0}^{n-1}\lambda_s^2(\overline{E}_\nu)+\sum_{\nu=0}^{n-2}\lambda_s(\overline{E}_\nu)\sum_{\mu=\nu+1}^{n-1}\lambda_s(\overline{E}_\mu)\left(\frac{2}{3A}\right)^{\mu-\nu}\right] \tag{8.1.10a}$$

which we can also write in the form

$$\overline{r_E^2}=2\sum_{\nu=0}^{n-1}\lambda_s(\overline{E}_\nu)\sum_{\mu=\nu}^{n-1}\lambda_s(\overline{E}_\mu)\left(\frac{2}{3A}\right)^{\mu-\nu}. \tag{8.1.10b}$$

Here we have approximated the average values of λ_s^2 and λ_s over the energy distribution after the ν-*th* collision by $\lambda_s^2(\overline{E}_\nu)$ and $\lambda_s(\overline{E}_\nu)$, respectively. For this to be a good approximation, A should be large and $\lambda_s(E)$ should vary slowly with energy. If both conditions are fulfilled, then we can further set

$$\sum_{\mu=\nu}^{n-1} \lambda_s(\bar{E}_\mu)\left(\frac{2}{3A}\right)^{\mu-\nu} \approx \lambda_s(\bar{E}_\nu)\sum_{\varkappa=0}^{\infty}\left(\frac{2}{3A}\right)^{\varkappa} = \frac{\lambda_s(\bar{E}_\nu)}{1-\frac{2}{3A}} \tag{8.1.11}$$

and obtain

$$\overline{r_E^2} = \frac{2}{1-\frac{2}{3A}} \sum_{\nu=0}^{n-1} \lambda_s^2(\bar{E}_\nu). \tag{8.1.12}$$

Finally, because of the large number of collisions we can replace the summation by an integration. The probability that a collision occurs in the energy interval $(E, E+dE)$ is $dE/\xi E$; thus

$$\overline{r_E^2} = \frac{2}{\xi\left(1-\frac{2}{3A}\right)} \int_E^{E_Q} \lambda_s^2(E')\frac{dE'}{E'}. \tag{8.1.13}$$

The contribution to this expression of the first flight of the neutron must still be taken into account. The sum in Eq. (8.1.12) contains a term $\frac{2\lambda_s^2(E_Q)}{1-2/3A}$ which is due to those neutrons that come directly from the source without making a collision. The contribution of these neutrons is not included in the integration in Eq. (8.1.13). We must therefore add a "first-flight correction"; doing so we get

$$\overline{r_E^2} = 2\lambda_s^2(E_Q) + \frac{2}{\xi\left(1-\frac{2}{3A}\right)} \int_E^{E_Q} \lambda_s^2(E')\frac{dE'}{E'}. \tag{8.1.14}$$

Notice here that the factor $\frac{1}{1-2/3A}$ has been omitted in the correction. The reason for this will become clear later (Sec. 8.2.5). Since the correction term is small in general and $2/3A \ll 1$, whether the factor $\frac{1}{1-2/3A}$ is included or not plays no important role in the calculation of $\overline{r_E^2}$. Similar considerations yield

$$\overline{r_E^{*2}} = 2\lambda_s^2(E_Q) + \frac{2}{\xi\left(1-\frac{2}{3A}\right)} \int_E^{E_Q} \lambda_s^2(E')\frac{dE'}{E'} + 2\lambda_s^2(E) \tag{8.1.15}$$

for the quantity $\overline{r_E^{*2}}$. Here a "last-flight correction" $2\lambda_s^2(E)$ occurs as well. We can understand the appearance of this term if we observe that in the calculation of $\overline{r_E^2}$ we summed only over the first $n-1$ collisions, since on the *n-th* collision the energy passed E; the energy is larger than E on the $n-1$-*st* flight path. In the calculation of $\overline{r_E^{*2}}$ we must obviously sum up to the $n+1$-*st* collision; for this reason a term $\overline{r_n^2}$ occurs in all the sums. We can easily convince ourselves that the correction term should not contain the factor $\frac{1}{1-2/3A}$. The expression $\overline{r_E^{*2}}/6 = \tau_E^*$ is occasionally called the flux age. Eq. (8.1.15) is important because in experimental determinations of the slowing-down length it is $\overline{r_E^{*2}}$ that is usually measured (cf. Chapter 16).

The formulas developed here hold rather accurately in graphite and represent a useful first approximation in other moderators.

For certain future applications, it will prove advantageous to calculate the mean squared slowing-down distance from a plane source in an infinite medium. Since in this case the slowing-down density depends only on the distance x from the source surface,

$$\overline{x_E^2} = \frac{\int\limits_0^\infty x^2 q(x, E)\, dx}{\int\limits_0^\infty q(x, E)\, dx}. \tag{8.1.16}$$

Since a plane source can be considered as made up of point sources, it must be possible to express $\overline{x_E^2}$ in terms of $\overline{r_E^2}$. Some simple geometric considerations show that

$$\overline{r_E^2} = 3\,\overline{x_E^2} \tag{8.1.17a}$$

while in general (with $s=1, 2, 3, \ldots$)

$$\overline{r_E^{2s}} = (2s+1)\,\overline{x_E^{2s}}. \tag{8.1.17b}$$

8.1.3. Formulation of an Exact Calculation of $\overline{r_E^2}$

Next we will show how $\overline{r_E^2}$ can be calculated by the so-called *moments method* (originally proposed by FERMI). This method can also be used to calculate the quantities $\overline{r_E^{2s}}$. In the discussion of this method, we shall study the P_N-approximation to the energy-dependent transport equation, which when $N=1$ is the point of departure for age theory and certain other approximations.

Let us consider an infinite medium containing at $x=0$ an infinite plane source that emits zero-lethargy neutrons. The one-dimensional transport equation then reads

$$\left.\begin{aligned} &\Sigma_t(u) F(x, \mathbf{\Omega}, u) + \cos\vartheta \frac{\partial F(x, \mathbf{\Omega}, u)}{\partial x} \\ &\quad = \int\limits_0^u \int\limits_{4\pi} \Sigma_s(\mathbf{\Omega}' \to \mathbf{\Omega}, u' \to u) F(x, \mathbf{\Omega}', u')\, d\Omega'\, du' + \frac{Q}{4\pi}\,\delta(u)\,\delta(x). \end{aligned}\right\} \tag{8.1.18}$$

We have assumed the source to emit isotropically, just as we did in Sec. 5.2. We proceed exactly as we did in Sec. 5.2.1, i.e., we expand $\Sigma_s(\mathbf{\Omega}' \to \mathbf{\Omega}, u' \to u)$ in Legendre polynomials of the scattering angle ϑ_0:

$$\Sigma_s(\mathbf{\Omega}' \to \mathbf{\Omega}, u' \to u) = \frac{1}{4\pi} \sum_l (2l+1)\, \Sigma_{sl}(u' \to u)\, P_l(\cos\vartheta_0) \tag{8.1.19a}$$

where

$$\Sigma_{sl}(u' \to u) = 2\pi \int\limits_0^\pi \Sigma_s(\mathbf{\Omega}' \to \mathbf{\Omega}, u' \to u)\, P_l(\cos\vartheta_0) \sin\vartheta_0\, d\vartheta_0. \tag{8.1.19b}$$

When the scattering is isotropic in the center-of-mass system, we can derive the cross section $\Sigma_s(\mathbf{\Omega}' \to \mathbf{\Omega}, u' \to u)$ from the results of Sec. 7.1: The probability that a neutron of lethargy u' has a lethargy u after one collision is

$$\begin{aligned} g(u' \to u) &= \frac{1}{1-\alpha}\, e^{-(u-u')} \quad &&\text{for} \quad u - \ln(1/\alpha) < u' < u, \\ g(u' \to u) &= 0 &&\text{otherwise.} \end{aligned}$$

It follows from Eqs. (7.1.6) and (7.1.7) that the cosine of the scattering angle of a collision in which the lethargy changes from u' to u is

$$\cos\vartheta_0 = \frac{A+1}{2} e^{-\frac{u-u'}{2}} - \frac{A-1}{2} e^{\frac{u-u'}{2}}. \tag{8.1.20}$$

Thus for $u - \ln(1/\alpha) < u' < u$

$$\left.\begin{aligned} \Sigma_s(\mathbf{\Omega}'\to\mathbf{\Omega}, u'\to u) &= \Sigma_s(u')\,\frac{1}{2\pi}\cdot\frac{1}{1-\alpha}\, e^{-(u-u')}\times \\ &\times \delta\left(\cos\vartheta_0 - \left[\frac{A+1}{2} e^{-\frac{u-u'}{2}} - \frac{A-1}{2} e^{\frac{u-u'}{2}}\right]\right). \end{aligned}\right\} \tag{8.1.21}$$

The factor $1/2\pi$ occurs because all azimuthal scattering angles are equally likely; the δ-function takes account of the fact that only one value of $\cos\vartheta_0$ corresponds to each lethargy change $u-u'$. From Eq. (8.1.19b) it follows that the first expansion coefficient is given by

$$\Sigma_{s0}(u'\to u) = \Sigma_s(u')\,\frac{e^{-(u-u')}}{1-\alpha} \quad \text{for} \quad u - \ln(1/\alpha) < u' < u, \tag{8.1.22a}$$

i.e., $\Sigma_{s0}(u'\to u)$ is the cross section for all scattering processes that transport a neutron from u' to u. The second expansion coefficient is given by

$$\left.\begin{aligned} \Sigma_{s1}(u'\to u) &= \Sigma_s(u')\,\frac{e^{-(u-u')}}{1-\alpha}\left(\frac{A+1}{2} e^{-\left(\frac{u-u'}{2}\right)} - \frac{A-1}{2} e^{\frac{u-u'}{2}}\right) \\ &\text{for} \quad u - \ln(1/\alpha) < u' < u. \end{aligned}\right\} \tag{8.1.22b}$$

The fact that Σ_{s0} and Σ_{s1} depend on $u-u'$ in a different manner is due to the fact that for an elastic collision the scattering angle and the energy change are uniquely connected. We obtain a connection with Eqs. (5.2.4) and (5.2.5) by integration over all final energies:

$$\int_{u'}^{u'+\ln(1/\alpha)} \Sigma_{s0}(u'\to u)\,du = \Sigma_s(u'), \tag{8.1.23a}$$

$$\int_{u'}^{u'+\ln(1/\alpha)} \Sigma_{s1}(u'\to u)\,du = \Sigma_s(u')\,\overline{\cos\vartheta_0} = \Sigma_s(u')\,\frac{2}{3A}. \tag{8.1.23b}$$

If we substitute Eq. (8.1.19a) in the transport equation and proceed exactly as in Sec. 5.2.1, we find

$$\left.\begin{aligned} \cos\vartheta\,\frac{\partial F(x,\mathbf{\Omega},u)}{\partial x} + \Sigma_t(u)\,F(x,\mathbf{\Omega},u) &= \frac{1}{2}\int_{u-\ln(1/\alpha)}^{u} \sum_{l=0}^{\infty} (2l+1)\,\Sigma_{sl}(u'\to u)\times \\ \times P_l(\cos\vartheta)\int_0^{\pi} F(x,\mathbf{\Omega}',u')\,P_l(\cos\vartheta')\sin\vartheta'\,d\vartheta'\,du' &+ \frac{Q}{4\pi}\,\delta(u)\,\delta(x). \end{aligned}\right\} \tag{8.1.24}$$

Next, the vector flux is expanded in Legendre polynomials

$$F(x,\mathbf{\Omega},u) = \frac{1}{4\pi}\sum_{l=0}^{N} (2l+1)\,F_l(x,u)\,P_l(\cos\vartheta) \tag{8.1.25a}$$

with

$$F_l(x,u) = 2\pi\int_0^{\pi} F(x,\mathbf{\Omega},u)\,P_l(\cos\vartheta)\sin\vartheta\,d\vartheta. \tag{8.1.25b}$$

If we multiply the transport Eq. (8.1.24) by $P_l(\cos\vartheta)$ and integrate over ϑ, then using Eqs. (8.1.25), we obtain the following set of $N+1$ integro-differential equations:

$$\left.\begin{aligned}
&\frac{\partial F_1(x,u)}{\partial x}+\Sigma_t(u)\,F_0(x,u)=\int\limits_{u-\ln(1/\alpha)}^{u}\Sigma_{s0}(u'\to u)\,F_0(x,u')\,du'+Q\,\delta(u)\,\delta(x),\\
&\frac{l+1}{2l+1}\,\frac{\partial F_{l+1}(x,u)}{\partial x}+\frac{l}{2l+1}\,\frac{\partial F_{l-1}(x,u)}{\partial x}+\Sigma_t F_l(x,u)\\
&\qquad\qquad=\int\limits_{u-\ln(1/\alpha)}^{u}\Sigma_{sl}(u'\to u)\,F_l(x,u')\,du',\\
&\qquad\qquad\qquad\qquad\qquad\qquad l=1\,\ldots\,N-1\\
&\frac{N}{2N+1}\,\frac{\partial F_{N-1}(x,u)}{\partial x}+\Sigma_t F_N(x,u)=\int\limits_{u-\ln(1/\alpha)}^{u}\Sigma_{sN}(u'\to u)\,F_N(x,u')\,du'.
\end{aligned}\right\}\tag{8.1.26}$$

Eqs. (8.1.26) obviously represent the generalization of Eqs. (5.2.46) to the energy-dependent case. They are the starting point for various approximate methods about which we shall learn in later sections. We begin here with the "moments" method.

We define the moments M_{nl} purely formally by

$$M_{nl}(u)=\int\limits_{-\infty}^{+\infty}\frac{x^n}{n!}\,F_l(x,u)\,dx.\tag{8.1.27}$$

Clearly $M_{00}(u)=\int\limits_{-\infty}^{+\infty}F_0(x,u)\,dx=\int\limits_{-\infty}^{+\infty}\Phi(x,u)\,dx$ is the space-integrated flux. Furthermore, $M_{20}(u)=\frac{1}{2}\int\limits_{-\infty}^{\infty}x^2\,\Phi(x,u)\,dx$; thus

$$\overline{x_u^{*\,2}}=\frac{1}{3}\,\overline{r_u^{*\,2}}=2\,\frac{M_{20}(u)}{M_{00}(u)}\tag{8.1.28a}$$

or generally (since M_{n0} vanishes for odd n)

$$\overline{x_u^{*\,2s}}=\frac{1}{2s+1}\,\overline{r_u^{*\,2s}}=(2s)!\,\frac{M_{(2s)0}(u)}{M_{00}(u)}.\tag{8.1.28b}$$

If we multiply Eqs. (8.1.26) by $x^n/n!$ and integrate over x from $-\infty$ to ∞, we obtain a system of coupled integral equations for the $M_{nl}(u)$. We can integrate these equations with the help of an electronic computing machine and from the moments so obtained determine the neutron flux distribution $\Phi(x,u)$ (cf. in this connection GOLDSTEIN or SPENCER and FANO); this procedure is one of the best and most general for the calculation of neutron attenuation in shields. However, we are principally interested here in the calculation of M_{00} and M_{20}, since from these moments we can immediately obtain the mean squared slowing-down distance. The equations for the first three moments are

$$\Sigma_t(u)\,M_{00}(u)=\int\limits_{u-\ln(1/\alpha)}^{u}\Sigma_{s0}(u'\to u)\,M_{00}(u')\,du'+Q\,\delta(u),\tag{8.1.29a}$$

$$\Sigma_t(u)\,M_{11}(u)=\int\limits_{u-\ln(1/\alpha)}^{u}\Sigma_{s1}(u'\to u)\,M_{11}(u')\,du'+\tfrac{1}{3}\,M_{00}(u),\tag{8.1.29b}$$

$$\Sigma_t(u)\,M_{20}(u)=\int\limits_{u-\ln(1/\alpha)}^{u}\Sigma_{s0}(u'\to u)\,M_{20}(u')\,du'+M_{11}(u).\tag{8.1.29c}$$

In the derivation of these equations integrals of the type $\int\limits_{-\infty}^{+\infty} \frac{x^2}{2} \frac{\partial F_1(x,u)}{\partial x} dx$ were integrated by parts:

$$\int\limits_{-\infty}^{+\infty} \frac{x^2}{2} \frac{\partial F_1(x,u)}{\partial x} dx = \frac{x^2}{2} F_1(x,u) \Big|_{-\infty}^{+\infty} - \int\limits_{-\infty}^{+\infty} x F_1(x,u)\, dx = -M_{11}(u).$$

The expression $\frac{x^2}{2} F_1(x,u) \Big|_{-\infty}^{\infty}$ vanishes because as $x \to \pm\infty$, F vanishes faster than any power of x. Furthermore, the fact that $\int\limits_{-\infty}^{\infty} F_l(x,u)\, dx = 0$ for $l \neq 0$ was also used.

Eq. (8.1.29a) for $M_{00}(u)$ corresponds to Eq. (5.1.16) for the flux in an infinite medium with homogeneously distributed sources; this becomes clear when we realize that $M_{00}(u)$ represents the flux integrated over all space and must therefore have the same energy dependence as the flux due to sources homogeneously distributed in an infinite medium.

By successive integration of Eqs. (8.1.29) we can obtain $M_{00}(u)$, $M_{11}(u)$, and $M_{20}(u)$ and thus also $\tau_u^* = M_{20}(u)/M_{00}(u)$. We shall not carry out these additional steps here because they result in rather cumbersome expressions; Marshak gives results for some special cases. In the general case (a mixture of various nuclei; cross sections that depend in different ways on energy; anisotropic scattering in the center-of-mass system), an analytic integration is not possible, and we must either integrate numerically or introduce approximations. We shall become acquainted with the results of such calculations in Chapter 16 during the discussion of measurements of the slowing-down length.

We can nevertheless still draw one important conclusion just from the form of Eqs. (8.1.29). In the first place, these equations are exact. Secondly, we would have been led to the same relations if we had started from a P_1-approximation for $F(x,u)$ $\left(F(x,u) = \frac{1}{4\pi} F_0(x,u) + \frac{3}{4\pi} F_1(x,u) \cos\vartheta\right)$, as we can easily check. It therefore follows that any theory of slowing down that is based on a P_1-approximation will give a correct value of $\overline{r_E^2}$, as long as no other approximations have been made.

8.2. Age Theory

8.2.1. The Energy-Dependent Diffusion Equation and Age Theory

In this section, we shall try to derive a simple differential equation for the energy- and space-dependent neutron flux in any medium. We start with a neutron balance for the neutrons in one cm³:

$$\frac{\partial q(\boldsymbol{r},u)}{\partial u} + \Sigma_a \Phi(\boldsymbol{r},u) + \operatorname{div} \boldsymbol{J}(\boldsymbol{r},u) = S(\boldsymbol{r},u). \tag{8.2.1}$$

The left-hand side represents the neutrons lost per cm³ and sec through slowing down, absorption, and diffusion; the right-hand side represents the sources. This equation is exact; it is a combination of our earlier balance Eqs. (7.2.10) and (6.1.8). According to Sec. 7.2.1, the slowing down density and the flux are related by the equation

$$q(u) = \int_{u-\ln(1/\alpha)}^{u} \Sigma_s(u')\,\Phi(u')\,\frac{e^{-(u-u')}-\alpha}{1-\alpha}\,du'. \tag{8.2.2}$$

We now assume that FICK's law relates the flux gradient and the current density:

$$\boldsymbol{J}(\boldsymbol{r}, u) = -D(u)\,\mathrm{grad}\,\Phi(\boldsymbol{r}, u). \tag{8.2.3a}$$

The diffusion constant is given by

$$D(u) = \frac{1}{3\Sigma_{tr}(u)} = \frac{1}{3\Sigma_s(u)(1-2/3A)}. \tag{8.2.3b}$$

Eq. (8.2.3a) is certainly of very limited validity; the conditions of its applicability will be investigated in Sec. 8.2.2. If we substitute Eq. (8.2.3a) in Eq. (8.2.1), we obtain

$$\frac{\partial q(\boldsymbol{r}, u)}{\partial u} + \Sigma_a\,\Phi(\boldsymbol{r}, u) - D(u)\,\nabla^2\Phi(\boldsymbol{r}, u) = S(\boldsymbol{r}, u). \tag{8.2.4}$$

Eq. (8.2.4) is an *energy-dependent diffusion equation* that is the starting point for a variety of approximate methods. In order to solve it, we must simplify the connection between the flux and slowing-down density given in Eq. (8.2.2).

In a "heavy" moderator like graphite the maximum energy loss is small. We can therefore approximate the quantity $\Sigma_s(u')\,\Phi(u')$ by its value at $u'=u$ and take it before the integral; doing this, we get

$$q(u) = \Sigma_s(u)\,\Phi(u) \int_{u-\ln(1/\alpha)}^{u} \frac{e^{-(u-u')}-\alpha}{1-\alpha}\,du' = \xi\,\Sigma_s\,\Phi(u), \tag{8.2.5}$$

i.e., the flux and slowing-down density are related exactly as in an infinite medium with homogeneously distributed sources. With the help of this relation, we can write Eq. (8.2.4) entirely in terms of either Φ or q. If we assume for simplicity that $\Sigma_a=0$ and limit ourselves to a source-free region, then

$$\frac{\partial q(\boldsymbol{r}, u)}{\partial u} = \frac{D(u)}{\xi\Sigma_s(u)}\,\nabla^2 q(\boldsymbol{r}, u). \tag{8.2.6}$$

If we introduce the FERMI age

$$\tau(u) = \int_0^u \frac{D(u')}{\xi\Sigma_s(u')}\,du' = \int_E^{E_Q} \frac{\lambda_s^2(E')}{3\xi(1-2/3A)}\,\frac{dE'}{E'} \tag{8.2.7}$$

in place of the energy variable, the celebrated *age equation* results:

$$\frac{\partial q}{\partial \tau} = \nabla^2 q. \tag{8.2.8}$$

This equation is formally identical with the differential equation of heat conduction if we identify q with the temperature and τ with the time. When $\Sigma_a \neq 0$ and in the presence of sources that emit neutrons of zero lethargy, i.e., neutrons with $\tau=0$, the age equation becomes

$$\frac{\partial q}{\partial \tau} = \nabla^2 q - \frac{\Sigma_a/\xi\Sigma_s}{D(u)/\xi\Sigma_s}\,q(\tau) + S(\boldsymbol{r})\,\delta(\tau). \tag{8.2.9a}$$

If we introduce $q^*(\boldsymbol{r}, \tau) = q(\boldsymbol{r}, \tau)\, p(\tau)$ where

$$p(\tau) = p(u) = e^{-\int_0^u \frac{\Sigma_a(u')}{\xi \Sigma_s} du'} \tag{8.2.9b}$$

the equation for q^* does not contain an absorption term; thus the absorption can be described in the usual way by a resonance escape probability p. In the following, and in particular in later applications, we shall only consider the case $\Sigma_a = 0$. Of course, it is clear that age theory is only valid when $\Sigma_a \ll \Sigma_s$.

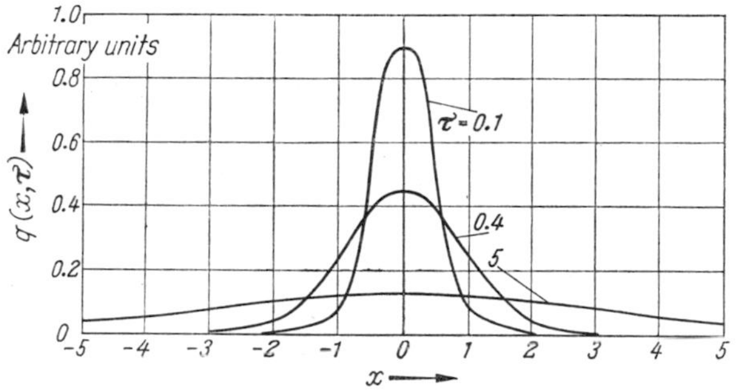

Fig. 8.2.1. The slowing-down density around a plane source as predicted by age theory. $\sqrt{\tau}$ is in the same units as x

The solution of Eq. (8.2.8) for a plane source emitting Q zero-lethargy neutrons per cm² and per sec into an infinite medium is (cf. Sec. 8.2.3)

$$q(x, \tau) = \frac{Q}{\sqrt{4\pi\tau}}\, e^{-\frac{x^2}{4\tau}}. \tag{8.2.10}$$

This solution is displayed in Fig. 8.2.1, where we can easily see how the slowing-down density, which always has a Gaussian shape, spreads out with increasing τ. The mean squared slowing-down distance is given by

$$\overline{x_E^2} = \frac{\overline{r_E^2}}{3} = \frac{\int_0^\infty x^2 \frac{Q}{\sqrt{4\pi\tau}}\, e^{-\frac{x^2}{4\tau}} dx}{\int_0^\infty \frac{Q}{\sqrt{4\pi\tau}}\, e^{-\frac{x^2}{4\tau}} dx} = 2\tau(E), \tag{8.2.11a}$$

i.e.,

$$\overline{r_E^2} = 6\tau(E) = \frac{2}{\xi(1 - 2/3A)} \int_E^{E_Q} \lambda_s^2(E')\, \frac{dE'}{E'}. \tag{8.2.11b}$$

Except for the first-flight correction this is identical with our earlier result [Eq. (8.1.14)]. It is characteristic of the age approximation that it does not include the contribution of the first flight; we shall return to this point in Sec. 8.2.5.

Confusion can easily arise in the use of the term age and the symbol τ. We shall denote as the *age* τ_E one-sixth of the mean squared slowing-down distance from a point source [Eq. (8.1.1)]. This definition is general and does not depend on whether age theory is valid or not. By *Fermi age* τ or $\tau(E)$ we shall mean the variable in the age equation; in a medium in which age theory is valid, $\tau(E) \approx \tau_E$. Finally, by *flux age* τ_E^* we shall mean one-sixth of the mean squared slowing-down distance determined from the flux distribution [Eq. (8.1.3)]. In the age approximation there is no difference between τ_E^* and τ_E. The name age indicates that we are dealing with a quantity that increases as the slowing-down process progresses; later we shall find that there is a direct connection between the Fermi age and the slowing-down time.

Before we go into any applications of age theory, we shall study the conditions of its validity.

8.2.2. Age Theory and the Transport Equation

In Sec. 5.2.1, we derived elementary diffusion theory from a P_1-approximation to the transport equation. We now attempt a P_1-approximation to the energy-dependent transport Eq. (8.1.24); thus we set

a) $$F(x, \mathbf{\Omega}, u) = \frac{1}{4\pi} F_0(x, u) + \frac{3}{4\pi} F_1(x, u) \cos\vartheta. \tag{8.2.12}$$

When $\Sigma_a = 0$, the following equations for F_0 and F_1 follow from Eq. (8.1.26):

$$\frac{\partial F_1(x, u)}{\partial x} + \Sigma_s F_0(x, u) = \int\limits_{u-\ln(1/\alpha)}^{u} \Sigma_{s0}(u' \to u) F_0(x, u') du' + Q\delta(u)\delta(x), \tag{8.2.13a}$$

$$\frac{1}{3}\frac{\partial F_0(x, u)}{\partial x} + \Sigma_s F_1(x, u) = \int\limits_{u-\ln(1/\alpha)}^{u} \Sigma_{s1}(u' \to u) F_1(x, u')\, du'. \tag{8.2.13b}$$

In contrast to the energy-independent case, the second equation does not yield FICK's law immediately, since the right-hand side contains the term

$$\int \Sigma_{s1}(u' \to u) F_1(x, u')\, du'.$$

Thus the P_1-approximation and diffusion theory are no longer the same in the energy-dependent case. We now make the following additional assumptions

b) $$\Sigma_s(u') F_0(u') = \Sigma_s(u) F_0(u) + (u' - u) \frac{\partial}{\partial u} [\Sigma_s(u) F_0(u)], \tag{8.2.14a}$$

c) $$\Sigma_s(u') F_1(u') = \Sigma_s(u) F_1(u), \tag{8.2.14b}$$

i.e., we assume that the collision density changes only a little in one collision interval. In this connection it is important to note that it is consistent to approximate $\Sigma_s(u') F_1(u')$ by only one term in its series expansion, since for the validity of the P_1-approximation we must assume that $F_1(x, u) \ll F_0(x, u)$. Eq. (8.2.14b), however, immediately transforms Eq. (8.2.13b) into FICK's law; with the help of Eq. (8.1.22b) we find that

$$F_1(x, u) = -\frac{1}{3\Sigma_s(1 - 2/3A)} \frac{\partial F_0(x, u)}{\partial x}. \tag{8.2.15a}$$

Using Eqs. (8.2.14a) and (8.1.22a), we obtain

$$\frac{\partial F_1(x, u)}{\partial x} = -\xi \frac{\partial [\Sigma_s(u) F_0(x, u)]}{\partial u} + Q\delta(u)\delta(x) \tag{8.2.15b}$$

from Eq. (8.2.13a). If we set $q(x, u) = \xi \Sigma_s(u) F_0(x, u)$, a combination of both these equations yields

$$\frac{\partial q(x,u)}{\partial u} = \frac{1}{3\xi \Sigma_s^2(u)(1-2/3A)} \frac{\partial^2 q(x,u)}{\partial x^2} + Q\,\delta(u)\,\delta(x) \tag{8.2.16}$$

which is essentially the same as Eq. (8.2.6). In order that this equation be valid, the approximations a, b, and c must obviously hold. We now investigate how well fulfilled they are by studying the flux distribution near a plane source; for this purpose, we write Eq. (8.2.10) in the form

$$\Sigma_s(u)\,F_0(u) = \frac{1}{\xi} \frac{Q}{\sqrt{4\pi\tau(u)}}\, e^{-\frac{x^2}{4\tau(u)}}. \tag{8.2.17a}$$

Approximation a is surely good if $F_1(x, u) \ll F_0(x, u)$; using Eqs. (8.2.15a) and (8.2.17a), this condition becomes

$$x \ll 6(1-2/3A)\,\Sigma_s(u)\,\tau(u). \tag{8.2.17b}$$

Age theory is thus only valid at relatively small distances from the source. This condition becomes somewhat clearer if we consider a medium in which Σ_s depends only weakly on u. In such a medium, $\tau = \frac{u}{3\xi(1-2/3A)\Sigma_s^2} = \frac{n}{3(1-2/3A)\Sigma_s^2}$, where n is the number of collisions necessary for moderation to the lethargy u. Eq. (8.2.17b) now becomes

$$x \ll 2n\lambda_s. \tag{8.2.17c}$$

This limitation is connected with the fact that at large distances from the source the neutron flux is predominantly determined by neutrons that have undergone only a few collisions, whose energy and direction have not changed much, and whose angular distribution is very anisotropic.

Approximation b presupposes that $\xi \frac{\partial}{\partial u}[\Sigma_s(u) F_0(x, u)] \ll \Sigma_s(u) F_0(x, u)$. Using Eqs. (8.2.17a) and (8.2.17b), we can rewrite this condition as

$$\xi \frac{\partial \tau}{\partial u} \ll \tau. \tag{8.2.17d}$$

Thus many collisions must occur during moderation, i.e., the moderating medium must be heavy. This condition is certainly not fulfilled in water, and we expect that age theory will be a poor approximation there.

Approximation c can only hold if $\xi \frac{\partial}{\partial u}[\Sigma_s(u) F_1(x, u)] \ll \Sigma_s(u) F_1(x, u)$; in view of the foregoing, this condition leads to

$$\xi \frac{d\lambda_s(u)}{du} \ll \lambda_s(u), \tag{8.2.17e}$$

i.e., the cross section can only change a little in one collision interval.

Furthermore the absorption must be very small ($\Sigma_a \ll \Sigma_s$); also we must be sufficiently far away (several λ_s) from any surfaces, since otherwise the angular distribution will be strongly anisotropic and condition a will no longer be fulfilled. In spite of these many restrictive conditions, age theory is applicable to a wide variety of slowing-down problems; in graphite and beryllium, the conditions derived here are well fulfilled. Next we shall study some methods of solving the age equation.

8.2.3. Application of the Age Equation to Simple Slowing-Down Problems

In this section, we shall calculate the slowing-down density for some simple arrangements of point and plane sources in homogeneous media. As a boundary condition at free surfaces, we require that $q(\boldsymbol{r}_G+\boldsymbol{n}d, \tau)=0$ for all τ, where $d=\frac{0.71}{\Sigma_s(1-2/3A)}$, exactly as in the monoenergetic theory. Here we must remember that the scattering cross section and thus the extrapolated endpoint depend on E and therefore on τ; taking this fact into strict account would lead to great mathematical complication, and throughout this section we shall use a suitable average value of d that is independent of energy. If we wish to treat this effect more accurately, we must use another description of the slowing-down process, the so-called multigroup method, into which we shall go further later. A similar complication occurs in the description of the slowing-down process in a medium composed of two or more regions that contain different moderators. In this case, the flux and current must be continuous at the interfaces at every lethargy. In such a problem it is no longer of advantage to use the age, since the relation between τ and u given by Eq. (8.2.7) is different in different media; in other words, as the neutron crosses an interface, its age can increase or decrease discontinuously. Even if we use Eq. (8.2.6), which is equivalent to the age equation, to describe slowing down in such a heterogeneous medium, a solution is hardly possible since a double infinity of boundary conditions must be satisfied at every interface. For such problems, multigroup methods are strongly to be preferred.

Plane Source in an Infinite Medium. The age equation takes the form

$$\frac{\partial q}{\partial \tau}=\frac{\partial^2 q}{\partial x^2}+Q\,\delta(x)\,\delta(\tau) \tag{8.2.18a}$$

with the boundary condition $q=0$ for $x=\pm\infty$. Here we can use to advantage the method of *Fourier transformation*, whose application to various problems is described in detail in the book of SNEDDON. The Fourier transform of $q(x, \tau)$ is defined by

$$f(\omega, \tau)=\int_{-\infty}^{+\infty} q(x, \tau)\, e^{-i\omega x}\, dx. \tag{8.2.18b}$$

The inverse transformation is given by

$$q(x, \tau)=\frac{1}{2\pi}\int_{-\infty}^{+\infty} f(\omega, \tau) e^{i\omega x}\, d\omega. \tag{8.2.18c}$$

It is easy to check that because q vanishes at infinity, the Fourier transform of $\partial q/\partial x$ is $i\omega f$ and the Fourier transform of $\partial^2 q/\partial x^2$ is $-\omega^2 f$. Thus Fourier transformation of Eq. (8.2.18a) leads to

$$\frac{\partial f(\omega, \tau)}{\partial \tau}+\omega^2 f=Q\,\delta(\tau) \tag{8.2.18d}$$

which has the solution

$$f(\omega, \tau)=Q\,e^{-\omega^2\tau} \tag{8.2.18e}$$

for $\tau>0$. Therefore

$$q(x, \tau)=\frac{Q}{2\pi}\int_{-\infty}^{+\infty} e^{-(\omega^2\tau-i\omega x)}\, d\omega. \tag{8.2.18f}$$

If we complete the square $\omega^2\tau - i\omega x = \left(\omega\sqrt{\tau} - \frac{ix}{2\sqrt{\tau}}\right)^2 + \frac{x^2}{4\tau}$ we get

$$q(x, \tau) = \frac{Q}{2\pi} e^{-\frac{x^2}{4\tau}} \int_{-\infty}^{+\infty} e^{-\left(\omega\sqrt{\tau} - \frac{ix}{2\sqrt{\tau}}\right)^2} d\omega. \tag{8.2.18g}$$

After the substitution $\omega\sqrt{\tau} - \frac{ix}{2\sqrt{\tau}} = u$, $d\omega = du/\sqrt{\tau}$ the integral can be carried out and gives

$$q(x, \tau) = \frac{Q}{\sqrt{4\pi\tau}} e^{-\frac{x^2}{4\tau}}. \tag{8.2.10}$$

This solution is well known in the theory of heat conduction. It has already been displayed in Fig. 8.2.1.

Point Source in an Infinite Medium.

$$\frac{\partial q}{\partial \tau} = \frac{\partial^2 q}{\partial x^2} + \frac{\partial^2 q}{\partial y^2} + \frac{\partial^2 q}{\partial z^2} + Q\,\delta(x)\,\delta(y)\,\delta(z)\,\delta(\tau). \tag{8.2.19a}$$

Here we define the Fourier transform

$$f(\boldsymbol{\omega}, \tau) = \iiint_{-\infty}^{+\infty} q(\boldsymbol{r}, \tau)\, e^{-i\boldsymbol{\omega}\cdot\boldsymbol{r}}\, dx\, dy\, dz, \tag{8.2.19b}$$

$$q(\boldsymbol{r}, \tau) = \frac{1}{(2\pi)^3} \iiint_{-\infty}^{+\infty} f(\boldsymbol{\omega}, \tau)\, e^{i\boldsymbol{\omega}\cdot\boldsymbol{r}}\, d\omega_1\, d\omega_2\, d\omega_3. \tag{8.1.19c}$$

Since the slowing-down density vanishes at infinity, Fourier transformation of Eq. (8.2.19a) leads to

$$\frac{\partial f(\boldsymbol{\omega}, \tau)}{\partial \tau} + \omega^2 f(\boldsymbol{\omega}, \tau) = Q\,\delta(\tau) \tag{8.2.19d}$$

which has the solution

$$f(\boldsymbol{\omega}, \tau) = Q e^{-\omega^2\tau} \tag{8.2.19e}$$

for $\tau > 0$. The inverse transformation is carried out exactly as in the case of a plane source and when we note that $\boldsymbol{\omega}^2 = \omega_1^2 + \omega_2^2 + \omega_3^2$ and $\boldsymbol{r}^2 = x^2 + y^2 + z^2$, yields

$$q(\boldsymbol{r}, \tau) = \frac{Q}{(4\pi\tau)^{\frac{3}{2}}} e^{-\frac{r^2}{4\tau}}. \tag{8.2.19f}$$

Thus

$$\overline{r_E^2} = \int_0^\infty r^2 \frac{e^{-\frac{r^2}{4\tau}}}{(4\pi\tau)^{\frac{3}{2}}} 4\pi r^2\, dr = 6\tau(E) \tag{8.2.19g}$$

while in general

$$\overline{r_E^{2s}} = \int_0^\infty r^{2s} \frac{e^{-\frac{r^2}{4\tau}}}{(4\pi\tau)^{\frac{3}{2}}} 4\pi r^2\, dr = \frac{(2s+1)!}{s!} \tau^s(E). \tag{8.2.19h}$$

Point Source at the Center of a Sphere. Let the effective radius of the sphere be R. Then $q(R, \tau)=0$ and[1]

$$\frac{\partial q}{\partial \tau}=\frac{\partial^2 q}{\partial r^2}+\frac{2}{r}\frac{\partial q}{\partial r}+\frac{Q}{4\pi r^2}\delta(r)\delta(\tau) \tag{8.2.20a}$$

or

$$\frac{\partial (rq)}{\partial \tau}=\frac{\partial^2 (rq)}{\partial r^2}+\frac{Q}{4\pi r}\delta(r)\delta(\tau). \tag{8.2.20b}$$

Next we develop rq in a Fourier series

$$rq=\sum_{l=1}^{\infty} A_l(\tau)\sin\left(\frac{l\pi r}{R}\right); \qquad A_l(\tau)=\frac{2}{R}\int_0^R rq\sin\left(\frac{l\pi r}{R}\right)dr. \tag{8.2.20c}$$

Fourier transformation of Eq. (8.2.20b) then yields

$$\frac{\partial A_l(\tau)}{\partial \tau}+\left(\frac{l\pi}{R}\right)^2 A_l(\tau)=\frac{Ql}{2R^2}\delta(\tau). \tag{8.2.20d}$$

This equation has the solution

$$A_l(\tau)=\frac{Ql}{2R^2}e^{-\left(\frac{l\pi}{R}\right)^2\tau} \tag{8.2.20e}$$

for $\tau>0$. Substituting Eq. (8.2.20e) into Eq. (8.2.20c) we obtain

$$q=\frac{Q}{2R^2 r}\sum_{l=1}^{\infty} l\sin\frac{l\pi r}{R}e^{-\left(\frac{l\pi}{R}\right)^2\tau}. \tag{8.2.20f}$$

The larger τ is, the better the sum on the right-hand side converges; when $\tau \gg (\pi/R)^2$, only the $l=1$ term contributes appreciably to the neutron distribution. When $R\to\infty$ the solution given in Eq. (8.2.20f) should reduce to that given in Eq. (8.2.19f). In this limit we can replace the summation by an integration by means of the substitutions $\omega=l\pi/R$, $d\omega=\pi/R$; Eq. (8.2.20f) then becomes

$$\lim_{R\to\infty} q(r,\tau)=\frac{Q}{2\pi^2 r}\int_0^\infty \sin(\omega r)\,e^{-\omega^2\tau}\,\omega\,d\omega. \tag{8.2.20g}$$

After the substitution $\sin\omega r=\dfrac{e^{i\omega r}-e^{-i\omega r}}{2}$, the integration can be carried out easily and yields Eq. (8.2.19f).

Point Source in an Infinitely Long Pile. Let the effective lengths of the pile's sides be a and b, and let the coordinate system be oriented as shown in Fig. 6.2.3. If the source is located at x_0, y_0, z_0, then

$$\frac{\partial q}{\partial \tau}=\nabla^2 q+Q\delta(x-x_0)\delta(y-y_0)\delta(z-z_0)\delta(\tau). \tag{8.2.21a}$$

Again we develop q in a Fourier series

$$q(x,y,z,\tau)=\sum_{l,m=1}^{\infty} A_{lm}(z,\tau)\sin\frac{l\pi x}{a}\sin\frac{m\pi y}{b} \tag{8.2.21b}$$

[1] When we transform to polar coordinates, $\delta(\boldsymbol{r})$ must be replaced by $\delta(r)/4\pi r^2$.

with the coefficients

$$A_{lm}(z,\tau)=\frac{4}{ab}\int_0^a\int_0^b q(x,y,z,\tau)\sin\frac{l\pi x}{a}\sin\frac{m\pi y}{b}\,dx\,dy. \qquad (8.2.21\,\text{c})$$

Fourier transformation of Eq. (8.2.21a) then gives

$$\left.\begin{aligned}&\frac{\partial A_{lm}(z,\tau)}{\partial\tau}+\pi^2\left(\frac{l^2}{a^2}+\frac{m^2}{b^2}\right)A_{lm}(z,\tau)\\&\quad=\frac{\partial^2 A_{lm}}{\partial z^2}+\frac{4}{ab}Q\,\delta(z-z_0)\sin\frac{l\pi x_0}{a}\sin\frac{m\pi y_0}{b}\,\delta(\tau).\end{aligned}\right\}\qquad(8.2.21\,\text{d})$$

The solution of this equation, which can be obtained by the method described in connection with the plane source, is

$$A_{lm}(z,\tau)=\frac{4Q}{ab}\,\frac{e^{-\pi^2\left(\frac{l^2}{a^2}+\frac{m^2}{b^2}\right)\tau}}{\sqrt{4\pi\tau}}\sin\frac{l\pi x_0}{a}\sin\frac{m\pi y_0}{b}\,e^{-\frac{(z-z_0)^2}{4\tau}}. \qquad (8.2.21\,\text{e})$$

Finally,

$$\left.\begin{aligned}q(x,y,z,\tau)=\frac{4Q}{ab\sqrt{4\pi\tau}}\sum_{l,m=1}^{\infty}&\left(e^{-\pi^2\left(\frac{l^2}{a^2}+\frac{m^2}{b^2}\right)\tau}\sin\frac{l\pi x_0}{a}\sin\frac{m\pi y_0}{b}\right)\times\\&\times\left(e^{-\frac{(z-z_0)^2}{4\tau}}\sin\frac{l\pi x}{a}\sin\frac{m\pi y}{b}\right).\end{aligned}\right\}\qquad(8.2.21\,\text{f})$$

Again the sum converges rapidly for large τ; when $\pi^2\left(\frac{1}{a^2}+\frac{1}{b^2}\right)\tau\gtrsim 1$, q is simply given by

$$q(x,y,z,\tau)\sim e^{-\frac{(z-z_0)^2}{4\tau}}\sin\frac{\pi x}{a}\sin\frac{\pi y}{b}.$$

Point Source in a Finite Pile. Let the effective lengths of the pile's sides be a, b, and c. The coordinate axes are oriented as shown in Fig. 6.2.3, and the source is located at x_0, y_0, z_0. By developing q in a Fourier series in all three coordinates, we can obtain the following solution

$$\left.\begin{aligned}q=\frac{8Q}{abc}\sum_{l,m,n=1}^{\infty}&\left(e^{-\pi^2\left(\frac{l^2}{a^2}+\frac{m^2}{b^2}+\frac{n^2}{c^2}\right)\tau}\sin\frac{l\pi x_0}{a}\sin\frac{m\pi y_0}{b}\sin\frac{n\pi z_0}{c}\right)\times\\&\times\left(\sin\frac{l\pi x}{a}\sin\frac{m\pi y}{b}\sin\frac{n\pi z}{c}\right).\end{aligned}\right\}\qquad(8.2.22)$$

8.2.4. The Physical Significance of the Fourier Transform of the Slowing-Down Density

In Sec. 8.2.3, we used the method of Fourier transformation as a mathematical tool to solve the age equation. Now we shall see that the Fourier transform of the slowing-down density due to a point source in an infinite medium has an important physical significance. This Fourier transform was introduced in Eq. (8.2.19b), which we shall first write in a more convenient form. Since in an infinite medium the slowing-down density depends only on the distance r from the source, we can use polar coordinates in the integration:

$$f(\omega,\tau)=\int_0^\infty\int_0^{2\pi}\int_0^{\pi} q(r,\tau)\,e^{-i\omega r\cos\vartheta}\,r^2\,dr\,d\varphi\sin\vartheta\,d\vartheta. \qquad (8.2.23)$$

The integrations over φ and ϑ are trivial; when they are carried out we obtain

$$f(\omega, \tau) = \int_0^\infty q(r, \tau) \frac{\sin \omega r}{\omega r} 4\pi r^2 \, dr. \tag{8.2.24}$$

This definition of the Fourier transformation is identical with that of Eq. (8.2.19b); there we wrote $f(\boldsymbol{\omega}, \tau)$, but it turned out later that f depended only on $\boldsymbol{\omega}^2 = \omega^2$. When we use polar coordinates in ω-space, Eq. (8.2.19c) gives the following expression for the inverse transformation

$$q(r, \tau) = \frac{1}{(2\pi)^3} \int_0^\infty f(\omega, \tau) \frac{\sin \omega r}{\omega r} 4\pi \omega^2 \, d\omega \tag{8.2.25}$$

[cf. also Eq. (8.2.20g)]. For the sake of simplicity, let us consider a non-absorbing medium and take the source to be of unit strength. Then $f(\omega, \tau) = e^{-\omega^2 \tau}$ results from Eq. (8.2.19e).

We can apparently write the general solution of the age equation in a finite medium containing arbitrary sources as follows:

$$q(\boldsymbol{r}, \tau) = \sum_n S_n \cdot R_n(\boldsymbol{r}) \cdot e^{-B_n^2 \tau}. \tag{8.2.26}$$

Here B_n^2 are the eigenvalues and R_n the eigenfunctions of the equation

$$\nabla^2 R(\boldsymbol{r}) + B^2 R(\boldsymbol{r}) = 0 \tag{8.2.27}$$

with the boundary condition $R = 0$ on the effective surface of the medium. The S_n are the coefficients in the expansion of the source distribution in these eigenfunctions:

$$q(\boldsymbol{r}, \tau = 0) = S(\boldsymbol{r}) = \sum_n S_n R_n(\boldsymbol{r}). \tag{8.2.28}$$

Our previous results for the finite pile [Eq. (8.2.22)] and the sphere [Eq. (8.2.20f)] are special cases of these general equations. Thus, we can describe the slowing-down process in a finite, homogeneous medium in a purely formal way as follows: The source distribution is composed of individual components $S_n \cdot R_n(\boldsymbol{r})$. During the slowing-down process, these individual components ("modes") diminish since the neutrons can leak out of the medium. The decrease of a component n during slowing down is given by the factor $e^{-B_n^2 \tau} = f(\omega = B_n, \tau)$. The Fourier transform of the slowing-down density due to a point source in an infinite medium $f(B_n, \tau)$ is thus the probability that a neutron of component n does not escape from the finite medium during moderation to Fermi age τ (non-escape probability).

It can now be shown that this result is much more general than the above considerations indicate (cf. the proof of the second fundamental theorem of reactor theory in Weinberg and Wigner): Let us consider an arbitrary homogeneous scattering medium, i.e., let us drop the assumptions of age theory. In this general case, the Fourier transform of the slowing-down density due to a point source in the infinite medium is defined by

$$f(\omega, E) = \int_0^\infty q(r, E) \frac{\sin \omega r}{\omega r} 4\pi r^2 \, dr, \tag{8.2.24a}$$

i.e., τ is replaced by the energy variable. Then, $f(\omega, E)$ is again the non-escape probability, i.e., in the finite medium with an arbitrary source distribution $S(\boldsymbol{r})$

$$q(\boldsymbol{r}, E) = \sum S_n R_n(\boldsymbol{r}) f(B_n, E) \tag{8.2.29}$$

holds; here R_n and B_n are eigenfunctions and eigenvalues of Eq. (8.2.27) with the boundary condition $R=0$ on the extrapolated surface of the medium. For Eq. (8.2.29) to hold rigorously there must be a single extrapolated surface on which the slowing-down density vanishes at all energies. Because of the energy dependence of the scattering cross section, this condition is generally not fulfilled in practice, and calculations are carried out with a suitable average of the extrapolated endpoint. Eq. (8.2.29) is then only valid at points more than several mean free paths from the surface; if the dimensions of the medium are only a few mean free paths, it is not applicable at all.

Eqs. (8.2.24a) and (8.2.29) are of great practical importance since they permit the calculation of the slowing-down density in any finite homogeneous medium without use of a specific slowing-down model. To this end, the slowing-down density due to a point source in the infinite medium is determined experimentally (cf. Chapter 16) and the Fourier transform according to Eq. (8.2.24a) is obtained by graphical or numerical integration. After normalization, the resulting non-escape probabilities are inserted into Eq. (8.2.29), which immediately yields $q(\boldsymbol{r}, E)$. As one can easily show, this procedure works even if the scattering medium is absorbing.

There is a simple connection, which we shall derive in the interest of completeness, between the Fourier transform of the slowing-down density and the moments $\overline{r_E^{2s}}$. If we develop the term $\sin \omega r$ in Eq. (8.2.24a) in a TAYLOR's series around $r=0$, we obtain

$$f(\omega, E) = \sum_{s=0}^{\infty} \frac{(-1)^s}{(2s+1)!} \overline{r_E^{2s}} \omega^{2s} = 1 - \frac{\overline{r_E^2}}{6} \omega^2 + \frac{\overline{r_E^4}}{120} \omega^4 - \cdots. \tag{8.2.30}$$

Here we have made use of the fact that the source is of unit strength and $\Sigma_a = 0$. This expression is exact, i.e., it does not depend on the assumptions of age theory. In age theory, on the other hand

$$f(\omega, \tau) = e^{-\omega^2 \tau} = \sum_{s=0}^{\infty} \frac{(-1)^s}{s!} \tau^s \omega^{2s}. \tag{8.2.31}$$

By comparing coefficients we find that

$$\overline{r_E^{2s}} = \frac{(2s+1)!}{s!} \tau^s(E) \tag{8.2.32}$$

a result that we derived directly earlier [Eq. (8.2.19h)].

8.2.5. Inclusion of the First Flight in Age Theory

In comparing the mean squared slowing-down distances calculated directly [Eq. (8.1.14)] and with the help of age theory [Eq. (8.2.11b)], we saw that age theory did not include the so-called "first-flight correction". This omission stems from the fact that age theory does not describe the energy spectrum of the neutrons properly. Let us consider the space-integrated neutron field in an infinite, non-absorbing medium containing a unit source at $u=0$. The collision density is given by

$$\psi(u)=\int\limits_{u-\ln(1/\alpha)}^{u}\psi(u')\frac{e^{-(u-u')}}{1-\alpha}\,du'+\delta(u) \tag{8.2.33}$$

[cf. Eq. (7.3.9a)]. The solution of this equation for $u>0$ is the Placzek function discussed in Sec. 7.3, whose asymptotic value is $1/\xi$. At $u=0$, the collision density has a δ-function singularity since all neutrons undergo their first collision at $u=0$.

In the age approximation the equation for the collision density is

$$\xi\frac{\partial\psi}{\partial u}=\delta(u) \tag{8.2.34}$$

[cf. Eqs. (8.2.14a) and (8.2.13a)]. The solution of this equation is $1/\xi$ for all $u>0$. Thus not only are the "Placzek wiggles" not described, but also the singularity at $u=0$. In other words, age theory does not include the "virgin" neutrons that have made no collision, and for this reason the first-flight correction is not included in Eq. (8.2.11b).

According to Flügge, age theory can be improved by simply taking into account the spatial distribution of the virgin neutrons. If $\Sigma_s(E_Q)$ is the scattering cross section of neutrons that have the source energy, then the probability that they make their first collision at a distance r from the source is

$$W(r)\,dr=\Sigma_s(E_Q)\frac{e^{-\Sigma_s(E_Q)r}}{4\pi r^2}4\pi r^2\,dr. \tag{8.2.35}$$

In describing the slowing-down density with the age equation, we now use $S(\boldsymbol{r})=Q\cdot W(|\boldsymbol{r}-\boldsymbol{r}_0|)$ as the source density instead of the original point source at $\boldsymbol{r}_0$. For a point source at the origin of an infinite medium there then results

$$q(r,\tau)=\frac{Q}{(4\pi\tau)^{\frac{3}{2}}}\Sigma_s(E_Q)\int\frac{e^{-\Sigma_s(E_Q)r'}}{4\pi r'^2}e^{-\frac{(\boldsymbol{r}-\boldsymbol{r}')^2}{4\tau}}\,d\boldsymbol{r}'. \tag{8.2.36}$$

Integration leads to the formula

$$q(r,\tau)=\frac{Q\Sigma_s(E_Q)}{8\pi\sqrt{\tau}\cdot r}e^{-\frac{r^2}{4\tau}}\int\limits_{\Sigma_s(E_Q)\sqrt{\tau}-\frac{r}{2\sqrt{\tau}}}^{\Sigma_s(E_Q)\sqrt{\tau}+\frac{r}{2\sqrt{\tau}}}t^2[1-\operatorname{erf}(t)]\,dt \tag{8.2.37}$$

where

$$\operatorname{erf}(t)=\frac{2}{\sqrt{\pi}}\int\limits_0^t e^{-y^2}\,dy.$$

If we calculate the mean squared slowing-down distance using Eq. (8.2.37), we obtain just Eq. (8.1.14) (without the factor $\frac{1}{1-2/3A}$ in the first-flight correction).

We might suppose that the variant of age theory just described is suitable at least in a rough first approximation to describe the slowing down of neutrons at large distances from the source, but that problem is actually more complicated, as we shall see in Chapter 16.

8.3. Other Approximate Methods of Calculating the Slowing-Down Density

8.3.1. The Selengut-Goertzel Method

A serious limitation of age theory is that it can only be applied to heavy moderators, for it presupposes that the collision density can be described over a single collision interval by only two terms in its Taylor series [cf. Eq. (8.2.14a)]. If a moderator contains hydrogen, a neutron can jump over an arbitrarily large lethargy interval in a single collision, and the simple Taylor series approximation is surely inapplicable. In the case of a hydrogenous mixture, SELENGUT and GOERTZEL recommend the following approximation: Let us assume, as we did in Sec. 8.2.1, that FICK's law still connects the current and the flux gradient, so that we shall be able to start from an energy-dependent diffusion equation:

$$\frac{\partial q(\boldsymbol{r}, u)}{\partial u} + \Sigma_a \Phi(\boldsymbol{r}, u) - D \nabla^2 \Phi(\boldsymbol{r}, u) = S(\boldsymbol{r}, u). \tag{8.2.4}$$

(Use of FICK's law means, as we have seen in Sec. 8.2.2, not only a P_1-approximation but also neglect of the correlation between the scattering angle and the energy change in a collision.) The slowing-down density is now decomposed into a part $q_H(\boldsymbol{r}, u)$ that accounts for the neutrons that have made their last collision with a hydrogen nucleus and a part $q_A(\boldsymbol{r}, u)$ that accounts for the neutrons that have made their last collision with a heavy nucleus:

$$q(\boldsymbol{r}, u) = q_H(\boldsymbol{r}, u) + q_A(\boldsymbol{r}, u).$$

Here

$$q_A(\boldsymbol{r}, u) = \int\limits_{u-\ln 1/\alpha}^{u} \Sigma_{sA} \Phi(\boldsymbol{r}, u') \frac{e^{-(u-u')} - \alpha}{1-\alpha} du' \approx \xi \Sigma_{sA} \Phi(\boldsymbol{r}, u), \tag{8.3.1}$$

i.e., we use the age approximation for moderation by the heavy nuclei [cf. Eq. (8.2.5)]. The moderation by hydrogen, on the other hand, is treated exactly (cf. Sec. 7.2.2):

$$q_H(\boldsymbol{r}, u) = \int\limits_0^u \Sigma_{sH} \Phi(\boldsymbol{r}, u') e^{-(u-u')} du'. \tag{8.3.2}$$

Eq. (8.2.4), together with Eqs. (8.3.1) and (8.3.2), forms the basis of the Selengut-Goertzel method. In the quite special case that we can neglect the contribution to the slowing-down density from collisions with heavy nuclei as well as the absorption of neutrons, we can combine Eqs. (8.2.4) and (8.3.2) into a single simple differential equation for q[1]:

$$\frac{\partial q}{\partial u} = \frac{D}{\Sigma_{0H}} \nabla^2 \left(q + \frac{\partial q}{\partial u}\right) + S(\boldsymbol{r}, u). \tag{8.3.3}$$

If we assume the neutrons are produced by a monoenergetic point source in an infinite medium, the Fourier transform of the slowing-down density satisfies the

[1] We can always do this, but in general the resulting equation is very complicated. We obtain Eq. (8.3.3) by differentiating Eq. (8.3.2) with respect to u, solving for $\Phi(\boldsymbol{r}, u)$, and substituting the latter into Eq. (8.2.4).

following equation:

$$\left.\begin{aligned} \frac{\partial f(\omega,u)}{\partial u} &= -\frac{D\omega^2}{\Sigma_{sH}}\left(f(\omega,u)+\frac{\partial f(\omega,u)}{\partial u}\right)+\delta(u) \\ \text{or} \qquad & \\ \frac{\partial f(\omega,u)}{\partial u}\left(1+\frac{D\omega^2}{\Sigma_{sH}}\right) &= -\frac{D\omega^2}{\Sigma_{sH}} f(\omega,u)+\delta(u). \end{aligned}\right\} \tag{8.3.4}$$

The solution of this equation is

$$f(\omega,u)=\frac{e^{-\omega^2\int\limits_0^u \frac{D(u')}{\omega^2 D(u')+\Sigma_{sH}}du'}}{\left(1+\frac{D\omega^2}{\Sigma_{sH}}\right)_{u=0}}. \tag{8.3.5}$$

In order to obtain the mean squared slowing-down distance from this expression, we must calculate $(\partial^2 f/\partial\omega^2)_{\omega^2=0}$ [cf. Sec. 8.2.4, Eq. (8.2.30)]:

$$\frac{\overline{r_E^2}}{6}=-\left(\frac{\partial^2 f}{\partial\omega^2}\right)_{\omega=0}=\left(\frac{D}{\Sigma_{sH}}\right)_{u=0}+\int\limits_0^u \frac{D(u')}{\Sigma_{sH}(u')}du'. \tag{8.3.6}$$

If we now remember that $D=1/3\,[\Sigma_{sA}(1-2/3A)+\frac{1}{3}\Sigma_{sH}]$ we then find that

$$\frac{\overline{r_E^2}}{6}=\tau_E=\frac{1}{\Sigma_{sH}[\Sigma_{sH}+3\Sigma_{sA}(1-2/3A)]_{u=0}}+\int\limits_0^u \frac{du'}{\Sigma_{sH}[\Sigma_{sH}+3\Sigma_{sA}(1-2/3A)]}. \tag{8.3.7}$$

The first term is a kind of first-flight correction; it is certainly too large since in the approximation being considered here only first collisions with protons are taken into account. Eq. (8.3.7) nevertheless is a very useful approximation for τ_E in water and other hydrogenous compounds.

While age theory represents a simple approximation for slowing down by heavy moderators ($A \gg 1$) and the Selengut-Goertzel method a simple approximation for slowing down by hydrogenous moderators, neither of these methods is particularly good in the region in between, e.g., for moderation by deuterium ($A=2$). In the latter case, one can proceed by the method of GOERTZEL and GREULING, which we studied in Sec. 7.3 in connection with space-independent slowing-down problems (cf. MACK and ZWEIFEL).

8.3.2. The Multigroup Method

In all the methods of calculating the spatial distribution of moderated neutrons heretofore discussed, the energy or the lethargy was treated as a continuous variable and we tried to find analytic solutions. A different kind of method that is also quite fruitful is to divide the energy or lethargy range into intervals or groups and imagine that the neutrons in each of these groups diffuse like monoenergetic neutrons; moderation by elastic or inelastic scattering is then a process that transports neutrons from one group to another. In this method, we must solve the diffusion problem in each group separately; however, we can apply all the approximate methods for the treatment of one-group transport problems hitherto developed, in particular the P_N- and S_N-methods.

The multigroup method is the only one with which we can treat slowing down in inhomogeneous media, a problem that has proved intractable when attacked with the comparatively simple age theory. Also, a practical description of the neutron field is possible only with the multigroup method in situations where the slowing down is predominantly due to inelastic scattering.

In the following, we shall restrict ourselves to the discussion of the so-called multigroup diffusion theory, i.e., that theory in which the energy-dependent diffusion equation is solved by subdivision of the energy range into groups; we shall furthermore restrict ourselves to a medium which slows neutrons down only by elastic scattering. We begin with the diffusion equation

$$\frac{\partial q(\mathbf{r}, u)}{\partial u} + \Sigma_a \Phi(\mathbf{r}, u) - D \nabla^2 \Phi(\mathbf{r}, u) = S(\mathbf{r}, u) \tag{8.2.4}$$

and divide the entire lethargy range $(0, u_{\max})$ into n lethargy intervals (groups). By integration of Eq. (8.2.4) over each of these groups, we obtain the following system of coupled differential equations:

$$q(\mathbf{r}, u_\nu) - q(\mathbf{r}, u_{\nu-1}) + \int_{u_{\nu-1}}^{u_\nu} \Sigma_a \Phi(\mathbf{r}, u)\, du - \int_{u_{\nu-1}}^{u_\nu} D \nabla^2 \Phi(\mathbf{r}, u)\, du = \int_{u_{\nu-1}}^{u_\nu} S(\mathbf{r}, u)\, du. \tag{8.3.8}$$

With the abbreviations

$$\left.\begin{aligned}
\Phi_\nu(\mathbf{r}) &= \int_{u_{\nu-1}}^{u_\nu} \Phi(\mathbf{r}, u)\, du, \\
S_\nu(\mathbf{r}) &= \int_{u_{\nu-1}}^{u_\nu} S(\mathbf{r}, u)\, du, \\
\Sigma_{a\nu} &= \frac{\int_{u_{\nu-1}}^{u_\nu} \Sigma_a \Phi(\mathbf{r}, u)\, du}{\int_{u_{\nu-1}}^{u_\nu} \Phi(\mathbf{r}, u)\, du}, \\
D_\nu &= \frac{\int_{u_{\nu-1}}^{u_\nu} D \nabla^2 \Phi(\mathbf{r}, u)\, du}{\int_{u_{\nu-1}}^{u_\nu} \nabla^2 \Phi(\mathbf{r}, u)\, du} \\
\text{and}\quad q_\nu(\mathbf{r}) &= q(\mathbf{r}, u_\nu)
\end{aligned}\right\} \tag{8.3.9}$$

we can write this system of differential equations as

$$\left.\begin{aligned}
-D_1 \nabla^2 \Phi_1 + \Sigma_{a1} \Phi_1 &= S_1 - q_1, \\
-D_\nu \nabla^2 \Phi_\nu + \Sigma_{a\nu} \Phi_\nu &= S_\nu + q_{\nu-1} - q_\nu \qquad \nu = 2, 3, \ldots, n.
\end{aligned}\right\} \tag{8.3.10}$$

In order to continue we must know the connection between the q_ν and the Φ_ν. In hydrogen this connection obviously has the form $q_\nu = \sum_{\mu=1}^{\nu} a_{\mu\nu} \Phi_\mu$ since neutrons from all the lethargy groups lying below u_ν can contribute to the slowing-down density at u_ν. However, if we use only a few, rather wide groups (e.g., $\Delta u = 3$), the probability that a neutron will jump over a group as the result of a collision with a proton is very small. For all other nuclei, this probability is zero, and we can set $q_\nu = \Sigma_{b\nu} \Phi_\nu$, where $\Sigma_{b\nu}$ is a "slowing-down" cross section.

In order to determine the group constants $\Sigma_{a\nu}$ and D_ν according to Eq. (8.3.9), we must know the variation of $\Phi(\mathbf{r}, u)$ with lethargy in each of the individual groups. Since as a rule we do not know this variation, we must make some assumption concerning it; this assumption introduces a certain arbitrariness into the method. This problem is particularly troublesome when the medium contains resonance absorbers that cause strong local flux depressions. We must then introduce effective cross sections like the resonance integrals dealt with in Chapter 7. We limit ourselves here to the simplest assumption, viz., that $\Phi(u)$ is constant in the individual groups, or in other words, that $\Phi(E) \sim 1/E$. Then

$$\left.\begin{aligned} D_\nu &= \frac{\int_{u_{\nu-1}}^{u_\nu} D(u)\,du}{u_\nu - u_{\nu-1}}, \\ \Sigma_{a\nu} &= \frac{\int_{u_{\nu-1}}^{u_\nu} \Sigma_a(u)\,du}{u_\nu - u_{\nu-1}}, \\ \Sigma_{s\nu} &= \frac{\int_{u_{\nu-1}}^{u_\nu} \Sigma_s(u)\,du}{u_\nu - u_{\nu-1}}. \end{aligned}\right\} \tag{8.3.11 a}$$

The number of collisions in group ν per cm³ and sec is $\Sigma_{s\nu}\Phi_\nu$. Since a neutron must make $N = \frac{u_\nu - u_{\nu-1}}{\xi}$ collisions before it is moderated from $u_{\nu-1}$ to u_ν, $q_\nu = \Sigma_{s\nu}\Phi_\nu/N$ and

$$\Sigma_{b\nu} = \frac{\Sigma_{s\nu}}{(u_\nu - u_{\nu-1})/\xi}. \tag{8.3.11 b}$$

Using Eqs. (8.3.11), we can now solve the group Eqs. (8.3.10). Let us now consider the special case of a monoenergetic unit point source at $\boldsymbol{r}_0 = 0$ in an infinite medium. If we assume $\Sigma_{a\nu} = 0$ and set $L_\nu^2 = \frac{D_\nu}{\Sigma_{b\nu}}$, then

$$\left.\begin{aligned} \Phi_1(\boldsymbol{r}) &= \frac{e^{-r/L_1}}{4\pi D_1 r}, \\ \Phi_2(\boldsymbol{r}) &= \int \frac{\Sigma_{b1}\,\Phi_1(\boldsymbol{r}_1)\,e^{-|\boldsymbol{r}-\boldsymbol{r}_1|/L_2}}{4\pi D_2 |\boldsymbol{r}-\boldsymbol{r}_1|}\,d\boldsymbol{r}_1 \end{aligned}\right\} \tag{8.3.12}$$

and in general

$$\Phi_\nu(\boldsymbol{r}) = \int \frac{\Sigma_{b\,\nu-1}\,\Phi_{\nu-1}(\boldsymbol{r}_{\nu-1})\,e^{-|\boldsymbol{r}-\boldsymbol{r}_{\nu-1}|/L_\nu}}{4\pi D_\nu |\boldsymbol{r}-\boldsymbol{r}_{\nu-1}|}\,d\boldsymbol{r}_{\nu-1}.$$

Here it is again advantageous to introduce the Fourier transform; in analogy with our former definitions of $f(\omega, \tau)$ and $f(\omega, E)$ we write here

$$f_\nu(\omega) = \int_0^\infty \underbrace{\Sigma_{b\nu}\,\Phi_\nu(r)}_{q_\nu(r)} \frac{\sin \omega r}{\omega r}\, 4\pi r^2\, dr. \tag{8.3.13}$$

In particular,

$$f_1(\omega) = \int_0^\infty \frac{\Sigma_{b1}}{D_1} \frac{e^{-r/L_1}}{r} \frac{\sin \omega r}{\omega r} r^2\, dr = \frac{1}{1+\omega^2 L_1^2}, \tag{8.3.14 a}$$

$$\left.\begin{aligned} f_2(\omega) &= \frac{1}{1+\omega^2 L_1^2}\,\frac{1}{1+\omega^2 L_2^2}, \\ f_n(\omega) &= \prod_{\nu=1}^{n} \frac{1}{1+\omega^2 L_\nu^2}. \end{aligned}\right\} \tag{8.3.14 b}$$

Eq. (8.3.14b) can easily be proven with the help of the convolution theorem for Fourier transforms (cf. WEINBERG and NODERER). Now we can calculate the neutron flux in each group by inverting the transforms in Eq. (8.3.14b):

$$\Sigma_{b\,\nu}\,\Phi_\nu(r) = \sum_{i=1}^{\nu} \frac{L_i^{2(\nu-1)}}{\prod\limits_{\mu \neq i}(L_i^2 - L_\mu^2)}\,\frac{e^{-r/L_i}}{4\pi L_i^2 r}\,. \tag{8.3.15}$$

In view of the fact that $\overline{r_\nu^2}/6 = -\left(\partial^2 f_\nu(\omega)/\partial\omega^2\right)_{\omega=0}$, the mean squared slowing-down distance is given by

$$\frac{\overline{r_\nu^2}}{6} = L_1^2 + L_2^2 + \cdots + L_\nu^2\,. \tag{8.3.16}$$

It is easy to see that in the limit $u_\nu - u_{\nu-1} \to 0$, $\nu \to \infty$, $\overline{r_\nu^2}/6$ approaches the Fermi age. Thus age theory is the limit of multigroup diffusion theory as the number of groups becomes infinite. The transition to this limit is naturally not valid when the medium contains hydrogen, for then the original assumption that no neutron could jump over a group is violated.

In practice one usually proceeds by (i) experimentally determining the slowing-down density near a point source in an infinite medium, (ii) Fourier transforming it, and (iii) trying to fit the experimental value of $f(\omega, E)$ thus determined with a product $\prod\limits_i \frac{1}{1+\omega^2 L_i^2}$ containing as few factors as possible. The inverted transform is frequently called the "synthetic" slowing-down density. In many cases, it turns out that a good approximation is achieved with relatively few groups. For many practical purposes, a simple two-group theory, in which all the fast neutrons are lumped together in one group (the thermal neutrons are the second group), is a sufficiently accurate first approximation. In this theory, we set $L_1 = \sqrt{\overline{r_E^2}/6}$ (with the energy E in the neighborhood of thermal energy), i.e., we demand that the synthetic slowing-down density correctly reproduce the measured value of $\overline{r_E^2}$. In this case, the treatment of heterogeneous media composed of different materials is particularly simple, since we need only require continuity of flux and current in each of the two groups at each interface.

8.3.3. The B_N-Approximation

We shall now describe another method for the treatment of slowing-down problems that resembles the P_N-method but yields much more accurate results for an equivalent expenditure of computational labor. Let us consider, as in the derivation of the moments method in Sec. 8.1.3, an infinite medium with a plane source of fast neutrons at $x=0$. Instead of the differential flux $F(x, \mathbf{\Omega}, u)$, let us consider its Fourier transform, which is defined by

$$f(\omega, \mathbf{\Omega}, u) = \int_{-\infty}^{+\infty} F(x, \mathbf{\Omega}, u)\, e^{-i\omega x}\, dx\,. \tag{8.3.17}$$

After multiplication by $e^{-i\omega x}$ and integration over all x, the transport Eq. (8.1.24) becomes

$$\left.\begin{aligned}\left(i\omega\cos\vartheta + \Sigma_t(u)\right) f(\omega, \mathbf{\Omega}, u) &= \frac{1}{2}\int_{u-\ln 1/\alpha}^{u} \sum_{l=0}^{\infty}(2l+1)\,\Sigma_{sl}(u'\to u)\,P_l(\cos\vartheta)\times\\ &\times\int_0^{\pi} f(\omega, \mathbf{\Omega}', u')\,P_l(\cos\vartheta')\sin\vartheta'\,d\vartheta'\,du' + \frac{Q}{4\pi}\,\delta(u)\,.\end{aligned}\right\} \tag{8.3.18}$$

Now, as in the treatment of the ordinary transport equation, $f(\omega, \mathbf{\Omega}, u)$ is expanded in Legendre polynomials and the expansion truncated after the N-*th* term:

$$f(\omega, \mathbf{\Omega}, u) = \frac{1}{4\pi} \sum_{l=0}^{N} (2l+1) f_l(\omega, u) P_l(\cos\vartheta), \tag{8.3.19a}$$

$$f_l(\omega, u) = \int_{-\infty}^{\infty} F_l(x, u) e^{-i\omega x} dx. \tag{8.3.19b}$$

Here the quantities $F_l(x, u)$ are the expansion coefficients introduced in Eq. (8.1.25). If we now multiply Eq. (8.3.18) by $P_l(\cos\vartheta)$ $(l=0, 1, \ldots, N)$ and integrate over solid angle, we obtain $N+1$ coupled differential equations that are identical with the Eq. (8.1.26) except that $f_l(\omega, u)$ replaces $F_l(x, u)$ everywhere. In other words, we have obtained a P_N-approximation for the Fourier transform of the differential flux that is in no way different from the P_N-approximation for the differential flux itself. The essence of the B_N-method is that we first divide the transport equation by $i\omega\cos\vartheta + \Sigma_t(u)$ and only then multiply by $P_l(\cos\vartheta)$ and integrate. We then obtain the following system of equations:

$$f_j(\omega, u) - \frac{1}{2} \sum_{l=0}^{N} (2l+1) \int_{u-\ln 1/\alpha}^{u} \Sigma_{sl}(u' \to u) f_l(\omega, u') A_{lj} \, du' + \frac{Q\,\delta(u) A_{0j}}{2} \tag{8.3.20a}$$

with

$$A_{lj} = \int_0^{\pi} \frac{P_l(\cos\vartheta) P_j(\cos\vartheta)}{i\omega\cos\vartheta + \Sigma_t(u)} \sin\vartheta \, d\vartheta. \tag{8.3.20b}$$

These coupled integral equations can be solved, for example, by the multigroup method; once we have determined $f_j(\omega, u)$ we can find $F_j(x, u)$ and thus $F(x, \mathbf{\Omega}, u)$ by inverting the Fourier transform

$$F_j(x, u) = \frac{1}{2\pi} \int_{-\infty}^{+\infty} f_j(\omega, u) e^{i\omega x} d\omega. \tag{8.3.21}$$

$f_0(\omega, u)$ yields the flux $\Phi(x, u)$ and all the moments $\overline{x_u^{2s}}$.

The significance of the B_N-approximation procedure can be made clear as follows: Let us assume that all the $\Sigma_{sl}(u' \to u)$ vanish except $\Sigma_{s0}(u' \to u)$, i.e., that the scattering is isotropic in the laboratory system. In this case, the angular distribution of $f(\omega, \mathbf{\Omega}, u)$ is simply const.$/\big(i\omega\cos\vartheta + \Sigma_t(u)\big)$ and only one of the Eqs. (8.3.20a) — namely, that with $j=0$ — is needed to calculate $f_0(\omega, u)$ and thus to solve the problem completely[1]. On the other hand, in a normal P_N-approximation many equations must be solved for an accurate calculation of the angle-dependent flux regardless of whether the scattering is isotropic or not. Thus in the B_N-approximation, we start from the angular distribution that the neutrons would have in the case of isotropic scattering and calculate the deviations from this distribution. It is clear that this procedure converges much faster than the P_N-method. A detailed decription and numerous applications of the B_N-approximation can be found in Bethe *et al.*

---

[1] In the case of isotropic scattering, once $f_0(\omega, u)$ has been found, all the other $f_j(\omega, u)$ can be obtained from the relation

$$f_j(\omega, u) = \tfrac{1}{2} \int_{u-\ln 1/\alpha}^{u} \Sigma_{s0}(u' \to u) f_0(\omega, u') A_{0j} \, du' + \tfrac{1}{2} Q\,\delta(u)\, A_{0j}.$$

8.4. The Distribution of Thermal Neutrons Arising from a Specified Fast Neutron Source Distribution

When a neutron reaches thermal energy at the end of the slowing-down process, it diffuses as a thermal neutron until it is captured or escapes through the surface of the medium. This transition from slowing down to thermal diffusion proceeds continuously; we shall discuss it and the general problem of "thermalization" in detail in Chapter 10. In first approximation, however, we can assume a discontinuous transition, i.e., with the methods developed in the last few sections we can calculate the slowing-down density $q_{\text{th}}(\boldsymbol{r}, E)$ at an energy E just above the thermal range and use this slowing-down density as the source density in describing the diffusion of thermal neutrons with elementary diffusion theory or some other better approximation. We shall illustrate this procedure in this section by working through several examples. In doing so we shall contradict in two places assumptions we previously made, viz., (i) that the slowing-down density could be calculated as through it arose from the elastic collision of neutrons with free atoms initially at rest (which is no longer true at energies below about 1 ev), and (ii) that the source of thermal neutrons emitted neutrons already in equilibrium with the thermal distribution (cf. Chapter 6). The errors occasioned by these inconsistencies are small, as a rule, and may often be compensated by simple corrections, as we shall see later.

As an elementary limiting case, let us consider first an infinite medium in which there is a homogeneously distributed source of fast neutrons. Let the absorption during the slowing-down process both here and in all subsequent examples be negligibly small. The slowing-down density q is then independent of both space and energy; accordingly, the thermal flux Φ_{th} is also space-independent and is related to q by

$$\Phi_{\text{th}} = \frac{q}{\Sigma_a}. \tag{8.4.1}$$

Remembering that the epithermal flux per unit lethargy $\Phi_{\text{epi}} = q/\xi\Sigma_s$ is constant between 1 ev and several kev because of the constancy of $\xi\Sigma_s$, we see that

$$\frac{\Phi_{\text{th}}}{\Phi_{\text{epi}}} = \frac{\xi\Sigma_s}{\Sigma_a}. \tag{8.4.2}$$

With a given epithermal flux, the thermal flux is larger the larger the moderating ratio (cf. Sec. 7.1.3). This relation is very useful for many purposes.

8.4.1. The Mean Squared Slowing-Down Distance and the Migration Area

Let us again consider a point source of fast neutrons in an infinite medium. Let $\Phi_{\text{th}}(r)$ be the thermal flux it produces. We now ask for the mean squared distance from the source at which a thermal neutron is absorbed. It is

$$\overline{r^2} = \frac{\int\limits_0^\infty r^2 \Sigma_a \Phi_{\text{th}}(r) 4\pi r^2 dr}{\int\limits_0^\infty \Sigma_a \Phi_{\text{th}}(r) 4\pi r^2 dr}. \tag{8.4.3}$$

In the case of a purely thermal source $\overline{r^2}=6L^2$, as we have already seen in Sec. 6.1.3. For a fast source we calculate $\overline{r^2}$ from the following relation:

$$\overline{r^2}=\overline{\boldsymbol{r}^2}=\overline{(\boldsymbol{r}_1+(\boldsymbol{r}-\boldsymbol{r}_1))^2}=\overline{\boldsymbol{r}_1^2}+\overline{(\boldsymbol{r}-\boldsymbol{r}_1)^2}+\overline{2\boldsymbol{r}_1\cdot(\boldsymbol{r}-\boldsymbol{r}_1)} \tag{8.4.4}$$

where $\boldsymbol{r}_1$ is the point at which the slowing-down process comes to an end and the neutron reaches thermal energy. However, $\overline{r_1^2}=\overline{r_{\text{th}}^2}=6\tau_{\text{th}}$ is the mean squared slowing-down length to thermal energy; furthermore, $\overline{(\boldsymbol{r}-\boldsymbol{r}_1)^2}=6L^2$ is the mean squared distance to absorption from a thermal point source. The mixed term in Eq. (8.4.4) vanishes upon averaging since there is no correlation between the directions of $\boldsymbol{r}_1$ and $\boldsymbol{r}-\boldsymbol{r}_1$. Finally then,

$$\overline{r^2}=6(\tau_{\text{th}}+L^2)=6M^2. \tag{8.4.5}$$

The quantity $M^2=L^2+\tau_{\text{th}}$ is called the *migration area*; occasionally $M=\sqrt{L^2+\tau_{\text{th}}}$ is called the *migration length.* Both are measures of the mean separation from a point source that a neutron achieves in the course of its life history.

8.4.2. Point Source in an Infinite Medium

We shall calculate $\Phi_{\text{th}}(r)$ for the following two forms of the slowing-down density:

Age theory:

$$q_{\text{th}}(r)=\frac{Q}{(4\pi\tau_{\text{th}})^{\frac{3}{2}}}\,e^{-r^2/4\tau_{\text{th}}}. \tag{8.4.6a}$$

Multigroup diffusion theory with one fast group:

$$q_{\text{th}}(r)=\frac{Q}{4\pi\tau_{\text{th}}}\,\frac{e^{-r/\sqrt{\tau_{\text{th}}}}}{r}. \tag{8.4.6b}$$

In both cases $\overline{r_{\text{th}}^2}=6\tau_{\text{th}}$.

We can calculate the thermal flux easily with the help of the *spherical-shell diffusion kernel* that was introduced in Sec. 6.2.2. The number of neutrons that become thermal in the spherical shell between r' and $r'+dr'$ is $4\pi r'^2\,q_{\text{th}}(r')\,dr'$. The thermal neutron flux is then given by

$$\Phi_{\text{th}}(r)=\int_0^\infty q_{\text{th}}(r')\,G_{ss}(r',r)\,4\pi\,r'^2\,dr' \tag{8.4.7}$$

with G_{ss} given by Eq. (6.2.10).

Age Theory:

$$\left.\begin{aligned}\Phi_{\text{th}}(r)=\frac{Q\,e^{\tau_{\text{th}}/L^2}}{8\pi Dr}&\left\{e^{-r/L}\left[1+\operatorname{erf}\left(\frac{r}{2\sqrt{\tau_{\text{th}}}}-\frac{\sqrt{\tau_{\text{th}}}}{L}\right)\right]\right.\\&\left.-e^{r/L}\left[1-\operatorname{erf}\left(\frac{r}{2\sqrt{\tau_{\text{th}}}}+\frac{\sqrt{\tau_{\text{th}}}}{L}\right)\right]\right\}\end{aligned}\right\} \tag{8.4.8a}$$

where $\operatorname{erf}(x)$ is the error integral [cf. Eq. (8.2.37)].

Group Theory:

$$\Phi_{\text{th}}(r)=\frac{QL^2}{4\pi Dr}\left[\frac{e^{-r/L}}{L^2-\tau_{\text{th}}}-\frac{e^{-r/\sqrt{\tau_{\text{th}}}}}{L^2-\tau_{\text{th}}}\right]. \tag{8.4.8b}$$

It is instructive to consider the behavior of the flux at large distances from the source. According to Eq. (8.4.8b), the flux at large distances is given by

$$\Phi_{\text{th}}(r)=\frac{Q L^2}{4\pi D(L^2-\tau_{\text{th}})}\,\frac{e^{-r/L}}{r}\qquad (L^2>\tau_{\text{th}}), \tag{8.4.9a}$$

$$\Phi_{\text{th}}(r)=\frac{Q L^2}{4\pi D(\tau_{\text{th}}-L^2)}\,\frac{e^{-r/\sqrt{\tau_{\text{th}}}}}{r}\qquad (\tau_{\text{th}}>L^2). \tag{8.4.9b}$$

Thus when $L^2>\tau_{\text{th}}$, the ratio of the thermal flux to the epithermal flux increases with increasing distance from the source; at very large distances from the source we have purely thermal neutrons. This is the case in graphite and heavy water, where the phenomenon is used in constructing "thermal columns" for reactors. On the other hand, when $L^2<\tau_{\text{th}}$ the behavior of the thermal neutrons even at large distances from the source is determined by the slowing-down density, and the ratio of thermal to epithermal flux approaches the constant value

$$\frac{\Phi_{\text{th}}}{\Phi_{\text{epi}}}=\frac{\xi\Sigma_s}{\Sigma_a}\,\frac{1}{1-L^2/\tau_{\text{th}}}. \tag{8.4.9c}$$

This is usually the case in ordinary water if the source neutrons have sufficiently high energies (a few Mev); clearly purely thermal neutrons cannot be produced.

With the help of age theory, we obtain

$$\Phi_{\text{th}}(r)=\frac{Q}{4\pi D}\,e^{\tau/L^2}\,\frac{e^{-r/L}}{r} \tag{8.4.10}$$

at large distances from the source whether $\tau>L^2$ or $<L^2$. In other words, in age theory, the thermal neutrons always predominate at large distances from the source. This conclusion is incorrect since age theory is not valid at large distances from the source.

8.4.3. Various Arrangements of Either Point or Plane Sources

We limit ourselves here to giving the thermal flux in several typical situations. The slowing-down density will be gotten from the solutions to the age equation obtained in Sec. 8.2.3. The thermal flux is calculated with the method of diffusion kernels, using the solutions of the diffusion equation obtained in Secs. 6.2 and 6.3. In finite media, a single effective surface that is the same for both thermal and non-thermal neutrons is assumed; such an assumption is only a very rough approximation.

Plane Source in an Infinite Medium:

$$\left.\begin{aligned}\Phi_{\text{th}}(z)=\frac{Q L e^{\tau_{\text{th}}/L^2}}{4D}&\left\{e^{-z/L}\left[1+\operatorname{erf}\left(\frac{z}{2\sqrt{\tau_{\text{th}}}}-\frac{\sqrt{\tau_{\text{th}}}}{L}\right)\right]+\right.\\&\left.+e^{z/L}\left[1-\operatorname{erf}\left(\frac{z}{2\sqrt{\tau_{\text{th}}}}+\frac{\sqrt{\tau_{\text{th}}}}{L}\right)\right]\right\}.\end{aligned}\right\} \tag{8.4.11}$$

Point Source at the Center of a Sphere of Radius R:

$$\Phi_{\text{th}}(r)=\frac{Q}{2\Sigma_a R^2}\sum_{l=1}^{\infty} l\,\frac{1}{1+\left(\frac{l^2\pi^2}{R^2}L^2\right)}\,e^{-\frac{l^2\pi^2}{R^2}\tau}\sin\frac{l\pi r}{R}. \tag{8.4.12}$$

Point Source at x_0, y_0, and z_0 in an Infinitely Long Pile:

$$\left.\begin{aligned}\Phi_{\text{th}}(x, y, z) &= \frac{Q e^{\tau/L^2}}{abD} \sum_{l,m=1}^{\infty} \left(\sin\frac{l\pi x_0}{a} \sin\frac{m\pi y_0}{b}\, e^{-\pi^2\left(\frac{l^2}{a^2}+\frac{m^2}{b^2}\right)\tau}\right) \times \\ &\times L_{lm}\left(e^{-\frac{z-z_0}{L_{lm}}}\left[1+\text{erf}\left(\frac{z-z_0}{2\sqrt{\tau}} - \frac{\sqrt{\tau}}{L_{lm}}\right)\right] + e^{\frac{z-z_0}{L_{lm}}}\left[1-\text{erf}\left(\frac{z-z_0}{2\sqrt{\tau}} + \frac{\sqrt{\tau}}{L_{lm}}\right)\right]\right) \times \\ &\times \sin\frac{l\pi x}{a} \sin\frac{m\pi y}{b}.\end{aligned}\right\} \quad (8.4.13)$$

Chapter 8: References

General

AMALDI, E.: loc. cit., especially § 71—80

DAVISON, B.: Neutron Transport Theory, Oxford: Clarendon Press, 1957, especially Part IV: Slowing-Down Problems.

GOLDSTEIN, H.: Fundamental Aspects of Reactor Shielding; Reading: Addison-Wesley, 1959, especially Chapter 6: Calculations of Fast Neutron Penetration.

MARSHAK, R. E.: The Slowing Down of Neutrons, Rev. Mod. Phys. **19**, 185 (1947).

SNEDDON, J. N.: Fourier Transforms, New York-Toronto-London: McGraw-Hill Co., 1951, especially Chapter VI: Slowing Down of Neutrons in Matter.

WEINBERG, A. M., and L. C. NODERER: Theory of Neutron Chain Reactions, AECD 3471 (1951), especially Chapter III: Slowing-Down of Neutrons.

WEINBERG, A. M., and E. P. WIGNER: The Physical Theory of Neutron Chain Reactors, Chicago: The University of Chicago Press 1958, especially Chapter XI: Diffusion and Thermalization of Fast Neutrons.

Special[1]

BLANCHARD, C. H.: Nucl. Sci. Eng. **3**, 161 (1953).
FERMI, E.: Ricerca Scientifica **7**, 13 (1936).
GOLDSTEIN, H., and J. CERTAINE: Nucl. Sci. Eng. **10**, 16 (1961).
HORWAY, G.: Phys. Rev. **50**, 897 (1936).
PLACZEK, G.: Phys. Rev. **69**, 423 (1946).
VOLKIN, H. C.: J. Appl. Phys. **26**, 127 (1955).
} Calculation of the Mean Squared Slowing-Down Length.

FERMI, E.: (Ed. J. G. BECKERLEY), AECD-2664 (1951), especially Chapter VI: The Slowing-Down of Neutrons.
FLÜGGE, S.: Phys. Z. **44**, 493 (1943).
WALLACE, P. R., and J. LECAINE: AECL-336 (1943).
} Age Theory.

HURWITZ, H., and P. F. ZWEIFEL: J. Appl. Phys. **26**, 923 (1955).
SIMON, A.: ORNL-2098 (1956).
} Selengut-Goertzel Approximation.

LEVINE, M. M. et al.: Nucl. Sci. Eng. **7**, 14 (1960).
MACK, R. J., and P. F. ZWEIFEL: Nucl. Sci. Eng. **7**, 144 (1960).
} Goertzel-Greuling Approximation.

BETHE, H. A., L. TONKS, and H. HURWITZ: Phys. Rev. **80**, 11 (1950) (B_N-Method).

EHRLICH, R., and H. HURWITZ: Nucleonics **12**, No 2, 23 (1954).
MANDL, M. E., and J. HOWLETT: Geneva 1955 P/430, Vol. 5, p. 433.
} Multigroup Method.

GOLDSTEIN, H., P. F. ZWEIFEL, and D. G. FOSTER: Geneva 1958 P/2375, Vol. 16, p. 379.
HURWITZ, H., and R. EHRLICH: Progr. Nucl. Energy, Ser. I, Vol. 1, p. 343 (1956).
WILKINS, J. E., R. L. HELLENS, and P. F. ZWEIFEL: Geneva 1955 P/597, Vol. 5, p. 62.
} Various Methods of Calculating Slowing Down in H_2O.

HOLTE, G.: Arkiv Fysik **2**, 523 (1951); **3**, 209 (1951); 8, 165 (1953).
SPENCER, L. V., and U. FANO: Phys. Rev. **81**, 464 (1951).
VERDE, M., and G. C. WICK: Phys. Rev. **71**, 852 (1947).
} Neutron Distribution at Large Distances from the Source.

[1] Cf. footnote on p. **53**.

9. Time Dependence of the Slowing-Down and Diffusion Processes

In this chapter, we shall study the behavior of neutron fields with non-stationary sources. In doing so we shall round out our discussion of the diffusion and slowing-down processes and in particular prepare ourselves to understand those important methods of measurement that employ non-stationary sources (these methods will be discussed in Chapter 18). First we shall consider the time-dependent slowing-down process in the absence of diffusion, then the time-dependent diffusion process in the absence of slowing down, and finally the space-time distribution during moderation, though only in the age approximation. We shall almost always assume a very short pulse of neutrons as the source $(S(t) \sim \delta(t))$. In practice, this is the most important case. Also, with suitable normalization the resulting $\Phi(E, t)$ can be considered as the probability that a neutron produced at time zero has the energy E at time t. Thus the life history of an "average" neutron can be read directly from the solution $\Phi(E, t)$. If we integrate the time-dependent solution over all t, we must again find the result we obtained earlier for a stationary source. In Sec. 9.4, we shall consider, in addition, the case in which the neutron sources are periodic in time.

9.1. Slowing Down in Infinite Media

For simplicity, let us assume that the medium is non-absorbing. The flux $\Phi(u, t)$ then satisfies the equation

$$\frac{1}{v}\frac{\partial \Phi(u,t)}{\partial t} = -\Sigma_s \Phi + \int_{u-\ln 1/\alpha}^{u} \Sigma_s \Phi(u') \frac{e^{-(u-u')}}{1-\alpha} du' + \delta(u)\,\delta(t). \tag{9.1.1}$$

Here we have assumed a monoenergetic, spatially homogeneous unit source and have supposed that the slowing down is due to elastic scattering that is isotropic in the C-system. A simple mathematical artifice that permits us to continue our treatment of this equation is to *Laplace transform* it with respect to t. The Laplace transform of $\Phi(u, t)$ is defined by

$$\overline{\Phi}(u,s) = \int_0^\infty \Phi(u,t)\, e^{-st}\, dt. \tag{9.1.2a}$$

The inverse transform is given by

$$\Phi(u,t) = \frac{1}{2\pi i} \int_{\sigma-i\infty}^{\sigma+i\infty} \overline{\Phi}(u,s)\, e^{st}\, ds. \tag{9.1.2b}$$

If we Laplace transform Eq. (9.1.1), it becomes

$$\left(\Sigma_s + \frac{s}{v}\right) \overline{\Phi}(u,s) = \int_{u-\ln 1/\alpha}^{u} \Sigma_s \overline{\Phi}(u', s) \frac{e^{-(u-u')}}{1-\alpha} du' + \delta(u). \tag{9.1.3}$$

Formally, this is the equation for the stationary neutron flux $\Phi(u)$ in a medium with the macroscopic absorption cross section s/v. We can therefore apply the results of Chapter 7 to the further treatment of Eq. (9.1.3).

9.1.1. Slowing Down in Hydrogen

According to Sec. 7.2.2, when there is a unit source at $u=0$ the neutron flux $\Phi(u)$ in hydrogen is given by

$$\Phi(u)=\left(\frac{\Sigma_s}{\Sigma_s+\Sigma_a}\right)_{u=0}\frac{e^{-\int\limits_0^u\frac{\Sigma_a}{\Sigma_a+\Sigma_s}du'}}{\Sigma_s+\Sigma_a}+\frac{\delta(u)}{\Sigma_s+\Sigma_a}. \tag{9.1.4a}$$

Thus in hydrogen the Laplace-transformed flux is given by

$$\bar{\Phi}(u,s)=\frac{\Sigma_s+\frac{s}{v_Q}}{\left(\Sigma_s+\frac{s}{v}\right)^3}\Sigma_s+\frac{\delta(u)}{\Sigma_s+\frac{s}{v_Q}}. \tag{9.1.4b}$$

Here we have assumed that Σ_s is constant in order to be able to carry out the integral in the exponent in Eq. (9.1.4a). v_Q is the velocity of the source neutrons ($u=0$), and v is the velocity of the neutrons with lethargy u. Thus $v=v_Q e^{-u/2}$. If we now invert the transform with the help of Eq. (9.1.2b), we find (cf. MARSHAK)

$$\Phi(u,t)=\frac{v}{2}(\Sigma_s vt)^2\left[1-\frac{v}{v_Q}+\frac{2}{\Sigma_s v_Q t}\right]e^{-\Sigma_s vt}+v_Q\,\delta(u)\,e^{-\Sigma_s v_Q t}. \tag{9.1.5}$$

Here the second term describes the decay of the uncollided flux. In order to better understand the first term, let us consider the "asymptotic" region of small energies, $v\ll v_Q$. Here $\Sigma_s v_Q t\gg 1$ so that when we introduce $\Phi(v,t)=\frac{2}{v}\Phi(u,t)$ we find

$$\Phi_{As}(v,t)=(\Sigma_s vt)^2 e^{-\Sigma_s vt}=x^2e^{-x}. \tag{9.1.6}$$

The asymptotic flux depends only on the dimensionless quantity $x=\Sigma_s vt$[1]. We can read directly from Eq. (9.1.6) either the velocity distribution $\Phi(v)$ at an arbitrary time t or the time distribution $\Phi(t)$ at an arbitrary velocity v. Fig. 9.1.1 illustrates the latter distribution for some velocities. Here Σ_s corresponds to the proton density of H_2O. The validity of the asymptotic solution presupposes that the sources have energies that are very much larger than the energy at which the slowing down is being considered. Furthermore, in order that Eqs. (9.1.4) and what follows it be valid, the scattering cross section must be independent of the energy. This is true in hydrogen between 1 and about 20,000 ev.

It is of interest to form an average from Eq. (9.1.6):

$$\int\limits_0^\infty \Phi_{As}(v,t)\,dt=\frac{2}{\Sigma_s v}. \tag{9.1.7a}$$

This equation has the following interpretation. In the stationary state, the flux from a unit source is $\Phi(E)=\frac{1}{\Sigma_s E}$ (cf. Sec. 7.2.2) or $\Phi(v)=\Phi(E)\frac{dE}{dv}=\frac{2}{\Sigma_s v}$.

---

[1] In reality, $\Phi(v,t)$ must have the dimensions cm^{-3}, whereas the right-hand side of Eq. (9.1.6) is obviously dimensionless. This comes about because we have specified the source density without the dimensions cm^{-3}.

When integrated over all time, the non-stationary solution gives the same value as the stationary solution with an equivalent source. This is to be expected. The average time $\overline{t_E}$ for moderation to an energy E (velocity v) is then given by

$$\overline{t_E} = \frac{\int_0^\infty t\,\Phi_{As}(v,t)\,dt}{\int_0^\infty \Phi_{As}(v,t)\,dt} = \frac{3}{\Sigma_s v}\,. \tag{9.1.7b}$$

Since we have used the asymptotic solution in this calculation of $\overline{t_E}$, it does not depend on the source energy. If we had calculated $\overline{t_E}$ from Eq. (9.1.5), we would have obtained for $v < v_Q$

$$\overline{t'_E} = \frac{3}{\Sigma_s v}\left[1 - \frac{v}{v_Q}\right] + \frac{2}{\Sigma_s v_Q}\,. \tag{9.1.7c}$$

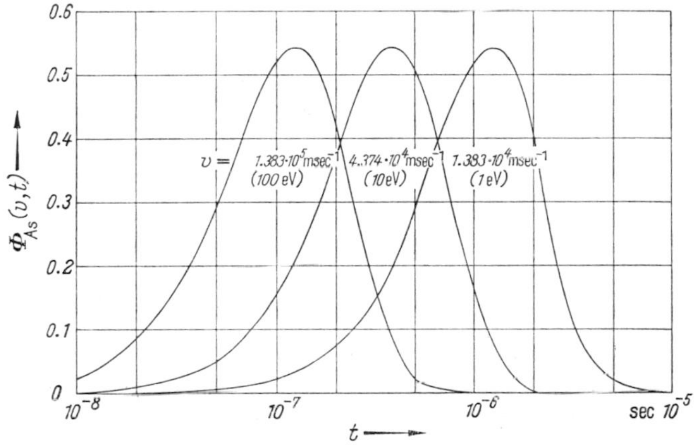

Fig. 9.1.1. The time distribution of the neutron flux in water at various velocities

When $v \ll v_Q$, the difference between $\overline{t'_E}$ and $\overline{t_E}$ is unimportant. According to Eq. (9.1.7b), $\overline{t_E}$ for moderation to 1 ev in water is 1.6 μsec; to 100 ev, 0.16 μsec; and to 10 kev, 0.016 μsec.

It is also of interest to form the quantity

$$\overline{t_E^2} = \frac{\int_0^\infty t^2\,\Phi_{As}(v,t)\,dt}{\int_0^\infty \Phi_{As}(v,t)\,dt} = \frac{12}{(\Sigma_s v)^2}\,. \tag{9.1.7d}$$

The quantity

$$\Delta^2 = \overline{t_E^2} - \overline{t_E}^2 = \frac{3}{(\Sigma_s v)^2} \tag{9.1.7e}$$

is a measure of the dispersion in time of the neutron flux at a fixed energy. In terms of $\overline{t_E}$, Δ is $\frac{\sqrt{3}}{\Sigma_s v} = \frac{\overline{t_E}}{\sqrt{3}}$. Knowledge of Δ is important for judging the resolving power of neutron spectrometers that employ pulsed sources and a hydrogeneous moderator (cf. Sec. 2.5.4).

In order to calculate the time variation of the average velocity, we must start from $n(v) = \Phi(v)/v$. Now[1]

$$\int_0^\infty n(v,t)\,dv = \int_0^\infty \frac{\Phi_{As}(v,t)}{v}\,dv = 1 \tag{9.1.8a}$$

since the source produces exactly one neutron per cm³. Furthermore,

$$\left.\begin{aligned} \bar{v}(t) &= \int_0^\infty \Phi_{As}(v,t)\,dv = \frac{2}{\Sigma_s t} \\ \overline{v^2}(t) &= \int_0^\infty v\,\Phi_{As}(v,t)\,dv = \frac{6}{(\Sigma_s t)^2} \\ \text{and}\qquad \overline{v^4}(t) &= \int_0^\infty v^3\,\Phi_{As}(v,t)\,dv = \frac{120}{(\Sigma_s t)^4}. \end{aligned}\right\} \tag{9.1.8b}$$

Thus

$$\bar{E}(t) = \frac{3m}{(\Sigma_s t)^2}, \tag{9.1.8c}$$

$\left(\text{so that for } H_2O,\ \bar{E} = \frac{1.8\,\text{ev}\cdot\mu\,\text{sec}^2}{t^2}\right)$ and

$$\overline{\left(\frac{\Delta E}{E}\right)^2} = \frac{\overline{E^2} - \bar{E}^2}{\bar{E}^2} = \frac{7}{3}. \tag{9.1.8d}$$

Clearly the energy spectrum during moderation in hydrogen is always very broad. This comes about because a neutron can lose an arbitrarily large fraction of its energy in a single collision.

9.1.2. Slowing Down in Heavy Moderators ($A \neq 1$)

To describe moderation by heavy moderators, we can again start with Eq. (9.1.3); but we immediately encounter the difficulty that there is no simple solution of the stationary slowing-down problem when $\Sigma_a \neq 0$. We must therefore use approximations, such as those with which we became acquainted in Sec. 7.3. Using the Goertzel-Greuling method, Koppel has recently obtained a very elegant approximate solution to the time-dependent slowing down problem to which we shall unfortunately be unable to devote any space here. Rather we shall limit ourselves to obtaining by the method of Marshak some useful average values [corresponding to those of Eqs. (9.1.7) and (9.1.8) for hydrogen] from which we shall be able to obtain all the important physical information we need.

For the following considerations it is best to start from the time-dependent slowing-down equation for the velocity-dependent density $n(v,t)$:

$$\frac{\partial n(v,t)}{\partial t} = -v\Sigma_s n(v,t) + \frac{2v}{1-\alpha}\int_v^{v/\sqrt{\alpha}} \Sigma_s n(v',t)\,\frac{dv'}{v'}. \tag{9.1.9}$$

[1] Here, we really should integrate only to v_Q and should use the complete solution rather than just the asymptotic solution. However, when $t \gg (\Sigma_s v_Q)^{-1}$ we can use the asymptotic solution and integrate to infinity.

We have omitted the source term here since we are seeking an asymptotic solution that is valid for velocities much smaller than the velocities of the source neutrons. We can introduce $x = v\Sigma_s t$ as a new variable in Eq. (9.1.9); the latter can then be written

$$\frac{d}{dx}\left[x\,n(x)\right] = -\,x\,n(x) + \frac{2x}{1-\alpha}\int_{x}^{x/\sqrt{\alpha}} n(x')\,\frac{dx'}{x'}\,. \tag{9.1.10}$$

Now we define the moments of n by

$$M_l = \int_0^\infty x^l\,n(x)\,dx\,. \tag{9.1.11}$$

If we multiply Eq. (9.1.10) by x^{l-1} and integrate, we obtain the following recursion formula for the moments:

$$M_l = M_{l-1}\left[\frac{l-1}{1-\dfrac{2}{l+1}\,\dfrac{1-\alpha^{\frac{l+1}{2}}}{1-\alpha}}\right]. \tag{9.1.12}$$

As normalization we take $M_0=1$. If we try to calculate M_1 from M_0 by means of this recursion formula, we encounter a slight difficulty, viz., that on the right-hand side of Eq. (9.1.12) there stands an indeterminate expression. However, if we set $l=1+\varepsilon$ and take the limit as $\varepsilon\to 0$, we find

$$M_1 = \frac{2}{\xi}\,M_0 = \frac{2}{\xi} \tag{9.1.13a}$$

and thus for $l \geqq 2$

$$M_l = \frac{2}{\xi}\prod_{\nu=2}^{l}\frac{\nu-1}{1-\dfrac{2}{\nu+1}\,\dfrac{1-\alpha^{\frac{\nu+1}{2}}}{1-\alpha}}\,. \tag{9.1.13b}$$

In case $A \gg 1$, we can approximate the recursion formula (9.1.12) by[1]

$$M_l = \frac{A+2}{1+(3-l)\dfrac{2}{3(A+2)}}\,M_{l-1}\,. \tag{9.1.14}$$

Thus if we take $M_0=1$ and neglect terms of higher order, we find that

$$\left.\begin{aligned} M_1 &= A+2/3 \quad [=2/\xi;\ \text{cf. Eq. (7.1.19)}]\\ M_2 &= A(A+2)\\ M_3 &= A^2(A+4)\\ M_4 &= A^3(A+20/3)\,. \end{aligned}\right\} \tag{9.1.15}$$

We shall next show the connection of these equations with Eqs. (9.1.7) and (9.1.8) for hydrogen. Let us consider the time dependence of the neutron flux at a fixed value of v. Since

$$n(x)\,dx = n(v,t)\,\frac{v}{t}\,dt$$

---

[1] Eq. (9.1.14) can be derived by replacing $\alpha^{\frac{m+1}{2}} = \left(1-\frac{4A}{(A+1)^2}\right)^{\frac{m+1}{2}}$ by the first three terms in its Taylor series.

we obtain

$$M_1 = \frac{2}{\xi} = v\Sigma_s \int_0^\infty n(v,t)\, v\, dt$$

or

$$\int_0^\infty v\, n(v,t)\, dt = \frac{2}{\xi \Sigma_s v}. \tag{9.1.16a}$$

This equation is analogous to Eq. (9.1.7a) and says that when integrated over all time the spectrum is the same as in the equivalent stationary case. Furthermore

$$\overline{t_E} = \frac{1}{v\Sigma_s}\frac{M_2}{M_1}, \tag{9.1.16b}$$

$$\overline{t_E^2} = \frac{1}{(v\Sigma_s)^2}\frac{M_3}{M_1}, \tag{9.1.16c}$$

$$\Delta^2 = \overline{t_E^2} - \overline{t}_E^2 = \frac{1}{(v\Sigma_s)^2}\frac{M_3 M_1 - M_2^2}{M_1^2}. \tag{9.1.16d}$$

With the help of Eqs. (9.1.15) we find for $A \gg 1$

$$\overline{t_E} = \frac{A+4/3}{\Sigma_s v} \approx \frac{A}{\Sigma_s v}, \tag{9.1.17a}$$

$$\overline{t_E^2} = \frac{A^2+10A/3}{(\Sigma_s v)^2} \approx \frac{A^2}{(\Sigma_s v)^2} \tag{9.1.17b}$$

and

$$\Delta^2 = \overline{t_E^2} - \overline{t}_E^2 = \frac{2A/3}{(\Sigma_s v)^2}. \tag{9.1.17c}$$

Thus $\Delta^2/\overline{t}_E^2 \approx 2/3A$. In other words, the relative dispersion in the time of moderation to a particular energy decreases with increasing mass number. This behavior originates in the fact that for large mass numbers the neutrons are slowed down by many collisions, each of which results in only a small energy loss. Owing to the large number of collisions, the statistical fluctuations in the energy loss per collision and the flight path between successive collisions cancel out to a large extent; in first approximation therefore we can describe the slowing-down process as a continuous decrease in a relatively sharply defined neutron energy.

Table 9.1.1. "*Asymptotic*" *Slowing-Down Times to 1 ev*

Substance	$\overline{t_E}$ (μsec)	Δ (μsec)
Be (1.78 g/cm³)	10.18	2.0
C (1.6 g/cm³)	25.40	4.7
Pb (11.34 g/cm³)	408.1	22.8

Table 9.1.1 gives values of $\overline{t}_E$ and $\Delta = (\overline{t_E^2} - \overline{t}_E^2)^{\frac{1}{2}}$ calculated with Eqs. (9.1.16b and d) for several heavy moderators.

The first few moments of the velocity distribution at a fixed time are given by

$$\overline{v}(t) = \frac{M_1}{\Sigma_s t} = \frac{2}{\xi \Sigma_s t}, \tag{9.1.18a}$$

$$\overline{v^2}(t) = \frac{M_2}{(\Sigma_s t)^2}, \tag{9.1.18b}$$

$$\overline{v^4}(t) = \frac{M_4}{(\Sigma_s t)^4}. \tag{9.1.18c}$$

When $A \gg 1$, the mean energy at a fixed time t is given by

$$\overline{E}(t) = \frac{m}{2}\,\frac{M_2}{(\Sigma_s t)^2} \approx \frac{m A^2}{2(\Sigma_s t)^2} = \frac{A^2}{(\Sigma_s t)^2}\,0.522\,\mu\mathrm{sec}^2\,\mathrm{cm}^{-2}\,\mathrm{ev} \qquad (9.1.19\,\mathrm{a})$$

and its dispersion is given by

$$\overline{\left(\frac{\Delta E}{E}\right)^2} = \frac{\overline{E^2} - \overline{E}^2}{\overline{E}^2} \approx \frac{8}{3A}\,. \qquad (9.1.19\,\mathrm{b})$$

Eq. (9.1.19b) shows that the energy spectrum during moderation in heavy substances is always very sharp. This sharpness is a result of the nearly continuous nature of the slowing-down process. For lead ($A = 207$), for example, $\sqrt{\overline{(\Delta E/E)^2}} = 11.4\%$. The so-called slowing-down-time spectrometer is an application of these ideas (cf. Sec. 16.3.2).

In concluding this section, let us note that Eq. (9.1.18a) can be derived in an elementary fashion. Let us consider the increase in lethargy that a neutron of velocity v (i.e. lethargy u) experiences in one second. Since the number of collisions per second is $v\Sigma_s$ and the average increase in lethargy per collision is ξ,

$$\frac{du}{dt} = \xi\,\Sigma_s v \qquad (9.1.20\,\mathrm{a})$$

or because $v = v_Q \cdot e^{-u/2}$

$$\frac{dv}{dt} = -\frac{\xi\,\Sigma_s}{2}\,v^2\,. \qquad (9.1.20\,\mathrm{b})$$

It then follows by integration that

$$t = \frac{2}{\xi\,\Sigma_s}\int_v^{v_Q}\frac{dv}{v^2} = \frac{2}{\xi\,\Sigma_s}\left(\frac{1}{v} - \frac{1}{v_Q}\right).$$

If $v \ll v_Q$, then

$$v(t) = \frac{2}{\xi\,\Sigma_s t}\,. \qquad (9.1.20\,\mathrm{c})$$

The derivation of this equation presupposes that the neutrons have a definite energy at every time, but as we have seen this assumption is only valid for very large A. For smaller A, Eq. (9.1.20c) is only correct when v represents the average velocity. Eq. (9.1.20c) is often used to calculate the slowing-down time, but as we can see from comparison with Eq. (9.1.17a) or (9.1.16b), this procedure is only correct for large A (for $A = 1$ the error is 50%).

9.2. The Diffusion of Monoenergetic Neutrons (Pulsed Sources)

9.2.1. The Time-Dependent Diffusion Equation

The neutron balance equation in a time-dependent, monoenergetic neutron field can be written

$$\frac{1}{v}\,\frac{\partial\Phi(\boldsymbol{r},t)}{\partial t} = -\operatorname{div}\boldsymbol{J}(\boldsymbol{r},t) - \Sigma_a\Phi(\boldsymbol{r},t) + S(\boldsymbol{r},t)\,. \qquad (9.2.1)$$

If we assume that the current and the density gradient are related by Fick's law, Eq. (9.2.1) immediately yields the time-dependent diffusion equation:

$$\frac{1}{v}\,\frac{\partial\Phi(\boldsymbol{r},t)}{\partial t} = D\nabla^2\Phi(\boldsymbol{r},t) - \Sigma_a\Phi(\boldsymbol{r},t) + S(\boldsymbol{r},t)\,. \qquad (9.2.2)$$

The same conditions are necessary for the validity of this equation as were necessary in Chapter 5 for the validity of the P_1-approximation, viz., boundary surfaces and localized sources must be far away and Σ_a must be $\ll \Sigma_s$. In Sec. 9.2.2, we shall see that there are still other restrictive conditions. In spite of these numerous limitations, Eq. (9.2.2) is of great practical importance. In particular, it forms the basis for the analysis of the non-stationary diffusion experiments discussed in Chapter 18. We shall next discuss some of these solutions for a pulsed source $(S(\boldsymbol{r}, t) = \delta(t) S(\boldsymbol{r}))$. We might surmise that the results for monoenergetic neutrons would also hold for *thermal* neutrons. In first approximation, this conclusion is correct, but we shall see in Sec. 10.3 that certain corrections must be applied when the results for monoenergetic neutrons are used for a pulsed thermal field.

When we are dealing with a pulsed source, we can remove the absorption term with the substitution

$$\Phi(\boldsymbol{r}, t) = \Phi_0(\boldsymbol{r}, t)\, e^{-v\Sigma_a t}. \tag{9.2.3}$$

Then

$$\frac{1}{v}\,\frac{\partial \Phi_0}{\partial t} = D\nabla^2 \Phi_0(\boldsymbol{r}, t) + S(\boldsymbol{r})\,\delta(t). \tag{9.2.4}$$

We can easily convince ourselves that in the important special case $\Sigma_a \sim 1/v$ the absorption can be taken into account by the substitution 9.2.3 even for energy-dependent problems. In particular, we could have taken a $1/v$-absorption into account in this way in Sec. 9.1. This possibility of splitting off the absorption always exists in pulsed neutron problems and does not depend on the use of diffusion theory.

In order to advance our treatment of Eq. (9.2.4), we may note with advantage that it is formally similar to the age equation, which was discussed in Sec. 8.2. If we make the substitutions $\tau = Dvt$ and $q = \Phi/v$, Eq. (9.2.4) becomes formally identical with the age equation, and we can use all the results obtained in Sec. 8.2.4. Some important cases are:

Plane Source in an Infinite Medium[1]

$$\Phi(x, t) = \frac{Qv}{\sqrt{4\pi Dvt}}\, e^{-\frac{x^2}{4Dvt} - v\Sigma_a t}. \tag{9.2.5}$$

Point Source in an Infinite Medium

$$\Phi(r, t) = \frac{Qv}{(4\pi Dvt)^{\frac{3}{2}}}\, e^{-\frac{r^2}{4Dvt} - v\Sigma_a t}. \tag{9.2.6}$$

It follows from Eq. (9.2.6) that the mean squared distance at time t is

$$\overline{r^2} = 6Dvt. \tag{9.2.7}$$

$\sqrt{\overline{r^2}}$ for thermal neutrons after 10 msec is 45 cm in H_2O and 110 cm in graphite.

Finite Media. Under the assumption that the linear dimensions of the medium are large compared to the mean free path, the time-dependent flux is given by

$$\Phi(\boldsymbol{r}, t) = \Sigma S_n v R_n(\boldsymbol{r})\, e^{-(v\Sigma_a + DvB_n^2)t}. \tag{9.2.8a}$$

Here B_n^2 are the eigenvalues and $R_n(\boldsymbol{r})$ the eigenfunctions of the equation

$$\nabla^2 \Phi + B^2 \Phi = 0 \tag{9.2.8b}$$

[1] In the following solutions the source strength Q does not have the dimensions sec^{-1} since Q is the total number of neutrons contained in a pulse.

with the boundary condition $\Phi=0$ on the effective surface of the medium. The S_n are the coefficients in the expansion of the source function in these eigenfunctions. The neutron flux is composed of an infinite number of "modes" that decay with the relaxation times $\frac{1}{v\Sigma_a+DvB_n^2}$. For very long times the "fundamental" mode predominates:

$$\Phi(\boldsymbol{r},t)\sim R_0(\boldsymbol{r})\,e^{-\alpha t}. \tag{9.2.8c}$$

Here

$$\alpha=v\Sigma_a+DvB^2=v\Sigma_a(1+L^2B^2) \tag{9.2.8d}$$

and R_0 is the lowest eigenfunction and B^2 the lowest eigenvalue. This lowest eigenvalue is frequently called the *buckling* or, more precisely, the *geometric buckling.*

The eigenfunctions and eigenvalues for a rectangular parallelepiped with the effective edges a, b, and c are given by

$$\left.\begin{aligned} R_{lmn}(x,y,z)&=\sin\frac{l\pi x}{a}\sin\frac{m\pi y}{b}\sin\frac{n\pi z}{c}\\ B_{lmn}^2&=\pi^2\left(\frac{l^2}{a^2}+\frac{m^2}{b^2}+\frac{n^2}{c^2}\right).\end{aligned}\right\} \tag{9.2.9a}$$

In particular

$$\left.\begin{aligned} R_0(x,y,z)&=\sin\frac{\pi x}{a}\sin\frac{\pi y}{b}\sin\frac{\pi z}{c}\\ B^2&=\pi^2\left(\frac{1}{a^2}+\frac{1}{b^2}+\frac{1}{c^2}\right).\end{aligned}\right\} \tag{9.2.9b}$$

The contribution of diffusion to the decay constant is obviously greater the smaller the linear dimensions of the medium; the characteristic unit of length in the medium is the diffusion length. According to Eq. (9.2.8d), when the linear dimensions of the medium are large compared to the diffusion length, $\alpha\approx v\Sigma_a$; however, this result has no practical significance because under the circumstances being considered, a very long time must elapse before the fundamental mode alone is present.

The eigenfunctions and eigenvalues of a cylinder of effective radius r_0 and height h are

$$\left.\begin{aligned} R_{lm}(r,z)&=I_0\left(\alpha_l\frac{r}{r_0}\right)\sin\frac{m\pi z}{h}\\ B_{lm}^2&=\left(\frac{\alpha_l}{r_0}\right)^2+\left(\frac{m\pi}{h}\right)^2\end{aligned}\right\} \tag{9.2.10}$$

if for simplicity we assume radial symmetry[1]. α_l is the *l-th* root of the Bessel function I_0 ($\alpha_0=2.405$, $\alpha_1=5.520$, $\alpha_2=8.654$, $\alpha_3=11.792$). SJÖSTRAND has summarized some additional solutions of Eq. (9.2.8b).

9.2.2. The Validity of the Time-Dependent Diffusion Equation

We can see immediately from Eq. (9.2.6) that the description of time-dependent neutron fields by the diffusion Eq. (9.2.4) cannot be entirely correct: For $t=0$, the neutron flux vanishes everywhere except the point $r=0$; this result is correct.

[1] If the sources are radially symmetric, only radially symmetric modes are excited.

For $t=\delta t>0$, on the other hand, $\Phi(r, t)$ vanishes nowhere, although it is small for all $r\neq 0$. In reality, however, $\Phi(r, t)$ must vanish outside the sphere $r=v\,\delta t$ because in the time δt no neutron can traverse a path longer than $v\,\delta t$. That the time-dependent diffusion equation does not describe this effect correctly comes about because FICK's law, which underlies Eq. (9.2.2), does not take into account the finite collision time. In order to see this clearly, let us remember the elementary derivation of FICK's law given at the beginning of Chapter 6: There we expressed the current in terms of the flux as follows

$$J\,dF=-\frac{dF}{4\pi}\Sigma_s\int\limits_0^\infty\int\limits_0^{2\pi}\int\limits_0^\pi \cos\vartheta\,\Phi(\boldsymbol{r})e^{-\Sigma_t r}\,dr\,d\varphi\,\sin\vartheta\,d\vartheta. \qquad (9.2.11\,\text{a})$$

Expansion of Φ in a Taylor series led immediately to FICK's law:

$$J=-\frac{1}{3\Sigma_s}\frac{\partial\Phi}{\partial z}. \qquad (9.2.11\,\text{b})$$

This procedure is correct in a stationary neutron field only if $\frac{1}{\Sigma_s}\frac{\partial\Phi}{\partial z}\ll\Phi$. In a non-stationary field, Eqs. (9.2.11) obviously no longer connect the flux and current at the same time since between the last collision of a neutron at the point $\boldsymbol{r}$ and its arrival at the surface element dF a time $t=r/v$ elapses. If we assume in first approximation that all the neutrons make their last collision at a distance of one mean free path from dF, the "retardation" time is $1/v\Sigma_s$ and FICK's law reads

$$J\left(z,\ t+\frac{1}{v\Sigma_s}\right)=-\frac{1}{3\Sigma_s}\frac{\partial\Phi(z,t)}{\partial z}. \qquad (9.2.12\,\text{a})$$

If we develop J in a power series in t and break it off after the second term, we find

$$J(z,\ t)+\frac{1}{v\Sigma_s}\frac{\partial}{\partial t}J(z,\ t)=-\frac{1}{3\Sigma_s}\frac{\partial\Phi(z,t)}{\partial z} \qquad (9.2.12\,\text{b})$$

or in the general three-dimensional case

$$\boldsymbol{J}+\frac{1}{v\Sigma_s}\frac{\partial\boldsymbol{J}}{\partial t}=-D\,\mathrm{grad}\,\Phi. \qquad (9.2.12\,\text{c})$$

If the change in $\boldsymbol{J}$ in one collision time can be neglected, we can drop the second term on the left-hand side and obtain FICK's law in its usual form[1].

We can now combine the balance Eqs. (9.2.1) and (9.2.12c) into a single equation by differentiating the first with respect to time and the second with respect to position. In a non-absorbing, source-free medium the result is

$$\frac{\partial^2\Phi}{\partial t^2}+\frac{v}{3D}\frac{\partial\Phi}{\partial t}=\frac{v^2}{3}\nabla^2\Phi. \qquad (9.2.13)$$

This equation has the form of the well-known telegrapher's equation of electrodynamics. It is a familiar property of the telegrapher's equation that it predicts the existence of a wave front which moves with the group velocity v_G for the propagation of any disturbance; outside this wave front the disturbance vanishes completely.

---

[1] Eq. (9.2.12) can also be derived by a P_1-approximation to the time-dependent transport equation.

In our case, $v_G^2 = v^2/3$, i.e., the velocity of propagation is $v/\sqrt{3}$. We might have expected that the velocity of propagation would be v, and in fact the factor $\sqrt{3}$ is a consequence of using the P_1-approximation, which does not describe the angular distribution of the advancing source neutrons very well. WEINBERG and NODERER have obtained a complete solution of Eq. (9.2.13) for an infinite medium, but we shall not reproduce it here. It has the wave-like form mentioned above, i.e., it is zero for $r > vt/\sqrt{3}$ and for $r \ll vt/\sqrt{3}$ becomes the solution to the time-dependent diffusion equation.

It is instructive to consider the time decay of a pulsed neutron field in a finite, non-absorbing medium, explicitly taking this collision-time effect into account. To begin with

$$\Phi(\boldsymbol{r}, t) = \Sigma S_n R_n(\boldsymbol{r}) T_n(t). \tag{9.2.14}$$

According to Eq. (9.2.13), $T_n(t)$ is given by

$$\frac{d^2 T_n}{dt^2} + \frac{v}{3D}\frac{dT_n}{dt} = -\frac{v^2}{3} B_n^2 T_n. \tag{9.2.15a}$$

If we set $T_n(t)$ equal to e^{-st}, the following characteristic equation for s results:

$$s^2 - \frac{v}{3D} s + \frac{v^2 B_n^2}{3} = 0, \tag{9.2.15b}$$

$$s_{1,2} = \frac{v}{6D} \pm \frac{v}{6D}\sqrt{1 - 12 D^2 B_n^2}. \tag{9.2.15c}$$

Thus two exponentials occur in the time decay of each mode. In general, a P_N-approximation leads to $N+1$ values of s. For large t, however, only the smallest value of s is significant. Eq. (9.2.15c) gives

$$s = \frac{v}{6D}\left(1 - \sqrt{1 - 12 D^2 B_n^2}\right) \approx D v B_n^2 + 3 D v D^2 B_n^4 + \cdots \tag{9.2.15d}$$

for this smallest value. Thus in first approximation we again obtain the result of elementary diffusion theory; however, correction terms, which are larger the larger B_n^2 is, do occur. We shall return to these transport corrections in another connection in Sec. 10.3.3. For $B_n^2 > \frac{1}{12 D^2} = \frac{3}{4}\Sigma_{tr}^2$, Eq. (9.2.15d) yields imaginary values of s. This might indicate that for such small systems no exponential time decay of the neutron flux exists. However, we must be aware that $B_n^2 > \frac{1}{12 D^2}$ holds only in very small systems, where the use of our simple diffusion theory — i.e., P_1-approximation — is not permitted. Much more laborious solutions of the transport equations have to be found in order to describe the neutron decay in small systems (cf. BOWDEN, KLADNIK, DAITCH and EBEOGLU) but we shall not discuss them here.

9.3. Age Theory and the Time-Dependent Diffusion Equation

In order to describe the time and space distribution of the neutrons during the slowing-down process, we must start with the balance equation

$$\frac{1}{v}\frac{\partial \Phi(\boldsymbol{r}, u, t)}{\partial t} = -\Sigma_a \Phi - \operatorname{div} \boldsymbol{J} - \frac{\partial q}{\partial u} + S(\boldsymbol{r}, u, t). \tag{9.3.1}$$

If we now introduce FICK's law in its elementary form $\boldsymbol{J} = -D \operatorname{grad} \Phi(u)$ and make the additional assumption that the moderation is due to elastic collisions with *heavy* nuclei, we can set $q(u) = \xi \Sigma_s \Phi$ as we did in Sec. 8.2 and obtain

$$\frac{1}{\xi \Sigma_s v} \frac{\partial q}{\partial t} = -\frac{\Sigma_a}{\xi \Sigma_s} q + \frac{D}{\xi \Sigma_s} \nabla^2 q - \frac{\partial q}{\partial u} + S(\boldsymbol{r}, u, t). \tag{9.3.2}$$

We restrict ourselves now in the interest of simplicity to a non-absorbing medium and a monoenergetic source $S(\boldsymbol{r}, u, t) = S(\boldsymbol{r}) \delta(u) f(t)$. If we transform the independent variable from lethargy to age $\tau = \int_0^u \frac{D(u')}{\xi \Sigma_s} du'$, we obtain

$$\frac{1}{D v} \frac{\partial q}{\partial t} + \frac{\partial q}{\partial \tau} = \nabla^2 q + S(\boldsymbol{r}) \delta(\tau) f(t). \tag{9.3.3}$$

This equation is clearly a generalization of the age equation to the case of non-stationary sources. It can easily be verified by substitution in Eq. (9.3.3) that the time dependence can be separated in the form

$$q(\boldsymbol{r}, t, \tau) = q_0(\boldsymbol{r}, \tau) f\left(t - \int_0^\tau \frac{d\tau'}{D v}\right) \tag{9.3.4a}$$

where $q_0(r, \tau)$ is the solution of the time-independent age equation

$$\frac{\partial q_0}{\partial \tau} = \nabla^2 q_0 + S(\boldsymbol{r}) \delta(\tau). \tag{9.3.4b}$$

We can easily see what this equation means if we assume a pulsed source $f(t) = \delta(t)$. Then $q(r, t, \tau) = q_0(r, \tau) \delta\left(t - \int_0^\tau \frac{d\tau'}{D v}\right)$, i.e., t and τ are not independent variables: to each time t after the beginning of the slowing-down process there corresponds a unique value of τ given by

$$t = \int_0^\tau \frac{d\tau'}{D v}; \qquad \tau = \int_0^t D v \, dt'. \tag{9.3.5}$$

This result is to be expected on the basis of what we learned in Sec. 9.1.2. There we saw that for $A \gg 1$ — and it is only in this case that age theory is valid at all — the neutrons formed a narrow energy group during moderation. Thus we can speak of a single sharply defined energy $E(t)$ or lethargy $u(t)$. To each time therefore there corresponds a unique lethargy and thus a unique Fermi age:

$$\tau = \int_0^u \frac{D}{\xi \Sigma_s} du' = \int_0^t \frac{D}{\xi \Sigma_s} \frac{du'}{dt'} dt' = \int_0^t D v \, dt'. \tag{9.3.6}$$

[Cf. Eqs. (8.2.7) and (9.1.20a.)] It is this direct connection between τ and t that suggests the nomenclature "age".

It is now understandable why the time-dependent diffusion equation discussed in Sec. 9.2.1 is very similar to the age equation: the time-dependent diffusion equation can immediately be transformed into the age equation by changing variables from time to age.

In order to describe the space and time distribution of the neutrons during moderation in a pulsed medium we thus need not seek any new solutions but can refer to the solutions of the stationary age equation developed in Sec. 8.2.

9.4. Diffusion of Monoenergetic Neutrons from a Harmonically Modulated Source

We consider last the space and time distribution of the neutrons in a medium in which there is a sinusoidally modulated neutron source. Such a source can be produced, for example, by periodic interruption of the neutron beam from a reactor. In the neighborhood of such a source "neutron waves" are produced from the nature of whose propagation we can draw conclusions about the properties of the medium. We shall return to the study of such neutron waves in Chapter 18.

The time-dependent diffusion equation for a point source $Q_0+\delta Q e^{i\omega t}$ [1] in an infinite medium takes the form

$$\frac{1}{v}\frac{\partial \Phi}{\partial t}=D\left(\frac{\partial^2\Phi}{\partial r^2}+\frac{2}{r}\frac{\partial\Phi}{\partial r}\right)-\Sigma_a\Phi+\frac{Q_0+\delta Q e^{i\omega t}}{4\pi r^2}\delta(r). \tag{9.4.1}$$

Its solution is

$$\Phi(r,t)=\Phi_0(r)+\delta\Phi_0(r)e^{i\omega t} \tag{9.4.2a}$$

with

$$\Phi_0(r)=\frac{Q_0}{4\pi D}\frac{e^{-r/L}}{r}, \tag{9.4.2b}$$

$$\delta\Phi_0(r)=\frac{\delta Q}{4\pi D}\frac{e^{-r/L_\omega}}{r}. \tag{9.4.2c}$$

L_ω is a "complex" diffusion length given by

$$\frac{1}{L_\omega^2}=\frac{\Sigma_a+i\omega/v}{D}=\frac{1}{L^2}+\frac{i\omega}{vD}. \tag{9.4.3a}$$

When $\omega/v\Sigma_a\gg 1$, i.e., when the period of oscillation is small compared to the mean lifetime against absorption,

$$\frac{1}{L_\omega}\approx\left(1+\frac{v\Sigma_a}{2\omega}\right)\sqrt{\frac{\omega}{2Dv}}+i\left(1-\frac{v\Sigma_a}{2\omega}\right)\sqrt{\frac{\omega}{2Dv}}=\varkappa(\omega)+ik(\omega). \tag{9.4.3b}$$

Thus

$$\delta\Phi_0(r)e^{i\omega t}=\frac{\delta Q e^{-\varkappa r}}{4\pi D r}e^{i(\omega t-kr)}. \tag{9.4.4}$$

The modulated part of the neutron flux is a damped wave with the depth of penetration

$$\frac{1}{\varkappa}=\frac{\sqrt{\frac{2Dv}{\omega}}}{1+\frac{v\Sigma_a}{2\omega}}\approx\sqrt{\frac{2Dv}{\omega}} \tag{9.4.5a}$$

the wavelength

$$\frac{2\pi}{k}=\frac{2\pi\sqrt{\frac{2Dv}{\omega}}}{1-\frac{v\Sigma_a}{2\omega}}\approx 2\pi\sqrt{\frac{2Dv}{\omega}} \tag{9.4.5b}$$

and the phase velocity

$$\frac{\omega}{k}=\frac{\sqrt{2Dv\omega}}{1-\frac{v\Sigma_a}{2\omega}}\approx\sqrt{2Dv\omega}. \tag{9.4.5c}$$

[1] δQ must always be $\leq Q_0$ since the source strength can never be negative.

The *response* is given by

$$\frac{|\delta \Phi_0(r)|}{\Phi_0(r)} = \frac{\delta Q}{Q_0} \frac{e^{-\varkappa r}}{e^{-r/L}}. \qquad (9.4.5\text{d})$$

We can easily see that at high frequencies the depth of penetration of the wave is small, causing the response to fall off rapidly with increasing r. At extremely low frequencies, as ω approaches zero, $\varkappa$ approaches $1/L$, and k approaches zero, i.e., the solution becomes quasi-stationary. The use of diffusion theory for the description of neutron waves is correct as long as $\omega \ll v\Sigma_s$, i.e., as long as the period of oscillation is long compared to the time between successive collisions.

Chapter 9: References

General

AMALDI, E.: Loc. cit. especially § 70 and § 124.
VON DARDEL, G. F.: The Interaction of Neutrons with Matter Studied with a Pulsed-Neutron Source, Trans. Roy. Inst. Technol. Stockholm **75**, 1954.
MARSHAK, R. E.: The Slowing Down of Neutrons, Rev. Mod. Phys. **19**, 185 (1947).
WEINBERG, A. M., and L. C. NODERER: Theory of Neutron Chain Reactions, AECD-3471 (1951), especially p. 1—82: The Time-Dependent Diffusion Equation.

Special

DYAD'KIN, G., and E. P. BATALINA: Atomnaya Energiya **10**, 5 (1961).
ORNSTEIN, L. S., and G. E. UHLENBECK: Physica **4**, 478 (1937).
} Time Dependence of the Slowing-Down Process in Hydrogen.

KAZARNOVSKY, M. V.: Thesis, Moscow 1955.
KOPPEL, J. U.: Nucl. Sci. Eng. **8**, 157 (1960).
ERIKSSON, K.-E.: Arkiv Fysik **16**, 1 (1959).
SVARTHOLM, N.: Trans. Chalmers Univ. Technol., Gothenburg **164**, 1955.
WALLER, I.: Geneva 1958 P/153 Vol. 16 p. 450.
} Time Dependence of the Slowing-Down Process in Media with $A \neq 1$.

SJÖSTRAND, N. G.: Nukleonik **1**, 89 (1958); Arkiv Fysik **13**, 229 (1958) (Solutions of $\nabla^2 \Phi + B^2 \Phi = 0$).
RAIEVSKI, V., and J. HOROWITZ: Compt. Rend. **238**, 1993 (1954) (Neutron Waves).

BOWDEN, R. L.: TID-18884 (1963).
DAITCH, P. B., and D. B. EBEOGLU: Nucl. Sci. Eng. **17**, 212 (1962).
KLADNIK, R.: Nukleonik **6**, 147 (1964).
SJÖSTRAND, N. G.: Arkiv Fysik **15**, 147 (1959).
WING, G. M.: An Introduction to Transport Theory; New York: John Wiley and Sons Inc., 1963.
} Transport Theory and Pulsed Moderators.

10. Thermalization of Neutrons

Our first discussion of thermal neutron fields in Sec. 5.3 was based on the idealizing assumption that the neutrons were in a true state of thermal equilibrium and thus had a Maxwellian energy distribution with the temperature of the medium. For reasons already mentioned there, the assumption of such an equilibrium state cannot be rigorously true. In this chapter we shall study the energy distribution at neutron energies below about 1 ev in detail. In doing so, we shall strive to connect the theory of the thermalization process with the theory of the slowing-down process that was previously developed for energies greater than about 1 ev on the basis of collisions with free, stationary nuclei.

The thermalization problem is especially difficult. In the first place, the elementary process, viz., the scattering of very slow neutrons by solid or liquid moderators, is complex and has not yet been fully explored. Secondly, even when the scattering law is known, the calculation of strongly space-dependent neutron spectra (the usual case) leads to difficult mathematical problems. In this chapter, we shall restrict ourselves to explaining the most important physical phenomena and to indicating the direction towards an exact treatment of neutron spectra. The subject is still very much in a state of flux and the presentation in this chapter must be considered preliminary.

In Sec. 10.1, the problem of inelastic scattering of slow neutrons is considered. Sec. 10.2 is concerned with the actual thermalization problem, i.e., with the determination of the spectrum $\Phi(E)$ or $\Phi(\boldsymbol{r}, E)$ in a medium with given sources. In Sec. 10.3 some properties of thermalized neutron fields, i.e., neutron fields that are asymptotically established far from the source (or in the case of pulsed sources, long after the pulse), are explained. The considerations of this section are important for the interpretation of the diffusion experiments on thermal neutron fields to be discussed in Chapters 17 and 18. Finally in Sec. 10.4, the question of how an already largely thermalized spectrum approaches the asymptotic distribution is discussed.

In the earliest period of thermalization physics, it was customary to approximate the thermal neutron spectrum by a Maxwell distribution with an effective temperature T: the main task was then to calculate the deviation of this temperature from the temperature T_0 of the moderator. There is no *a priori* physical basis for such an approximation, and it turns out that in many cases the neutron temperature concept leads to quantitatively false results. The temperature concept is nevertheless useful for qualitative purposes and we shall use it repeatedly (Secs. 10.2.1, 10.2.3, 10.3.1, and 10.4.1) to explain the various phenomena of neutron thermalization.

10.1. The Scattering of Slow Neutrons

The scattering process is characterized by the differential scattering cross section $\sigma_s(E'\to E, \boldsymbol{\Omega}'\to\boldsymbol{\Omega})$. In an isotropic substance $\sigma_s(E'\to E, \boldsymbol{\Omega}'\to\boldsymbol{\Omega}) = \frac{1}{2\pi}\sigma_s(E'\to E, \cos\vartheta_0)$ where ϑ_0 is the scattering angle. Frequently for thermalization problems knowledge of $\sigma_s(E'\to E) = \int_{-1}^{+1} \sigma_s(E'\to E, \cos\vartheta_0)\, d\cos\vartheta_0$ is sufficient. Whereas for energies greater than 1 ev we can calculate $\sigma_s(E'\to E)$ from the total scattering cross section with the help of the laws of elastic collision (cf. Sec. 7.1), in the thermal energy range we must take the thermal motion and the chemical binding of the scattering atoms into account. We shall treat this problem in two steps: First we shall neglect chemical binding but take into account the thermal motion of the atoms, i.e., we shall calculate the scattering from a hypothetical ideal gas with the density and temperature of the actual thermalizing medium. Next we shall indicate how the effects of chemical binding in real media can be taken into account. Finally we shall discuss various experimental methods for studying the scattering of slow neutrons.

10.1.1. Calculation of $\sigma_s(E'\to E)$ for an Ideal, Monatomic Gas

The following derivation is due to WIGNER and WILKINS. Let the moderating gas consist of atoms of mass $M = A m_n$ and let these atoms have the energy-independent scattering cross section σ_{sf}. Furthermore, let the gas atoms have a Maxwell velocity distribution:

$$P(V)\,dV = \left(\frac{M}{2\pi kT_0}\right)^{\frac{3}{2}} e^{-\frac{MV^2}{2kT_0}}\, 4\pi\, V^2\, dV. \tag{10.1.1}$$

We first consider collisions between neutrons of velocity v' and atoms with a particular velocity V. Suppose that before the collision the directions of the neutron and the gas atom make an angle ε, ($\cos\varepsilon = \mu$). Then the relative velocity is

$$v_{\text{rel}} = \sqrt{v'^2 + V^2 - 2v'V\mu}\,.$$

If the atomic density of the gas is N atoms/cm³, there are $N P(V) dV$ gas atoms with a velocity between V and $V+dV$ per cm³. Since all directions are equally likely for both the neutron and the gas atom, the probability of a collision angle between ε and $\varepsilon + d\varepsilon$ is $d\mu/2$. Thus the number of such collisions per cm³ and sec is

$$d\psi = v_{\text{rel}}\cdot\sigma_{sf}\cdot N\cdot P(V)\,dV\cdot\frac{d\mu}{2}\,. \tag{10.1.2a}$$

We can also write $d\psi$ in the form

$$d\psi = v'\,d\sigma_{v'V\mu}\cdot N \tag{10.1.2b}$$

where $d\sigma_{v'V\mu}$ is the microscopic cross section for a neutron of velocity v' to collide with a gas atom whose velocity V is inclined at an angle ε to the direction of the neutron. Obviously

$$d\sigma_{v'V\mu} = \frac{v_{\text{rel}}\cdot\sigma_{sf}}{v'}\,P(V)\,dV\,\frac{d\mu}{2}\,. \tag{10.1.3}$$

We now wish to find the probability $g(v'\to v)\,dv$ that the neutron falls into the velocity interval $(v, v+dv)$ after such a collision. We determine it by considering the collision in the neutron-gas-atom center-of-mass system. In this system, the neutron has the velocity $\frac{A}{A+1}v_{\text{rel}}$ before the collision. This velocity only changes its direction in a collision; thus the velocity in the laboratory system after the collision is given by

$$v = \sqrt{v_m^2 + \left(\frac{A}{A+1}\right)^2 v_{\text{rel}}^2 + 2v_m\frac{A}{A+1}v_{\text{rel}}\cos\psi}\,.$$

Here v_m is the velocity of the center of mass and ψ is the scattering angle in the center-of-mass system. v_m is given by

$$v_m = \frac{\sqrt{v'^2 + A^2V^2 + 2Av'V\mu}}{A+1}\,.$$

Assuming that the scattering in the center-of-mass system is isotropic, we find that

$$\left.\begin{aligned} g(v'\to v) &= 0, & v &< v_{\min}\\ &= \frac{2v}{v_{\max}^2 - v_{\min}^2}, & v_{\min} &< v < v_{\max}\\ &= 0, & v &> v_{\max}. \end{aligned}\right\} \tag{10.1.4}$$

Here $v_{\max}$ and $v_{\min}$ are respectively the largest and smallest velocities that the neutron can have after the collision, viz.,

$$v_{\max}=v_m+\frac{A}{A+1}v_{\mathrm{rel}}, \qquad v_{\min}=v_m-\frac{A}{A+1}v_{\mathrm{rel}}.$$

If we now combine Eqs. (10.1.3) and (10.1.4) and integrate over all directions and velocities V, we obtain the scattering cross section for processes in which a neutron of velocity v' is scattered into the velocity interval $(v, v+dv)$:

$$\sigma_s(v'\to v)\,dv=\frac{1}{2}\cdot\frac{1}{v'}\cdot\int_0^\infty dV\int_{-1}^{+1} v_{\mathrm{rel}}\,d\mu\,\sigma_{sf}\,P(V)\,g(v'\to v)\,dv. \qquad (10.1.5)$$

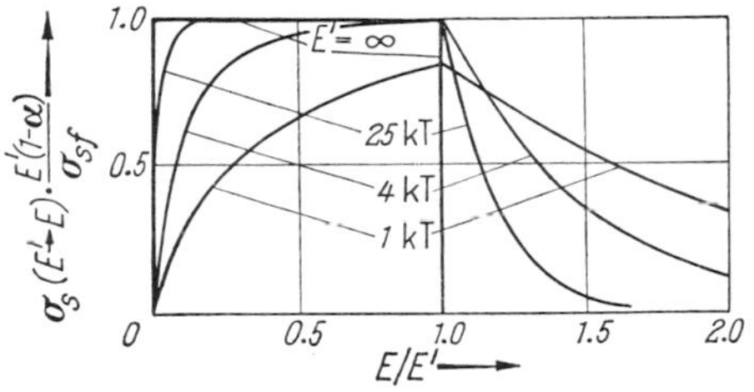

Fig. 10.1.1. The differential scattering cross section of a gas of thermal protons

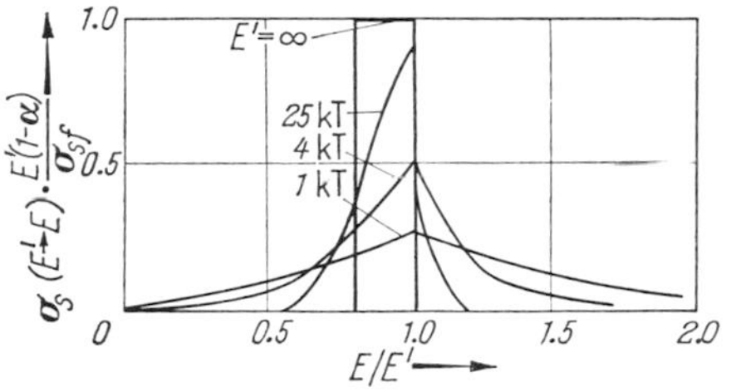

Fig. 10.1.2. The differential scattering cross section of a gas of thermal Oxygen nuclei

If we set $\eta=\frac{A+1}{2\sqrt{A}}$ and $\varrho=\frac{A-1}{2\sqrt{A}}$, the result of carrying out the integration can be written

$$\left.\begin{aligned}\sigma_s(v'\to v)=\eta^2\,\frac{\sigma_{sf}}{v'^2}\,v\Big\{&\mathrm{erf}\Big[\sqrt{\frac{m_n}{2kT_0}}\,(\eta v-\varrho v')\Big]\pm\mathrm{erf}\Big[\sqrt{\frac{m_n}{2kT_0}}\,(\eta v+\varrho v')\Big]+\\&+\exp\Big[\frac{m_n}{2kT_0}\,(v'^2-v^2)\Big]\cdot\Big(\mathrm{erf}\Big[\sqrt{\frac{m_n}{2kT_0}}\,(\eta v'-\varrho v)\Big]\mp\\&\mp\mathrm{erf}\Big[\sqrt{\frac{m_n}{2kT_0}}\,(\eta v'+\varrho v)\Big]\Big)\Big\}.\end{aligned}\right\} \qquad (10.1.6)$$

The upper sign holds for $v'>v$, the lower for $v'<v$. $\sigma_s(E'\to E)$ is then given by

$$\left.\begin{aligned}\sigma_s(E'\to E)=\frac{\eta^2}{2}\,\frac{\sigma_{sf}}{E'}\,e^{E'/kT_0}\Big\{&e^{-E/kT_0}\,\mathrm{erf}\Big[\eta\sqrt{\frac{E'}{kT_0}}-\varrho\sqrt{\frac{E}{kT_0}}\Big]+\\&+e^{-E'/kT_0}\,\mathrm{erf}\Big[\eta\sqrt{\frac{E}{kT_0}}-\varrho\sqrt{\frac{E'}{kT_0}}\Big]-\Big|e^{-E/kT_0}\,\mathrm{erf}\Big[\eta\sqrt{\frac{E'}{kT_0}}+\varrho\sqrt{\frac{E}{kT_0}}\Big]-\\&-e^{-E'/kT_0}\,\mathrm{erf}\Big[\eta\sqrt{\frac{E}{kT_0}}+\varrho\sqrt{\frac{E'}{kT_0}}\Big]\Big|\Big\}.\end{aligned}\right\} \qquad (10.1.7)$$

As an illustration of Eq. (10.1.7), $\sigma_s(E'\to E)\cdot\frac{E'(1-\alpha)}{\sigma_{sf}}$ for various values of E'/kT_0 is plotted in Fig. 10.1.1 for $A=1$ and in Fig. 10.1.2 for $A=16$. It is clear that as E'/kT_0 approaches infinity, $\sigma_s(E'\to E)$ approaches the behavior found in Sec. 7.1; in this case the thermal velocities of the scattering nuclei can be neglected compared to the velocity of the neutron. At smaller neutron energies the thermal motion of the scattering atoms becomes noticeable; in fact, collisions in which the neutron gains energy become possible, and the probability for collisions with an energy loss becomes smaller.

The *total scattering cross section* $\sigma_s(E') = \int_0^\infty \sigma_s(E' \to E)\, dE$ can be obtained directly from Eq. (10.1.3) by integrating over V and μ. It is

$$\left.\begin{aligned} \sigma_s(E') &= \frac{\overline{v_{\text{rel}}}^{V,\mu}}{v'} \sigma_{sf} \\ &= \sigma_{sf} \frac{1}{\beta^2 \sqrt{\pi}} \psi(\beta) \end{aligned}\right\} \tag{10.1.8}$$

where $\overline{v_{\text{rel}}}^{V,\mu}$ is the relative velocity averaged over V and μ, $\beta^2 = AE'/kT_0$, and

$$\psi(\beta) = \beta e^{-\beta^2} + (2\beta^2 + 1) \frac{\sqrt{\pi}}{2} \operatorname{erf}(\beta). \tag{10.1.9}$$

The function $\psi(\beta)$, known in the kinetic theory of gases, has been tabulated by JEANS[1]. Fig. 10.1.3 shows $\sigma_s(E')/\sigma_{sf}$ as a function of β. The increase at small β, i.e., at energies $E' < kT_0/A$, comes about because a very slow neutron is frequently struck during its flight by the more quickly moving gas atoms (the so-called bumping effect). For $A \gg 1$, $\sigma_s(E') \approx \sigma_{sf}$ since the increase of $\sigma_s(E')$ first begins at energies below $E' = kT_0/A$.

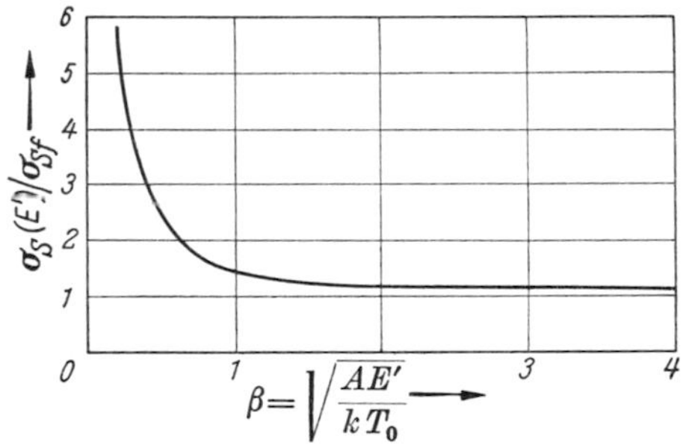

Fig. 10.1.3. The energy-dependent total scattering cross section of a gas of thermal nuclei

We can also calculate the *angular distribution* of the neutrons scattered by a gas of atoms. If we continue to assume isotropic scattering in the center-of-mass system, the average cosine of the scattering angle in the laboratory system is given by

$$\overline{\cos \vartheta_0} = \frac{2}{3A} \operatorname{erf}(\beta) + \frac{1}{3\beta^2} \left\{ \operatorname{erf}(\beta) - \frac{2}{\sqrt{\pi}} \beta e^{-\beta^2} \right\}. \tag{10.1.10}$$

When $E' \gg kT_0/A$, the angular distribution is the same as for the scattering of neutrons on stationary atoms. However, as E' falls, $\overline{\cos \vartheta_0}$ decreases monotonically, and for $E' \ll kT_0/A$, $\overline{\cos \vartheta_0} \approx 0$, i.e., the scattering is also isotropic in the laboratory system.

We shall now investigate the differential cross section $\sigma_s(E' \to E)$ somewhat more closely. We begin by noting that the expression in the braces in Eq. (10.1.7) remains unchanged when E and E' are permuted. Thus

$$E' e^{-E'/kT_0} \sigma_s(E' \to E) = E e^{-E/kT_0} \sigma_s(E \to E'). \tag{10.1.11}$$

In order to make the physical significance of Eq. (10.1.11) clear, let us remember that in a state of true thermodynamic equilibrium (which would exist in an infinite, non-absorbing medium), the neutron flux has a Maxwell distribution $\Phi(E) \sim E e^{-E/kT_0}$ of energies. Eq. (10.1.11) then says that in equilibrium as many neutrons make transitions from the energy E to the energy E' as make transitions from the energy E' to the energy E. This assertion is the generally valid *principle of detailed balance*, well known from statistical mechanics. Eq.

[1] JEANS, SIR JAMES, The Kinetic Theory of Gases. Cambridge: University Press, 1948.

(10.1.11) holds for the scattering cross section of any arbitrary scatterer and in particular for the chemically bound systems to be discussed later. We shall see later that the treatment of most thermalization problems is strongly influenced by this relation.

Particularly useful for some applications are the moments

$$\overline{\sigma_s(\Delta E)^\nu} = \int_0^\infty \sigma_s(E' \to E)(E' - E)^\nu \, dE \tag{10.1.12}$$

which contain information about the energy loss per collision. We can calculate the $\overline{\sigma_s(\Delta E)^\nu}$ by integration using Eq. (10.1.7), but the resulting formulas are complicated (cf. von Dardel). However, when $A \gg 1$ and $E' \gg kT_0/A$, these formulas give very simple expressions for the first few moments:

$$\overline{\sigma_s(\Delta E)} = \int_0^\infty \sigma_s(E' \to E)(E' - E)\, dE = \frac{2}{A}\sigma_{sf}(E' - 2kT_0), \tag{10.1.13a}$$

$$\overline{\sigma_s(\Delta E)^2} = \int_0^\infty \sigma_s(E' \to E)(E' - E)^2\, dE = \frac{4}{A}\sigma_{sf} E' \cdot kT_0. \tag{10.1.13b}$$

The higher moments are of higher order in $1/A$. Eq. (10.1.13a) for the average energy loss in a collision is particularly instructive: In a collision with a stationary heavy nucleus, $\overline{\Delta E} = 2E'/A$ (cf. Sec. 7.1). When $E' \gg kT_0$, Eq. (10.1.13a) leads to the same result. As the neutron energy decreases, so does the energy loss per collision; it vanishes when $E' = 2kT_0$, and when $E' < 2kT_0$ the neutron gains energy.

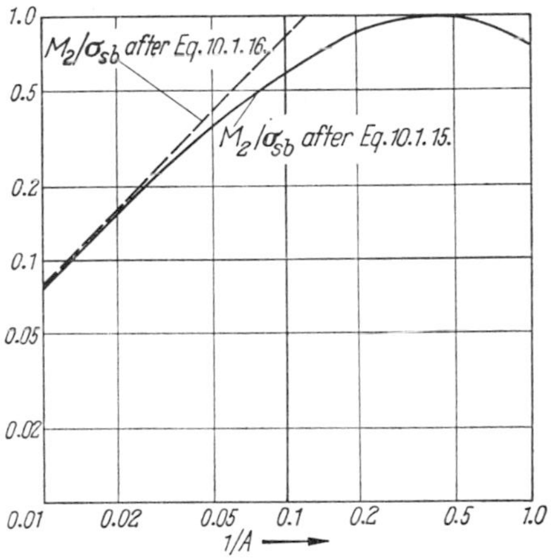

Fig. 10.1.4. The mean squared energy loss in an equilibrium Maxwellian neutron spectrum for a gas of thermal nuclei. —— Heavy gas approximation

Besides $\overline{\sigma_s(\Delta E)}$ and $\overline{\sigma_s(\Delta E)^2}$, the quantity

$$\left.\begin{aligned} M_2 = \frac{1}{(kT_0)^2}\int_0^\infty\int_0^\infty M(E') \times \\ \times (E' - E)^2 \sigma_s(E' \to E)\, dE'\, dE \end{aligned}\right\} \tag{10.1.14}$$

is also important. Here $M(E)$ is a Maxwell distribution $\frac{E}{(kT_0)^2} e^{-E/kT_0}$ normalized to one neutron per cm² per second. M_2 is a measure of the mean squared energy exchanged by collision in one second between a thermal neutron and the scattering atoms[1]. For arbitrary A,

$$M_2 = 8\frac{A^{\frac{1}{2}}}{(A+1)^{\frac{3}{2}}}\sigma_{sf} = \frac{8\sigma_{sb}}{A\left(1+\frac{1}{A}\right)^{\frac{3}{2}}} \tag{10.1.15}$$

[1] One might try to characterize the energy exchange by the linear quantity $M_1 = \frac{1}{kT_0}\int_0^\infty\int_0^\infty M(E')(E' - E)\sigma_s(E' \to E)\, dE'\, dE$. However, M_1 vanishes, as is physically clear and can easily be proved using Eq. (10.1.11).

and for $A \gg 1$,

$$M_2 = \frac{8}{A}\,\sigma_{sf} \approx \frac{8}{A}\,\sigma_{sb}\,. \tag{10.1.16}$$

Fig. 10.1.4 shows curves of M_2/σ_{sb} as a function of $1/A$ calculated from Eqs. (10.1.15) and (10.1.16).

10.1.2. The General Nature of Neutron Scattering on Chemically Bound Atoms

The model developed in Sec. 10.1.1 is unrealistic (though nonetheless useful, as we shall see later). At the densities at which they occur in solids or liquids, the binding between the scattering atoms can no longer be neglected, and particularly not at neutron energies that are small compared to the binding energy. If the binding were completely rigid, slow neutrons could not exchange any energy by collisions since the atoms would have an infinite effective mass. In other words, scattering would be elastic in the laboratory system. This however is not the case; instead the neutrons exchange energy with the "internal" degrees of freedom of the scatterer. These degrees of freedom are lattice vibrations in the case of solids and molecular rotations and vibrations, as well as some more or less hindered translations, in the case of molecular liquids like H_2O.

In these latter cases, the differential scattering cross section can only be calculated correctly using quantum mechanics and then only when the participating states of the scattering substances are known. In the following, let us consider scattering by a substance that consists of only one type of atom. Let the scattering be *purely incoherent*, i.e., let there be no interference effects at all. The Born approximation then yields the following approximation for the differential cross section[1]:

$$\sigma_s(E' \to E, \cos\vartheta_0) = \frac{\sigma_{sb}}{2}\,\frac{K}{K'} \sum_{a,b} p_a(T_0)\,\overline{|\langle\psi_b|\,e^{i\boldsymbol{\varkappa}\cdot\boldsymbol{r}}\,|\psi_a\rangle|^2}\;\delta(E_b - E_a + E - E')\,. \tag{10.1.17}$$

Here K' and K are respectively the wave numbers of the incident and outgoing neutrons; $K = p/\hbar = \sqrt{2mE}/\hbar$, etc. σ_{sb} is the (total) scattering cross section of the rigidly bound atoms, viz., $\left(1+\frac{1}{A}\right)^2 \sigma_{sf}$ (cf. Sec. 1.4). $\boldsymbol{\varkappa}$ gives the change in wave number in a single collision, $\boldsymbol{\varkappa} = \boldsymbol{K}' - \boldsymbol{K}$. According to the law of cosines

$$\varkappa^2 = K'^2 + K^2 - 2KK'\cos\vartheta_0\,. \tag{10.1.18}$$

The matrix element $|\langle\psi_b|e^{i\boldsymbol{\varkappa}\cdot\boldsymbol{r}}|\psi_a\rangle|^2$ gives the probability that in a collision in which the neutron wave number changes by $\boldsymbol{\varkappa}$, i.e., in which a momentum $\hbar\boldsymbol{\varkappa}$ is taken up, the scattering system goes from the state a to the state b. It is summed over all initial and final states, and the population of the initial state is weighted according to Boltzmann's factor $p_a(T_0) = e^{-E_a/kT_0}/\sum_a e^{-E_a/kT_0}$. The δ-function guarantees that only states b will contribute to the sum for which the condition $E_b - E_a = E' - E$ is fulfilled. Finally, the overbar indicates that an average is to be taken over all orientations of the scattering substance with respect to the in-

[1] Cf., e.g., E. Amaldi, loc. cit., p. 401 ff.

cident neutron beam. After this average, the scattering cross section can only depend on the magnitude but not on the direction of $\boldsymbol{\varkappa}$.

If we replace the δ-function in Eq. (10.1.17) by its representation as a Fourier integral

$$\delta(E_b - E_a + E - E') = \frac{1}{2\pi\hbar}\int_{-\infty}^{+\infty} e^{i(E_b - E_a + E - E')t/\hbar}\, dt$$

and further set

$$e^{iE_b t/\hbar}\,\psi_b = e^{iHt/\hbar}\psi_b \tag{10.1.19}$$

where H is the Hamiltonian of the system, Eq. (10.1.17) becomes[1]

$$\sigma_s(E' \to E, \cos\vartheta_0) = \frac{\sigma_{sb}}{4\pi\hbar}\sqrt{\frac{E}{E'}}\int_{-\infty}^{+\infty} e^{i(E-E')t/\hbar}\,\chi(\varkappa, t)\, dt \tag{10.1.20a}$$

with

$$\chi(\varkappa, t) = \sum_a p_a(T_0)\,\overline{\langle\psi_a|\, e^{-i\boldsymbol{\varkappa}\cdot\boldsymbol{r}}\, e^{iHt/\hbar}\, e^{i\boldsymbol{\varkappa}\cdot\boldsymbol{r}}\, e^{-iHt/\hbar}|\,\psi_a\rangle}. \tag{10.1.20b}$$

If we identify t with the time, we can consider the quantity

$$e^{iHt/\hbar}\, e^{i\boldsymbol{\varkappa}\cdot\boldsymbol{r}}\, e^{-iHt/\hbar} = e^{i\boldsymbol{\varkappa}\cdot\boldsymbol{r}(t)} \tag{10.1.21}$$

as a Heisenberg operator. Setting $F = e^{i\boldsymbol{\varkappa}\cdot\boldsymbol{r}}$, we can finally write

$$\chi(\varkappa, t) = \sum_a p_a(T_0)\,\overline{\langle\psi_a|\, F^*(0)\, F(t)\,|\,\psi_a\rangle}, \tag{10.1.20c}$$

Eq. (10.1.20c), which was derived by WICK (cf. also ZEMACH and GLAUBER), is the starting point of various approximate calculations of the inelastic scattering cross section. A survey of the various methods of making such calculations, as well as numerous citations from the literature, can be found in NELKIN. We shall restrict ourselves here to quoting results for several important special cases.

For an *isotropic harmonic oscillator*[2] with the frequency ω

$$\chi(\varkappa, t) = \exp\left\{\frac{\varkappa^2}{2M}\,\frac{\hbar}{\omega}\left[(\overline{n}+1)(e^{i\omega t} - 1) + \overline{n}(e^{-i\omega t} - 1)\right]\right\}. \tag{10.1.22}$$

Here

$$\overline{n} = \frac{\sum_{n=0}^{\infty} n\, e^{-\frac{\hbar\omega}{kT_0}\left(n+\frac{1}{2}\right)}}{\sum_{n=0}^{\infty} e^{-\frac{\hbar\omega}{kT_0}\left(n+\frac{1}{2}\right)}} = \frac{1}{e^{\frac{\hbar\omega}{kT_0}} - 1}$$

is the average occupation number of the oscillator and $M = A\, m_n$ is the mass of the scattering atom. Substitution of Eq. (10.1.22) into (10.1.20a) then gives the differential scattering cross section. Unfortunately the time integration cannot

[1] In deriving Eq. (10.1.20), use has been made of the closure property, viz., $\sum_b \langle\psi_a|\mathfrak{A}|\psi_b\rangle\langle\psi_b|\mathfrak{B}|\psi_a\rangle = \langle\psi_a|\mathfrak{A}\mathfrak{B}|\psi_a\rangle$; here $\mathfrak{A} = e^{-i\varkappa r} e^{i\frac{Ht}{\hbar}}$, $\mathfrak{B} = e^{i\varkappa r} e^{-i\frac{Ht}{\hbar}}$.

[2] i.e., when the scattering atom is bound in a fixed harmonic oscillator potential. This case is practically realized in an Einstein solid or in molecules in which the scattering atoms are bound to a heavy, slightly scattering rump.

be carried out in closed form, but a series expansion of $\chi(\varkappa, t)$, which leads to a series expansion of σ, is possible. Setting $y=e^{-\frac{\hbar\omega}{2kT_0}-i\omega t}$ we have

$$\chi(\varkappa, t)=\exp\left\{-\frac{\varkappa^2}{2M}\frac{\hbar}{\omega}\coth\frac{\hbar\omega}{2kT_0}\right\}\exp\left\{\frac{\varkappa^2\hbar}{4M\omega\sinh\frac{\hbar\omega}{2kT_0}}\left(y+\frac{1}{y}\right)\right\}. \quad (10.1.23\text{a})$$

The second factor in Eq. (10.1.23a) has the form of the generating function of the Bessel functions, so that

$$\left.\begin{aligned}\chi(\varkappa, t)=\exp&\left\{-\frac{\varkappa^2}{2M}\frac{\hbar}{\omega}\coth\frac{\hbar\omega}{2kT_0}\right\}\times\\ &\times\sum_{n=-\infty}^{+\infty}e^{-in\omega t-\frac{n\hbar\omega}{2kT_0}}I_n\left(\frac{\varkappa^2\hbar}{2M\omega\sinh\frac{\hbar\omega}{2kT_0}}\right).\end{aligned}\right\} \quad (10.1.23\text{b})$$

Substitution in Eq. (10.1.20a) and term-by-term integration gives

$$\left.\begin{aligned}\sigma_s(E'\to E, \cos\vartheta_0)=\frac{\sigma_{sb}}{2}\sqrt{\frac{E}{E'}}\exp&\left\{-\frac{\varkappa^2}{2M}\frac{\hbar}{\omega}\coth\frac{\hbar\omega}{2kT_0}\right\}\times\\ \times\sum_{n=-\infty}^{+\infty}e^{-\frac{n\hbar\omega}{2kT_0}}I_n&\left(\frac{\varkappa^2\hbar}{2M\omega\sinh\frac{\hbar\omega}{2kT_0}}\right)\delta(E'-E+n\hbar\omega).\end{aligned}\right\} \quad (10.1.24)$$

We see immediately that the individual summands on the right-hand side correspond to processes in which n "phonons" are either transferred from the neutron to the oscillator or from the oscillator to the neutron[1]. The term with $n=0$ describes elastic scattering, i.e., scattering in which the quantum state of the oscillator does not change. The exponential function before the sum plays the role played by the Debye-Waller factor in the theory of X-ray scattering. To see this let us consider elastic scattering, for which $E'=E$, $K'=K$, and $\varkappa^2=2K^2(1-\cos\vartheta_0)=\left(\frac{4\pi\sin\frac{\vartheta_0}{2}}{\lambda}\right)^2$ (where $\lambda=2\pi/K$ is the de Broglie wavelength), and note that the "average amplitude of oscillation" of a quantum mechanical oscillator of temperature T_0 is

$$\overline{u^2}=\frac{\hbar^2}{6M}\cdot\frac{1}{\hbar\omega}\coth\frac{\hbar\omega}{2kT_0}. \quad (10.1.25)$$

Then — when we disregard the factor $I_0\left(\frac{\varkappa^2\hbar}{2M\omega\sinh\frac{\hbar\omega}{2kT_0}}\right)$, which can frequently be set equal to unity — we find that

$$\sigma_{el}(\cos\vartheta_0)=\sigma_s(E\to E', \cos\vartheta_0)=\frac{\sigma_{sb}}{2}\exp\left\{-3\overline{u^2}\left(\frac{4\pi\sin\frac{\vartheta_0}{2}}{\lambda}\right)^2\right\}. \quad (10.1.26)$$

The characteristic angular distribution described by Eq. (10.1.26) arises because the neutrons are not scattered from a point scattering center but rather from a "thermal cloud" that is smeared out over a region of radius $\overline{u^2}$.

In the limiting case $E'\gg\hbar\omega$ and $kT_0\gg\hbar\omega$, the cross section should approach the scattering cross section of the monatomic gas derived in Sec. 10.1.1. However,

[1] For this reason, the expansion is called the "phonon" expansion.

we cannot show that this is the case using Eq. (10.1.24) since many terms then contribute to the cross section, i.e., the phonon expansion converges poorly.

We can arrive at a more convenient representation of the cross section by developing the function $\chi(\varkappa, t)$ in Eq. (10.1.22) in powers of its exponent. This procedure leads to

$$\chi(\varkappa, t)=\sum_{n=0}^{\infty}\frac{1}{n!}\left[\frac{\varkappa^2\hbar}{2M\omega}\{(\bar{n}+1)(e^{i\omega t}-1)+\bar{n}(e^{-i\omega t}-1)\}\right]^n \tag{10.1.26a}$$

and

$$\left.\begin{aligned}\sigma_s(E'\to E, \cos\vartheta_0)=\frac{\sigma_{sb}}{4\pi\hbar}\sqrt{\frac{E}{E'}}\sum_{n=0}^{\infty}\frac{1}{n!}\int_{-\infty}^{+\infty}e^{i(E-E')t/\hbar}\times\\ \times\left[\frac{\varkappa^2\hbar}{2M\omega}\{(\bar{n}+1)(e^{i\omega t}-1)+\bar{n}(e^{-i\omega t}-1)\}\right]^n dt.\end{aligned}\right\} \tag{10.1.26b}$$

The individual terms on the right-hand side are of the order of magnitude of $1/A^n$. In our discussion of scattering by a monatomic gas, we saw that in the case $A\gg 1$ many expressions simplified considerably. If we restrict ourselves to heavy nuclei, we can truncate the expansion of $\chi(\varkappa, t)$ after the second term ("heavy crystal approximation") and obtain

$$\left.\begin{aligned}\sigma_s(E'\to E, \cos\vartheta_0)=\frac{\sigma_{sb}}{2}\sqrt{\frac{E}{E'}}\left[\left(1-\frac{\varkappa^2}{2M}\frac{\hbar}{\omega}(2\bar{n}+1)\right)\delta(E-E')+\right.\\ \left.+\frac{\varkappa^2}{2M}\frac{\hbar}{\omega}\{(\bar{n}+1)\delta(E'-E-\hbar\omega)+\bar{n}\delta(E'-E+\hbar\omega)\}\right].\end{aligned}\right\} \tag{10.1.27a}$$

Then

$$\left.\begin{aligned}\sigma_s(E'\to E)=\int_{-1}^{+1}\sigma_s(E'\to E,\cos\vartheta_0)\,d\cos\vartheta_0=\sigma_{sb}\sqrt{\frac{E}{E'}}\left[\left(1-\frac{2E}{A\hbar\omega}\coth\frac{\hbar\omega}{2kT_0}\right)\times\right.\\ \times\delta(E-E')+\frac{E+E'}{2A\hbar\omega\sinh\frac{\hbar\omega}{2kT_0}}\times\\ \left.\times\left\{e^{\frac{\hbar\omega}{2kT_0}}\delta(E'-E-\hbar\omega)+e^{\frac{-\hbar\omega}{2kT_0}}\delta(E'-E+\hbar\omega)\right\}\right]\end{aligned}\right\} \tag{10.1.27b}$$

and

$$\left.\begin{aligned}\sigma_s(E')=\int_0^{\infty}\sigma_s(E'\to E)\,dE=\sigma_{sb}\left[\underbrace{\left(1-\frac{2E'}{A\hbar\omega}\coth\frac{\hbar\omega}{2kT_0}\right)}_{\text{elastic processes}}+\right.\\ \left.+\underbrace{\sqrt{1-\frac{\hbar\omega}{E'}}\frac{2E'-\hbar\omega}{A\hbar\omega}\frac{e^{\frac{\hbar\omega}{2kT_0}}}{2\sinh\frac{\hbar\omega}{2kT_0}}}_{\text{processes with energy loss}}+\underbrace{\sqrt{1+\frac{\hbar\omega}{E'}}\frac{2E'+\hbar\omega}{A\hbar\omega}\frac{e^{-\frac{\hbar\omega}{2kT_0}}}{2\sinh\frac{\hbar\omega}{2kT_0}}}_{\text{processes with energy gain}}\right].\end{aligned}\right\} \tag{10.1.27c}$$

When $E'\gg\hbar\omega$, Eq. (10.1.27c) becomes

$$\sigma_s(E')=\sigma_{sb}\left(1-\frac{2}{A}\right). \tag{10.1.27d}$$

This corresponds to the result for the heavy gas on page 184. There we found $\sigma_s(E')=\sigma_{sf}$, but $\sigma_{sf}=\frac{\sigma_{sb}}{\left(1+\frac{1}{A}\right)^2}\approx\sigma_{sb}\left(1-\frac{2}{A}\right)+$ higher terms in $\frac{1}{A}$, which we can consistently neglect here.

In analogy with what we did in Sec. 10.1.1, we can now calculate the averages

$$\overline{\sigma_s(\Delta E)^\nu}=\int_0^\infty \sigma_s(E'\to E)(E'-E)^\nu\, dE. \tag{10.1.28a}$$

Using Eq. (10.1.27b), we find that

$$\left.\begin{aligned}\int_0^\infty \sigma_s(E'\to E)(E'-E)\,dE=\sigma_{sb}\bigg[\frac{1}{2A\sinh(\hbar\omega/2kT_0)}\times\\ \times\bigg\{\sqrt{1-\frac{\hbar\omega}{E'}}\,(2E'-\hbar\omega)\,e^{\frac{\hbar\omega}{2kT_0}}-\sqrt{1+\frac{\hbar\omega}{E'}}\,(2E'+\hbar\omega)\,e^{-\frac{\hbar\omega}{2kT_0}}\bigg\}\bigg].\end{aligned}\right\} \tag{10.1.28b}$$

When $E'\gg\hbar\omega$, it follows that

$$\overline{\sigma_s(\Delta E)}=\frac{2}{A}\,\sigma_{sb}\left[E'-\hbar\omega\coth\frac{\hbar\omega}{2kT_0}\right], \tag{10.1.28c}$$

$$\overline{(\sigma_s\Delta E)}=\begin{cases}\dfrac{2}{A}\,\sigma_{sb}[E'-\hbar\omega] & \hbar\omega\gg kT_0\\[2mm] \dfrac{2}{A}\,\sigma_{sb}[E'-2kT_0] & \hbar\omega\ll kT_0.\end{cases} \tag{10.1.28d}$$

The last result again corresponds to a result for the heavy gas, Eq. (10.1.13a). We can see clearly how the chemical binding gradually becomes ineffective at high energies. Finally let us calculate M_2:

$$\left.\begin{aligned}M_2&=\frac{1}{(kT_0)^2}\int_0^\infty\int_0^\infty M(E')(E'-E)^2\sigma_s(E'\to E)\,dE\,dE'\\ &=\sigma_{sb}\frac{(\hbar\omega)^3K_2\left(\dfrac{\hbar\omega}{2kT_0}\right)}{2A\,(kT_0)^3\sinh\dfrac{\hbar\omega}{2kT_0}}.\end{aligned}\right\} \tag{10.1.29}$$

(In this connection, cf. PUROHIT). Here K_2 is the modified Bessel function of the second kind and second order. For $\hbar\omega\ll kT_0$, $K_2\left(\frac{\hbar\omega}{2kT_0}\right)=8\left(\frac{kT_0}{\hbar\omega}\right)^2$ and $M_2=8\sigma_{sb}/A$; this result is analogous to Eq. (10.1.16). Fig. 10.1.5 shows $M_2A/8\sigma_{sb}$ as a function of $\frac{\Theta}{T_0}=\frac{\hbar\omega}{kT_0}$ (Θ is the Einstein temperature). This figure shows particularly clearly the effect of chemical binding on the thermalizing power of a moderating substance. When $T_0>\Theta$, binding effects are barely palpable. When T_0 falls below Θ, M_2 decreases rapidly. This comes about because on the one hand energy loss processes are improbable since only a few neutrons in the Maxwell distribution have energies in excess of $k\Theta$, and because on the other hand energy gain processes are also improbable owing to the weak thermal excitation of the moderator.

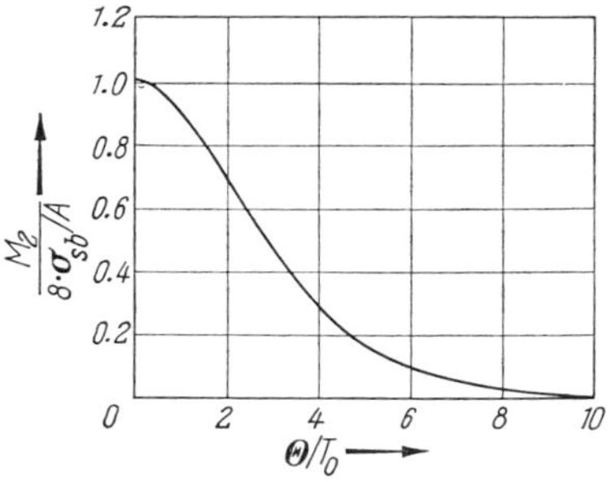

Fig. 10.1.5. The mean squared energy loss in an equilibrium Maxwellian neutron spectrum for a heavy Einstein crystal (relative to its heavy gas value)

Next we shall study scattering by a solid body. In the simple case that the lattice vibrations are harmonic and that there is only one atom in each unit cell, the function $\chi(\varkappa, t)$ is given by

$$\chi(\varkappa, t) = \exp\left\{\frac{\varkappa^2}{2M}\int_0^{\omega_{\max}} p(\omega)\frac{\hbar}{\omega}\left[(\bar{n}+1)(e^{i\omega t}-1)+\bar{n}(e^{-i\omega t}-1)\right]d\omega\right\}. \tag{10.1.30}$$

This expression differs from that of Eq. (10.1.22) only in that the exponent is averaged over the normalized distribution of lattice vibrations. According to the frequently used Debye model, this distribution is

$$p(\omega) = 3\left(\frac{\hbar}{k\Theta_D}\right)^3 \omega^2, \qquad 0 \leqq \omega \leqq \frac{k\Theta_D}{\hbar}. \tag{10.1.31}$$

Here Θ_D is the Debye temperature. We can also expand Eq. (10.1.30) in a phonon expansion, but we shall be hampered by the same difficulties as in the case of the Einstein crystal (cf. SJÖLANDER in this connection). It is nonetheless instructive to calculate the cross section for processes in which a *single* phonon is absorbed or emitted. We find[1]

$$\left.\begin{aligned}\sigma_{+1}(E' \to E, \cos\vartheta_0) &= \frac{\sigma_{sb}}{2}\sqrt{\frac{E}{E'}}\, e^{-2W}\frac{\varkappa^2}{4M\omega_0}\, p(\omega_0)\left[\coth\frac{\hbar\omega_0}{2kT_0} - 1\right];\\ &\qquad E > E'; \quad \hbar\omega_0 = E - E',\end{aligned}\right\} \tag{10.1.32a}$$

$$\left.\begin{aligned}\sigma_{-1}(E' \to E, \cos\vartheta_0) &= \frac{\sigma_{sb}}{2}\sqrt{\frac{E}{E'}}\, e^{-2W}\frac{\varkappa^2}{4M\omega_0}\, p(\omega_0)\left[\coth\frac{\hbar\omega_0}{2kT_0} + 1\right];\\ &\qquad E < E'; \quad \hbar\omega_0 = E' - E.\end{aligned}\right\} \tag{10.1.32b}$$

Here $e^{-2W} = \exp\left\{-\frac{\varkappa^2}{2M}\int_0^{\omega_{\max}} p(\omega)\frac{\hbar}{\omega}\coth\frac{\hbar\omega}{2kT_0}\,d\omega\right\}$ is the Debye-Waller factor. Eq. (10.1.32a) shows that for very low incident energies the spectrum of the scattered neutrons gives a picture of the spectrum of the lattice vibrations (albeit distorted by the temperature factor). Insofar as we can keep the contribution of multiphonon processes small, we can determine the spectrum of the lattice vibrations from measurements on the inelastically scattered neutrons.

For the following let us postulate the Debye model. Expansion of Eq. (10.1.30) in powers of the exponent of $\chi(\varkappa, t)$ and truncation of the series after the second term ("heavy crystal approximation") leads to

$$\left.\begin{aligned}\sigma_s(E' \to E, \cos\vartheta_0) &= \frac{\sigma_{sb}}{2}\sqrt{\frac{E}{E'}}\left\{\left[1 - \frac{3\varkappa^2\hbar^4}{2M(k\Theta_D)^3}\int_0^{k\Theta_D/\hbar}\omega(2\bar{n}+1)\,d\omega\right]\times\right.\\ &\quad\left.\times\,\delta(E-E') + \frac{3\varkappa^2\hbar^2}{2M(k\Theta_D)^3}\left[\frac{E'-E}{1-e^{-(E'-E)/kT_0}}\right]\right\}\\ &\text{for}\quad |E-E'| < k\Theta_D; \quad \sigma_s(E' \to E, \cos\vartheta_0) = 0 \text{ otherwise}\end{aligned}\right\} \tag{10.1.33a}$$

from which it follows that

$$\left.\begin{aligned}\sigma_s(E' \to E) &= \sigma_{sb}\left\{\left[1 - \frac{12E'}{A\,k\Theta_D}\,\Phi(T_0/\Theta_D)\right]\delta(E-E') + \right.\\ &\quad\left. + \frac{3}{A}\sqrt{\frac{E}{E'}}\,\frac{E+E'}{(k\Theta_D)^3}\,\frac{E'-E}{1-e^{-(E'-E)/kT_0}}\right\}\\ &\text{for}\quad |E'-E| \leqq k\Theta_D; \ \sigma_s(E' \to E) = 0 \text{ otherwise.}\end{aligned}\right\} \tag{10.1.33b}$$

[1] The phonon expansion is carried out in the following way: first the Debye-Waller factor is factored out, and then the remaining expression is expanded in powers of its exponent.

The function $\Phi(T_0/\Theta_D)=\frac{1}{4}+\frac{T_0^2}{\Theta_D^2}\int\limits_0^{\Theta_D/T_0}\frac{x\,dx}{e^x-1}$ has been tabulated by BLACKMANN[1]. We can again calculate $\sigma_s(E')$, $\overline{\sigma_s(\Delta E)}$, etc. from Eq. (10.1.33b), but the resulting expressions are very complicated. Finally, shown in Fig. 10.1.6, is

$$M_2=\sigma_{sb}\frac{3\hbar^6}{2A(k\Theta_D)^3(kT_0)^3}\int\limits_0^{k\Theta_D/\hbar}\frac{\omega^5 K_2\left(\frac{\hbar\omega}{2kT_0}\right)d\omega}{\sinh\left(\frac{\hbar\omega}{2kT_0}\right)} \qquad (10.1.33\text{c})$$

as a function of Θ_D/T_0 [cf. also Eqs. (10.1.29) and (10.1.31)]. For $\Theta_D/T_0<1$, the effect of the binding is again hardly noticeable, while for smaller temperatures the solid gradually "freezes up".

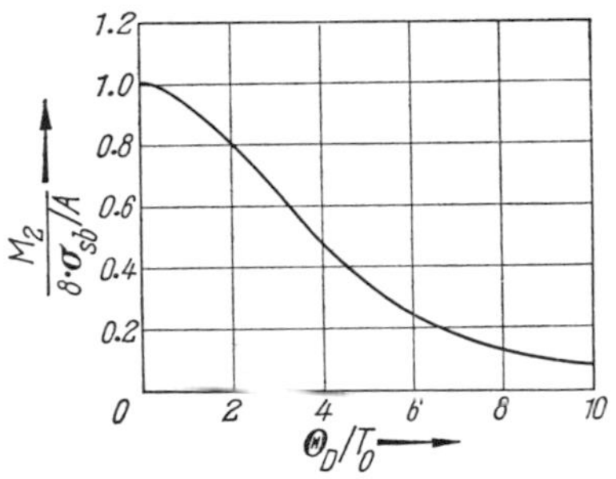

Fig. 10.1.6. The mean squared energy loss in an equilibrium Maxwellian neutron spectrum for a heavy Debye crystal (relative to its heavy gas value)

We can also use the quantum mechanical formalism developed here for a very short derivation of the scattering cross section for a monatomic gas, Eq. (10.1.7). In this case, the wave functions appearing in Eq. (10.1.20b) describe plane waves, and the Hamiltonian is that of free particles. Then (cf. ZEMACH and GLAUBER)

$$\chi(\varkappa,t)=\exp\left\{\frac{\hbar\varkappa^2}{2M}\left(it-\frac{kT_0}{\hbar}t^2\right)\right\} \qquad (10.1.34\text{a})$$

and with the help of Eq. (10.1.20a)

$$\left.\begin{aligned}&\sigma(E'\to E,\cos\vartheta_0)\\&\quad=\frac{\sigma_{sb}}{4\pi}\sqrt{\frac{E}{E'}}\sqrt{\frac{2\pi M}{\hbar^2\varkappa^2 kT_0}}\exp\left\{-\frac{M}{2kT_0\hbar^2\varkappa^2}\left[(E'-E)-\frac{\hbar^2\varkappa^2}{2M}\right]^2\right\}.\end{aligned}\right\} \qquad (10.1.34\text{b})$$

At this point, we need merely to integrate over all scattering angles ϑ_0 to obtain Eq. (10.1.7). It is instructive to compare Eqs. (10.1.34a) and (10.1.22). If we carry out in the latter an expansion of $e^{i\omega t}$ and $e^{-i\omega t}$ in powers of the time t, we obtain

$$\chi(\varkappa,t)=\exp\left\{\frac{\hbar\varkappa^2}{2M}\left[it-\frac{(\bar n+\frac{1}{2})\hbar\omega}{\hbar}t^2+\text{higher powers of } t\right]\right\}. \qquad (10.1.35\text{a})$$

In the term linear in t, both exponents agree. In the Fourier transformation that leads to the cross section, small t corresponds to large energy transfers $E'-E$. This means, however, that the differential cross sections for both bound and free atoms tend to the same value in the limit of large energy transfers. This limiting value is

$$\left.\begin{aligned}\sigma_s(E'\to E,\cos\vartheta_0)&=\frac{\sigma_{sb}}{2}\sqrt{\frac{E}{E'}}\frac{1}{2\pi\hbar}\int\limits_{-\infty}^{+\infty}\exp\left\{it\left[\frac{\hbar\varkappa^2}{2M}+\frac{E-E'}{\hbar}\right]\right\}dt\\&=\frac{\sigma_{sb}}{2}\sqrt{\frac{E}{E'}}\,\delta\left(E-E'+\frac{1}{A}\left[E'+E-2\sqrt{EE'}\cos\vartheta_0\right]\right)\end{aligned}\right\} \qquad (10.1.35\text{b})$$

which becomes

$$\sigma_s(E'\to E)=\frac{\sigma_{sb}A}{4E'}=\frac{\sigma_{sf}}{E'}\frac{(A+1)^2}{4A}=\frac{\sigma_{sf}}{(1-\alpha)E'} \quad \text{for} \quad \alpha E'<E<E' \qquad (10.1.35\text{c})$$

[1] M. BLACKMANN, "The Specific Heat of Solids", Handbuch der Physik, VII/1, 377 (1957).

after integration over all scattering angles ϑ. Here α has the same meaning as in Sec. 7.1.1; Eq. (10.1.35c) is then the result given there for scattering from free, stationary nuclei. We see now that the term in Eq. (10.1.34a) quadratic in t expresses modifications of the scattering cross section caused by the thermal motion of the scattering atoms. As $T_0 \to 0$, this so-called Doppler term vanishes. A Doppler term also appears in the case of the oscillator, but instead of the thermal energy kT_0 of the free particles, the mean thermal energy of the oscillators appears. The quadratic term (as well as the higher terms) is characteristically determined by the chemical binding. We denote as the Doppler approximation that approximation in which only the terms linear and quadratic in t (the latter, however, reflecting the nature of the chemical binding) are used in the calculation of σ. (In this connection, cf. PUROHIT and RAJAGOPAL.)

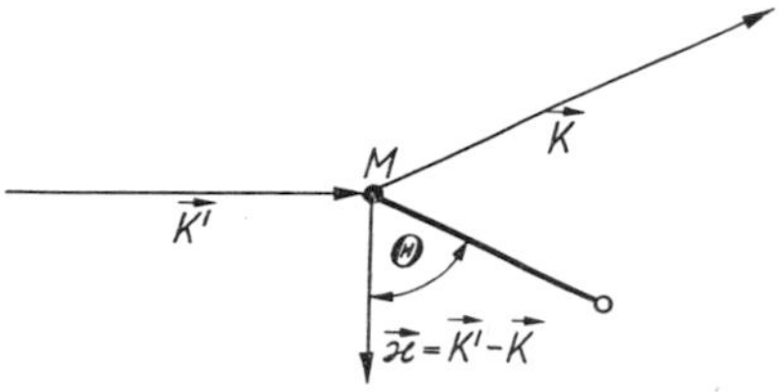

Fig. 10.1.7. Neutron scattering by a rigid rotator

The validity of the *principle of detailed balance* for the scattering systems considered here still remains to be investigated. In order that this principle hold [cf. Eq. (10.1.11)], we must clearly have

$$e^{-\frac{E}{kT_0}} \int_{-\infty}^{+\infty} \chi(\varkappa, t)\, e^{i\frac{(E'-E)}{\hbar}t}\, dt = e^{-\frac{E'}{kT_0}} \int_{-\infty}^{+\infty} \chi(\varkappa, t)\, e^{i\frac{(E-E')}{\hbar}t}\, dt \qquad (10.1.36\text{a})$$

or

$$\int_{-\infty}^{+\infty} \chi(\varkappa, t)\, e^{i\frac{(E'-E)}{\hbar}t}\, dt = \int_{-\infty}^{+\infty} \chi(\varkappa, t)\, e^{i\frac{(E-E')}{\hbar}\left(t-\frac{i\hbar}{kT_0}\right)}\, dt. \qquad (10.1.36\text{b})$$

Eq. (10.1.36b) will be true if

$$\chi(\varkappa, t) = \chi\left(\varkappa, -\left(t - \frac{i\hbar}{kT_0}\right)\right). \qquad (10.1.36\text{c})$$

We can easily convince ourselves using Eqs. (10.1.22), (10.1.30), and (10.1.34a) that this condition is actually fulfilled by $\chi(\varkappa, t)$. Eq. (10.1.36c) holds for every $\chi(\varkappa, t)$ defined by Eq. (10.1.20b).

Finally, let us consider neutron scattering by a free rotator. This problem has been studied by ZEMACH and GLAUBER as well as by VOLKIN; its exact treatment leads to very complicated expressions and allows the calculation of the cross section only in special cases. However, at sufficiently high neutron energies and temperatures[1] it is possible to use a simple semiclassical model. Let us consider the collision, depicted in Fig. 10.1.7, of a neutron with an atomic nucleus of mass M that is connected by a rigid link to a fixed point in such a way that it can rotate about this point in an arbitrary fashion ("rigid, free rotator"). Let the momentum of the neutron change from $\boldsymbol{p}'$ to $\boldsymbol{p}$ in the course of the collision, i.e., let the rotator take up the recoil momentum $\boldsymbol{p}' - \boldsymbol{p} = \hbar\boldsymbol{\varkappa}$. Furthermore, let the rotator be at rest before the collision. In the course of the collision, the rotator will receive an amount of energy ΔE; however, only components of $\hbar\boldsymbol{\varkappa}$ per-

[1] kT_0 as well as E' should be large compared to the energy of the lowest rotational level. In H_2O vapor, e.g., this level lies at about 0.0001 ev so that the condition is fulfilled over a wide range of neutron energies.

pendicular to the link joining the rotator to the fixed point are effective in accelerating it. Thus

$$\Delta E = E' - E = \frac{\hbar^2 \varkappa^2}{2M} \sin^2 \Theta . \tag{10.1.37a}$$

If the nucleus were completely free, ΔE would equal $\hbar^2\varkappa^2/2M$. In view of the relation between the energy and the momentum transferred, the rotator behaves as if it were a free nucleus whose "effective mass" depends on the angle Θ between $\varkappa$ and the axis of the link. If we average over all angles we obtain

$$\overline{\Delta E} = \frac{\hbar^2 \varkappa^2}{2M} \cdot \frac{1}{2} \int_{-1}^{+1} \sin^2 \Theta \, d \cos \Theta = \frac{\hbar^2 \varkappa^2}{2 M_{\text{eff}}} \tag{10.1.37b}$$

with $M_{\text{eff}} = 3M/2$. Averaged over all directions of $\boldsymbol{\varkappa}$, the rotator behaves like a free nucleus of mass M_{eff}. For polyatomic molecules, the calculation of the effective mass is of course more complicated. To calculate the cross section, we can now use Eqs. (10.1.34a)ff., replacing M everywhere by M_{eff}. This simple approximate procedure originated in work of Sachs and Teller and of Brown and St. John; its validity has been investigated in some detail by Krieger and Nelkin, among others. However, since free rotation does not occur for the moderating substances important in practice, we shall not go into this problem any more deeply here.

10.1.3. Neutron Scattering in Water, Beryllium, and Graphite

Now with the help of the formalism developed in the last section, we shall try to determine what can be said about the scattering of neutrons by real moderators. We must expect right from the start that the best we shall be able to achieve in this way is a very rough description of the scattering process since the actual motions of the atoms in solids and liquids are much more complicated than those in the models developed in Sec. 10.1.2.

Water. The free water molecule can execute oscillations with quantum energies of 0.198, 0.474, and 0.488 ev. Its rotational spectrum is very complicated and has states with quantum energies down to 10^{-4} ev. When the water molecule is not free but interacts with others in a sample of liquid water, the vibrational energies remain essentially the same. On the other hand, the rotation of the water molecules is strongly hindered by their strong dipole interaction with their neighbors. Instead of rotation, a band of torsional oscillation appears, whose quantum energies (according to measurements of Raman spectra) are centered roughly at 60 mev. Thus for $E' \gtrsim 60$ mev a neutron can transfer energy to the torsional oscillations, and for $E' > 0.2$ ev it can transfer energy to the vibrational motions. The inverse processes are rare since at room temperature both the vibrational degrees of freedom and the torsional oscillations are only slightly excited. For this reason, when $E' < 60$ mev the neutron can only exchange energy with the translatory motion of the water molecule as a whole. These latter motions are also hindered, as we know, for example, from measurements of the specific heat; but in the calculation of the scattering cross section they can be considered free with an accuracy adequate for thermalization calculations. Starting with these ideas, Nelkin carried out a calculation of the scattering for water. He set

$$\chi(\varkappa, t) = \chi_T(\varkappa, t)\, \chi_R(\varkappa, t)\, \chi_V(\varkappa, t). \tag{10.1.38a}$$

Here

$$\chi_T(\varkappa, t) = \exp\left\{\frac{\hbar^2 \varkappa^2}{2M}\left(it - \frac{kT_0}{\hbar}\, t^2\right)\right\} \tag{10.1.38b}$$

is the translatory part with $M = 18\, m_p$. χ_R describes the torsional oscillation: it is treated as an ordinary oscillation with an effective mass $m_R = 2.32\, m_p$ (KRIEGER and NELKIN). Then

$$\chi_R(\varkappa, t) = \exp\left\{\frac{\hbar \varkappa^2}{2 m_R \omega}\left[(\bar{n}+1)\,(e^{i\omega t}-1) + \bar{n}\,(e^{-i\omega t}-1)\right]\right\} \tag{10.1.38c}$$

with $\hbar\omega = 0.060$ ev. Finally, the vibrational part is given by[1]

$$\chi_V(\varkappa, t) = \exp\left[\frac{\hbar \varkappa^2}{2 m_V}\left\{\frac{1}{3\omega_1}\,(e^{i\omega_1 t}-1) + \frac{2}{3\omega_2}\,(e^{i\omega_2 t}-1)\right\}\right] \tag{10.1.38d}$$

where $\hbar\omega_1 = 0.20$ ev, and the two vibrational states at 0.474 and 0.488 ev have been combined into a single state at $\hbar\omega_2 = 0.481$ ev. The mass m_V is determined by the following considerations. For very large energy transfers, the scattering cross section should go over into that for free protons, i.e.,

$$\lim_{t\to 0} \chi(\varkappa, t) = \exp\left\{\frac{\hbar \varkappa^2}{2 m_p}\, it\right\}. \tag{10.1.38e}$$

If we expand the arguments of the exponentials in Eqs. (10.1.38c and d) in powers of t and combine Eqs. (10.1.38b, c, and d), we obtain

$$\chi(\varkappa, t) = \exp\left\{\frac{\hbar \varkappa^2}{2}\left[\frac{1}{M} + \frac{1}{m_R} + \frac{1}{m_V}\right] it + \cdots\right\}. \tag{10.1.38f}$$

Thus $1/m_p$ must equal $1/M + 1/m_R + 1/m_V$, from which it follows that $m_V = 1.95\, m_p$.

The Fourier transformation that leads from Eq. (10.1.38a) to the scattering cross section cannot be carried out explicitly in general; however, in particular ranges of incident neutron energy the expressions for $\chi(\varkappa, t)$ can be still further simplified. Thus for incident energies under about 0.20 ev, only elastic interactions with the vibrational degrees of freedom need be taken into account, and $\chi_V(\varkappa, t)$ reduces to a Debye-Waller factor. If we introduce a phonon expansion in this region, then we obtain

$$\left.\begin{aligned}\sigma_s(E' \to E, \cos\vartheta_0) = \frac{\sigma_{sb}}{2}\sqrt{\frac{E}{E'}}\sqrt{\frac{M}{2\pi k T_0 \hbar^2 \varkappa^2}}\exp\left\{-\frac{\hbar \varkappa^2}{2}\, p\right\} \cdot \sum_{n=-\infty}^{+\infty} e^{-\frac{n\hbar\omega}{2kT_0}} \times \\ \times I_n\left(\frac{\varkappa^2 \hbar}{2 m_R \omega \sinh \frac{\hbar\omega}{2kT_0}}\right)\exp\left\{-\frac{M}{2kT_0 \hbar^2 \varkappa^2}\left(E - E' - n\hbar\omega + \frac{\hbar^2 \varkappa^2}{2M}\right)^2\right\}\end{aligned}\right\} \tag{10.1.38g}$$

where

$$p = \frac{1}{3 m_V}\left[\frac{1}{\omega_1} + \frac{2}{\omega_2}\right] + \frac{\coth \frac{\hbar\omega}{2kT_0}}{m_R \cdot \omega}.$$

[Cf. Eqs. (10.1.24) and (10.1.34b).] For $E < 0.1$ ev, i.e., in the true thermal energy range, only the terms with $n = 0$ and $n = \pm 1$ are important. In other words, only

[1] Since $kT_0 \ll \hbar\omega_1$ and $\hbar\omega_2$, $\bar{n} = 0$ was assumed.

processes need be taken into account in which either no energy at all is exchanged with the torsional oscillations or only one quantum is emitted or absorbed. Above 0.2 ev and in particular above 0.48 ev, we must take into account the loss of energy to the vibrational modes; however, $\chi_T(\varkappa, t)$ may then be simplified by a Doppler approximation. We shall not give the resulting expressions for $\sigma_s(E' \to E)$ here.

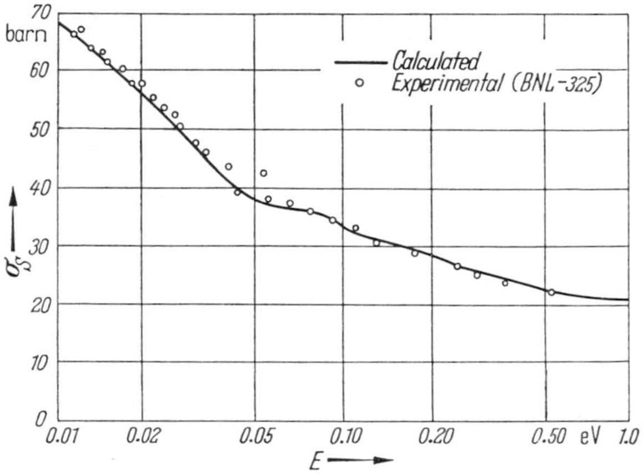

Fig. 10.1.8. A comparison of the measured H_2O total scattering cross section and calculations based on Eq. (10.1.38g)

In Fig. 10.1.8, measured values of the total scattering cross section are compared with calculations based on the models developed here. The agreement is good. The theory also reproduces quite faithfully the average cosine of the scattering angle averaged over all secondary energies as a function of the incident neutron's energy (cf. Fig. 10.1.9). The model works exceptionally well in the calculation of integral thermalization effects, as we shall see later[1].

It seems plausible that the model developed here could also be used to describe neutron scattering in *heavy water*, but there is the following objection to this procedure. The formalism developed in Sec. 10.1.2 holds for purely incoherent scattering, i.e., scattering in which all interference phenomena are neglected. Its application to water is justified only because the protons mainly scatter incoherently[2] ($\sigma_{sb} = 4\sigma_{sf} = 81.5$ barn, $\sigma_{coh} = 1.79$ barn). On the other hand, the scattering in deuterium is largely coherent ($\sigma_{sf} = 7.6$ barn, $\sigma_{coh} = 5.4$ barn), and we may expect that interference effects play an important role in scattering. A calculation of the scattering cross section taking this effect into account has not yet been carried out[3]. However, HONECK has shown (taking into account the difference of mass and scattering cross section) that a modified Nelkin model correctly describes various integral thermalization effects that have been investigated in D_2O.

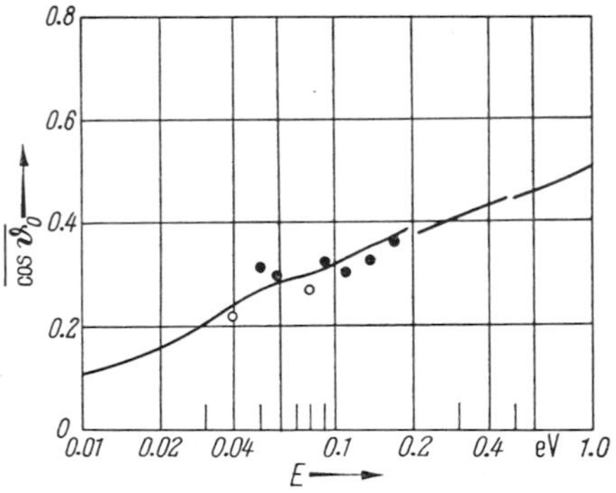

Fig. 10.1.9. The average cosine of the scattering angle for slow neutron scattering on H_2O. ● ○ Measurements; ——— calculation with Eq. (10.1.38g)

Crystalline Moderators. Both beryllium and graphite predominantly scatter neutrons coherently. For Be, $\sigma_{sb} \approx \sigma_{coh} = 7.53$ barn while for C, $\sigma_{sb} \approx \sigma_{coh} =$

---

[1] The model yields $M_2 = 91$ barns per H_2O molecule at room temperature.

[2] This is no longer true at very low temperatures, where partial spin correlations are possible.

[3] Note added in proof: Cf., however, D. BUTLER, Proc. Phys. Soc. 81, Part 2, No. 520, 276 (1963).

5.50 barn. The "incoherent approximation", i.e., the neglect of interference effects in the calculation of the scattering cross section, will clearly fail completely in the calculation of the cross section for elastic scattering processes. In Sec. 1.4.3 we showed that in coherently scattering crystalline moderators the total cross section exhibits a "cut-off" behavior, which comes about because coherent elastic scattering is no longer possible for neutron wavelengths that are larger than twice the largest separation between lattice planes. Naturally, calculations based on the incoherent approximation cannot reproduce this behavior. Also the continuous angular distribution predicted by the incoherent calculations is completely wrong since the actual angular distribution of scattering from a polycrystal exhibits sharp maxima at discrete scattering angles (in the sense of Debye-Scherrer rings). Only at neutron energies so high that reflections of many orders on essentially every conceivable set of lattice planes take place (about 0.05 ev in Be and 0.03 ev in graphite) can the cross section and angular distribution for elastic scattering be calculated in the incoherent approximation with satisfactory accuracy. Interference effects also occur in inelastic scattering and have been treated among others by PLACZEK and VAN HOVE and by SINGWI and KOTHARI. MARSHALL and STUART have calculated the energy-dependent cross sections of magnesium and aluminum for processes in which one or more phonons are emitted or absorbed including interference effects and have shown that neglect of the interference effects causes only small errors (several %). From these calculations and from some qualitative arguments, one can conclude that in thermalization calculations the inelastic scattering cross section can be described with sufficient accuracy by the incoherent approximation developed in Sec. 10.1.2. As we shall see later, elastic scattering plays no role in space-independent thermalization problems and therefore we introduce no additional errors by treating it in the incoherent approximation. However, if the thermalization process is coupled with the transport of neutrons, elastic scattering processes play an important role, and we must supply elastic scattering cross sections either from measurements or from special calculations. The elastic scattering cross section for beryllium has been calculated by BHANDARI and that of graphite by KUBCHANDANI *et al.*

To calculate the scattering cross section in the incoherent approximation, we start from Eq. (10.1.30). The main difficulties in thermalization calculations for solids are the choice of the frequency distribution function on one hand and the inversion of the Fourier transform, which leads from the function $\chi(\varkappa, t)$ to the cross section on the other. In *beryllium* a Debye spectrum $p(\omega)$ with $\Theta_D=1000$ °K approximates the actual spectrum of the lattice vibrations rather well. The inelastic scattering cross section in the "heavy crystal approximation" is then given by Eqs. (10.1.33a, b, and c). As NELKIN has shown, the scattering cross sections calculated in this way reproduce quite well the values measured with cold neutrons, even though this approximation does not seem justified because of beryllium's small mass ($A=9$). Because of the strongly anisotropic structure of *graphite*, its spectrum of lattice vibrations cannot be approximated by a Debye distribution. In fact, we must distinguish between the spectrum of vibrations parallel to the lattice planes of the graphite and the spectrum of vibrations perpendicular to them. The first extends to $\hbar\omega_{\|}/k=2600$ °K, the second to $\hbar\omega_{\perp}/k=1600$ °K. In a very rough first approximation, we can set

$$\chi(\varkappa, t) = \tfrac{1}{3}\chi_\perp(\varkappa, t) + \tfrac{2}{3}\chi_\parallel(\varkappa, t) \qquad (10.1.39\,\text{a})$$

where $\chi_\perp$ and $\chi_\parallel$ are calculated with Debye spectra corresponding to Θ_D of 1600 °K and 2600 °K, respectively. A much more satisfactory model has recently been developed by PARKS (cf. also TAKAHASHI). PARKS starts from

$$\chi(\varkappa, t) = \exp\left\{\frac{\varkappa^2\hbar}{2M}\int_0^\infty\int_0^1 \frac{p(\omega, l)}{\omega}\left[(\bar{n}+1)(e^{i\omega t}-1)+\bar{n}(e^{-i\omega t}-1)\right] dl\, d\omega\right\} \qquad (10.1.39\,\text{b})$$

with

$$p(\omega, l) = l^2 p_\perp(\omega) + (1-l^2)\, p_\parallel(\omega). \qquad (10.1.39\,\text{c})$$

Here l is the cosine of the angle between the direction of $\boldsymbol{\varkappa}$ and the normal to the lattice planes, and $p_\perp(\omega)$ and $p_\parallel(\omega)$ are the frequency distribution functions for vibrations perpendicular and parallel to the lattice planes, respectively. $p_\perp(\omega)$ and $p_\parallel(\omega)$ are taken from the theoretical calculations of YOSHIMORI and KITANO[1]. In order to calculate the cross section from the $\chi(\varkappa, t)$ specified above, a phonon expansion was carried out. A great many terms must be included in this expansion, and the resulting expressions are complex and can only be manipulated with large electronic computing machines. We shall become familiar with some numerical results later[2].

In the discussion of neutron scattering, we shall restrict ourselves to the above cases. Calculations of the scattering cross section of BeO have been done by BHANDARI *et al.*, of zirconium hydride by MILLER *et al.*, and of various organic moderators by BOFFI *et al.* and by GOLDMANN and FEDERIGHI.

10.1.4. Experimental Investigation of Slow Neutron Scattering

Because of the great difficulty of theoretically calculating differential scattering cross sections, it seems clear that we must measure them. Over and above their special necessity in neutron physics, such studies hold great interest for the physics of the solid and liquid states since they enable us to draw conclusions about the internal dynamics of solids and liquids. This aspect of slow neutron scattering, however, is outside the scope of this book[3]. The principle of the measurements is the following. A scattering sample which is small enough to exclude multiple scattering is bombarded with monoenergetic neutrons in the range 0—0.5 ev and the intensity of the scattered neutrons measured as a function of angle and energy. We must clearly always construct two neutron spectrometers, and in principle all the methods of making monochromatic neutrons discussed in Sec. 2.6.3 can be employed. The best suited for these investigations is the *time-of-flight method*, which is explained in Fig. 10.1.10. A neutron beam from a reactor first encounters a "pulsing monochromator". Here neutrons in a narrow energy band around E' are sorted out, and the continuous beam is chopped into short pulses. The neutrons are then scattered and can be detected by a large number

[1] A. YOSHIMORI and Y. KITANO, J. Phys. Soc. Japan **11**, 352 (1956).

[2] PARKS obtains $M_2 = 1.068$ barn in room temperature graphite.

[3] Cf., e.g., Inelastic Scattering of Neutrons in Solids and Liquids, Proc. of the 1962 Chalk River Conference; Vol. I and II. Vienna: International Atomic Energy Agency, 1963.

of detectors placed at various scattering angles ϑ_0. Each detector is connected with a time analyzer; thus detection is equivalent to an energy measurement. As the pulsing monochromator we frequently use a *double chopper*, i.e., two phase-synchronized choppers located one behind the other in which phase shift, angular velocity, and mutual distance define the energy E' (EGELSTAFF, BRUGGER). Less frequently used is the *rotating crystal spectrometer* (Fig. 10.1.10) (BROCKHOUSE, GLÄSER). In this instrument, a small, rapidly rotating single crystal is located between suitable entrance and exit collimators. If reflections of order higher than the first and reflections from undesired lattice planes can be suppressed, the reflection condition is fulfilled only twice per revolution; thus pulses of monochromatic neutrons leave the exit collimator. We shall not go into the numerous technical problems associated with such neutron spectrometers any more deeply here. The measurements yield the energy and angle distribution of the scattered neutrons in relative units. In order to obtain the cross section $\sigma_s(E'\to E, \cos\vartheta_0)$ from these data, we normalize them using the value of the elastic scattering cross section of *vanadium* ($\sigma_{sb}=5.13\pm0.02$ barns). Here we can neglect coherent effects and calculate the angular distribution of the elastically scattered neutrons exactly using the Debye-Waller factor.

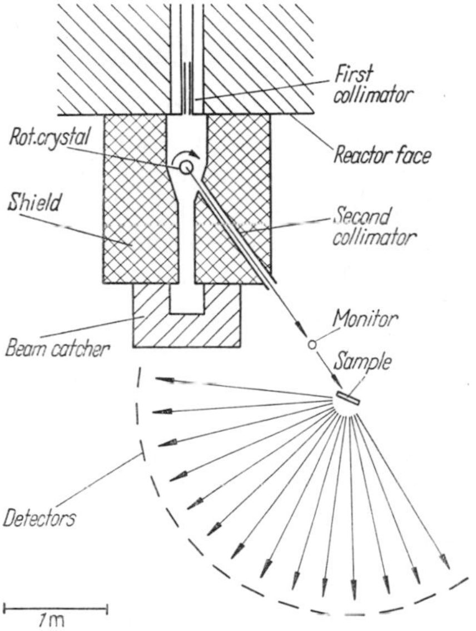

Fig. 10.1.10. A typical set-up for slow neutron inelastic scattering measurements

In the evaluation of such scattering measurements, it is customary to introduce an even function $S(\varkappa, \hbar\omega)$ defined in the following way:

$$\sigma_s(E'\to E, \cos\vartheta_0)=2\pi\sqrt{\frac{E}{E'}}\,e^{-\frac{\hbar\omega}{2kT_0}}\,S(\varkappa,\hbar\omega). \tag{10.1.40}$$

Here $\hbar\omega=E'-E$. In the defining Eq. (10.1.40), we make use of the fact that aside from the factor $\sqrt{\frac{E}{E'}}$, $\sigma_s(E'\to E, \cos\vartheta_0)$ depends only on the magnitude of the momentum transfer $\hbar\varkappa$ [cf. Eq. (10.1.17) and the related discussion]. Furthermore, *every* cross section formed according to Eq. (10.1.40) (that is with an arbitrary even function $S(\varkappa, \hbar\omega)$) satisfies the principle of detailed balance. Thus the measured cross section, which is a function of the three parameters, E', E, and $\cos\vartheta_0$, can be expressed in terms of the two-parameter function $S(\varkappa, \hbar\omega)$, whose behavior is easier to grasp. What is more, since for given values of $\varkappa$ and $\hbar\omega$, S can be determined in various ways (i.e., by more than one suitable combination of E', E, and $\cos\vartheta_0$), a check on the consistency of the method of measurement is possible. A serious problem in evaluating scattering measurements arises out of the fact that for want of intensity at very small and very large neutron energies, the S-function can only be studied in a restricted range of $\varkappa$ and ω.

With the aid of simple assumptions about the scattering process, however, the experimentally determined S-values can be extrapolated over a much wider range of $\varkappa$ and ω. This has been done by EGELSTAFF and SCHOFIELD among others.

BRUGGER has compiled a summary of all known $S(\varkappa, \hbar\omega)$-measurements. We can use these results either as numerical input data in thermalization problems or to check and improve the theoretical models for calculating differential cross sections.

10.2. The Calculation of Stationary Neutron Spectra

In practice, the most important problem in thermalization physics is the calculation of the space- and energy-dependent neutron flux $\Phi(\boldsymbol{r}, E)$ in a finite, absorbing, and sometimes even inhomogeneous medium due to specified sources of fast neutrons. To solve this problem, we must start from the general transport Eq. (5.1.12) and seek a solution based on the assumption of either a theoretically calculated or an empirical scattering law. Even when we apply very far-reaching approximations, a simple analytic treatment is no longer possible in the general space-dependent case; we can only obtain numerical results by the use of large electronic calculating machines.

The space-independent case (an infinite medium with homogeneously distributed sources) presents a much simpler situation. Here the simple balance equation

$$\Sigma_a(E)\,\Phi(E) = L\Phi(E) + S(E) \tag{10.2.1}$$

with the thermalization "operator"

$$L\Phi = \int_0^\infty \Sigma_s(E' \to E)\,\Phi(E')\,dE' - \Sigma_s(E)\,\Phi(E) \tag{10.2.2a}$$

applies. L gives the excess of neutrons scattered into the energy E over those scattered out. In view of the principle of detailed balance, the equation

$$LM = 0 \tag{10.2.2b}$$

holds for a Maxwell spectrum $M(E) = \frac{E}{(kT_0)^2}\, e^{-E/kT_0}$. Furthermore, for any physically significant $\Phi(E)$

$$\int_0^\infty L\Phi(E)\,dE = 0. \tag{10.2.2c}$$

$S(E)$ represents the source density in Eq. (10.2.1). Since we are interested in the spectrum at energies that are small compared to the large energies of the source neutrons, we can take the source energy to be infinite. We need not then include the source term explicitly in Eq. (10.2.1) but must require instead that the flux is properly normalized, i.e., that $\int_0^\infty \Sigma_a(E)\,\Phi(E)\,dE = S$. Although the space-independent case does not occur rigorously in practice, its treatment is still of great interest because it provides insight into the mechanism of thermalization and because in some cases it is possible to reduce space-dependent problems to space-independent ones. For this reason, we first treat space-independent spectra in detail; following this, we append a short discussion of space-dependent spectra in which we limit ourselves to qualitative considerations.

10.2.1. The Space-Independent Spectrum in a Heavy Gas Moderator

A closed analytic solution of Eq. (10.2.1) cannot be found for any of the scattering laws introduced in Sec. 10.1. However, two simple limiting cases of Eq. 10.2.1 can be reduced to differential equations of the second order whose further treatment is simpler than that of the original integral equation. The first case concerns a gas of protons (WIGNER and WILKINS) and will not be treated further here. The second case, which we shall investigate in detail, is that of a thermal gas of heavy nuclei.

To begin with, let us introduce the quantity $\psi(E)=\Phi(E)/M(E)$ into Eq. (10.2.1). With the help of the law of detailed balance the latter becomes

$$\Sigma_a(E)\psi(E)=\int_0^\infty \psi(E')\Sigma_s(E\to E')\,dE'-\Sigma_s(E)\psi(E). \tag{10.2.3a}$$

Now let us develop $\psi(E')$ in a Taylor series around the point $E'=E$; then

$$\Sigma_a(E)\psi(E)=\sum_{n=1}^{\infty}\frac{(-1)^n}{n!}\frac{d^n\psi(E)}{dE^n}\overline{\Sigma_s(\Delta E)^n}. \tag{10.2.3b}$$

Here $\overline{\Sigma_s(\Delta E)^n}=N\overline{\sigma_s(\Delta E)^n}$ are the moments of the scattering cross section defined in Eq. (10.1.12). In Sec. 10.1.1 it was shown that for a heavy gas the first two moments are of the order of magnitude of $1/A$ while all the other moments are of higher order. If we take $\overline{\sigma_s\Delta E}$ and $\overline{\sigma_s(\Delta E)^2}$ from Eq. (10.1.12), then in leading order

$$\Sigma_a(E)\psi(E)=\xi\Sigma_s\left[(2kT_0-E)\frac{d\psi}{dE}+EkT_0\frac{d^2\psi}{dE^2}\right]. \tag{10.2.3c}$$

Here we have introduced the approximation $\xi=2/A$. The flux $\Phi(E)$ is then given by

$$\Sigma_a(E)\Phi(E)=\frac{d}{dE}\xi\Sigma_s\left\{(E-kT_0)\Phi+EkT_0\frac{d\Phi}{dE}\right\}, \tag{10.2.4}$$

i.e., for a heavy gas the thermalization operator is a simple second-order differential operator. We can easily check that Eq. (10.2.2b) holds for this operator.

Let us next consider Eq. (10.2.4) in the limiting case $E\gg kT_0$! We can then set $kT_0=0$ and in the absence of absorption obtain

$$\frac{d}{dE}\left(\xi\Sigma_s E\Phi(E)\right)=0 \tag{10.2.5a}$$

which has the solution

$$\Phi(E)=\frac{\text{const.}}{\xi\Sigma_s E}. \tag{10.2.5b}$$

This is the $1/E$-behavior of the epithermal flux familiar from the theory of slowing down. The constant follows from the requirement that $\lim_{E\to\infty}\xi\Sigma_s E\Phi(E)$ be equal to the source density.

For all the succeeding considerations of this subsection, we assume $1/v$-absorption, viz., $\Sigma_a(E)=\Sigma_a(kT_0)\sqrt{kT_0/E}$. Fig. 10.2.1 shows the values of $E\Phi(E)$ obtained by HURWITZ *et al.* for various values of $\Sigma_a(kT_0)/\xi\Sigma_s$ as functions of $\sqrt{E/kT_0}$. The values were obtained by numerical solution of Eq. (10.2.4) (in this connection cf. also COHEN). In such a solution, we must take as a boundary condition $\Phi(E=0)=0$; the solution still contains a constant factor as a free parameter which can be determined from the requirement $\int_0^\infty \Sigma_a(E)\,\Phi(E)\,dE$ = source density. Under certain circumstances, another normalization may be convenient. We can see from Fig. 10.2.1 that only in the presence of weak absorption $\left(\Sigma_a(kT_0) \ll \xi\Sigma_s\right)$ does a "thermal maximum" exist. In the presence of strong absorption, most of the neutrons are absorbed before they become thermal.

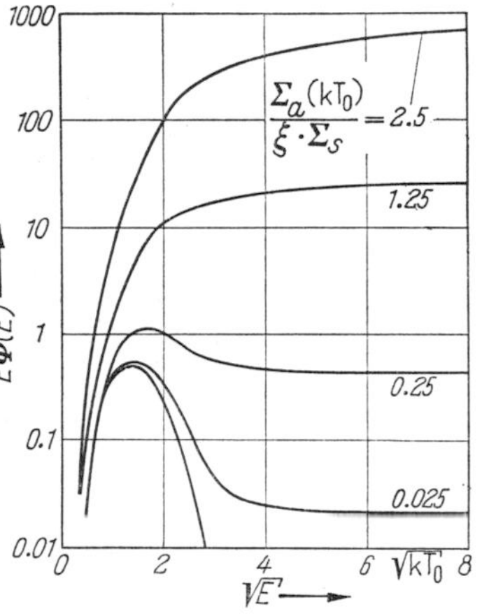

Fig. 10.2.1. The neutron spectrum in a thermal heavy gas moderator

Weak absorption. The case of weak absorption is of great interest since in many media $\Sigma_a(kT_0) \ll \xi\Sigma_s$. Here an instructive approximate solution of Eq. (10.2.4) is possible. Let us set

$$\Phi(E) = M(E) + F(E), \qquad (10.2.6\text{a})$$

i.e., we split the flux into a Maxwellian component and a perturbation which clearly vanishes when $\Sigma_a = 0$. (We shall return to the question of normalization later.) If we substitute this form in Eq. (10.2.4), we find

$$\Sigma_a(E)\,M(E) + \Sigma_a(E)\,F(E) = LF. \qquad (10.2.6\text{b})$$

As long as the absorption is small, the term on the left-hand side involving $F(E)$ is a small perturbation, and we can neglect it in first approximation. Then

$$\Sigma_a M(E) = LF = \frac{d}{dE}\left[\xi\Sigma_s\left\{(E - kT_0)\,F(E) + E\,kT_0\frac{dF}{dE}\right\}\right]. \qquad (10.2.6\text{c})$$

In the case of $1/v$-absorption, integration of this inhomogeneous differential equation yields

$$\left.\begin{aligned} F(E) = \frac{\Sigma_a(kT_0)}{\xi\Sigma_s}\,M(E)\int_0^{E/kT_0}\frac{e^y}{y^2}\int_0^y\sqrt{x}\,e^{-x}\,dx\,dy + \\ + \left\{a\,M(E) + b\left[M(E)\,Ei^*\left(\frac{E}{kT_0}\right)\right]\right\}.^1 \end{aligned}\right\} \qquad (10.2.6\text{d})$$

The term in braces is the general solution of the homogeneous equation $LF = 0$ and contains two free constants, a and b. Since $F(E=0)$ must equal zero, $b = 0$. We can determine the constant a from a neutron balance. Clearly,

$$\lim_{E\to\infty}\xi\Sigma_s E\,\Phi(E) = \int_0^\infty \Sigma_a(E)\,\Phi(E)\,dE, \qquad (20.2.7\text{a})$$

i.e., the number of neutrons absorbed per cm³ per sec must equal the source density. Now, on the one hand

$$\left.\begin{aligned} \lim_{E\to\infty}\xi\Sigma_s E\,\Phi(E) &= \lim_{E\to\infty}\xi\Sigma_s E\,F(E) \\ &= \lim_{E\to\infty}\Sigma_a(kT_0)\,E\,M(E)\int_0^{E/kT_0}\frac{e^y}{y^2}\int_0^y\sqrt{x}\,e^{-x}\,dx\,dy = \frac{\sqrt{\pi}}{2}\,\Sigma_a(kT_0) \end{aligned}\right\} \qquad (10.2.7\text{b})$$

---

[1] Cf. footnote on p. 246.

while on the other hand for $1/v$-absorption

$$\left.\begin{aligned}\int_0^\infty \Sigma_a[M(E)+F(E)]\,dE &= \frac{\sqrt{\pi}}{2}\,\Sigma_a(kT_0)(1+a)+\\ &+\frac{\Sigma_a(kT_0)}{\xi\Sigma_s}\int_0^\infty \Sigma_a(E)M(E)\int_0^{E/kT_0}\frac{e^y}{y^2}\int_0^y \sqrt{x}\,e^{-x}\,dx\,dy\,dE.\end{aligned}\right\}\qquad(10.2.7\,\mathrm{c})$$

Now (cf. COHEN)

$$\left.\begin{aligned}&\int_0^\infty \Sigma_a(E)M(E)\int_0^{E/kT_0}\frac{e^y}{y^2}\int_0^y \sqrt{x}\,e^{-x}\,dx\,dy\,dE\\ &=\Sigma_a(kT_0)\int_0^\infty \sqrt{z}\,e^{-z}\int_0^z\frac{e^y}{y^2}\int_0^y\sqrt{x}\,e^{-x}\,dx\,dy\,dz=\Sigma_a(kT_0)\times 0.7989\times 2\sqrt{\pi}.\end{aligned}\right\}\qquad(10.2.7\,\mathrm{d})$$

If follows by equating Eqs. (10.2.7b and c) that

$$a=-\frac{\Sigma_a(kT_0)}{\xi\Sigma_s}\times 0.7989\times 4.$$

Thus we finally obtain

$$\Phi(E)=M(E)+\frac{\Sigma_a(kT_0)}{\xi\Sigma_s}\,M(E)\left[\int_0^{E/kT_0}\frac{e^y}{y^2}\int_0^y\sqrt{x}\,e^{-x}\,dx\,dy-4\times 0.7989\right]\qquad(10.2.7\,\mathrm{e})$$

or

$$\Phi(E)=M(E)+\frac{\frac{\sqrt{\pi}}{2}\,\Sigma_a(kT_0)}{\xi\Sigma_s}\,F_1(E).\qquad(10.2.8\,\mathrm{a})$$

Here we have introduced the function

$$F_1(E)=M(E)\left[\frac{2}{\sqrt{\pi}}\int_0^{E/kT_0}\frac{e^y}{y^2}\int_0^y\sqrt{x}\,e^{-x}\,dx\,dy-\frac{8}{\sqrt{\pi}}\times 0.7989\right]\qquad(10.2.8\,\mathrm{b})$$

which for high energies approaches $1/E$. Fig. 10.2.2 shows $M(E)$ and $F_1(E)$ as functions of E/kT_0.

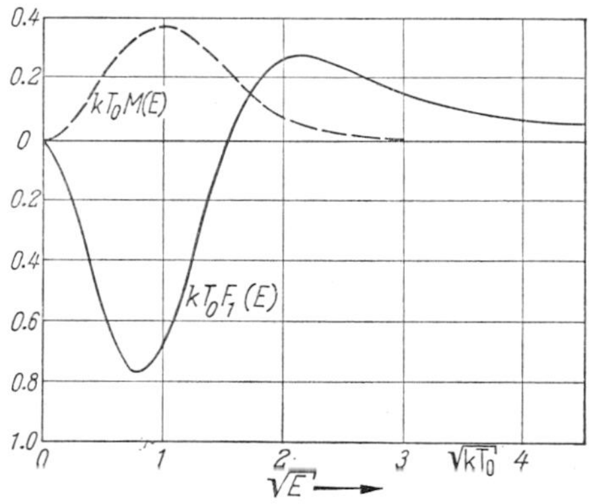

Fig. 10.2.2. The Horowitz-Tretiakoff function $F_1(E)$ in a thermal heavy gas moderator

This elegant representation of the spectrum is due to HOROWITZ and TRETIAKOFF. It applies to a heavy gas moderator for weak absorption ($\Sigma_a(kT_0)/\xi\Sigma_s$ must be <0.1). We can compare it with the result of the elementary calculation given in Sec. 8.4. There, completely neglecting thermalization effects, we divided the spectrum into a $1/E$-part for energies $\gg kT_0$ and a Maxwell spectrum for thermal energies. A neutron balance showed that the ratio between the thermal flux and the epithermal flux per unit lethargy is given by the moderating ratio $\xi\Sigma_s\Big/\frac{\sqrt{\pi}}{2}\Sigma_a(kT_0)$. This result remains unaffected here; however, for low energies the function $F_1(E)$ describes the transition between the two limiting behaviors. The representation of $\Phi(E)$ by Eq. (10.2.8a) has still another interesting property: In order to discuss it let us form the total *neutron density*

$$n = \int_0^\infty \Phi(E)\,\frac{1}{v}\,dE = \int_0^\infty \frac{1}{v}\,M(E)\,dE + \frac{\frac{\sqrt{\pi}}{2}\,\Sigma_a(kT_0)}{\xi\,\Sigma_s}\int_0^\infty \frac{1}{v}\,F_1(E)\,dE\,. \qquad (10.2.88\text{c})$$

With the help of Eqs. (10.2.8b) and (10.2.7d) we can easily convince ourselves that $\int_0^\infty \frac{1}{v}\,F_1(E)\,dE$ vanishes. In other words, the perturbing function does not

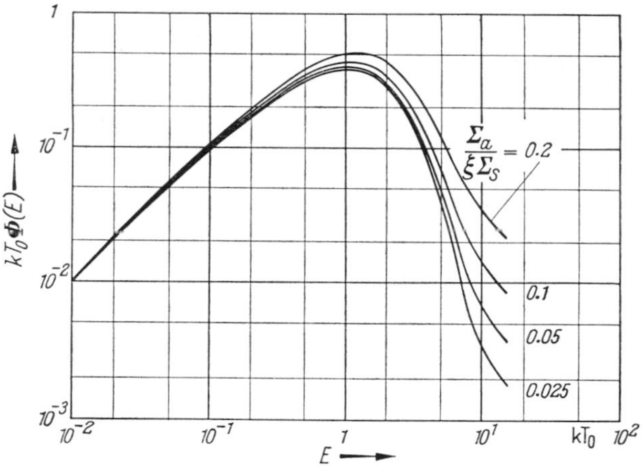

Fig. 10.2.3. A log-log plot of neutron spectra calculated in a weakly absorbing heavy gas moderator

contribute to the neutron density, and the total density is normalized to the same value as the Maxwellian portion. In our work, $n = \int_0^\infty \frac{1}{v}\,M(E)\,dE = \sqrt{\frac{\pi m_n}{8kT_0}}$; in the work of HOROWITZ and TRETIAKOFF n is normalized to unity, i.e., our Eq. (10.2.8a) must be multiplied by the factor $\sqrt{\frac{8kT_0}{\pi m_n}}$ in order to agree with the work of these authors. Which normalization factor we use is unimportant as long as the source density is not specified. However, if a source density $S\,(\text{cm}^{-3}\text{sec}^{-1})$ is given, the normalization is fixed and we have

$$\Phi(E) = \frac{S}{\frac{\sqrt{\pi}}{2}\,\Sigma_a(kT_0)}\left[M(E) + \frac{\frac{\sqrt{\pi}}{2}\,\Sigma_a(kT_0)}{\xi\,\Sigma_s}\,F_1(E)\right]. \qquad (10.2.8\text{d})$$

The Effective Neutron Temperature. Next we shall discuss the concept, useful for practical purposes, of the effective neutron temperature. For this purpose let us consider in Fig. 10.2.3 a log-log plot of the calculated spectra $\Phi(E)$ versus E for various small values of $\Sigma_a(kT_0)/\xi\,\Sigma_s$. It turns out that in the thermal range the various spectra have very similar shapes but with increasing absorption shift more and more to the right, i.e., more and more to higher energies. This suggests that when $\Sigma_a \neq 0$ the thermal spectrum can be described by a Maxwell distribution whose temperature — the neutron temperature — is higher than the moderator temperature. There is no good physical foundation for such a supposition, and it can only be justified by its success. In Fig. 10.2.4, $\ln \frac{\Phi(E)}{E}$ is plotted against E/kT_0 for the spectra shown in Fig. 10.2.3. If the perturbed spectra were Maxwell distributions, the plots would be straight lines from whose slopes the neutron temperatures could be determined[1]. We see that in fact in the energy region $E < 5\,kT_0$ and for $\Sigma_a(kT_0)/\xi\Sigma_s < 0.2$ straight lines actually do occur. Fig. 10.2.5 shows the ratio $(T-T_0)/T_0$ determined from straight-line fits to the spectra as a function of $\Sigma_a(kT_0)/\xi\Sigma_s$. The broken line shown in the figure corresponds to the equation

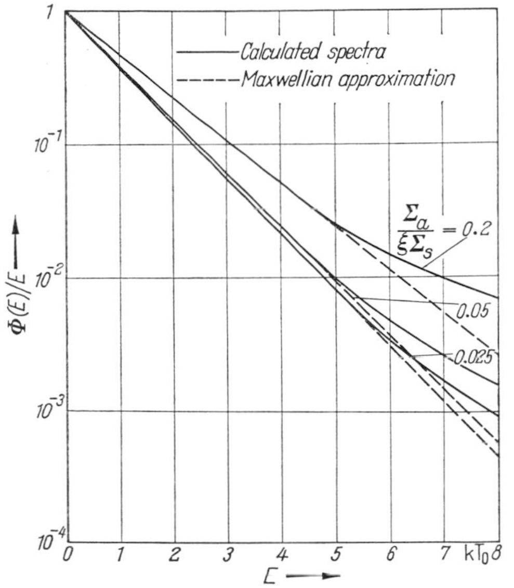

Fig. 10.2.4. The fitting of neutron spectra calculated in a weakly absorbing heavy gas moderator by shifted Maxwell distributions

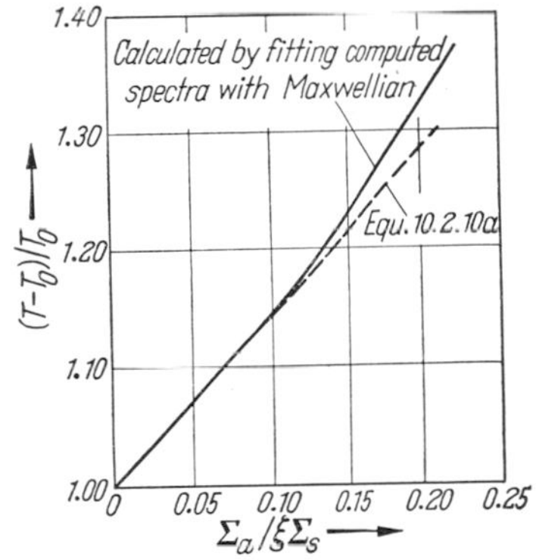

Fig. 10.2.5. The neutron temperatures obtained by fitting the calculated heavy gas spectra with shifted Maxwell distributions

$$T = T_0\left[1 + 1.46\,\frac{\Sigma_a(kT_0)}{\xi\Sigma_s}\right] = T_0\left[1 + 0.73\,A\,\frac{\Sigma_a(kT_0)}{\Sigma_s}\right] \qquad (10.2.10\text{a})$$

which we can consider as a recipe for calculating the neutron temperature in a heavy gas moderator. We can easily obtain other recipes if we perform the rather arbitrary fitting of the spectra by Maxwell distributions in another way; for example, Cohen finds

$$T = T_0\left[1 + 0.6\,A\,\frac{\Sigma_a(kT_0)}{\Sigma_s}\right] \qquad (10.2.10\text{b})$$

while Coveyou *et al.* obtain[2]

$$T = T_0\left[1 + 0.91\,A\,\frac{\Sigma_a(kT_0)}{\Sigma_s}\right]. \qquad (10.2.10\text{c})$$

[1] For if $\Phi(E) \sim \frac{E}{(kT)^2}\,e^{-E/kT}$, $\ln\frac{\Phi(E)}{E} = -\frac{E}{kT} + \ln\left(\frac{1}{kT}\right)^2$.

[2] Coveyou *et al.* start from spectra that were calculated by the Monte Carlo method for a gaseous moderator of arbitrary mass. Thus Eq. (10.2.10c) is not restricted to a heavy gas moderator.

Using the average thermal absorption cross section $\overline{\Sigma}_a = \frac{\sqrt{\pi}}{2}\Sigma_a(kT) \approx \frac{\sqrt{\pi}}{2}\Sigma_a(kT_0)$ and the fact that $\overline{\Sigma}_a/\xi\Sigma_s = \Phi_{\text{epi}}/\Phi_{\text{th}}$ (see below), we can write these relations in the form

$$T - T_0 = \frac{\Phi_{\text{epi}}}{\Phi_{\text{th}}}\,\Theta \tag{10.2.10d}$$

where according to Eq. (10.2.10a), $\Theta = 4\cdot 0.73\cdot T_0/\sqrt{\pi} = 1.65\,T_0$. In this form we can see very clearly the cause of the temperature rise, which is greater the larger the ratio of the epithermal to the thermal flux, i.e., the larger the ratio of the "influx" to the "population".

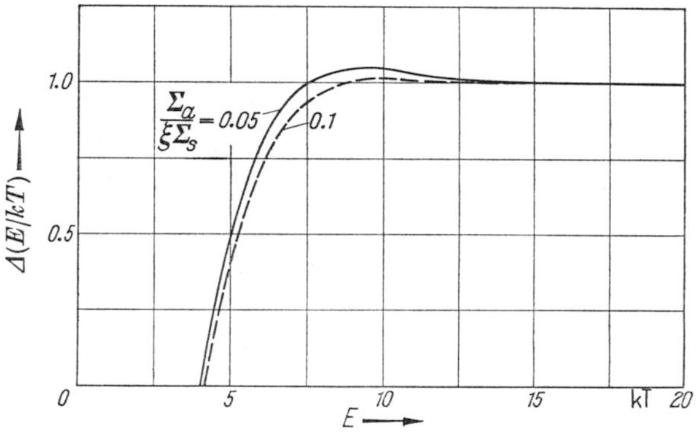

Fig. 10.2.6. The joining function in a thermal heavy gas moderator

Next we turn to the question of what shape the epithermal flux has when the thermal flux is represented by a Maxwell distribution shifted in temperature. We therefore subtract from the spectra shown in Fig. 10.2.3 the Maxwell distributions fitted to them by means of Fig. 10.2.4. The resulting function $F_2(E)$ is normalized so that for $E \gg kT_0$, $F_2(E) \approx 1/E$. Fig. 10.2.6 shows $E\cdot F_2(E)$ as a function of E/kT (not E/kT_0!). The figure shows that $F_2(E)$ depends on the absorption; in a rough approximation sufficient for many practical purposes, we can introduce an average $F_2(E) = \Delta\left(\frac{E}{kT}\right)/E$. $\Delta\left(\frac{E}{kT}\right)$ is called the "joining function"; as we can see from Fig. 10.2.6, it vanishes for $E < 4\,kT$, goes through a maximum at $E \approx 8\,kT$, and approaches the value 1 for $E > 15\,kT$. We can now write the total spectrum in the form

$$\Phi(E) = \frac{E}{(kT)^2}\,e^{-E/kT} + \lambda\,\frac{\Delta(E/kT)}{E}. \tag{10.2.10e}$$

The value of λ follows from a neutron balance: $\xi\Sigma_s\lambda = \int_0^\infty \Sigma_a(E)\,\Phi(E)\,dE = \frac{\sqrt{\pi}}{2}\Sigma_a(kT) + \lambda\int_0^\infty \Sigma_a(E)\,\Delta(E/kT)\,dE/E$, i.e.[1],

$$\lambda = \frac{\frac{\sqrt{\pi}}{2}\Sigma_a(kT)}{\xi\Sigma_s}\cdot\frac{1}{1 - \frac{\Sigma_a(kT)}{\xi\Sigma_s}}. \tag{10.2.10f}$$

When $\Sigma_a \ll \xi\Sigma_s$ we can set the second factor equal to one and obtain the simple result that the ratio of the Maxwell flux to the epithermal flux per unit lethargy is equal to the moderating ratio.

[1] The joining function shown in Fig. 10.2.6 approximately satisfies the following relation for $1/v$-absorption: $\int_0^\infty \Sigma_a(E)\,\frac{\Delta(E/kT)}{E}\,dE \approx \Sigma_a(kT)$.

As we shall see later, the approximate representation of the spectrum by Eq. (10.2.10e) is frequently possible in real moderators. It turns out that the "joining function" $\Delta(E/kT)$ depends rather weakly on the properties of the moderator. Thus in many practical cases, we can describe the spectrum entirely by means of the two parameters λ and T. This practice is especially customary in the evaluation of probe measurements (cf. Chapters 11 and 12). This description of the spectrum is certainly not very exact, and in fact the temperature concept fails completely when the absorption cross section has resonances near thermal energy or if $\Sigma_a(kT_0)$ is not $\ll \xi\Sigma_s$.

10.2.2. The Space-Independent Spectrum in Real Moderators

If we wish to calculate the neutron spectrum in an actual moderator, we must solve Eq. (10.2.1) using a scattering law which correctly describes the chemical binding. In principle, the following possibilities are available.

(a) Direct Numerical Integration of Eq. (10.2.1). We can transform Eq. (10.2.1) into a system of linear equations by dividing the continuous energy range into a number of discrete groups in the manner of multigroup theory. The number of groups should be large (25 to 50) in order that the energy variation be well approximated. The system of equations can then be solved on an electronic computing machine. An advantage of this method is the possibility of using an absorption cross section with any arbitrary energy variation and any arbitrary scattering law, including one that may have been determined experimentally. A disadvantage is the considerable expenditure of effort involved in the calculation, for the spectrum must be calculated anew for each value of Σ_a. In addition, we lose any physical "feel" we might have had for the connection between the spectrum and the scattering law.

(b) A Modified Horowitz-Tretiakoff Method. If the absorption is weak and follows the $1/v$-law[1], we can set

$$\Phi(E) = M(E) + \frac{\frac{\sqrt{\pi}}{2}\Sigma_a(kT_0)}{\xi\Sigma_s} F_1(E) \tag{10.2.11a}$$

just as we did in Sec. 10.2.1. Here $F_1(E)$ is determined by

$$\Sigma_a(E)\,M(E) = \frac{\frac{\sqrt{\pi}}{2}\Sigma_a(kT_0)}{\xi\Sigma_s} L F_1(E) \tag{10.2.11b}$$

with the condition

$$\int_0^\infty \frac{F_1(E)}{v}\,dE = 0. \tag{10.2.11c}$$

$F_1(E)$ must now be determined by numerical solution of Eq. (10.2.11b) using a measured or calculated scattering law; this leads to an expenditure of computational effort similar to that in method (a). However, we only need to carry out such a calculation once since there is a single, characteristic $F_1(E)$ for each

[1] The latter condition is not necessary, but the determination of $F_1(E)$ becomes much more complicated when $\Sigma_a \neq 1/v$.

moderator[1]. Shown in Fig. 10.2.7 is $F_1(E)$ as calculated for graphite compared with the heavy gas approximation. There are marked differences, especially at low energies; because of the strong binding, the spectrum in graphite is much harder than in a heavy gas. Unfortunately, there are no calculations of $F_1(E)$ for other moderators. Moreover, there has still been no successful attempt to extract $F_1(E)$ from experimentally measured spectra.

(c) The Method of CORNGOLD. As CORNGOLD has shown, the general solution of Eq. (10.2.1) can be written in the form of a power series

$$\Phi(E) = \frac{1}{\xi \Sigma_s E} \sum_{n=0,1,2,\dots} a_n \left(\frac{kT_0}{E}\right)^{n/2} \tag{10.2.12}$$

where the coefficients a_n depend on the absorption cross section and on certain integrals of the function $\chi(\varkappa, t)$ that characterizes the scattering law. A deeper discussion of this physically very elegant but formally rather difficult method is beyond the scope of this text (cf. also PARKS). It is particularly well suited to representing the spectrum in the transition region between the thermal and epithermal ranges.

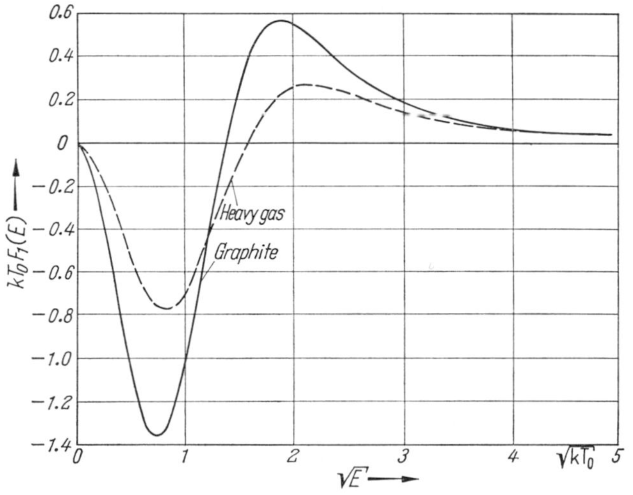

Fig. 10.2.7. The Horowitz-Tretiakoff function $F_1(E)$ in room-temperature graphite

(d) Semi-Empirical Methods. If the scattering law for a moderator is not known or is only known poorly, or if the mathematical aids to computing the spectrum are not available, we can try to modify some of the methods introduced in Sec. 10.2.1 with the aid of experimental data in order to at least arrive at a rough description of the spectrum. Such procedures were frequently used in the dawn of thermalization physics.

The most elementary of such methods uses the concept of effective neutron temperature. The spectrum is represented by Eq. (10.2.10e) (with an experimentally determined joining function or with that of the heavy gas model); the value of λ can be taken from Eq. (10.2.10f)[2]. The relation between T and T_0 is determined from experiment. Frequently the concept of "effective" mass is also introduced. The difference between the neutron temperature and the moderator temperature in a heavy gas is proportional to the mass number A [cf. Eqs. 10.2.10a—c)]. If there is appreciable chemical binding, the temperature dif-

[1] This holds of course only as long as $\Sigma_a(kT_0) < 0.1$.

[2] The slowing-down power $\xi\Sigma_s$ must be given its "plateau" value, i.e., the value it has above 1 ev.

ference can under certain circumstances be larger since the thermalization process is hindered. The "effective" mass is then that value of A that we must substitute into Eq. (10.2.10a) (or b or c) in order to obtain the actual temperature difference. It turns out that the effective mass can be many times larger than the true mass.

Another semiempirical approximation procedure is based on the thermalization operator of the heavy gas [cf. Eq. (10.2.4)]. Using an energy-dependent slowing-down power, we can construct a more general operator

$$L = \frac{d}{dE} (\xi \Sigma_s)(E) \left[(E - kT_0) + kT_0 E \frac{d}{dE} \right] \tag{10.2.13}$$

which obviously satisfies the principle of detailed balance $LM = 0$. $\xi \Sigma_s$ must go over into the plateau value for $E > 1$ ev, but for smaller energies it can behave arbitrarily. We can try to approximate the binding effects by a suitably chosen decrease of the slowing-down power with decreasing energy. We can then determine the function $F_1(E)$ for the resulting operator. (Cf. CADILHAC as well as SCHAEFER and ALLSOPP.)

A comparison between calculated and experimentally observed spectra and temperatures follows in Chapter 15.

10.2.3. Basic Facts about Space-Dependent Neutron Spectra

For the following qualitative considerations we shall assume the validity of elementary diffusion theory. Further we shall use the results of Chapter 8 to the following extent, viz., we shall assume that the slowing-down density is known either from measurement or from calculation above a cut-off energy $E_m \gg kT_0$ [1] and use it as the source density for the thermal neutrons. Then for $E < E_m$ in a homogeneous medium

$$\left.\begin{aligned} &-D\nabla^2 \Phi(\mathbf{r}, E) + \Sigma_a(E)\,\Phi(\mathbf{r}, E) \\ &\quad = \int_0^{E_m} \Sigma_s(E' \to E)\,\Phi(\mathbf{r}, E')\,dE' - \Sigma_s(E)\,\Phi(\mathbf{r}, E) + q(\mathbf{r}, E_m)\,f(E). \end{aligned}\right\} \tag{10.2.14}$$

Here $f(E)$ in the energy distribution of the source neutrons that have made their last collision above E_m. $f(E)$ is normalized so that $\int_0^{E_m} f(E)\,dE = 1$; in a heavy moderator $f(E) \approx \delta(E - E_m)$.

Let us now try to separate the variables in Eq. (10.2.14) with the substitution

$$\Phi(\mathbf{r}, E) = \Phi_{th}(\mathbf{r})\,\Phi(E)$$

where

$$\int_0^{E_m} \Phi(E)\,dE = 1. \tag{10.2.15}$$

In other words, let us try to represent the thermal spectrum as the product of a normalized, space-independent spectral function and the total thermal flux.

[1] The cut-off energy should be high enough that $\Sigma_s(E' \to E) \approx 0$ for $E' < E_m < E$.

This procedure leads to

$$\left.\begin{aligned}&-D(E)\frac{\nabla^2\Phi_{th}}{\Phi_{th}}\Phi(E)+\Sigma_a(E)\Phi(E)\\&\quad=\int_0^{E_m}\Sigma_s(E'\to E)\Phi(E')\,dE'-\Sigma_s(E)\Phi(E)+\frac{q(\boldsymbol{r},E_m)}{\Phi_{th}(\boldsymbol{r})}f(E)\end{aligned}\right\}\tag{10.2.16}$$

and

$$-\overline{D}\nabla^2\Phi_{th}+\overline{\Sigma}_a\Phi_{th}=q(\boldsymbol{r},E_m)\tag{10.2.17}$$

with

$$\overline{D}=\int_0^{E_m}D(E)\Phi(E)\,dE\tag{10.2.18a}$$

and

$$\overline{\Sigma}_a=\int_0^{E_m}\Sigma_a(E)\Phi(E)\,dE.\tag{10.2.18b}$$

It is immediately clear from Eq. (10.2.14) that $\Phi(E)$ is space-independent if and only if the flux curvature $\nabla^2\Phi_{th}/\Phi_{th}$ and $q(\boldsymbol{r},E_m)/\Phi_{th}(\boldsymbol{r})$ are space-independent. This is always the case if there is local equilibrium between the thermal neutrons and their sources. Such equilibrium typically occurs in the inner regions of a homogeneous reactor or at large distances from a source of fast neutrons in a hydrogenous moderator. If these conditions are fulfilled, we can, at least in principle, specify $\Phi(E)$ immediately: The spectrum $\Phi(E)$ is the same as that in an infinite medium with homogeneously distributed sources and the effective absorption cross section

$$\Sigma_a^{\text{eff}}(E)=\Sigma_a(E)+D(E)B^2\qquad\left(\text{with } B^2=-\frac{\nabla^2\Phi_{th}}{\Phi_{th}}\right).$$

Unfortunately, most of the methods and results given in Secs. 10.2.1 and 10.2.2 are valid only for $1/v$-absorption, while because of the appearance of $D(E)$, $\Sigma_a^{\text{eff}}(E)$ certainly does not follow the $1/v$-law. As a rule, therefore, multigroup methods are used for the calculation of $\Phi(E)$. De Sobrino and Clark have solved Eq. (10.2.16) numerically by series expansion for the case of a heavy gas moderator with $1/v$-absorption and an energy-independent diffusion coefficient. Fig. 10.2.8 shows some of their results for various values of the parameter $DB^2/\xi\Sigma_s$. When $A\Sigma_a(kT_0)/\Sigma_s\ll 1$ and $ADB^2/\Sigma_s\ll 1$, these spectra in the thermal range can again be approximated by Maxwell distributions with shifted temperatures given by

$$\frac{T-T_0}{T_0}=0.73\frac{A\Sigma_a(kT_0)}{\Sigma_s}+0.70\frac{ADB^2}{\Sigma_s}.\tag{10.2.19}$$

The calculations of de Sobrino and Clark were carried out only for $B^2>0$ (i.e., only for cases in which there was a net leakage of neutrons out of each volume element). Eq. (10.2.19) is also valid, however, for negative B^2, i.e., for cases in which there is a net influx of neutrons into each volume element. The first summand on the right-hand side is the same as that which applies in an infinite medium; in addition, there occurs an increase or decrease of the temperature according to the sign of B^2. It is instructive to bring Eq. (10.2.19) into the form of Eq. (10.2.10d): Setting $\Sigma_a^{\text{eff}} = \frac{\sqrt{\pi}}{2}\,\Sigma_a(kT) + DB^2$ and $\Sigma_a^{\text{eff}}/\xi\Sigma_s = \Phi_{\text{epi}}/\Phi_{th}$ we obtain

$$T - T_0 = \frac{\Phi_{\text{epi}}}{\Phi_{th}}\,\Theta - \frac{1}{8}\,DB^2\,\frac{AT_0}{\Sigma_s}\,. \tag{10.2.20}$$

Here the first term, which always vastly exceeds the second, is identical with the right-hand side of Eq. (10.2.10d), i.e., the temperature increase is determined just as in an infinite medium by the ratio of the "influx" to the "population". The correction term comes about because the energy dependence of the diffusion coefficient is different from $1/v$ and represents an additional temperature increase or decrease, depending on the sign of B^2. This additional effect is identical with the diffusion cooling or diffusion heating effect in a thermalized neutron field (cf. Sec. 10.3[1]). When $B^2 = 0$, all diffusion effects disappear, and Eq. (10.2.20) becomes Eq. (10.2.10d).

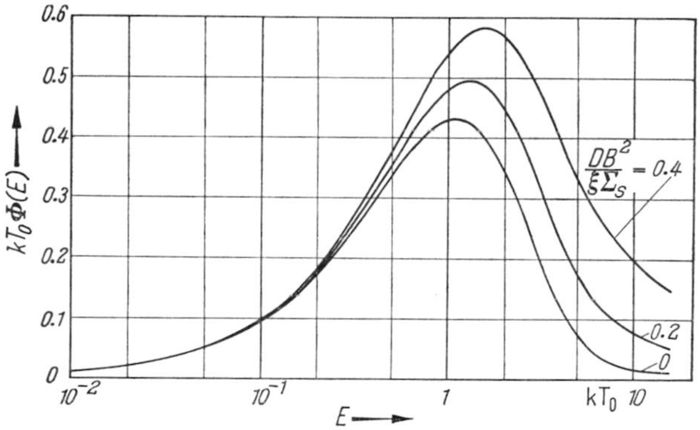

Fig. 10.2.8. The neutron spectrum in a thermal heavy gas moderator at $\frac{\Sigma_a(kT_0)}{\xi\Sigma_s} = 0.1$ for different amounts of leakage

The formalism developed here is approximately valid even when the space-energy distribution is not rigorously separable if the spectrum varies slowly with position. In this case, we must introduce a "local" effective absorption cross section

$$\Sigma_a^{\text{eff}}(\boldsymbol{r}, E) = \Sigma_a(E) - D(E)\,\frac{\nabla^2 \int \Phi(\boldsymbol{r}, E)\,dE}{\int \Phi(\boldsymbol{r}, E)\,dE}\,.$$

The "local" flux curvature must be obtained either from calculations or from measurements. This method is certainly not suitable for a clean analytic calculation of space-dependent spectra, but it can be used to obtain preliminary qualitative information about a spectrum. Eq. (10.2.20) is also very useful in the analysis of spectrum and temperature measurements in estimating the change in the spectrum caused by diffusion effects (cf. Chapter 15).

Because of their great significance in the theory of heterogeneous reactors, many methods have been developed for calculating space-dependent spectra. Some authors try to treat the problem largely analytically. They may, for example, start from a Fourier transformation of the energy-dependent diffusion Eq. (10.2.14) and solve the resulting equation in the thermal range by expansion in eigenfunctions of the thermalization operator (cf. Sec. 10.3). However, the inverse transformation is difficult and leads to complicated expressions. In contrast to these few, most authors employ multigroup methods, which *ab initio* require the

[1] There it will be shown that the temperature change caused by diffusion is generally given by $\frac{T - T_0}{T_0} = \frac{\nabla^2\Phi_{th}}{\Phi_{th}}\,\frac{\bar{D}}{NM_2}\left(1 + 2\,\frac{d\ln\bar{D}}{d\ln T}\right)$; the second term in Eq. (10.2.20) is the value appropriate to the special case of constant D and a heavy gas moderator.

use of electronic computing machines. In these methods, the thermal range is divided into a large number of energy groups (up to 70), and the energy-dependent transport equation splits into a system of energy-independent transport equations. Various approximations (P_1-, P_3-, S_4-, and S_8-approximations) are used to solve these systems. Some authors have also treated the integral form of the transport equation numerically. In every case, the computational labor is enormous and is larger, the greater the number of groups. A detailed discussion of these various methods, which we cannot undertake here, can be found in HONECK.

10.3. Some Properties of Thermalized Neutron Fields

Neutrons emitted by a fast neutron source lose energy in successive collisions with the atoms of the moderating substance and eventually arrive in the neighborhood of the thermal range. In this range, they ultimately achieve an asymptotic state in which their average energy no longer changes from collision to collision. We call neutrons in this asymptotic state "thermalized" and shall investigate their properties in some detail in this section.

The spectra we dealt with in Sec. 10.2 do not represent thermalized neutrons; rather, because of the presence of stationary sources of non-thermal neutrons, they are averages over the spectra of neutrons in all stages of slowing down and thermalization. On the other hand, there are two important classes of experiments in which pure thermalized neutron fields do occur, viz., the stationary diffusion experiments to be discussed in Chapter 17 and the pulsed neutron measurements to be discussed in Chapter 18. In the stationary measurements, the neutron distribution is studied at such large distances from the source that only thermalized neutrons are present[1]. In an infinite medium in plane geometry, for example, the flux is given by

$$\Phi(x, E) \sim \Phi(E) e^{-x/L}. \tag{10.3.1}$$

In pulsed neutron experiments, we shoot a short burst of fast neutrons into a system, wait until all the neutrons are thermalized, and observe the decay of the resulting neutron field, which follows the law

$$\Phi(\mathbf{r}, E, t) \sim R(\mathbf{r}) \Phi(E) e^{-\alpha t}. \tag{10.3.2}$$

In earlier sections, we became familiar with certain relations derived on the basis of one-group theory between L and α and the diffusion and absorption properties of the medium. Next we shall study the changes in these relations caused by thermalization effects and by the changing properties of the spectrum $\Phi(E)$. To this end, we shall use in Sec. 10.3.1 a greatly simplified model in which the space dependence of the flux is given by elementary diffusion theory and its energy dependence is handled by means of the concept of effective neutron temperature. In Sec. 10.3.2 the concept of effective neutron temperature will be

[1] Because of the sharp decrease of the neutron-proton cross section above 0.1 Mev, a purely thermal field is never reached in hydrogenous media with high-energy sources. There the primary neutrons from the source always determine the spreading out of the distribution so that one must either use sources of lower energy (e.g., (Sb-Be) sources) or employ a cadmium difference method (cf. Sec. 17.1.1).

abandoned, and in Sec. 10.3.3 the diffusion approximation will be given up. Thus the treatment of thermalized neutron fields will essentially be exact, except for the question, to be discussed in Sec. 10.3.4, of separating the space and energy variables in finite media.

10.3.1. Elementary Treatment of Thermalized Neutron Fields: Diffusion Cooling and Diffusion Heating

At first let us consider stationary and non-stationary fields together. In the framework of elementary diffusion theory, the energy-, space-, and time-dependent flux is given by

$$\frac{1}{v}\frac{\partial \Phi(E,\boldsymbol{r},t)}{\partial t} = -\Sigma_a(E)\,\Phi(E,\boldsymbol{r},t) + D(E)\,\nabla^2\Phi(E,\boldsymbol{r},t) + L\Phi. \tag{10.3.3}$$

Here L is again the thermalization operator

$$L\Phi = \int_0^\infty \Sigma_s(E'\to E)\,\Phi(E',\boldsymbol{r},t)\,dE' - \Sigma_s(E)\,\Phi(E,\boldsymbol{r},t).$$

No source term has been included in Eq. (10.3.3). In the thermalized field the flux must be separable as follows: $\Phi(E,\boldsymbol{r},t) = \Phi(E)\cdot\Phi(\boldsymbol{r},t)$. Following integration over all energies, Eq. (10.3.3) becomes

$$\frac{1}{\bar{v}}\frac{\partial \Phi(\boldsymbol{r},t)}{\partial t} = -\bar{\Sigma}_a\,\Phi(\boldsymbol{r},t) + \bar{D}\,\nabla^2\Phi(\boldsymbol{r},t) \tag{10.3.4}$$

where $\bar{\Sigma}_a$ and $\bar{D}$ are the usual averages over the spectrum $\Phi(E)$. Eq. (10.3.4) is a balance equation for the number $n(\boldsymbol{r},t) = \Phi(\boldsymbol{r},t)/\bar{v}$ of neutrons in one cm³. We can also set up a balance equation for the energy density $n(\boldsymbol{r},t)\cdot\bar{E} = \int_0^\infty \frac{1}{v}\,\Phi(E,r,t)\cdot E\,dE$. To do this, let us multiply Eq. (10.3.3) with E and integrate over all energies. There then results

$$\left.\begin{aligned}\frac{\bar{E}}{\bar{v}}\frac{\partial \Phi(\boldsymbol{r},t)}{\partial t} &= -\bar{\Sigma}_a E_a\,\Phi(\boldsymbol{r},t) + \bar{D}\,E_D\,\nabla^2\Phi(\boldsymbol{r},t) + \\ &\quad + \Phi(\boldsymbol{r},t)\int_0^\infty E\left\{\int_0^\infty \Sigma_s(E'\to E)\,\Phi(E')\,dE' - \Sigma_s(E)\,\Phi(E)\right\}dE\end{aligned}\right\} \tag{10.3.5}$$

where

$$E_a = \frac{\int_0^\infty E\,\Sigma_a(E)\,\Phi(E)\,dE}{\int_0^\infty \Sigma_a(E)\,\Phi(E)\,dE}. \tag{10.3.6}$$

For $1/v$-absorption, which we now assume, $E_a = \bar{E}$. Furthermore,

$$E_D = \frac{\int_0^\infty E\,D(E)\,\Phi(E)\,dE}{\int_0^\infty D(E)\,\Phi(E)\,dE}. \tag{10.3.7}$$

Clearly

$$\int_0^\infty E\left[\int_0^\infty \Sigma_s(E'\to E)\,\Phi(E')\,dE' - \Sigma_s(E)\,\Phi(E)\right]dE = \int_0^\infty\int_0^\infty (E-E')\,\Sigma_s(E'\to E)\,\Phi(E')\,dE\,dE'.$$

If we now multiply Eq. (10.3.4) with $\bar{E}$ and subtract it from Eq. (10.3.5), we obtain after some simple rearrangement

$$D\frac{\nabla^2\Phi}{\Phi}(E_D-\bar{E})=\int_0^\infty\int_0^\infty(E'-E)\Sigma_s(E'\to E)\Phi(E')\,dE'\,dE. \tag{10.3.8}$$

When $\nabla^2\Phi/\Phi=0$, i.e., when there is no diffusion, the left-hand side of Eq. (10.3.8) vanishes. The right-hand side must then also vanish, and $\Phi(E)$ must equal $M(E)=\frac{E}{(kT_0)^2}e^{-E/kT_0}$ (cf. the footnote on p. 185), i.e., the spectrum of the thermalized neutrons must be a true equilibrium spectrum. Since E_D never equals $\bar{E}$ — otherwise, $D(E)$ would have to be proportional to $1/v$ — the left-hand side does not vanish if $\nabla^2\Phi/\Phi$ is not zero. Depending on the sign of $\nabla^2\Phi/\Phi$, each volume element either loses or gains energy through diffusion. This energy must be gained or lost through collisions with the atoms of the moderator, i.e., $\Phi(E)$ must depart from the equilibrium spectrum.

Aside from the diffusion approximation, Eq. (10.3.8) is exact. Now we make the assumption that the spectrum can be represented by a Maxwell distribution with a shifted temperature $T\neq T_0$, viz., $\Phi(E)=\frac{E}{(kT)^2}e^{-E/kT}$. Then $\bar{E}=\frac{3}{2}kT$ and

$$E_D=\frac{\int_0^\infty E\cdot D(E)\cdot\frac{E}{(kT)^2}e^{-E/kT}\,dE}{\int_0^\infty D(E)\frac{E}{(kT)^2}e^{-E/kT}\,dE}=2kT+kT^2\frac{d\ln\bar{D}}{dT}. \tag{10.3.9}$$

(Because $\bar{D}=\int_0^\infty D(E)\frac{E}{(kT)^2}e^{-E/kT}\,dE$, $\frac{d\bar{D}}{dT}=-\frac{2\bar{D}}{T}+\frac{1}{kT^2}\int_0^\infty E\cdot D(E)\cdot\frac{E}{(kT)^2}\times$ $\times e^{-E/kT}dE$; the differentiation is with respect to the *neutron temperature*.) Furthermore, when $|T-T_0|/T_0\ll 1$ — and we shall only consider this case of slight deviation from true equilibrium — it follows that

$$\Phi(E)=\frac{E}{(kT)^2}e^{-E/kT}=M(E)+\frac{T-T_0}{T_0}\left(\frac{E}{kT_0}-2\right)M(E). \tag{10.3.10}$$

This equation is obtained by expanding $\Phi(E)$ in a Taylor series around $T=T_0$ and truncating the series after the second term. Then[1]

$$\left.\begin{aligned}&\int_0^\infty\int_0^\infty(E'-E)\Sigma_s(E'\to E)\Phi(E')\,dE'\,dE\\&\quad=\frac{T-T_0}{kT_0^2}\iint E'(E'-E)\Sigma_s(E'\to E)M(E')\,dE'\,dE\\&\quad=\frac{1}{2}k(T-T_0)\cdot N\cdot M_2\end{aligned}\right\} \tag{10.3.11}$$

[1] Since $M_2=\frac{1}{(kT_0)^2}\iint(E'-E)^2M(E')\sigma_s(E'\to E)\,dE'\,dE-\frac{1}{(kT_0)^2}\iint(E'^2+E^2-2EE')\times$ $\times M(E')\sigma_s(E'\to E)\,dE'\,dE=\frac{2}{(kT_0)^2}\iint E'(E'-E)M(E')\sigma_s(E'\to E)\,dE'\,dE$. The last step follows from the principle of detailed balance.

(with N=the number of atoms per cm³), i.e., the average energy absorbed or given up in a collision is proportional to the product of the difference between the neutron temperature and the moderator temperature and the mean squared energy loss introduced in Sec. 10.1 [in Eq. (10.1.14)]. Combining Eqs. (10.3.11), (10.3.9), and (10.3.8) and noting that $\bar{E}=3kT/2$, we obtain

$$\bar{D}\cdot\frac{\nabla^2\Phi}{\Phi}\cdot\frac{kT}{2}\left(1+2\frac{d\ln\bar{D}}{d\ln T}\right)=\frac{1}{2}k(T-T_0)NM_2 \qquad (10.3.12\text{a})$$

or, if we set $T\approx T_0$ on the left-hand side and solve for $T-T_0$,

$$\frac{T-T_0}{T_0}=\bar{D}\frac{\nabla^2\Phi}{\Phi}\frac{1+2\frac{d\ln\bar{D}}{d\ln T}}{NM_2}. \qquad (10.3.12\text{b})$$

Non-Stationary Case. Here $\Phi(\boldsymbol{r},t)=R(\boldsymbol{r})\cdot e^{-\alpha t}$, where $R(\boldsymbol{r})$ is the lowest eigenfunction of $\nabla^2R+B^2R=0$ subject to the boundary condition that the flux vanish on the effective surface of the medium. It then follows from Eq. (10.3.12b) that

$$\frac{T-T_0}{T_0}=-\bar{D}B^2\frac{1+2\frac{d\ln\bar{D}}{d\ln T}}{NM_2}, \qquad (10.3.13)$$

i.e., *diffusion cooling* occurs. The physical reason for this diffusion cooling is that the more energetic neutrons preferentially leak through the surface of the medium during the decay of the pulsed neutron field. This preferential leakage leads to a cooling of the spectrum which is more pronounced the weaker the energetic coupling of the neutron gas to the moderator. The decay constant α is given by Eq. (10.3.4) as

$$\alpha=\bar{v}\bar{\Sigma}_a+\bar{D}\bar{v}B^2. \qquad (10.3.14)$$

When the absorption obeys the $1/v$-law, the first term on the right-hand side does not depend on the spectrum: $\bar{v}\bar{\Sigma}_a=\overline{v\Sigma_a}=v_0\Sigma_a(v_0)$. In what follows this absorption term will frequently be denoted by α_0. $\bar{D}\cdot\bar{v}$ depends on the spectrum and through Eq. (10.3.13) also on B^2; for

$$\bar{D}\bar{v}(T)=(\bar{D}\bar{v})(T_0)+(T-T_0)\frac{d\bar{D}\bar{v}}{dT} \qquad (10.3.15\text{a})$$

so that if we set $(\bar{D}\bar{v})(T_0)=D_0$ and $T_0\bar{D}\cdot\frac{d\bar{D}\bar{v}}{dT}\cdot\frac{1+2\frac{d\ln\bar{D}}{d\ln T}}{NM_2}=C$ we can combine Eqs. (10.3.15a) and (10.3.13) and obtain

$$\bar{D}\bar{v}=D_0-CB^2. \qquad (10.3.15\text{b})$$

Thus finally,

$$\alpha=\alpha_0+D_0B^2-CB^4. \qquad (10.3.16)$$

In contrast to the linear relation between α and B^2 predicted by simple one-group theory, now as a consequence of the diffusion cooling effect there is a parabolic relation[1]. The downward curvature of the α-vs.-B^2 curve has been observed experimentally. C is called the "diffusion cooling constant." It is given by

$$C=\frac{\bar{v}\bar{D}^2}{2NM_2}\left(1+2\frac{d\ln\bar{D}}{d\ln T}\right)^2. \qquad (10.3.17\text{a})$$

[1] Naturally, higher terms in B^2 also appear, but for small B^2 they can be neglected.

or with $\bar{D}\bar{v} \approx D_0$

$$C = \frac{2\left(\frac{d\,D_0}{d \ln T}\right)^2}{\bar{v} N M_2}. \tag{10.3.17b}$$

The quantity $c = C/D_0$ is also used frequently; it is given by

$$c = \frac{\bar{D}}{2 N M_2}\left(1 + 2\,\frac{d \ln \bar{D}}{d \ln T}\right)^2. \tag{10.3.17c}$$

In graphite and beryllium $D(E)$ is nearly constant so that $\frac{d \ln \bar{D}}{d \ln T} = 0$. Then C is simply $\frac{\bar{v}\bar{D}^2}{2 N M_2} = \frac{D_0^2}{2 \bar{v} N M_2}$ and c is $\frac{\bar{D}}{2 N M_2}$. Some calculated values of C are given in Table 10.3.1; obviously they depend sensitively on the thermalization model used.

Table 10.3.1. *Calculated Values for C According to the Method of Effective Neutron Temperatures (at 20 °C)*

Substance	$\frac{d \ln \bar{D}}{d \ln T}$	$\bar{D}$ (cm)	C (cm^{-4} sec^{-1})	M_2 from
H_2O	$\frac{1}{2}$	0.144	3400	Nelkin Model
			2650	Eq. (10.1.15): $A=1$
			4250	Eq. (10.1.15): $A=18$
Be (1.85 g/cm³) . .	0	0.5	9×10^4	Eq. (10.1.33c)
			5.6×10^4	Eq. (10.1.15): $A=9$
Graphite (1.6 g/cm³)	0	0.86	10.8×10^5	Parks Model
			4×10^5	Eq. (10.1.15): $A=12$

Stationary Case. Here $\Phi(\boldsymbol{r}, t) = R(\boldsymbol{r})$ with $\nabla^2 R - R/L^2 = 0$. The temperature is given by

$$\left.\begin{aligned} \frac{T - T_0}{T_0} &= \frac{\bar{D}}{L^2}\,\frac{1 + 2\,\frac{d \ln \bar{D}}{d \ln T}}{N M_2} \\ &= \Sigma_a\,\frac{1 + 2\,\frac{d \ln \bar{D}}{d \ln T}}{N M_2}. \end{aligned}\right\} \tag{10.3.18}$$

In other words, *diffusion heating* occurs; it does so because the average energy of the neutrons streaming into each volume element is larger than that of the neutrons being absorbed there. The diffusion length is given by

$$L^2 = \frac{\bar{D}}{\Sigma_a} = \frac{\bar{D}\bar{v}}{v\,\Sigma_a(v)}. \tag{10.3.19a}$$

In analogy with Eq. (10.3.15b)

$$\bar{D}\bar{v} = D_0 + C\,\frac{1}{L^2} \tag{10.3.19b}$$

and therefore

$$L^2 = \frac{D_0}{v\,\Sigma_a(v)} + \frac{C}{v\,\Sigma_a(v)\,L^2}. \tag{10.3.19c}$$

With $L_0^2 = D_0/v\,\Sigma_a(v)$ and $C/v\,\Sigma_a(v)\,L^2 \approx C/D_0 = c$

$$L^2 = L_0^2 + c. \tag{10.3.19d}$$

Frequently, one also writes

$$\frac{1}{L^2} = \frac{1}{L_0^2}\left(1 - \frac{c}{L_0^2}\right). \tag{10.3.19e}$$

10.3.2. Treatment of the Diffusion Cooling Effect by the Method of Laguerre Polynomials

The neutron temperature model is certainly suitable for achieving qualitative understanding of diffusion effects in thermalized fields, but unfortunately involves the assumption that $\Phi(E)$ can always be described by a Maxwell distribution. We shall now free ourselves of this assumption. In illustrating how we do this, we shall only consider the case of diffusion cooling in a pulsed neutron field; the treatment of diffusion heating is entirely analogous.

Let us again start with Eq. (10.3.3) and again set $\Phi(E, \boldsymbol{r}, t) = \Phi(E) \cdot R(\boldsymbol{r}) \cdot e^{-\alpha t}$; in view of the fact that $\nabla^2 R/R = -B^2$,

$$\frac{\alpha}{v}\Phi(E) = D(E)\, B^2\, \Phi(E) - L\Phi. \qquad (10.3.20)$$

In the interest of brevity we have assumed $\Sigma_a = 0$; since a $1/v$-absorption does not influence the spectrum in a pulsed field but merely increases the decay constant by the amount $\alpha_0 = \overline{v\Sigma_a(v)}$, this assumption occasions no loss of generality. Eq. (10.3.20) defines an eigenvalue problem; there exists an entire spectrum of eigenvalues α_ν and eigenfunctions $\Phi_\nu(E)$. Here we are only interested in the lowest eigenvalue and its associated energy spectrum; in Sec. 10.4 we shall return to the question of the higher eigensolutions, which determine the approach to the thermalized state.

In order to solve Eq. (10.3.20), we attempt to transform the integral equation into a system of linear equations. This can be done, e.g., by the multigroup method, i.e., by dividing the continuous energy interval into a number n of discrete energy groups. More elegant and much more tractable, however, is the development of $\Phi(E)$ in a series of orthogonal functions. Let us set

$$\Phi(E) = \sum_{k=0}^{\infty} A_k \cdot L_k^{(1)}\left(\frac{E}{kT_0}\right) \cdot M(E) \qquad (10.3.21)$$

where the A_k are constants and the $L_k^{(1)}$ are the associated Laguerre polynomials of the first kind[1]. The latter are given by

$$L_k^{(1)}(x) = \sum_{\nu=0}^{k} \binom{k+1}{\nu+1} \frac{(-x)^\nu}{\nu!} \qquad (10.3.22\text{a})$$

and in particular

$$\left.\begin{aligned} L_0^{(1)}(x) &= 1 \\ L_1^{(1)}(x) &= 2 - x \\ L_2^{(1)}(x) &= 3 - 3x + x^2/2 . \end{aligned}\right\} \qquad (10.3.22\text{b})$$

Furthermore, they satisfy the following orthogonality relation:

$$\int_0^\infty x e^{-x} L_k^{(1)}(x)\, L_i^{(1)}(x)\, dx = (k+1)\,\delta_{ik}. \qquad (10.3.22\text{c})$$

Because of this orthogonality relation, the $L_k^{(1)}$ are particularly well suited to treating thermalized neutron fields; and, as a matter of fact, the equations take

[1] Cf., e.g., "Higher Transcendental Functions", Vol. 1, p. 179, McGraw-Hill Book Co., New York, 1953.

a very simple form in the case of a heavy gas moderator. If we substitute Eq. (10.3.21) into Eq. (10.3.20), multiply by $L_i^{(1)}(E/kT_0)$, and integrate over the energy, we obtain

$$\alpha \sum_{k=0}^{\infty} V_{ik} A_k = B^2 \sum_{k=0}^{\infty} D_{ik} A_k - \sum_{k=0}^{\infty} L_{ik} A_k . \tag{10.3.23}$$

These equations represent a homogeneous, linear system of equations for the A_k. In order that there be a non-trivial solution, the determinant of the coefficients must vanish:

$$|\alpha V_{ik} - B^2 D_{ik} + L_{ik}| = 0 . \tag{10.3.24}$$

Eq. (10.3.24) is an algebraic equation that determines the eigenvalues. Before we solve it, we must evaluate the matrix elements V_{ik}, D_{ik}, and L_{ik}. To begin with,

$$\left.\begin{aligned} V_{ik} &= \int_0^{\infty} L_i^{(1)}\left(\frac{E}{kT_0}\right) \frac{1}{v} M(E) L_k^{(1)}\left(\frac{E}{kT_0}\right) dE \\ &= \frac{1}{v_T} \int_0^{\infty} L_i^{(1)}(x) \sqrt{x}\, e^{-x} L_k^{(1)}(x)\, dx \end{aligned}\right\} \tag{10.3.25a}$$

with $v_T = \sqrt{\frac{2kT_0}{m_n}}$. It then follows that

$$V_{00} = \frac{1}{v_T} \frac{\sqrt{\pi}}{2}; \quad V_{01} = V_{10} = \frac{1}{2} V_{00}; \quad V_{11} = \frac{7}{8} V_{00}. \tag{10.3.25b}$$

Häfele and Dresner have shown that in general for $i \leq k$

$$V_{ik} = \frac{1}{\pi v_T} \sum_{l=0}^{i} \frac{\Gamma(i-l+\frac{1}{2})\, \Gamma(k-l+\frac{1}{2}) \Gamma(l+\frac{3}{2})}{(i-l)!\,(k-l)!\, l!} . \tag{10.3.25c}$$

Secondly,

$$D_{ik} = \int_0^{\infty} L_i^{(1)}\left(\frac{E}{kT_0}\right) D(E) M(E) L_k^{(1)}\left(\frac{E}{kT_0}\right) dE . \tag{10.3.25d}$$

In the important special case that $D(E)$ is constant, which we shall be considering from now on, we find using Eq. (10.3.22c) that

$$D_{ik} = (k+1) D \delta_{ik} . \tag{10.3.25e}$$

Finally,

$$L_{ik} = \int_0^{\infty} L_i\left(\frac{E}{kT_0}\right) \left\{ L M(E) L_k\left(\frac{E}{kT_0}\right)\right\} dE . \tag{10.3.25f}$$

In the special case of a heavy gas moderator, the L_{ik} take a particularly simple form. In this case, L is given by

$$L\Phi = \frac{d}{dE} \xi \Sigma_s \left[(E - kT_0)\, \Phi(E) + E\, kT_0 \frac{d\Phi}{dE}\right]$$

[cf. Eq. (10.2.4)]. Now

$$\frac{d}{dx}\left[(x-1) + x\frac{d}{dx}\right] x e^{-x} L_k^{(1)}(x) = -k L_k^{(1)}(x)\, x e^{-x} , \tag{10.3.25g}$$

i.e., the functions $L_k^{(1)}(E/kT_0)\cdot M(E)$ are eigenfunctions of the heavy gas thermalization operator corresponding to the eigenvalue $-k\xi\Sigma_s$. Using Eq. (10.3.22c), we therefore obtain

$$L_{ik}=-k(k+1)\xi\Sigma_s\delta_{ik} \tag{10.3.25h}$$

for the heavy gas moderator. It can easily be shown that even for an arbitrary scattering law all the L_{0k} and L_{k0} vanish; furthermore

$$L_{11}=-\frac{NM_2}{2} \tag{10.3.25i}$$

where N is the number of scattering atoms per cm³ and M_2 is the mean squared energy loss (cf. Sec. 10.1). General expressions for the L_{ik} have been given by PUROHIT and by TAKAHASHI; the higher moments of the scattering kernel appear in these expressions.

In order to solve Eq. (10.3.24), we shall break off the expansion of $\Phi(E)/M(E)$ after the *l-th* term. Eq. (10.3.24) is then of *l-th* order in α. For $l=1$ and 2, we find, respectively, that

$$\left.\begin{aligned}&\alpha V_{00}-B^2D_{00}=0\\&\alpha=\frac{D_{00}B^2}{V_{00}}=D\frac{2}{\sqrt{\pi}}v_T B^2=D_0B^2\end{aligned}\right\}l=1 \tag{10.3.26a}$$

and

$$\left.\begin{aligned}&\begin{vmatrix}\alpha V_{00}-D_{00}B^2 & \alpha V_{01}\\ \alpha V_{01} & \alpha V_{11}-B^2D_{11}+L_{11}\end{vmatrix}=0\\&\alpha=\frac{D_{00}}{V_{00}}B^2-\frac{D_{00}^2}{V_{00}L_{11}}\left(\frac{V_{01}}{V_{00}}\right)^2B^4\end{aligned}\right\}l=2 \tag{10.3.26b}$$ [1]

or

$$\alpha=D_0B^2-CB^4$$

with

$$\left.\begin{aligned}C&=\frac{D_{00}^2}{V_{00}L_{11}}\left(\frac{V_{01}}{V_{00}}\right)^2\\&=\frac{D^2\frac{2}{\sqrt{\pi}}v_T}{2NM_2}\end{aligned}\right\}l=2. \tag{10.3.26c}$$

This result for C is identical with that of Eq. (10.3.17a) (for constant D), which was derived from the neutron temperature model. We can easily understand this result with the help of Eq. (10.3.10); as long as we consider only terms up to order B^2, the $l=2$ approximation is identical with the representation of the perturbed spectrum by a Maxwell distribution with a shifted temperature.

We can now investigate what errors we make with these approximations by taking more terms in the expansion of $\Phi(E)/M(E)$. For a heavy gas,

$$C^{(l=2)}=\frac{AD^2\frac{2}{\sqrt{\pi}}v_T}{16\Sigma_s} \tag{10.3.27a}$$

[1] Here terms of order higher than B^4 have been neglected.

since $M_2 = 8\sigma_s/A$. According to Häfele and Dresner, for arbitrary l

$$C^{(l)} = \frac{A D^2 \frac{2}{\sqrt{\pi}} v_T}{2\Sigma_s} \cdot \sum_{\nu=1}^{l-1} \frac{1}{\nu(\nu+1)} \left[\frac{(2\nu-1)!!}{(2\nu)!!}\right]^2. \tag{10.3.27b}$$ [1]

Table 10.3.2 shows $C^{(l)} \Big/ \frac{A D^2 \frac{2}{\sqrt{\pi}} v_T}{2\Sigma_s}$ as a function of l. We see that $C^{(l)}$ converges slowly; its asymptotic value $C^{(\infty)}$ is very clase to

$$C^{(\infty)} = \frac{A D^2 \frac{2}{\sqrt{\pi}} v_T}{12\Sigma_s} \tag{10.3.27c}$$

and is some 33% higher than the value given by the $l=2$ approximation. Eq. (10.3.27c) can also be obtained without the use of the Laguerre polynomial approximation by direct integration of Eq. (10.3.20) (cf. Hurwitz and Nelkin as well as Beckurts).

Table 10.3.2. $\dfrac{C^{(l)}}{\left(\dfrac{A D^2 \frac{2}{\sqrt{\pi}} v_T}{2\Sigma_s}\right)}$ *according to* Eq. (10.3.27b)

$l=2$	3	4	5	6	7	8	9	10	11
0.1250	0.1484	0.1566	0.1603	0.1623	0.1635	0.1643	0.1649	0.1652	0.1655

$l=12$	13	14	15	16
0.1657	0.1659	0.1660	0.1661	0.1662

Takahashi has studied the higher-order approximations to C for crystalline moderators. He used the heavy crystal approximation and assumed D to be constant. He showed that one must go to $l=20$ to get good convergence of the value of C. Takahashi found that $C^{(l=20)} = 2.48\, C^{(l=2)}$ for *graphite* and $C^{(l=20)} = 3.36\, C^{(l=2)}$ for *beryllium*. With these adjustments, we can use the values of $C^{(l=2)}$ in Table 10.3.1. These results also show that in crystalline moderators with strong binding the neutron temperature model is a very poor approximation.

Finally, we wish to note briefly another procedure for calculating the diffusion cooling constant. The solution of Eq. (10.3.20) can be written generally in the form

$$\left.\begin{aligned} \Phi(E) &= M(E) + B^2 \Phi_2(E) + B^4 \Phi_4(E) + \cdots \\ \alpha &= D_0 B^2 - C B^4 + F B^6 + \cdots . \end{aligned}\right\} \tag{10.3.28}$$

If we insert these expressions in Eq. (10.3.20) and note that the coefficients of the various powers of B must separately vanish, we find that

$$\left(\frac{D_0}{v} - D(E)\right) M(E) + L\Phi_2(E) = 0, \tag{10.3.29a}$$

$$-\frac{C}{v} M(E) + \frac{D_0}{v} \Phi_2(E) - D(E)\Phi_2(E) + L\Phi_4(E) = 0 \tag{10.3.29b}$$

[1] $x!! = x(x-2)(x-4) \ldots (2)$ or (1).

etc. If we integrate both of these equations over all energy, it follows that

$$D_0 = \frac{\int_0^\infty D(E)\,M(E)\,dE}{\int_0^\infty \frac{1}{v}\,M(E)\,dE}, \tag{10.3.29c}$$

$$C = \frac{\int_0^\infty \left(\frac{D_0}{v} - D(E)\right)\Phi_2(E)\,dE}{\int_0^\infty \frac{1}{v}\,M(E)\,dE} \tag{10.3.29d}$$

since $\int_0^\infty L\Phi\,dE \equiv 0$. In this case, we do not need to solve an eigenvalue problem; instead we must determine $\Phi_2(E)$ from Eq. (10.3.29a) and then the diffusion cooling constant from Eq. (10.3.29d). For a heavy gas, we can easily integrate Eq. (10.3.29a); when we do so we again obtain Eq. (10.3.27c) for C. For other scattering laws, we can solve Eq. (10.3.29a) by the multigroup method or by expansion in Laguerre polynomials. These methods are also suitable for determining the higher terms (F etc.) in the $\alpha(B^2)$-relation.

10.3.3. Transport Theory of the Thermalized Neutron Field

Now we must investigate the errors that arise from treating thermalized neutron fields with elementary diffusion theory. We start from the transport equation in a source-free, isotropically scattering medium in plane geometry:

$$\left.\begin{aligned} &\frac{1}{v}\,\frac{\partial F(E,\mu,x,t)}{\partial t} \\ &\quad = -\Sigma_t(E)\cdot F - \mu\frac{\partial F}{\partial x} + \frac{1}{2}\int_0^\infty\int_{-1}^{+1} \Sigma_s(E'\to E)\,F(E',\mu',x,t)\,dE'\,d\mu'. \end{aligned}\right\} \tag{10.3.30a}$$

(The assumption of isotropic scattering will be dropped later.) In the stationary case, in an infinite medium, we can write for the thermalized neutron field $F(E,\mu,x,t) = F(E,\mu)\,e^{-\varkappa x}$. Then

$$(\Sigma_t(E) - \varkappa\mu)\,F(E,\mu) = \frac{1}{2}\int_0^\infty\int_{-1}^{+1} \Sigma_s(E'\to E)\,F(E',\mu')\,dE'\,d\mu'. \tag{10.3.30b}$$

In the non-stationary case, $F(E,\mu,x,t) = F(E,\mu,x)\,e^{-\alpha t}$ and

$$\left(\Sigma_t(E) - \frac{\alpha}{v}\right)F(E,\mu,x) = -\mu\frac{\partial F}{\partial x} + \frac{1}{2}\int_0^\infty\int_{-1}^{+1} \Sigma_s(E'\to E)\,F(E',\mu',x)\,dE'\,d\mu'. \tag{10.3.30c}$$

These two equations are eigenvalue problems for the relaxation constant $\varkappa = 1/L$ and the decay constant α, respectively. In Eq. (10.3.30c) boundary conditions must be specified on the surface of the medium. Before we proceed with the solution of these equations, let us consider an elementary argument concerning the existence of an upper limit for the eigenvalues. $F(E, \mu)$ and $F(E, \mu, x)$ are always $\geqq 0$. Furthermore, the in-scattering integral $\int \Sigma_s(E' \to E) F \, dE' \, d\mu'$ is always $\geqq 0$. It therefore follows from Eq. (10.3.30b) that $\Sigma_t(E) - \varkappa\mu \geqq 0$ or since $\mu \leqq 1$ that

$$\varkappa \leqq \min\left(\Sigma_t(E)\right). \tag{10.3.30d}$$

Thus the relaxation constant can never be larger than the minimum value of the total cross section over the entire energy range. It follows similarly from Eq. (10.3.30c) that

$$\alpha \leqq \min\left(v \Sigma_t(E)\right) \tag{10.3.30e}$$

provided that we first assume space independence ($\partial F/\partial x = 0$). NELKIN has shown that Eq. (10.3.30e) also holds when $\partial F/\partial x \neq 0$. The intuitive grounds for these limitations are as follows. The presence of an asymptotic spectrum in either space or time means that all groups of neutrons, no matter what their energies, decay equally rapidly in either space or time, respectively. The intensity of neutrons with energy E cannot decay faster than $e^{-\Sigma_t(E)x}$ in the stationary case and $e^{-v\Sigma_t(E)t}$ in the non-stationary case, for $1/\Sigma_t(E)$ is the mean free path between collisions and $1/v\Sigma_t(E)$ is the mean collision time. Thus the neutrons for which Σ_t or $v\Sigma_t$ has a minimum limit the rate of decay. ("The grade school class cannot proceed faster than the slowest pupil." — CORNGOLD.) In H_2O and D_2O, the minimum value of $\Sigma_t(E)$ occurs at $E = \infty$, and in crystalline moderators immediately below the Bragg cutoff energy. The minimum of $v\Sigma_t(E)$ always occurs at $E = 0$. Table 10.3.3 shows values of $\frac{1}{(\Sigma_s)_{\min}}$ and $\frac{1}{(v\Sigma_s)_{\min}}$ for some moderators.

Table 10.3.3. *Lower Limits for the Relaxation Time and Relaxation Length for Various Moderators at Room Temperature*

	$\frac{1}{(v\Sigma_s)_{\min}}$ [μsec]	$\frac{1}{(\Sigma_s)_{\min}}$ [cm]
Graphite (1.6 g/cm³)	380	23
Beryllium (1.85 g/cm³)	260	18
H_2O	≈ 3	0.6
D_2O	≈ 30	2.8

If the absorption is increased[1] (in the stationary case) or the geometric dimensions of the medium decreased (in the pulsed case), $\varkappa$ or α will increase, respectively. They approach the limits specified by Eqs. (10.3.30d and e). There arises the question of whether these limiting values are achieved asymptotically, i.e., in the limit of infinite Σ_a or B^2, or for finite $\Sigma_a\left((\Sigma_a)_c\right)$ or $B^2\left((B^2)_c\right)$. If the latter is the case, then for $\Sigma_a > (\Sigma_a)_c$ or $B^2 > (B^2)_c$ there will be no solution to Eqs. (10.3.30b and c). In other words, under these extreme circumstances there exists no asymptotic spectrum of thermalized neutrons. There are indications from experiments and calculations that indeed the latter is the case but we are still far from a complete understanding of the behaviour of the neutron field under such extreme circumstances. As a rule, for weak absorption or in moderators that are not too small we are well under the critical limits, and we shall now try to treat Eqs. (10.3.30b and c) further in these practical cases.

[1] e.g., by poisoning the medium with boron (cf. Sec. 17.2).

If we introduce $\Phi(E)=\int_{-1}^{+1} F(E,\mu)\,d\mu$ into Eq. (10.3.30b), we obtain a result similar to that of Sec. 5.2.2, viz.,

$$\Phi(E)=\frac{1}{2\varkappa}\ln\left(\frac{\Sigma_t(E)+\varkappa}{\Sigma_t(E)-\varkappa}\right)\int_0^\infty \Sigma_s(E'\to E)\,\Phi(E')\,dE'. \tag{10.3.31a}$$

A reduction of Eq. (10.3.30c) for the non-stationary case to a simple integral equation is not possible because of the space dependence. We can, however, carry out a Fourier transform. Let us introduce $F(E,\mu,B)$, the Fourier transform of $F(E,\mu,x)$, defined by

$$F(E,\mu,B)=\int_{-\infty}^{+\infty} F(E,\mu,x)\,e^{-iBx}dx. \tag{10.3.31b}$$

It satisfies the equation

$$\left.\begin{aligned}&\left(\Sigma_t(E)-\frac{\alpha}{v}\right)F(E,\mu,B)\\&\quad=iB\mu F(E,\mu,B)+\frac{1}{2}\int_0^\infty\int_{-1}^{+1}\Sigma_s(E'\to E)\,F(E',\mu',B)\,dE'\,d\mu'.\end{aligned}\right\} \tag{10.3.31c}$$

If we now introduce $\Phi(E,B)=\int_{-1}^{+1} F(E,\mu,B)\,d\mu$ and proceed exactly as we did in Sec. 5.2.2, we find that

$$\Phi(E,B)=\frac{1}{B}\arctan\left\{\frac{B}{\Sigma_t(E)-\frac{\alpha}{v}}\right\}\int_0^\infty \Sigma_s(E'\to E)\,\Phi(E',B)\,dE'. \tag{10.3.31d}$$

We can solve the integral Eqs. (10.3.31 a and d) numerically and thereby determine the eigenvalues $\varkappa$ and α and the spectra corresponding to them as functions of the absorption and B^2, respectively. Such numerical calculations have been done by Honeck. According to Nelkin, however, an analytic solution in the form of a power series is also possible, and we shall now find such a solution for Eq. (10.3.31d) for the non-stationary case. We shall neglect the absorption ($1/v$-absorption only increases the decay constant by an amount $\alpha_0=\Sigma_a(v_0)\,v_0$ and does not affect the spectrum) and write Eq. (10.3.31d) in the form

$$\left\{B/\arctan\left(\frac{B}{\Sigma_s(E)-\frac{\alpha}{v}}\right)-\Sigma_s(E)\right\}\Phi(E,B)=L\Phi \tag{10.3.32a}$$

where L is again the usual thermalization operator. Now let us set

$$\Phi(E,B)=M(E)+B^2\Phi_2(E)+B^4\Phi_4(E)+\cdots \tag{10.3.32b}$$

and

$$\alpha=D_0B^2-CB^4+FB^6+\cdots. \tag{10.3.32c}$$

If we substitute these equations in Eq. (10.3.32a), expand $B/\arctan\left(\frac{B}{\Sigma_s(E)-\frac{\alpha}{v}}\right)$, and equate the coefficients of like powers of B, we get

$$\left(\frac{1}{3\Sigma_s(E)}-\frac{D_0}{v}\right)M(E)=L\Phi_2(E), \tag{10.3.32d}$$

$$\left(\frac{1}{3\Sigma_s(E)}-\frac{D_0}{v}\right)\Phi_2(E)+\left(\frac{C}{v}+\frac{1}{3\Sigma_s^2(E)}\cdot\left\{\frac{D_0}{v}-\frac{4}{15\Sigma_s(E)}\right\}\right)M(E)=L\Phi_4(E). \tag{10.3.32e}$$

Eq. (10.3.32d) is identical with the result (10.3.29a) of elementary diffusion theory; in particular

$$D_0 = \frac{\int\limits_0^\infty \frac{1}{3\Sigma_s(E)} M(E)\, dE}{\int\limits_0^\infty \frac{1}{v} M(E)\, dE} \qquad (10.3.33\text{a})$$

corresponding to Eq. (10.3.29c). Integration of Eq. (10.3.32e) over all energies leads to

$$C = \underbrace{\frac{\int\limits_0^\infty \left(\frac{D_0}{v} - \frac{1}{3\Sigma_s(E)}\right) \Phi_2(E)\, dE}{\int\limits_0^\infty \frac{1}{v} M(E)\, dE}}_{C_D} + \underbrace{\frac{\int\limits_0^\infty \frac{1}{3\Sigma_s^2(E)} \left(\frac{4}{15\Sigma_s(E)} - \frac{D_0}{v}\right) M(E)\, dE}{\int\limits_0^\infty \frac{1}{v} M(E)\, dE}}_{C_t}. \qquad (10.3.33\text{b})$$

Here the first term is identical with the result of Eq. (10.3.29d). The second term obviously has nothing to do with diffusion cooling since it does not contain Φ_2. Instead, it represents a purely transport theoretic correction to the results of elementary diffusion theory. The reasons for the appearance of such corrections were already discussed in Sec. 9.2; the above result, however, is different from — and more accurate than — Eq. (9.2.15d). For energy-independent cross sections, C_t is simply $-D_0/15\,\Sigma_s^2$. In a heavy gas, on the other hand, $C_D = \frac{A D_0}{36\,\Sigma_s^2}$ with the result that for $A \gg 1$, $C_D \gg C_t$.

Analogous considerations are also possible in the case of anisotropic scattering if we set $\Sigma_s(E' \to E, \mathbf{\Omega}' \to \mathbf{\Omega}) = \frac{1}{4\pi}\,\Sigma_{s0}(E' \to E) + \frac{3}{4\pi}\,\Sigma_{s1}(E' \to E)\cos\vartheta_0 + \cdots$. We shall not write down here the very complicated formulas that result but shall instead give some numerical results of Honeck, who, in addition to the direct numerical eigenvalue calculations mentioned above, also carried out calculations of the first few coefficients in the $\alpha(B^2)$-expansion for several substances (cf. Table 10.3.4). In these calculations, the anisotropy of scattering was taken into account by inclusion of N terms in the expansion of $\Sigma_s(E' \to E, \cos\vartheta_0)$ in Legendre polynomials $P_l(\cos\vartheta_0)$. These results will be compared with experimental results later. Note the small contribution of C_t to the diffusion cooling constant C.

Results analogous to those found here also hold in the stationary case.

Table 10.3.4. *Diffusion Parameters at 20 °C According to* Honeck

Substance	D_0 (cm² sec⁻¹)	$C = C_D + C_t$ (cm⁴ sec⁻¹)	C_t/C_D	F (cm⁶ sec⁻¹)	Model
H_2O ($N=3$)	3.746×10^4	2.878×10^4	-0.057	2.7×10^2	Nelkin Kernel
D_2O ($N=2$)	2.069×10^5	4.852×10^5	-0.104	3.73×10^5	Modified Nelkin Kernel
Graphite (1.6 g/cm³) ($N=1$)	2.178×10^5	2.457×10^6	-0.006	-8.3×10^7	Parks Kernel

10.3.4. Remarks on the Milne Problem for Thermal Neutrons

In the treatment of pulsed, thermalized neutron fields, we have hitherto tacitly made the assumption that a separation of the space and energy dependences was possible, i.e., that $\Phi(E, \boldsymbol{r})$ could be set equal to $\Phi(E) \cdot R(\boldsymbol{r})$, where $R(\boldsymbol{r})$ is the lowest eigenfunction of $\nabla^2 R + B^2 R = 0$. [This assumption is equivalent to the Fourier transformation of Eq. (10.3.31 b).] Now we shall investigate the validity of this assumption.

Let us first consider the general Milne problem for thermal neutrons. Let the half-space $x > 0$ be filled with a non-absorbing material of temperature T_0. Let there be a source at $x = \infty$ that emits neutrons with an equilibrium energy distribution. Let the half-space $x < 0$ be empty (vacuum). The asymptotic flux distribution at large distances from the surface then has the form

$$\Phi(E, x) \sim (x + d) M(E) \qquad (10.3.34)$$

with an energy-independent extrapolated endpoint d. For monoenergetic neutrons $d = 0.71\,\lambda_{tr}$ (Sec. 5.2.4). In media with only weakly energy-dependent scattering cross sections, the behavior of the thermalized neutron field will not differ much from that of a monoenergetic field, and $d \approx 0.71\,\bar{\lambda}_{tr}$. On the other hand, in the case of a strongly energy-dependent scattering cross section (like that of H_2O), the spectrum will be strongly distorted near the surface owing to the different leakage probabilities for neutrons of different energies, and the extrapolation length will differ in value from $0.71\,\bar{\lambda}_{tr}$.

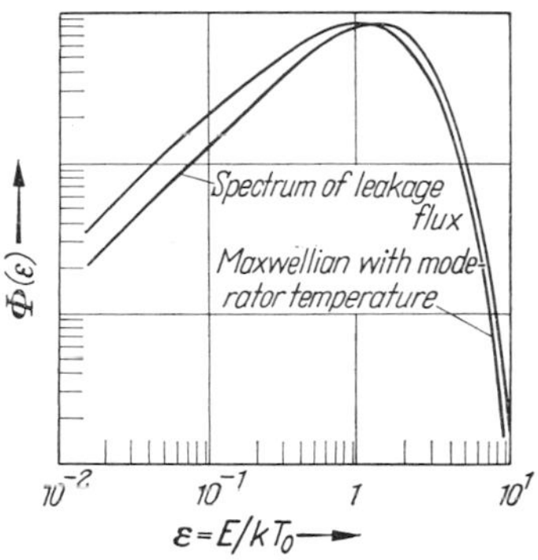

Fig. 10.3.1. The neutron spectrum leaking from a free water surface calculated using a P_1-approximation and the Nelkin model

The mathematical treatment of the energy-dependent Milne problem is very difficult and will not be given here (cf. Kladnik, Nelkin, Conkie, as well as Kiefhaber). Kiefhaber performed his calculations for water in the P_1-approximation using the Nelkin model. His results show that in the absence of absorption Eq. (10.3.34) holds very accurately for distances > 1 cm from the surface; d is 9% greater than the extrapolated endpoint obtained by averaging $1/\Sigma_{tr}(E)$ over a Maxwell distribution. For distances < 1 cm from the surface, the spectrum deviates from the Maxwell distribution. The deviation becomes stronger as we approach the surface; at the surface, the spectrum is some 20% "hotter" than in the interior (cf. Fig. 10.3.1). Unfortunately, there are no similar calculations for other moderators (except gaseous moderators). In beryllium and graphite[1] these edge effects will probably be less marked than in water since the scattering cross sections depend only slightly on energy.

We can conclude from these results that in the interior of a pulsed block of moderator whose linear dimensions are large compared to the average transport mean free path there is a space-independent spectrum of thermalized neutrons.

[1] However, at low temperatures, "cold" neutrons, i.e., neutrons with energies below the Bragg cut-off energy, which have very long mean free paths, lead to very large edge perturbations.

This conclusion is verified by calculations of GELBARD and DAVIS, who numerically calculated $\Phi(E, x)$ and α for slabs of water by a multigroup method in the P_3-approximation without any additional assumptions about separability. GELBARD and DAVIS calculated the extrapolated endpoint d as a function of the slab thickness a by assigning to each slab a B^2 such that the directly calculated value of α equals $\alpha_0+D_0B^2-CB^4+\cdots$. Since $B^2=\left(\frac{\pi}{a+2d}\right)^2$, $d=\frac{1}{2}\left(\frac{\pi}{B}-a\right)$. Fig. 10.3.2 shows a curve of $d(B^2)$ derived from these calculations. For $B^2=0$, $d\approx 0.76\,\bar{\lambda}_{tr}$, in good agreement with the result of KIEFHABER. The curve of $d(B^2)$ for spheres is very similar to the curve shown in Fig. 10.3.2. Fig. 10.3.3 shows $-\frac{\nabla^2\int\Phi(E, x)\,dE}{\int\Phi(E, x)\,dE}$ as a function of x for a slab 5.1 cm thick from these

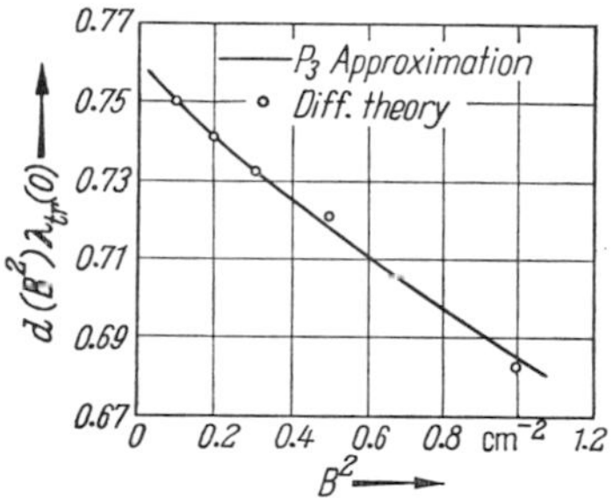

Fig. 10.3.2. The calculated extrapolated endpoint for H_2O slabs as a function of their buckling

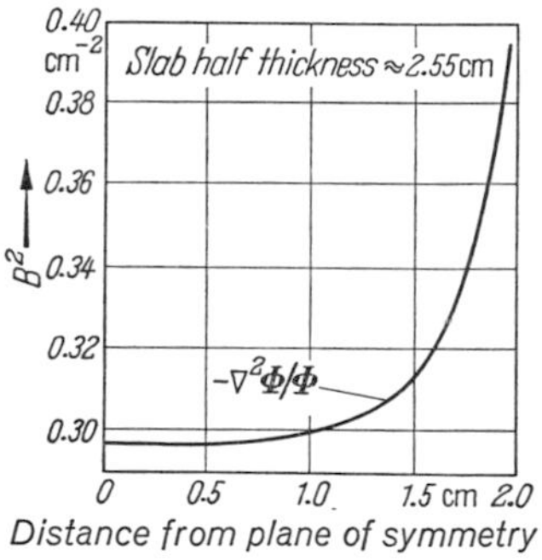

Fig. 10.3.3. The calculated local buckling $B^2=-\frac{\nabla^2\Phi}{\Phi}$ as a function of position in a pulsed slab

calculations. If $\Phi(E, x)$ were rigorously separable and $\nabla^2R+B^2R=0$, this quantity should be a constant. In the interior region of the slab, this is actually the case. In contrast, near the surface the local flux buckling depends strongly on position. In this region, the space and energy dependences are no longer separable.

A detailed study of these problems can be found in WILLIAMS.

10.4. The Approach to Equilibrium in Pulsed Neutron Fields

In this section we shall study how an incompletely thermalized neutron field approaches the equilibrium distribution. We shall limit ourselves to considering the approach to equilibrium in time. These considerations are important for interpreting some of the experiments to be discussed in Chapter 18. In addition, they give good insight into the mechanism of neutron thermalization. Discussions of the spatial approach to equilibrium may be found in KOTTWITZ, SELENGUT, KUSCER, and others.

Mathematically, the question of the approach to equilibrium will lead us to consider the higher eigenvalues and eigenfunctions of Eq. (10.3.30c); the asymptotic state treated in Sec. 10.3 corresponds to the lowest eigenvalue. However, before we embark on this formal treatment, the essence of the approach to equilibrium will be studied by means of the simple temperature concept.

10.4.1. Elementary Treatment of the Approach to Equilibrium

We limit ourselves in the following to processes in infinite, non-absorbing media. As in Sec. 10.3.1, we set up a neutron balance:

$$\frac{1}{v}\,\frac{\partial \Phi(E,t)}{\partial t} = L\Phi + \delta(t)\,\delta(E-E_Q). \tag{10.4.1a}$$

Eq. (10.4.1a) describes the thermalization of a pulse of neutrons of energy E_Q shot into an infinite medium at time $t=0$. It follows by integration over all energies that $\frac{d}{dt}\int\limits_0^\infty \frac{1}{v}\,\Phi(E,t)\,dE=0$ for $t>0$, i.e., that the total density n of neutrons is constant in time. With the source normalization chosen above, it is equal to one. If we multiply Eq. (10.4.1a) by E and integrate over all E, it then follows that for $t>0$,

$$-n\frac{d\overline{E}}{dt} = \int\limits_0^\infty\int\limits_0^\infty (E'-E)\,\Phi(E',t)\,\Sigma_s(E'\to E)\,dE'\,dE. \tag{10.4.1b}$$

Here we have transformed the integral $\int E\,L\,\Phi\,dE$ exactly as in Sec. 10.3.1. Eq. (10.4.1b) says that the energy given up per cm³ and sec via collisions equals the decrease per sec in the mean energy per cm³. Eq. (10.4.1b) is exact. When $\Phi(E,t)=M(E)$, the right-hand side vanishes; in this case we are dealing with the asymptotic state, for which $\overline{E}=3\,kT_0/2$.

Next, we again make the assumption that in the neighborhood of this equilibrium state the spectrum can be represented by a Maxwell distribution with a shifted temperature, which in this case must depend on the time. Because of the normalization to constant neutron density, we set

$$\Phi(E,t) = n\overline{v}\,\frac{E}{[kT(t)]^2}\,e^{-E/kT(t)}. \tag{10.4.2a}$$

If we substitute this expression into Eq. (10.4.1b) and rewrite the right-hand side with the help of Eqs. (10.3.10) and (10.3.11), we obtain the following result when $T(t)-T_0 \ll T_0$:

$$-\frac{dT(t)}{dt} = \frac{\overline{v}\,N\,M_2}{3}\left(T(t)-T_0\right). \tag{10.4.2b}$$

If we replace $\overline{v}=\sqrt{\frac{8\,kT}{\pi m_n}}$ by $\sqrt{\frac{8\,kT_0}{\pi m_n}}$ on the right-hand side, then with the abbreviation

$$t_T = \frac{3}{\sqrt{\frac{8\,kT_0}{\pi m_n}}\cdot N\,M_2} \tag{10.4.2c}$$

it follows that

$$T(t)-T_0 \sim e^{-t/t_T}. \tag{10.4.2d}$$

The neutron temperature thus approaches equilibrium exponentially with the time constant $1/t_T$. We shall call the quantity t_T defined by Eq. (10.4.2c) the "temperature relaxation time"; the quantity $\gamma = 1/t_T$ is frequently used in the literature. Table 10.4.1 contains some values of t_T.

Table 10.4.1. *The Temperature Relaxation Time t_T at 20 °C*

Substance	t_T (μsec)	M_2 from
H_2O	4	Nelkin Model
	3.2	Eq. (10.1.15): $A=1$
	5.05	Eq. (10.1.15): $A=18$
Be (1.85 g/cm³) . .	34	Eq. (10.1.33c)
	22	Eq. (10.1.15): $A=9$
Graphite (1.6 g/cm³)	140	Parks Model
	52	Eq. (10.1.15): $A=12$

By comparing Eqs. (10.4.2c) and (10.3.17b), we obtain the important relation

$$C = \frac{2}{3}\left(\frac{d \ln D_0}{d \ln T}\right)^2 t_T \qquad (10.4.3\text{a})$$

between the diffusion cooling constant and the temperature relaxation time. For an energy-independent D, we have simply

$$C = \frac{D_0^2 t_T}{6}. \qquad (10.4.3\text{b})$$

Next, we shall calculate the temperature relaxation in a finite medium. In this case, we find by a straightforward generalization of Eqs. (10.3.12a) and (10.4.2b) with $\nabla^2 \Phi/\Phi = -B^2$ that

$$-\frac{dT(t)}{dt} = \frac{1}{t_T}\left(T(t) - T_0\right) + \bar{D}\bar{v} B^2 \frac{T(t)}{3}\left(1 + 2\frac{d \ln \bar{D}}{d \ln T}\right). \qquad (10.4.4\text{a})$$

This equation leads to

$$T(t) - T_0\left(1 - \bar{D} B^2 \frac{1 + 2\dfrac{d \ln \bar{D}}{d \ln T}}{N M_2}\right) \sim e^{-\left(\frac{1}{t_T} + \frac{D_0 B^2}{3}\left[1 + 2\frac{d \ln \bar{D}}{d \ln T}\right]\right) t}. \qquad (10.4.4\text{b})$$

The temperature reached at the end of the thermalization process is lower than T_0 because of the diffusion cooling effect [cf. Eq. (10.3.13)]. Owing to the preferential leakage of the more energetic neutrons, thermalization proceeds faster than in an infinite medium.

10.4.2. Treatment of Time-Dependent Thermalization as an Eigenvalue Problem

After injection of a neutron pulse into an infinite, non-absorbing medium, the following equation holds [Eq. (10.4.1a) for $t > 0$]:

$$\frac{1}{v}\frac{\partial \Phi(E, t)}{\partial t} = L\Phi.$$

A general solution of this equation should be possible in the form

$$\Phi(E, t) = \sum_n B_n \Phi_n(E) e^{-\alpha_n t} \qquad (10.4.5\text{a})$$

where α_n and $\Phi_n(E)$ are respectively the eigenvalues and eigenfunctions of the equation

$$\frac{\alpha}{v}\Phi(E) = -L\Phi. \qquad (10.4.5\text{b})$$

In order that the solution can be written in the form of Eq. (10.4.5a), the $\Phi_n(E)$ must form a complete, orthogonal system of functions. We can easily prove the orthogonality of the eigenfunctions of Eq. (10.4.5b)[1]. However, the question of the completeness is very complex. The discrete eigenvalues obey the condition $\alpha_n \leq \min(v\Sigma_s(v))$ (cf. Sec. 10.3.3), and we can show that in the range $\min(v\Sigma_s(v)) < \alpha < \max(v\Sigma_s(v))$ there exists a continuum of eigenvalues and eigenfunctions (cf. CORNGOLD, MICHAEL and WOLLMANN as well as KOPPEL). Instead of Eq. (10.4.5a), therefore, we must write

$$\Phi(E, t) = \sum_n B_n \Phi_n(E) e^{-\alpha_n t} + \int_{\min(v\Sigma_s)}^{\max(v\Sigma_s)} B(\alpha) F_\alpha(E) e^{-\alpha t} \, d\alpha \tag{10.4.5c}$$

and only if we include both the discrete and continuous eigenfunctions is the system complete. However, we are not interested here in a complete solution $\Phi(E, t)$ valid for all times after the pulse but rather wish only to know how the spectrum approaches the equilibrium state long after the pulse. For this reason, we only need to know the lowest two eigensolutions of Eq. (10.4.5b). The equilibrium state corresponds to the eigenvalue $\alpha = 0$ and the eigenfunction $M(E)$; for long times we then have

$$\left.\begin{aligned} \Phi(E, t) = B_0 M(E) + B_1 \Phi_1(E) e^{-\alpha_1 t} + \\ + \text{ additional terms decaying even more rapidly with time.} \end{aligned}\right\} \tag{10.4.5d}$$

Now we shall try to calculate α_1 and $\Phi_1(E)$. $1/\alpha_1$ is clearly a characteristic measure of the duration of the thermalization process; we shall call $t_{th} = 1/\alpha_1$ the "thermalization time". By use of the form (10.4.5d) we are tacitly making the assumption that there is in fact at least one other discrete eigenvalue besides $\alpha = 0$ below the limit $\min(v\Sigma_s)$; this assertion has not yet been proven in general. It will turn out, indeed, that the values of α_1 calculated below will be of the same order of magnitude as the critical value in graphite and even beyond it in beryllium[2].

We again use the Laguerre-polynomial method for the calculations. Let us substitute

$$\Phi(E) = \sum_{k=0}^{\infty} A_k L_k^{(1)}\left(\frac{E}{kT_0}\right) M(E) \tag{10.3.21}$$

into Eq. (10.4.5b), multiply by $L_i^{(1)}\left(\frac{E}{kT_0}\right)$, and integrate over all energy. The result is a linear, homogeneous system of equations for the A_k; the condition that there be a non-trivial solution is that the determinant of the coefficients vanish:

$$|\alpha V_{ik} + L_{ik}| = 0. \tag{10.4.6a}$$

In the $l=2$ approximation, Eq. (10.4.6a) becomes

$$\begin{vmatrix} V_{00} & \alpha V_{01} \\ \alpha V_{10} & \alpha V_{11} + L_{11} \end{vmatrix} = 0 \tag{10.4.6b}$$

---

[1] When $n \neq m$, $\int_0^\infty \Phi_n^+(E) \frac{1}{v} \Phi_m(E) \, dE = 0$. Here $\Phi_n^+(E)$ is an eigenfunction of the equation adjoint to Eq. (10.4.5b); it is given by $\Phi_n(E)/M(E)$.

[2] It should be noted here that for the heavy gas model $\min(v\Sigma_s) = \infty$, i.e., the collision time is zero. For this reason, the system of discrete eigenfunctions is complete for this model. The heavy gas model is obtained by letting $A \to \infty$ while keeping the quantity $\xi\Sigma_s = 2\Sigma_s/A$ finite. Thus no conclusions about time-dependent thermalization phenomena should be drawn from the heavy gas model.

when we factor out the trivial root $\alpha=0$. Thus

$$\left.\begin{aligned}\alpha_1&=\frac{V_{00}L_{11}}{V_{00}V_{11}-V_{01}^2}\\&=\frac{2}{\sqrt{\pi}}v_T\frac{NM_2}{3},\\t_{th}&=\frac{1}{\alpha_1}=\frac{3}{\sqrt{\frac{8kT_0}{\pi m_n}}NM_2}.\end{aligned}\right\}\tag{10.4.6c}$$

In the $l=2$ approximation, t_{th} is identical with the temperature relaxation time introduced in Eq. (10.4.2c)! Using Eqs. (10.3.10) and (10.4.2a), we can easily convince ourselves that the temperature approximation is again identical with the approximation of only using two Laguerre polynomials, just as it was in the description of diffusion cooling.

By keeping more Laguerre polynomials, TAKAHASHI investigated how good an approximation the $l=2$ case really is. It turns out that good convergence of α_1 is not reached until $l=20$. TAKAHASHI finds that

$$\alpha_1^{(l=20)}=\frac{\alpha_1^{(l=2)}}{1.28}\quad\text{for a heavy gas,}$$

$$\alpha_1^{(l=20)}=\frac{\alpha_1^{(l=2)}}{3.0}\quad\text{for graphite at 20 °C,}$$

$$\alpha_1^{(l=20)}=\frac{\alpha_1^{(l=2)}}{4.4}\quad\text{for beryllium at 20 °C.}$$

The $l=2$ approximation thus considerably overestimates α_1, especially in media with strong binding. This means that the effective neutron temperature concept is unsuitable for the quantitative treatment of time-dependent thermalization processes. In order to obtain the right thermalization times, we must multiply the temperature relaxation times given in Table 10.4.1 by the factors specified above. For graphite we thus obtain $t_{th}\approx 420$ μsec from the Parks model, a value which is just above the critical limit, while for beryllium $t_{th}\approx 150$ μsec follows from the Debye model. However, according to Eq. (10.3.30c) and Table 10.3.3, t_{th} for beryllium should be no less than 260 μsec. Thus the calculations of TAKAHASHI seem not to be applicable to this case, and indeed a calculation of CORNGOLD and SHAPIRO shows that in beryllium no discrete mode above the fundamental one exists. The factor has not been calculated for H_2O, but the value 1.28 appropriate to the heavy gas model may be a useful approximation in this case.

As we can see by comparison of Eqs. (10.4.9c) and (10.3.26c), when the diffusion constant is independent of energy, the following relation exists between C and t_{th} in the $l=2$ approximation:

$$C=\frac{D_0 t_{th}}{6}.\tag{10.4.7}$$

This equation corresponds to Eq. (10.4.3b) of the temperature model. In passing from $l=2$ to $l=20$, both C and t_{th} increase, and by roughly the same factor, irrespective of the model. Thus Eq. (10.4.7) remains approximately valid. TAKAHASHI has shown without reference to any special thermalization model that Eq. (10.4.7) is always approximately true.

We can also calculate α_1 for *finite* media; PUROHIT has done this in the $l=2$ approximation. An increase with increasing B^2 is evident as is to be expected from the qualitative considerations of the temperature model (cf. Sec. 10.4.1).

Chapter 10: References

General

AMALDI, E.: loc. cit., especially § 95—112.

CORNGOLD, N. (ed.): Proceedings of the Brookhaven Conference on Neutron Thermalization, BNL-719 (1962).

POOLE, M. J., M. S. NELKIN, and R. S. STONE: The Theory and Measurement of Reactor Spectra, Progr. Nucl. Energy Ser. I, Vol. 2, p. 91 (1958).

KOTHARI, L. S., and V. P. DUGGAL: Scattering of Thermal Neutrons from Solids and their Thermalization Near Equilibrium, Advances in Nuclear Science and Technology, Vol. 2, p. 186 (1964).

Special[1]

Neutron Scattering by a Gas of Atomic Nuclei.

COHEN, E. R.: Geneva 1955, P/611, Vol. 5, p. 531.
DARDEL, G. F. VON: Phys. Rev. 94, 1272 (1954).
WIGNER, E. P., and J. E. WILKINS: AECD-2275 (1944).
WILKINS, J. E.: CP-2482 (1944).

Fundamentals of Inelastic Neutron Scattering by Chemically Bound Atoms.

BROWN, H. D., and D. S. ST. JOHN: DP-33 (1954).
FERMI, E.: Ricerca Sci. 1, 13 (1936).
JANIK, J. A., and A. KOWALSKA: Institute of Nuclear Physics Cracow, Report No. 255 (1963).
KRIEGER, T. J., and M. NELKIN: Phys. Rev. 106, 290 (1957).
NELKIN, M.: GA-1689 (1960).
PLACZEK, G.: Phys. Rev. 86, 377 (1952).
PUROHIT, S. N.: BNL 719, 203 (1962).
PUROHIT, S. N., and A. K. RAJAGOPAL: BNL 719, 238 (1962).
SACHS, R. G., and E. TELLER: Phys. Rev. 60, 18 (1941).
SJÖLANDER, A.: Arkiv Fysik 14, 315 (1959).
VOLKIN, H. C.: Phys. Rev. 113, 866 (1959); 117, 1029 (1960).
WICK, G. C.: Phys. Rev. 94, 1223 (1954).
ZEMACH, A. C., and R. J. GLAUBER: Phys. Rev. 101, 118 (1956).

Inelastic Neutron Scattering by H_2O, D_2O, and Other Hydrogenous Compounds.

BOFFI, V. C., V. C. MOLINARI, and D. E. PARKS: BNL 719, 69 (1962).
GOLDMANN, D. T., and F. D. FEDERIGHI: BNL 719, 100 (1962).
GÖSSMANN, G.: Nukleonik 4, 110 (1962).
HONECK, H.: Trans. Am. Nucl. Soc. 5, 1, 47 (1962).
MILLER, J., R. L. BREHM, and W. J. ROBERTS: NAA-SR-7140 (1962).
NELKIN, M.: Phys. Rev. 119, 741 (1960).
SPRINGER, T.: Nukleonik 3, 110 (1961).

Inelastic Neutron Scattering by Crystalline Solids.

BHANDARI, R. C.: J. Nucl. Energy A 6, 104 (1957).
BHANDARI, R. C., L. S. KOTHARI, and K. S. SINGWI: J. Nucl. Energy A 7, 45 (1958).
KOTHARI, L. S., and K. S. SINGWI: Solid State Physics 8, 109 (1959).
KHUBCHANDANI, P. G., L. S. KOTHARI, and K. S. SINGWI: Phys. Rev. 110, 70 (1958).
MARSHALL, W., and R. STUART: Inelastic Scattering of Neutrons in Solids and Liquids, p. 75, Vienna: International Atomic Energy Agency, 1961.
NELKIN, M.: Nucl. Sci. Eng. 2, 373 (1957).
PARKS, D. E.: GA-2438 (1961); Nucl. Sci. Eng. 13, 306 (1962).
PLACZEK, G., and L. VAN HOVE: Phys. Rev. 93, 1207 (1954).
TAKAHASHI, H.: BNL-719, 1299 (1962).

Apparatus for Studying Inelastic Neutron Scattering.

BROCKHOUSE, B. N.: Bull. Am. Phys. Soc. 3, 233 (1958).
BRUGGER, R. M., and J. E. EVANS: Nucl. Instr. Methods 12, 75 (1961).
EGELSTAFF, P. A., J. COCKING, and J. ALEXANDER: CRRP-1078 (1960).
GLÄSER, W.: Neutron Time of Flight Methods, p. 301, Brussels: Euratom 1961.

[1] Cf. footnote on p. 53.

Brugger, R. M.: BNL-719, 3 (1962).
Egelstaff, P.: Nucl. Sci. Eng. **12**, 250 (1962).
Egelstaff, P., and P. Schofield: Nucl. Sci. Eng. **12**, 260 (1962).
Haywood, S. C., and I. M. Thorson: BNL-719, 26 (1962).
Sinclair, R. N.: Inelastic Scattering of Neutrons in Solids and Liquids, Vol. 2, p. 199, Vienna: International Atomic Energy Agency, 1963.

Measurements of Inelastic Neutron Scattering.

Cohen, E. R.: Nucl. Sci. Eng. **2**, 227 (1957).
Coveyou, R. F., R. R. Bate, and R. K. Osborn: ORNL-1958 (1955).
Horowitz, J., and O. Tretiakoff: EANDC-(E)-14 (1960).
Hurwitz, H., M. S. Nelkin, and G. I. Habetler: Nucl. Sci. Eng. **1**, 280 (1956).
Wigner, E. P., and J. E. Wilkins: AECD-2275 (1944).
Wilkins, J. E.: CP-2481 (1944).

Space-Independent Neutron Spectra in Gaseous Moderators.

Cadilhac, M., et al.: BNL-719, 439 (1962).
Corngold, N.: Ann. Phys. **6**, 368 (1959); **11**, 338 (1960).
Parks, D. E.: Nucl. Sci. Eng. **9**, 430 (1961).
Schaefer, G. W., and K. Allsopp: BNL-719, 614 (1962).

Space-Independent Spectra in Moderators with Chemical Binding.

Honeck, H.: BNL-719, RD-1 (1962); BNL-821 (T-319) (1963).
Marchuk, G. I., et al.: BNL-719, 706 (1962).
Michael, P.: Nucl. Sci. Eng. **8**, 426 (1960).
De Sobrino, L., and M. Clark: Nucl. Sci. Eng. **10**, 388 (1961).

Space-Dependent Neutron Spectra.

Beckurts, K. H.: Nucl. Sci. Eng. **2**, 516 (1957).
Beckurts, K. H.: Z. Naturforsch. **12**a, 956 (1957).
von Dardel, G. F.: Phys. Rev. **94**, 1272 (1954).
Häfele, W., and L. Dresner: Nucl. Sci. Eng. **7**, 304 (1960).
Hurwitz, H., and M. S. Nelkin: Nucl. Sci. Eng. **3**, 1 (1958).
Nelkin, M. S.: J. Nucl. Energy A **8**, 48 (1958).
Purohit, S. N.: Nucl. Sci. Eng. **9**, 157 (1961).
Singwi, K. S.: Arkiv Fysik **16**, 385 (1960).
Takahashi, H.: BNL-719, 1299 (1962).

Theory of the Diffusion Cooling Effect.

Honeck, H.: BNL-719, 1186 (1962).
Nelkin, M.: Nucl. Sci. Eng. **7**, 210 (1960).
Nelkin, M.: Physica **29**, 261 (1963); GA-3122, 3298 (1962).
Sjöstrand, N. G.: Arkiv Fysik **15**, 147 (1959).

Transport Theory of the Thermal Neutron Field.

Conkie, W. R.: Nucl. Sci. Eng. **7**, 295 (1960).
Gelbard, E. M., and J. A. Davis: Nucl. Sci. Eng. **13**, 237 (1962).
Kiefhaber, E.: Nucl. Sci. Eng. **18**, 404 (1964).
Kladnik, R.: BNL-719, 1211 (1962).
Kladnik, R., and I. Kuscer: Nucl. Sci. Eng. **13**, 149 (1962).
Williams, M. M. R.: J. Nucl. Energy A & B **17**, 55 (1963).

Space-Energy Separability in the Thermalized Field; the Milne Problem.

Corngold, N., P. Michael, and W. Wollmann: Nucl. Sci. Eng. **15**, 13 (1963); BNL-719, 1103 (1962).
Corngold, N., and C. S. Shapiro: Nucl. Sci. Eng. (in press).
Koppel, J. U.: Nucl. Sci. Eng. **16**, 101 (1963); GA-2988 (1962); BNL-719, 1232 (1962).
Purohit, S. N.: Nucl. Sci. Eng. **9**, 157 (1961).

Time-Dependent Thermalization.

Bennett, R. A., and R. E. Heineman: Nucl. Sci. Eng. **8**, 294 (1963).
Kottwitz, D. A.: Nucl. Sci. Eng. **7**, 345 (1960).
Kuscer, I.: J. Nucl. Energy A & B **17**, 49 (1963).
de Ladonchamps, J. R. L., and L. M. Grossman: Nucl. Sci. Eng. **12**, 238 (1962).
Selengut, D. S.: Nucl. Sci. Eng. **9**, 94 (1961).
Takahashi, H.: BNL-719, 1299 (1962).
Weiss, Z., and W. Suwalski: Nukleonika **6**, 243, 443, 691, 704 (1961).

Neutron Diffusion in the Neighborhood of a Temperature Discontinuity; the Spatial Approach to Equilibrium.

Part III

The Determination of Flux and Spectrum in a Neutron Field

11. Measurement of the Thermal Neutron Flux with Probes

The neutron detectors discussed in the third chapter are not always suitable for determining the intensity of a stationary neutron field, but we can always use to advantage the so-called "radioactive indicators", i.e., we can always activate a probe in the neutron field and then count its radioactivity. In this way, we can achieve very precise relative and absolute measurements of neutron intensities. Since there are many probe substances whose activation cross sections depend in various different ways on the neutron energy, we can also derive information about the energy distribution in a neutron field from probe measurements.

In this chapter, we shall examine probe measurements in purely thermal neutron fields. In the process, questions will be answered which are important for all probe measurements. The following chapters will then be devoted to the measurement of the epithermal neutron flux with resonance probes and the separation of thermal and epithermal activation (Chapter 12) as well as the use of threshold detectors for the detection of fast neutrons (Chapter 13). Chapters 14 and 15 also touch in places on the use of probes.

11.1. General Facts about Probes

11.1.1. Activation and Activity

We shall begin by specifying the geometric forms that will interest us hereafter. In practice, the most widely used probes are *foils*, i.e., probes made in the form of a thin, disc-shaped foil with a surface area in the range 0.1—10 cm^2. Probes in the form of long *tapes* or *wires* are occasionally used to measure flux profiles. For foils and tapes, activation and activity are usually expressed per cm^2 of area, while for wires they are expressed per cm of length.

The *activation* C is the number of radioactive atoms formed by neutron capture per second and cm^2 of probe area. C depends on the intensity of the neutron field $F(\boldsymbol{r}, \boldsymbol{\Omega}, E)$, the thickness d of the probe, and its cross sections $\Sigma_{act}(E)$, $\Sigma_a(E)$, and $\Sigma_s(E)$. We shall learn the general connection between these quantities and C in Sec. 11.2.1. For $\big(\Sigma_a(E)+\Sigma_s(E)\big)\, d \ll 1$, i.e., for thin probes, the neutron field traverses the probe practically unattenuated, and

$$C = \Phi \overline{\Sigma}_{\text{act}} d. \tag{11.1.1}$$

Here Φ is the thermal flux[1] and $\bar{\Sigma}_{act}$ the "average" activation cross section in the sense of Sec. 5.3.2. The activation of thin probes is thus proportional to the flux or to the density; if $\Sigma_{act}(v)$ obeys the $1/v$-law, then $C = n\bar{v}\Sigma_{act}(\bar{v})d$ and we can obtain the neutron density from the activation without knowledge of the spectrum.

The *total activation* $B(t)$ is the number of radioactive atoms present per cm² of probe area. If λ is the decay constant, then

$$B(t) = \frac{C}{\lambda}(1 - e^{-\lambda t}) \tag{11.1.2}$$

for a constant activation that began at time $t = 0$. $B(t)$ reaches the saturation value C/λ after an "infinitely long" irradiation. After an irradiation time of one half-life $T_{\frac{1}{2}} = 0.693/\lambda$, 50% of the saturation value is reached, after ten half-lives, 99.9%. After the end of the irradiation, $B(t)$ falls off according to the law

$$B(t_2) = \frac{C}{\lambda}(1 - e^{-\lambda t_1})e^{-\lambda t_2} \tag{11.1.3a}$$

where t_1 is the irradiation time and t_2 is the elapsed time since the end of the irradiation.

Finally, the *activity* A is the number of atoms decaying per sec and cm² of probe surface. $A = \lambda B$ and $A(t_2) = C(1 - e^{-\lambda t_1})e^{-\lambda t_2}$ or

$$AT = C \tag{11.1.3b}$$

with the *time factor*

$$T = \frac{e^{\lambda t_2}}{1 - e^{-\lambda t_1}}. \tag{11.1.3c}$$

In order to make different probe measurements comparable, we customarily calculate the activation by multiplying the measured activity by the appropriate time factor. Note that A is the "true" probe activity whereas simple counting gives an "apparent" activity that is smaller than the true activity owing to the self-absorption of the emitted radiation and to geometrical (solid-angle) effects.

11.1.2. Materials for Thermal Probes

Table 11.1.1 contains some useful data on substances that can be used for thermal probes. The first three columns contain the number of atoms per gram of the substance, its absorption cross section, and its scattering cross section. In addition, the table specifies the activable isotopes, their abundances in the element, their activation cross sections, the isotopes produced, and the latter's half-lives. The principal activity is given in italics; knowledge of the other activities is important since in order to evaluate the probe measurements we must know how long to wait for the short-lived activities to die out and how much background to subtract to account for the long-lived activities.

Manganese has the special advantage of being a pure $1/v$-absorber in the thermal energy range. In practice, it is mainly used in the form of manganese-nickel foils containing about 90 wt-% of manganese ($\varrho = 7.6$ g/cm³). Neutron capture by Ni^{64} leads to a perturbing activity which coincidentally has the same half-

[1] More precisely: the average of the thermal flux over the probe surface.

life as Mn^{56} but only contributes 0.01% as much as the principal activity. Manganese-nickel foils are very stable and can be rolled as thin as *ca.* 6 mg/cm² ≙ 7.9×10^{-3} mm. The absorption and activation coefficients[1] are $\mu_a=0.135$ and $\mu_{ac}=0.130$ cm²/g. Such foils are suitable for measurements in fluxes greater than 10^4 n/cm²/sec. Fig. 11.1.1 shows the decay scheme of manganese. β- as well as γ-radiation occurs, and both can be used for counting. Manganese is also used in the manganese bath method of integrating neutron fields (cf. Sec. 14.3.3).

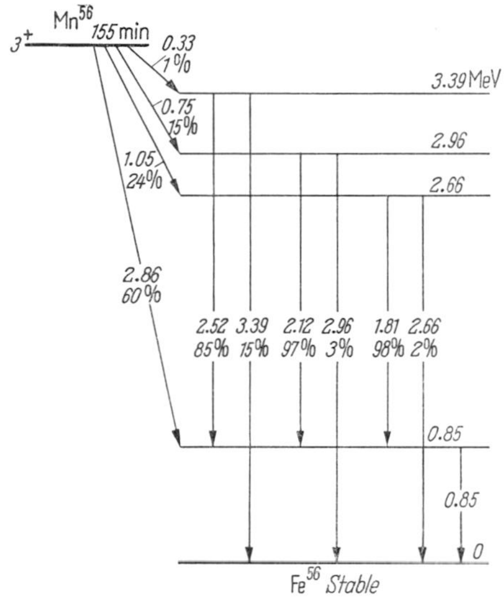

Fig. 11.1.1. The decay scheme of Mn^{56}

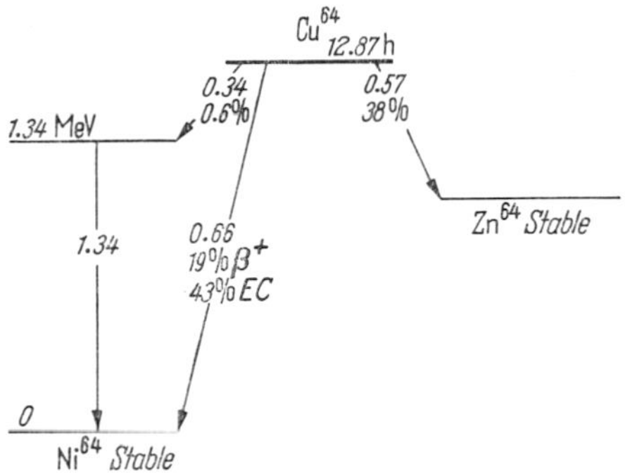

Fig. 11.1.2. The decay scheme of Cu^{64}

Table 11.1.1. *Substances for Thermal Probes*

Element	$N/\varrho\times10^{-22}$ (g⁻¹)	σ_a (barn) ($v_0=2200$ m/sec)	σ_s (barn)	Isotope (abundance in %)	σ_{act} (barn) ($v_0=2200$ m/sec)	Radioactive Isotope (Half-Life)
$_{25}$Mn	1.097	13.2±0.1	2.3±0.3	Mn^{55} (100)	13.2±0.1	*Mn^{56} (2.58 h)*
$_{27}$Co	1.022	37.1±1.0	7±1	Co^{59} (100)	16.9±1.5 20.2±1.9	Co^{60m} (10.4 min) (10.4 min→5.28 y (99%)) *Co^{60} (5.28 y)*
$_{29}$Cu	0.948	3.81±0.03	7.2±0.6	Cu^{63} (69.1) Cu^{65} (30.9)	4.41±0.20 1.8±0.4	*Cu^{64} (12.87 h)* Cu^{66} (5.14 min)
$_{47}$Ag	0.558	64.5±0.6 g(0.0253 ev)= 1.0045	6±1	Ag^{107} (51.35) Ag^{109} (48.65)	45±4 3.2±0.4 105±15	*Ag^{108} (2.3 min)* Ag^{110m} (253 d) Ag^{110} (24 sec)
$_{49}$In	0.525	194±2 g(0.0253 ev)= 1.0194	2.2±0.5	In^{113} (4.23) In^{115} (95.77)	56±12 2.0±0.6 160±2 42±1	In^{114m_1} (49 d) In^{114} (72 sec) *In^{116m_1} (54.12 min)* In^{116} (14.1 sec)
$_{66}$Dy	0.371	940±20	100±20	Dy^{164} (28.18)	2000±200 800±100	Dy^{165m} (1.3 min) *Dy^{165} (140 min)*
$_{79}$Au	0.3055	98.5±0.4 g(0.0253 ev)= 1.0052	9.3±1.0	Au^{197} (100)	98.5±0.4	*Au^{198} (2.695 d)*

[1] In dealing with probes it is customary to measure the thickness in terms of the surface mass loading $\delta=\varrho d$ (g/cm²) rather than in cm; then the mass absorption coefficient $\mu_a=\Sigma_a/\varrho$ (cm²/g) appears in place of the macroscopic absorption coefficient Σ_a.

Because of its long half-life and its small activation cross section, *cobalt* is used for the measurement and frequently for the long-time integration[1] of extremely high neutron fluxes (10^{10}—10^{13} n/cm²/sec). It is used principally in the form of wires. The irradiated probes are counted most conveniently by means of their γ-radiation, since Co^{60} emits only very weak β-radiation.

Copper like manganese has the advantage of being a pure $1/v$-absorber. Because of their longer half-life and smaller activation cross section, copper probes are less sensitive than manganese-nickel foils (flux range above 10^5 n/cm²/sec). Pure metallic copper ($\varrho = 8.90$ g/cm³) can easily be made into thin foils or tapes with good mechanical properties and surface loadings as low as 5 mg/cm². $\mu_a = 0.0361$ cm²/g and $\mu_{act} = 0.0289$ cm²/g for

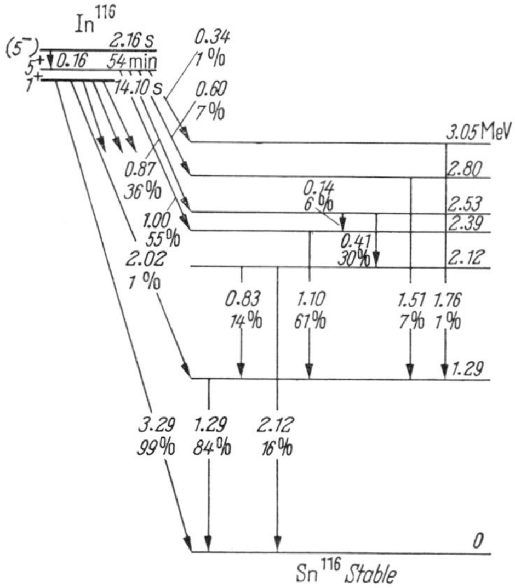

Fig. 11.1.3. The decay scheme of In^{116}, $In^{116\,m_1}$ and $In^{116\,m_2}$

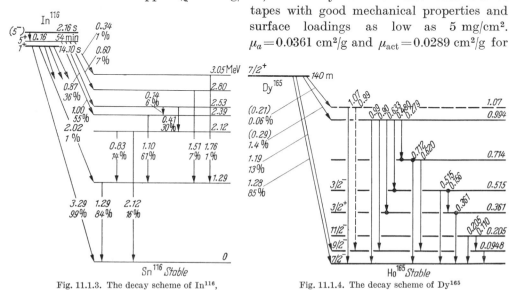

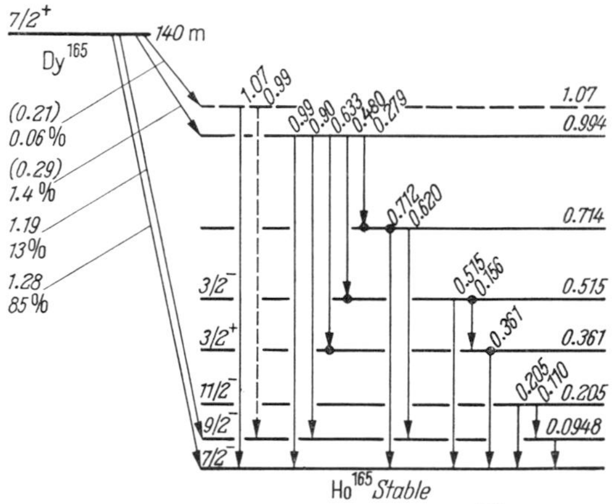

Fig. 11.1.4. The decay scheme of Dy^{165}

the 12.87-hour activity. Fig. 11.1.2 shows the decay scheme of Cu^{64}. The probes are best counted by means of the β^-- and β^+-radiation or by means of the positron annihilation radiation. γ—γ coincidence methods may be used here with advantage.

Because of its short half-life and high activation cross section, *silver* — and also *rhodium* — is often used in demonstration experiments involving neutron activation; we have included it in Table 11.1.1 only for the sake of completeness.

The 54-minute activity of *indium* is frequently used for the determination of low fluxes. The cross section does not follow the $1/v$-law; in fact, there are resonances in the evrange which can lead to a strong epithermal activation of the probe (cf. Sec. 12.1.2). Besides the activities given in the table, a perturbing 4.5-hour activity is produced by an isomer of In^{115} that can be excited by the inelastic scattering of fast neutrons (cf. Sec. 13.1.2). Durable foils can be manufactured out of indium metal ($\varrho = 7.28$ g/cm³) with thicknesses down to 10 mg/cm²; "thinner" foils can be made by evaporating indium onto suitable backings or

[1] A long-time integration is necessary to find the total dose received by a sample irradiated in a reactor.

alloying it with tin. For indium metal, $\mu_a = 1.019\ cm^2/g$ and $\mu_{act}(54\ min) = 0.8045\ cm^2/g$. Fig. 11.1.3 shows the decay scheme of In^{116}; obviously the activity can be counted by means of either the β- or the γ-radiation.

Dysprosium is also suitable for the measurement of low neutron fluxes. The activation cross section deviates slightly from the $1/v$-law; compared to indium, the contribution of epithermal activation is smaller. Dysprosium is sometimes used in the form of dysprosium metal, in the form of dysprosium-aluminum alloys or in the form of dysprosium oxide Dy_2O_3, which is deposited on an aluminum backing after being mixed with a binder. For Dy_2O_3, $\mu_a = 1.586\ cm^2/g$ and $\mu_{act} = 0.381\ cm^2/g$ for the 140-minute activity. We can easily see from the decay scheme of Dy^{165} given in Fig. 11.1.4 that dysprosium is only suitable for β-counting.

Gold is very well suited for precision measurements, particularly absolute measurements. Metallic gold foils ($\varrho = 19.32\ g/cm^3$) with thicknesses down to

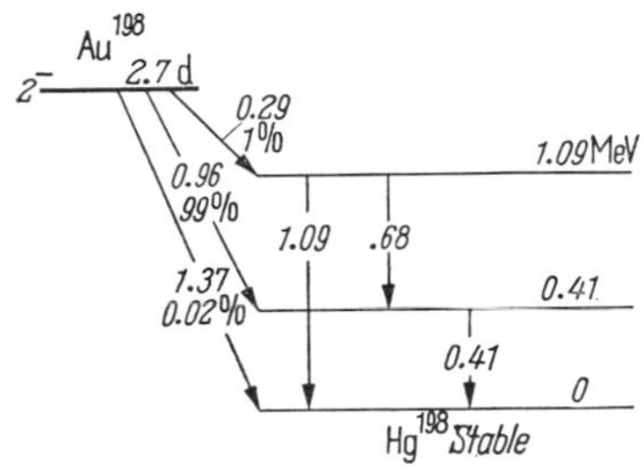

Fig. 11.1.5. The decay scheme of Au^{198}

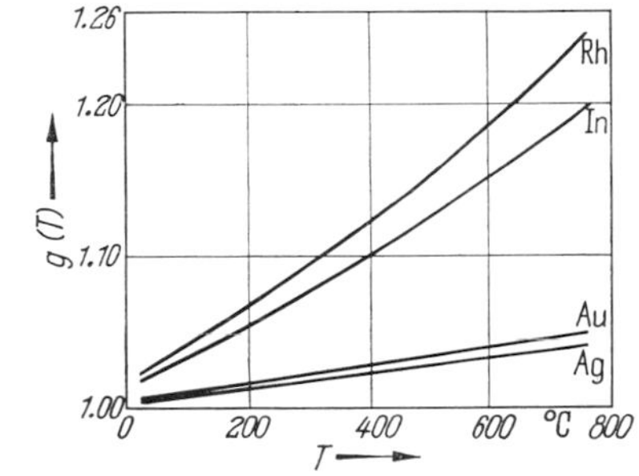

Fig. 11.1.6. The temperature-dependent g-factors of several foil substances [cf. Eqs. (5.3.19—21)]

$5\ mg/cm^2$ can easily be rolled; thin foils can also be made by evaporation. $\mu_a = \mu_{act} = 0.3008\ cm^2/g$. Fig. 11.1.5 shows the decay scheme of gold; β- as well as γ-counting is possible[1].

Fig. 11.1.6 shows the neutron-temperature-dependent g-factors of several probe substances. Data on the cross sections in the epithermal range follow in Sec. 12.1.2.

11.1.3. The Determination of the Probe Activity

We must distinguish between the problems of determining the absolute and the relative probe activity. The former is necessary when we want to determine the neutron flux absolutely; for most investigations, however, it is only the relative variation of the flux that interests us, and a relative activity measurement, i.e., one in which the activity of one probe is referred to that of another or of a standard suffices. We shall next consider such relative measurements.

If the activity is measured by means of γ-radiation, it is best to use a scintillation counter whose photopeak matches one of the characteristic γ-lines of the activity. Since the energy of the γ-rays is usually quite high (Table 11.1.2), their self-absorption in the probe material will be small if the probe is not too thick. The "apparent" activity, i.e., the number of γ-rays leaving one side of the foil

[1] Au^{198} has an absorption cross section of 30,000 barns. With very high neutron fluxes errors can occur owing to the loss of activated nuclei by neutron capture.

per cm² and per second, is thus equal to half of the true activity (aside from the fact that each decay may not produce exactly one γ-ray). The measurement thus yields a counting rate proportional to the true activity. Many probe substances do not emit any useable γ-rays. In this case, we must determine their activity

Table 11.1.2. *Data for γ-Counting of Probe Activities*

Substance	$T_{\frac{1}{2}}$	γ-Energy (Mev)	Number of γ-Rays per Decay	Total Absorption Coefficient μ_γ [cm²/g] in the Probe Substance[1]
$_{25}$Mn	2.58 h (Mn^{56})	2.12 1.81 0.845	0.145 0.235 0.989	0.0415 (Mn) 0.0405 (Ni) 0.0450 (Mn) 0.0435 (Ni) 0.0656 (Mn) 0.0645 (Ni)
$_{27}$Co	5.28 y (Co^{60})	1.333 1.173	1.00 0.99	0.0512 0.0548
$_{49}$In	54.12 min ($In^{116\,m_1}$)	2.12 1.76 1.51 1.29 1.10 0.83 0.41 0.14	0.16 0.02 0.07 0.84 0.61 0.14 0.30 0.06	0.0400 0.0425 0.0458 0.0496 0.0540 0.0635 0.0970 0.6650
$_{79}$Au	2.695 d (Au^{198})	0.412	0.99	0.192

[1] If the probe is a foil of thickness δ that has been homogeneously activated, the average probability that a γ-ray escapes is $\varphi_0(\mu_\gamma\delta)/2\mu_\gamma\delta$. The total absorption coefficient has been used here because Compton-scattered γ-rays suffer an energy loss and thus can no longer be detected if a window discriminator is used.

Table 11.1.3. *Data for β-Counting of Probes*

Substance	$T_{\frac{1}{2}}$	$E_\beta$max (Mev)	$\alpha_0\left[\frac{cm^2}{g}\right]$ for Aluminum	$\bar{\alpha}\left[\frac{cm^2}{g}\right]$ for the Probe Substance
$_{25}$Mn	2.58 h (Mn^{56})	2.86 (60%) 1.05 (24%) 0.75 (16%)	11.6±1.0	7.5
$_{29}$Cu	12.87 h (Cu^{64})	0.573 (38%) 0.656 (19%) (β^+)	34.9±1.5	35
$_{49}$In	54.12 min ($In^{116\,m_1}$)	1.00 (51%) 0.87 (28%) 0.60 (21%)	25.0+2.0	22
$_{66}$Dy	140 min (Dy^{165})	1.28 (85%) 1.19 (13%) 0.29 (1.4%)	12.5 [Eq. (11.1.5)]	≈12.5
$_{79}$Au	2.695 d (Au^{198})	0.957 (99.0%) 0.295 (1%)	20.0±1.0	25

by means of their β-radiation. Even with substances which emit γ-rays, β-counting is sometimes preferred because it requires a much smaller expenditure of apparatus. Here, however, the apparent activity is smaller than half the true activity because of the strong β-self-absorption. In particular, the apparent activity may depend on the foil orientation in the neutron field, as we shall show in Sec. 11.2.5. β-self-absorption can be characterized approximately by an exponential law of attenuation: If a β-emitter whose surface density is negligibly small is covered by a layer of thickness x, the probability $w(x)$ that a decay electron penetrates the layer is given by

$$w(x) = e^{-\alpha x}. \tag{11.1.4}$$

This law holds quite accurately for substances that have a simple decay scheme (gold); for substances with a complex β-spectrum, such as indium, it is necessary to take the attenuation of each individual component i into account through its own separate mass absorption coefficient α_i. For modest accuracies, however,

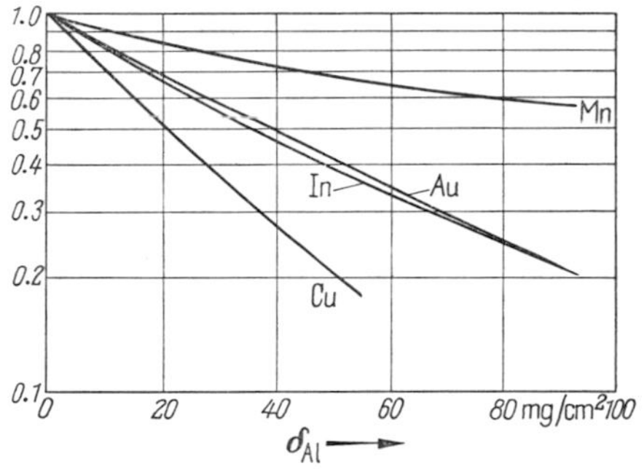

Fig. 11.1.7. Absorption curves for β-rays emitted from thin detector foils as a function of Aluminum absorber thickness. The curves represent smooth interpolations of MEISTER'S experimental values

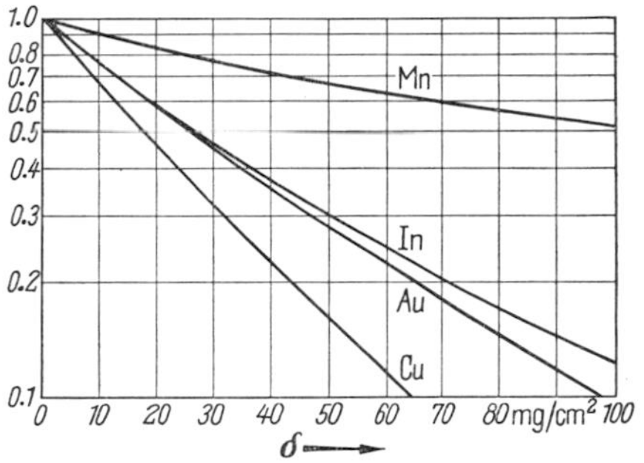

Fig. 11.1.8. Absorption curves for β-rays emitted from thin detector foils as a function of the thickness of the absorber (which is the same material as the β-emitting foil). The curves represent smooth interpolations of MEISTER'S experimental values

use of an average α is usually sufficient. Table 11.1.3 contains data on the β-energies as well as the mass absorption coefficients of some probe substances. The fourth column contains mass absorption coefficients α_0 determined from the initial portion of an absorption curve measured for aluminum (cf. Fig. 11.1.7). α_0 is given rather accurately by the following empirical expression (GLEASON):

$$\alpha_0 = 17.0\, E_{\beta\max}^{-1.43}\ [\mathrm{cm^2/g}]\quad (E_{\beta\max}\ \text{in Mev}). \tag{11.1.5}$$

When the spectrum is complex, α_0 for each component must be calculated from Eq. (11.1.5) and then all the values averaged. The fifth column gives an average absorption coefficient that is chosen so that the measured absorption curve (with the probe substance as absorber) is fit as well as possible by an exponential function $e^{-\bar{\alpha} x}$ in the range 0—100 mg/cm² (cf. Fig. 11.1.8). We see that $\bar{\alpha}$ and α_0 do not differ strongly; therefore, Eq. (11.1.5) can be used to estimate $\bar{\alpha}$ for substances where no measurements have been reported.

11.2. The Theory of the Activity of Neutron-Detecting Foils

In order to take full advantage of the high accuracy attainable with probe measurements, we must know the theory of foils that will be developed in this and the next section. The theory will be developed in two steps: First we shall discuss the relationship between the activation or activity and the neutron intensity incident on the foil. In this discussion, it will be assumed that the neutron field incident on the foil is the same as that which was present before the foil was introduced. In reality, each foil perturbs the neutron field, and it is necessary to apply *foil corrections*. We shall go into this latter problem in detail in Sec. 11.3.

11.2.1. Activation and the Neutron Flux

Let us first consider a monoenergetic neutron field. Let the vector flux be represented as usual by a series of Legendre polynomials in $\cos\vartheta$, where ϑ is the angle to the field axis:

$$F(\boldsymbol{r}, \boldsymbol{\Omega}) = \frac{1}{4\pi} \sum_{l=0}^{\infty} (2l+1)\, F_l(\boldsymbol{r})\, P_l(\cos\vartheta). \tag{11.2.1}$$

Let ψ be the angle between the field axis and the normal to the foil (cf. Fig. 11.2.1). Then in a system of polar coordinates around the normal to the foil, we have

$$\left.\begin{aligned} F(\boldsymbol{r}, \boldsymbol{\Omega}^*) = \frac{1}{4\pi} \sum_{l=0}^{\infty} (2l+1)\, F_l(\boldsymbol{r}) \Big[& P_l(\cos\vartheta^*)\, P_l(\cos\psi) + \\ & + 2 \sum_{m=1}^{l} \frac{(l-m)!}{(l+m)!} P_l^m(\cos\vartheta^*)\, P_l^m(\cos\psi) \cos^m\varphi^* \Big]. \end{aligned}\right\} \tag{11.2.2}$$

We assume that the $F_l(\boldsymbol{r})$ do not change appreciably along the foil surface, with the result that we can henceforth omit the argument $\boldsymbol{r}$. Furthermore, we assume that (a) no scattering occurs in the foil ($\mu_t = \mu_a$), and (b) no neutrons enter the foil through its edges (or, in other words, the radius of the foil is very large compared to its thickness).

The number of neutrons that are absorbed per cm² per sec at depths between x and $x + dx$ is then

$$\left.\begin{aligned} P_A(x)\,dx = & \left(\int_0^{\pi/2} \int_0^{2\pi} F(\boldsymbol{\Omega}^*)\, e^{-\frac{\mu_a(\delta - x)}{\cos\vartheta^*}} \sin\vartheta^*\, d\vartheta^*\, d\varphi^* \right) \mu_a\, dx + \\ & + \left(\int_{\pi/2}^{\pi} \int_0^{2\pi} F(\boldsymbol{\Omega}^*)\, e^{-\frac{\mu_a x}{|\cos\vartheta^*|}} \sin\vartheta^*\, d\vartheta^*\, d\varphi^* \right) \mu_a\, dx. \end{aligned}\right\} \tag{11.2.3}$$

The first term on the right-hand side describes the contribution to the absorption of the neutrons incident "from below", the second term the contribution of those incident "from above". If we substitute the value of $F(\boldsymbol{\Omega}^*)$ from Eq. (11.2.2), the terms containing the associated Legendre polynomials vanish upon integration over φ^* and we obtain

$$P_A(x) = \tfrac{1}{2} \sum_{l=0}^{\infty} (2l+1)\, F_l\, g_l(\mu_a, x, \delta)\, P_l(\cos\psi) \tag{11.2.4a}$$

with

$$\left.\begin{aligned} g_l(\mu_a, x, \delta) = \mu_a \Bigg[& \int_0^{\pi/2} e^{-\frac{\mu_a(\delta - x)}{\cos\vartheta^*}} P_l(\cos\vartheta^*) \sin\vartheta^*\, d\vartheta^* + \\ & + \int_{\pi/2}^{\pi} e^{-\frac{\mu_a x}{|\cos\vartheta^*|}} P_l(\cos\vartheta^*) \sin\vartheta^*\, d\vartheta^* \Bigg]. \end{aligned}\right\} \tag{11.2.4b}$$

In particular, if $\cos\vartheta^* = t$,

$$\left.\begin{aligned} g_0(\mu_a, x, \delta) &= \mu_a \int_0^1 \left[e^{-\frac{\mu_a(\delta-x)}{t}} + e^{-\frac{\mu_a x}{t}}\right] dt \\ &= \mu_a\left[E_2(\mu_a\{\delta-x\}) + E_2(\mu_a x)\right], \end{aligned}\right\} \quad (11.2.4c)$$

$$\left.\begin{aligned} g_1(\mu_a, x, \delta) &= \mu_a \int_0^1 t\left[e^{-\frac{\mu_a(\delta-x)}{t}} - e^{-\frac{\mu_a x}{t}}\right] dt \\ &= \mu_a\left[E_3(\mu_a\{\delta-x\}) - E_3(\mu_a x)\right], \end{aligned}\right\} \quad (11.2.4d)$$

$$\left.\begin{aligned} g_2(\mu_a, x, \delta) &= \mu_a \int_0^1 \frac{1}{2}(3t^2-1)\left[e^{-\frac{\mu_a(\delta-x)}{t}} + e^{-\frac{\mu_a x}{t}}\right] dt \\ &= \frac{\mu_a}{2}\left[3E_4(\mu_a\{\delta-x\}) + 3E_4(\mu_a x) - E_2(\mu_a\{\delta-x\}) - E_2(\mu_a x)\right]. \end{aligned}\right\} \quad (11.2.4e)$$

The functions

$$E_n(x) = \int_1^\infty \frac{e^{-ux}}{u^n}\, du = \int_0^1 t^{n-2} e^{-\frac{x}{t}}\, dt \quad (11.2.4f)$$

have been tabulated by PLACZEK among others (cf. Appendix III).

Fig. 11.2.2 shows g_0, g_1, and g_2. All g_l with even l remain invariant under the transformation $x \to \delta - x$; on the other hand, g_l with odd l change their signs. The activation C is obtained from $P_A(x)$ by multiplication with μ_{act}/μ_a and integration over the foil thickness:

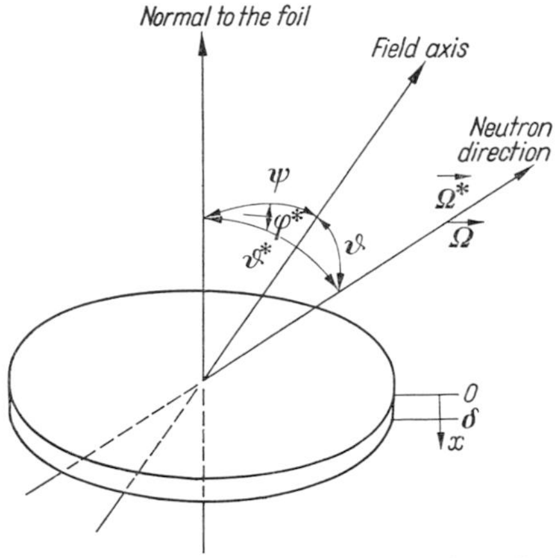

Fig. 11.2.1. The relation between ψ, ϑ, ϑ^* and φ^* (see text)

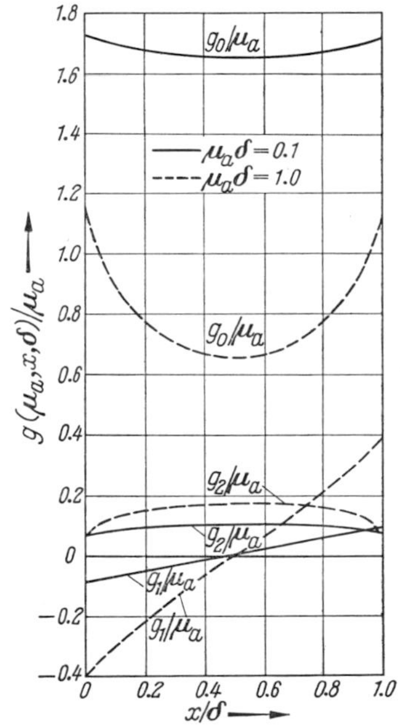

Fig. 11.2.2. The functions $g_l(\mu_a, x, \delta)$ for $l = 0, 1, 2$ and for $\mu_a\delta = 0.1$ and 1.0

$$\left.\begin{aligned} C &= \frac{\mu_{act}}{\mu_a} \int_0^\delta P_A(x)\, dx \\ &= \frac{\mu_{act}}{\mu_a} \frac{1}{2} \sum_{l=0}^{\infty} (2l+1) F_l \varphi_l(\mu_a\delta) P_l(\cos\psi). \end{aligned}\right\} \quad (11.2.5a)$$

Here

$$\left.\begin{aligned}\varphi_l(\mu_a\delta) &= \int_0^\delta g_l(\mu_a, x, \delta)\,dx\\ &= \mu_a \int_0^\delta \left[\int_0^{\pi/2} e^{-\frac{\mu_a(\delta-x)}{\cos\vartheta^*}} P_l(\cos\vartheta^*)\sin\vartheta^*\,d\vartheta^* + \right.\\ &\quad \left. + \int_{\pi/2}^{\pi} e^{-\frac{\mu_a x}{|\cos\vartheta^*|}} P_l(\cos\vartheta^*)\sin\vartheta^*\,d\vartheta^*\right] dx.\end{aligned}\right\} \qquad (11.2.5\text{b})$$

We can see immediately that all the $\varphi_l(\mu_a\delta)$ with odd l vanish. For even l on the other hand,

$$\varphi_l(\mu_a\delta) = 2\int_0^1 P_l(t)\left(1 - e^{-\frac{\mu_a\delta}{t}}\right) t\,dt \qquad (11.2.5\text{c})$$

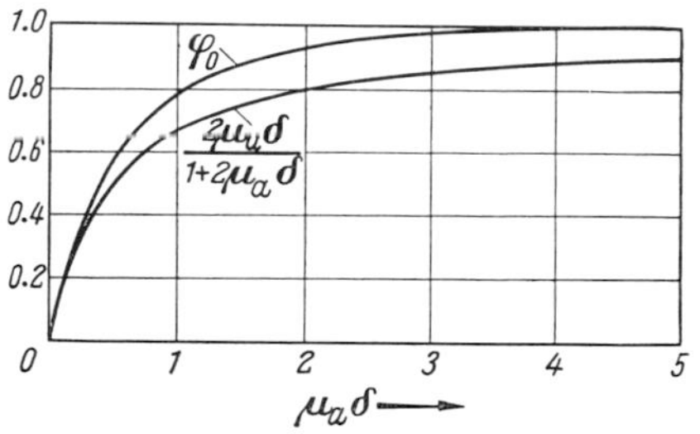

Fig. 11.2.3. The absorption probability $\varphi_0(\mu_a\delta)$ of a foil in an isotropic neutron field and a simple rational approximation to it

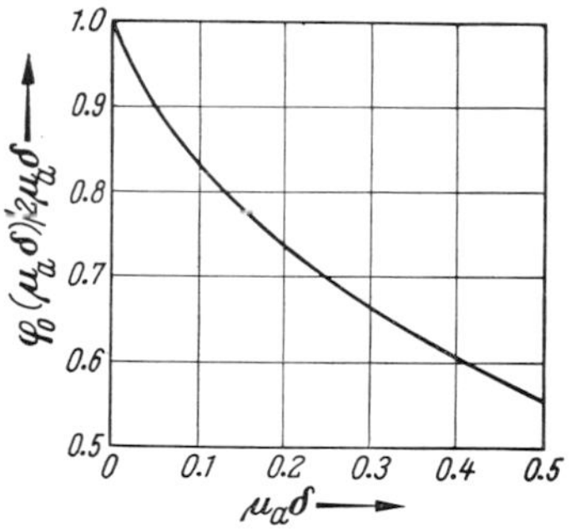

Fig. 11.2.4. The self-shielding factor $\varphi_0(\mu_a\delta)/2\mu_a\delta$ of an absorbing foil in an isotropic neutron field

where $\cos\vartheta^* = t$. In particular,

$$\varphi_0(\mu_a\delta) = 1 - 2E_3(\mu_a\delta), \qquad (11.2.5\text{d})$$

$$\varphi_2(\mu_a\delta) = \tfrac{1}{4} + E_3(\mu_a\delta) - 3E_5(\mu_a\delta). \qquad (11.2.5\text{e})$$

In the following, we shall limit ourselves to neutron fields whose vector flux is described by only the first two terms in the expansion in Legendre polynomials. Then, simply,

$$C = \frac{\mu_{\text{act}}}{\mu_a}\,\frac{\Phi(\boldsymbol{r})}{2}\,\varphi_0(\mu_a\delta), \qquad (11.2.6)$$

i.e., the activation is independent of the foil orientation and proportional to the flux. The current term affects the local distribution of the activation in the foil — cf. Fig. 11.2.2 — but not the total activation integrated over the foil thickness. Since $\Phi(\boldsymbol{r})/2$ is the number of neutrons falling on both sides of the foil per cm^2, $\varphi_0(\mu_a\delta)$ can obviously be identified as the absorption probability of the foil in an isotropic neutron field. For an extremely thick foil ($\mu_a\delta \gg 1$) we have $\varphi_0(\mu_a\delta) = 1$; thus $C = \frac{\mu_{\text{act}}}{\mu_a}\frac{\Phi(\boldsymbol{r})}{2}$. On the other hand, for an extremely thin foil, $\varphi_0(\mu_a\delta) = 2\mu_a\delta$ and

$$C = \frac{\mu_{\text{act}}}{\mu_a}\,\frac{\Phi(\boldsymbol{r})}{2}\,2\mu_a\delta = \mu_{\text{act}}\,\delta\,\Phi(\boldsymbol{r}) = \Sigma_{\text{act}}\,d\,\Phi(\boldsymbol{r}). \qquad (11.2.7)$$

[Cf. Eq. (11.1.1).] Fig. 11.2.3 shows $\varphi_0(\mu_a\delta)$ and the simple approximation $2\mu_a\delta/(1+2\mu_a\delta)$ to it, which we shall occasionally use later.

Frequently, Eq. (11.2.6) is written in the form

$$C=\mu_{\text{act}}\,\delta\,\frac{\varphi_0(\mu_a\delta)}{2\mu_a\delta}\,\Phi(\boldsymbol{r}). \tag{11.2.8}$$

$\frac{\varphi_0(\mu_a\delta)}{2\mu_a\delta}\,\Phi(\boldsymbol{r})$ is the average flux in the foil; Fig. 11.2.4 shows the self-shielding factor[1] $\frac{\varphi_0(\mu_a\delta)}{2\mu_a\delta}$ as a function of $\mu_a\delta$.

11.2.2. The Effect of Scattering in the Foil

Next we shall investigate the errors introduced into the calculation of the activation by the neglect of scattering in the foil. Since in the energy range under consideration the scattering cross sections of most substances used in practice are much less than their absorption cross sections, it is enough to consider only the first and second collisions in the foil. Since in this case also the current term does not contribute to the activation, let us consider the activation in an isotropic neutron field. The number of neutrons which suffer their first collision at depths between x and $x+dx$ is [cf. Eqs. (11.2.4a—c)]

$$P_t(x)\,dx=\frac{\Phi}{2}\left[E_2(\mu_t\{\delta-x\})+E_2(\mu_t x)\right]\mu_t\,dx. \tag{11.2.9}$$

Here $\mu_t=\mu_a+\mu_s$ is the total collision coefficient. Of these collisions, the fraction μ_{act}/μ_t leads immediately to activation. The fraction μ_s/μ_t represents scattered neutrons, whose history we must follow further. In so doing, let us assume that the scattering is isotropic. The probability that a neutron isotropically scattered at x escapes from the foil without making another collision is $\frac{1}{2}\left[E_2(\mu_t\{\delta-x\})+E_2(\mu_t x)\right]$; therefore the number of neutrons which make a second collision in the foil is

$$\frac{\mu_s}{\mu_t}\int_0^\delta P_t(x)\left(1-\frac{1}{2}\left[E_2(\mu_t\{\delta-x\})+E_2(\mu_t x)\right]\right)dx=\frac{\Phi(\boldsymbol{r})}{2}\,\mu_s\delta\,\chi(\mu_t\delta). \tag{11.2.10a}$$

The function

$$\left.\begin{aligned}\chi(\mu_t\delta)=\frac{1}{2\mu_t\delta}\int_0^\delta&\left[E_2(\mu_t\{\delta-x\})+E_2(\mu_t x)\right]\times\\ &\times\left[2-\{E_2(\mu_t\{\delta-x\})+E_2(\mu_t x)\}\right]\mu_t\,dx\end{aligned}\right\} \tag{11.2.10b}$$

is shown in Fig. 11.2.5. Since the fraction μ_{act}/μ_t of the second collisions also leads to activation, the activation of the foil resulting from the first and second collisions is

$$\left.\begin{aligned}C&=\frac{\mu_{\text{act}}}{\mu_t}\left[\int_0^\delta P_t(x)\,dx+\frac{\Phi(\boldsymbol{r})}{2}\,\mu_s\delta\,\chi(\mu_t\delta)\right]\\ &=\frac{\mu_{\text{act}}}{\mu_t}\cdot\frac{\Phi(\boldsymbol{r})}{2}\left[\varphi_0(\mu_t\delta)+\mu_s\delta\,\chi(\mu_t\delta)\right].\end{aligned}\right\} \tag{11.2.11}$$

[1] This self-shielding factor gives the ratio of the average flux $\bar{\Phi}$ in the foil to the unperturbed incident flux Φ and *not* to the flux Φ_s on the foil surface. Φ_s is smaller than Φ: $\Phi_s=\frac{\Phi}{2}\left[1+E_2(\mu_a\delta)\right]$ [cf. Eqs. (11.2.4a and b)].

In Fig. 11.2.6 is shown the quantity $C/\mu_{\text{act}}\,\delta\,\Phi$ calculated according to Eqs. (11.2.11) and (11.2.8) for various foil thicknesses and for $\mu_s/\mu_t = 0.01$ and 0.1. We see that the effect of scattering is small: in most cases scattering can be neglected and the activation calculated according to Eq. (11.2.8), even when μ_s is not negligible compared to μ_a. The physical reason for the small role played by scattering processes is the following: The average path length in the foil substance of normally incident neutrons is increased by scattering while that of obliquely incident neutrons is decreased. In first approximation, the two effects cancel each other.

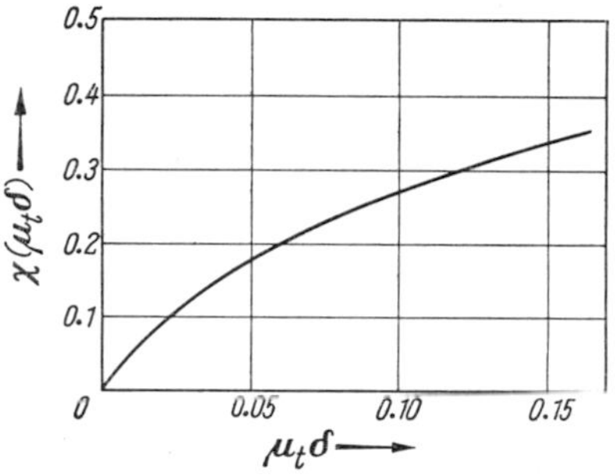

Fig. 11.2.5. The function $\chi(\mu_t\delta)$

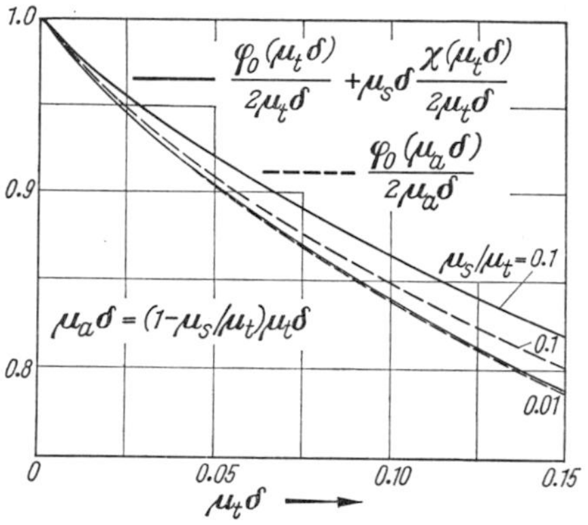

Fig. 11.2.6. A comparison of self-shielding factors that do and do not into account scattering processes in the foil

In all the considerations to follow, we shall neglect scattering in the foil and start from Eq. (11.2.6) for the activation.

11.2.3. Activation in a Thermal Neutron Field

Our previous results, particularly Eq. (11.2.6), must now be averaged over the Maxwell distribution of neutron energies. For thin foils, naturally, we simply have

$$C = \Phi_{th}\,\overline{\mu_{\text{act}}}\,\delta = \Phi_{th}\,g(T)\sqrt{\frac{293.6^\circ}{T}}\,\frac{\sqrt{\pi}}{2}\,\Sigma_{\text{act}}(0.0253\text{ ev})\,d. \qquad (11.2.12)$$

The temperature-dependent g-factors of several foil substances were shown in Fig. 11.1.6. For thick foils, we shall restrict ourselves to cases where the ratio μ_{act}/μ_a does not depend on the neutron energy. Then

$$C = \frac{\mu_{\text{act}}}{\mu_a}\,\frac{\Phi_{th}}{2}\,\overline{\varphi_0(\mu_a\delta)} \qquad (11.2.13\text{a})$$

with

$$\overline{\varphi_0(\mu_a\delta)} = \int_0^\infty \frac{E}{kT}\,e^{-\frac{E}{kT}}\,\varphi_0(\mu_a(E)\,\delta)\,\frac{dE}{kT}. \qquad (11.2.13\text{b})$$

Fig. 11.2.7 shows $\overline{\varphi_0(\mu_a\delta)}$ as a function of $\mu_a(kT)\,\delta$ in the important special case of $1/v$-absorption $(\mu_a(E)\sim 1/\sqrt{E})$. This curve was obtained by numerical integration of Eq. (11.2.13b); the function $\overline{\varphi_0(\mu_a\delta)}$ has been tabulated by Martinez among others. Also shown in the figure is the approximation

$$\overline{\varphi_0(\mu_a \delta)} \approx \frac{\sqrt{\pi}}{2} \varphi_0(\mu_a(kT)\delta) \quad (11.2.13\text{c})$$

which reproduces the correct value to better than 0.5% for $\mu_a(kT)\delta < 0.5$.

Fig. 11.2.8 shows $\overline{\frac{\mu_{\text{act}}}{\mu_t} \varphi_0(\mu_t \delta)}$ for gold and indium foils (absorption cross section $\neq 1/v$) as a function of the thickness for $T = 293.6$ °K; these curves were obtained by numerical integration.

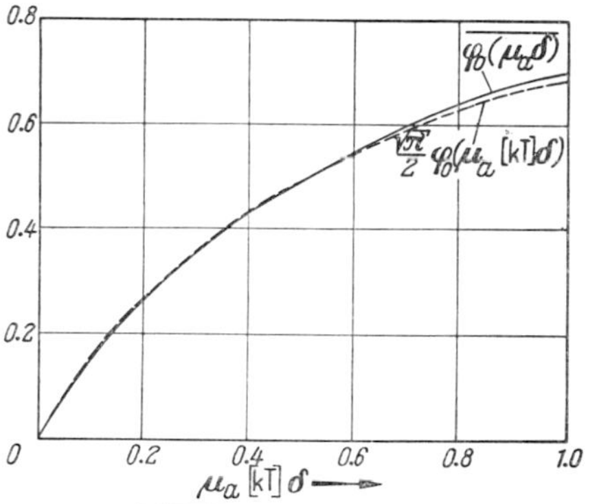

Fig. 11.2.7. $\overline{\varphi_0(\mu_a \delta)}$ for $1/v$-absorber foils in a Maxwellian spectrum. The dotted line represents a simple approximation

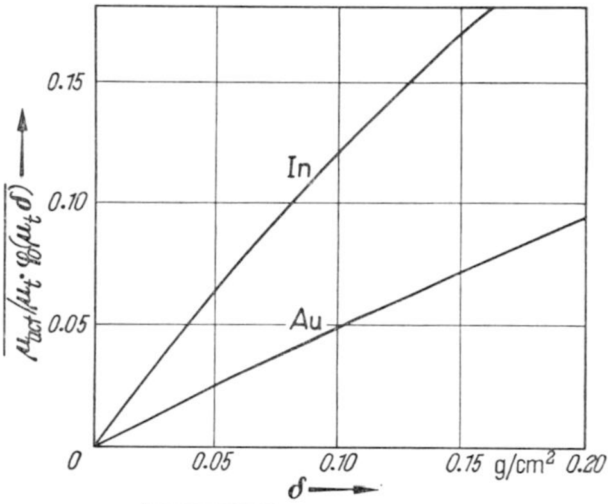

Fig. 11.2.8. $\overline{\frac{\mu_{act}}{\mu_t} \varphi_0(\mu_t \delta)}$ for Gold and Indium foils in a Maxwellian spectrum with $T = 293.6$ °K

11.2.4. Activation by Neutrons Incident on the Edges of the Foil

Up to this point, our considerations hold rigorously only for infinite foils since the activation due to neutrons which enter the edges of the foil has been neglected. The contribution of such neutrons to the activation of disc-shaped foils has been carefully investigated by HANNA. In a monoenergetic isotropic neutron field, HANNA writes

$$\tilde{C} = C(1 + \varepsilon) \quad (11.2.14)$$

where C is the activation of the infinite foil and

$$\varepsilon = \frac{2\mu_a \delta}{\varphi_0(\mu_a \delta)} \frac{2}{\pi n} [I(\mu_a \delta) - \Delta(n \mu_a \delta)], \quad (11.2.15)$$

$n = 2R/d$ (R = the foil radius, d = the foil thickness) and

$$\Delta(n\mu_a \delta) = \int_0^1 E_2(n \mu_a \delta \sin \varphi)\, d \cos \varphi, \quad (11.2.16\text{a})$$

$$I(\mu_a \delta) = \frac{2}{\mu_a \delta} \int_0^{\pi/2} \left[1 + e^{-\frac{\mu_a \delta}{\cos \vartheta}} - \frac{2}{\mu_a \delta}\left(1 - e^{-\frac{\mu_a \delta}{\cos \vartheta}}\right)\right] \sin^2 \vartheta \, d\vartheta. \quad (11.2.16\text{b})$$

HANNA finds $\Delta(1) = 0.2258$; $\Delta(2) = 0.0881$ and $\Delta(4) = 0.0235$. For foils that are not too thick, $I(\mu_a \delta)$ is given approximately by

$$I(\mu_a \delta) = 1 - \frac{\pi}{6} \mu_a \delta. \quad (11.2.16\text{c})$$

In practical cases, we can usually neglect $\Delta(n\mu_a\delta)$ compared to $I(\mu_a\delta)$ in Eq. (11.2.15) and obtain

$$\varepsilon \approx \frac{2\mu_a\delta}{\varphi_0(\mu_a\delta)} \frac{d}{\pi R} \cdot \left(1 - \frac{\pi}{6}\mu_a\delta\right). \tag{11.2.17}$$

For a gold foil with $d=0.25$ mm and $R=6$ mm, $\varepsilon=0.015$; thus the correction for neutrons entering the foil through its edges is indeed small, but in precision measurements it cannot be neglected. According to Sec. 11.2.3, in a thermal neutron field ε must be replaced by

$$\left.\begin{aligned} \varepsilon &= \frac{2\bar{\mu}_a\delta}{\varphi_0(\mu_a\delta)} \frac{d}{\pi R}\left(1 - \frac{\pi}{6}\bar{\mu}_a\delta\right) \\ &\approx \frac{2\mu_a(kT)\delta}{\varphi_0(\mu_a(kT)\delta)} \frac{d}{\pi R}\left(1 - \frac{\sqrt{\pi}}{3}\mu_a(kT)\delta\right). \end{aligned}\right\} \tag{11.2.18}$$

11.2.5. Calculation of the Apparent Activity Taking Account of β-Self-Absorption

Since for thick foils the activation is not uniform over the thickness x — cf. Fig. 11.2.2 — and since in addition the probability that a decay electron leaves the foil depends on the distance to the foil surface, the apparent β-activity A^* of thick foils is no longer proportional to their activation. However, A^* can easily be calculated on the basis of the exponential law of attenuation for β-radiation introduced in Sec. 11.1.3.

If we take the vector flux to be the same as that used in Sec. 11.2.1 and neglect scattering in the foil and neutrons incident on the edges of the foil just as we did there, then in a monoenergetic field

$$A_+^* = \frac{1}{2} \frac{\mu_{\text{act}}}{\mu_a} \frac{1}{2T} \sum_{l=0}^{\infty} (2l+1) F_l(\mathbf{r})\, \varphi_l(\nu, \beta)\, P_l(\cos\psi) \tag{11.2.19}$$

gives the apparent activity of the upper (in the sense of Fig. 11.2.1) surface[1]. Here T is the time factor, $\nu=\mu_a\delta$, and $\beta=\alpha\delta$. The functions $\varphi_l(\nu, \beta)$ are given by

$$\varphi_l(\nu, \beta) = \int_0^\delta e^{-\alpha x}\, g_l(\mu_a, x, \delta)\, dx \tag{11.2.20}$$

with $g_l(\mu_a, x, \delta)$ given by Eq. (11.2.4). The integration gives

$$\left.\begin{aligned} \varphi_0(\nu, \beta) &= \frac{\nu}{\beta}(1 - e^{-\beta})(1 + E_1(\nu)) \\ &\quad - \frac{\nu^2}{\beta^2}\left[E_1(\nu+\beta) + \ln\frac{\nu+\beta}{\beta} + e^{-\beta}\left\{E_1(\nu-\beta) - \ln\frac{|\nu-\beta|}{\beta}\right\}\right] + \\ &\quad + \frac{\nu^2}{\beta^2}(1 + e^{-\beta})E_2(\nu), \end{aligned}\right\} \tag{11.2.21 a}^2$$

$$\varphi_1(\nu, \beta) = \frac{\nu}{\beta}\left[\varphi_0(\nu, \beta) - \frac{1}{2}(1 + e^{-\beta})(1 - 2E_3(\nu))\right]. \tag{11.2.21 b}$$

[1] Eq. (11.2.19) follows immediately from multiplication of the value of $P_A(x)$ given in Eq. (11.2.4a) by $e^{-\alpha x}(\mu_{\text{act}}/\mu_a)$ and integration over the foil tickness. In addition, the time factor must be included as well as a factor of $\frac{1}{2}$ that accounts for the fact that half of the decay electrons are emitted upwards and half downwards.

[2] For $\beta > \nu$, $E_1(\nu - \beta)$ must be replaced by $-Ei^*(\beta - \nu)$ where $Ei^*(x) = \int_{-\infty}^{x} \frac{e^u}{u}\, du$ is tabulated in JAHNKE-EMDE-LÖSCH, Tables of Higher Functions, Stuttgart: B. G. Teubner 1960.

$\varphi_0(\nu, \beta)$ and $\varphi_1(\nu, \beta)$ for gold and indium foils in a thermal neutron field are shown in Figs. 11.2.9 and 11.2.10. Tabular values for many ν and β can be found in Appendix IV. The apparent activity of the lower (in the sense of Fig. 11.2.1) surface is given by

$$A_-^* = \frac{1}{2} \frac{\mu_{\text{act}}}{\mu_a} \frac{1}{2T} \sum_{l=0}^{\infty} (2l+1) F_l(\boldsymbol{r}) (-1)^l \varphi_l(\nu, \beta) P_l(\cos\psi). \tag{11.2.22a}$$

We can draw the following important conclusions from these formulas:

a) In an isotropic neutron field, the apparent activity is always proportional to the flux Φ. However, the proportionality constant does not increase indefinitely with increasing foil thickness, but instead goes through a maximum, which in the case of indium lies in the neighborhood of $\delta = 120$ mg/cm². It is impractical to use foils for flux measurement thicker than that which corresponds to this maximum sensitivity. The decrease in foil sensitivity at larger thickness comes about because in β-counting only a thin superficial region (of thickness $\approx 1/\alpha$) is effective and the probability that neutrons "from behind" reach this region decreases with increasing foil thickness.

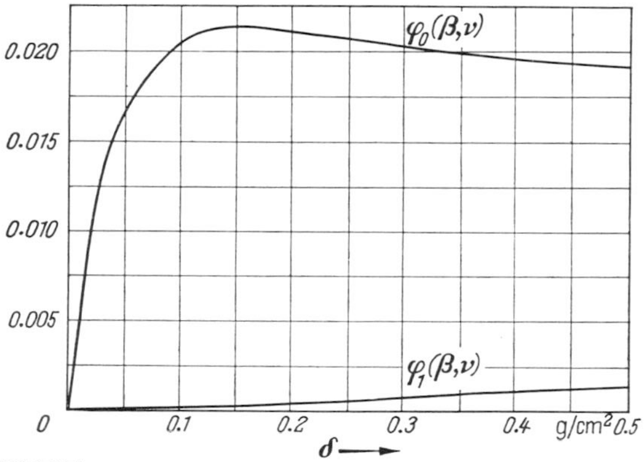

Fig. 11.2.9. $\varphi_0(\nu, \beta)$ and $\varphi_1(\nu, \beta)$ for Gold foils

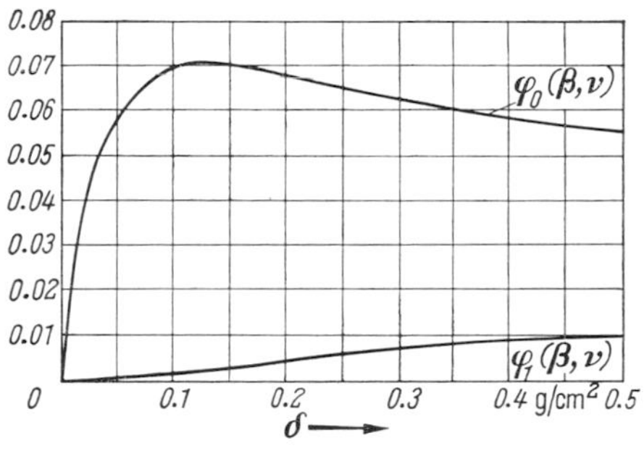

Fig. 11.2.10. $\varphi_0(\nu, \beta)$ and $\varphi_1(\nu, \beta)$ for Indium foils

b) Even in weakly anisotropic neutron fields that can be described by only the first two terms in the expansion of the vector flux in Legendre polynomials, the apparent activities A_+^* and A_-^* show some dependence on the orientation angle ψ. This dependence is more strongly marked the thicker the foil; for very thin foils it can be neglected. In principle, this dependence should make it possible to determine the neutron current, but the effect is too small for exact measurements.

c) As a rule, we are only interested in knowing the neutron flux and must therefore eliminate the current term from the apparent activity measured with a thick foil. This can easily be done by counting both the upper and lower sides of the foil and adding the results; when we neglect F_2 and higher even terms, this procedure gives

$$A_+^* + A_-^* = \frac{\mu_{\text{act}}}{\mu_a} \cdot \frac{1}{T} \cdot \frac{\Phi(\boldsymbol{r})}{2} \cdot \varphi_0(\nu, \beta) \tag{11.2.22b}$$

since all the odd terms cancel. As $\beta \to 0$, $\varphi_0(\nu, \beta)$ approaches $\varphi_0(\mu_a \delta)$ and we obtain $A_+^* + A_-^* = C/T$ which is the same as Eq. (11.1.3).

It should be noted in conclusion that Eqs. (11.2.19—22) were derived with the help of the exponential law of attenuation for β-radiation. While this law holds very accurately for the attenuation of β-radiation in thinly deposited absorbers, its integration to determine the apparent activity of thicker foils is problematic since in such foils electron backscatter processes occur that are difficult to describe. These effects increase the activity over what would be calculated using the theory of this section. The results of this section thus have only qualitative significance; for absolute measurements the β-self-absorption must be determined experimentally. We shall return to this point in Sec. 14.1.3.

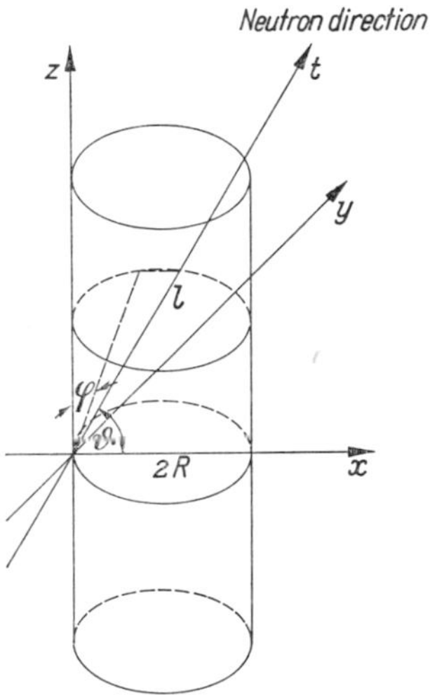

Fig. 11.2.11. Neutron absorption by a cylindrical probe (see text)

11.2.6. Activation of a Cylindrical Probe

Occasionally, we use probes in the form of long tapes or wires for the measurement of flux profiles. Whereas we can calculate the activity and activation of tape probes with the formulas already developed — except for the inconsiderable correction for neutrons that enter the edges of the foil — new considerations are necessary for wires.

Let us calculate the activation per cm of length of an infinite cylinder of radius R in an isotropic neutron field and let us neglect scattering by the probe substance. Let us begin by considering an incident neutron with the flight direction (ϑ, φ) (cf. Fig. 11.2.11). If this neutron is not absorbed, it will traverse a flight path of length $l = \frac{2R\cos\vartheta}{1-\sin^2\vartheta\cos^2\varphi}$ before it leaves the cylinder[1]. The absorption probability is then $1 - e^{-\Sigma_a l(\vartheta, \varphi)}$ and

$$C = 2\pi R \frac{\Sigma_{\text{act}}}{\Sigma_a} \frac{\Phi}{4\pi} \int_0^{2\pi} \int_0^{\pi/2} (1 - e^{-\Sigma_a l(\vartheta, \varphi)}) \cos\vartheta \sin\vartheta \, d\vartheta \, d\varphi. \qquad (11.2.23)$$

Here the factor $2\pi R$ takes into account the fact that to 1 cm of cylinder length correspond $2\pi R$ cm^2 of surface; we have assumed that the incident flux does not vary over the circumference. Now we write

$$C = \pi R \frac{\Sigma_{\text{act}}}{\Sigma_a} \frac{\Phi}{2} \chi_0(\Sigma_a R) \qquad (11.2.24)$$

with

$$\chi_0(\Sigma_a R) = \frac{1}{\pi} \int_0^{2\pi} \int_0^{\pi/2} \left(1 - e^{-\frac{2\Sigma_a R\cos\vartheta}{1-\sin^2\vartheta\cos^2\varphi}}\right) \cos\vartheta \sin\vartheta \, d\vartheta \, d\varphi. \qquad (11.2.25)$$

[1] This relation for l can be obtained as follows: In cartesian coordinates, $(x-R)^2 + y^2 = R^2$ is the equation for the surface of the cylinder and $x = t\cos\vartheta$, $y = t\sin\vartheta\sin\varphi$, $z = t\sin\vartheta\cos\varphi$ are the equations for the neutron trajectory. One point of intersection of the line and the cylinder is at $x = y = z = t = 0$, and the other at $t = l$. Thus $l^2 = x^2 + y^2 + z^2 = 2Rx + z^2 = 2Rl\cos\vartheta + l^2\sin^2\vartheta\cos^2\varphi$ and $l = \frac{2R\cos\vartheta}{1-\sin^2\vartheta\cos^2\varphi}$.

Clearly $\chi_0(\Sigma_a R)$ is the absorption probability of neutrons with an isotropic distribution of velocities incident on the infinite cylinder (since $\pi R\Phi/2$ is the number of neutrons incident per sec on a 1-cm length of the cylinder). Then

$$\chi_0(\Sigma_a R) = 1 - \frac{1}{\pi}\int\limits_0^{2\pi}\int\limits_0^{\pi/2} e^{-\frac{2\Sigma_a R\cos\vartheta}{1-\sin^2\vartheta\cos^2\varphi}}\cos\vartheta\sin\vartheta\,d\vartheta\,d\varphi. \qquad (11.2.26\text{a})$$

By introduction of the variables $y = \frac{\cos\vartheta}{1-\sin^2\vartheta\cos^2\varphi}$ and $x = \frac{\cos\vartheta}{\sqrt{1-\sin^2\vartheta\cos^2\varphi}}$ we obtain

$$\chi_0(\Sigma_a R) = 1 - \frac{4}{\pi}\int\limits_0^1 \frac{x^4\,dx}{\sqrt{1-x^2}}\int\limits_x^\infty \frac{e^{-2\Sigma_a R y}\,dy}{y^3\sqrt{y^2-x^2}}. \qquad (11.2.26\text{b})$$

Integration yields (cf. CASE, DE HOFFMANN, and PLACZEK)

$$\left.\begin{aligned}\chi_0(\Sigma_a R) = \frac{4}{3}(\Sigma_a R)^2\Big\{2\big[\Sigma_a R\{K_1(\Sigma_a R)I_1(\Sigma_a R) + {}\\ + K_0(\Sigma_a R)I_0(\Sigma_a R)\} - 1\big] + \frac{K_1(\Sigma_a R)I_1(\Sigma_a R)}{\Sigma_a R} - {}\\ - K_0(\Sigma_a R)I_1(\Sigma_a R) + K_1(\Sigma_a R)I_0(\Sigma_a R)\Big\}\end{aligned}\right\} \qquad (11.2.27)$$

where I and K are modified Bessel functions of the first and second kinds, respectively. $\chi_0(\Sigma_a R)$ is shown in Fig. 11.2.12. The function approaches unity for large values of $\Sigma_a R$; for small $\Sigma_a R$, i.e., for thin cylinders, $\chi_0(\Sigma_a R) \approx 2\Sigma_a R$ and thus $C = \pi R^2 \Sigma_{act}\Phi$. The quantity $1-\chi_0(\Sigma_a R)/2\Sigma_a R$ has been tabulated by CASE, DE HOFFMANN, and PLACZEK.

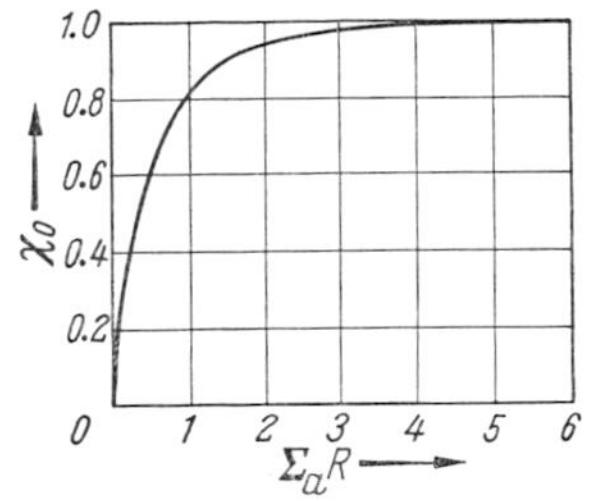

Fig. 11.2.12. The absorption probability of a cylindrical probe

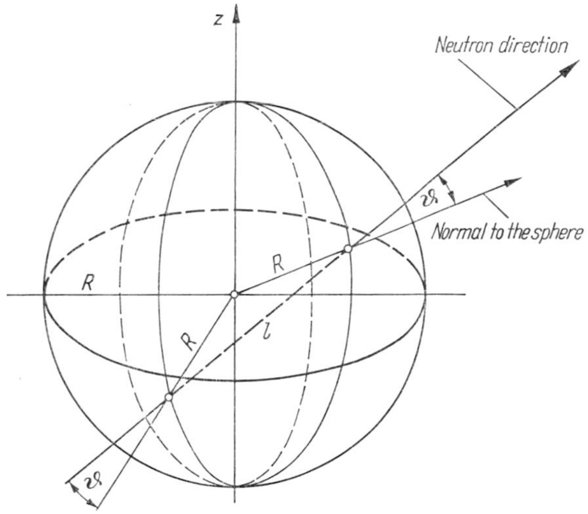

Fig. 11.2.13. Neutron absorption by a spherical probe (see text)

Since we already know the absorption probability for slabs and cylinders, for the sake of completeness we shall now give it also for spheres. We can derive the following formula for the activation of a "sphere probe" of radius R in an isotropic neutron field in a manner similar to that used above for cylinders:

$$C = 4\pi R^2\frac{\Sigma_{act}}{\Sigma_a}\frac{\Phi}{4\pi}\int\limits_0^{2\pi}\int\limits_0^{\pi/2}(1-e^{-\Sigma_a l(\vartheta)})\cos\vartheta\sin\vartheta\,d\vartheta\,d\varphi. \qquad (11.2.28)$$

Here C is the total activation and not the activation per cm². Now $l=2R\cos\vartheta$ (cf. Fig. 11.2.13) and therefore

$$C=\pi R^2\frac{\Sigma_{\text{act}}}{\Sigma_a}\Phi\,\psi_0(\Sigma_a R) \tag{11.2.29}$$

with the absorption probability

$$\psi_0(\Sigma_a R)=2\int_0^{\pi/2}(1-e^{-2\Sigma_a R\cos\vartheta})\cos\vartheta\sin\vartheta\,d\vartheta \tag{11.2.30a}$$

$$=1+\frac{e^{-2\Sigma_a R}}{\Sigma_a R}-\frac{1-e^{-2\Sigma_a R}}{2\cdot(\Sigma_a R)^2}. \tag{11.2.30b}$$

Fig. 11.2.14 shows $\psi_0(\Sigma_a R)$. For $\Sigma_a R\ll 1$, $\psi_0(\Sigma_a R)\approx\frac{4}{3}\Sigma_a R$ and $C=\frac{4}{3}\pi R^3\Sigma_a\Phi$.

In Fig. 11.2.15 we compare the absorption probabilities of slabs, cylinders,

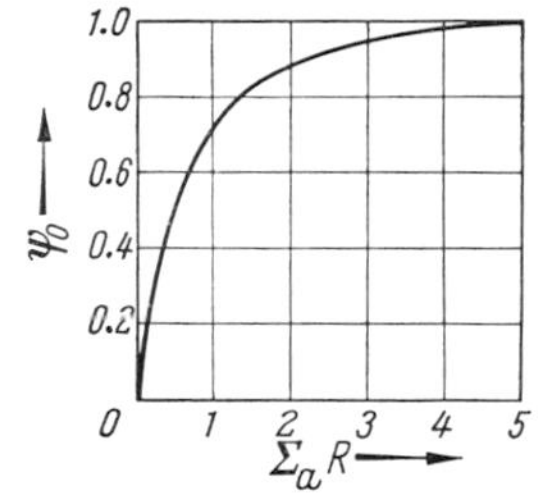

Fig. 11.2.14. The absorption probability of a spherical probe

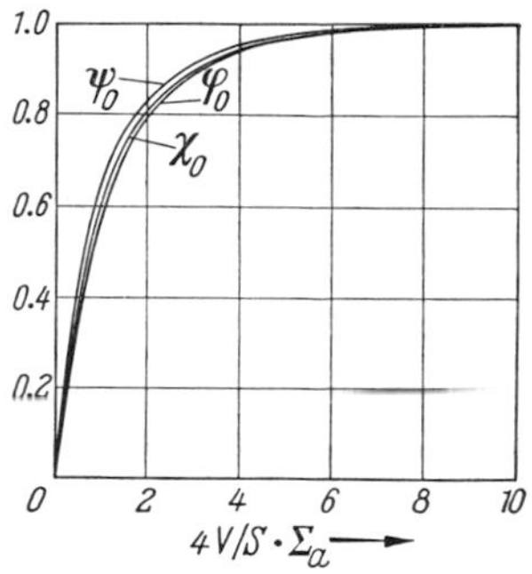

Fig. 11.2.15. A comparison of the absorption probabilities for slab, cylinder, and sphere. $\frac{4V}{S}=2d$ for slabs, $2R$ for cylinders, and $\frac{4}{3}R$ for spheres

and spheres [given by Eqs. (11.2.5d), (11.2.27), and (11.2.30b), respectively]. The abscissa is the quantity $\frac{4V}{S}\Sigma_a$ (V=volume, S=surface); $\frac{4V}{S}$ is $2d$ for slabs, $2R$ for cylinders and $\frac{4}{3}R$ for spheres. It turns out that with this choice of abscissa the three curves do not differ from one another very much.

11.3. The Theory of Foil Perturbations

In Sec. 11.2, the probe activation or activity was related to the incident neutron flux. It was assumed that the introduction of the probe did not change the flux. This would be so if there were no "backscattering", i.e., if each neutron on its way from birth to absorption would only cross the surface of the probe once. In practice, however, a considerable backscattering probability exists, and neutrons that are absorbed in the probe are absent in the backscattered flux. The actual activation C is therefore smaller than that activation C_0 that would occur in the unperturbed field. Let us set

$$\frac{C_0-C}{C}=\varkappa_c \tag{11.3.1}$$

where $\varkappa_c$ is the "activation correction". $\varkappa_c$ is a function of the properties of both the probe and the surrounding medium. In Secs. 11.3.1—11.3.3 we shall study various means of calculating it[1].

[1] In Secs. 11.3.1 and 11.3.2, various calculations in monoenergetic fields will be considered; in Sec. 11.3.2, the various results will be compared with one another. Activation corrections in thermal neutron fields will be considered in Sec. 11.3.3.

The following simple considerations show the qualitative connection between the backscattering probability, the absorption in the probe, and the activation correction. $N_0 = F\frac{\Phi}{2}$ neutrons cross a surface F per second in a neutron field. Of these, $N_0^{(1)}$ cross for the first time and $N_0^{(2)} = N_0 - N_0^{(1)}$ for the second, third, etc. time. If p is the backscattering probability, i.e., the probability that a neutron that has crossed the surface once will cross it again after one or more collisions, then

$$N_0^{(2)} = N_0^{(1)} p + N_0^{(1)} p^2 + \cdots = N_0^{(1)} \frac{p}{1-p}. \tag{11.3.2a}$$

Therefore

$$N_0 = \frac{N_0^{(1)}}{1-p}. \tag{11.3.2b}$$

Let us now replace the surface being considered by a real foil. The number of neutrons that are incident on the foil per second is $N = N_0^{(1)} + N^{(2)}$. The number of neutrons incident for the first time is naturally the same as without the foil. In contrast, $N^{(2)} \leq N_0^{(2)}$ since only a fraction of the neutrons incident on the foil can penetrate it and become available for backscattering. If we assume that the distribution of directions of the incident neutrons is always isotropic, the probability of penetrating the foil is $1 - \varphi_0(\mu_a \delta)$. Thus

$$\left.\begin{aligned} N^{(2)} &= N_0^{(1)} [1 - \varphi_0(\mu_a\delta)] p + N_0^{(1)} [1 - \varphi_0(\mu_a\delta)]^2 p^2 + \cdots \\ &= N_0^{(1)} \frac{[1 - \varphi_0(\mu_a\delta)] p}{1 - [1 - \varphi_0(\mu_a\delta)] p} \end{aligned}\right\} \tag{11.3.2c}$$

or

$$N = \frac{N_0^{(1)}}{1 - [1 - \varphi_0(\mu_a\delta)] p}. \tag{11.3.2d}$$

It then follows that

$$\frac{N}{N_0} = \frac{1}{1 + \frac{p}{1-p} \varphi_0(\mu_a\delta)} \tag{11.3.2e}$$

or with $C/C_0 \approx N/N_0$ that

$$\varkappa_c = \frac{p}{1-p} \varphi_0(\mu_a\delta). \tag{11.3.2f}$$

The activation correction is proportional to the absorption in the foil and is larger the larger the backscattering probability.

Besides the perturbation of the activation, which we must know when we make absolute measurements[1] or when we wish to compare flux measurements made in different media, we occasionally wish to know the density or flux perturbations. If Φ_0 represents the unperturbed flux, the flux $\Phi(\boldsymbol{r})$ in the neighborhood of the foil is smaller than Φ_0 owing to the additional absorption. We call

$$\varkappa_\Phi(\boldsymbol{r}) = \frac{\Phi_0 - \Phi(\boldsymbol{r})}{\overline{\Phi(\boldsymbol{r}_s)}} \tag{11.3.3}$$

the "flux correction". $\overline{\Phi(\boldsymbol{r}_s)}$ is the average flux on the foil surface. We must know $\varkappa_\Phi(\boldsymbol{r})$ in order to estimate the mutual influence of several foils exposed simultaneously. In Sec. 11.3.4, we shall familiarize ourselves with some results

[1] We determine $C_0 = C(1 + \varkappa_c)$ from the experimental value of C and a value of $\varkappa_c$ calculated by the methods of this section.

for $\varkappa_\Phi(\boldsymbol{r})$; the flux correction $\varkappa_\Phi(0)$ on the foil surface, on the other hand, can again be estimated from simple considerations similar to those we used above for the activation correction. The flux at the upper surface of a foil is composed of neutrons from above that are directly incident on the foil and neutrons from below that have penetrated the foil. According to the considerations given above, the contribution of the former to the flux is $\frac{1}{2}\frac{\Phi_0}{1+\varkappa_c}$, that of the latter $\frac{1}{2}\frac{\Phi_0}{1+\varkappa_c}\left(1-\varphi_0(\mu_a\delta)\right)$. The flux at the foil surface is thus given by[1]

$$\Phi(\boldsymbol{r}_s)=\frac{\Phi_0}{1+\varkappa_c}\left(1-\frac{1}{2}\varphi_0(\mu_a\delta)\right), \tag{11.3.4a}$$

$$\frac{\Phi_0}{\Phi(\boldsymbol{r}_s)}=1+\varkappa_\Phi(0)=\frac{1+\varkappa_c}{1-\frac{1}{2}\varphi_0(\mu_a\delta)}, \tag{11.3.4b}$$

$$\varkappa_\Phi(0)\approx\varkappa_c+\tfrac{1}{2}\varphi_0(\mu_a\delta). \tag{11.3.4c}$$

Because of the "shadowing effect" of the foil, the flux correction at the foil surface is always greater than the activation correction; in fact there is even a flux perturbation when the backscattering probability vanishes.

11.3.1. Calculation of the Activation Correction with Elementary Diffusion Theory

Let us consider a disc-shaped foil of radius R large compared to the transport mean free path in the surrounding medium. In the interest of simplicity, let us take this surrrounding medium to be infinite. In the absence of the foil there is a homogeneous neutron flux Φ_0 everywhere. If we denote the perturbed neutron flux by $\Phi(\boldsymbol{r})$, then $\Phi(\boldsymbol{r})=\Phi_0-\Delta\Phi(\boldsymbol{r})$. According to elementary diffusion theory, the flux perturbation $\Delta\Phi(\boldsymbol{r})$ is given by

$$\Delta\Phi(\boldsymbol{r})=\frac{1}{4\pi D}\int Q(\boldsymbol{r}')\frac{e^{-\frac{|\boldsymbol{r}-\boldsymbol{r}'|}{L}}}{|\boldsymbol{r}-\boldsymbol{r}'|}\,dS' \tag{11.3.5}$$

where $Q(\boldsymbol{r}')\,dS'$ is the number of neutrons absorbed per second by the surface element dS' of the foil. The integration extends over the entire foil (cf. Fig. 11.3.1). Instead of $Q(\boldsymbol{r}')$ we can also use the "local" activation $C(\boldsymbol{r}')=(\mu_{\text{act}}/\mu_a)\,Q(\boldsymbol{r}')$; in terms of $C(\boldsymbol{r}')$, C is given by $C=\int C(\boldsymbol{r}')\,dS'/\pi R^2$. We must now express $C(\boldsymbol{r}')$ in terms of the perturbed neutron field at the foil surface. We take as the vector flux of the perturbed neutron field the expression

$$F(\boldsymbol{r},\boldsymbol{\Omega})=\frac{1}{4\pi}F_0(\boldsymbol{r})+\frac{3}{4\pi}F_1(\boldsymbol{r})\cos\vartheta.$$

The angle ψ between the field axis and the normal to the foil now depends on position on the foil surface (cf. Fig. 11.3.2). Since everything is symmetric in the foil surface, we can limit our considerations to one side. Neglecting neutrons that enter the foil through its edges and ignoring scattering in the foil, we have

$$C(\boldsymbol{r})=\frac{\mu_{\text{act}}}{\mu_a}\left[\frac{F_0(\boldsymbol{r})}{2}\varphi_0(\mu_a\delta)+\frac{3}{2}F_1(\boldsymbol{r})\,\varphi_1^*(\mu_a\delta)\cos\psi\right] \tag{11.3.6}$$

[1] More precisely, $\Phi(\boldsymbol{r}_s)=\frac{\Phi_0}{1+\varkappa_c}\cdot\frac{1}{2}\left(1+E_2(\mu_a\delta)\right)$ (cf. footnote on p. 243).

where

$$\left.\begin{aligned}\varphi_1^*(\mu_a\delta) &= 2\int_0^1 t^2(1-e^{-\mu_a\delta/t})\,dt\\ &= \frac{2}{3} - 2E_4(\mu_a\delta)\\ &= \frac{2}{3}(1-e^{-\mu_a\delta}) + \frac{\mu_a\delta}{3}(1-\varphi_0(\mu_a\delta)).\end{aligned}\right\} \quad (11.3.7\text{a})$$

When $\mu_a\delta < 0.3$,

$$\varphi_1^*(\mu_a\delta) \approx \frac{1}{2}\varphi_0(\mu_a\delta). \quad (11.3.7\text{b})$$

In contrast to Sec. 11.2.1, here the term containing F_1 does *not* vanish since the current is directed at the foil from both sides. If we now note that $F_1(\boldsymbol{r})\cos\psi$ is

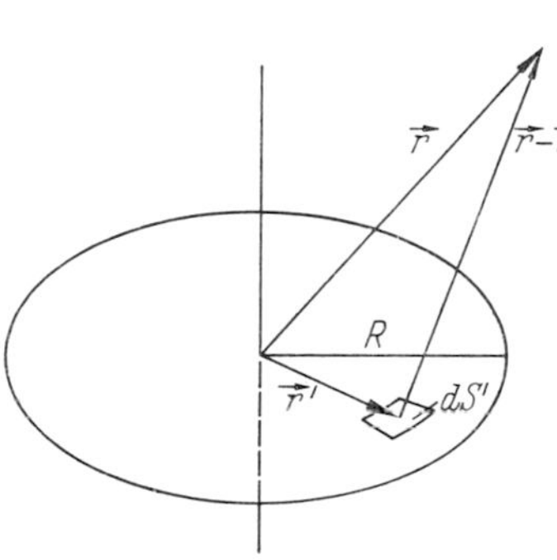

Fig. 11.3.1. Calculation of the flux perturbation around a foil (see text)

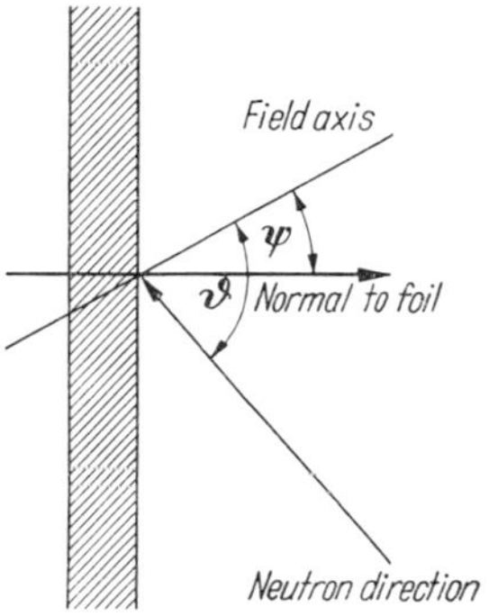

Fig. 11.3.2. Geometry of the perturbed neutron field

the normal component of the current directed at the foil and that the total normal current $2F_1(\boldsymbol{r})\cos\psi$ directed at the foil from both sides equals the number $Q(\boldsymbol{r})$ of neutrons absorbed per unit surface, we can write

$$C(\boldsymbol{r}) = \frac{\mu_{\text{act}}}{\mu_a}\cdot\frac{F_0(\boldsymbol{r})}{2}\cdot\varphi_0(\mu_a\delta) + \frac{3}{4}C(\boldsymbol{r})\varphi_1^*(\mu_a\delta). \quad (11.3.8)$$

If we now express $F_0(\boldsymbol{r})$ in terms of Φ_0 and $\Delta\Phi(\boldsymbol{r})$, then it follows with the help of Eq. (11.3.5) that

$$\left.\begin{aligned}C(\boldsymbol{r}) = \frac{\mu_{\text{act}}}{\mu_a}\frac{\Phi_0}{2}\varphi_0(\mu_a\delta) - \frac{1}{8\pi D}\int\frac{C(\boldsymbol{r}')e^{-|\boldsymbol{r}-\boldsymbol{r}'|/L}}{|\boldsymbol{r}-\boldsymbol{r}'|}dS'\,\varphi_0(\mu_a\delta) +\\ + \frac{3}{4}C(\boldsymbol{r})\varphi_1^*(\mu_a\delta).\end{aligned}\right\} \quad (11.3.9)$$

Now $(\mu_{\text{act}}/\mu_a)\dfrac{\Phi_0}{2}\varphi_0(\mu_a\delta)$ is the activation C_0 that would occur in the unperturbed neutron field. Therefore the activation depression is given by

$$C_0 - C(\boldsymbol{r}) = \frac{1}{8\pi D}\int\frac{C(\boldsymbol{r}')e^{-|\boldsymbol{r}-\boldsymbol{r}'|L}}{|\boldsymbol{r}-\boldsymbol{r}'|}dS'\,\varphi_0(\mu_a\delta) - \frac{3}{4}C(\boldsymbol{r})\varphi_1^*(\mu_a\delta). \quad (11.3.10\text{a})$$

By integration over the foil surface and division by C we obtain the activation correction

$$\varkappa_c = \frac{C_0 - C}{C} = \frac{1}{8\pi D}\frac{\iint\frac{C(\boldsymbol{r}')e^{-|\boldsymbol{r}-\boldsymbol{r}'|/L}}{|\boldsymbol{r}-\boldsymbol{r}'|}dS\,dS'}{\int C(\boldsymbol{r})\,dS}\varphi_0(\mu_a\delta) - \frac{3}{4}\varphi_1^*(\mu_a\delta). \quad (11.3.10\text{b})$$

We can obtain an approximate solution to Eq. (11.3.10b) by neglecting the position dependence of the activation on the foil surface; then

$$\left.\begin{aligned}\varkappa_c = \frac{1}{8\pi D}\,\varphi_0(\mu_a\delta)\,\frac{1}{\pi R^2}\int_0^R 2\pi r\,dr\int_0^R r'\,dr'\int_0^{2\pi} d\varphi\,\frac{e^{-\sqrt{r^2+r'^2-2rr'\cos\varphi}/L}}{\sqrt{r^2+r'^2-2rr'\cos\varphi}}\\ -\frac{3}{4}\,\varphi_1^*(\mu_a\delta).\end{aligned}\right\}\qquad(11.3.10\text{c})$$

In general, the integral cannot be carried out explicitly, and for this reason we consider two limiting cases. The first is the case $R \gg L$. Here because of the rapid decay of the exponential, the integral

$$\int_0^R r'\,dr'\int_0^{2\pi} d\varphi\,\frac{e^{-\sqrt{r^2+r'^2-2rr'\cos\varphi}/L}}{\sqrt{r^2+r'^2-2rr'\cos\varphi}}$$

is independent of position r on the foil surface and can be replaced by its value $2\pi L(1-e^{-R/L}) \approx 2\pi L$ at $r=0$. Then

$$\varkappa_c = \frac{1}{4}\,\frac{L}{D}\,\varphi_0(\mu_a\delta) - \frac{3}{4}\,\varphi_1^*(\mu_a\delta). \qquad (11.3.11\text{a})$$

The second limiting case is the case $R \ll L$. Here the exponential function can be set equal to unity; since

$$\frac{1}{\pi R^2}\int_0^R 2\pi r\,dr\int_0^R r'\,dr'\int_0^{2\pi}\frac{d\varphi}{\sqrt{r^2+r'^2-2rr'\cos\varphi}} = 2\pi R W$$

with $W = \frac{8}{3\pi} = 0.85$, $\varkappa_c$ becomes

$$\varkappa_c = \frac{1}{4}\,\frac{R}{D}\,W\,\varphi_0(\mu_a\delta) - \frac{3}{4}\,\varphi_1^*(\mu_a\delta). \qquad (11.3.11\text{b})$$

An interpolation formula that has the correct values in the limits $R \gg L$ and $R \ll L$ and is probably a good representation in between is

$$\varkappa_c = \frac{3}{4}\left\{\frac{L}{\lambda_{tr}}\,(1-e^{-WR/L})\,\varphi_0(\mu_a\delta) - \varphi_1^*(\mu_a\delta)\right\}. \qquad (11.3.11\text{c})$$

We have replaced D by $\lambda_{tr}/3$. For practical values of $\mu_a\delta$ we can apply Eq. (11.3.7b) to transform this result into

$$\varkappa_c = \frac{3}{4}\left\{\frac{L}{\lambda_{tr}}\,(1-e^{-WR/L}) - \frac{1}{2}\right\}\varphi_0(\mu_a\delta). \qquad (11.3.11\text{d})$$

As R approaches zero, Eqs. (11.3.11b, c and d) predict a negative foil correction, which is naturally nonsense. This absurdity occurs because our results have been derived with the help of elementary diffusion theory, which is inapplicable in this limiting case. It will be shown later that Eq. (11.3.11d) is quite accurate if the $\frac{1}{2}$ in the braces is dropped. Further calculations of foil perturbations based on elementary diffusion theory can be found in Bothe, in Tittle, and in Vigon and Wirtz.

11.3.2. Transport-Theoretic Treatment of the Activation Correction

Skyrme has calculated the activation correction for disc-shaped foils by means of an approximate solution of the transport equation (first-order perturbation theory). He finds that in an isotropically scattering medium[1]

$$\varkappa_c = \frac{3}{4}\left\{\frac{L}{\lambda}\, S\left(\frac{2R}{L}\right) - \frac{2}{3}\, K\left(\frac{2R}{\lambda}, \gamma\right)\right\} \varphi_0(\mu_a \delta). \tag{11.3.12}$$

Here $\lambda = 1/\Sigma_t$ is the mean free path in the medium, γ is the ratio λ/λ_s, R is the foil radius, and L is the diffusion length. The functions $S(x)$ and $K(x, \gamma)$ are shown in Figs. 11.3.3 and 11.3.4. For small values of x,

$$S(x) = \frac{4}{3\pi}\, x - \frac{1}{8}\, x^2 + \frac{4}{45\pi}\, x^3 - \frac{1}{192}\, x^4 + \cdots \tag{11.3.13a}$$

while for large x,

$$S(x) = 1 - \frac{4}{\pi}\left(\frac{1}{x} + \frac{1}{x^3} + \cdots\right). \tag{11.3.13b}$$

The term $\frac{2}{3} K(x, \gamma)$ is always < 0.1 and as a rule represents a negligibly small correction.

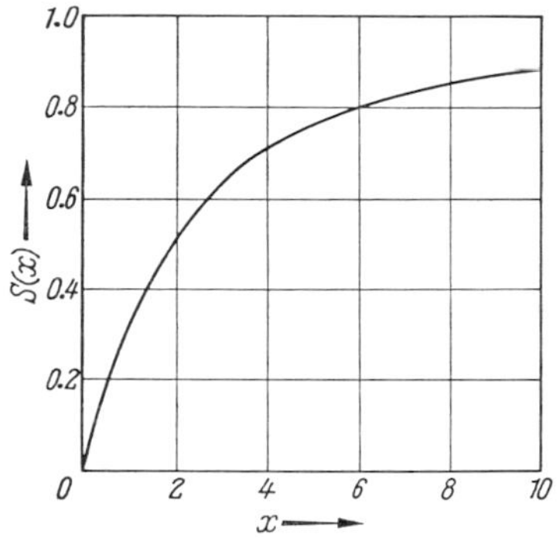

Fig. 11.3.3. The Skyrme function $S(x)$ $\left(\text{where } x = \frac{2R}{L}\right)$

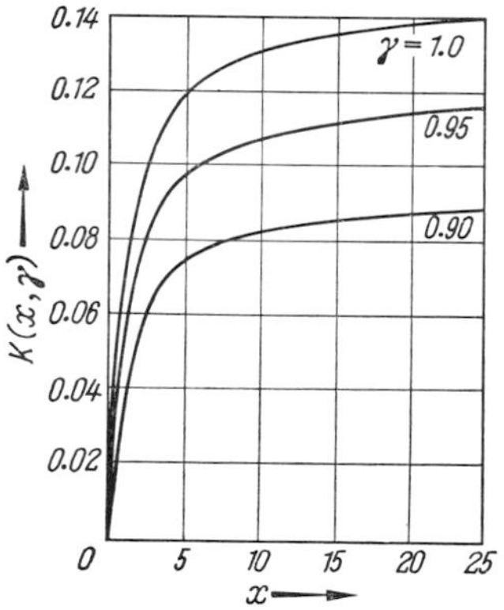

Fig. 11.3.4.
The function $K(x, \gamma)$ $\left(\text{where } x = \frac{2R}{\lambda}, \gamma = \frac{\lambda}{\lambda_s}\right)$

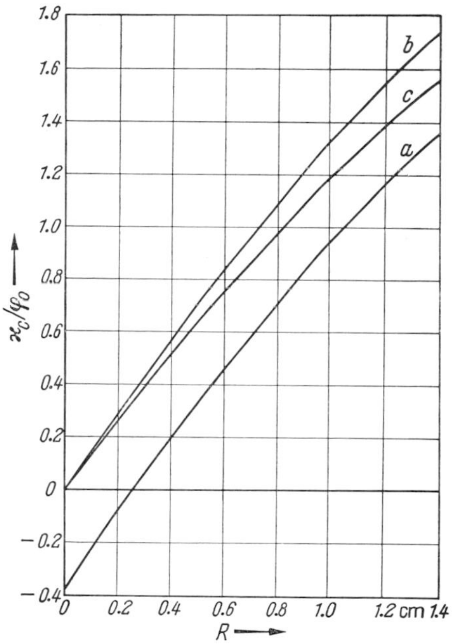

Fig. 11.3.5. A comparison of foil perturbations in water as predicted by elementary diffusion theory and by the Skyrme theory. Curve a: Eq. (11.3.11d). Curve b: Eq. (11.3.11d), additive term $\frac{1}{2}$ dropped. Curve c: Eq. (11.3.12)

In Fig. 11.3.5, we compare values of $\varkappa_c/\varphi_0(\mu_a \delta)$ appropriate to water ($L = 2.76$ cm, $\lambda_{tr} = 0.43$ cm) that have been calculated with Eqs. (11.3.11d) and (11.3.12). In this comparison, λ was replaced by λ_{tr} in Eq. (11.3.12). We see that diffusion theory always gives smaller values than SKYRME's theory, but that the variation of $\varkappa_c/\varphi_0(\mu_a \delta)$ with increasing R/λ_{tr} is very similar in both

[1] SKYRME's results had another form in his original work; here we use the notation suggested by RITCHIE and ELDRIDGE.

theories. This is easily understood if we note that for the range of foil sizes important in practice $2R/L \lesssim 1$. Thus according to Eq. (11.3.11d)

$$\frac{\varkappa_c}{\varphi_0(\mu_a\delta)} = \frac{2}{\pi}\frac{R}{\lambda_{tr}}\left[1 - \frac{32}{9\pi^2}\frac{R}{L} + \cdots\right] - \frac{3}{8} \tag{11.3.14a}$$

while according to Eq. (11.3.12)

$$\frac{\varkappa_c}{\varphi_0(\mu_a\delta)} = \frac{2}{\pi}\frac{R}{\lambda_{tr}}\left[1 - \frac{3\pi}{16}\frac{R}{L} + \cdots\right] - \frac{1}{2}K. \tag{11.3.14b}$$

The first terms (i.e., the expressions $\frac{2}{\pi}\frac{R}{\lambda_{tr}}[1 - \cdots]$) differ only slightly for small values of R/L. They represent the "asymptotic" contribution of widely separated foil elements to the activation perturbation, which is reproduced quite well by diffusion theory. The essential difference is in the non-asymptotic terms

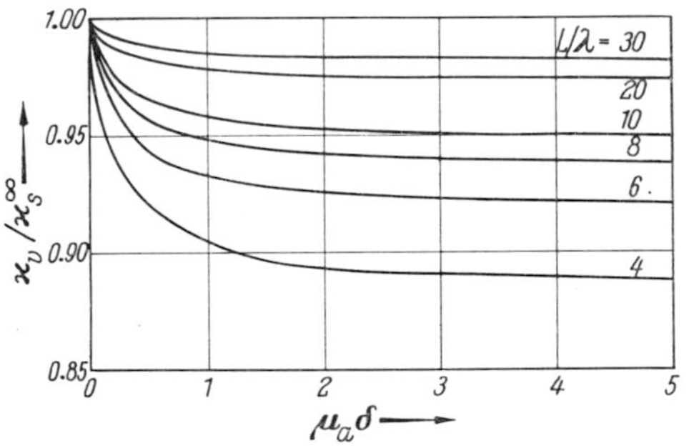

Fig. 11.3.6. The ratio of the Ritchie-Eldridge to the Skyrme foil perturbations for infinite foils in an isotropically scattering medium

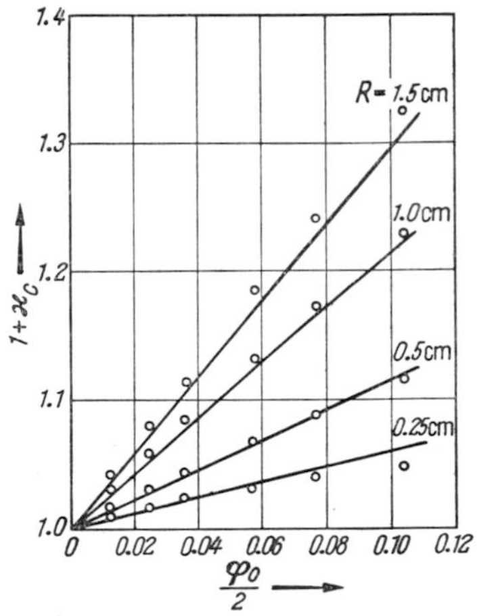

Fig. 11.3.7. Perturbations of gold foils in water. o o Numerical calculations by DALTON and OSBORN; —— Eq. (11.3.15)

$\frac{3}{8}$ and $\frac{1}{2}K$. We obtain quite good agreement between the result of diffusion theory and Eq. (11.3.12) if we drop the additive term $\frac{1}{2}$ in Eq. (11.3.11d). The corresponding curve is also shown in Fig. 11.3.5.

The validity of the SKYRME theory for large foil radii has been studied by RITCHIE and EDRIDGE, who obtained a very exact variational solution to the transport equation for an infinite foil in an isotropically scattering medium. In Fig. 11.3.6 we show $\varkappa_v/\varkappa_s^\infty$, the ratio of the Ritchie-Eldridge activation correction to that obtained from Eq. (11.3.12) by setting $R=\infty$. From this curve we see that $\varkappa_s^\infty$ is always somewhat greater than $\varkappa_v$, particularly in strongly absorbing media and for thick foils. In practical cases ($L/\lambda > 6$, $\mu_a\delta \lesssim 0.3$), however, the difference is smaller than 5%. When the foil radius is finite RITCHIE and ELDRIDGE recommend in place of Eq. (11.3.12) the relation

$$\varkappa_c = \frac{3}{4}\left\{\frac{L}{\lambda} S\left(\frac{2R}{L}\right) - \frac{2}{3} K\left(\frac{2R}{\lambda}, \gamma\right)\right\} \frac{\varkappa_v}{\varkappa_s^\infty} \varphi_0(\mu_a\delta) \tag{11.3.15}$$

which gives the right result in the limiting case $R \to \infty$. This so-called "modified Skyrme formula" is very widely used; the difference between it and Eq. (11.3.12) is very small.

DALTON and OSBORN have carried out an independent transport-theoretic investigation of probe perturbations and probe activation. These authors start from the integral form of the transport equation in an infinite medium (cf. Sec. 5.1.3) in which the flux before the introduction of the detector is Φ_0. They then calculate numerically the flux $\bar{\Phi}$ averaged over the entire probe volume. For a foil, $C = \bar{\Phi}\mu_{\text{act}}\,\delta$. In this treatment, the self-shielding effect that was discussed in Sec. 11.2.1 and the activation perturbation are taken into account simultaneously. If we intend to compare the results of DALTON and OSBORN with those already obtained in this section, we must express $\bar{\Phi}/\Phi_0$ as follows:

$$\frac{\bar{\Phi}}{\Phi_0} = \frac{\varphi_0(\mu_a\delta)}{2\mu_a\delta}\,\frac{1+\varepsilon}{1+\varkappa_c}. \tag{11.3.16}$$

In Fig. 11.3.7 we show $1+\varkappa_c$ for gold foils of various sizes in water (from HANNA); the points have been obtained with the help of Eq. (11.3.16) from flux ratios calculated by DALTON and OSBORN. The solid curve was calculated according to the modified Skyrme theory [Eq. (11.3.15)]. The same values of the scattering and absorption cross sections of gold and water were used in both calculations. In the Dalton-Osborn calculations, the anisotropy of scattering was taken into account approximately; in analogy, $\lambda_{tr} = \lambda_s/(1-\bar{\mu})$ was used instead of λ in the modified Skyrme calculations. With the exception of the smallest foil radii, where a maximum deviation of 20% occurs, the agreement between the modified Skyrme and the Dalton-Osborn calculations is satisfactory. We conclude from these considerations that in the range $R \gtrsim \lambda_{tr}$, i.e., in the range of practical foil sizes in hydrogenous media, the modified Skyrme theory with $\lambda = \lambda_{tr}$ is a useful approximation; this assertion holds for the time being only in a monoenergetic field. Since Eqs. (11.3.12) and (11.3.15) differ only slightly, we can also use the simple Skyrme theory or the result (11.3.11d) of elementary diffusion theory, although in the latter case we must drop the additive term $\frac{1}{2}$.

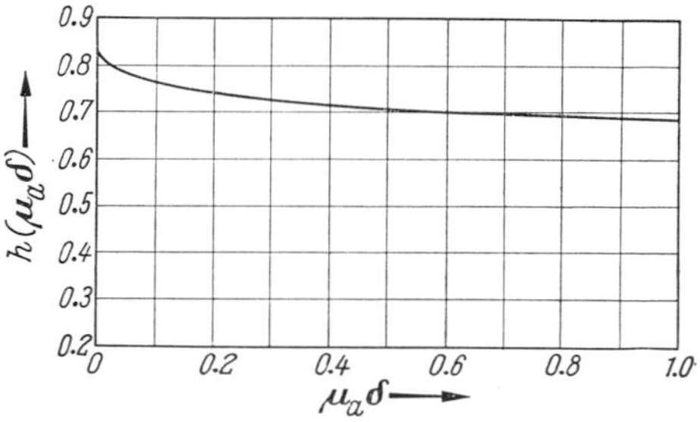

Fig. 11.3.8. The function $h(\mu_a\delta)$

In the range $R \ll \lambda_{tr}$, which is the range of practical foil sizes in graphite, beryllium, and D_2O, there has been as yet no comparison of the modified Skyrme theory with numerically calculated Dalton-Osborn values. However, in this range a complete solution of the transport equation is not necessary for the calculation of the activation correction. Rather consideration of individual collisions suffices, as BOTHE has already shown. According to MEISTER, when $R \ll \lambda_s$ in an isotropically scattering medium

$$\varkappa_c = \frac{2}{\pi}\cdot\frac{R}{\lambda_s}\,h(\mu_a\delta)\,\varphi_0(\mu_a\delta). \tag{11.3.17}$$

The function $h(\mu_a\delta)$ is plotted in Fig. 11.3.8; its value at the origin is $h(0) = 0.828$. When $R \ll \lambda_{tr}$, $\varkappa_c/\varphi_0(\mu_a\delta)$ depends much more strongly on the foil thickness than in the Ritchie-Eldridge calculation for infinite foils. In Fig. 11.3.9 we compare $\varkappa_c$

calculated according to the modified Skyrme theory and according to the Meister theory for foils of various sizes in graphite. The Meister theory always gives the smaller values of $\varkappa_c$, but for thin foils and when $R/\lambda_{tr} \to 0$ both calculations agree. As we shall see later, Meister's theory reproduces the measured foil perturbations better than the modified Skyrme calculations when $R \ll \lambda_{tr}$.

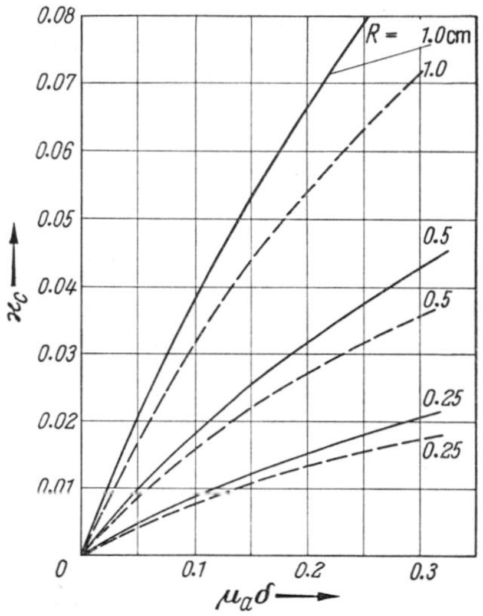

Fig. 11.3.9. A comparison of the modified Skyrme theory and Meister's theory for foil perturbations in graphite. ——— SKYRME [Eq. (11.3.15)]; ----- MEISTER [Eq. (11.3.17)]

Comparison with the diffusion-theoretic result for $\varkappa_c$ in the range of small R/λ_{tr} is not meaningful since when $R \ll \lambda_{tr}$ elementary diffusion theory no longer applies.

11.3.3. Activation Correction in the Thermal Field

Up to this point, all our results held only in a monoenergetic field, and we must now investigate how to average them over the velocity distribution of the neutrons. In order to arrive at an exact answer to this question, we must solve the transport equation of the foil perturbation problem taking into account energy exchange between the thermal neutrons and the scattering substance. There has been no such treatment to date, and we must be content with some more qualitative arguments.

In Eqs. (11.3.11d, 12, 15, and 17), $\varkappa_c$ is the product of a flux perturbation and the absorption probability $\varphi_0(\mu_a\delta)$. This flux perturbation is normalized to (negative) unit source strength; its sources are the absorption processes in the probe. Let $\Phi_0(E)$, the neutron spectrum in the unperturbed field before introduction of the foil, be a Maxwell distribution $M(E)$ with the moderator temperature. The energy distribution of the sources of the flux perturbation is the same as the energy distribution of the absorption processes in the foil; as a rule, it is not a Maxwell distribution. For example, in thin foils with $1/v$-absorption it is $\sim M(E)/\sqrt{E}$. Which energy spectrum the flux perturbation has depends on the thermalization properties of the moderating medium as well as on the number of collisions a neutron experiences before it is scattered back into the foil. If this number is large and if the medium thermalizes neutrons efficiently, the flux perturbation will have a Maxwell energy distribution; and instead of $\varphi_0(\mu_a\delta)$ we must use the average value $\overline{\varphi_0(\mu_a\delta)}$ given in Eq. (11.2.13b) for the absorption probability. On the other hand, if the backscattering takes place in a single collision ($R \ll \lambda_{tr}$) and if the energy coupling between the neutron and the moderator is weak, the spectrum of the flux perturbation incident on the foil as a rule is softer than a Maxwell distribution and the absorption probability larger than $\overline{\varphi_0(\mu_a\delta)}$. HANNA has shown that for negligible rethermalization and a thin $1/v$-absorber, the absorption probability is $\frac{4}{\pi}\,\overline{\varphi_0(\mu_a\delta)}$; in all practical cases, the absorption probability should lie between $\overline{\varphi_0(\mu_a\delta)}$ and $\frac{4}{\pi}\,\overline{\varphi_0(\mu_a\delta)}$.

Since the difference between these two limiting cases is not very large and since the flux perturbation in most practical cases is largely rethermalized, we suggest that the value $\overline{\varphi_0(\mu_a \delta)}$ given by Eq. (11.2.13b) always be used for the absorption probability. For $1/v$-absorbers that are not too thick $\overline{\varphi_0(\mu_a \delta)} = \frac{\sqrt{\pi}}{2}\, \varphi_0\big(\mu_a(kT)\,\delta\big)$; in this case also $\overline{h(\mu_a \delta)\, \varphi_0(\mu_a \delta)} = \frac{\sqrt{\pi}}{2}\, h\big(\mu_a(kT)\,\delta\big)\, \varphi_0\big(\mu_a(kT)\,\delta\big)$, which is worth noting when MEISTER's Eq. (13.3.17) is being used.

An additional question is how to average the value of $\varkappa_c/\varphi_0(\mu_a \delta)$, or what is the same question, what value of the transport mean free path to use in the formulas for this quantity. Since λ_{tr} depends weakly on energy in graphite and beryllium, we know without any detailed investigation that we can use the usual thermal average of λ_{tr} in these substances. In hydrogenous moderators the energy dependence of λ_{tr} is much stronger; in water in particular $\lambda_{tr}(E) \sim \sqrt{E}$ (cf. Sec. 17.1.4). Now for practical foil measurements in hydrogenous media, $R \gg \lambda_{tr}$, i.e., most of the backscattered neutrons have experienced several collisions and are correctly described by diffusion theory. In addition, the energy coupling between neutrons and hydrogenous media is very strong, and neutrons from the source of the perturbation are very rapidly thermalized. Therefore, one can use for this case also the usual thermal average of the transport mean free path in the calculation of $\varkappa_c/\varphi_0(\mu_a \delta)$.

11.3.4. Calculation of the Flux Perturbation

Let us consider a disc-shaped foil in an infinite medium in which the flux has the homogeneous value Φ_0 before the introduction of the foil. Let us introduce polar coordinates r, z and try to calculate the flux correction

$$\varkappa_\Phi(r=0, z) = \frac{\Phi_0 - \Phi(r=0, z)}{\Phi(r, z=0)} = \frac{\Delta\Phi(r=0, z)}{\Phi(r, z=0)}$$

on the foil axis. Using Eq. (11.3.5), which gives the diffusion-theoretic value of $\Delta\Phi$, we have

$$\Delta\Phi(r=0, z) = \frac{1}{2D} \int_0^R \frac{e^{-\sqrt{r'^2+z^2}/L}}{\sqrt{r'^2+z^2}}\, Q(r')\, r'\, dr'. \tag{11.3.18}$$

According to Eq. (11.3.8),

$$Q(r') = \frac{\Phi(r', z=0)}{2}\, \frac{\varphi_0(\mu_a \delta)}{1 - \frac{3}{4}\varphi_1^*(\mu_a \delta)}$$

and therefore

$$\varkappa_\Phi(r=0, z) = \frac{1}{4D}\, \frac{\varphi_0(\mu_a \delta)}{1 - \frac{3}{4}\varphi_1^*(\mu_a \delta)}\, \frac{\int_0^R \Phi(r', z=0)\, \frac{e^{-\sqrt{r'^2+z^2}/L}}{\sqrt{r'^2+z^2}}\, r'\, dr'}{\Phi(r, z=0)}. \tag{11.3.19}$$

Exactly as we did in the calculation of the activation correction, we now replace the position-dependent flux on the foil surface by its average value; by integration we then obtain

$$\varkappa_\Phi(r=0, z) = \frac{3}{4}\, \frac{L}{\lambda_{tr}} \left[e^{-z/L} - e^{-\frac{1}{L}\sqrt{R^2+z^2}}\right] \frac{\varphi_0(\mu_a \delta)}{1 - \frac{3}{4}\varphi_1^*(\mu_a \delta)}. \tag{11.3.20}$$

When $R \gg L$,

$$\varkappa_\Phi = \frac{3}{4} \frac{L}{\lambda_{tr}} e^{-z/L} \frac{\varphi_0(\mu_a \delta)}{1-\frac{3}{4}\varphi_1^*(\mu_a\delta)} \tag{11.3.21}$$

and when $R \ll L$, $z \gg R$,

$$\varkappa_\Phi = \frac{3}{8} \frac{R^2}{\lambda_{tr}} \frac{e^{-z/L}}{z} \frac{\varphi_0(\mu_a \delta)}{1-\frac{3}{4}\varphi_1^*(\mu_a\delta)}\,. \tag{11.3.22}$$

The flux perturbation at the foil surface is given by

$$\varkappa_\Phi(r=0,\, z=0) = \frac{3}{4} \frac{L}{\lambda_{tr}} (1-e^{-R/L}) \frac{\varphi_0(\mu_a \delta)}{1-\frac{3}{4}\varphi_1^*(\mu_a\delta)}\,. \tag{11.3.23}$$

If we compare this formula with Eq. (11.3.11d), which was also derived with elementary diffusion theory, we find that for small $\mu_a\delta$ and with $W=1$,

$$\varkappa_\Phi(r=0,\, z=0) = \varkappa_c + \tfrac{3}{8}\varphi_0(\mu_a\delta)\,.$$

This result is analogous to Eq. (11.3.4c), which was derived at the beginning of Sec. 11.3 in an entirely different way.

11.3.5. Activation Perturbation of Tape Probes

All the considerations advanced so far in this section hold for disc-shaped foils, which are the kind most frequently used in practice. Occasionally square foils are also used; for them the foil perturbation is probably about the same as that of a disc-shaped foil of the same area, i.e., if a is the length of the foil's side, we use the formula for a disc foil with $R = a/\sqrt{\pi}$. For tape probes that are much longer in one direction than in the other, such a simple adaptation is not possible, and we shall try to estimate the perturbation using elementary diffusion theory. Let a be the width of the probe; let its length be large compared to L. Let the neutron field be constant everywhere before the introduction of the probe; afterwards let us assume that it does not vary in the long direction of the foil. The calculation of the activation correction then proceeds exactly as in Sec. 11.3.1 except that in place of Eq. (11.3.10b) we obtain

$$\varkappa_c = \frac{C_0 - C}{C} = \frac{\frac{1}{4\pi D}\int_0^a\int_0^a C(x') K_0\left(\frac{|x-x'|}{L}\right) dx\, dx'}{\int_0^a C(x)\, dx}\, \varphi_0(\mu_a\delta) - \frac{3}{4}\varphi_1^*(\mu_a\delta)\,. \tag{11.3.24}$$

Here $\frac{1}{2\pi} K_0\left(\frac{|x-x'|}{L}\right)$ is the diffusion kernel of a line source (cf. Sec. 6.2.2). If we now again replace the position-dependent activation by its average over the probe, we obtain

$$\varkappa_c = \frac{1}{4\pi a D}\int_0^a\int_0^a K_0\left(\frac{|x-x'|}{L}\right) dx\, dx'\, \varphi_0(\mu_a\delta) - \frac{3}{4}\varphi_1^*(\mu_a\delta)\,. \tag{11.3.25}$$

Now we assume further that the width of the tape is small compared to the diffusion length; then $K_0\left(\frac{|x-x'|}{L}\right) \approx \ln \frac{2L}{\gamma|x-x'|}$, where $\gamma = 0.577$ is Euler's constant, and the integration leads to

$$\varkappa_c = \frac{3a}{4\pi\lambda_{tr}}\left(\ln\frac{2L}{\gamma a} - \frac{1}{2}\right)\varphi_0(\mu_a\delta) - \frac{3}{4}\varphi_1^*(\mu_a\delta)\,. \tag{11.3.26}$$

No transport-theoretic calculation of the perturbation exists in this case. In analogy with the situation for disc-shaped foils, the formula derived from diffusion theory ought to give the right value for the perturbation if the term $\frac{3}{4}\varphi_1^*(\mu_a\delta)$ in Eq. (11.3.26) is dropped. Eq. (11.3.26) should also permit rough estimation of $\varkappa_c$ for an infinite cylinder (wire probe) if a is replaced by πR and $\varphi_0(\mu_a\delta)$ by $\chi_0(\Sigma_a R)$.

11.4. Experimental Studies of Foil Perturbations

11.4.1. Methods of Measuring the Activation Correction

The only procedure for the direct determination of $\varkappa_c$ is the cavity method (MEISTER; cf. also BOTHE). Here the perturbed activation C in a medium is compared with the activation C_0 of the same foil at a point where there is no activation perturbation. This unperturbed case is realized by irradiating the foil inside a cavity in the medium whose dimensions are large compared with those of the foil. The backscattering probability is then vanishingly small and thus in view of Eq. (11.3.2f) so is $\varkappa_c$. The quantities C and C_0 must be referred to the same unperturbed flux by use of a reference foil whose perturbing effect is known. An effect similar to that which is achieved with a cavity can be achieved with a medium having an extremely small scattering cross section, e.g., Al ($\lambda_s \approx 10$ cm); however, it is difficult to produce a pure thermal neutron field in such a medium. An advantage of the cavity method is that β-counting can be used to determine the foil activity. Since the distribution of activation processes over the foil depth changes very little from the unperturbed to the perturbed case, the β-self-absorption factors are nearly equal and cancel out in the expression $(C-C_0)/C_0$ for $\varkappa_c$.

Nevertheless, most experiments for the determination of $\varkappa_c$ are done indirectly and without the use of a cavity. In these experiments, a series of foils of varying thickness are successively irradiated at the same place in a medium and the specific activity, which is proportional to the specific activation $C^* = C/\delta$, is determined. These measurements determine the ratio $C^*/C^*(\delta=0) = \overline{\Phi}/\Phi_0$ from which $\varkappa_c$ can be calculated using Eq. (11.3.16). $C^*(\delta=0)$ is the specific activity of a foil for which there is no perturbation — neither self-shielding nor an activation perturbation. $C^*(\delta=0)$ is obtained either by measurement with extremely thin foils ($\mu_a\delta<10^{-3}$) or by extrapolation of the values of $C^*(\delta)$ measured for the thicker foils to zero thickness. RANDALL and WALKER have given a special procedure for the determination of $C^*(0)$.

In contrast to the cavity method, in this method we must measure the true activity, i.e., we must count the foil's γ-activity or eliminate the β-self-absorption by calculation or even better by measurement (cf. Sec. 14.1.3).

In making clean measurements of the activation correction, we must so arrange the irradiation apparatus that the unperturbed flux Φ_0 does not vary noticeably over the face of the foil. The neutron field should also be isotropic; in case we have to, we can always satisfy this requirement by suitable rotation of the probe. The neutron spectrum of the unperturbed field should be a Maxwell distribution with a temperature as close as possible to the moderator temperature. Sometimes, it may be necessary to check this point by a neutron temperature measurement (cf. Sec. 15.3). If the spectrum contains an epithermal component, it must be eliminated using the cadmium difference method (cf. Sec. 12.2).

11.4.2. Activation Perturbation in Graphite ($R \ll \lambda_{tr}$)

For $R \ll \lambda_{tr}$ the activation correction is very small compared to the self-shielding correction; for example, for a 0.2-mm-thick gold foil with $R=1$ cm in graphite, $\varphi_0(\mu_a \delta)/2\mu_a \delta \approx 0.8$ but $(1+\varkappa_c)^{-1} \approx 0.97$. Determination of $\varkappa_c$ from measured values of $\overline{\Phi}/\Phi_0$ by means of Eq. (11.3.16) is therefore not very exact (cf., e.g., KLEMA and RITCHIE, THOMPSON, and GALLAGHER); the cavity method probably gives much better values. Fig. 11.4.1 shows values of $\varkappa_c$ for various indium foils measured by MEISTER with this method. For comparison, calculated values obtained from MEISTER's theory [Eq. (11.3.17)] are also plotted[1]. The agreement of the measured values and Eq. (11.3.17) is well within the limits of experimental error, and we conclude therefrom that Eq. (11.3.17) is a serviceable approximation for the determination of $\varkappa_c$ in the case $R \ll \lambda_{tr}$.

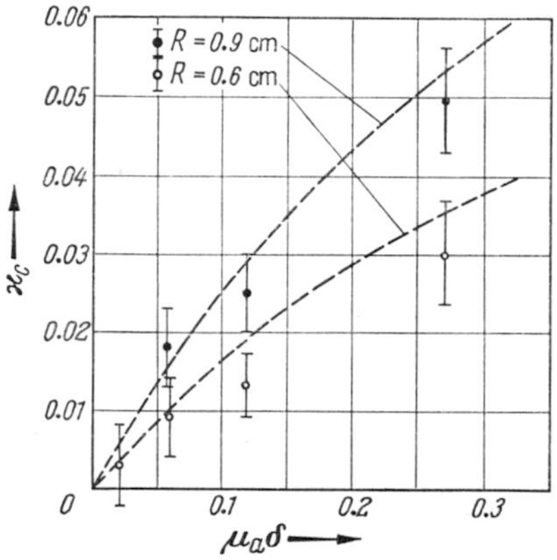

Fig. 11.4.1. Indium foil perturbations as measured by MEISTER in graphite. — — — Eq. (11.3.17)

11.4.3. Activation Perturbation in Water and Paraffin ($R \gtrsim \lambda_{tr}$)

Fig. 11.4.2 shows MEISTER's measured values of $\varkappa_c$ for indium foils in paraffin (cavity method). The curves represent extrapolations done according to the formula $\varkappa_c = \text{const.}\ \overline{\varphi_0(\mu_a \delta)}$. The values of $\varkappa_c/\overline{\varphi_0(\mu_a \delta)}$ determined for various foil radii are compared in Fig. 11.4.3 with calculations done with the Skyrme formula [Eq. (11.3.12)], which in this range of parameters is practically identical with the modified Skyrme formula[2]. Agreement is very good, and we can conclude that the Skyrme formula can be used for calculating the activation correction when $R \gtrsim \lambda_{tr}$. We can also use, as Fig. 11.4.3 shows, the simpler formula

$$\varkappa_c = \frac{3}{4}\,\frac{L}{\lambda_{tr}}\,(1 - e^{-0.85\,R/L})\,\overline{\varphi_0(\mu_a \delta)} \tag{11.4.1}$$

which arises from dropping the non-asymptotic term $-\frac{1}{2}$ in Eq. (11.3.11d).

Very careful studies of the activation correction in water by measurement of $\overline{\Phi}/\Phi_0$ have been carried out by STELZER, by HASNAIN *et al.*, by ZOBEL, and by WALKER, RANDALL, and STINSON. These measurements pretty well corroborate MEISTER's results, i.e., they show that $\varkappa_c$ is accurately given by the SKYRME or modified Skyrme theories as long as the thermal average $\overline{\varphi_0(\mu_a \delta)}$ of the absorption

[1] The calculations were performed with the values $\lambda_{tr} = 2.6$ cm for graphite and σ_a(2200 m/sec)$=190$ barns for indium. A neutron temperature measurement at the foil position gave the result $T = 320$ °K and $\mu_a(kT)$ was evaluated at this temperature.

[2] $\lambda_{tr} = 0.325$ cm and $L = 2.15$ cm for paraffin.

probability is used and the thermally averaged transport mean free path λ_{tr} is used for λ. As an example, we show in Fig. 11.4.4 values of $\bar{\Phi}/\Phi_0$ for indium foils ($R=1$ cm) in H_2O taken from the work of WALKER, RANDALL, and STINSON. The solid curve was calculated from Eq. (11.3.16) using values of $\varkappa_c$ obtained from the modified Skyrme theory [Eq. (11.3.15)]. The agreement is exceptional.

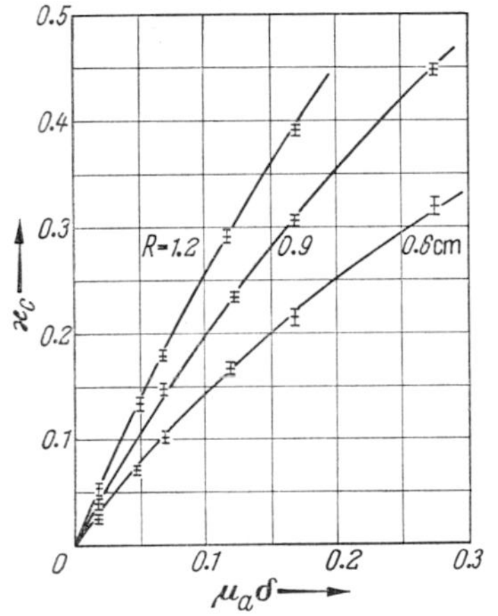

Fig. 11.4.2. Indium foil perturbations as measured by MEISTER in paraffin. ——— Extrapolations according to $\varkappa_c = \text{const} \cdot \varphi_0(\mu_a \delta)$

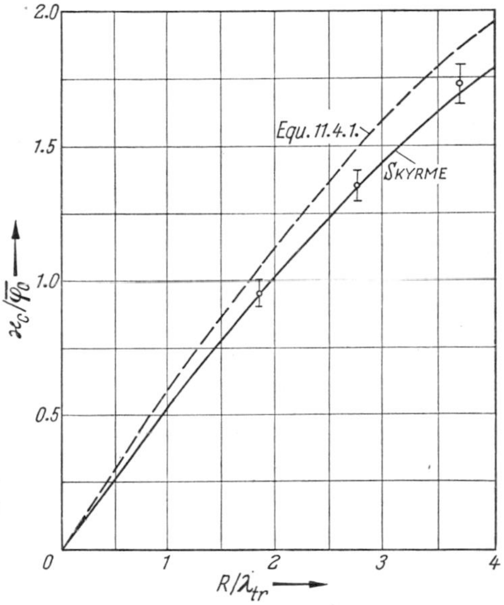

Fig. 11.4.3. $\varkappa_c/\varphi_0(\mu_a\delta)$ for Indium foils in paraffin taken from MEISTER's experiments in comparison to the Skyrme theory and Eq. (11.4.1)

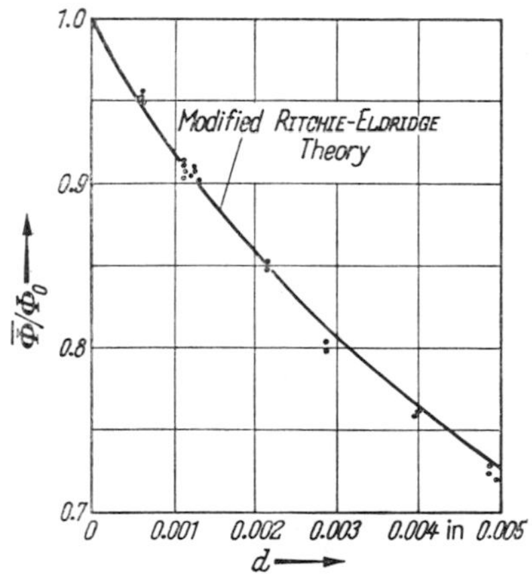

Fig. 11.4.4. $\bar{\Phi}/\Phi_0$ for Indium foils ($R=1$ cm) in water. ..... Experiments of WALKER, RANDALL and STINSON; ——— Eq. (11.3.16) with $\varkappa_c$ from Eq. (11.3.15)

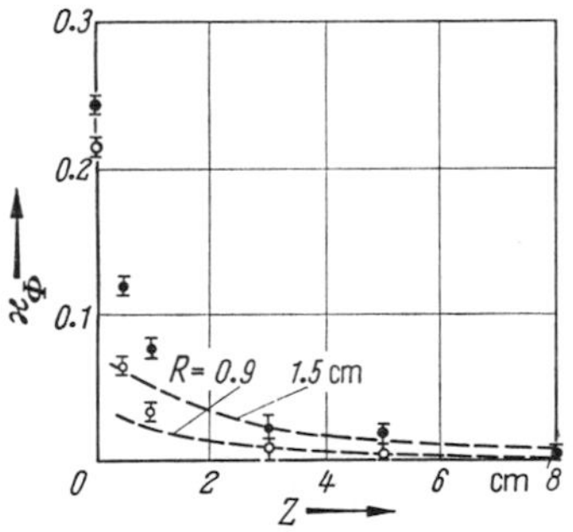

Fig. 11.4.5. The flux perturbation in the vicinity of Indium foils ($\mu_a\delta \approx 0.12$) in graphite. Experiments on foils with $R=1.5$ resp. 0.9 cm; — — — Eq. (11.3.20)

11.4.4. Measurement of the Flux Perturbation

Fig. 11.4.5 shows the flux perturbation in the neighborhood of indium foils in graphite according to MEISTER. The perturbing field was measured with very small dysprosium foils whose own contribution to the perturbation was negligible. Also plotted is the flux perturbation $\varkappa_\Phi(z)$ calculated with elementary diffusion theory [Eq. (11.3.20)]. We see immediately that near the foil Eq. (11.3.20) predicts values for the flux perturbation that are much too small; this comes about because for $z < \lambda_s$ elementary diffusion theory is not applicable. For larger distances z from the foil, Eq. (11.3.20) is quite accurate. This result is of practical significance, for in practice Eq. (11.3.20) is used only to insure that the distance between foils is large enough to make their mutual influence negligibly small. According to Fig. 11.4.5, diffusion theory is sufficiently accurate for this purpose. Using Eq. (11.3.20), we find that for indium foils with surface loadings of 100 mg/cm^2 and radii of 1 cm at a distance from each other of ≈ 15 cm in graphite and ≈ 8 cm in H_2O, $\varkappa_\Phi(z)$ is $< 10^{-3}$.

Chapter 11: References

TITTLE, C. W.: Nucleonics 8, No. 6, 6 (1951); ibid. **9**, No. 1, 60 (1951) (General Facts About Thermal Probes).

HUGHES, D. J., and R. B. SCHWARTZ: BNL-325 (1958).
WESTCOTT, C. H.: AECL-1101 (1960).
} Activation Cross Sections and g-Factors.

STROMINGER, D., J. M. HOLLANDER, and G. T. SEABORG: Rev. Mod. Phys. **30**, 585 (1958) (Decay Schemes, β- and γ-Energies, and Half-Lives).

GLEASON, G. I., J. D. TAYLOR, and D. L. TABERN: Nucleonics 8, No. 5, 12 (1951).
MEISTER, H.: Z. Naturforsch. **13**a, 722 (1958).
} β-Self-Absorption

DAVISSON, C. M., and R. D. EVANS: Rev. Mod. Phys. **24**, 79 (1952) (γ-Absorption Coefficients).

BOTHE, W.: Z. Phys. **120**, 457 (1943).
CASE, K. M., F. DE HOFFMANN, and G. PLACZEK: Introduction to the Theory of Neutron Diffusion, Los Alamos Scientific Laboratory (1953).
HANNA, G. C.: Nucl. Sci. Eng. **15**, 325 (1963).
MARTINEZ, J. S.: UCRL-6526 (1961).
VIGON, M. A.: Z. Naturforsch. **8**a, 727 (1953).
} Theory of Foil Activation.

PLACZEK, G.: NRC 1547 (1951) (E_n Functions).

BOTHE, W.: Z. Physik **120**, 437 (1943).
CORINALDESI, E.: Nuovo Cim. **3**, 131 (1946).
DALTON, G. R., and R. K. OSBORN: Nucl. Sci. Eng. **9**, 198 (1961).
DALTON, G. R.: Nucl. Sci. Eng. **13**, 190 (1962).
MEISTER, H.: Z. Naturforsch. **11**a, 347 (1956).
MEISTER, H.: Ibid., **11**a, 579 (1956).
OSBORN, R. K.: Nucl. Sci. Eng. **15**, 245 (1963).
RITCHIE, R. H., and H. B. ELDRIDGE: Nucl. Sci. Eng. 8, 300 (1960).
SKYRME, T. H. R.: UKAEA-Report MS 91 (1944), reprinted 1961.
VIGON, M., and K. WIRTZ: Z. Naturforsch. **9**a, 286 (1954).
} The Theory of Foil Perturbations.

GALLAGHER, T. L.: Nucl. Sci. Eng. **3**, 110 (1957).

HASNAIN, S. A., T. MUSTAFA, and T. V. BLOSSER: ORNL-3193 (1961).
KLEMA, E. D., and R. H. RITCHIE: Phys. Rev. **87**, 167 (1952).
MEISTER, H.: Z. Naturforsch. **10**a, 669 (1955); Ibid. **11**a, 356 (1956).
SOLA, A.: Nucleonics **18** (3), 78 (1960).
STELZER, K.: Nukleonik **1**, 10 (1958).
THOMPSON, M. W.: J. Nuclear Energy **2**, 286 (1955).
DE TROYER, A., and G. C. TAVERNIER: Bull. Acad. R. Belg., Cl. sci., Sér. 5, **39**, 880 (1953).
TRUBEY, D. K., T. V. BLOSSER, and G. M. ESTABROOK: ORNL-2842 (1959).
WALKER, J. V., J. D. RANDALL, and R. C. STINSON jr.: Nucl. Sci. Eng. **15**, 309 (1963).
ZOBEL, W.: ORNL-3407 (1963).
} Measurements of Foil Perturbations.

12. Activation by Epithermal Neutrons

The considerations begun in Chapter 11 will now be extended to the practically important case in which the neutron field contains both thermal and epithermal neutrons. The very same activity that can be excited by absorption of thermal neutrons in the probe substance can also be excited by absorption of epithermal neutrons, and it is necessary to separate these two parts by additional measurements. In doing so we simultaneously obtain information about the epithermal neutron flux.

We shall first discuss in Sec. 12.1 activation by epithermal neutrons without consideration of thermal activation; next in Sec. 12.2 we shall discuss the separation of the two parts of the activation. The related problem of measuring resonance integrals will be discussed in Sec. 12.3.

12.1. Activation by Epithermal Neutrons

Let us consider a foil in a homogeneous neutron field whose energy distribution is given by

$$\Phi(E)\,dE = \Phi_{\text{epi}}(E)\,\frac{dE}{E}$$

and let $\Phi_{\text{epi}}(E)$ vary slowly with energy. If we neglect scattering in the foil, the activation (disregarding the foil perturbation) is given by

$$C = \frac{1}{2}\int_{E_C}^{E_{\max}} \Phi_{\text{epi}}(E)\,\frac{\mu_{\text{act}}(E)}{\mu_a(E)}\,\varphi_0\big(\mu_a(E)\,\delta\big)\,\frac{dE}{E}. \qquad (12.1.1)$$

The exact value of the lower limit will be specified later; the energy $E_{\max}$ is the highest energy appearing in the spectrum and can frequently be taken as infinite without appreciable error. In general, the energy integral is difficult to carry out exactly, but in many cases, as we shall show, it can be evaluated with the help of simple approximate procedures.

12.1.1. Approximate Calculation of the Activation

In calculating C, we make the following assumptions:

a) $\Phi_{\text{epi}}(E)$ is constant, i.e., the neutrons have a pure $1/E$-spectrum.

b) The absorption coefficient $\mu_a(E)$ of the foil substance may be expressed in the form

$$\mu_a(E) = \mu_a(kT)\sqrt{\frac{kT}{E}} + \sum_i \frac{\mu_{a_0}^i}{1+\left(\dfrac{E-E_R^i}{\Gamma^i/2}\right)^2}, \qquad (12.1.2)$$

i.e., it may be divided into a $1/v$-part and a part that is the sum of symmetric Breit-Wigner line shapes arising from the various resonances. This decomposition is generally possible when the resonance energies are large, i.e., when $E_R^i \gg \Gamma^i$. For indium (energy of the first resonance $= 1.46$ ev; $\Gamma = 75$ mev), however, this representation is inexact.

c) The φ_0-function may be replaced by the rational approximation introduced in Sec. 11.2.1, viz.,

$$\varphi_0(\mu_a\delta)=\frac{2\mu_a\delta}{1+2\mu_a\delta}$$

and it may further be written

$$\frac{2\mu_a\delta}{1+2\mu_a\delta}\approx 2\mu_a(kT)\sqrt{\frac{kT}{E}}\,\delta+2\sum_i\frac{\mu^i_{a_0}\delta}{1+\left(\frac{E-E^i_R}{\Gamma^i/2}\right)^2+2\mu^i_{a_0}\delta}. \tag{12.1.3}$$

In deriving this last equation, we have assumed that we could neglect the self-shielding of the $1/v$-part of the cross section $\left(\text{since } \mu_a(kT)\sqrt{\frac{kT}{E}}\,\delta \text{ is usually} \ll 1 \text{ when } E>E_C\right)$ and separate the approximate φ_0-function into contributions from the individual resonances, which are assumed to be well separated.

With these assumptions, we have

$$C=\Phi_{\text{epi}}\left[\lambda_0\int\limits_{E_C}^{E_{\max}}\mu_a(kT)\sqrt{\frac{kT}{E}}\,\delta\frac{dE}{E}+\sum_i\lambda_i\int\frac{\mu^i_{a_0}\delta}{1+\left(\frac{E-E^i_R}{\Gamma^i/2}\right)^2+2\mu^i_{a_0}\delta}\frac{dE}{E}\right], \tag{12.1.4}$$

$\lambda_i\,(i=1, 2, 3, \ldots)$ specifies the fraction of the absorptions in the *i-th* resonance that lead to the activity being considered. For example, in In^{115}, only 79% of the captures in the main resonance at 1.46 ev and 65% of the captures in the resonance at 3.68 ev lead to the 54-min activity of In^{116}; the rest lead to the 14-sec activity. λ_0 refers to the $1/v$-absorption.

The integration over the $1/v$-absorption in Eq. (12.1.4) is easily carried out and when we assume $E_{\max}=\infty$ yields

$$\left.\begin{aligned}C_{1/v}&=2\lambda_0\Phi_{\text{epi}}\mu_a(kT)\sqrt{\frac{kT}{E_C}}\,\delta=2\lambda_0\Phi_{\text{epi}}\,\mu_a(E_C)\,\delta\\&=2\Phi_{\text{epi}}N\sigma^{1/v}_{\text{act}}(E_C)\,d.\end{aligned}\right\} \tag{12.1.5}$$

We have again introduced $d=\delta/\varrho$ and $\sigma=\mu\varrho/N$ and have furthermore written $\sigma^{1/v}_{\text{act}}=\lambda_0\sigma^{1/v}_a$. The contribution of a single resonance is given by

$$\left.\begin{aligned}C&=\Phi_{\text{epi}}\lambda_i\mu^i_{a_0}\delta\int\frac{1}{1+\left(\frac{E-E^i_R}{\Gamma^i/2}\right)^2+2\mu^i_{a_0}\delta}\frac{dE}{E}\\&=\Phi_{\text{epi}}\lambda_i N I^i_{\text{eff}}\,d\end{aligned}\right\} \tag{12.1.6}$$

with

$$I^i_{\text{eff}}=\sigma^i_{a_0}\int\frac{1}{1+\left(\frac{E-E^i_R}{\Gamma^i/2}\right)^2+2Nd\sigma^i_{a_0}}\frac{dE}{E}. \tag{12.1.7}$$

Here I^i_{eff} is identical with the effective resonance integral of a symmetric resonance that was introduced in Sec. 7.4.2, except that σ_p is replaced by $\frac{1}{2Nd}$. Therefore

$$I^i_{\text{eff}}=\frac{I^i_\infty}{\sqrt{1+2Nd\sigma^i_{a_0}}} \tag{12.1.8a}$$

with

$$I^i_\infty=\frac{\pi}{2}\sigma^i_{a_0}\frac{\Gamma^i}{E^i_R}. \tag{12.1.8b}$$

Finally, we obtain

$$C = \Phi_{\text{epi}} N I_{\text{eff act}} d \tag{12.1.9a}$$

with

$$I_{\text{eff act}} = 2\sigma_{\text{act}}^{1/v}(E_C) + \sum_i \frac{\lambda_i I_\infty^i}{\sqrt{1+2Nd\sigma_{a_0}^i}}. \tag{12.1.9b}$$

We can see from Eq. (12.1.9b) that the activation of the foil by the $1/v$-part of the absorption is proportional to the foil's volume. On the other hand, owing to the considerable self-shielding [Eq. (12.1.8a)], the resonances contribute more and more weakly to the activation the thicker the foil is. For thick foils, $2Nd\sigma_{a_0}^i \gg 1$, and the resonance part of $I_{\text{eff act}}$ is proportional to $1/\sqrt{d}$, i.e., the resonance activation of the foil is proportional to the root of its thickness. For "infinitely" thin foils, $Nd\sigma_{a_0}^i \ll 1$, and $I_{\text{eff act}}$ approaches[1]

$$I_{\text{act}} = 2\sigma_{\text{act}}^{1/v}(E_C) + \sum_i \lambda_i I_\infty^i = \int_{E_C}^{\infty} \sigma_{\text{act}}(E)\frac{dE}{E}. \tag{12.1.9c}$$

Here the second expression on the right-hand side is the exact form of I_{act}^∞ while the first expression holds only in the approximation that a decomposition into a $1/v$-part and a series of symmetric Breit-Wigner terms is possible.

Frequently, it is convenient to introduce the "epithermal self-shielding factor"

$$G_{\text{epi}} = \frac{I_{\text{eff act}}}{I_{\text{act}}}. \tag{12.1.10}$$

Fig. 12.1.1 shows BROSES's values of $G_{\text{epi}}(d)$ for gold (lower energy limit $E_C = 0.68$ ev; cf. Sec. 12.2.2). In this figure we compare measured values, an elementary calculation done by the methods developed in this section, and an exact calculation. In an exact calculation, we may no longer neglect scattering and moderation in the foil, we may no longer decompose the cross section as we did in Eq. (12.1.2), we must use a better representation of the φ_0-function, and we must take the Doppler effect into account. In attempting such a calculation, we can fall back on the methods developed for the calculation of resonance absorption in heterogeneous reactors (in this connection see DRESNER or ADLER and NORDHEIM). Fig. 12.1.1 shows that the elementary calculation reproduces the experimentally observed behavior quite well. Similar experiments on indium foils have been performed by TRUBEY, BLOSSER and ESTABROOK and by BROSE and KNOCHE, among others.

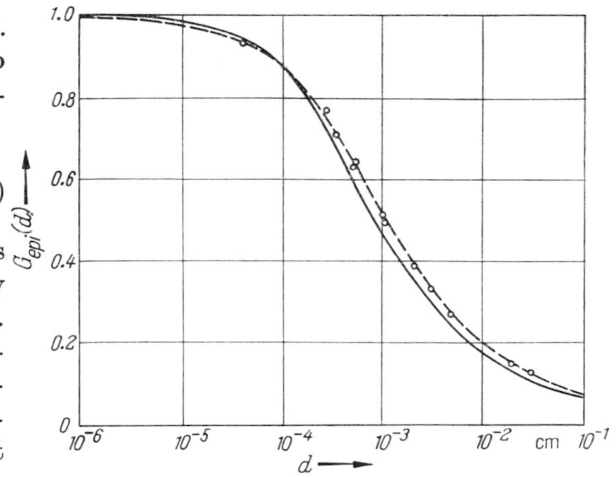

Fig. 12.1.1. The epithermal self-shielding factor G_{epi} for Gold foils as a function of their thickness. o o Measured values; — — — accurate calculation; ——— elementary calculation using Eq. (12.1.9)

[1] For a more precise definition of the resonance integral see Sec. 12.3.1.

It should be added that TRUBEY *et al.* have performed a much more accurate calculation of the epithermal self-shielding factor of a purely absorbing foil than we did above. They start from

$$G_{\text{epi}} = \frac{\frac{1}{2}\int \varphi_0(\mu_a(E)\,\delta)\,\frac{dE}{E}}{\int \mu_a(E)\,\delta\,\frac{dE}{E}}\,. \tag{12.1.11a}$$

Instead of replacing $\varphi_0(\mu_a\delta)$ by its simple rational approximation, its exact form $1-2E_3(\mu_a\delta)$ is retained. Assuming a single resonance and neglecting the $1/v$-part, one obtains after integrating the denominator and putting $x=\frac{E-E_R}{\Gamma/2}$

$$G_{\text{epi}} = \frac{1}{\pi\,\mu_{a_0}\delta}\cdot\int_{-\infty}^{+\infty}\left[\frac{1}{2}-E_3\left(\frac{\mu_{a_0}\delta}{1+x^2}\right)\right]dx\,. \tag{12.1.11b}$$

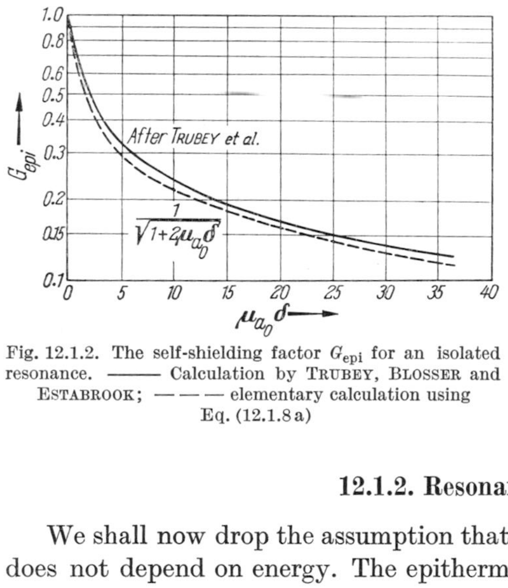

Fig. 12.1.2. The self-shielding factor G_{epi} for an isolated resonance. ——— Calculation by TRUBEY, BLOSSER and ESTABROOK; — — — elementary calculation using Eq. (12.1.8a)

TRUBEY *et al.* have shown that Eq. (12.1.11b) yields

$$G_{\text{epi}} \approx 1+\frac{\mu_{a_0}\cdot\ln(\mu_{a_0}\delta)}{4}-0.3274\,\mu_{a_0}\delta \quad \text{for} \quad \mu_{a_0}\delta \ll 1,$$

$$G_{\text{epi}} \approx \frac{4}{3\sqrt{\pi}}\cdot\frac{1}{\sqrt{\mu_{a_0}\delta}} \quad \text{for} \quad \mu_{a_0}\delta \gg 1.$$

For intermediate values of $\mu_{a_0}\delta$, they have performed the integration numerically. Their result is shown in Fig. 12.1.2 together with our simple approximation $G_{\text{epi}}=1/\sqrt{1+2\mu_{a_0}\delta}$ [cf. Eq. (12.1.8a)].

12.1.2. Resonance Probes

We shall now drop the assumption that the epithermal flux per unit lethargy does not depend on energy. The epithermal foil activation is then given by

$$C = N d\left[\int_{E_C}^{\infty}\sigma_{\text{act}}^{1/v}(E)\,\Phi_{\text{epi}}(E)\,\frac{dE}{E}+\sum_i \Phi_{\text{epi}}(E_i)\,\lambda_i I_{\text{eff}}^i\right]. \tag{12.1.12}$$

Table 12.1.1. *Resonance Detectors (cf. also Fig. 12.1.3a—c)*

Substance	Isotope	$T_{\frac{1}{2}}$	Resonance Energy [ev]	$\int_{0.55\,\text{ev}}^{\infty}\sigma_{\text{act}}(E)\,dE$ [barn]	$\frac{\lambda_i I_\infty^i}{\int_{0.55\,\text{ev}}^{\infty}\sigma_{\text{act}}(E)\,dE}$
In	In^{115}	54.12 min	1.457	2700	~0.96
Au	Au^{197}	2.695 d	4.905	1150	~0.95
W	W^{186}	24 h	18.8	400	~0.98
La	La^{139}	40.2 h	73.5	11	~0.97
Mn	Mn^{55}	2.58 h	337	15.7	~0.88

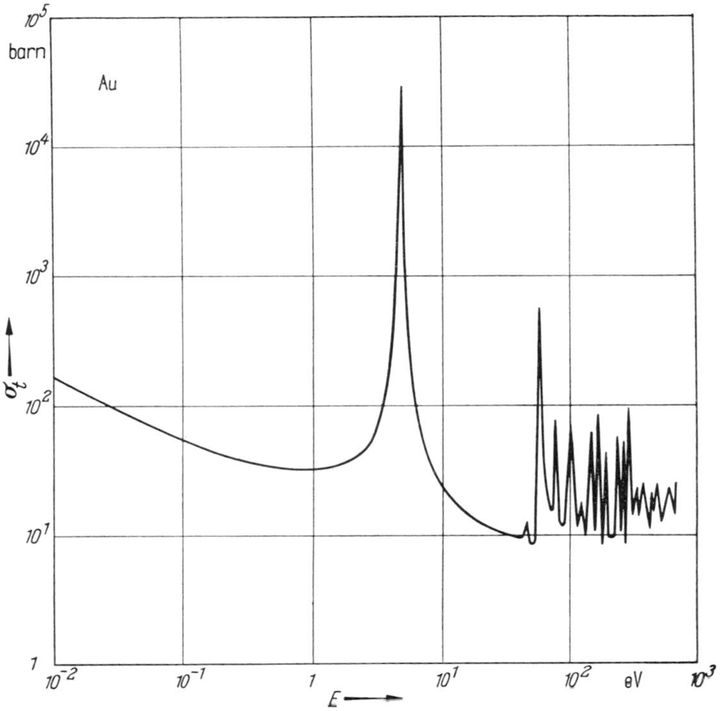

Fig. 12.1.3 a

Fig. 12.1.3a—c. The energy-dependent cross sections of some resonance detector materials. The cross sections of In and Mn were shown in Figs. 1.3.3 and 1.3.2

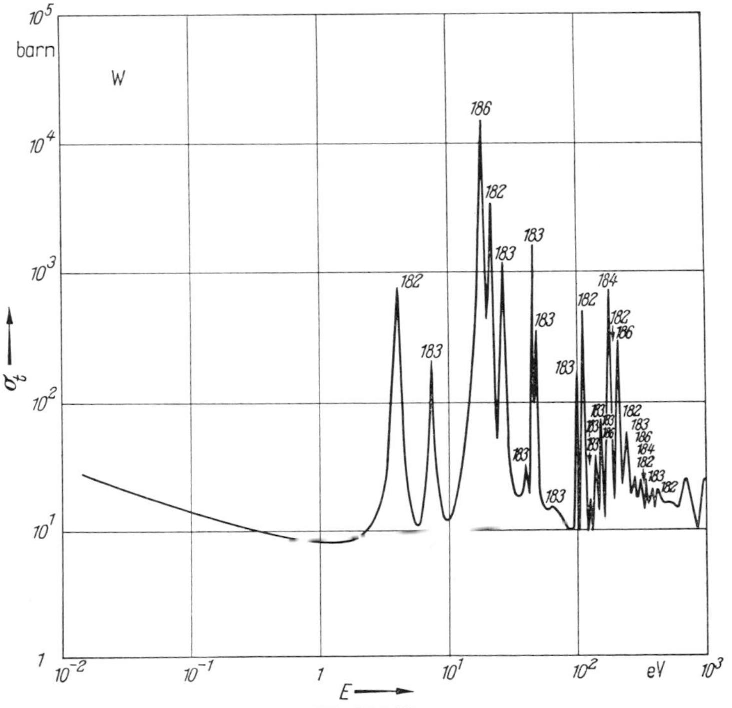

Fig. 12.1.3 b

Frequently, a probe substance has such a large main resonance that the capture processes in this resonance are responsible for the main part of the activation. Then

$$C = N d \Phi_{\mathrm{epi}}(E_i) \lambda_i I_{\mathrm{eff}}^i, \tag{12.1.13}$$

i.e., the activation is proportional to the flux at a particular resonance energy. Such resonance foils are frequently used (under a cadmium cover in order to suppress thermal activation) to make measurements on epithermal neutron fields whose energy distributions deviate from $1/E$. Table 12.1.1 contains some useful data on resonance detectors[1]. The last column gives the fraction of the activation of an infinitely thin foil caused by capture in the main resonance; in all cases it is quite near one. However, the situation becomes less favorable when we consider foils of finite thickness. In these foils, the resonances are self-shielded and indeed more strongly the higher the resonance. $\lambda_i I_{\mathrm{eff}}^i / I_{\mathrm{eff\,act}}$ therefore decreases with increasing foil thickness, as can easily be seen for several foil substances in Fig. 12.1.4. For this reason, we must make a resonance foil as thin as possible or apply corrections, for whose calculation $\Phi_{\mathrm{epi}}(E)$ must at least be roughly known. The sandwich method, which we shall discuss in Sec. 12.1.3, offers a way out of these difficulties. Resonance detectors are unsuitable for absolute measurements because the cross sections at the resonances are not known accurately enough. However, they can be calibrated by activation in a $1/E$-flux of known intensity. In such measurements we must not forget that the cadmium cover used for shielding the foil from thermal neutrons also absorbs some of the resonance neutrons. However, this resonance absorption can be ignored if the same foil and cover are used both for calibration and measurement.

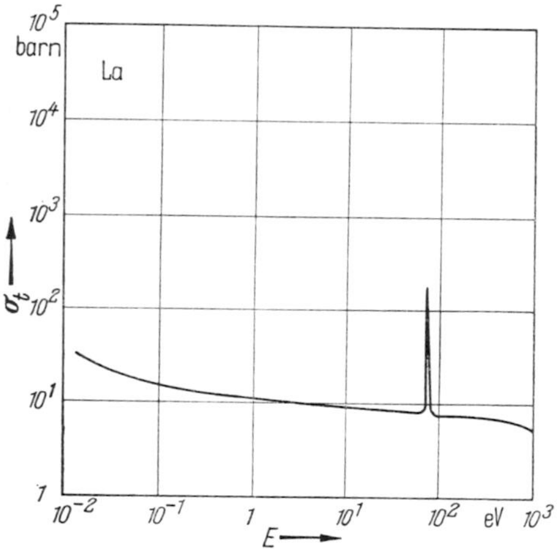

Fig. 12.1.3 c

The activation perturbation in resonance foil measurements is very small. The reason for this is that the average energy loss of a neutron in a collision with an atom of the surrounding moderator is usually larger than the width of the resonance. Thus a neutron that has once traversed the foil with the resonance energy has a very small probability of being scattered back to the foil with nearly the same energy; thus it cannot cause a perturbation by its absence.

12.1.3. The Sandwich Method

We can even make resonance detectors out of thick foils if we make a "sandwich" of three layers of the same material. The two outer layers predominantly absorb neutrons with energies near the main resonance. The inner layer is activated nearly as strongly as the outer ones by the $1/v$-part of the absorption and the

[1] Further substances that are occasionally used as resonance detectors are Co^{59} (132 ev), Na^{23} (2950 ev), Cl^{37} (26 kev), and F^{19} (27 kev).

minor resonances. The fraction of the difference in activation between an outer and inner layer due to the main resonance is thus greater than in the case of a single foil. Fig. 12.1.5 shows the fraction of the activation difference due to the main resonance for sandwiches of identical foils of several probe substances. Note the considerable improvement compared to the individual foils (Fig. 12.1.4).

Fig. 12.1.6 shows the construction of a typical foil sandwich. In order to avoid activation of the inner foil by "unshielded" neutrons incident on the edges

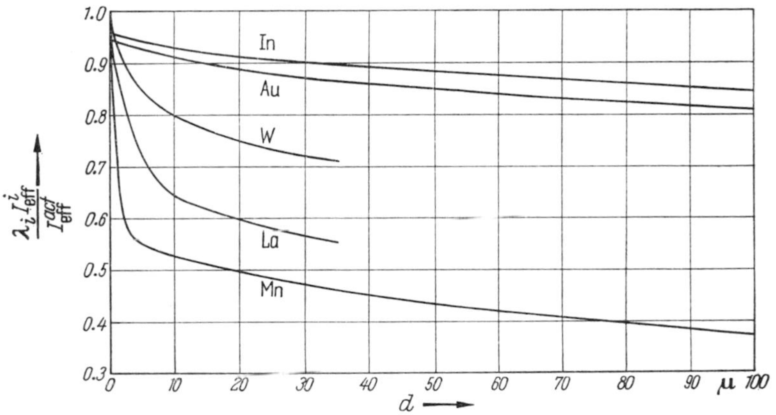

Fig. 12.1.4. The contribution of the main resonance to the activation of various resonance detectors as a function of their thickness

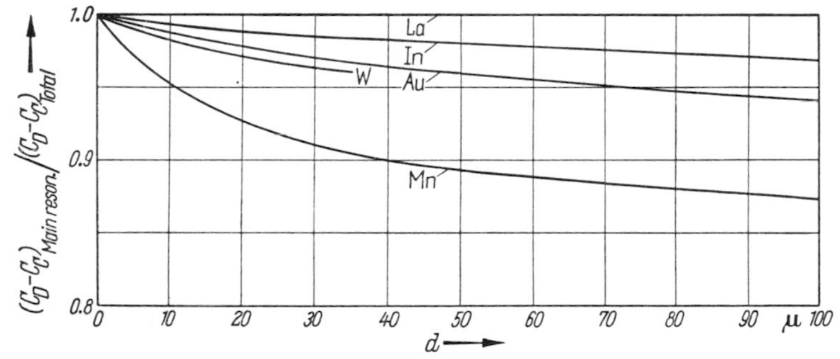

Fig. 12.1.5. The contribution of the main resonance to the activation difference of various sandwich resonance detectors as a function of their thickness

of the sandwich, a shielding ring about 5 mm thick must be used. The foil is irradiated under a cadmium cover. If the foils are so thin that the thermal activation is homogeneous, it will drop out on subtraction, and the sandwich can be irradiated without the cadmium cover. This procedure is not very accurate, however, since in it we form the small difference between two large quantities.

Fig. 12.1.7 shows the activation ratio C_D/C_C (the ratio of the activation of an outer foil to that of an inner foil) for identical foils in a $1/E$-spectrum; this ratio is important in choosing the dimensions of the foil sandwich. In order that $C_D - C_C$ be determined as accurately as possible, C_D/C_C must be as large as possible. It is clear that in gold, indium, and tungsten it is easy to achieve large ratios. In lanthanum and manganese the differences are small, and C_D and C_C must be determined very accurately.

It will be instructive now, as well as in some future discussions, to consider the theory of activation of a foil sandwich. Let us therefore consider a foil of thickness δ and absorption and activation coefficient μ_a that is surrounded on both sides by cover foils (δ', μ_a'). The activation of the inner foil in a monoenergetic, isotropic neutron flux Φ is then given by

$$C_C = \frac{\Phi}{2} \cdot 2 \int_0^{\pi/2} e^{-\frac{\mu_a' \delta'}{\cos\vartheta^*}} \left[1 - e^{-\frac{\mu_a \delta}{\cos\vartheta^*}}\right] \cos\vartheta^* \sin\vartheta^* \, d\vartheta^* . \qquad (12.1.14\text{a})$$

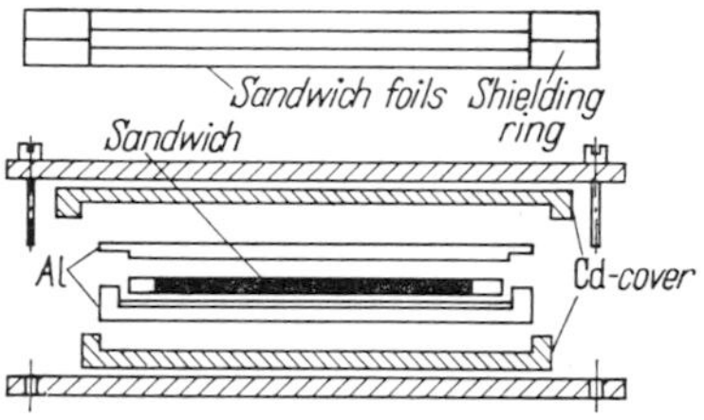

Fig. 12.1.6. The construction of a typical sandwich detector

(Cf. Sec. 11.2.1.) The integration yields

$$\left.\begin{aligned} C_C &= \Phi\left[E_3(\mu_a'\delta') - E_3(\mu_a'\delta' + \mu_a\delta)\right] \\ &= \frac{\Phi}{2}\left[\varphi_0(\mu_a'\delta' + \mu_a\delta) - \varphi_0(\mu_a'\delta')\right]. \end{aligned}\right\} \qquad (12.1.14\text{b})$$

For identical foils of the same material ($\mu_a = \mu_a'$, $\delta = \delta'$),

$$C_C = \frac{\Phi}{2}\left[\varphi_0(2\mu_a\delta) - \varphi_0(\mu_a\delta)\right]. \qquad (12.1.15)$$

We can easily obtains the activity of an outer foil if we note that the activation of the entire sandwich of identical foils is given by

$$C_C + 2C_D = \frac{\Phi}{2}\varphi_0(3\mu_a\delta). \qquad (12.1.16)$$

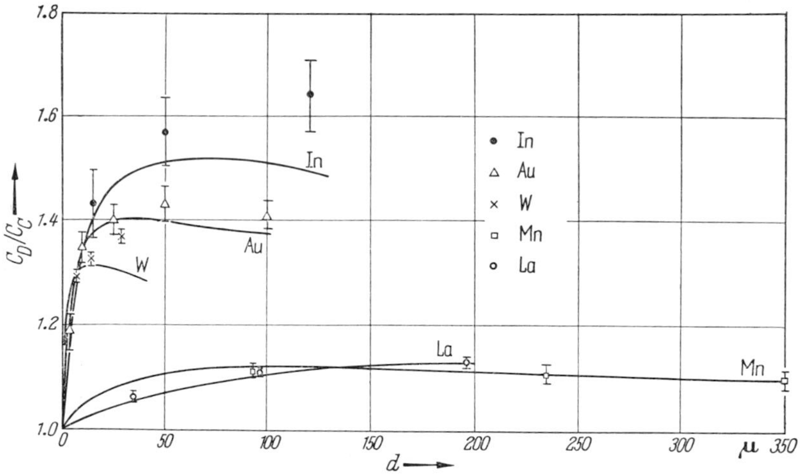

Fig. 12.1.7. The ratio of the activation of an outer foil of a sandwich detector to that of the inner foil as a function of their thickness. ——— Calculated values ; ● △ × □ ○ measurements by EHRET

Then

$$C_D = \frac{\Phi}{4}\left[\varphi_0(3\mu_a\delta) - \varphi_0(2\mu_a\delta) + \varphi_0(\mu_a\delta)\right]. \qquad (12.1.17)$$

We can now integrate C_D and C_C over the $1/E$-spectrum by the methods of Sec. 12.1.1 and form the ratio C_D/C_C. The curves in Fig. 12.1.7 were obtained in this way (with $E_C = 0.68$ ev).

12.2. Simultaneous Thermal and Epithermal Foil Activation

12.2.1. The Cadmium Difference Method

Let us now make the following assumption for the energy distribution of the neutron flux:

$$\Phi(E)=\Phi_{th}\frac{E}{(kT)^2}e^{-E/kT}+\Phi_{\text{epi}}\frac{\Delta(E/kT)}{E}. \tag{12.2.1}$$

Here Φ_{th} and Φ_{epi} are constants, T is the neutron temperature, and $\Delta(E/kT)$ is the "joining" function introduced in Sec. 10.2.1. Then if we neglect scattering in the foil[1], its activation is given by

$$\left.\begin{aligned}C=&\frac{\Phi_{th}}{2}\overline{\frac{\mu_{\text{act}}}{\mu_a}\varphi_0(\mu_a\delta)}+\frac{\Phi_{\text{epi}}}{2}\times\\&\times\int_0^\infty\frac{\mu_{\text{act}}}{\mu_a}\varphi_0\big(\mu_a(E)\,\delta\big)\frac{\Delta(E/kT)}{E}dE\\&=C_{th}+C_{\text{epi}}.\end{aligned}\right\} \tag{12.2.2}$$

In order to separate the thermal and epithermal activation, we now irradiate the foil under a completely sealed cover of from 0.5- to 1.5-mm-thick cadmium. The absorption cross section of cadmium (cf. Fig. 12.2.1) is very high at low energies and drops sharply in the vicinity of 0.5 ev; thus in a rough first approximation we can assume that the cadmium cover captures all the thermal neutrons and transmits all the epithermal neutrons. Thus under cadmium we only measure the epithermal activation; more precisely [cf. Eq. (12.1.14b)], with a cadmium thickness δ^{CD}

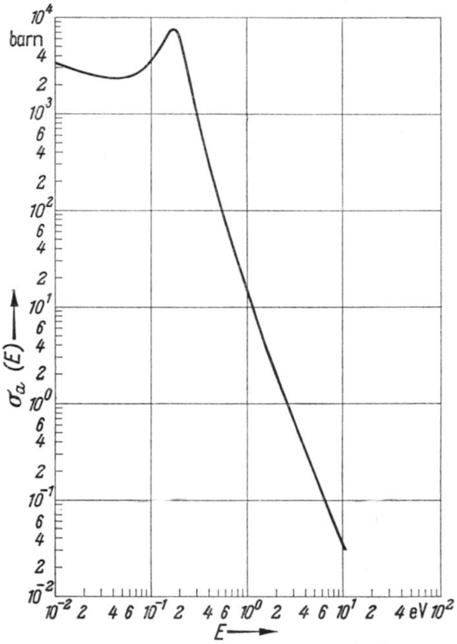

Fig. 12.2.1. The absorption cross section of Cadmium as a function of neutron energy

$$\left.\begin{aligned}C^{CD}=\frac{\Phi_{\text{epi}}}{2}\int_0^\infty\frac{\mu_{\text{act}}}{\mu_a}\big[\varphi_0\big(\mu_a^{CD}(E)\,\delta^{CD}+\mu_a(E)\,\delta\big)-\\-\varphi_0\big(\mu_a^{CD}(E)\,\delta^{CD}\big)\big]\frac{\Delta(E/kT)}{E}dE=\frac{C_{\text{epi}}}{F_{CD}}.\end{aligned}\right\} \tag{12.2.3}$$

Here we have introduced a cadmium correction factor

$$F_{CD}=\frac{\displaystyle\int_0^\infty\frac{\mu_{\text{act}}}{\mu_a}\varphi_0(\mu_a(E)\,\delta)\,\Delta(E/kT)\frac{dE}{E}}{\displaystyle\int_0^\infty\frac{\mu_{\text{act}}}{\mu_a}[\varphi_0(\mu_a^{CD}(E)\,\delta^{CD}+\mu_a(E)\,\delta)-\varphi_0(\mu^{CD}(E)\,\delta^{CD})]\,\Delta(E/kT)\frac{dE}{E}} \tag{12.2.4}$$

[1] In addition, we neglect the (thermal) activation perturbation both here and in what follows.

which takes into account the fact that cadmium captures neutrons between the "cut-off" energy of the epithermal spectrum (≈ 0.1 ev) and the "cadmium cut-off" energy that will be defined later. F_{CD} depends on the thickness of the cadmium cover, the thickness of the foil, the foil substance, and the joining function, which may depend on the moderator substance and under certain circumstances also on the neutron temperature. Figs. 12.2.2 and 3 show correction factors that were calculated for gold and indium foils using JOHANSSON'S joining function (cf. also Sec. 12.2.2).

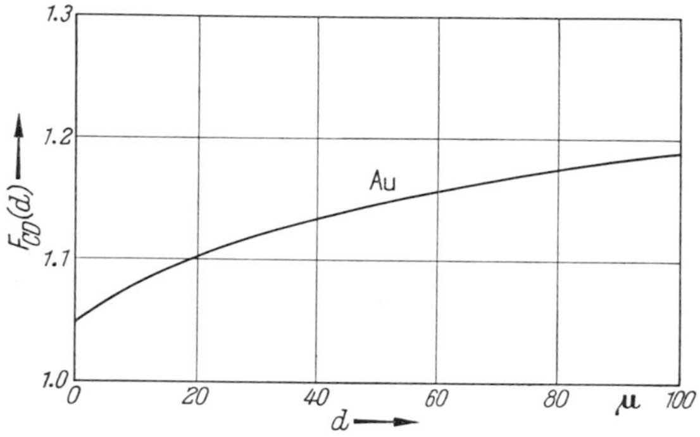

Fig. 12.2.2. The Cadmium correction factor for Gold foils in an isotropic neutron field as a function of Gold foil thickness. Cd filter thickness 1 mm

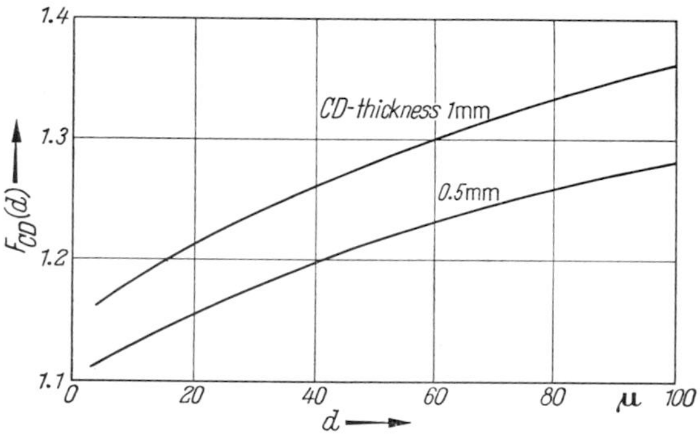

Fig. 12.2.3. The Cadmium correction factor for Indium foils in an isotropic neutron field as a function of Indium foil thickness. Cd filter thickness 0.5 and 1 mm

In order to obtain the purely thermal activation from the total activation of a foil, we must subtract the epicadmium activation multiplied by the cadmium correction factor:

$$C_{th} = C - F_{CD} C^{CD}. \qquad (12.2.5)$$

All the considerations of the eleventh chapter now hold for C_{th}. Since γ-counting measures an activity proportional to the activation, in this case all our results, particularly the correction factors shown in Figs. 12.2.2 and 3, also hold for the activity. In β-counting, electron self-absorption must be taken into account; however, this factor enters the numerator and the denominator of Eq. (12.2.4) in nearly the same way and in first approximation cancels out.

12.2.2. Calculation of F_{CD} for a Thin $1/v$-Absorber

Let the $1/v$-absorber be so thin that self-shielding in the foil can be neglected. In addition, let all absorption processes lead to activation. We show next that the "thermal transparency" T_{th}^{CD} of a sufficiently thick cadmium cover can be neglected. We begin by noting that

$$\left.\begin{aligned} C_{th}^{CD} &= \frac{\Phi_{th}}{2} \int_0^\infty \left[\varphi_0\big(\mu_a^{CD}(E)\,\delta^{CD} + \mu_a(E)\,\delta\big) - \varphi_0\big(\mu_a^{CD}(E)\,\delta^{CD}\big)\right] \frac{E}{kT}\, e^{-E/kT}\, \frac{dE}{kT} \\ &= \Phi_{th} \int_0^\infty \mu_a(E)\,\delta \cdot E_2\big(\mu_a^{CD}(E)\,\delta^{CD}\big)\, \frac{E}{kT}\, e^{-E/kT}\, \frac{dE}{kT} \end{aligned}\right\} \qquad (12.2.6)$$

(since $\mu_a(E)\,\delta \ll 1$). Thus

$$\left.\begin{aligned} T_{th}^{CD} = \frac{C_{th}^{CD}}{C_{th}} &= \frac{\int_0^\infty E_2(\mu_a^{CD}(E)\,\delta^{CD})\sqrt{\frac{E}{kT}}\,e^{-E/kT}\,\frac{dE}{kT}}{\int_0^\infty \sqrt{\frac{E}{kT}}\,e^{-E/kT}\,\frac{dE}{kT}} \\ &= \frac{2}{\sqrt{\pi}}\int_0^\infty E_2\big(\mu_a^{CD}(E)\,\delta^{CD}\big)\sqrt{\frac{E}{kT}}\,e^{-E/kT}\,\frac{dE}{kT}. \end{aligned}\right\} \qquad (12.2.7)$$

Here we set $\mu_a(E)=\mu_a(kT)\sqrt{\frac{kT}{E}}$; $\mu_a(kT)$ then drops out. If we assume the Breit-Wigner formula for the absorption cross section of cadmium, Eq. (12.2.7) can be integrated. Fig. 12.2.4 shows T_{th}^{CD} as a function of the thickness of the cadmium cover for a neutron temperature $T=300$°K. We see that for cover thicknesses >0.5 mm, cadmium is opaque to thermal neutrons; thus the neglect of C_{th}^{CD} in Eq. (12.2.3) is justified. In the following, we shall only admit cadmium thicknesses >0.5 mm.

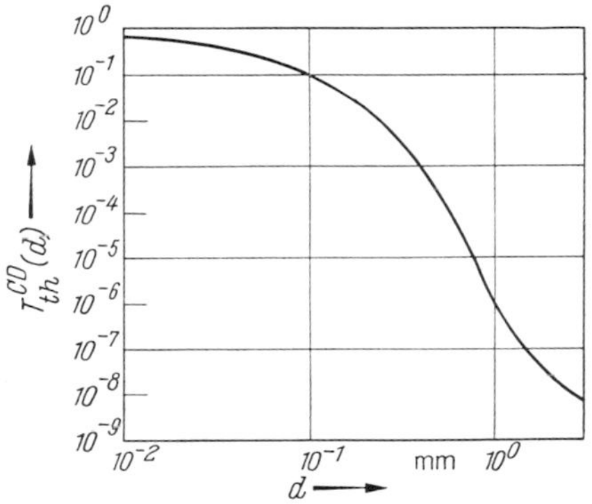

Fig. 12.2.4. The transmission of an isotropic thermal neutron field ($T=300$ °K) through Cadmium covers as a function of Cadmium thickness

In order to determine F_{CD} we next form the integral

$$C_{\text{epi}}^{CD} = C^{CD} = \Phi_{\text{epi}}\int_0^\infty \mu_a(E)\,\delta\,E_2\big(\mu_a^{CD}(E)\,\delta^{CD}\big)\,\frac{\Delta(E/kT)}{E}\,dE. \qquad (12.2.8)$$

If the cadmium cross section were an ideal step function, i.e., if it were extremely high below a cut-off energy E_{CD} and zero above it, then C^{CD} would be given by

$$C_{CD} = \Phi_{\text{epi}}\int_{E_{CD}}^\infty \mu_a(E)\,\delta\,\frac{dE}{E}. \qquad (12.2.9)$$

It has been assumed here that the joining function $\Delta(E/kT)$ is already equal to one for $E \geqq E_{CD}$. Now the cadmium cross section only approximates a step function. Nevertheless, it is customary to use Eq. (12.2.9) to define a cadmium cut-off energy. In other words, we require

$$\int_{E_{CD}}^\infty \mu_a(E)\,\frac{dE}{E} = \int_0^\infty \mu_a(E)\,E_2\big(\mu_a^{CD}(E)\,\delta^{CD}\big)\,\frac{\Delta(E/kT)}{E}\,dE \qquad (12.2.10\text{a})$$

and determine E_{CD} as a function of the cadmium thickness δ^{CD} and the joining function $\Delta(E/kT)$. If $\mu_a(E)=\mu_a(kT)\sqrt{\frac{kT}{E}}$, the integration on the left-hand side can be carried out, and we obtain

$$E_{CD} = \frac{4}{\left(\int_0^\infty \Delta(E/kT)\,E_2(\mu_a^{CD}(E)\,\delta^{CD})\,\frac{dE}{E^{\frac{3}{2}}}\right)^2}. \qquad (12.2.10\text{b})$$

The main contribution to the integral in the denominator comes (for cadmium that is not too thin) from the energy range above 0.3 ev; in this range, all suggested variants of the joining function are very close to one, and E_{CD} therefore depends only slightly on the choice of joining function. Fig. 12.2.5 shows E_{CD} as a function of the cadmium thickness calculated using the joining function suggested by JOHANSSON *et al.* for D_2O at $T=300$ °K. When $d=1$ mm, $E_{CD}=0.68$ ev, which we shall frequently use as a standard value. Then for our $1/v$-absorber we simply have

$$C^{CD}=\Phi_{\text{epi}}\, 2\mu_a(E_{CD})\,\delta\,. \tag{12.2.11}$$

In addition, we must calculate

$$C_{\text{epi}}=\Phi_{\text{epi}}\int\limits_0^\infty \mu_a(E)\,\delta\,\frac{\Delta(E/kT)}{E}\,dE\,. \tag{12.2.12}$$

The function $\Delta(E/kT)$ also approximates a step function, and in this case too we introduce a cut-off energy E_{ET} defined in the following way:

$$\int\limits_{E_{ET}}^\infty \mu_a(E)\,\frac{dE}{E}=\int\limits_0^\infty \mu_a(E)\,\frac{\Delta(E/kT)}{E}\,dE\,. \tag{12.2.13a}$$

If we set $\mu_a(E)=\mu_a(kT)\sqrt{\frac{kT}{E}}$, we can carry out the left-hand integration and obtain

$$E_{ET}=\frac{4}{\left(\int\limits_0^\infty \Delta(E/kT)\,\frac{dE}{E^{\frac{3}{2}}}\right)^2}\,. \tag{12.2.13b}$$

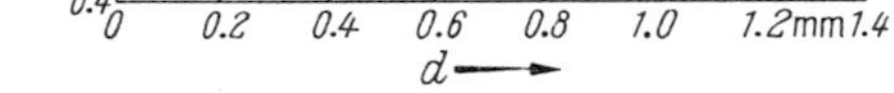

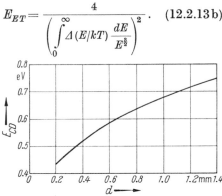

Fig. 12.2.5. The Cadmium cut-off energy for a thin $1/v$-absorber in an isotropic neutron field as a function of the Cadmium filter thickness

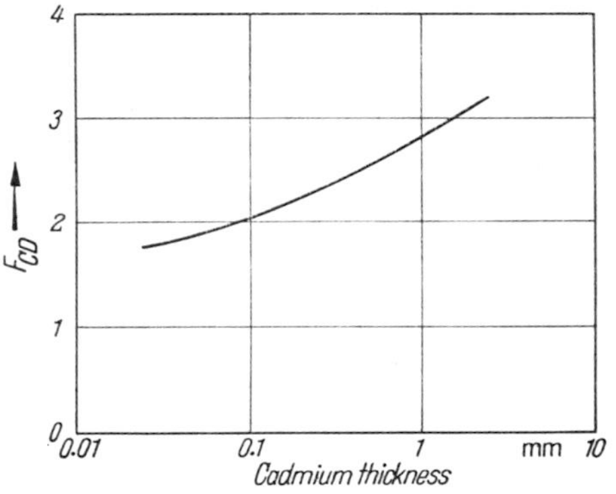

Fig. 12.2.6. The Cadmium correction factor for a thin $1/v$-foil in an isotropic neutron flux as a function of the Cadmium thickness

Since the joining function only depends on E/kT, we can introduce the new variable $x=E/kT$ in the integral and obtain

$$E_{ET}=\frac{4kT}{\left(\int\limits_0^\infty \Delta(x)\,x^{-\frac{3}{2}}\,dx\right)^2}=\mu\, kT\,. \tag{12.2.13c}$$

$\mu\approx 3.6$ (± 0.4) for the joining function in D_2O (according to JOHANSSON *et al.*) and 3.4 (± 0.3) in graphite (according to COATES). Then, for a $1/v$-absorber

$$C_{\text{epi}}=\Phi_{\text{epi}}\, 2\mu_a(E_{ET})\,\delta \tag{12.2.14}$$

and the cadmium correction factor for a thin $1/v$-absorber is given by

$$F_{CD} = \frac{C_{\text{epi}}}{C^{CD}} = \sqrt{\frac{E_{CD}}{\mu\, kT}}\,. \tag{12.2.15}$$

Thus, for example, when the cadmium cover is 1 mm thick and $T=293.6$ °K, $F_{CD} \approx 2.75$. This factor, which is shown in Fig. 12.2.6 as a function of cadmium thickness, is quite a bit larger than the correction factors shown in Figs. 12.2.2 and 3 for gold and indium. The difference arises from the fact that in the latter substances the epithermal activation is largely due to the resonances, which all lie at such high energies that they are only slightly affected by the absorption in cadmium.

Some of the results of this section can be applied to thin foils made of substances whose cross sections can be decomposed into a $1/v$-part and a resonance part as long as the first resonance lies at a high energy where cadmium no longer absorbs neutrons. Then

$$C_{\text{epi}} = \Phi_{\text{epi}} \int_{E_{ET}}^{\infty} \mu_{\text{act}}(E)\, \frac{dE}{E} \tag{12.2.16a}$$

and

$$C^{CD} = \Phi_{\text{epi}} \int_{E_{CD}}^{\infty} \mu_{\text{act}}(E)\, \frac{dE}{E}\,. \tag{12.2.16b}$$

Thus

$$F_{CD} = \frac{\int\limits_{E_{ET}}^{\infty} \sigma_{\text{act}}(E)\, \frac{dE}{E}}{\int\limits_{E_{CD}}^{\infty} \sigma_{\text{act}}\, \frac{dE}{E}} = 1 + \frac{\int\limits_{E_{ET}}^{E_{CD}} \sigma_{\text{act}}(E)\, \frac{dE}{E}}{\int\limits_{E_{CD}}^{\infty} \sigma_{\text{act}}(E)\, \frac{dE}{E}} \tag{12.2.16c}$$

with E_{ET} and E_{CD} given respectively by Eqs. (12.2.13c) and (12.2.10b). Eq. (12.2.16c) is a very good approximation for gold just as it is for dysprosium. For indium, on the other hand, E_{CD} and E_{ET} must be redefined since they are affected by the first resonance at 1.46 ev.

12.2.3. The Cadmium Ratio and the Determination of the Epithermal Flux

The cadmium ratio is defined as the ratio of the activity of the bare foil to its activity when entirely enclosed in cadmium:

$$R_{CD} = \frac{C}{C^{CD}} = \frac{C_{th} + C_{\text{epi}}}{\frac{1}{F_{CD}} \cdot C_{\text{epi}}}\,. \tag{12.2.17a}$$

Obviously

$$\left.\begin{aligned} R_{CD} - F_{CD} &= \frac{C_{th}}{\frac{1}{F_{CD}}\, C_{\text{epi}}} \\ &= \frac{\Phi_{th}\, \overline{\frac{\mu_{\text{act}}}{\mu_a}\, \varphi_0(\mu_a \delta)}}{\Phi_{\text{epi}} \int\limits_0^{\infty} \frac{\mu_{\text{act}}}{\mu_a} \left[\varphi_0(\mu_a^{CD}(E)\, \delta^{CD} + \mu_a(E)\, \delta) - \varphi_0(\mu_a^{CD}(E)\, \delta^{CD})\right] \frac{\Delta(E/kT)}{E}\, dE}\,. \end{aligned}\right\} \tag{12.2.17b}$$

For a thin foil,

$$R_{CD} - F_{CD} = \frac{\Phi_{th}}{\Phi_{\text{epi}}} \frac{\frac{\sqrt{\pi}}{2} \cdot g(T) \sqrt{\frac{293.6^\circ}{T}}\, \sigma_{\text{act}}(2200 \text{ m/sec})}{\int\limits_{E_{CD}}^{\infty} \sigma_{\text{act}}(E)\, \frac{dE}{E}}. \tag{12.2.18a}$$

In particular, for a thin $1/v$-absorber,

$$R_{CD} - F_{CD} = \frac{\Phi_{th}}{\Phi_{\text{epi}}} \cdot \frac{\sqrt{\pi}}{4} \sqrt{\frac{E_{CD}}{kT}}. \tag{12.2.18b}$$

Thus we can determine the epithermal flux from a measurement of the cadmium ratio if Φ_{th} is known. If we use a thin $1/v$-detector, such as a BF_3-counter, we

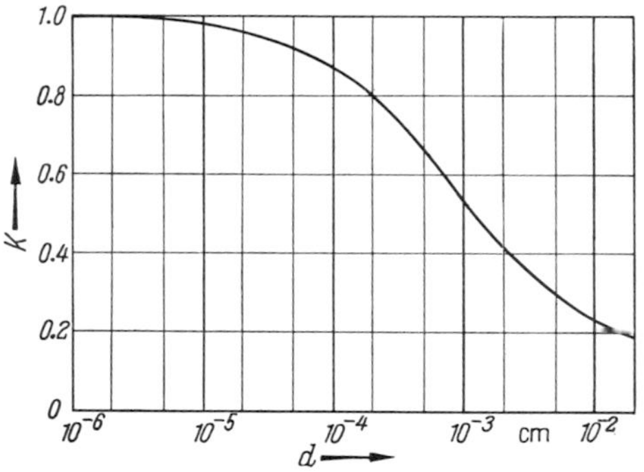

Fig. 12.2.7. Calculated values of the correction factor K [cf. Eq. (12.2.19)] for Gold foils as a function of their thickness. Cd cover thickness 1 mm

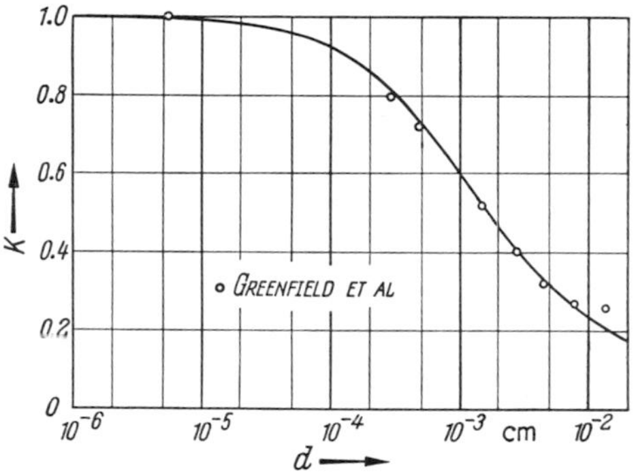

Fig. 12.2.8. ——— Calculated values of the correction factor K [cf. Eq. (12.2.19)] for Indium foils as a function of their thickness. Cd cover thickness 1 mm. ooo Experimental results by GREENFIELD *et al.* for K_β. A Cd cover thickness of 0.5 mm was used in these experiments

need no additional data (save values of E_{CD} and F_{CD}, which we obtain using the methods of Sec. 12.2.2). If we use thin foils of another material, we must know the epicadmium resonance activation integral and the thermal activation cross section. Values for gold and indium are given in Table 12.1.1 and in Sec. 11.1.2; except in the case of gold these data are not very accurate. When thick foils are used, we must be careful to take into account the self-shielding of the thermal and above all of the epithermal neutrons. These effects can be described by introduction of a correction factor K, which we define as follows:

$$\left.\begin{aligned} K &= \frac{(R_{CD} - F_{CD})_{\delta=0}}{(R_{CD} - F_{CD})_{\delta}} \\ &= \frac{\frac{\sqrt{\pi}}{2}\, g(T) \sqrt{\frac{293.6^\circ}{T}}\, \sigma_{\text{act}}(2200 \text{ m/sec})}{\overline{\frac{\mu_{\text{act}}}{\mu_a}\, \varphi_0(\mu_a \delta)}} \times \\ &\times \frac{\int\limits_0^{\infty} \frac{\mu_{\text{act}}}{\mu_a} \left[\varphi_0(\mu_a^{CD} \delta^{CD} + \mu_a \delta) - \varphi_0(\mu_a^{CD} \delta^{CD})\right] \frac{\Delta(E/kT)}{E}\, dE}{\int\limits_{E_{CD}}^{\infty} \sigma_{\text{act}}(E)\, \frac{dE}{E}}. \end{aligned}\right\} \tag{12.2.19}$$

For a given foil substance, K is a function of δ^{CD}, δ, and T and also depends on the choice of the joining function. We can also define a factor K_β, which includes the effects of β-self absorption. We can measure K by determining the cadmium ratios of foils of various thicknesses in the same field, and we can calculate it by the methods of Secs. 11.2 and 12.2.1. Figs. 12.2.7 and 12.2.8 show some measured and calculated values of K and K_β for gold and indium foils. Using these data, we determine $\Phi_{th}/\Phi_{\text{epi}}$ as follows: First we measure C and C^{CD} with a thick foil. Then we determine $R_{CD}-F_{CD}$ and multiply it with either K or K_β. Eq. (12.2.18a) now holds for the resulting quantity, and if the cross sections are known, the ratio $\Phi_{th}/\Phi_{\text{epi}}$ can be determined.

12.2.4. The Two-Foil Method

If we attempt to separate the thermal and epithermal activation of a foil in a multiplying medium the following difficulty may arise. The cadmium perturbs the thermal flux in the neighbourhood of the foil and in so doing also affects the source of fast neutrons (via the fission rate). Thus the epithermal flux being measured in the presence of cadmium is affected in a way that is difficult to describe, especially when a large number of cadmium-covered foils are simultaneously exposed. The two-foil method offers a way out of this difficulty. In this method, two foils of different materials are fastened together in a package and simultaneously irradiated. Foil 1 is made of a material with a small resonance integral (Cu, Mn), while foil 2 is made of a material whose resonance integral is large compared to its thermal cross section (e.g., gold). Then

$$\left.\begin{aligned} C_1&=a_{11}\Phi_{th}+a_{12}\Phi_{\text{epi}}\\ C_2&=a_{21}\Phi_{th}+a_{22}\Phi_{\text{epi}}.\end{aligned}\right\}\tag{12.2.20a}$$

Here the a_{ik} are constants which depend on the cross sections of the foil substances, the foil thicknesses, the neutron temperature, and the joining function. If the a_{ik} are known, Φ_{th} and Φ_{epi} can be determined by separately counting C_1 and C_2:

$$\left.\begin{aligned} \Phi_{th}&=b_{11}C_1+b_{12}C_2\\ \Phi_{\text{epi}}&=b_{21}C_1+b_{22}C_2\end{aligned}\right\}\tag{12.2.20b}$$

with

$$b_{11}=\frac{a_{22}}{a_{22}a_{11}-a_{12}a_{21}},\qquad b_{12}=\frac{-a_{12}}{a_{22}a_{11}-a_{12}a_{21}},\tag{12.2.20c}$$

etc. The a_{ik} and b_{ik} can be calculated from the specifications of the foils, but the recommended procedure is to determine them experimentally by measurements with and without cadmium in a standard spectrum for which Φ_{th} and Φ_{epi} are known. The standard spectrum should resemble the spectrum of the field being investigated since the a_{ik} and b_{ik} depend on the neutron temperature and the joining function. The accuracy with which Φ_{th} and Φ_{epi} can be determined by this method depends on the spectrum and as a rule is less than the accuracy of the cadmium difference method.

12.2.5. Thin Foils: Description According to Westcott's Convention

The reaction rate ψ [cm^{-3} sec^{-1}] in a foil that is so thin that self-shielding effects can be neglected is given by

$$\psi=N d\left[\Phi_{th}\frac{\sqrt{\pi}}{2}g(T)\sqrt{\frac{293.6^\circ}{T}}\,\sigma_a(2200\text{ m/sec})+\Phi_{\text{epi}}\int_0^\infty\sigma_a(E)\,\frac{\Delta(E/kT)}{E}\,dE\right].\tag{12.2.21}$$

Following a suggestion of WESTCOTT, we can represent ψ in a considerably simpler and more elegant way if we make use of a somewhat different formal description of the spectrum than we have hitherto been using. According to WESTCOTT, we write

$$\psi = N \cdot d \cdot n \cdot \hat{\sigma}_a \cdot v_0 \tag{12.2.22a}$$

(with $v_0 = 2200$ m/sec). Here n is the total neutron density

$$n = \int_0^\infty n(v)\,dv = \int_0^\infty \frac{\Phi(E)}{v}\,dE \tag{12.2.22b}$$

and $\hat{\sigma}_a$ is an effective cross section defined by

$$n v_0 \hat{\sigma}_a = \int_0^\infty n(v)\,v\,\sigma_a(v)\,dv = \int_0^\infty \Phi(E)\,\sigma_a(E)\,dE. \tag{12.2.22c}$$

To aid us in calculating $\hat{\sigma}_a$, let us introduce some auxiliary quantities; let n_{th} be the thermal density

$$n_{th} = \frac{\Phi_{th}}{\bar{v}} = \frac{\sqrt{\pi}}{2}\,\frac{\Phi_{th}}{v_T} \tag{12.2.23a}$$

and let n_{epi} be the epithermal density

$$n_{\text{epi}} = \Phi_{\text{epi}} \int_0^\infty \frac{\Delta(E/kT)}{E v}\,dE = \frac{2}{\sqrt{\mu}}\,\frac{\Phi_{\text{epi}}}{v_T}. \tag{12.2.23b}$$

[Cf. Eq. (12.2.13c).] Obviously, $n = n_{th} + n_{\text{epi}}$. Introducing the abbreviations $f = n_{\text{epi}}/n$, $1 - f = n_{th}/n$ and using Eq. (12.2.21), we find that

$$\left.\begin{aligned} \hat{\sigma}_a &= (1-f)\,g(T)\,\sigma_a(v_0) + f\,\frac{\sqrt{\mu}\,v_T}{2 v_0} \int_0^\infty \frac{\Delta(E/kT)}{E}\,\sigma_a(E)\,dE \\ &= g(T)\,\sigma_a(v_0) + f\left[\frac{\sqrt{\mu}\,v_T}{2 v_0} \int_0^\infty \frac{\Delta(E/kT)}{E}\,\sigma_a(E)\,dE - g(T)\,\sigma_a(v_0)\right]. \end{aligned}\right\} \tag{12.2.24a}$$

Because $\int_0^\infty \frac{\Delta(E/kT)}{vE}\,dE = \frac{2}{\sqrt{\mu}}\,\frac{1}{v_T}$, we can also write the term in the square brackets in the form

$$\frac{\sqrt{\mu}\,v_T}{2 v_0} \int_0^\infty \left(\sigma_a(E) - g(T)\,\sigma_a(v_0)\,\frac{v_0}{v}\right) \frac{\Delta(E/kT)}{E}\,dE. \tag{12.2.24b}$$

The expression in the braces defines an *excess resonance integral*

$$I' = \int_0^\infty \left(\sigma_a(E) - g(T)\,\sigma_a(v_0)\,\frac{v_0}{v}\right) \frac{\Delta(E/kT)}{E}\,dE. \tag{12.2.25}$$

In terms of it,

$$\hat{\sigma}_a = g(T)\,\sigma_a(v_0) + \frac{\sqrt{\mu}}{2}\,\frac{v_T}{v_0}\,f I'.$$

Two additional auxiliary quantities will now be introduced. First, let the excess resonance integral I' be expressed by

$$s = \frac{1}{\sigma_a(v_0)} \frac{2}{\sqrt{\pi}} \sqrt{\frac{T}{293.6^\circ}}\, I'. \tag{12.2.26}$$

The quantity s is dimensionless and depends on the neutron temperature. Second, let us introduce a new spectral index

$$r = f \frac{\sqrt{\pi\mu}}{4}. \tag{12.2.27 a}$$

From the definition of f and Eqs. (12.2.23a and b) it follows that

$$r = \frac{\sqrt{\pi\mu}/4}{1 + \frac{\sqrt{\pi\mu}}{4} \frac{\Phi_{th}}{\Phi_{\text{epi}}}}. \tag{12.2.27 b}$$

$\Phi_{th}/\Phi_{\text{epi}}$ is $\gg 1$ for "soft" spectra, and thus $r \approx \Phi_{\text{epi}}/\Phi_{th}$. Finally, $\hat{\sigma}_a$ can be expressed in terms of r and s as follows:

$$\hat{\sigma}_a = \sigma_a(v_0)\,[g + r s]. \tag{12.2.28}$$

Detailed tables of g and s as functions of the neutron temperature can be found in Westcott (cf. also Campbell and Freemantle). For a pure $1/v$-absorber, $g=1$ and $s=0$; thus $\hat{\sigma}_a = \sigma_a(v_0)$. In a pure thermal neutron field, $r=0$ and thus $\sigma_a = \sigma_a(v_0)\cdot g$. For most nuclides, the values of s and I' do not depend on the joining function, for as long as there are no resonances at very low energies the integrand in Eq. (12.2.25) practically vanishes in the region where $\Delta(E/kT)$ changes rapidly. Lutetium and plutonium are noteworthy exceptions to this rule; we shall come back to them later (Sec. 15.2).

It is instructive to rewrite the formulas for the cadmium ratio in the language of the Westcott convention. The activation of a thin $1/v$-detector under cadmium is given by (cf. also Sec. 12.2.2),

$$C^{CD} = N d \sigma_a(v_0)\, \Phi_{\text{epi}} \int_{E_{CD}}^{\infty} \frac{v_0}{v} \frac{dE}{E} \tag{12.2.29 a}$$

or after integration $\left(\text{using } \Phi_{\text{epi}} = \frac{2}{\sqrt{\pi}} v_T n r\right)$

$$C^{CD} = N d n \sigma_a(v_0)\, v_0 \frac{4}{\sqrt{\pi}} r \sqrt{\frac{kT}{E_{CD}}}. \tag{12.2.29 b}$$

Because $C = N d \cdot n \sigma_a(v_0)\, v_0$ it follows that

$$R_{CD} = \frac{C}{C^{CD}} = \frac{1}{r} \cdot \frac{\sqrt{\pi}}{4} \sqrt{\frac{E_{CD}}{kT}}. \tag{12.2.29 c}$$

Thus we can determine r from the cadmium ratio of a $1/v$-absorber just as we can determine $\Phi_{th}/\Phi_{\text{epi}}$ [cf. Eq. (12.2.18b)].

The activation of a substance whose cross section follows the $1/v$-law at low energies ($g \approx 1$) but which has resonances at higher energies (>1 ev) is given by

$$C^{CD} = N d \sigma_a(v_0)\, n v_0 \left[\frac{4}{\sqrt{\pi}} r \sqrt{\frac{kT}{E_{CD}}} + r s\right]. \tag{12.2.30 a}$$

Thus

$$R_{CD}=\frac{1+rs}{r\left[s+\frac{4}{\sqrt{\pi}}\sqrt{\frac{kT}{E_{CD}}}\right]}. \tag{12.2.30b}$$

The Westcott convention is particularly well suited to the evaluation of neutron temperature measurements with foils (Sec. 15.2) and to the description of resonance integral measurements by the two-spectrum method (Sec. 12.3.3). Its generalization to thick foils is problematic.

12.3. The Measurement of Resonance Integrals

Many of the questions dealt with in this chapter have significance far beyond the theory of foil activation: In the approximation in which the neutron field is considered to be made up of a thermal Maxwell spectrum and an epithermal $1/E$-spectrum, all reaction rates can be expressed in terms of average thermal cross sections and resonance integrals, no matter whether absorption, fission, activation, or scattering is involved. There is therefore a considerable interest in the direct measurement of resonance integrals both at infinite dilution and in situations where self-shielding plays an important role. In this section, we shall familiarize ourselves with the most important methods of measuring the resonance integrals of thin foils. However, first we must make our definition of the resonance integral more precise.

12.3.1. Precise Definition of the Resonance Integral

According to Goldstein *et al.*, the epicadmium resonance integral of a substance is *defined* by the relation

$$I_x=\int_{0.55\,\text{ev}}^{2\,\text{Mev}}\sigma_x(E)\frac{dE}{E} \tag{12.3.1}$$

where $x=a$ means absorption, $x=f$ means fission, $x=s$ means scattering, and $x=$act means activation. We must be careful to distinguish these quantities from the quantities I_x^{exp} that we actually *measure* in a reaction-rate experiment with an infinitely thin, cadmium-covered foil, viz.,

$$I_x^{\text{exp}}=\int_{E_{CD}}^{E_{\max}}\sigma_x(E)\frac{f(E)}{E}\,dE. \tag{12.3.2}$$

Here the notation $f(E)$ indicates that the spectrum with which the actual measurements are carried out may deviate from the ideal $1/E$-behavior. E_{CD} depends on the cadmium thickness, the absorber substance, and the geometry of the neutron field or the arrangement of the irradiation apparatus. We have already calculated E_{CD} for a thin foil with a $1/v$-cross section in an isotropic neutron field in Sec. 12.2.3. The values of Fig. 12.3.1 apply for a neutron beam normally incident on a foil with a $1/v$-cross section. For other geometries see Halperin *et al.* In general, E_{CD} will be different from the 0.55 ev required above. The upper limiting energy $E_{\max}$ will also generally be different from 2 Mev, but as a rule this difference has no appreciable effect on I_x. In any case, we must nearly always apply calculated corrections to the measured resonance integrals.

In addition, we can define an *excess resonance integral* I'_x [cf. Eq. (12.2.25)]

$$I'_x = \int_0^{2\,\text{Mev}} \left(\sigma_x(E) - g_x(T)\sigma_x(v_0)\frac{v_0}{v}\right)\frac{\Delta(E/kT)}{E}\,dE. \tag{12.3.3}$$

In contrast to the definition of the epicadmium integral, this definition is not unique since it contains the joining function, which depends on the neutron field. As we previously saw, for many substances the exact form of $\Delta(E/kT)$ plays no role. Here, too, we must distinguish between the defined and directly measured values.

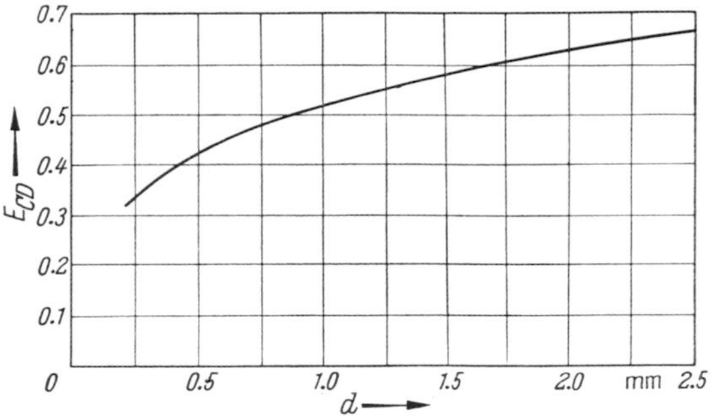

Fig. 12.3.1. The Cadmium cut-off energy (in ev) for a thin $1/v$-absorber in a monodirectional neutron beam as a function of the Cadmium filter thickness

The symbols I_x and I'_x introduced here should not be confused with our earlier notation I^k_{eff} and I^k_∞, which represented the resonance absorption integral of a symmetric Breit-Wigner resonance with and without self-shielding, respectively.

12.3.2. Determination of the Resonance Integral from the Cadmium Ratio

Determination of the cadmium ratio offers a simple way of comparing resonance integrals. The reaction rates of a thin foil with and without a cadmium cover are determined, and from them the cadmium ratio is formed. For a thin foil, we have

$$R_{CD} - F_{CD} = \frac{\Phi_{th}}{\Phi_{\text{epi}}}\,\frac{\frac{\sqrt{\pi}}{2}\,g_x(T)\sqrt{\frac{293.6^\circ}{T}}\,\sigma_x(v_0)}{I_x^{\text{exp}}}. \tag{12.3.4}$$

F_{CD} can now be calculated with Eq. (12.2.16c) (with σ_x replacing σ_{act}) with the help of a preliminary estimate of the resonance integral. If one measures the cadmium ratio of a standard substance for which the resonance integral is already known then

$$\frac{(R_{CD} - F_{CD})_{\text{standard}}}{R_{CD} - F_{CD}} = \frac{I_x^{\text{exp}}}{I_{\text{standard}}}\;\frac{[g(T)\,\sigma(v_0)]_{\text{standard}}}{g_x(T)\,\sigma_x(v_0)} \tag{12.3.5}$$

and one can determine I_x^{exp} as long as the thermal cross section and the g-factor are known with sufficient accuracy. In order to obtain I_x from I_x^{exp}, we must calculate the contribution of the energy range between 0.55 ev and E_{CD}. In doing this, we must fall back on knowledge of the energy variation of $\sigma_x(E)$ obtained from differential measurements. Sometimes a correction must also be made for the upper limiting energy of the spectrum, but as a rule such a correction is unnecessary. Finally, under some circumstances deviations from the $1/E$-spectrum must be taken into account (cf. HARDY *et al.*).

The measurements can be carried out either in a neutron field or in a neutron beam. The reaction rates that are measured are usually activation rates, but by use of small fission chambers fission rates can be measured, and by observation of the prompt γ-radiation capture rates can be measured.

The 2.695-day activity of gold $\left(\sigma_a(E)=\sigma_{\mathrm{act}}(E)\right)$ usually serves as a standard in such measurements. By comparison with the cadmium ratio of a boron counter, which does not depend on the cross section since $\sigma\sim 1/v$ [cf. Eq. (12.2.18b)], JIRLOW and JOHANSSON found that $\int_{0.5\,\mathrm{ev}}^{\infty} \sigma_a(E)\frac{dE}{E}=1535\pm 40$ barns. A calculation based on very carefully measured Breit-Wigner parameters gave $I_a=1566$ barns, in satisfactory agreement with the measured value.

Numerous results of resonance integral experiments can be found in Appendix II.

In the older literature, Eq. (12.3.5) was frequently writen with F_{CD} equal to one, i.e., the contribution of the neutrons between the thermal cut-off energy E_{ET} and the cadmium cut-off energy E_{CD} to the reaction rate of the bare foil was neglected. This is incorrect, particularly when the cadmium ratio being measured is not very large compared to one. We may only set F_{CD} equal to one if we define the "effective" thermal cross section in a suitable way (cf. HALPERIN *et al.*).

12.3.3. Other Methods of Measuring Resonance Integrals

In the *two-spectrum* method, the reaction rate of an (infinitely thin) foil is measured in two spectra of different hardnesses. For example, the reaction rate in a neutron field with $r=0$ is given by

$$R_0=n_0 v_0 \sigma(v_0) g \tag{12.3.6a}$$

while in a field with $r\neq 0$, it is given by

$$R_1=n_1 v_0 \sigma(v_0)\,[g+rs]. \tag{12.3.6b}$$

If we measure n_1/n_0 with a thin BF_3-counter, we obtain

$$\frac{R_1 n_0}{R_0 n_1}=1+\frac{rs}{g} \tag{12.3.6c}$$

from which we can find s if r has already been determined by measurement of the cadmium ratio of a $1/v$-detector. With the help of Eq. (12.2.26), I' can be determined from s. This method has the advantage of being free of the details of the cadmium absorber. It fails for substances that have an extremely small value of s.

Many measurements of resonance absorption integrals have been carried out with the pile oscillator method.

SPYVAK has described a transmission method for the measurement of resonance integrals. A neutron beam penetrates a thin foil of the resonance absorber and enters a detector whose detection efficiency does not depend on neutron energy. The transmission (the ratio of the counting rate Z^+ with the foil to the counting rate Z^0 without it) is then given by

$$\frac{Z^+}{Z^0}=1-Nd\int \sigma_t(E)\frac{dE}{E}=1-Nd\,I^{\mathrm{exp}}. \tag{12.3.7}$$

In the arrangement used by Spyvak, the neutrons scattered in the foil could still be detected; therefore, the transmission measurements yielded the resonance absorption integral I_a^{exp}.

Chapter 12: References[1]

Adler, F. T., and L. W. Nordheim: GA-377 (1958).
Dresner, L.: Resonance Absorption in Nuclear Reactors. Oxford: Pergamon Press 1960.
} Calculation of the Effective Resonance Integral.

Brose, M.: Dissertation, Karlsruhe 1962. Cf. also Nukleonik 6, 134 (1964).
Brose, M., and M. Knoche: Unpublished Karlsruhe report (1963).
Trubey, D. K., T. V. Blosser, and G. M. Estabrook: ORNL-2842, 204 (1959).
} Measurement and Calculation of Epithermal Self-Shielding Factors.

Coates, M. S.: Brussels: Euratom 1961. Neutron Time-of-Flight Methods, p. 233.
Johansson, E., E. Lampa, and N. G. Sjöstrand: Arkiv Fysik 18, 513 (1960).
} Measurement of the Joining Function $\Delta(E/kT)$.

Bigham, R. M., and C. B. Pearce: AECL-1228 (1961).
Ehret, G.: Atompraxis 7, 393 (1961).
Golubev, V. I., et al.: Atomnaya Energiya 11, 522 (1961).
Stewart, H. B., and G. B. Gavin: Chapter 6 in The Physics of Intermediate Spectrum Reactors, Ed. J. Stehn, USAEC 1959.
} Resonance Detectors; the Sandwich Method.

Dayton, I. E., and W. G. Pettus: Nucleonics 15, No. 12, 86 (1957).
Farinelli, U.: Neutron Dosimetry Vol. I, p. 195. Vienna: International Atomic Energy Agency, 1963.
Hickman, G. D., and W. B. Leng: Nucl. Sci. Eng. 12, 523 (1962).
Stoughton, R. W., J. Halperin, and M. P. Lietzke: Nucl. Sci. Eng. 6, 441 (1959).
Stoughton, R. W., and J. Halperin: Nucl. Sci. Eng. 15, 314 (1963).
} Effective Cadmium Cut-Off Energies.

Martin, D. H.: Nucleonics 13, No. 3, 52 (1955) (Cadmium Correction Factors).

Fastrup, B.: Riso Report No. 1, 1959; J. Nucl. Energy A 11, 143 (1963).
Fastrup, B., and J. Olsen: Riso Report No. 43, 1962; cf. also Neutron Dosimetry, Vol. I, p. 227. Vienna: International Atomic Energy Agency, 1963.
Greenfield, M. A., R. L. Koontz, and A. A. Jarett: Nucl. Sci. Eng. 2, 246 (1957).
Männer, W., and T. Springer: Nukleonik 1, 337 (1959).
} The Cadmium Ratio and the Epithermal Neutron Flux.

Campbell, C. G., and R. C. Freemantle: AERE, RP/R 2031 (1957).
Nisle, R. G.: Neutron Dosimetry, Vol. I, p. 111. Vienna: International Atomic Energy Agency, 1963.
Stoughton, R. W., and J. Halperin: Nucl. Sci. Eng. 6, 100 (1959).
Westcott, C. H., W. H. Walker, and T. K. Alexander: Geneva 1958 P/202, Vol. 16, p. 70.
Westcott, C. H.: AECL 1101 (1960).
} Conventions for Reaction Rates in Mixed Neutron Fields.

Goldstein, H., *et al.*: EANDC-12, 1961 (Definition of the Resonance Integral).

Hardy, J., D. Klein, and G. G. Smith: Nucl. Sci. Eng. 9, 341 (1961).
Harris, S. P., C. O. Muehlhause, and G. E. Thomas: Phys. Rev. 79, 11 (1950).
Jirlow, K., and E. Johansson: J. Nucl. Energy A 11, 101 (1960).
Johnston, F. J., J. Halperin, and R. W. Stoughton: J. Nucl. Energy A 11, 95 (1960).
Macklin, R. L., and H. S. Pomerance: Progr. Nucl. Energy Ser. I, Vol. 1, 179 (1956).
Spivak, P. E., *et al.*: Geneva 1955 P/659, Vol. 5, p. 91.
Tatersall, R. B., *et al.*: J. Nucl. Energy A 12, 32 (1960).
} Measurement of Resonance Integrals.

[1] Cf. footnote on p. 53.

13. Threshold Detectors for Fast Neutrons

Certain nuclear reactions, such as $(n, 2n)$ reactions, inelastic scattering, and some (n, p) and (n, α) reactions, occur only when the neutrons have energies above a particular threshold energy. In many cases, these reactions lead to a radioactivity which can be measured after the irradiation. With the help of threshold detectors in the form of foils, we can use these reactions to determine the flux, the energy spectrum, or the dose of fast neutrons. Such foils are of special significance in measurements in bulk media (reactors); as a rule, the detectors considered in Chapter 3 are preferred in measurements in free neutron beams.

13.1. General Facts about Threshold Detectors

The activation of a disc-shaped threshold detector foil that is not too thick is given by

$$C = N d \int_0^\infty \sigma_{\text{act}}(E)\, \Phi(E)\, dE, \tag{13.1.1}$$

for the cross sections at high energies are so small that (for reasonable foil thicknesses) scattering and self-shielding can be neglected. The central difficulty in all threshold detector measurements is that the energy dependences of the fast neutron flux $\Phi(E)$ and the cross section $\sigma_{\text{act}}(E)$ are usually quite complicated. The energy dependence of the activation cross section is known in some cases, but with the exception of the special case of monoenergetic neutrons from an accelerator, the energy dependence of the flux is not known exactly. One must therefore try to obtain additional information about the spectrum. We shall discuss the problem of evaluating threshold detector measurements in detail in Sec. 13.2, but to begin with we shall use the simple concept of the "effective threshold energy".

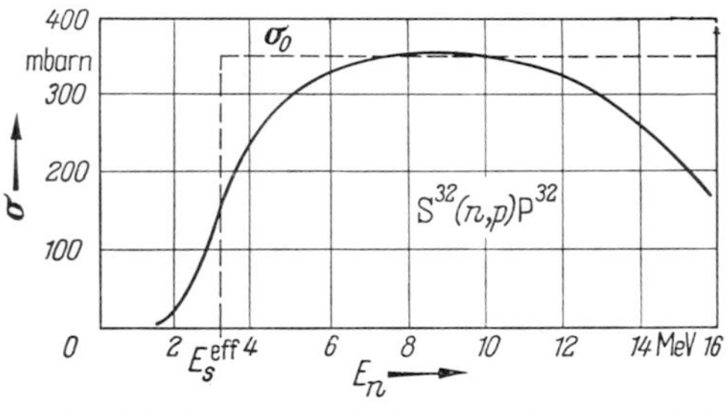

Fig. 13.1.1. The cross section for the reaction $S^{32}(n, p)P^{32}$ as a function of neutron energy. In this and the following figures, the smooth curve represents an interpolation between many different measurements

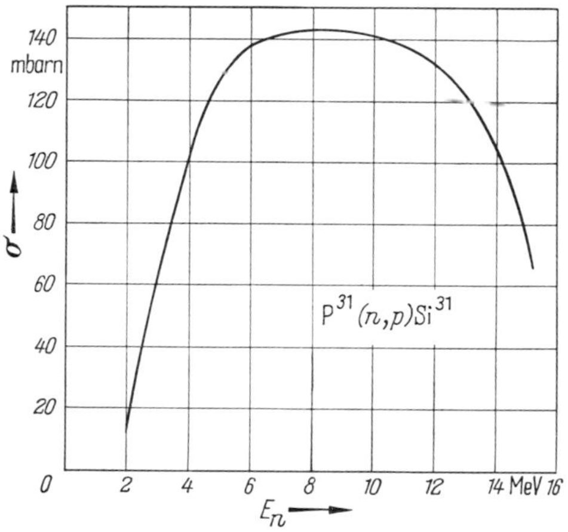

Fig. 13.1.2. The cross section for the reaction $P^{31}(n, p)Si^{31}$ as a function of neutron energy

13.1.1. Definition of the "Effective Threshold Energy"

The activation cross section of an ideal threshold detector should have the following kind of step-function behavior, namely, it should be zero below the threshold energy E_s and equal to σ_0 above E_s. The activation would then be proportional to $\sigma_0 \int_{E_s}^{\infty} \Phi(E)\, dE$, i.e., the measurement would then yield the integrated flux above E_s. From measurements with various substances, we could then obtain the integrated flux as a function of energy, and by differentiation finally obtain $\Phi(E)$. Now the actual cross sections exhibit this step-function behavior only in a very rough approximation [cf. Fig. 13.1.1 for the reaction $S^{32}(n, p)P^{32}$]. However, we can define an effective threshold energy E_s^{eff} in the following way. Let σ_0 be the "plateau" value the cross section reaches at high energies. Then let us set

$$\left.\begin{aligned} &\int_0^{\infty} \sigma_{\text{act}}(E)\, \Phi(E)\, dE \\ &\quad = \sigma_0 \int_{E_s^{\text{eff}}}^{\infty} \Phi(E)\, dE, \end{aligned}\right\} \qquad (13.1.2)$$

i.e., let us choose the effective threshold energy to give the true reaction rate. This definition of E_s^{eff} clearly depends on the form of $\Phi(E)$; in other words, E_s^{eff} is not a universal constant. However, as a rough first approximation the neutron spectrum in reactors is a fission neutron spectrum above 0.5 Mev, and it is conventional to define E_s^{eff} for all substances using the neutron spectrum $N(E)$ of thermal fission of U^{235} for $\Phi(E)$ in Eq. (13.1.2). Values of E_s^{eff} obtained in this way are given in Tables 13.1.1—3, which also contain various other data on threshold

Table 13.1.1. *Substances for (n, p) Threshold Detectors*

Element	Target Nucleus (Abundance)	Reaction Product (Half-Life)	E_s [Mev]	σ_0 [mbarn]	E_s^{eff} [Mev]	$\bar{\sigma}$ [mbarn]	Practical Detector Substances	Emitted Radiation	Minor Activities (Threshold Energy [Mev] and Half-Life)
P	P^{31} (100%)	Si^{31} (157 min)	0.716	140	3.0	30	Phosphor Containing Glasses	β^-	P^{30} (12.7; 2.6 min) Al^{28} (2.0; 2.3 min) Al^{29} (13.6; 6.6 min) P^{32} (thermal; 14.2 d)
S	S^{32} (95%)	P^{32} (14.2 d)	0.954	350	3.2	66	Tablets of Flowers of Sulfur	β^-	P^{33} (8.7; 25 d) Si^{31} (1.32; 157 min) S^{35} (thermal; 87 d)
Ni	Ni^{58} (68%)	Co^{58} (72 d)	— ($Q=0.399$)	600	3.45	100	Metallic Ni	γ, (EC), β^+	Ni^{57} (12.0; 36 h) Co^{61} (0.5; 1.65 h) Ni^{56} (thermal; 2.56 h)
Fe	Fe^{54} (5.82%)	Mn^{54} (290 d)	— ($Q=0.094$)	400	3.75	53	Metallic Fe	γ, (EC)	Cr^{51} (—; 27.8 d) Mn^{56} (2.95; 154.5 min)
Al	Al^{27} (100%)	Mg^{27} (9.393 min)	1.89	80	5.30	3.5	Metallic Al	γ, β^-	Al^{28} (thermal; 2.3 min) Na^{24} (3.26; 14.97 h)
Fe	Fe^{56} (91.3%)	Mn^{56} (154.5 min)	2.95	130	7.70	0.97	Metallic Fe	γ, β^-	Cr^{51} (—; 27.8 d) Mn^{54} (—; 290 d)
Mg	Mg^{24} (78.8%)	Na^{24} (14.97 h)	4.9	200	8.00	1.2	Metallic Mg	γ, β^-	—

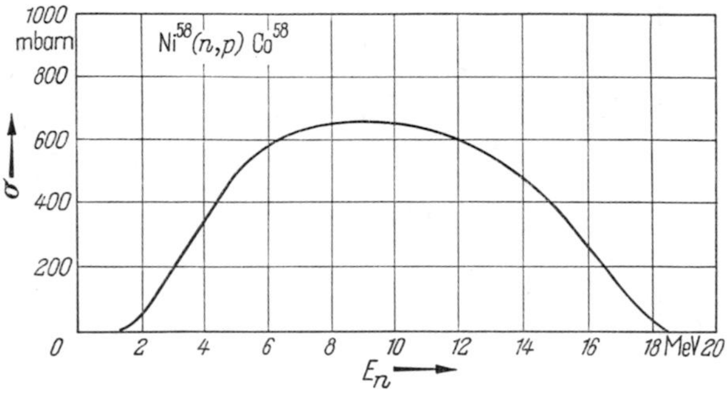

Fig. 13.1.3. The cross section for the reaction $Ni^{58}(n, p)Co^{58}$ as a function of neutron energy

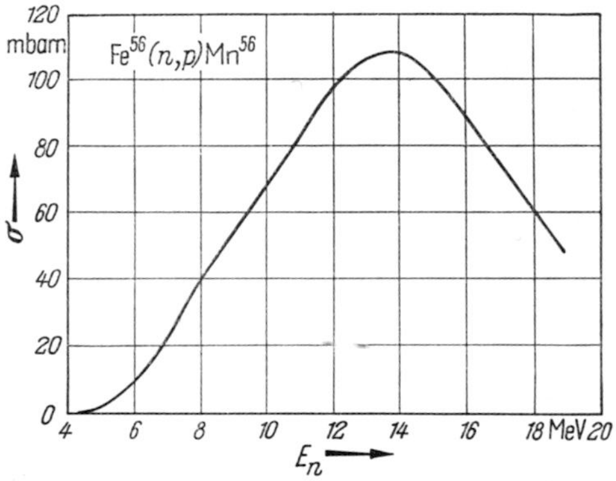

Fig. 13.1.4. The cross section for the reaction $Fe^{56}(n, p)Mn^{56}$ as a function of neutron energy

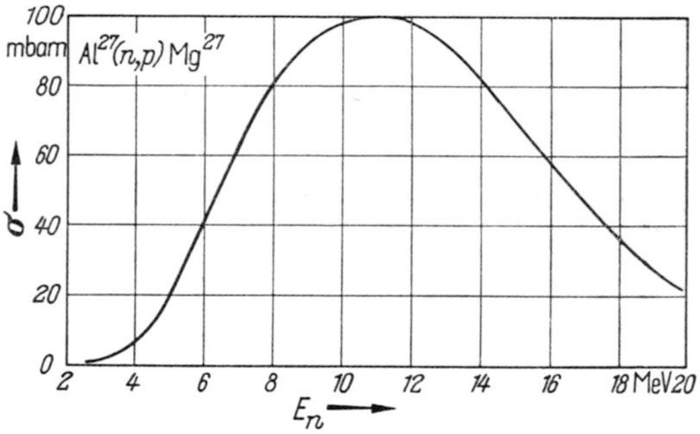

Fig. 13.1.5. The cross section for the reaction $Al^{27}(n, p)Mg^{27}$ as a function of neutron energy

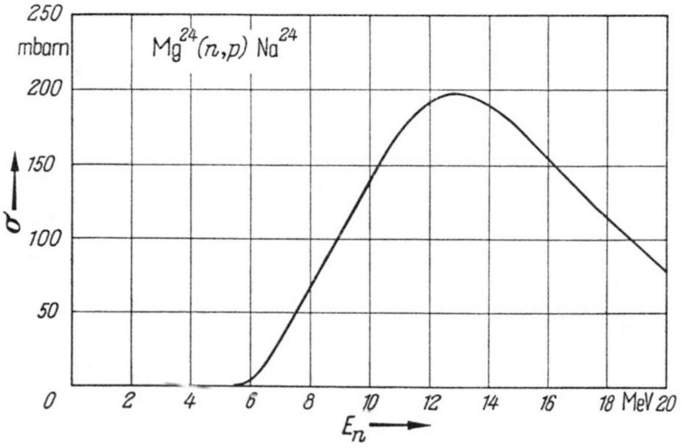

Fig. 13.1.6. Thee ross section for the reaction $Mg^{24}(n, p)Na^{24}$ as a function of neutron energy

detectors. σ_0 is the plateau value, or in those cases where no plateau exists the maximum value, of the activation cross section. These values lose their meaning when the foil is irradiated in a spectrum that differs strongly from a fission neutron spectrum, a situation that arises in many practical cases. Nevertheless, they are quite useful since they specify the energy range in which the foil may be used. Later we shall show how the concept of the effective threshold energy may be modified.

Also given in the tables is $\bar{\sigma}$, the activation cross section averaged over the fission spectrum $N(E)$; $\bar{\sigma}$ can be calculated from $\sigma_{act}(E)$ or directly measured by activation in the flux behind a converter plate. The following relation exists between $\bar{\sigma}$, σ_0, E_s^{eff}, and $\sigma_{act}(E)$:

$$\left.\begin{aligned} &\int_0^\infty N(E)\,\sigma_{act}(E)\,dE \\ &\quad=\sigma_0\int_{E_s^{eff}}^\infty N(E)\,dE \\ &\quad=\bar{\sigma}\int_0^\infty N(E)\,dE\,. \end{aligned}\right\} \qquad (13.1.3)$$

13.1.2. Substances for Threshold Detectors

Table 13.1.1 contains useful data on (n, p) threshold detectors; the substances are arranged according to increasing threshold energy E_s^{eff}. The fourth

Table 13.1.2. *Substances for (n, n'), (n, α), and $(n, 2n)$ Threshold Detectors*

Element	Target Nucleus (Abundance) Type of Reaction	Reaction Product (Half-Life)	E_s [Mev]	σ_0 [mbarn]	E_s^{eff} [Mev]	$\bar{\sigma}$ [mbarn]	Practical Detector Substances	Emitted Radiation	Minor Activities (Threshold [Mev] and Half Life)
Rh	Rh^{103} (100%) (n, n')	Rh^{103m} (57 min)	0.040	1500	0.9	1093	Metallic Rh	γ, e^-	Rh^{104m} (thermal; 4.4 min) Rh^{102} (9.5; 210 d)
In	In^{115} (95.8%) (n, n')	In^{115m} (4.5 h)	0.335	350	1.65	171	Metallic In	γ, e^-	In^{116m_1} (thermal; 54.1 min) In^{113m} (0.393; 1.7 h) Ag^{113} (9.5; 5.3 h) Ag^{112} (—; 3.4 h)
Al	Al^{27} (100%) (n, α)	Na^{24} (14.97 h)	3.26	120	8.15	0.61	Metallic Al	γ, β^-	Mg^{27} (1.89; 9.39 min)
Cu	Cu^{65} (30.91%) $(n, 2n)$	Cu^{64} (12.8 h)	10.1	1000	11.7	0.31	Metallic Cu	γ, β^+, β^-	Co^{61} (10.5; 1.65 h) Ni^{65} (—; 2.56 h) Cu^{64} (thermal; 12.8 h)
Cu	Cu^{63} (69.09%) $(n, 2n)$	Cu^{62} (9.8 min)	10.9	800	13.2	0.073	Metallic Cu	β^+	Co^{60m} (—; 10.5 min) Cu^{66} (thermal; 5.1 min) Co^{62} (—; 13.9 min) Ni^{65} (—; 2.56 h) Co^{61} (—; 1.65 h)
Ni	Ni^{58} (68.0%) $(n, 2n)$	Ni^{57} (37 h)	12.0	80	14	4×10^{-3}	Metallic Ni	γ, β^+	Ni^{56} (thermal; 2.56 h) Co^{61} (0.5; 1.65 h)

column gives the (true) threshold energy at which the process being considered becomes energetically possible. Owing to the presence of the Coulomb barrier, E_s is nearly always very much smaller than E_s^{eff}. The values of σ_0 in the fifth column

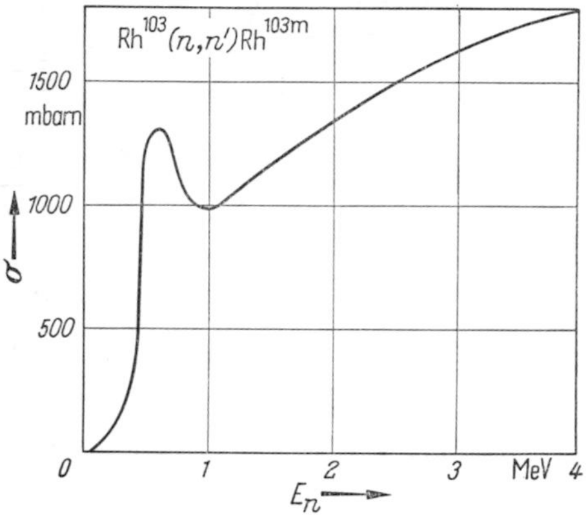

Fig. 13.1.7. The cross section for the reaction $Rh^{103}(n, n')Rh^{103m}$ as a function of neutron energy

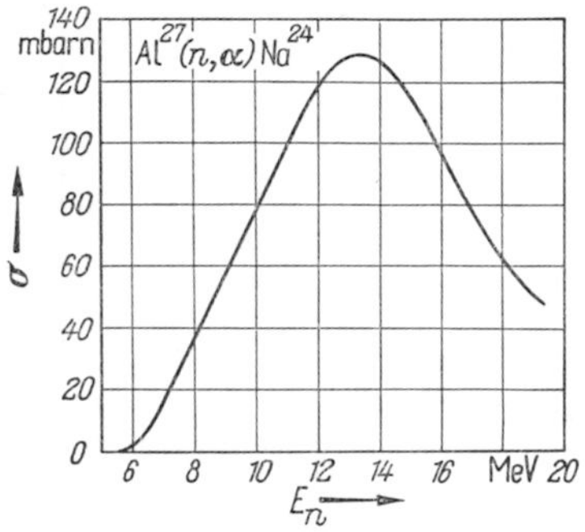

Fig. 13.1.9. The cross section for the reaction $Al^{27}(n, \alpha)Na^{24}$ as a function of neutron energy

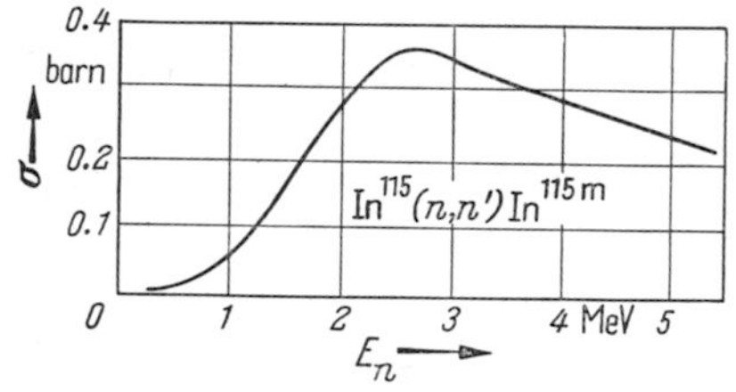

Fig. 13.1.8. The cross section for the reaction $In^{115}(n, n')In^{115m}$ as a function of neutron energy

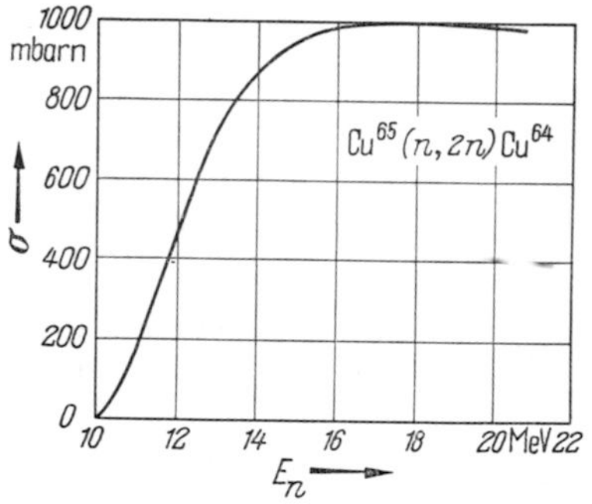

Fig. 13.1.10. The cross section for the reaction $Cu^{65}(n, 2n)Cu^{64}$ as a function of neutron energy

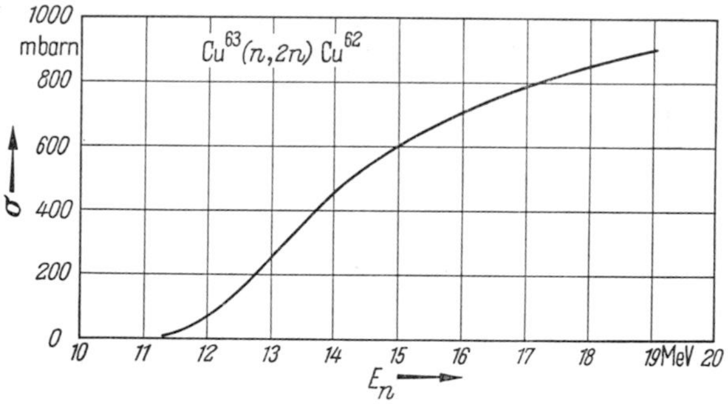

Fig. 13.1.11. The cross section for the reaction $Cu^{63}(n, 2n)Cu^{62}$ as a function of neutron energy

were taken from the curves shown in Figs. 13.1.1—6[1]. With these values and the values of $\bar{\sigma}$ in the seventh column, which were taken from a tabulation of Beaugé, E_s^{eff} was calculated with the help of Eq. (13.1.3). Eq. (2.6.1) was used for the fission spectrum. In addition, the table contains data on those minor activities

[1] $\sigma(E)$ for the reaction $Fe^{54}(n, p)Mn^{54}$ is not shown since measured values are available only at 2 and 14 Mev. σ_0 has been taken to be σ(14 Mev) in Table 13.1.1.

that occur after irradiation of the natural element whose half-lives are not negligibly small compared to that of the main activity. The foils are irradiated under cadmium covers in order to suppress activation by thermal neutrons. Several foil substances have particularly long half-lives ($Fe^{54}(n, p)Mn^{54}$, $Ni^{58}(n, p)Co^{58}$) and for this reason are well suited to the time integration of the flux necessary in determinations of the neutron dose.

Table 13.1.2 contains the corresponding data for additional threshold detectors that are based on (n, n'), (n, α), and $(n, 2n)$ reactions. These detectors are complementary to the (n, p) detectors insofar as the effective threshold energy is concerned because they allow measurements with either very much larger or very much smaller thresholds. Figs. 13.1.7—11 show the cross sections of the relevant reactions as functions of the neutron energy. The values of $\bar{\sigma}$ again come from the tabulation of BEAUGÉ; E_s^{eff} and σ_0 were obtained in the same way as for the (n, p) detectors.

Neither the energy-dependent cross sections in Figs. 13.1.7—11 nor the average values $\bar{\sigma}$ in Tables 13.1.1 and 2 are particularly accurate; in some cases individual measurements deviate from one another by as much as 50%. The values of $\bar{\sigma}$ for the $S^{32}(n, p)P^{32}$ and $P^{31}(n, p)Si^{31}$ reactions are probably among the most trustworthy. Measurements and tabulations of $\bar{\sigma}$-values can be found in MELLISH, in ROCHLIN, in PASSELL and HEATH, and in BEAUGÉ. LISKIEN and PAULSEN, BAYHURST and PRESTWOOD, and BEAUGÉ give information on the energy-dependent cross sections. BYERLY has tabulated the properties of many other substances that can be used as threshold detectors.

13.1.3. Fission Chambers as Threshold Detectors

As we saw in the discussion of fission cross sections in the first and third chapters, many of these cross sections have a very marked step structure with thresholds in the neighborhood of 0.3—1.3 Mev; this suggests that we use fission detectors similarly to the (n, p), (n, α), $(n, 2n)$, and (n, n') activation detectors for the measurement of fast neutron fluxes. Table 13.1.3 gives some useful data on fission detectors. The cross sections as functions of energy have already been shown in Fig. 3.3.1.

Table 13.1.3. *Fission Detectors as Threshold Detectors*

Isotope	E_s[Mev]	σ_0 [mbarn]	E_s^{eff} [Mev]	$\bar{\sigma}$ [mbarn]
U^{234}	≈ 0.3	1500	0.62	1200
Np^{237}	≈ 0.4	1500	0.87	1100
U^{236}	≈ 0.7	850	1.25	520
Th^{232}	≈ 1.3	140	1.40	28
U^{238}	≈ 1.3	606	1.55	310

As a rule, we determine the fission rate in a fission chamber. For this purpose, particularly small chambers which can easily be introduced into a neutron field have been developed. Multiple fission chambers which make it possible to simultaneously determine the fission rate in several substances have been built. In the construction of such chambers, care must be taken that the fissionable material be present in extremely pure form. For example, if a U^{238} deposit contains only 0.7% of U^{235}, upon irradiation in a typical thermal reactor about 99% of all fissions occur in U^{235}.

One can also determine the fission rate from the activity of the fission products. To do so, one can either count the fission foils directly after irradiation or surround them during irradiation with so-called "catcher foils" and count the fission product activity in the latter after irradiation (cf. KÖHLER and ROMANOS).

13.2. Evaluation of Threshold Detector Measurements

There are three different groups of methods for evaluating threshold measurements. There are "mathematical" methods in which one tries to determine an unknown neutron spectrum from measurements with several detectors (Sec. 13.2.1). There are "semiempirical" methods, in which the spectrum is also determined but with the help of additional assumptions about its form. Finally, there are the cases in which the neutron spectrum is already known from calculation. In these cases, measurements with threshold detectors serve to verify the calculated spectral distribution and eventually to fix the absolute value of the fast flux (cf. KÖHLER). Frequently in the investigation of fast reactor systems use is made of threshold detectors and especially of the spectral indicators discussed in Sec. 13.1.3; we shall not go into this application of threshold detectors here.

The methods of determining unknown spectra that we plan to explain in some detail in what follows have only been carefully worked out in a few cases and because of the large uncertainties in the cross sections probably only give exact results in favorable cases.

13.2.1. "Mathematical" Methods

(a) The Multigroup Method. The activation of a threshold detector k is proportional to $A_k = \int_0^\infty \sigma^k \Phi(E)\,dE$. If the energy range is divided into a series of energy groups, then

$$A_k = \sum_{j=1}^{N} \sigma_j^k \Phi_j \tag{13.2.1}$$

where Φ_j is the flux integrated over the j-th group and σ_j^k is a suitable average value of σ^k in the same group. If we now expose M threshold detectors ($k=1, 2, 3, \ldots M$), Eq. (13.2.1) becomes a system of M linear equations which can be solved for the unknown Φ_j:

$$\Phi_j = \sum_{k=1}^{M} S_j^k A_k. \tag{13.2.2}$$

Here S_j^k is the inverse of the matrix σ_j^k. This inverse exists, as is well known, only if the determinant of σ_j^k does not vanish, i.e., only if the cross sections are linearly independent of one another. If the σ_j^k are known — they can be calculated if we know $\sigma^k(E)$, but only if we make some plausible assumption regarding the energy variation of the flux in the j-th group — then we also know the S_j^k and can thus calculate the behavior of the flux from the A_k. Small errors in the S_j^k and the A_k can produce rather large errors in the group fluxes; in fact high accuracy is probably only attainable by making the number M of threshold detectors quite a bit larger than the number N of energy groups and determining the Φ_j ($j=1 \ldots N$) to best reproduce the A_k ($k=1 \ldots M$) in the least-squares sense.

Both FISCHER and DIETRICH have tried to apply the multigroup method; cf. also UTHE.

(b) HARTMANN'S Method. In this method, the flux $\Phi(E)$ is approximated in terms of a series of auxiliary functions $\varphi_n(E)$:

$$\Phi'(E)=\sum_{n=1}^{N} a_n \varphi_n(E). \tag{13.2.3}$$

In order that $\Phi(E)$ be approximated as well as possible,

$$\int [\Phi(E)-\Phi'(E)]^2 dE = \text{minimum} \tag{13.2.4}$$

should hold. If we differentiate with respect to a_k and set the resulting expression equal to zero, we find

$$\int \Phi(E)\varphi_k(E)\,dE=\sum_{n=1}^{N} a_n \int \varphi_n(E)\varphi_k(E)\,dE. \tag{13.2.5}$$

If we now assume that the auxiliary functions $\varphi_k(E)$ are the activation cross sections of the various threshold detectors being used (or a suitable linear combination thereof), then it follows with $A_k=\int \Phi(E)\sigma^k(E)\,dE$ and $\varphi_{nk}=\int \sigma^n(E)\sigma^k(E)\,dE$ that

$$A_k=\sum_{n=1}^{N} \varphi_{nk} a_n \tag{13.2.6}$$

and the a_n can be calculated from the A_k by inversion. One respect in which this method is unsatisfactory is that the development of $\Phi(E)$ in functions with the energy dependences of the cross sections has little physical sense; as DIETRICH has shown, under some circumstances we may actually obtain approximation functions $\Phi'(E)$ that assume negative values in places (cf. Fig. 13.2.2).

13.2.2. Semiempirical Methods

All the methods to be discussed below serve to determine unknown spectra assuming, however, that something is already known about their basic form. Without exception, they are more accurate than the purely mathematical methods.

(a) The Method of Effective Threshold Energy. If we assume that the spectrum to be investigated deviates only slightly from a fission neutron spectrum, then by using the effective threshold energies introduced in Sec. 13.1.1 we can immediately determine the integral spectrum:

$$A_k=\sigma_0^k \int_{E_{sk}^{\text{eff}}}^{\infty} \Phi(E)\,dE=\sigma_0^k F(E_{sk}^{\text{eff}}). \tag{13.2.7}$$

If we use many threshold detectors we obtain $F(E)$ and by differentiation $\Phi(E)$.

GRUNDL and USNER have suggested a useful modification of the concept of effective threshold energy. These authors start from the connection between σ_0 and E_s^{eff} given in Eq. (13.1.2). Whereas in the usual methods, σ_0 is set equal to the plateau value of $\sigma(E)$ and E_s^{eff} is calculated from $\bar{\sigma}$, GRUNDL and USNER recommend that we determine σ_0 and E_s^{eff} in such a way that σ_0 varies as little as possible when the spectrum deviates slightly from a fission neutron spectrum; however, no change will be made in Eq. (13.1.2). In order to carry out the variations of the neutron spectrum in a simple way, GRUNDL and USNER chose the form

$$N(E)=\frac{2\beta^{3/2}}{\sqrt{\pi}}\sqrt{E}\,e^{-\beta E} \tag{13.2.8}$$

for $N(E)$ (E in Mev). When $\beta = 0.77$, this form reproduces the experimentally observed spectrum of the fission neutrons about as well as Eq. (2.6.1). We illustrate the method by means of Fig. 13.2.1. Here σ_0 — for the reaction $Al^{27}(n, p)Mg^{27}$ — was calculated as a function of β for various values of the parameter E_s^{eff} by means of the relation

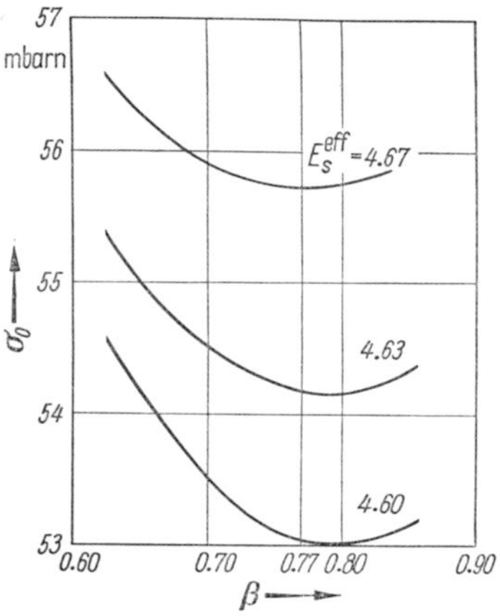

Fig. 13.2.1. σ_0 for the reaction $Al^{27}(n, p)Mg^{27}$ as a function of the spectral index β for various values of the effective threshold energy E_s^{eff} (Mev)

$$\sigma_0(\beta) = \frac{\int\limits_0^\infty \sigma(E)\sqrt{E}\,e^{-\beta E}\,dE}{\int\limits_{E_s^{eff}}^\infty \sqrt{E}\,e^{-\beta E}\,dE}. \tag{13.2.9}$$

For each value of E_s^{eff} we obtain a curve of $\sigma_0(\beta)$. Out of this family, we seek that curve for which $\frac{d\sigma_0}{d\beta} = 0$ when $\beta = 0.77$; the value of $\sigma_0(\beta = 0.77)$ on this curve is the value of σ_0 being sought. Table 13.2.1 gives values of E_s^{eff} and σ_0 found in this way by GRUNDL and USNER. These values are more useful than those given in the earlier tables since they apply even when the spectrum to be studied is no longer strictly a fission spectrum.

(b) UTHE's Method. In this procedure, the neutron spectrum is written in the form

$$\Phi'(E) = N(E) \sum b_n E^n, \tag{13.2.10}$$

i.e., as the product of a fission neutron spectrum and a polynomial in E. The b_n are chosen so that the A_k determined with the threshold detectors are best approximated in the least-squares sense:

$$\sum_{k=1}^{N} \left(A_k - \sum b_n \sigma_{nk}\right)^2 = \text{minimum} \tag{13.2.11a}$$

with

$$\sigma_{nk} = \int \sigma_k(E) N(E) E^n dE. \tag{13.2.11b}$$

Table 13.2.1. *Effective Threshold Energies According to* GRUNDL *and* USNER

Reaction	E_s^{eff} [Mev]	σ_0 [mbarn]
P^{31} (n, p)	2.71	118.9
Al^{27} (n, p)	4.67	55.7
Fe^{56} (n, p)	6.33	52.4
Al^{27} (n, α)	7.25	58.7
Cu^{63} $(n, 2n)$	12.77	694
S^{32} (n, p)	2.78	272.0
U^{238} (n, f)	1.67	650
Np^{237} (n, f)	0.80	533

Differentiating Eq. (13.2.11a) with respect to each of the b_n and setting the resulting expressions equal to zero, we obtain a system of linear equations for the determination of the b_n from the A_k. Since Eq. (13.2.8) probably already can represent the spectrum of the fast neutrons in a reactor with only a few b_n, this method should give good results if we use sufficiently many threshold detectors and if the cross sections are well known.

(c) DIETRICH's Method. In the special case of a water-moderated reactor, we may assume that each neutron produced in fission loses so much energy in one collision that after the collision it can no longer contribute to the activation of the threshold detector. Then

$$\Phi(E) \sim \frac{N(E)}{\Sigma_t(E)} \tag{13.2.12a}$$

with $$\Sigma_t(E) = \Sigma_0 + \Sigma_{sH}(E). \quad (13.2.12\text{b})$$

Here $\Sigma_{sH}(E)$ is the strongly energy-dependent scattering cross section of hydrogen; the absorption cross section and the inelastic scattering cross sections of all other materials present in the reactor are combined in Σ_0. In first approximation, the energy dependence of Σ_0 can be neglected. In the energy range from 2 to 12 Mev, $\Sigma_{sH}(E)$ varies approximately as $E^{-0.75}$. Thus the energy dependence of the flux is given by

$$\Phi(E) \sim \frac{N(E)}{1+\alpha E^{-0.75}} \quad (13.2.13)$$

where α is a parameter that can be determined by measurement with threshold detectors. This determination consists of calculating the A_k with the flux given in Eq. (13.2.13) and changing α until the relative deviation of the measured and calculated values is a minimum. In the core of the Danish swimming pool reactor DR-2, DIETRICH found that $\alpha = 11$; in his determination, the activities measured with P^{31}, S^{32}, Al^{27}, Fe^{56} (n, p), and Al^{27} (n, α) were reproduced rather well ($\approx 5\%$). It should be possible to apply this method to other systems.

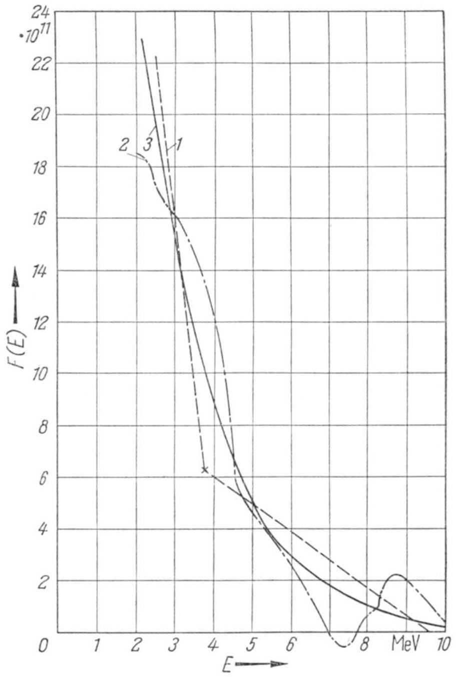

Fig. 13.2.2. The integral spectrum $F(E) = \int_E^\infty \Phi(E)\,dE$ in the Danish swimming pool reactor DR 2 observed using threshold detectors. Curve 1: Multigroup method evaluation. Curve 2: Evaluation by HARTMANN's method. Curve 3: Evaluation by DIETRICH's method

In his measurements in the core of DR-2, DIETRICH made a comparison of various methods of evaluation. Fig. 13.2.2 shows the integral flux $F(E) = \int_E^\infty \Phi(E)\,dE$ obtained from measurements with the five threshold detectors mentioned above. Curve 1 was obtained with the multigroup method, curve 2 with HARTMANN's method, and curve 3 with the last method described (with α set equal to 11).

Chapter 13: References[1]

General Facts about Threshold Detectors.

BYERLY, P. R.: In Fast Neutron Physics, part I, New York: Interscience Publishers, 1960, Chapter IV C.

GRUNDL, I., and A. USNER: Nucl. Sci. Eng. 8, 598 (1960).

HUGHES, D. J.: Pile Neutron Research, Cambridge: Addison-Weseley, 1953, especially Chapter 4, Fast Neutron Research.

KÖHLER, W.: Atomkernenergie 9, 81 (1964).

Neutron Dosimetry, Proceedings of the 1962 Harwell Symposium, Vol. 1 and 2. Vienna: International Atomic Energy Agency, 1963.

PASSELL, T. O., and R. L. HEATH: Nucl. Sci. Eng. 10, 308 (1961).

ROCHLIN, R. S.: Nucleonics 17, No. 1, 54 (1959).

[1] Cf. footnote on p. 53.

BEAUGÉ, R.: Sections Efficaces Pour les Detecteurs de Neutrons Par Activation, Recommandees par le Groupe de Dosimetrie d'Euratom. CEA-Report (1963).
MELLISH, C. E.: Nucleonics **19**, No. 3, 114 (1961).
ROY, I. C., and I. I. HAWTON: AECL-1181 (1960).
Effective Cross Sections.

BAYHURST, B. P., and R. J. PRESTWOOD: LA-2493 (1960).
HUGHES, D. J., and R. B. SCHWARTZ: BNL-325 (1958).
LISKIEN, H., and A. PAULSEN: EANDC (E) 28 (1961).
Energy Dependence of the Cross Sections of Detector Substances.

AGER-HANNSSEN, H., and I. M. DØDERLEIN: Geneva 1958 P/566, Vol. 14, p. 455.
DIERCKX, R.: in Neutron Dosimetry, Vol. I, p. 325; Vienna: International Atomic Energy Agency 1963, cf. also CEN-Rapport R 2128 (1961).
DIETRICH, O. W., and I. THOMAS: in Physics of Fast and Intermediate Reactors, Vol. 1, p. 377; Vienna: International Atomic Energy Agency, 1962.
FISCHER, G. I.: Nucl. Sci. Eng. **7**, 355 (1960).
HARTMANN, S. R.: WADC-TR-57/375 (1957).
KÖHLER, W.: FRM-Berichte No. **27** (1960), No. **30** (1960), No. **39** (1961), No. **43** (1962), No. **52** (1963), No **56** (1963).
UTHE, P.M.: WADC-TR-57/3 (1957).
TRICE, I. B.: CF-55-10-140 (1955), Oak Ridge National Laboratory.
Application of Threshold Detectors; Procedures for Evaluation.

HURST, G. S., et al.: Rev. Sci. Inst. **27**, 153 (1956).
KÖHLER, W., and J. ROMANOS: Nukleonik **5**, 159 (1963).
REINHARDT, P. W., and F. J. DAVIS: Health Phys. **1**, 169 (1958).
Fission Chambers as Threshold Detectors.

14. Standardization of Neutron Measurements

A great many investigations in neutron physics require precise absolute measurements. These absolute measurements are of two kinds: the absolute determination of source strengths and the absolute measurement of fluxes in bulk media or in free neutron beams. Absolute source strength measurements are necessary in all investigations of neutron-producing nuclear reactions, especially $\bar{\nu}$- and η-measurements on fissionable substances. Absolute flux measurements are a necessity in many cross section measurements (cf. Chapter 4); they are also the basis of neutron dosimetry, particularly in nuclear reactors. The problems of absolute flux and source strength measurement are naturally quite intimately connected.

In Secs. 14.1 and 14.2 we shall treat methods of absolute flux measurement, in Sec. 14.3 methods of determining source strengths, and in Sec. 14.4 the important comparison methods that permit us to compare arbitrary source strengths or neutron fluxes with standards.

14.1. Absolute Measurement of Thermal Neutron Fluxes with Probes

Neutron probes are particularly well suited for the absolute measurement of thermal neutron fluxes. The main problem in their use is the absolute measurement of their activity. In this section we shall discuss some suitable methods for the absolute measurement of activity. The applicability of these methods is naturally not limited to absolute flux measurements; in fact, they form the basis for all measurements of activation cross sections (cf. Sec. 4.3).

If we once know the absolute value of the thermal flux in a medium, we can easily determine the absolute value of Φ_{epi} by the cadmium ratio method. We can also absolutely determine the fast flux from probe measurements, but only under those restrictions (explained in detail in Chapter 13) that arise from the difficulty of determining the spectrum.

14.1.1. Probe Substances for Absolute Measurements

Let us suppose that the flux spectrum in a given medium is that of Eq. (12.2.1). If we determine the activation of a foil in this field and eliminate the epithermal activation by the cadmium difference method as explained in Sec. 12.2.1, then the remaining thermal activation is given by (cf. Chapter 11),

$$C_{th} = \frac{\Phi_{th}\overline{\left(\frac{\mu_{\text{act}}}{\mu_a}\varphi_0(\mu_a\delta)\right)}(1+\varepsilon)}{2(1+\varkappa_c)}. \tag{14.1.1a}$$

Thus we obtain the thermal flux from the activity $A_{th} = C_{th}/T$ (T = time factor) using the equation

$$\Phi_{th} = \frac{2A_{th}T(1+\varkappa_c)}{\overline{\left(\frac{\mu_{\text{act}}}{\mu_a}\varphi_0(\mu_a\delta)\right)}(1+\varepsilon)}. \tag{14.1.1b}$$

For thin foils, $\mu_a\delta \ll 1$ and we have

$$\Phi_{th} = \frac{2A_{th}T(1+\varkappa_c)}{\sqrt{\pi}\,g(T)\sqrt{\frac{293.6^\circ}{T}}\,(\mu_{\text{act}}(v_0)\,\delta)(1+\varepsilon)}. \tag{14.1.1c}$$

Thus Φ_{th} can be absolutely determined if A_{th} can be absolutely counted, if the activation correction $\varkappa_c$ is known, and if $\mu_a(E)$ and $\mu_{\text{act}}(E)$, or in the case of a thin foil the thermal activation cross section and the g-factor, are known. In addition, the neutron temperature must be known[1].

We ought to demand from an ideal probe substance that its decay scheme be well suited to absolute counting, that its activation cross section be very accurately known and deviate at most slightly from the $1/v$-law ($g \approx 1$), and that it have a conveniently large half-life. Of all the probe substances discussed in Sec. 11.1.2, gold comes closest to filling these requirements: Its activation cross section ($\sigma_{\text{act}} = \sigma_a$) for $v_0 = 2200$ m/sec ($\sigma_{\text{act}} = 98.5 \pm 0.4$ barn) is very well known; its g-factor (Fig. 11.1.6) is always close to one; its half-life is favorable, and its simple decay scheme permits precise absolute counting, particularly by coincidence methods. Indium has occasionally been suggested for absolute measurements, but its activation cross section is less certain and its complex decay scheme less suitable for absolute counting. Sodium ($\sigma_{\text{act}}(v_0) = 0.531 \pm 0.008$ barn, $g = 1$, $T_{\frac{1}{2}} = 14.97$ h), cobalt, and manganese have occasionally been used for absolute measurements, but always with an accuracy less than that attainable with gold. According to Grimeland, the sodium activity may be counted quite simply by irradiating a NaI(Tl)-crystal in a neutron field, waiting for the iodine activity ($T_{\frac{1}{2}} = 24.09$ min) to decay, and then counting the scintillations of the crystal with

[1] The neutron temperature only enters the neutron density $n = \Phi_{th}/\bar{v}$ through $g(T)$ and $\varkappa_c$. For an extremely thin $1/v$-absorber, $\varkappa_c = 0$ and $g(T) = 1$; thus we can determine n without knowing the temperature.

a photomultiplier. Since each decay of a sodium nucleus produces one scintillation, no corrections for absorption or counter geometry are necessary.

14.1.2. The Absolute Determination of Probe Activities

For the absolute determination of foil activities, we can fall back on the procedures developed in nuclear physics for the absolute calibration of radioactive β- and γ-emitters. The most important procedures are:

(a) 4π β-Counting. (Cf. HOUTERMANNS *et al.*, PATE and YAFFE). The foil to be counted is placed with as few mechanical supports as possible in the interior of a 4π-counter consisting of two half-cylindrical chambers (cf. Fig. 14.1.1). An arrangement is sought in which all the β-particles leaving the probe produce pulses. Most frequently employed are methane flow counters, which are characterized by exceptional stability and a good detection efficiency ($\approx 1.00 \pm 0.01$ according to PATE and YAFFE). We shall not go into the various possible kinds of counter construction here. The numbers N_1 and N_2 of events counted per second in chambers 1 and 2, respectively, as well as the coincidence rate N_{12}, are recorded. After correction for dead time and background, the quantity

$$N_\beta = N_1 + N_2 - N_{12} \qquad (14.1.2)$$

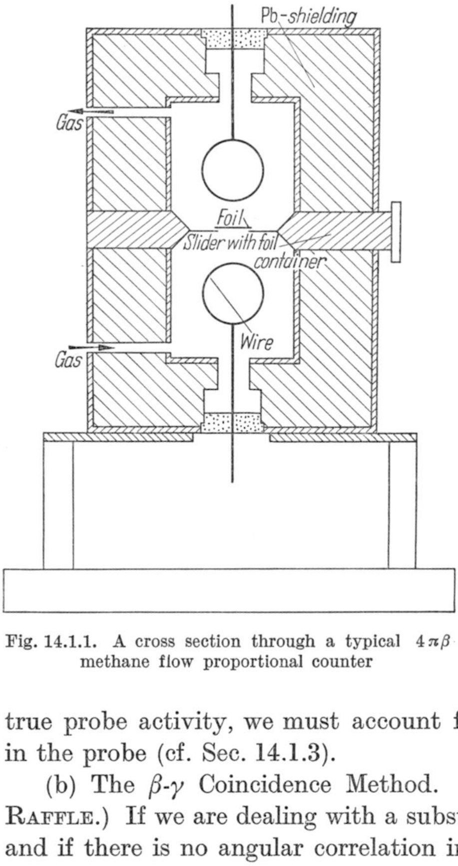

Fig. 14.1.1. A cross section through a typical $4\pi\beta$ methane flow proportional counter

is equal to the number of particles leaving the probe per second[1] (not including some corrections of higher order). If we wish to go from N_β to the true probe activity, we must account for the self-absorption of the β-radiation in the probe (cf. Sec. 14.1.3).

(b) The β-γ Coincidence Method. (Cf. LARRSSON, v. PLANTA and HUBER, RAFFLE.) If we are dealing with a substance that has a simple β-γ decay scheme and if there is no angular correlation in the β-γ cascade, we may use both a β-sensitive counter (stilbene or anthracene crystal, proportional counter) and one which responds only to γ-rays (NaI) and record the coincidence rate. If η_β is the detection efficiency of the β-counter (defined as the number of counts per decay in the probe) and η_γ that of the γ-counter, and if $F \cdot A$ (F = foil area) is the true decay rate in the foil, then after correction for background and dead time and after elimination of accidental coincidences the β-counting rate is given by $N_\beta = \eta_\beta FA$, the γ-counting rate by $N_\beta = \eta_\beta FA$, and the coincidence rate by $N_{\beta\gamma} = \eta_\beta \eta_\gamma FA$.

[1] It is not necessary to record N_1, N_2, and N_{12} separately in the determination of N_β. If we connect the outputs of the two chambers in parallel and without any delay between them, the resulting count rate is obviously N_β.

Thus the true decay rate is given by

$$F \cdot A = \frac{N_\beta N_\gamma}{N_{\beta\gamma}}. \tag{14.1.3}$$

Apparently, the β-self-absorption does not appear in this result. In order that Eq. (14.1.3) hold, the following conditions must be fulfilled in addition to those mentioned above:

1. No conversion electrons may appear.
2. The β-counter may not respond to γ-rays nor the γ-counter to β-particles.
3. The specific activity of the foil must be homogenous.
4. The sensitivity for β-radiation may not depend on position.
5. The sensitivity for γ-radiation may not depend on position. (Only two of conditions 3—5 need be fulfilled.)

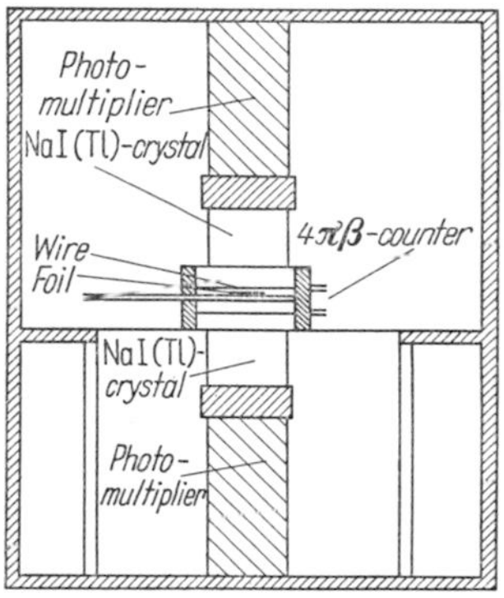

Fig. 14.1.2. A cross section through a typical $4\pi\beta-\gamma$ coincidence apparatus

As a rule, these conditions and those mentioned initially are either not fulfilled or only fulfilled approximately, and it is necessary to apply corrections to Eq. (14.1.3). The appearance of these corrections limits the accuracy attainable with the β-γ coincidence method; even with gold, FA can hardly be determined to better than 1%.

(c) The $4\pi\,\beta$-γ Coincidence Method (Campion, Wolf). The accuracy of the β-γ coincidence method can be considerably increased and its realm of application simultaneously extended to substances with complex decay schemes if the 4π-counter described in (a) is used for the β-counting (cf. Figs. 14.1.2 and 14.1.3). The sources of error of the ordinary β γ coincidence method discussed in (b) do not have such a strong effect here, especially for thin foils. Campion gives a detailed discussion of this method and the corrections it requires (also cf. Pönitz). With this method using gold a determination of FA of metallic foils with an accuracy of 0.3% is probably possible.

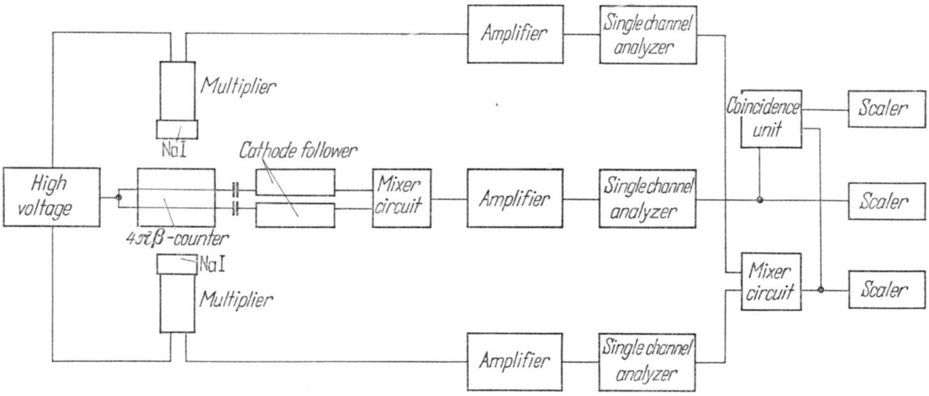

Fig. 14.1.3. The electronic block diagram of a $4\pi\beta-\gamma$ coincidence apparatus

(d) Activation Measurement with a $4\pi\,\beta$-Liquid Scintillator (STEYN, BELCHER). β-self-absorption can be gotten around by dissolving the radioactive substance and mixing it with a liquid scintillator. It is then possible in principle to count each decay.

14.1.3. Electron Self Absorption in 4π β-Counting

Of the methods described in Sec. 14.1.2, that requiring the least expenditure of apparatus and the most convenient in execution is $4\pi\,\beta$-counting. With the exception of the difficult liquid scintillator method, it is the only one that can be

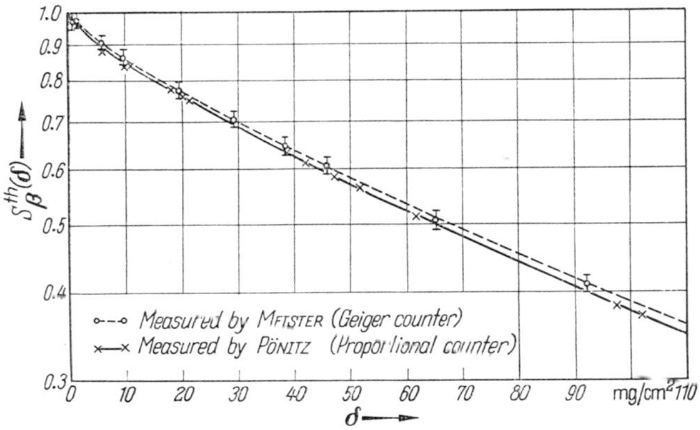

Fig. 14.1.4. The effective self-absorption factor $S_\beta^{\text{th}}(\delta)$ for thermally activated Gold foils

used with weak activities such as occur in low neutron fluxes. In this method, however, the self-absorption of the electrons must be taken into account. This is done by introduction of an effective self-absorption factor S_β defined by the relation[1]

$$N_\beta = S_\beta \cdot FA. \qquad (14.1.4)$$

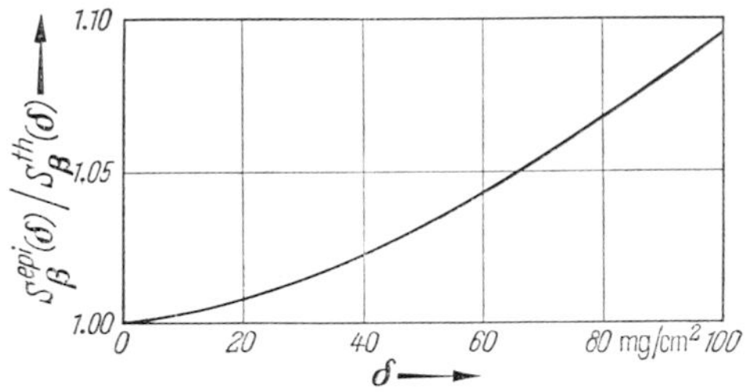

Fig. 14.1.5. $\frac{S_\beta^{\text{epi}}(\delta)}{S_\beta^{\text{th}}(\delta)}$ for Gold foils as measured by PÖNITZ

The expression "effective" denotes the fact that S_β not only takes into account the β-self-absorption but also any other deviation of N_β from the number of electrons leaving the foil per second. S_β is a function of the foil thickness; it can be experimentally determined by determining N_β for foils of various thicknesses in a $4\pi\,\beta$-counter and the true decay rate FA, for example, by the 4π β-γ coincidence method (for which the necessary corrections must be calculated). One can also determine S_β without a coincidence apparatus by the following extrapolation method: One activates a set of foils of various thicknesses in the same neutron

[1] As was shown in Sec. 11.2.5, under certain circumstances the apparent β-activity of a foil depends on its orientation. However, this is not true for $4\pi\beta$-counting because when both sides of the foil are counted the effect of the current term vanishes.

field and calculates their true activity $A=C/T$ with the methods of Chapter 11. One then extrapolates their specific activities to zero foil thickness, for which S_β should be one.

Since the self-absorption factor depends on the depth distribution of the foil activation, S_β is different in thermal and epithermal activation, and under some circumstances the two values must be separately determined.

Fig. 14.1.4 shows $S_\beta(\delta)$ for thermally activated gold foils. The measured values of MEISTER were obtained with the extrapolation method by use of a 4π β-counter operated in the Geiger region. The values of PÖNITZ shown in Fig. 14.1.4 were obtained from carefully corrected 4π β-γ coincidence measurements carried out with a methane flow proportional counter. PÖNITZ has by the same method determined the self-absorption factor for epithermal activation of gold foils; $S_\beta^{\mathrm{epi}}/S_\beta^{th}$ from his measurements is shown in Fig. 14.1.5. Figs. 14.1.6–8 show the "thermal" $S_\beta(\delta)$ of indium, copper, and manganese (Mn-Ni foils) found by MEISTER with a 4π Geiger counter by the extrapolation method.

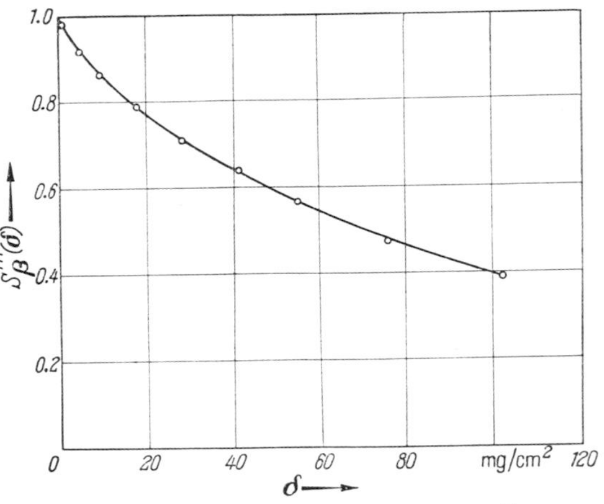

Fig. 14.1.6. The effective self-absorption factor $S_\beta^{\mathrm{th}}(\delta)$ for thermally activated In foils (54-min activity)

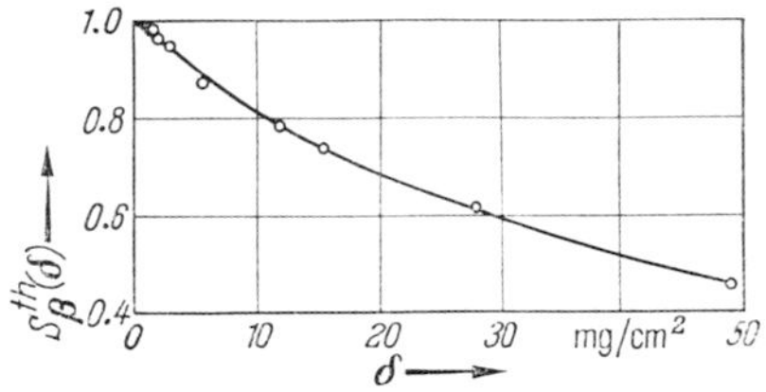

Fig. 14.1.7. The effective self-absorption factor $S_\beta^{\mathrm{th}}(\delta)$ for thermally activated Cu foils (12.8h activity)

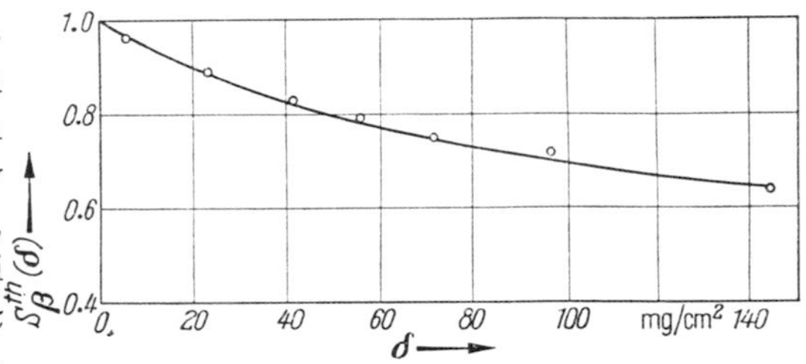

Fig. 14.1.8 The effective self-absorption factor $S_\beta^{\mathrm{th}}(\delta)$ for thermally activated Mn—Ni foils

14.2. Other Methods of Absolute Flux Measurement

In the following survey of other methods of flux measurement, we shall be mainly concerned with free neutron beams from reactors and accelerators. Let us only deal with monoenergetic neutrons, i.e., either let the source be monoenergetic or let there be an energy definition via time-of-flight. In principle, any of the neutron detectors discussed in Chapter 3 can be used for absolute flux measurement as long as the detector geometry is simple, the cross section of the detection reaction sufficiently well known, and absolute counting of the events in the detector possible. The energy-dependent cross section of the detection reaction usually represents the main problem: in order to determine this cross section we must measure the reaction rate of the detector substance in a known flux; in order to measure this flux, we must know the cross section of another detector substance, etc. Thus in making absolute flux measurements we must try to use detector substances whose cross sections have been determined by methods that give absolute σ-values directly. The main such method is the transmission method.

In the energy range from 100 kev to 20 Mev, we can make absolute flux measurements by means of neutron-proton scattering, the cross section for which is known from transmission experiments and semiempirical interpolation to about 1% (cf. Fig. 1.4.2). We may assume that up to about 10 Mev the scattering in the C-system is isotropic; at higher energies the anisotropy known from scattering experiments must be taken into account (Gammel). Suitable devices for counting are the proton recoil counter discussed in Sec. 3.2.1, some of the counter telescopes discussed in Sec. 3.2.2, or nuclear photoplates. The accuracy of such flux measurements is about 3—5%. Since the fission cross section of U^{235} is known to about 4—7% in the energy range from 100 kev to 10 Mev — from comparison measurements with the neutron-proton scattering cross section — and since fission counting is easier to perform than recoil proton counting, fission counters with U^{235} are frequently used for absolute neutron flux measurements in this energy range.

In the energy range 0—10 kev, one can make absolute flux measurements by means of the $B^{10}(n, \alpha)Li^7$ reaction. The cross section has been measured by transmission (with a correction for the scattering cross section, whose influence increases with increasing energy) and is known to a few percent; within this accuracy the cross section follows a $1/v$-law. B^{10} is used in BF_3-counters, in boron-coated counters or in boron-loaded glasses. In the latter, corrections for neutron scattering may sometimes be necessary. When the neutron energy is less than 10 kev, a sufficiently thick slab of B^{10} will absorb all neutrons incident on it and we can determine the flux without knowledge of the $B^{10}(n, \alpha)Li^7$ cross section if we can determine the reaction rate by absolute counting of the 478-kev γ-ray of Li^{7*}. For this determination, knowledge of the energy-dependent branching ratio of the $B^{10}(n, \alpha)Li^7$ reaction is necessary.

No good techniques have been developed so far for absolute flux measurements in the energy range 10—100 kev. Though the neutron-proton scattering cross section is well-known in this energy region, recoil proton detectors cannot be used since it is hardly possible to count recoil protons of such low energy. Therefore, one must use detectors based on the reactions $B^{10}(n, \alpha)Li^7$, $He^3(n, p)H^3$ or $Li^6(n, \alpha)H^3$, but unfortunately the cross sections for these reactions are not very well known in the energy range 10—100 kev owing to the difficulty of measuring them in a transmission experiment ($\sigma_s \gtrsim \sigma_a$). A thorough discussion of the problems of flux and cross section measurements in this energy range can be found in Batchelor.

Besides the procedures discussed here for direct flux measurement, there exists in many cases an indirect possibility, viz.: if the energy spectrum and the strength of a neutron source are known, then as long as there is no backscattering the flux is also known in the immediate neighborhood of the source. We can then compare the known flux produced in this way with an unknown by means of a suitable detector, e.g., a long counter. Use of the so-called standard pile, which has occasionally been employed in the absolute determination of thermal and epithermal neutron fluxes, can be counted among the indirect methods. A standard pile is a graphite block some $2 \times 2 \times 3$ m along the edges in which a calibrated neutron source has been placed. The thermal and epithermal fluxes are then calculated with the help of age and diffusion theory. An unknown flux is determined by comparing the foil activity it produces to that produced by the flux in the standard pile. In making the comparison, the foil correction must be taken into account.

14.3. Determination of the Strength of Neutron Sources

The following different methods for the absolute calibration of neutron sources may be distinguished:

(a) The Method of Associated Particles. Here we make use of the fact that in some neutron-producing reactions a charged particle is emitted either promptly or after some delay; if we succeed in counting these particles absolutely, we know the source strength.

(b) Direct Flux Measurement. If the spectrum and the angular distribution of the emitted neutrons is known, the source strength can be determined from the flux in the neighborhood of the source as long as we are careful to suppress any backscattering. This procedure is the inverse of the indirect method of determining the neutron flux that was mentioned earlier.

(c) The Method of Total Absorption. The source is placed in a very large absorbing medium. Then all the neutrons emitted by the source are slowed down and finally absorbed in the surrounding medium; the source strength follows from a spatial integration of the absorption rate.

In Sec. 14.3.1, we shall deal with the method of associated particles. In Secs. 14.3.2 and 14.3.3, we shall discuss various ways of applying the total absorption method. The procedure described under (b) requires no further explanation; in fact, it is not particularly accurate and is also only applicable in special cases.

14.3.1. The Method of Associated Particles

In the reaction $H^3(d, n)He^4$, an α-particle appears simultaneously with the neutron. Its kinetic energy depends on the direction of emission of the emitted neutron and on the deuteron energy; for $E_d=0$, $E_\alpha=3.49$ Mev in all directions (cf. Sec. 2.1.2). We can easily detect the α-particles emitted in a limited range of solid angles from a tritium target with a scintillation counter (preferably ZnS), a proportional counter, or a solid state counter. Fig. 14.3.1 shows a typical counting apparatus used by Larsson. At low deuteron energies ($E_d < 200$ kev), the angular distribution of the neutrons is quite isotropic in the C-system and approximately so in the L-system. If $\Delta\Omega$ is the solid angle subtended by the α-counter, i.e., if *all* the particles emitted from the target in this angular region are counted, then the count rate of the α-counter is simply given by

$$N_\alpha \approx \frac{\Delta\Omega}{4\pi} Q \tag{14.3.1}$$

and the source strength can easily be determined. When higher accuracy is necessary and particularly at higher deuteron energies, we must take the angular distribution of the neutrons and the α-particles into account. Thus to determine the source strength from N_α, the (deuteron-energy-dependent) angular distribution of the neutrons in the C-system must be employed; fortunately, it is very accurately known.

The reaction $H^2(d, n)He^3$ can also be absolutely counted by this method. However, we must take into account the fact that when deuterium is bombarded with deuterons the reaction $H^2(d, p)H^3$ can also occur; at low deuteron energies the cross section of this reaction is about the same as that of the $H^2(d, n)He^3$ reaction. The $H^3(p, n)He^3$ reaction can also be absolutely counted. Since the emission of the neutron and the helium nucleus in all these reactions occurs simultaneously, the α-particle or He^3 nucleus can also be used as a time mark; the attainable precision is about 5×10^{-9} sec.

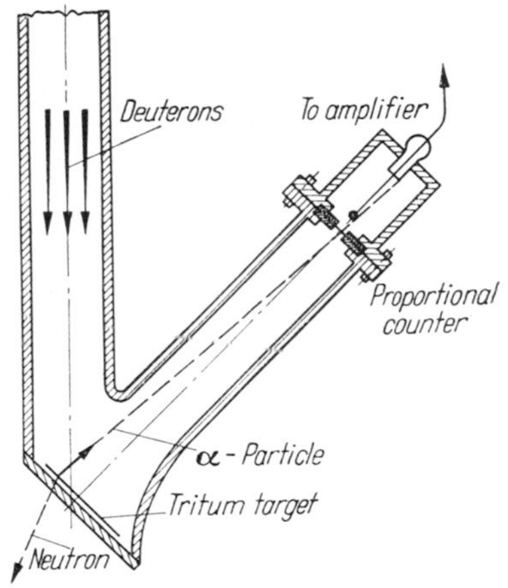

Fig. 14.3.1. A typical set-up for associated particle counting

The accuracy with which source strengths may be determined in this way is about 2% in the case of the $H^3(d, n)He^4$ reaction. LARSSON gives more details.

According to RICHMOND and GARDNER, the absolute strength of a neutron source based on the photodisintegration of the deuteron can be determined by absolutely counting the photoprotons arising from the reaction $H^2(\gamma, n)H^1$. These authors used a ThC'' preparation as the γ-emitter ($E_\gamma = 2.62$ Mev), and placed it between two deuterium-filled ionization chambers. In each (γ, n) process, one neutron and one proton with a kinetic energy of 190 kev are produced. It is quite easy to detect the proton. The source strength is very small.

Various neutron-producing reactions lead to radioactive residual nuclei, and it is possible to determine the source strength after irradiation by absolutely counting the radioactivity. As a rule, to do this it is necessary to destroy the source. An important example is the (α, n) reaction on F^{19} (Sec. 2.2.2), which leads to the well-known positron emitter Na^{22}. The $Li^7(p, n)Be^7$ reaction leads to an electron capture activity in Be^7 ($T_{\frac{1}{2}} = 53.4$ d) that can best be detected by means of the 478-kev γ-radiation of Li^7. Finally, we can determine the source strength of the reaction $Be^9(\gamma, n)Be^8$ by quantitatively determining the amount of helium formed by the α-decay of Be^8.

14.3.2. The Water-Bath Gold Method

This method is the most usual and far and away the most accurate for the absolute determination of radioactive source strengths. The source is placed in the middle of a vessel filled with distilled water that is so large that only a negligible fraction of the neutrons escape from it[1]. Then all the neutrons emitted from the source are slowed down and ultimately absorbed in the water. Thus

$$Q = \iint \Sigma_a(E)\, \Phi(\boldsymbol{r}, E)\, d\boldsymbol{r}\, dE. \tag{14.3.2}$$

The integration extends over the entire volume including the source. $\Sigma_a(E)$ is either the absorption cross section of the water or of the source substance.

If we base our considerations on the thermal flux, take the absorption of neutrons during moderation into account by means of a resonance escape probability p, and correct for the neutron absorption in the source by a factor K_Q, then

$$p\,Q = K_Q \int \Sigma_{a_{th}} \Phi_{th}(r)\, 4\pi r^2\, dr. \tag{14.3.3}$$

We have assumed here that the source radiates isotropically so that Φ_{th} will only depend on the distance r from the source. If this is not the case, we can force the results to agree with those for an isotropic source by rotating the source during the measurement. $\Sigma_{a_{th}}$ is the absorption cross section of water for thermal neutrons $\left(\Sigma_{a_{th}} = \frac{\sqrt{\pi}}{2}\sqrt{\frac{293.6^\circ}{T}}\,\Sigma_a(v_0)\right)$. $\Sigma_a(v_0)$ is very accurately known [$\sigma_a(v_0)$ for hydrogen is 327 ± 2 mbarns; the contribution of the oxygen can be neglected].

Since we can absolutely determine the thermal neutron flux very accurately ($\approx 0.5\%$) by cadmium difference measurements on gold foils, Eq. (14.3.3) permits a simple determination of the source strength. In making this determination, however, we must carry out a flux integration over the water volume. This integration is most simply done by measuring Φ_{th} as a function of r and graphically integrating the function $4\pi r^2 \Phi(r)$. However, we can also move the foil through the water bath during the irradiation in such a way that its activity at the end of the irradiation directly determines the flux integral (cf. CURTISS or HAGE). The correction factor K_Q can be calculated with the help of diffusion theory from the specifications of the source; for a typical (Ra—Be) source, $K_Q \approx 1.025$.

Taking the resonance escape probability into account is more difficult. p is composed of a factor p_1 that describes the (n, α) process in O^{16} above 3.6 Mev and also the (n, p) process above 10 Mev, and a factor p_2 that accounts for the epithermal absorption in hydrogen. p_2 can be calculated using Eq. (7.3.21), or it can be derived from the epithermal absorption rate using the measured cadmium ratio; it is about 0.985. p_1 depends on the spectrum of the source, and in the case of a (Ra—Be) source, for example, cannot be calculated with sufficient accuracy because the contribution of the neutrons above 3.6 Mev to the total source strength is not accurately known (on account of the uncertainties in the spectrum for $E < 1$ Mev; cf. Sec. 2.2.1). We can determine p_1 experimentally by successively introducing the same source into a water and a paraffin bath and measuring the flux integral with a resonance probe. Since the paraffin contains no oxygen, the value of p_1 is one there, and the ratio of the resonance flux integrals should give p_1. Since $\xi \Sigma_s$ is different in H_2O and paraffin, we must make an additional normalization measurement using an (Sb—Be) source, for which p_1 is always one. In this way, DE TROYER and TAVERNIER found $p_1 = 0.978$ for a (Ra—Be) source[2]; PÖNITZ

[1] If a (Ra—Be) source is placed in the center of a sphere with a 1-m radius, only 0.5% of all the neutrons escape.

[2] This value comes from a reevaluation of the measurements of DE TROYER and TAVERNIER; the original value was 0.975.

found $p_1 = 0.982 \pm 0.004$. p_1 decreases rapidly with increasing source energy because of the rise of the cross sections for the (n, α) and (n, p) reactions in O^{16}; for 14-Mev neutrons, $p_1 = 0.826$ according to LARSSON. It is obviously no longer sensible to use a water bath with high-energy sources, and it is much better to use paraffin or another organic liquid that contains no oxygen. Since with increasing neutron energy the requirement of "infinite size" leads to ever greater dimensions of the bath, a paraffin bath for 14-Mev neutrons is quite expensive.

Sources of error in the water-bath gold method arise from (i) the absolute determination of Φ_{th} ($\approx 0.5\%$), (ii) the flux integration ($\approx 0.5\%$), (iii) the correction factors p_1, p_2, and K_Q (about 0.3% each for a typical (Ra—Be) source), and (iv) from the absorption cross section of hydrogen ($\approx 0.6\%$); the total error is about 1.5%. The scatter of the various independent source strength determinations is of the same order of magnitude (cf. Sec. 14.5).

14.3.3. Other Total Absorption Methods

(a) The Manganese-Bath Method. The source is placed in the middle of a sufficiently large vessel that contains an aqueous solution of $MnSO_4$. A large fraction of the moderated neutrons are then captured by the manganese, leading to the well-known 2.58-hour activity of Mn^{56}. At the end of the irradiation, the solution is carefully stirred to uniformly distribute the activity, a definite volume is taken out, and the activity of the Mn^{56} it contains absolutely counted. If we could make the concentration of the manganese in the water so high that all the neutrons would be absorbed in the manganese, this method would obviously yield a result independent of any knowledge of the cross sections. In practice only about 50% of the neutrons are absorbed in the manganese, and the absorption cross sections of Mn and H_2O must be used in the evaluation. In addition, neutron absorption in the sulfur cannot entirely be neglected. The accuracy of absolute determination of source strengths by this method is about 2%. This error is mainly due to the difficulty of absolutely counting the manganese activity. The manganese-bath method is much more accurate when used as a comparison method (cf. Sec. 14.4).

(b) The Boron-Bath Method. Here the absorption of the neutrons emitted from the source occurs in an aqueous solution of a boron compound, e.g., boric acid. If N_B (N_H) is the number of boron (hydrogen) atoms per cm³, then neglecting resonance absorption and absorption in the source we have

$$\left.\begin{aligned} Q &= (N_B \sigma_B + N_H \sigma_H) \int_0^\infty 4\pi\, r^2\, \Phi(r)\, dr \\ &= N_B \sigma_B \left(1 + \frac{N_H \sigma_H}{N_B \sigma_B}\right) \int_0^\infty 4\pi\, r^2\, \Phi(r)\, dr. \end{aligned}\right\} \tag{14.3.4}$$

Now we determine $\Phi(r)$ with a small boron counter. If V_z is the sensitive volume and N_z the number of boron atoms per cm³ of counter volume, the counting rate is given by

$$Z(r) = \Phi(r) N_z V_z \sigma_B. \tag{14.3.5}$$

If we substitute this in Eq. (14.3.4), we obtain

$$Q = \frac{N_B}{N_z V_z} \left(1 + \frac{N_H \sigma_H}{N_B \sigma_B}\right) \int_0^\infty 4\pi\, r^2 Z(r)\, dr. \tag{14.3.6}$$

The cross sections of boron and hydrogen only enter here through the correction term $N_H \sigma_H / N_B \sigma_B$, which for high boron concentration is small compared to one. By variation of N_H/N_B, we can make the method entirely independent of the cross sections. The achievable accuracy is about 2%.

(c) Total Absorption in a Liquid Scintillator. The neutron source is placed in the interior of a very large liquid scintillator in which a cadmium salt has been dissolved. The neutrons are moderated by collisions with the atoms of the scintillating substance and are finally captured by the cadmium. Capture by the cadmium initiates a complicated γ-cascade with a total energy of about 9 Mev. The scintillations produced by these γ-rays are detected by photomultipliers on the surface of the tank. An advantage of this method is that there is a prompt ($\lesssim 20\ \mu$sec) response to the neutrons so that coincidence methods are possible. The coincidence technique is frequently used, for example, in measurements of the average fission multiplicity $\bar{\nu}$. A disadvantage of this method is the extreme γ-sensitivity of the tank; cosmic radiation alone produces a high background counting rate, and it is necessary to suppress low-energy γ-rays by use of an integral discriminator. In this way, however, we lose some of the γ-rays arising from true capture events, i.e., the detection efficiency of the apparatus is no longer 100%. Thus we must either calibrate the apparatus or calculate the detection efficiency with the help of a Monte Carlo calculation.

14.4. Methods for the Comparison of Source Strengths

If we possess a standard neutron source that has been calibrated by one of the methods discussed in Sec. 14.3, we can refer the strength of an arbitrary source to it as long as there is some suitable comparison technique. Such a comparison is simple when the source to be compared with the standard has the same energy spectrum as the latter and when the angular distributions of the emitted neutrons also agree. Then all we have to do is simply to compare the fluxes arising from the two sources with an arbitrary detector in a reproducible way. If the energy spectra of the sources to be compared do not agree, the comparison can be carried out under some circumstances by making flux measurements with the long counter discussed in Sec. 3.1.1; however, in this procedure, deviations in the two angular distributions must be carefully taken into account by an integration over all angles. Even more satisfactory are comparison methods in which not only is the detection sensitivity independent of the neutron energy, but the neutrons emitted from the source are also counted in a 4π-geometry. In principle, all total absorption methods fall into this category. The manganese bath method discussed in Sec. 14.3.3 is very frequently used for source comparison. Since the main source of error — the absolute counting of the manganese activity — drops out, the method is extremely accurate ($\approx 0.5\%$) when used for comparison measurements. (The error estimate does not include errors which may arise from the absorption of energetic neutrons in oxygen. However, for neutron energies less than 3.6 Mev, these latter errors do not occur.) The scintillator tank method also serves very frequently for comparison measurements; the tank is calibrated once and then used for source strength measurements.

In addition, much specific comparison apparatus has been developed that operates in a 4π-geometry with a detection efficiency that is nearly independent of the spectrum. We shall become more familiar with some of this apparatus in the following.

(a) The Boron Pile. Fig. 14.4.1 shows an apparatus developed by Colvin and Sowerby that is mainly used to measure the fission neutron multiplicity $\bar{\nu}$. It consists of 240 BF_3-counters in a graphite cube 220 cm on a side. This sensitive zone is surrounded by a graphite reflector which in turn is surrounded by an external neutron shield. The source to be investigated is placed in the middle of the boron pile.

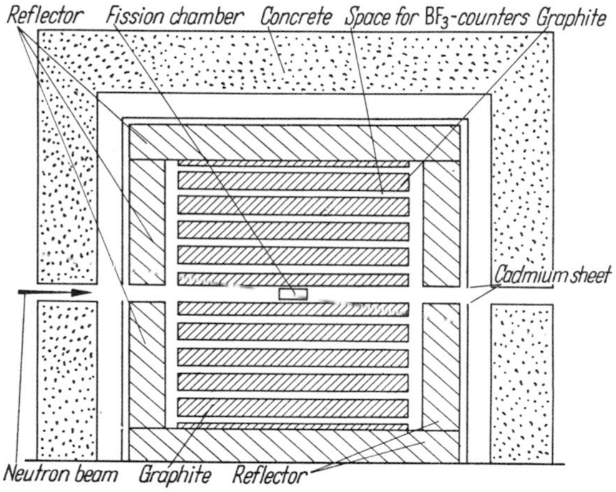

Fig. 14.4.1. The Boron pile

The pile is calibrated by absolute counting of the photoprotons from the reaction $H^2(\gamma, n)H^1$. For this calibration, an ionization chamber containing deuterated methane (CD_4) is bombarded by γ-rays of various energies [obtained from (p, γ) reactions in light nuclei]. The sensitivity determined in this way is practically independent of the neutron energy. The sensitivity depends somewhat on the direction of the emitted neutron; this dependence comes from the strongly anisotropic construction of the boron pile. This directional sensitivity and the slight deviations of the sensitivity curve from flattness must be taken into account in measurements on other sources, but we only need to know the spectrum of the source to be investigated very roughly.

In using the boron pile to measure $\bar{\nu}$, one places the fissionable material in the center of the boron pile in the form of a fission chamber and bombards it with an external neutron beam. Each time a fission process takes place, the pulse from the fission chamber opens an electronic gate and allows the signals from the BF_3-counters to be recorded in a counter for about 15 msec. In addition, the number of pulses from the fission chamber is recorded. $\bar{\nu}$ follows from the ratio of the number of BF_3-signals to the number of fission chamber signals after correction for lost counts, background, and the energy dependence of the sensitivity of the boron pile. The sensitivity of the method to background is considerably reduced by use of the electronic gate technique; similar methods are used in $\bar{\nu}$-measurements in liquid scintillators.

The accuracy with which $\bar{\nu}$ can be determined in a boron pile is about 0.5%.

(b) Macklin's Sphere. Fig. 14.4.2 shows a simple arrangement devised by Macklin of eight BF_3-counters in a graphite sphere. The neutron source is introduced into the middle of the sphere through a 10×10 cm canal. MACKLIN calculated the sensitivity of this apparatus with the help of age and diffusion theory; it is constant to within about 1% of its value between 1 kev and 1 Mev and is somewhere in the neighborhood of 0.035. However, it falls off rapidly for

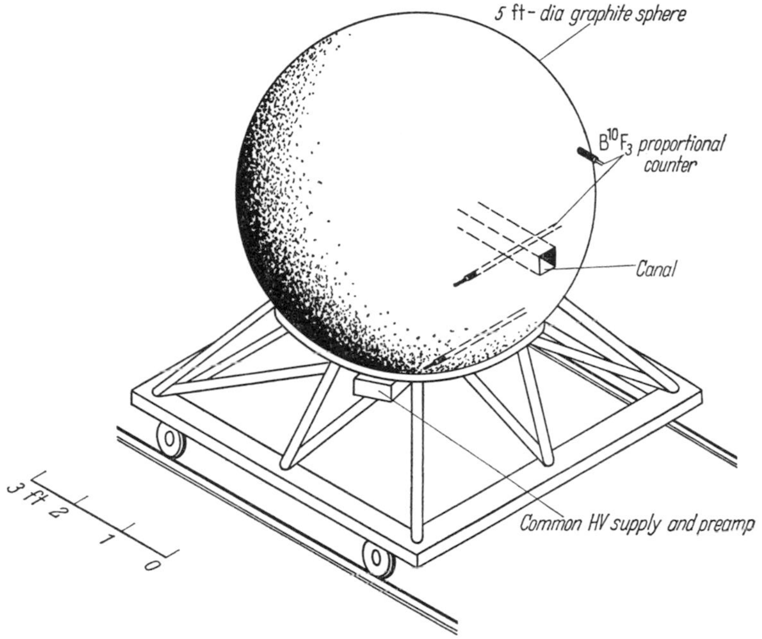

Fig. 14.4.2. A graphite sphere neutron detector (MACKLIN)

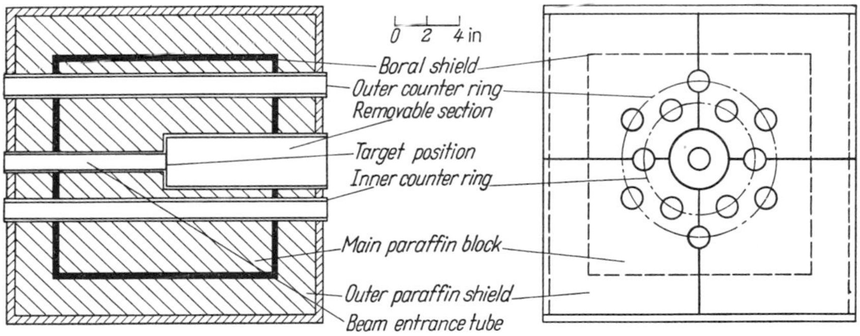

Fig. 14.4.3. A neutron detector for source strength comparisons (MARION *et al.*)

higher energies. MACKLIN was able to show experimentally that the sensitivity does not depend on the angular distribution and that the leakage of neutrons out of the canal left open for the introduction of the source has no appreciable effect on the sensitivity. This apparatus is particularly suitable for the study of neutron-producing nuclear reactions with accelerators.

(c) Other Apparatus. Various arrangements of BF_3-counters in water or paraffin have been used. As an example, we show in Fig. 14.4.3 a paraffin detector investigated by MARION *et al.* A total of 12 BF_3-counters are arranged in two concentric rings in the middle of which is located a canal for introducing the neutron source (accelerator target). The sensitivity of this apparatus is about 0.1 and it is constant within about 5% of its value in the energy range from 100 kev to 2.5 Mev. Compared to graphite detectors, we can easily attain a higher sensitivity with smaller dimensions, but because of the shorter diffusion length we cannot build detectors whose sensitivity plateau reaches into the low kev region.

(d) Use of a Subcritical or Critical Reactor. According to EGGLER and WATTENBERG, one can compare different neutron sources by placing them in a slightly subcritical reactor and then comparing the resulting thermal fluxes. In order that the comparison yield a result independent of the source spectrum, the reactor must be so large that differences in the escape probability of the various fast neutrons do not come into consideration. This method is hardly used any more. If one introduces a neutron source into an exactly critical reactor, the latter will show an increase in power that is linear in time and whose magnitude depends on the excess neutron production. We can compensate the excess neutron production by introduction of a negative source, i.e., an absorber. In steady state, the absorption rate is equal to the source strength. If the absorption rate can be determined by absolute counting of the absorber activity, the source strength is also known. Upon these principles depends the method of LITTLER. The accuracy of this method is limited by the fact that the additional neutrons are emitted at high energy but absorbed at thermal energy, and thus corrections must be applied for resonance absorption and leakage.

14.5. A Comparison of Various Source-Strength Measurements on (Ra—Be) Sources

In order to check the various procedures for determining source strengths and also in order to obtain an internationally recognized standard of source strength, comparisons have frequently been made of the sources absolutely calibrated by

Table 14.5.1. *International Comparison Measurements on (Ra—Be) Sources*

Laboratory	Strength of the Primary Source (n/sec) According to Own Measurement	Strength of the Same Source as Measured by Other Laboratories			
		PTB	ABA	UKAEA	UM
NRC Canada	3.21×10^6 $\pm1.5\%$	3.227×10^6 $\pm2.2\%$	3.18×10^6 $\pm2.2\%$		3.234×10^6 $\pm2.2\%$
PTB Germany	1.95×10^6 $\pm2\%$		1.937×10^6 $\pm2.2\%$		1.975×10^6 $\pm2\%$
AB Atomenergi Sweden	2.65×10^6 $\pm2\%$	2.658×10^6 $\pm2\%$			2.689×10^6 $\pm2\%$
UKAEA Harwell England	9.08×10^6 $\pm1.7\%$		9.15×10^6 $\pm2.2\%$		
Union Minière Belgium	7.88×10^6 $\pm2\%$	7.88×10^6 $\pm2\%$			
Univ. Basel Switzerland	1.52×10^6 $\pm2.8\%$		1.516×10^6 $\pm2.2\%$	1.560×10^6 $\pm1.7\%$	
USSR	5.96×10^6 $\pm3\%$		6.00×10^6 $\pm2.2\%$		

various laboratories. Such investigations have mainly been made with (Ra—Be) sources, which because of their long half-life and good yield are particularly suitable. In these investigations, particular attention must be paid to the increase in time of the source strength due to the buildup of polonium. This increase is described by

$$Q(t_1) = Q(t_0)\,\frac{1+0.14\,(1-e^{-0.0357\,(t_0+t_1)})}{1+0.14\,(1-e^{-0.0357\,t_0})}\,. \qquad (14.5.1)$$

Here $Q(t_0)$ is the source strength determined at the time of calibration t_0 (years) — reckoned from the time of radium extraction — and $Q(t_1)$ is the source strength t_1 years after calibration. This formula of course only holds when the initial increase of the source strength (Sec. 2.2.1) is ended. Table 14.5.1 contains the results of some international comparison measurements (as of February 1962). The agreement is generally satisfactory; we can moreover conclude that source strength measurements can be carried out today with an accuracy of *ca.* 1.5%.

Chapter 14: References

General

HUGHES, D. J.: Pile Neutron Research, Cambridge: Addison-Wesley, 1953; especially p. 72ff.: Neutron Standardization.

LARSSON, K. E.: J. Nucl. Energy **6**, 322 (1958).

Neutron Dosimetry, Proceedings of the 1962 Harwell Symposium, Vol. 1 and 2. Vienna: International Atomic Energy Agency, 1963.

RICHMOND, R.: Progr. Nucl. Energy, Ser. I, Vol. 2, p. 165 (1958).

WATTENBERG, A.: The Standardization of Neutron Measurements. Ann. Rev. Nucl. Sci. **3**, 119 (1953).

Special[1]

GRIMELAND, B.: Phys. Rev. **86**, 937 (1952) (Absolute Flux Measurement with NaI).

COHEN, R.: Ann. Phys. (Paris) XII, **7**, 185 (1952).
HOUTERMANS, F. G., L. MEYER-SCHÜTZMEISTER u. D. H. VINCENT: Z. Physik **134**, 1 (1952).
PATE, B. D., and L. YAFFE: Can. J. Chem. **33**, 15 (1955).
— 4π-Counter.

MOSBURG, E. R., and W. M. MURPHEY: J. Nucl. Energy A & B **14**, 25 (1961).
PUTMAN, J. L.: Brit. J. Radiol. **23**, 46 (1950).
RAFFLE, J. F.: J. Nucl. Energy A **10**, 8 (1959).
— β-γ Coincidence Method.

CAMPION, P. J.: Int. J. Appl. Radiation Isotopes **4**, 232 (1958/59).
WOLF, G.: Nukleonik **2**, 255 (1961).
— 4π β-γ Coincidence Method.

BELCHER, E. H.: J. Sci. Instr. **30**, 286 (1953).
STEYN, J.: in Metrology of Radionuclides, p. 279; Vienna: International Atomic Energy Agency, 1960.
— 4π β-Scintillation Counting.

MEISTER, H.: Z. Naturforsch. **13**a, 722 (1958).
PÖNITZ, W.: KFK-180 (1963).
— Self-Absorption in 4π β-Counting.

COCKING, S. J., and J. F. RAFFLE: J. Nucl. Energy **3**, 70 (1956).
SEREN, L., H. N. FRIEDLANDER, and S. H. TURKEL: Phys. Rev. **72**, 888 (1947).
STOUGHTON, R. W., and J. HALPERIN: Nucl. Sci. Eng. **6**, 100 (1959).
WOLF, G.: Nukleonik **2**, 255 (1961).
— Determination of Activation Cross Sections for Thermal Neutrons.

GAMMEL, J. L.: In MARION and FOWLER, loc. cit., Vol. II, Chapter V. T.
PERRY, J. E.: In MARION and FOWLER, loc. cit. Vol. I, Chapter IV. B.
— Absolute Flux Measurement with Recoil Protons.

[1] Cf. footnote on p. 53.

BATCHELOR, R. (editor): EANDC-33 U (1963) (Absolute Flux Measurement in the Energy Range 10—100 kev).
DEBRUE, J.: CEN-Report R 1860 (1960) (The Standard Pile).

BARSCHALL, H. H., L. ROSEN, and R. F. TASCHEK: Rev. Mod. Phys. **24**, 1 (1952). LARSSON, K. E.: Arkiv Fysik **9**, 293 (1954). MARIN, P. C.: Nucl. Instrum. Methods **5**, 26 (1959). RICHMOND, R., and B. J. GARDNER: AERE-R/R 2097 (1957).	Source Strength Measurement by the Method of Associated Particles.
LARSSON, K. E.: Arkiv Fysik **7**, 323 (1954). PLANTA, C., and P. HUBER: Helv. Phys. Acta **29**, 375 (1956). TROYER, A. DE, et G. C. TAVERNIER: Bull. Acad. R. Belg., Cl. sci., Sér. 5, **40**, 150.	The Water-Bath Gold Method.
CURTISS, L. F.: Introduction to Neutron Physics. Princeton: D. von Nostrand Company Inc. (1959). HAGE, W.: Nukleonik **2**, 73 (1960). WALKER, R. L.: MDDC-414 (1946).	Flux Integration with Mechanically Moved Foils.
AXTON, E. J., and P. CROSS: J. Nucl. Energy A&B **15**, 22 (1961). FISHER, G. J.: Phys. Rev. **108**, 99 (1957). JUREN, J. DE, and J. CHIN: J. Res. Nat. Bur. Stand. **55**, 311 (1955). WALKER, R. L.: MDDC 414 (1946).	The Manganese Bath Method.

O'NEAL, R. D., and G. SCHARFF-GOLDHABER: Phys. Rev. **69**, 368 (1946) (The Boron Bath Method).

DIVEN, B. C., *et al.*: Phys. Rev. **101**, 1012 (1956). REINES, F., *et al.*: Rev. Sci. Instr. **25**, 1061 (1954).	Liquid Scintillators for Source-Strength Measurements.
COLVIN, D. W., and M. G. SOWERBY: Geneva 1958 P/52, Vol. 16, p. 121. COLVIN, D. W., and M. G. SOWERBY: EANDC(UK)-3 (1960). LITTLER, D. J.: Proc. Phys. Soc. (London) A **64**, 638 (1951). MACKLIN, R. L.: Nucl. Instrum. Methods **1**, 335 (1957). MARION, J. B., *et al.*: Nucl. Instrum. Methods 8, 297 (1960). WATTENBERG, A., and C. EGGLER: AECD 3002 (1950).	Methods of Source-Strength Comparison.

15. Investigation of the Energy Distribution of Slow Neutrons

In this chapter, we shall consider experiments for determining the stationary spectra of very slow ($E < 10$ ev) neutrons in bulk media. The main part of the chapter is centered around the discussion of various methods of measurements (Secs. 15.1 and 15.2); in the last short section we summarize some important results for various moderator substances and relate them to the general theory of Chapter 10.

We must distinguish between *differential* and *integral* methods of measurement. The former are based on the detailed energy analysis by time-of-flight or with a crystal spectrometer of a neutron beam extracted from the medium. They immediately yield the spectrum $\Phi(E)$. In contrast, we only obtain from integral measurements certain averages over the spectrum for whose further evaluation we need additional knowledge of the spectrum. For example, we can determine the neutron temperature if we know that the thermal part of the spectrum can be approximated by a Maxwell distribution.

With regard to the physical information they supply, the differential methods are therefore far superior to the integral methods. However, they require far more apparatus and are limited in their applicability to relatively simple systems.

For example, it is usually not possible to extract from a reactor lattice a neutron beam that is representative of the entire neutron field. On the other hand, integral probe measurements can nearly always be carried out. Today, therefore, we use differential methods for experiments on "clean" systems; these experiments are directed toward revealing the basic physical character of the spectrum $\Phi(E)$ and toward aiding the development of a general theory. Integral measurements are then used to check the theory for special systems that are not always accessible to the differential measurements.

Before the introduction of sufficiently reliable differential methods, integral measurements yielded the only usable information on thermalization in a stationary neutron field.

15.1. Observation of the Differential Spectrum by the Time-of-Flight Method

A differential spectrum can be studied with a crystal spectrometer or by the time-of-flight method. Although many very careful investigations have been carried out with crystal spectrometers, this method has not prevailed over the time-of-flight method, chiefly because of the difficulty of calculating the energy dependence of the spectrometer's sensitivity. We shall limit ourselves therefore in the following to discussing the time-of-flight method. First, however, we shall discuss a problem common to all differential methods, viz., the extraction of a representative neutron beam from a medium.

15.1.1. The Problems of Beam Extraction

In order to obtain a neutron beam for the spectral analysis of a neutron field, we use an extraction canal that reaches from the surface of the medium to the point $\boldsymbol{r}$ at which the spectrum is to be observed. Such a canal generally has the form of a thimble, as shown in Fig. 15.1.1. Let us denote by $J(E)$ [$\text{cm}^{-2}\,\text{sec}^{-1}\,\text{ev}^{-1}$] the energy-dependent current density of the beam so obtained. What we wish to have is the energy spectrum $\Phi(\boldsymbol{r}, E)$ of the flux at the point $\boldsymbol{r}$. We shall now determine the connection between $\Phi(\boldsymbol{r}, E)$ and $J(E)$ under the assumption that the perturbation of the neutron field resulting from the introduction of the extraction canal can be neglected. Following that, we shall estimate this perturbation.

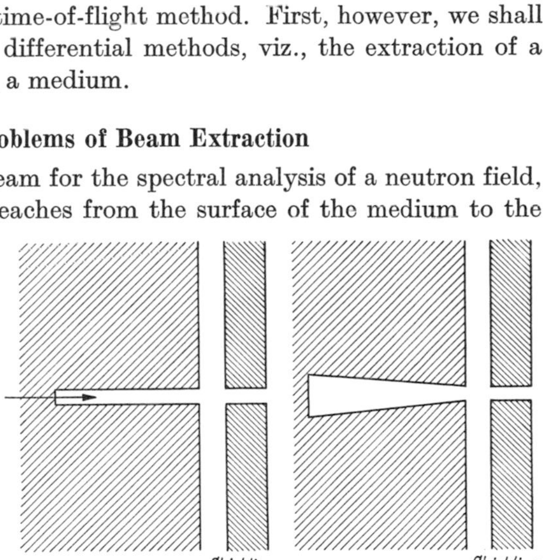

Fig. 15.1.1. Typical extraction channels for neutron spectrum measurements. In a liquid moderator, a thimble made of a weakly absorbing substance like aluminum must be used

If $\boldsymbol{\Omega}$ is the direction of the extraction canal, then obviously $J(E) \sim F(\boldsymbol{r}, \boldsymbol{\Omega}, E)$ (F is the differential vector flux). The beam spectrum is thus representative of the spectrum of the vector flux in the beam direction and not of the angle-integrated flux $\Phi(\boldsymbol{r}, E)$. We must therefore either integrate over all directions $\boldsymbol{\Omega}$ or find a simple relation between $\Phi(\boldsymbol{r}, E)$ and $F(\boldsymbol{r}, \boldsymbol{\Omega}, E)$.

The latter relation takes a particularly simple form in an isotropic neutron field. There $F(\boldsymbol{r}, \boldsymbol{\Omega}, E)$ is simply $\Phi(\boldsymbol{r}, E)/4\pi$ and $J(E) \sim \Phi(\boldsymbol{r}, E)$. We should thus take the beam whenever possible out of a region of isotropic flux, i.e., a region of vanishing flux gradient. If there is a weak dependence of $\Phi(\boldsymbol{r}, E)$ on position, we can take the connection between $\Phi(\boldsymbol{r}, E)$ and $F(\boldsymbol{r}, \boldsymbol{\Omega}, E)$ from diffusion theory:

$$F(\boldsymbol{r}, \boldsymbol{\Omega}, E) = \frac{1}{4\pi}\left[\Phi(\boldsymbol{r}, E) + 3D(E)\boldsymbol{\Omega}\cdot\nabla\Phi(\boldsymbol{r}, E)\right]. \tag{15.1.1}$$

Thus we have

$$\left.\begin{aligned} J(E) &\sim \Phi(\boldsymbol{r}, E) + 3D(E)\frac{\partial\Phi(\boldsymbol{r}, E)}{\partial z} \\ &\sim \Phi(\boldsymbol{r}, E)\left[1 + 3D(E)\frac{\partial \ln\Phi(\boldsymbol{r}, E)}{\partial z}\right] \end{aligned}\right\} \tag{15.1.2}$$

or approximately

$$\Phi(\boldsymbol{r}, E) \sim J(E)\left[1 - 3D(E)\frac{\partial \ln\Phi(\boldsymbol{r}, E)}{\partial z}\right]. \tag{15.1.3}$$

Here $\partial/\partial z$ signifies differentiation in the direction of the extraction canal (and in fact in the direction medium $\to$ canal) at the point $\boldsymbol{r}$ of observation. Thus we must apply a correction to the measured $J(E)$ for whose calculation $D(E)$ and the flux gradient must be known. Under the assumption that the flux gradient does not depend on energy, it can be determined from probe measurements; $D(E)$ must be calculated approximately[1].

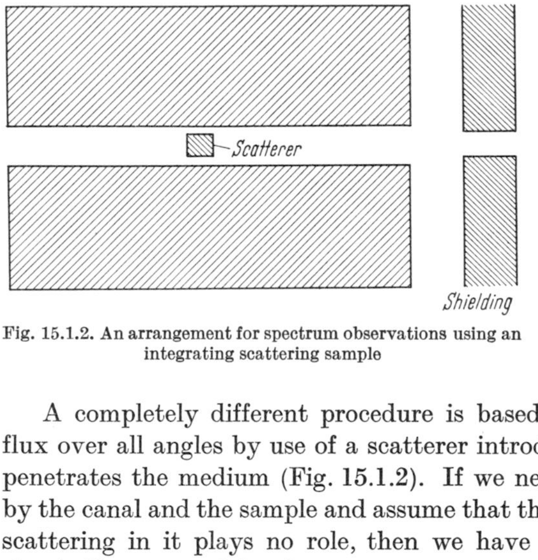

Fig. 15.1.2. An arrangement for spectrum observations using an integrating scattering sample

This procedure only gives reliable values of $\Phi(\boldsymbol{r}, E)$ when the corrections are small (less than about 20%).

A completely different procedure is based on the integration of the vector flux over all angles by use of a scatterer introduced into a canal that completely penetrates the medium (Fig. 15.1.2). If we neglect the perturbation of the field by the canal and the sample and assume that the scatterer is so small that multiple scattering in it plays no role, then we have for the spectrum of the neutrons elastically scattered by the scatterer

$$J(E) \sim \int F(\boldsymbol{r}, \boldsymbol{\Omega}', E)\,\Sigma_s(\boldsymbol{\Omega}' \to \boldsymbol{\Omega}, E))\, d\Omega' \tag{15.1.4}$$

or for isotropic scattering

$$J(E) \sim \Sigma_s(E) \int F(\boldsymbol{r}, \boldsymbol{\Omega}', E)\, d\Omega' = \Sigma_s(E)\,\Phi(\boldsymbol{r}, E).$$

An additional contribution due to inelastically scattered neutrons may also appear. In an ideal scatterer, this contribution should be negligibly small; in addition,

[1] For H_2O, where the energy dependence of the diffusion constant is particularly strong, we can use the values of $\overline{\cos\vartheta_0}\,(E)$ and $\sigma_t(E)$ given in Figs. 10.1.9 and 10.1.8. In graphite and beryllium, D is nearly constant above the cut-off energy.

the elastic scattering should be as isotropic as possible, the absorption cross section as small as possible, and the scattering cross section as energy-independent as possible. Zirconium metal approaches these specifications (BEYSTER *et al.*); lead, graphite, and D_2O have also been used (JOHANSSON *et al.*).

The question of what *perturbations* a neutron spectrum suffers owing to the introduction of an extraction canal has not as yet been theoretically investigated in detail. In this connection, one must carefully distinguish between homogeneous and heterogeneous systems according to whether the spectrum changes slowly or rapidly from point to point inside the system. In a heterogeneous system, the introduction of an extraction canal can perturb the spectrum to be studied very strongly since neutrons from other regions, i.e., neutrons with an entirely different energy distribution, can penetrate to the front surface of the canal. In homogeneous systems, this does not occur. We shall assume that other spectrum perturbations remain small as long as the diameter of the extraction canal is small compared to the mean free path of the neutrons. Experiments on large graphite and water systems (POOLE, BEYSTER) show that this is indeed the case and that furthermore the observed energy distribution $J(E)$ hardly changes when the diameter of the extraction canal is increased to about 5 cm. One comes to this conclusion on one hand by comparison with spectra that have been taken with extremely small canal diameters (for which there is surely no perturbation) and on the other from integral measurements in which one compares the activation ratio Lu^{176}/Lu^{175} (Sec. 15.2.2) at the point $\boldsymbol{r}$ with and without a canal.

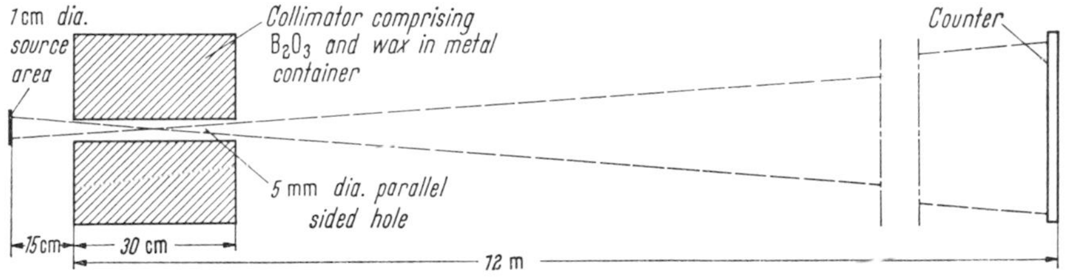

Fig. 15.1.3. The collimator arrangement used by POOLE for neutron spectrum measurements (cf. also Fig. 15.1.9)

By means of Fig. 15.1.3, we next call attention to the important problem of the *collimator* that must be used whenever we extract a beam. It should guarantee that only those neutrons that enter the flight tube through the front surface of the extraction canal can reach the detector. It must be made of a material that strongly absorbs slow neutrons, such as boron carbide (B_4C), boron metal, or even better, boron-10 metal, and must have a form that is well adapted to the optical conditions.

15.1.2. The Chopper Method

Fig. 15.1.4 explains the principle of a chopper for the measurement of spectra. The beam leaving the medium is chopped into short pulses by a rotating interrupter; the pulses traverse a flight tube at whose end is a detector. The energy distribution of $J(E)$ can be determined from the time distribution of detector signals, providing that

(a) the resolving power of the system is sufficiently high. In other words, the uncertainty Δt in the flight time (caused both by the chopper and the detector) should be small compared to the flight time of the neutrons along a flight tube of length l. We shall return to the general connection between the energy spectrum and the flight-time distribution with finite resolving power again in Sec. 15.1.4; suffice it now to say that in practice $\Delta t/l$ is most often around 10 μsec/m (thus e.g., $\Delta t = 50$ μsec and $l = 5$ m).

(b) the transmission function of the chopper and the efficiency of the detector are accurately known either from calculations or from measurements.

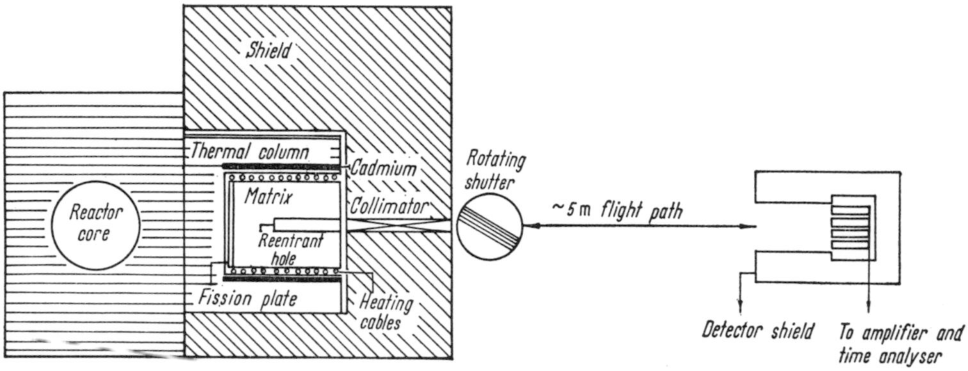

Fig. 15.1.4. The chopper set-up used by STONE and SLOVACEK for slow neutron spectrum measurements

(c) all perturbing effects are either suppressed or very carefully taken into account. For this purpose, the detector, or even better, the entire apparatus, must be shielded against stray neutrons. In addition, we must avoid, wherever possible, scattering or absorption of neutrons on their way to the detector. The perturbing air scattering that occurs on long flight paths can be suppressed by use of an evacuated flight tubes.

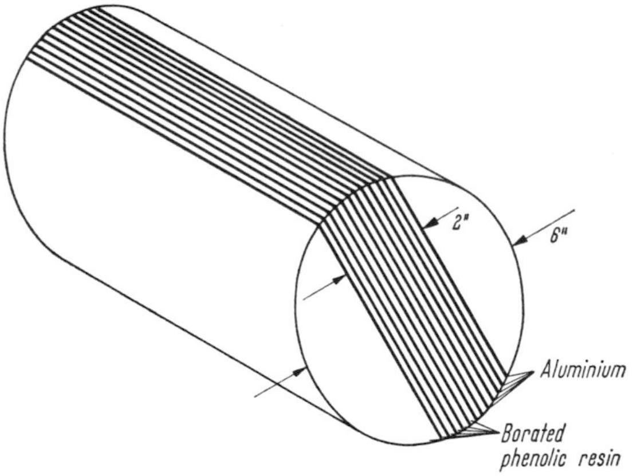

Fig. 15.1.5. A typical rotor for slow-neutron spectrum measurements

The detector of choice is the BF_3-counter, many of which can be connected together in the form of a counter bank in order to obtain as large a sensitive area as possible (cf. Fig. 15.1.4), To increase the sensitivity, particularly at higher neutron energies, several counter banks can be arranged one behind the other[1]. The energy-dependent detection sensitivity of a counter bank can be calculated from its dimensions, the gas pressure, and the cross section of the $B^{10}(n, \alpha)Li^7$ reaction; in doing this calculation we must average over the surface of the bank. Neutron absorption in the walls of the counters must also be taken into account.

[1] M. FORTE describes a simple electronic delay device for compensating for the time-of-flight difference between the individual banks in the Proceedings of the Conference on Neutron Time-of-Flight Methods, Brussels: Euratom 1961, p. 457.

One can also easily determine the sensitivity experimentally by comparison with an extremely thin counter filled with natural boron, whose sensitivity accurately follows the $1/v$-law.

Fig. 15.1.5 shows a rotor used for the analysis of neutron spectra. Since the spectra to be studied usually contain epithermal neutrons, cadmium cannot be used to make the walls defining the slits; instead we must use a substance that absorbs neutrons as strongly as possible at all energies. Plastics loaded with boron or suitable alloys, such as K-Monel (30% Cu, 70% Ni), are customary. The form of the slits is very important in determining the transmission function of the system. Let us discuss this point further for the case of a rotor of radius R with a plane slit of width h (cf. Fig. 15.1.6). Let the rotor turn at a constant angular velocity ω in the counterclockwise direction. At $t=0$ $\left(\text{and also at } t=\pm\frac{n\pi}{\omega},\ n=1, 2, 3, \ldots\right)$, let the slit be parallel to the direction of incidence. Let the parallel incident beam be trimmed to the slit width h by an entrance collimator. Let h be $\ll R$. Then according to STONE and SLOVACEK, the chopping action of this arrangement is the same as that of a slit at $x=0$ with a time-dependent slit width $h\cdot W(v, t)$. $W(v, t)$, the fraction of the transmitted neutrons, is in the velocity range $4\omega R^2/h \leqq v \leqq \infty$:

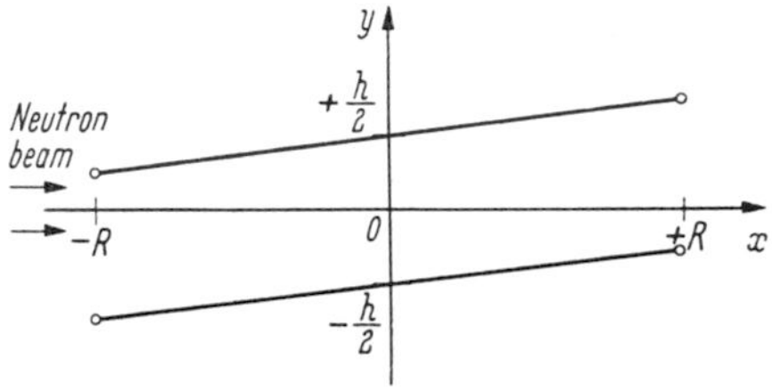

Fig. 15.1.6. Derivation of the chopper transmission function

$$\left.\begin{aligned}
&W(v,t)=0, && t\leqq -h/2\omega R,\\
&W(v,t)=1+\frac{2\omega R}{h}t, && -\frac{h}{2\omega R}\leqq t\leqq -\frac{2R}{v},\\
&W(v,t)=1-\frac{v\omega}{4h}\left(\frac{2R}{v}-t\right)^2, && -\frac{2R}{v}\leqq t\leqq 0,\\
&W(v,t)=1-\frac{v\omega}{4h}\left(\frac{2R}{v}+t\right)^2, && 0\leqq t\leqq \frac{2R}{v},\\
&W(v,t)=1-\frac{2\omega R}{h}t, && \frac{2R}{v}\leqq t\leqq \frac{h}{2\omega R},\\
&W(v,t)=0, && \frac{h}{2\omega R}\leqq t,
\end{aligned}\right\}\qquad (15.1.5)$$

in the velocity range $\omega R^2/h \leqq v \leqq 4\omega R^2/h$:

$$\left.\begin{aligned}
&W(v,t)=0, && t\leqq -\sqrt{\frac{4h}{\omega v}}+\frac{2R}{v},\\
&W(v,t)=1-\frac{\omega v}{4h}\left(\frac{2R}{v}-t\right)^2, && -\sqrt{\frac{4h}{\omega v}}+\frac{2R}{v}\leqq t\leqq 0,\\
&W(v,t)=1-\frac{\omega v}{4h}\left(\frac{2R}{v}+t\right)^2, && 0\leqq t\leqq \sqrt{\frac{4h}{\omega v}}-\frac{2R}{v}.\\
&W(v,t)=0, && \sqrt{\frac{4h}{\omega v}}-\frac{2R}{v}\leqq t.
\end{aligned}\right\}\qquad (15.1.6)$$

For $v<\omega R^2/h$, $W(v, t)$ is always equal to zero. Furthermore, $W(v, t)$ is periodic in time with a period equal to half the period of rotation, i.e., $W(v, t)=$

$W\left(v, t \pm \frac{n\pi}{\omega}\right)$. Figs. 15.1.7a and b show $W(v, t)$ for two typical values of v. We see that in the limiting case $v \to \infty$, $W(v, t)$ becomes a simple triangle function with a base width $h/\omega R$. At smaller neutron velocities, the finite neutron time of flight makes itself felt: neutrons that enter the slit at a particular slit orientation and that could go all the way through if the rotor were stationary in fact collide with the moving rotor wall. In particular, neutrons with the velocity $v < \frac{\omega R^2}{h} = v_{c0}$ cannot get through the rotor at all since the time $2R/v$ that they need to traverse the rotor is greater than the maximum time $2h/\omega R$ the rotor is open. The following

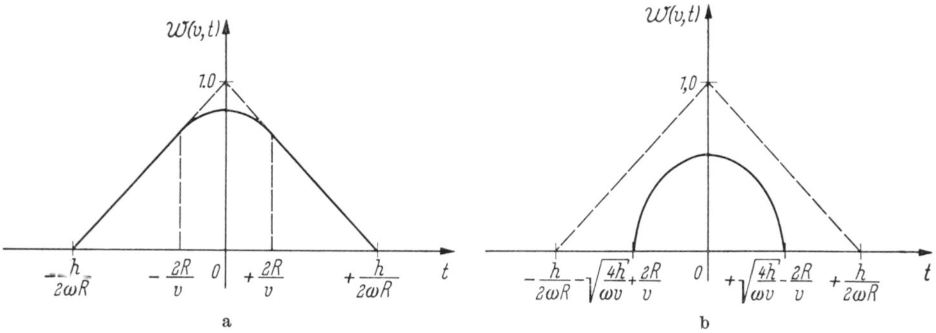

Fig. 15.1.7a and b. Typical burst shapes from a flat plate chopper. a: $v > 4\omega R^2/h$; b: $v < 4\omega R^2/h$

obvious relation exists between the base width $2 \cdot \Delta t$ (which is important in determining the resolving power), the limiting velocity v_{c0}, and the rotor radius R:

$$R = 2 \cdot v_{c0} \Delta t. \tag{15.1.7}$$

The choice of design specifications is sharply limited by this relation. For example, if we wish to have $v_{c0} = 500$ m/sec and $2\Delta t = 100$ μsec, then it follows that $R = 5$ cm. Thus $h/\omega = 5 \times 10^{-4}$ cm sec and $\omega = 400$ sec^{-1} for $h = 2$ mm. We can decrease v_{c0} by choosing curved slits. Calculations for curved slits can be found in Marseguerra and Pauli.

More important than the function $W(v, t)$ is the transmission function $T(v)$, which is defined as the time-averaged probability of penetration for a neutron of velocity v. By integrating Eqs. (15.1.5) and (15.1.6) between the limits $-\frac{\pi}{2\omega} \leqq t \leqq \frac{\pi}{2\omega}$ and multiplying by ω/π, we find in the velocity range $4\omega R^2/h \leqq v \leqq \infty$:

$$T(v) = \frac{h}{2\pi R}\left[1 - \frac{8}{3}\frac{v_{c0}^2}{v^2}\right] \tag{15.1.8a}$$

in the velocity range $\omega R^2/h \leqq v \leqq 4\omega R^2/h$:

$$T(v) = \frac{h}{2\pi R}\left[\frac{8}{3}\frac{v_{c0}^2}{v^2} - 8\frac{v_{c0}}{v} + \frac{16}{3}\sqrt{\frac{v_{c0}}{v}}\right]. \tag{15.1.8b}$$

Frequently we introduce the variable $\beta = v_{c0}/v$ and form $\tau(\beta)$, the transmission relative to the transmission $\frac{h}{2\pi R}$ for neutrons with infinite velocity:

$$\tau(\beta) = 1 - \tfrac{8}{3}\beta^2, \qquad 0 \leqq \beta \leqq \tfrac{1}{4}, \tag{15.1.9a}$$

$$\tau(\beta) = \tfrac{8}{3}\beta^2 - 8\beta + \tfrac{16}{3}\sqrt{\beta}, \qquad \tfrac{1}{4} \leqq \beta \leqq 1. \tag{15.1.9b}$$

Fig. 15.1.8 shows $\tau(\beta)$ according to Eq. (15.1.9). This function enters directly as a factor in the energy distribution observed in a chopper experiment and must be carefully corrected for. The expressions for $\tau(\beta)$ are more complicated in the case of curved slits. The transmission function can be determined experimentally (cf. JOHANSSON *et al.*).

Chopper experiments for the determination of neutron spectra are described in more detail by STONE and SLOVACEK, by MOSTOVOI, by JOHANSSON, LAMPA

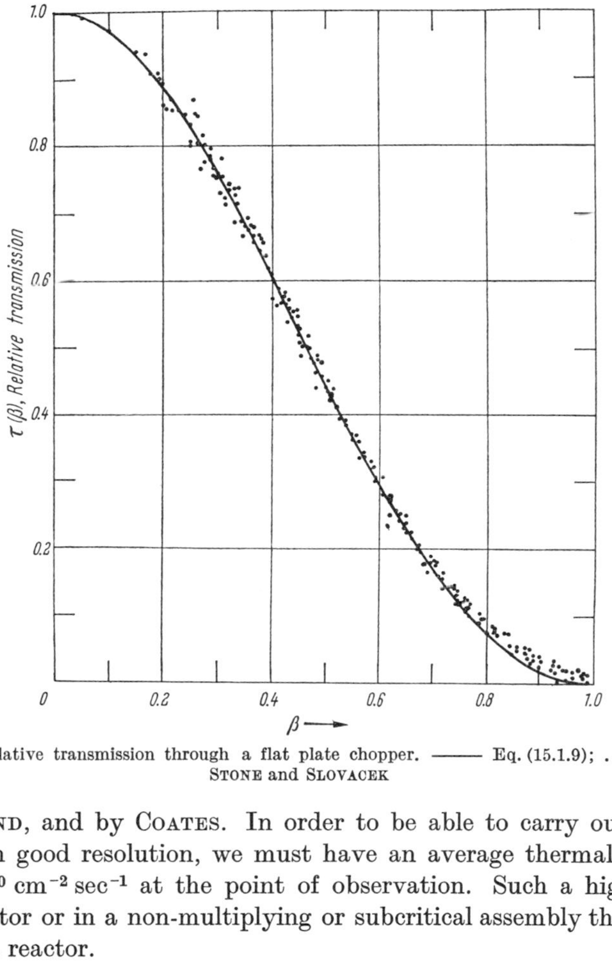

Fig. 15.1.8. The relative transmission through a flat plate chopper. ——— Eq. (15.1.9); ..... measured by STONE and SLOVACEK

and SJÖSTRAND, and by COATES. In order to be able to carry out a clean experiment with good resolution, we must have an average thermal flux of from 5×10^8 to 10^{10} cm^{-2} sec^{-1} at the point of observation. Such a high flux exists only in a reactor or in a non-multiplying or subcritical assembly that is fed with neutrons by a reactor.

15.1.3. The Method of Pulsed Sources

Fig. 15.1.9 shows another very useful arrangement for the time-of-flight measurement of neutron spectra. A pulsed source periodically emits short bursts of neutrons into the system being studied; the neutrons leaving the system after each pulse traverse a flight tube and are finally observed by the detectors. If we take care that the decay time of the neutron field after each pulse is short compared to the neutron flight time, we can obtain the energy spectrum from the flight-time distribution, just as in a chopper experiment.

In principle, the same considerations govern the design of the detectors and the flight tube here as in the chopper method. In non-multiplying media, to which we shall restrict ourselves here, the decay time in the epithermal range, i.e., for energies above 0.5 ev, is of the order of magnitude of the slowing-down time to the particular energy. On the other hand, in the thermal range, the decay is exponential with the time constant $\alpha = \overline{v\Sigma_a} + D_0 B^2$. $1/\alpha$ is usually much larger than the slowing-down time and is therefore decisive for the resolving power. In an infinite sample of pure water, for example, $1/\alpha$ is of the order of magnitude of 200 μsec compared to slowing-down times in the epithermal region of a few μsec. If we strive again for a resolving power of a least 10 μsec/m, the length of the

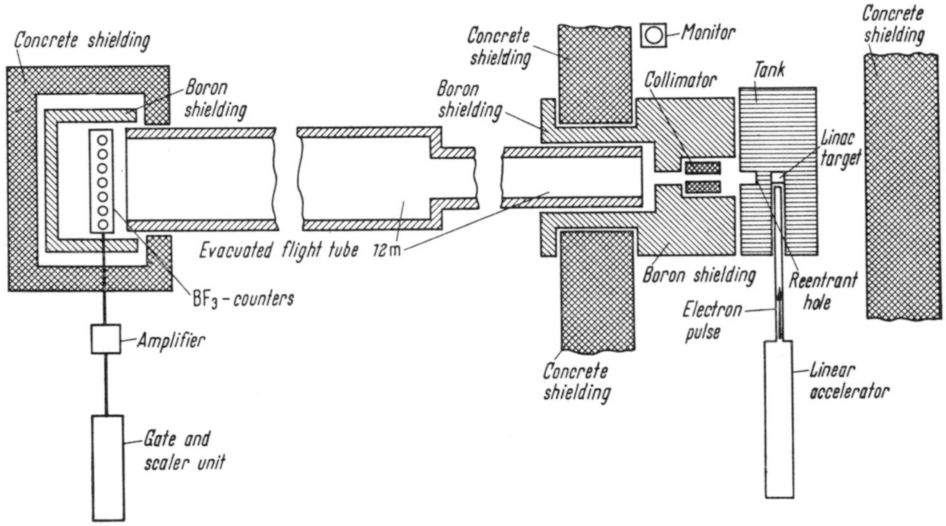

Fig. 15.1.9. The linear accelerator facility used by POOLE for slow-neutron spectrum measurements

flight path must be about 20 m in the case of pure water. The resolution in the epithermal region is then very much better (< 1 μsec/m). In graphite, the slowing-down time to 0.5 ev is about 25 μsec; the thermal lifetime can be as much as several milliseconds according to size and absorption. Thus flight paths of several hundred meters are necessary; such long flight paths are very costly and lead to large neutron losses. However, we can drastically reduce the thermal lifetime by strong poisoning of the moderator, i.e., by mixing it with an absorber. Similar considerations hold for D_2O and beryllium systems. In multiplying media the relaxation time is independent of the neutron energy and, for strong multiplication, is very large; here also long flight paths are necessary. In practice, flight-path lengths up to 60 m are used, i.e., only systems with relaxation times < 0.5 msec are considered; for longer-lived systems, the chopper method is preferable. A more precise discussion of resolution effects in the two methods follows in Sec. 15.1.4.

The particular advantages of the pulsed-source method compared to the chopper method are (i) no correction for the chopper transmission is necessary, and (ii) a much smaller average flux is necessary in the medium. Let us compare, for example, a chopper and a pulsed-source experiment with equally long flight paths and the same beam cross section. In the chopper experiment, only about 1% of the neutrons leaving the extraction canal are used; the other 99% are intercepted by the closed interruptor. In the pulsed-source experiment, all the neutrons are used; for the same counting rate, the average flux can be smaller by a factor of 100.

Poole and Beyster *et al.* have undertaken experiments with large linear accelerators. Reichardt has done some measurements using a pulsed $T(d, n)He^4$ neutron source.

15.1.4. The Energy Resolution in Time-of-Flight Measurements

Let $J(E)$ be the energy distribution of the neutron beam leaving the medium, and let $\varepsilon(E)$ be the energy-dependent detector sensitivity. Let any other corrections, such as those for the transmission function of the chopper in a chopper experiment or for absorption processes in the flight tube, be accounted for by a single correction factor $T(E)$. In a time-of-flight experiment, the time-dependent counting rate $Z(t)$ is observed; we now ask for the connection between $Z(t)$ and $J(E)$. Let

$$\tilde{J}(E) = J(E)\, T(E)\, \varepsilon(E). \tag{15.1.10}$$

Then in the case of ideal resolution, i.e., when there is no uncertainty at all in the flight time, we have

$$Z_0(t) = \text{const.} \int \tilde{J}(E)\, \delta\left(\frac{l}{\sqrt{2E/m}} - t\right) dE. \tag{15.1.11a}$$

Here the constant contains irrelevant factors such as the solid angle and the repetition frequency. The δ-function expresses the fact that only those neutrons contribute to $Z_0(t)$ whose flight time $\frac{l}{\sqrt{2E/m}}$ is equal to t. Here l is the length of the flight path, which in a chopper experiment is measured from the middle of the rotor to the detector and in a pulsed-source experiment from the point of observation to the detector. From Eq. (15.1.11a) it follows that

$$\left.\begin{aligned} Z_0(t) &= \text{const.}\, \tilde{J}\left(E = \frac{m}{2}\left(\frac{l}{t}\right)^2\right) \left|\frac{dE}{dt}\right| \\ &= \text{const.}\, \tilde{J}\left(E = \frac{m}{2}\left(\frac{l}{t}\right)^2\right) \frac{m l^2}{t^3} \end{aligned}\right\} \tag{15.1.11b}$$

i.e., that $Z_0(t)|dt| = J(E)|dE|$. This latter conclusion is immediately obvious and could have been used instead of Eq. (15.1.11a) at the beginning of our considerations. The use of the δ-function is better if we intend ultimately to go over to the more important practical case in which the uncertainty in the flight time can no longer be neglected. Then in place of the δ-function there appears a normalized time resolution function $R(\lambda)$ $\left(\text{with} \int_{-\infty}^{+\infty} R(\lambda)\, d\lambda = 1\right)$. $R(\lambda)$ is determined on the one hand by the time-dependent transmission of the chopper in a chopper experiment or the time dependence of $\Phi(\boldsymbol{r}, E, t)$ in a pulsed-source experiment and on the other hand by the channel width of the time analyzer connected to the detectors. As a rule, enough channels are available to allow them to be made very narrow and thereby to eliminate the effect of channel width on the resolution function; therefore, in the following we shall neglect this effect. A difficulty now arises because as a rule $R(\lambda)$ depends on the neutron energy. In a chopper experiment with plane slits, the time-dependent transmission depends on velocity through Eqs. (15.1.5) and (15.1.6); in a pulsed-source experiment, the decay time in the epithermal range depends on the neutron energy. However, for the following somewhat qualitative considerations we shall assume that $R(\lambda)$ is energy independent. In the case of the chopper, let

$$R(\lambda)=\frac{1}{\Delta t}-\frac{|\lambda|}{(\Delta t)^2},\ |\lambda|\leqq\Delta t;\ R(\lambda)=0 \text{ otherwise (cf. Fig. 15.1.10a).} \quad (15.1.12\text{a})$$

Eq. (15.1.12a) represents a triangle function with a base width $2\Delta t=h/R\omega$. $\lambda=0$ corresponds to the zero time-point at which the chopper slits are parallel to the incident beam. In the case of the pulsed-source experiment, let

$$\left.\begin{aligned} R(\lambda)&=0, & \lambda&<-1/\alpha \\ R(\lambda)&=\alpha e^{-(1+\alpha\lambda)}, & \lambda&>-1/\alpha. \end{aligned}\right\} \quad (15.1.12\text{b})$$

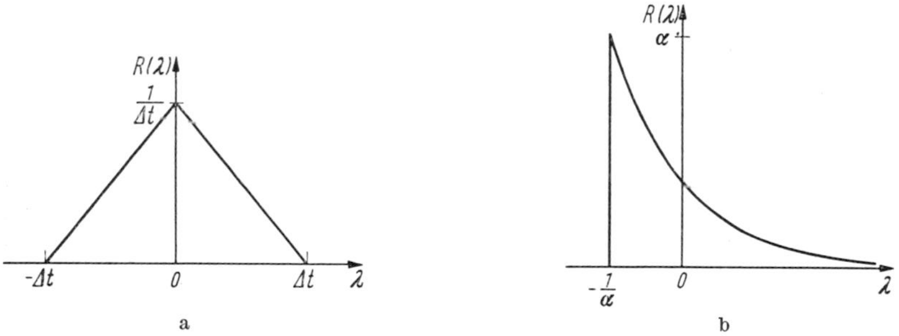

Fig. 15.1.10a and b. Idealized resolution functions for a: chopper; b: pulsed assembly

(Cf. Fig. 15.1.10b.) Here it has been assumed that after the injection of the pulse, the neutron flux $\Phi(\boldsymbol{r}, E, t)$ decays purely exponentially at all energies; this is certainly not the case in a non-multiplying medium in the epithermal range. For reasons to become obvious later, the zero time-point is not coincident with the injection of the pulse, but is shifted from it by one average lifetime $1/\alpha$.

We now have

$$Z(t)=\text{const.}\int_0^\infty \tilde{J}(E)\,R\left(\frac{l}{\sqrt{2E/m}}-t\right)dE \quad (15.1.13\text{a})$$

or using Eq. (15.1.11b)

$$Z(t)=\int_{-\infty}^{+\infty} Z_0(t+\lambda)\,R(\lambda)\,d\lambda. \quad (15.1.13\text{b})$$

Eq. (15.1.13b) connects the actually observed flight-time distribution $Z(t)$ with the flight-time distribution $Z_0(t)$ that would be observed in the case of ideal resolution. If we develop $Z_0(t)$ in a Taylor series, it follows because of the normalization of the resolution function that

$$Z(t)=Z_0(t)+\sum_{n=1}^{\infty}\frac{d^n}{dt^n}Z_0(t)\cdot\frac{1}{n!}\int_{-\infty}^{+\infty}\lambda^n R(\lambda)\,d\lambda. \quad (15.1.14\text{a})$$

We can easily convince ourselves by means of Eq. (15.1.12a) that for the chopper because of the symmetry of $R(\lambda)$ all odd moments of the resolution function, in particular the first, vanish. In the case of the pulsed source, the first moment vanishes because the zero time-point was shifted with respect to the time of pulse injection by one mean lifetime. Thus when we can neglect terms of order higher than the second, we have

$$Z(t)=Z_0(t)+Z_0''(t)\cdot\tfrac{1}{2}\int_{-\infty}^{+\infty}\lambda^2 R(\lambda)\,d\lambda, \tag{15.1.14b}$$

or

$$Z_0(t)=Z(t)\left[1-\frac{Z''(t)}{Z(t)}\cdot\frac{1}{2}\int_{-\infty}^{+\infty}\lambda^2 R(\lambda)\,d\lambda\right]. \tag{15.1.14c}$$

In order to determine an energy spectrum from $Z(t)$ we must proceed in the following way. We calculate $Z''(t)/Z(t)$ from the measured $Z(t)$[1] and using the second moment of the resolution function calculate $Z_0(t)$. $\tilde{J}(E)$ then follows from Eq. (15.1.11b) and $J(E)$ from Eq. (15.1.10). In order to go from here to the flux $\Phi(\boldsymbol{r}, E)$, it is necessary under certain circumstances to apply the corrections discussed in Sec. 15.1.1.

If we set

$$\tfrac{1}{2}\int_{-\infty}^{+\infty}\lambda^2 R(\lambda)\,d\lambda=\tfrac{1}{2}\Delta^2 \tag{15.1.15a}$$

by way of abbreviation, we obtain from Eq. (15.1.12a) for a chopper with a triangle resolution function of half-width Δt

$$\Delta^2=\frac{1}{6}(\Delta t)^2;\quad \Delta=\frac{\Delta t}{\sqrt{6}} \tag{15.1.15b}$$

and for a pulsed assembly with a decay time $1/\alpha$

$$\Delta^2=\left(\frac{1}{\alpha}\right)^2;\quad \Delta=\frac{1}{\alpha}. \tag{15.1.15c}$$

For a rectangular resolution function with the base width $\Delta t'$

$$\Delta^2=\frac{(\Delta t')^2}{3};\quad \Delta=\frac{\Delta t'}{\sqrt{3}}. \tag{15.1.15d}$$

If several resolution functions are superposed, then $\Delta^2=\Delta_1^2+\Delta_2^2+\cdots$. For example, if in a pulsed-source experiment we observe the time-of-flight distribution with channels of width $\Delta t'$, then $\Delta=\sqrt{\left(\frac{1}{\alpha}\right)^2+\frac{1}{3}(\Delta t')^2}$, and we can neglect the effect of finite channel width as long as $\Delta t'\lesssim\frac{1}{2\alpha}$.

If we know Δ, we can estimate what errors arise in a measurement owing to neglect of the resolution correction. Conversely, we can start with a given spectrum, require that the error due to neglect of the resolution correction be less, say, than 1%, and from this requirement calculate the necessary size of Δ. Fig. 15.1.11 shows the values of Δ/l (in μsec/m) that may not be exceeded in order that

[1] In order to calculate $Z''(t)/Z(t)$, $Z(t)$ must be very accurately known. In practice, we fit a smooth curve to the measured $Z(t)$ in some narrow range and then differentiate the fitted curve.

$\frac{Z(t)-Z_0(t)}{Z_0(t)}$ be $< 10^{-2}$, i.e., in order that the errors in the uncorrected spectrum, be less than 1% (POOLE). The assumptions underlying the calculations were black detectors, $\varepsilon(E)=1$, $T(E)=1$, and $J(E)\sim M(E)$ or $J(E)\sim 1/E$. Δ/l naturally depends on the energy and decreases with increasing energy. The upper end of the Maxwellian region is particularly critical since the spectrum changes rapidly there. It is seen that a $\Delta/l \approx 5-10$ μsec/m is sufficient to allow measurements to be carried out at energies below 10 ev to 1% or better. The resolution achieved in some actual experiments is also indicated in the figure.

Additional consideration of the resolution correction can be found in STONE and SLOVACEK as well as in the works of POOLE and of BEYSTER.

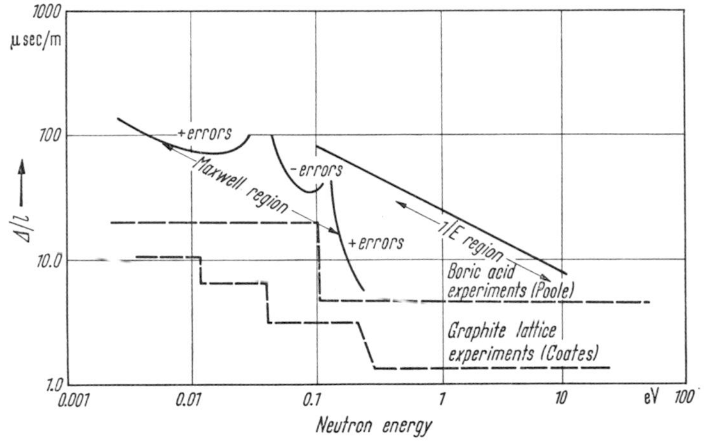

Fig. 15.1.11. ——— The value of Δ/l which should not be exceeded in order to obtain 1% accuracy in time-of-flight measurements of neutron spectra. — — — Δ/l in actual experiments

15.2. Investigation of the Spectrum by Integral Methods

Integral methods are based on the measurement of the effective cross sections of probes in the neutron field. These measurements can either be transmission measurements (attenuation of the probe activity by superposition of filters) or activation measurements with probes whose activation cross sections have some characteristic energy dependence. Such measurements can be interpreted by comparing the measured effective cross section with a value obtained by averaging the known differential cross section over the theoretically calculated neutron spectrum. Somewhat more direct but not always applicable is the evaluation of the probe measurements in terms of effective neutron temperatures. According to Sec. 10.2.1 we can frequently approximate the spectrum of slow neutrons with the form

$$\Phi(E) = \Phi_{th} \cdot \frac{E}{(kT)^2} e^{-E/kT} + \Phi_{\text{epi}} \cdot \frac{\Delta(E/kT)}{E}. \tag{15.2.1}$$

The spectrum is then fully characterized by the parameters $\Phi_{\text{epi}}/\Phi_{th}$ and T (or when the Westcott convention is used by r and T; cf. Sec. 12.2.5). $\Phi_{\text{epi}}/\Phi_{th}$ (or r) can be determined from the cadmium ratio by the methods described in Sec. 12.2; we shall now show that the temperature T can be derived from probe measurements.

15.2.1. Temperature Determination by the Transmission Method

Fig. 15.2.1 explains the principle of a transmission measurement on a free neutron beam. A collimated neutron beam with a Maxwellian energy distribution $J(E) \sim \frac{E}{(kT)^2} e^{-E/kT}$ traverses an absorber foil of thickness d and a $1/v$-cross section, $\Sigma_a(E) = \Sigma_a(kT_0)\sqrt{\frac{kT}{E}}\sqrt{\frac{T_0}{T}}$, and then impinges on a detector. Let the cross section of the detector substance obey the $1/v$-law and let the detector be so thin that its sensitivity likewise varies as $1/v$. We then have for the ratio of the counting rate Z^+ *with* the absorber to the counting rate Z^0 *without* it the following expression:

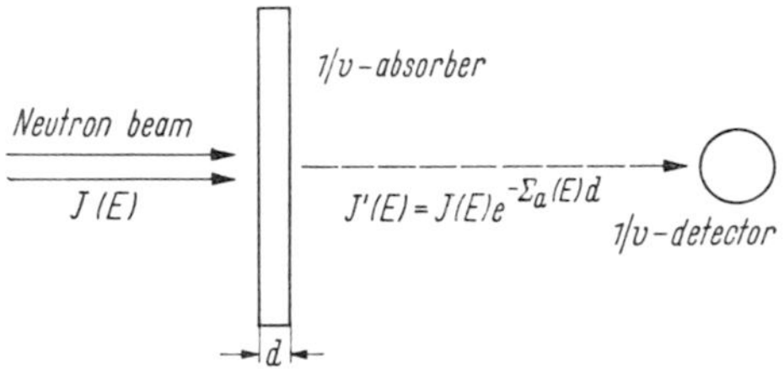

Fig. 15.2.1. A schematic arrangement for neutron temperature measurements by the transmission method

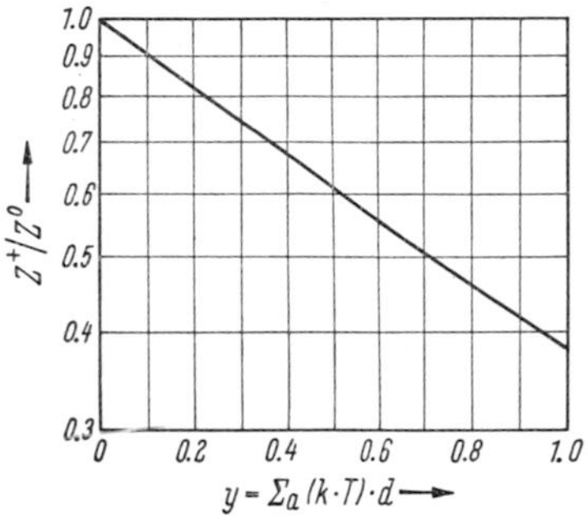

Fig. 15.2.2. The transmission of a monodirectional beam of thermal neutrons through a $1/v$-absorber foil

$$\frac{Z^+}{Z^0} = \frac{\int\limits_0^\infty e^{-\Sigma_a(E)d} J(E) \frac{dE}{\sqrt{E}}}{\int\limits_0^\infty J(E) \frac{dE}{\sqrt{E}}} \tag{15.2.2a}$$

or with $\sqrt{\frac{T_0}{T}}\,\Sigma_a(kT_0)d = y$ and $\frac{E}{kT} = x$

$$\frac{Z^+}{Z^0} = \frac{\int\limits_0^\infty e^{-y/\sqrt{x}}\sqrt{x}\,e^{-x}\,dx}{\int\limits_0^\infty \sqrt{x}\,e^{-x}\,dx} = \frac{2}{\sqrt{\pi}}\int\limits_0^\infty e^{-y/\sqrt{x}}\sqrt{x}\,e^{-x}\,dx. \tag{15.2.2b}$$

Fig. 15.2.2 shows Z^+/Z^0 as a function of y according to Zahn. The decrease is approximately exponential, but with a slowly decreasing decay constant. This decrease comes from the hardening of the Maxwell spectrum by the $1/v$-absorption in the foil. For thin absorber foils, $\Sigma_a(kT_0)d \ll 1$, and we can expand the exponential function in Eq. (15.2.2b) and obtain

$$\frac{Z^+}{Z^0} = 1 - \frac{2}{\sqrt{\pi}} y. \tag{15.2.2c}$$

For very thick absorber foils $\left(\Sigma_a(kT_0)d \gg 1\right)$ Laporte gives

$$\frac{Z^+}{Z_0} = \frac{4}{\sqrt{3}}\left(\frac{y}{2}\right)^{\frac{2}{3}} e^{-3(y/2)^{\frac{2}{3}}}\left[1 + \frac{17}{36}\left(\frac{2}{y}\right)^{\frac{2}{3}} - \cdots\right]. \tag{15.2.2d}$$

In each case Z^+/Z^0 is a unique function of y. y follows uniquely from a transmission measurement, and thus if d and $\Sigma_a(kT_0)$ are known, we can find T. Such measurements have been carried out by FERMI and MARSHALL as well as by HUGHES, WALLACE, and HOLTZMANN.

A variant of the transmission method of interest to us has been described by BRANCH and further developed by KÜCHLE. In this variant, we use an indium foil as detector and enclose it with two gold foils. A typical sandwich arrangement is shown in Fig. 15.2.3. Here again a shielding ring is used to avoid activation of the inner foil by laterally incident neutrons. According to Sec. 11.2.3, if we neglect scattering in the foil, the activation of the bare inner foil in a Maxwell spectrum is given by

$$C^0 = \frac{1}{2}\int_0^\infty \frac{E}{(kT)^2}\, e^{-E/kT}\, \frac{\mu_{\mathrm{act}}(E)}{\mu_a(E)}\, \varphi_0\big(\mu_a(E)\,\delta\big)\, dE. \tag{15.2.3}$$

If the foil is covered on both sides by covers whose absorption coefficient is

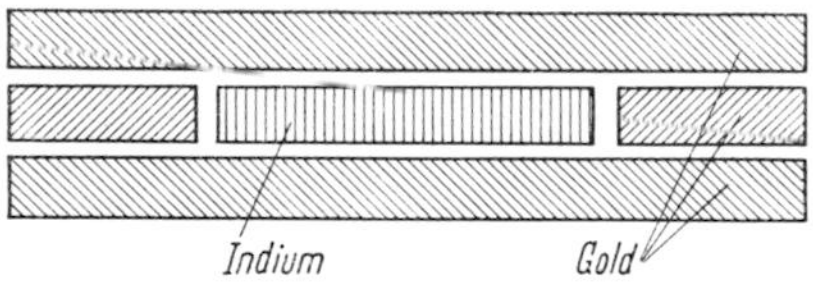

Fig. 15.2.3. A foil sandwich for neutron temperature measurements

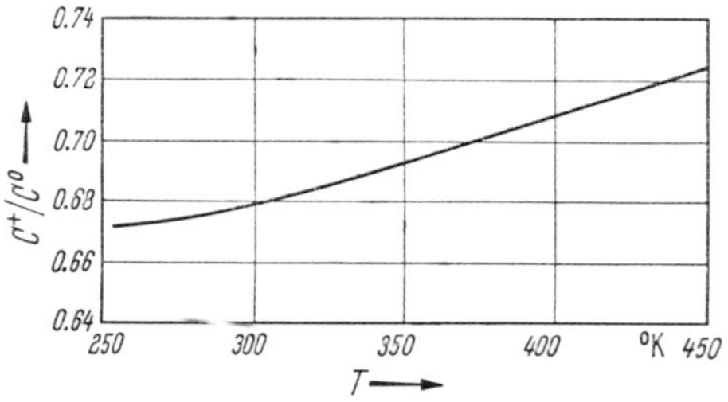

Fig. 15.2.4. C^+/C^0 for an Indium foil ($\delta = 65.7$ mg/cm²) covered by gold foils ($\delta' = 360$ mg/cm²) as a function of the neutron temperature

$\mu_a'(E)$ and whose thickness is δ', then again neglecting scattering[1], its activation according to Sec. 12.1.3 is

$$C^+ = \frac{1}{2}\int_0^\infty \frac{E}{(kT)^2}\, e^{-E/kT} \left\{\frac{\mu_{\mathrm{act}}(E)}{\mu_a(E)} \left[\varphi_0\big(\mu_a(E)\,\delta + \mu_a'(E)\,\delta'\big) - \varphi_0\big(\mu_a'(E)\,\delta'\big)\right]\right\} dE. \tag{15.2.4}$$

Thus for given μ_{act}, μ_a, μ_a', δ, and δ' the activation ratio C^+/C^0 is a function only of the temperature. Fig. 15.2.4 shows C^+/C^0 according to KÜCHLE for a particular sandwich arrangement. One obtains C^+/C^0 by counting the covered and bare foils, respectively, after irradiation; we must correct for any differences due to different counting and irradiation times by means of the time factor. The sensitivity of the counting apparatus does not matter because one is interested in the ratio of two counting rates. However, one must be very careful that the observed counting rates are proportional to the true activity of the foil (Sec. 11.2.5). One must therefore either count γ-rays or average the β-counting rate over both sides of the foil.

As a rule, the thermal neutron fields to be studied also contain epithermal neutrons. We must then again separate thermal and epithermal activation by

[1] The influence of scattering in the cover foils on C^+ has been studied in detail for a gold absorber by KÜCHLE. It turns out to be negligibly small; for on one hand, some nearly normally incident neutrons cannot reach the foil owing to backscattering, while on the other hand, some obliquely incident neutrons that would otherwise remain in the absorber are scattered into the foil.

means of the cadmium difference method. To do this we determine C^+ and C^0 once without and once with the entire sandwich tightly enclosed in a cadmium shell (0.5—1 mm thick) and thereby obtain four values C_0^+, C_{CD}^+, C_0^0, and C_{CD}^0. Then we form

$$C_{th}^+ = C_0^+ - F_{CD}^+ C_{CD}^+ \tag{15.2.5a}$$

and

$$C_{th}^0 = C_0^0 - F_{CD}^0 C_{CD}^0 . \tag{15.2.5b}$$

F_{CD}^0 and F_{CD}^+ are the cadmium correction factors (cf. Sec. 12.2.1). F_{CD}^0 is identical with the correction factor for a bare foil; F_{CD}^+ is that of the sandwich. The measurements and calculations of KÜCHLE show that for the foils used by him, $F_{CD}^+ \approx F_{CD}^0$; this conclusion is probably valid for all conceivable practical sandwiches. With $F_{CD}^+ \approx F_{CD}^0 = F_{CD}$, we can employ the values given in Fig. 12.2.3.

In this method, the activation perturbation has a very serious effect on the temperature measurement since a foil surrounded by an absorber causes a much larger flux depression than a bare one. Thus the measured value of C_{th}^+/C_{th}^0 must be corrected. This can be done with the help of the formulas given in Sec. 11.3. However, if, for example, we use gold absorbers each with 360 mg/cm² and an indium foil of 65.7 mg/cm², the entire package has a $\mu_a \delta$ of 0.28, and we have overshot the range in which there are reliable measurements of the activation perturbation. The perturbation is greater the smaller the transport mean free path in the surrounding medium compared to the foil thickness, and probably excludes a temperature determination in H_2O by this method. We can avoid the perturbation, however, by placing the foil package in a sufficiently large cavity. There, there is no field perturbation and the measured transmission values need not be corrected.

The accuracy with which the neutron temperature can be determined with this method is about 10°. In Sec. 15.3, we shall return to some results of such measurements.

15.2.2. Temperature Determination with Lutetium Foils

We can obtain information about the neutron spectrum without transmission measurements from the activation of individual foils as long as we use a foil substance whose activation cross section deviates from the $1/v$-law. Whereas the activation of a foil with a $1/v$-cross section is always proportional to the density and independent of the spectrum, that of a foil with an energy-independent cross section, for example, is proportional to the flux; from the ratio of the two, we can determine the average neutron velocity.

Lutetium is a suitable foil substance for such measurements. Natural lutetium consists of 97.40% Lu^{175} and 2.60% Lu^{176}. Lu^{175} is stable, while Lu^{176} decays with a half-life of $2.2 \cdot 10^{10}$ years. Neutron capture in Lu^{175} leads to an isomeric state of Lu^{176} which decays with a half-life of 3.684 h (cf. Fig. 15.2.5a). Neutron capture in Lu^{176} leads to Lu^{177} which decays with a half-life of 6.8 d (cf. Fig. 15.2.5b for its decay scheme). Whereas the activation cross section of Lu^{175} follows the $1/v$-law in the thermal range, that of Lu^{176} deviates from it sharply because of a strong resonance at 0.142 ev. Therefore, if we irradiate a lutetium foil in a neutron field and thereafter count the $Lu^{176\,m}$ and Lu^{177} activities separately, we obtain a useful spectral index from which we can determine the neutron temperature. We can also ignore the Lu^{175} activation and instead use a manganese or copper foil as a $1/v$-reference detector. In this case, we do not need to determine both of the lutetium activities separately; we simply wait long enough after the irradiation for the $Lu^{176\,m}$ activity to decay. The works of SCHMID and STINSON as well as of BURKART contain numerous practical hints for working with lutetium foils.

We shall show in the following how we can determine the neutron temperature from such measurements. For this purpose, we shall use the formalism of WEST-

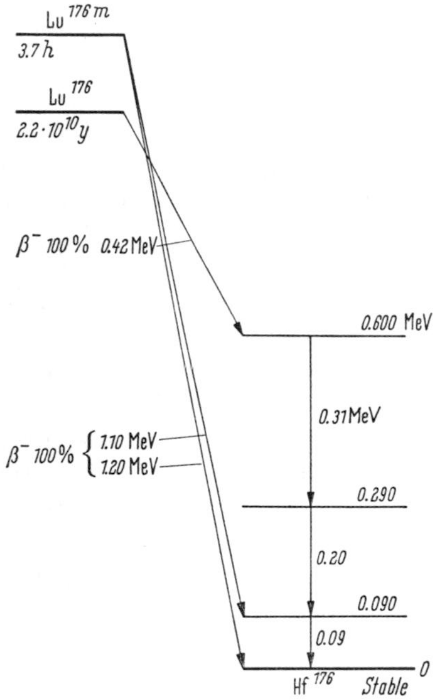

Fig. 15.2.5a. The decay scheme of Lu^{176} and $Lu^{176\,m}$

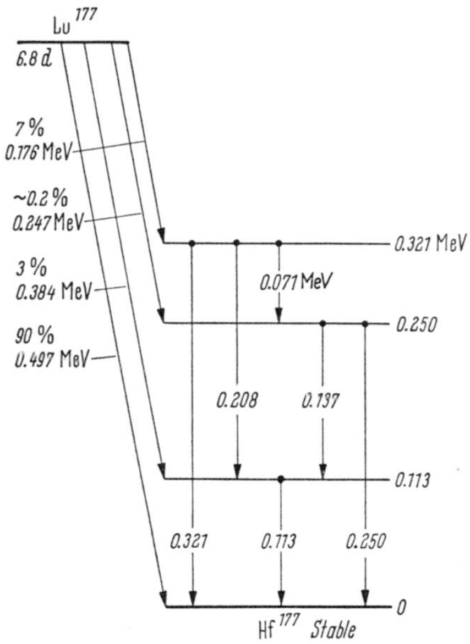

Fig. 15.2.5b. The decay scheme of Lu^{177}

COTT introduced in Sec. 12.2.5; our considerations therefore hold only for thin foils. Let the neutron spectrum be characterized by the parameters r and T. Let us consider the activation of a lutetium foil or of a sandwich composed of a lutetium foil and a $1/v$-reference detector. The quantities $g(T)$ and $s(T)$ refer to the excitation of the Lu^{177} activity; the quantity $s^{1/v}(T)$ refers to the $1/v$-reference detector (i.e., either the $Lu^{176\,m}$ activity or the activity of the copper or manganese foil[1]). Then we have

$$\left.\begin{aligned} C &= K[g(T)+r s(T)] \\ C^{1/v} &= K^{1/v}[1+r s^{1/v}(T)]. \end{aligned}\right\} \qquad (15.2.6)$$

Here K and $K^{1/v}$ are constants that contain the neutron density, the absolute value of the activation cross section, and the foil loading. Let us now form the activation ratio

$$\frac{C}{C^{1/v}} = \frac{K}{K^{1/v}}\,\frac{g(T)+r s(T)}{1+r s^{1/v}(T)} = \tilde{K}\,\frac{g(T)+r s(T)}{1+r s^{1/v}(T)}\,. \qquad (15.2.7)$$

[1] If the cross section of the latter follows the $1/v$-law rigorously at all energies, then $s=0$; however, this is not the case for copper, manganese, and Lu^{175}.

Here on the left is the (measurable) quantity $C/C^{1/v}$; on the right are the unknown quantities $\widetilde{K}$, r, and T. $\widetilde{K}$ can be eliminated by additional measurements in a known neutron spectrum. The neutron field in the thermal column of a reactor is particularly suitable for this purpose. In the thermal column there is certainly an equilibrium spectrum; then $r=0$ and T is equal to the moderator temperature T_{TC}. Thus

$$\left(\frac{C}{C^{1/v}}\right)_{TC}=\widetilde{K}\,g(T_{TC}). \tag{15.2.8}$$

Since T_{TC} is known, $g(T_{TC})$ can be taken from the tables of WESTCOTT (cf. also Fig. 15.2.6) and thus $\widetilde{K}$ can be determined. Then

$$\left(\frac{C}{C^{1/v}}\right)^{*}=\frac{g(T)+r s(T)}{1+r s^{1/v}(T)} \tag{15.2.9}$$

where the star indicates that the measured activation ratio has been divided by the $\widetilde{K}$-value found in the additional measurement.

Eq. (15.2.9) still does not permit a unique determination of T since the right-hand side still contains the unknown quantity r. We must therefore measure the cadmium ratio of the $1/v$-reference foil. According to Sec. 12.2.5, we have [Eq. (12.2.30b)]

$$R_{CD}=\frac{C^{1/v}}{C^{1/v}_{CD}}=\frac{1+r s(T)}{r\left[s(T)+\frac{4}{\sqrt{\pi}}\sqrt{\frac{kT}{E_{CD}}}\right]}. \tag{15.2.10}$$

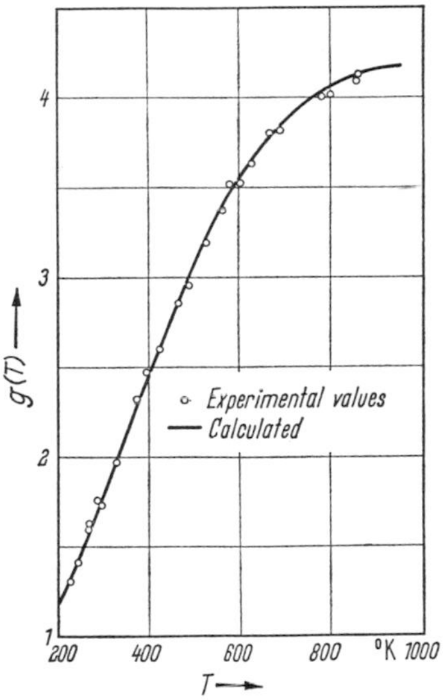

Fig. 15.2.6. The temperature-dependent g-factor for Lu^{176}

Here E_{CD} is the cadmium cut-off energy, which depends on the cadmium thickness (cf. Fig. 12.2.5). We now have two independent equations for the determination of the two parameters r and T which can be solved graphically, for example. However, the following iteration procedure is simpler. We start with a first estimate of T and determine a value of r from the measured cadmium ratio using Eq. (15.2.10); from this value of r and the measured activation ratio $(C/C^{1/v})^*$ we can calculate a new value of T, with which the value of r may be improved, etc.

Fig. 15.2.6 shows the g-factor of Lu^{176} as a function of T. The values of g were calculated with Eq. (5.3.20); the cross section was represented by a superposition of individual Breit-Wigner resonances. Table 15.2.1 gives the parameters used for the first few resonances. Fig. 15.2.6 also contains some g-factors experimentally determinod (by measurements on a heated thermal column) by SCHMID and STINSON. We see that for $T<500$ °C, g depends very sensitively on the temperature. Fig. 15.2.7a shows $s(T)$ for Lu^{176}. The values were calculated with Eqs. (12.2.25a) and (12.2.26) from the energy-dependent

Table 15.2.1. *The Resonance Parameters of Lu^{176}*

E_R [ev]	$2g\Gamma_n$[1] [mev]	Γ_γ [mev]
0.142	0.08742 ± 0.00035	58.8 ± 0.2
1.57	0.40 ± 0.06	≈ 60
4.35	0.48 ± 0.04	≈ 60
6.13	1.46 ± 0.14	≈ 60
9.77	1.47 ± 0.13	≈ 60

[1] The g-factor is actually unknown, but because $I=7$, g is nearly $\frac{1}{2}$ and $2g\Gamma_n \approx \Gamma_n$.

cross section. $s(T)$ for Lu^{176} depends somewhat on the joining function since the first resonance lies in the neighborhood of the transition range between the thermal and epithermal regions. However the differences, even for extreme assumptions about the joining function, are not very great.

For Cu, Mn, and Lu^{175}, $g(T)$ is always 1. Figs. 15.2.7a and b show $s(T)$ for these three substances. We see from the figure that $s(T)$ for Lu^{175} is rather large; moreover, its values are only poorly known. Therefore, the determination of r from the cadmium ratio of Lu^{175} is not at all precise, and it is preferable to use Cu or Mn as the $1/v$-reference detector.

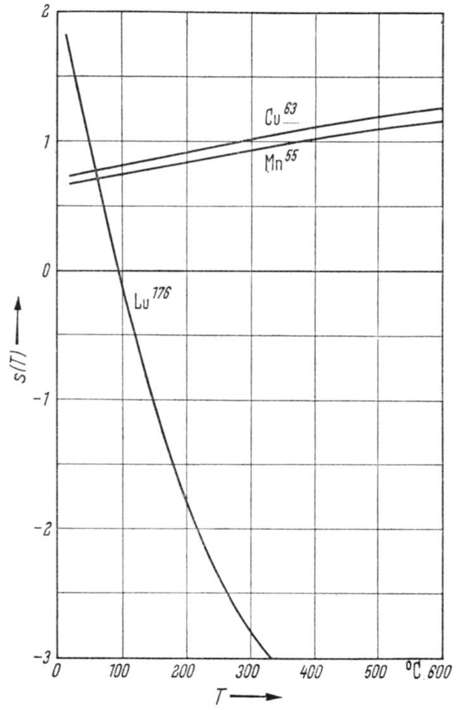

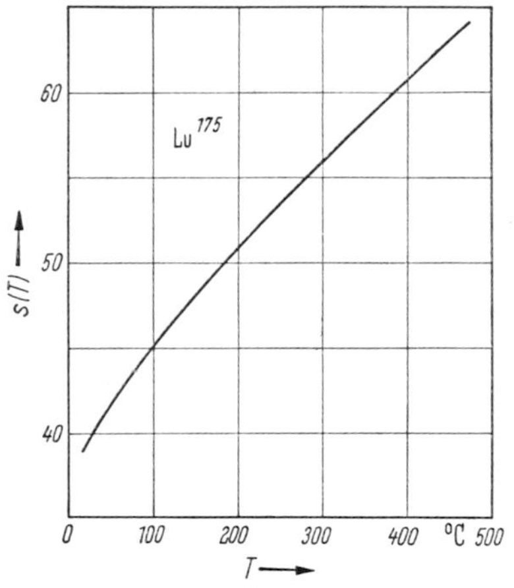

Fig. 15.2.7a and b. $s(T)$ values (according to WESTCOTT) for substances used in neutron temperature measurement. a: Lu^{176}, Cu^{63}, Mn^{55}; b: Lu^{175}

The accuracy with which we can determine the neutron temperature with this method is about $\pm 5°$. It is thus more accurate (and also somewhat simpler) than the transmission method of Sec. 15.2.1. A disadvantage is that because of their strong β-self-absorption, Lu foils must be very thin. In order to obtain adequate statistics, we must irradiate them in a flux of at least 10^5 cm^{-2} sec^{-1} (as compared to 10^3 cm^{-2} sec^{-1} for the transmission method). We shall return to the results of such measurements in Sec. 15.3.

15.2.3. Other Methods of Determining the Neutron Temperature

Instead of Lu^{176}, we can also use Eu^{151} for the activation measurement described in Sec. 15.2.2. Natural europium consists of 47.77% Eu^{151} and 52.23% Eu^{153}. Neutron capture in Eu^{151} leads to a 9.2-h activity (the decay scheme is given in Fig. 15.2.8); the cross section for this activation deviates strongly from the $1/v$-law and in the thermal range has practically a $1/E$-behavior. In contrast with lutetium, the g factor decreases strongly with temperature. Fig. 15.2.9 shows $g(T)$ for Eu^{151}; unfortunately, these values are only poorly known. The evaluation of the measurements is done exactly as for lutetium. The work of SPRINGER gives more details.

We recommend the combination of several detector substances, e.g., Lu, Eu, and Cu, or Lu, Pu, and Mn, since we may then judge how well the spectrum can

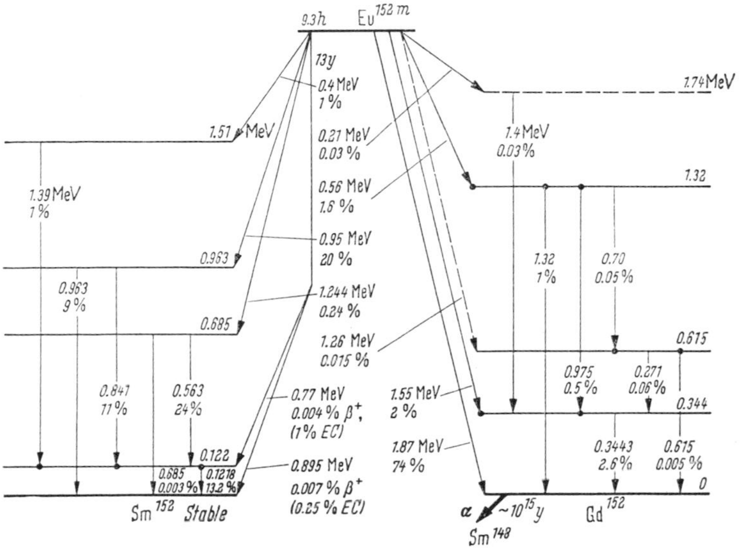

Fig. 15.2.8. The decay scheme of Eu^{152m}

be described by a Maxwell distribution by how well the various temperature measurements agree.

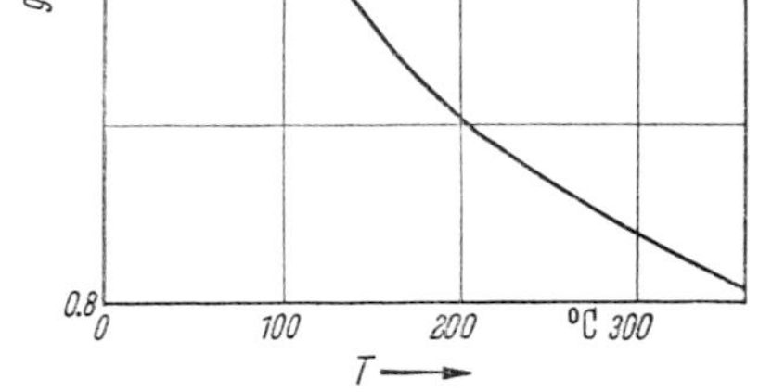

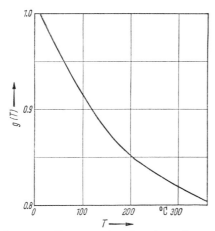

Fig. 15.2.9. The temperature-dependent g-factor for the reaction $Eu^{151}(n, \gamma)Eu^{152m}$

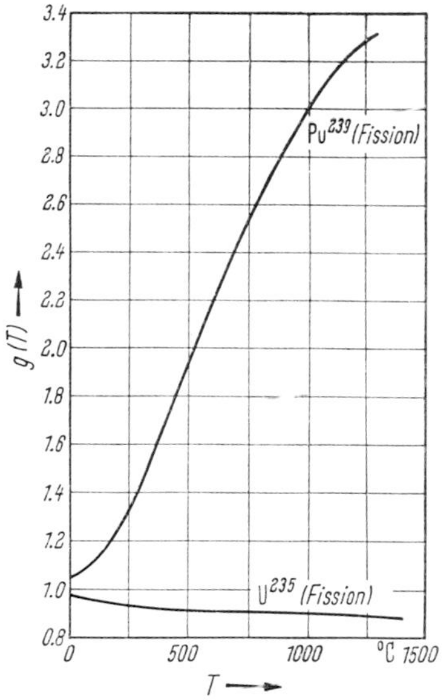

Fig. 15.2.10. The temperature-dependent g-factor for thermal neutron fission of Pu^{239} and U^{235}

Because of a strong resonance at 0.297 ev, the *fission cross section* of *plutonium-239* also shows a sharp deviation from the $1/v$-law. Figs. 15.2.10—12 show $g(T)$ and $s(T)$ for Pu^{239} and U^{235}. We see that particularly at high temperatures $g(T)$ for Pu^{239} increases very much with increasing temperature. Thus we can determine the neutron temperature by simultaneous fission-rate measurements on Pu^{239} and U^{235}. The measurements are evaluated in principle exactly as in the case of lutetium. Investigations with fission chambers have been carried out by CAMPBELL, POOLE, and FREEMANTLE. STINSON, SCHMID, and HEINEMANN describe a procedure in which U^{235} and Pu^{239} are irradiated in a neutron field and their fission rates determined thereafter by counting their fission product activities. A particular advantage of fission-rate measurements in the study of reactor spectra is that — quite independently of the spectrum evaluation procedure — we immediately obtain important technical data.

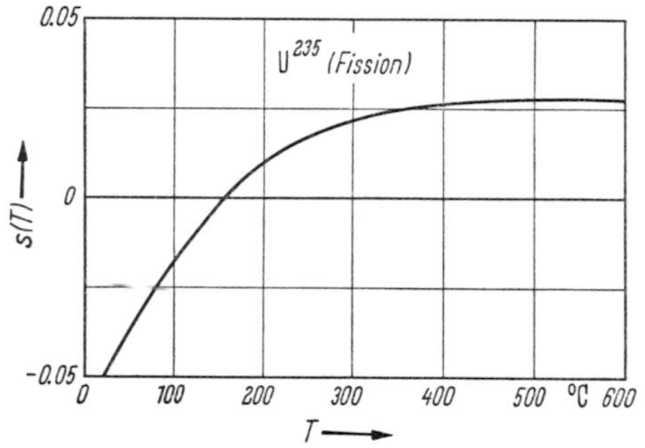

Fig. 15.2.11. $s(T)$ values for U^{235} fission

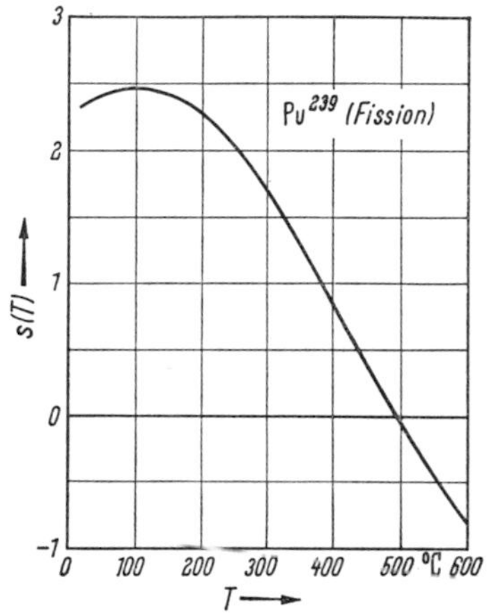

Fig. 15.2.12. $s(T)$ values for Pu^{239} fission

Fig. 15.2.13 indicates how we can use cadmium to measure the temperature of a free neutron beam. Since an 0.5-mm-thick cadmium sheet is "black" for all thermal neutrons, the capture rate in it is proportional to the flux; in contrast the capture rate in a $1/v$-detector (a thin BF_3-counter) is proportional to the density. The average neutron velocity and thus the neutron temperature follows from the ratio of the two capture rates. Unfortunately, neutron capture in cadmium does not lead to activation; however, the capture rate in cadmium can be determined from the intensity of the capture γ-radiation. In principle, we can use any "black" detector, e.g., a thick Li^6I crystal or a high-pressure BF_3-counter, instead of cadmium.

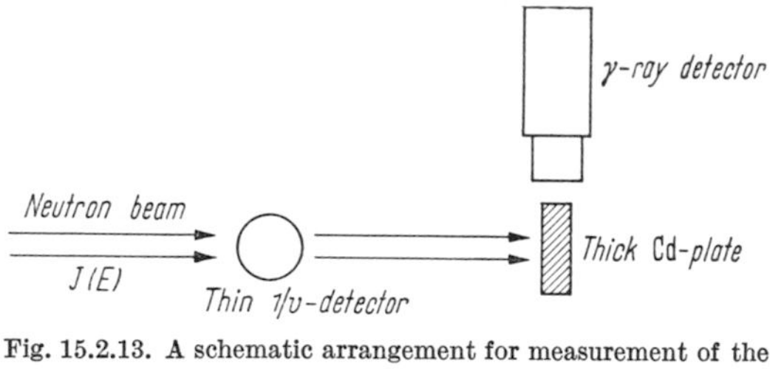

Fig. 15.2.13. A schematic arrangement for measurement of the average neutron velocity

We mention in this connection a procedure given by ANDERSON and later improved by GAVIN for determining the neutron temperature in a reactor. If we put an absorber substance in a reactor, its multiplication factor falls by an amount that is proportional to the number of additional neutron absorptions per second. By comparing the effect induced by cadmium and by a $1/v$-absorber, such as boron, Mn, or Cu, we can determine the average neutron velocity and the neutron temperature just as we did above.

15.3. Results of Measurements of Stationary Spectra

15.3.1. Measurements on Water

Measurements of neutron spectra in water and in aqueous solutions of various absorbers have been carried out by POOLE, by BEYSTER and coworkers, by REICHARDT (all by the method of pulsed sources), and by STONE and SLOVACEK (by the chopper method). Measurements on water are easily carried out because the

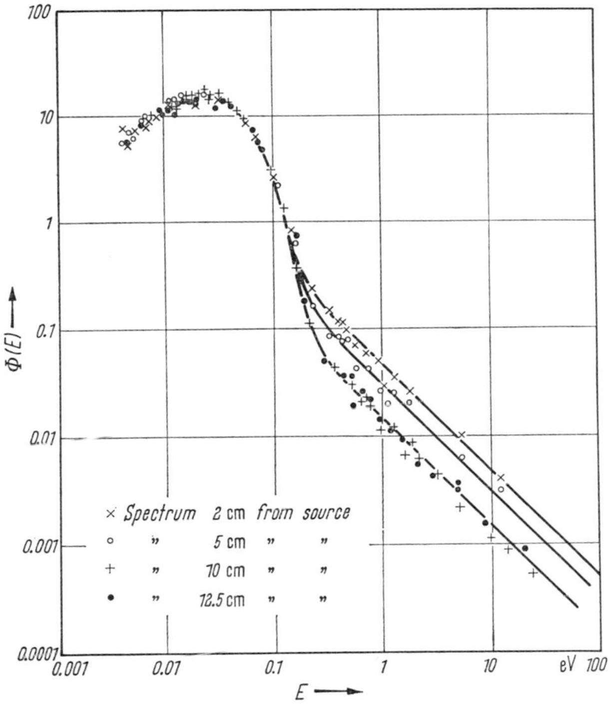

Fig. 15.3.1. Neutron spectra in pure water at various distances from the source

diffusion length (2.76 cm in pure water, smaller in absorbing solutions) is small compared to the relaxation length λ with which the fast neutron intensity falls off. As a result, an equilibrium state is rapidly established in the neighborhood of the source; the spectrum in the adjacent equilibrium region then differs only slightly from that in an infinite medium with homogeneously distributed sources and is easily accessible to theoretical calculation. In Fig. 15.3.1 are shown the spectra observed by POOLE at distances of 2, 5, 10, and 12.5 cm from the source. At source distances > 10 cm, the spectrum no longer changes.

Fig. 15.3.2 shows spectra in boric acid solutions of various concentrations taken by BEYSTER. For comparison, infinite-medium spectra calculated with the Nelkin model are also shown. The agreement of theory and experiment is good, except for pure H_2O. The Nelkin model also describes the measurements well in the case of samarium solutions (cf. Fig. 15.3.3). Note the minima in the spectra caused by resonance absorption. On the other hand, the Wigner-Wilkins model (neutron scattering on free protons) describes the measurements less well.

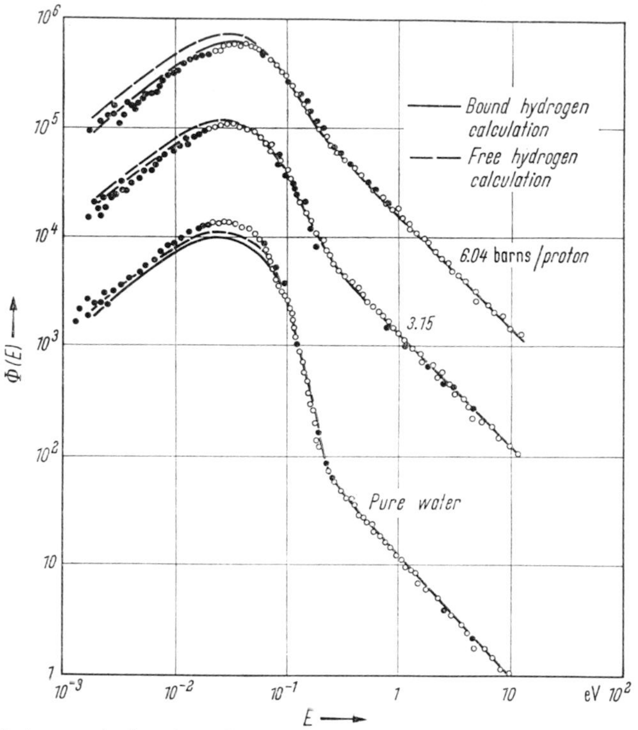

Fig. 15.3.2. Neutron spectra in water and aqueous boric acid solutions. ○● Measurements by BEYSTER; ——— calculated with Nelkin model; — — — calculated for a free proton gas

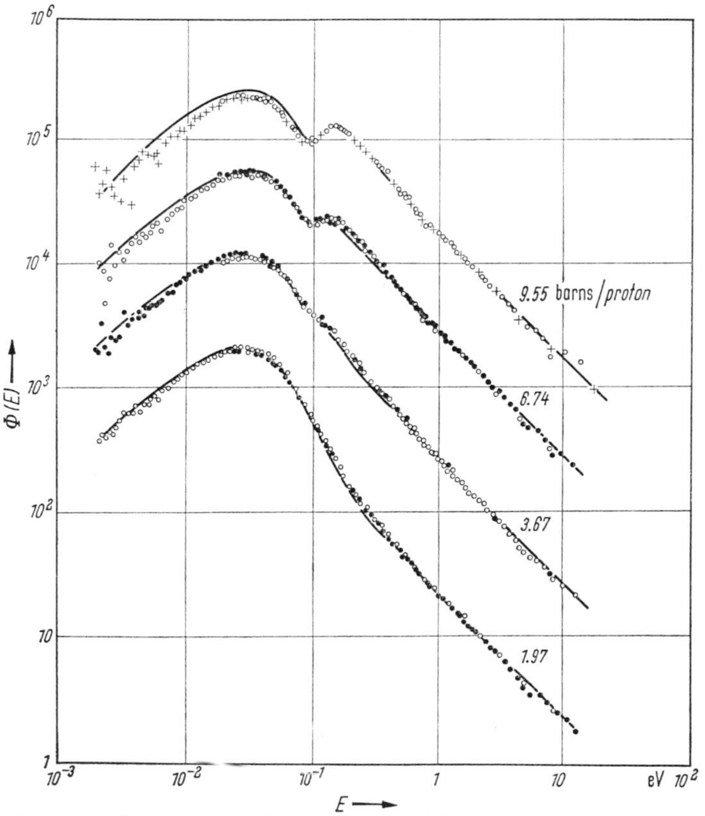

Fig. 15.3.3. Neutron spectra in aqueous samarium solutions. ○●+ Measurements by BEYSTER; ——— calculated with Nelkin model

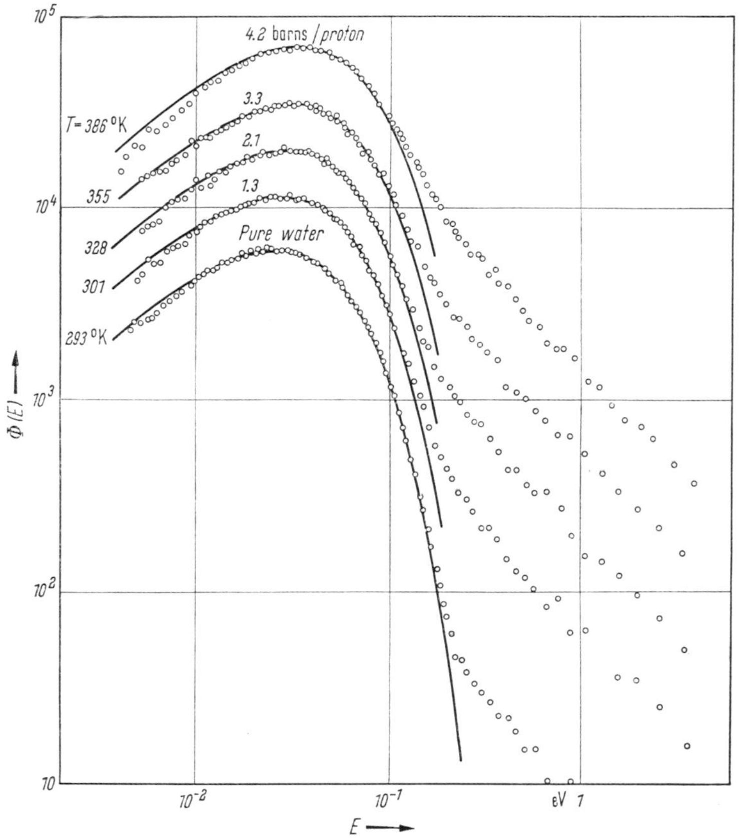

Fig. 15.3.4. Neutron spectra in water ($T_0 = 293$ °K) and aqueous boric acid solutions. ooo Measurements by BURKART and REICHARDT; —— Maxwell distributions fitted to thermal parts of spectra. The neutron temperatures thereby obtained are given at the left of each curve

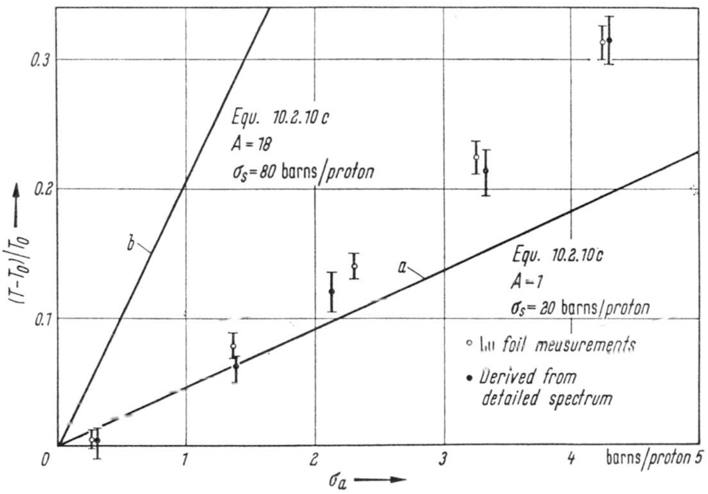

Fig. 15.3.5. Neutron temperatures in aqueous boric acid solutions. $T_0 = 293$ °K

Poole and later Reichardt and Burkart have tried to resolve the spectra measured in boric acid solutions into a Maxwell distribution and a $1/E$-part. Fig. 15.3.4 shows that representation of the thermal spectral region as a Maxwell distribution is in fact quite possible; Fig. 15.3.5 shows neutron temperatures determined in this way as a function of the amount of absorber present. Reichardt and Burkart have also measured neutron temperatures with lutetium foils; these measurements agree well with the values derived from differential measurements. Line "a" in Fig. 15.3.5 shows the neutron temperature to be expected according to Eq. (10.2.10c) from a free proton gas ($A=1$, $\sigma_s=20$ barn/H-atom). Line "b" has been calculated with Eq. (10.2.10c) for a gas of protons rigidly bound to

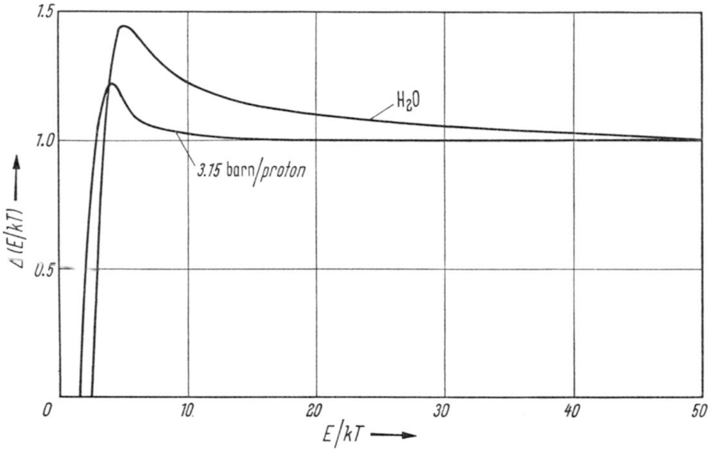

Fig. 15.3.6. The joining functions for water and for an aqueous boric acid solution. $T_0 = 293$ °K

oxygen atoms ($A=18$, $\sigma_s=80$ barn/H-atom). The measured temperature increases lie between the two limits, as was to be expected. Finally, Fig. 15.3.6 shows the joining functions for pure water and for a boric acid solution with an absorption cross section of 3.15 barn/H-atom; these joining functions were obtained by subtracting the fitted Maxwell distributions from the measured spectra. The two joining functions are clearly different, indicating a weakness of the effective neutron temperature concept.

15.3.2. Measurements on Graphite

Systematic investigation of the spectrum in graphite is much more difficult than in the case of H_2O. First of all, it is not possible to poison graphite systems homogeneously. The effective absorption cross section can only be increased by constructing the system of alternating thin graphite sheets and absorber plates of copper, boron steel, etc. Furthermore, because of the large diffusion and slowing-down lengths in graphite systems, no equilibrium is reached between the primary neutron source and the thermal flux, i.e., the spectrum in the neighborhood of a fast source is always strongly influenced by diffusion phenomena; and we must — if we wish to compare the calculated and measured spectra — fall back on the space-dependent theory.

The only systematic comparison between theory and experiment was carried out by PARKS, BEYSTER, and WIKNER. Fig. 15.3.7 shows the experimental arrangement. A graphite cube 60 cm on an edge is heterogeneously poisoned by boron steel plates. The thermal absorption cross section per carbon atom is about 0.4 barn ($\Sigma_a/\xi\Sigma_s = 0.53$). The entire apparatus is insulated and can be heated to *ca.* 500 °C. Adjacent faces are used for neutron injection and beam

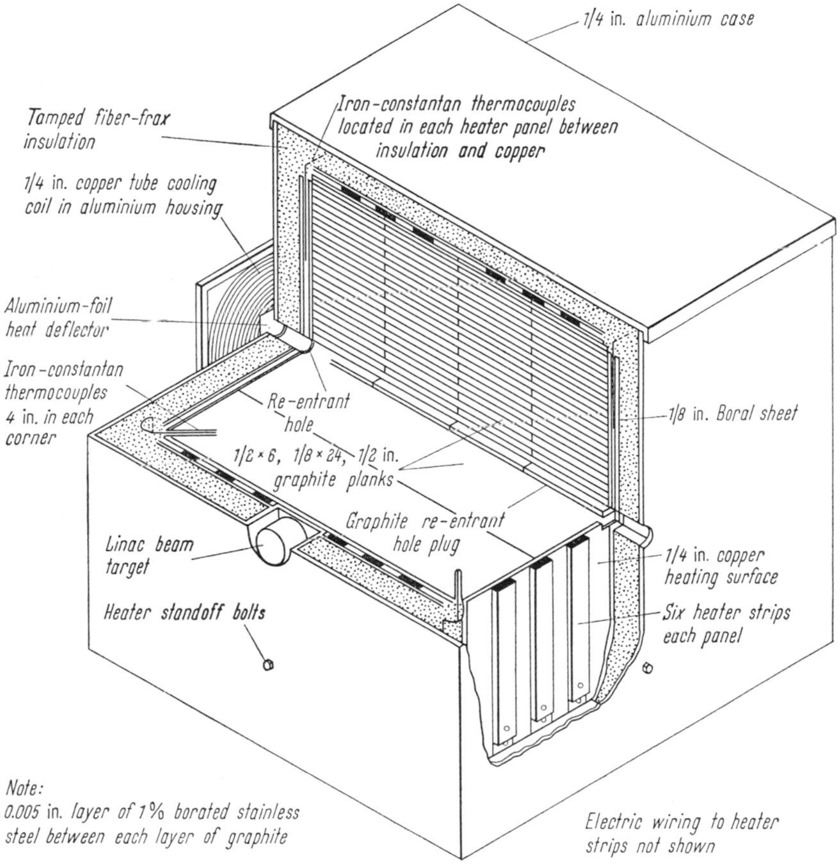

Fig. 15.3.7. The arrangement used by PARKS, BEYSTER and WIKNER for neutron spectrum measurements in graphite

extraction. The measurements are done by the pulsed-source method with a flight path 60 m long. Since the relaxation time of thermal neutrons in such a strongly poisoned system is of the order of 200 μsec, the energy resolution is adequate. Figs. 15.3.8a—e show the spectra measured at various temperatures in comparison with calculations done with the Parks model (Sec. 10.1.3) and with a free gas model (using mass 12). In the calculations, diffusion effects were taken into account by use of an effective absorption cross section (Sec. 10.2.3); the local flux curvature was determined from foil measurements. PARKS' model reproduces the measured spectra quite well. The gas model is bad at low temperatures, but becomes much better at higher graphite temperatures since the chemical binding is then less noticeable.

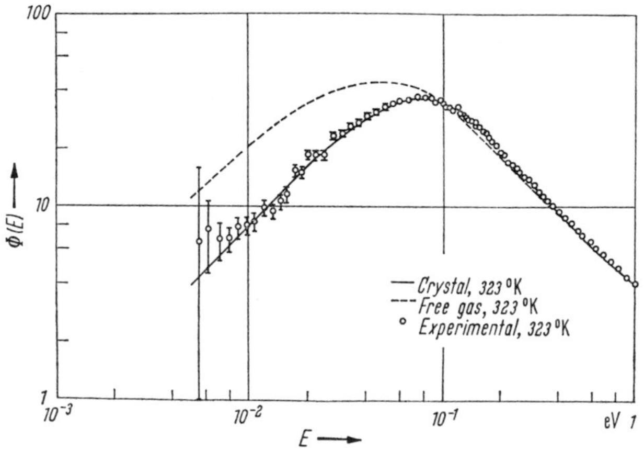

Fig. 15.3.8a

Fig. 15.3.8a–e. Neutron spectra in poisoned graphite. Graphite temperature: a $T_0 = 323$ °K; b $T_0 = 422$ °K; c $T_0 = 588$ °K; d $T_0 = 700$ °K; e $T_0 = 810$ °K

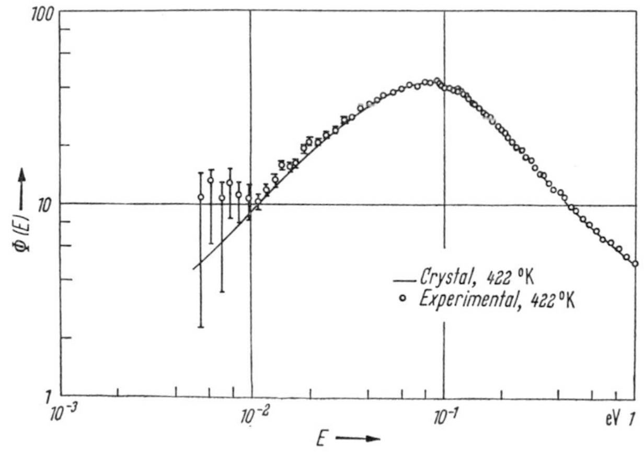

Fig. 15.3.8b

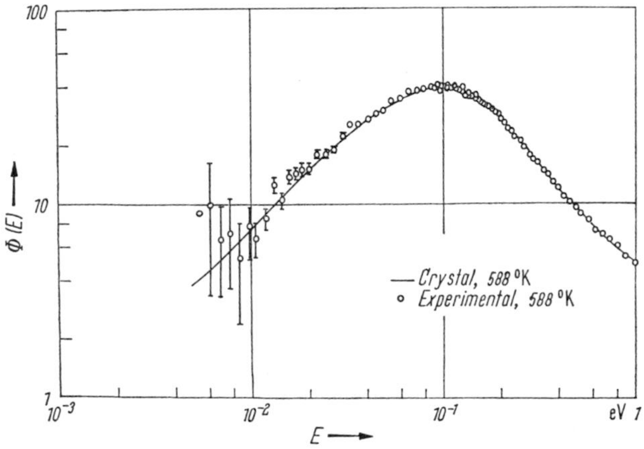

Fig. 15.3.8c

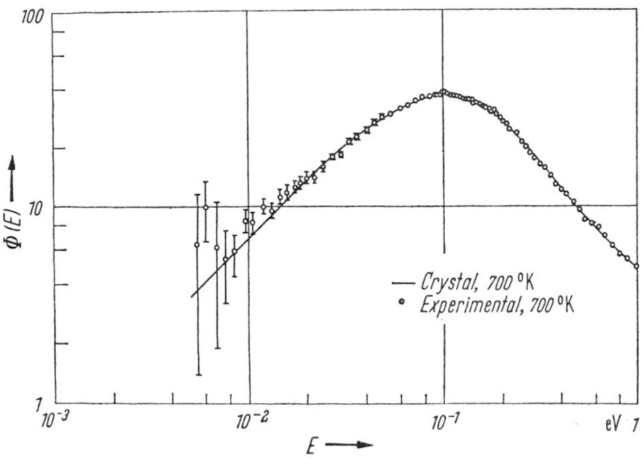

Fig. 15.3.8 d

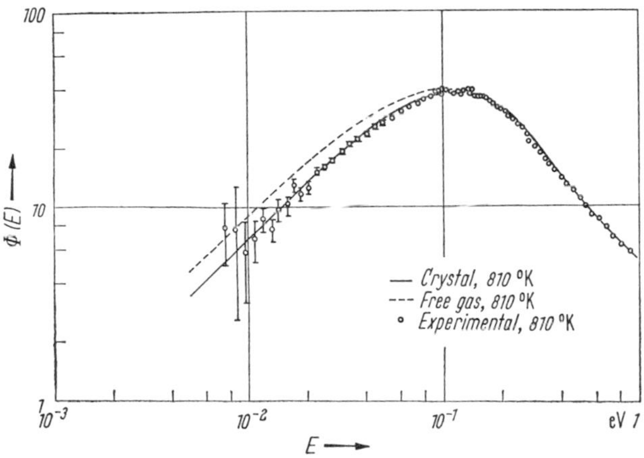

Fig. 15.3.8 e

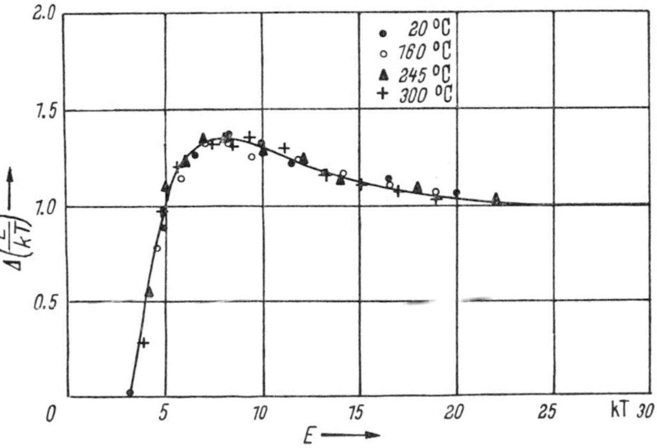

Fig. 15.3.9. The joining function in graphite as measured by COATES

Neutron temperature measurements have been carried out in graphite systems by COATES, by KÜCHLE, and by CAMPBELL, POOLE, and FREEMANTLE. COATES determined the spectrum in a subcritical assembly of graphite and natural uranium at various temperatures, and resolved it according to Eq. (10.2.10d) into a Maxwell distribution and an epithermal part. He used the relation

$$T = T_0 \left[1 + 0.5\, A_{\text{eff}} \frac{\Sigma_a(kT)}{\Sigma_s}\right] \tag{15.3.1}$$

for the neutron temperature[1], and found $A_{\text{eff}}=30$ ($T_0=300$ °K), $A_{\text{eff}}=26$ ($T_0=430$ °K), $A_{\text{eff}}=23$ ($T_0=520$ °K), and $A=19$ ($T_0=590$ °K). Fig. 15.3.9 shows the joining function derived from his measurements; the measurements at different graphite temperatures lead to the same function.

KÜCHLE determined the neutron temperature in the neighborhood of a (Ra—Be) source in graphite by the transmission method. He found that $T - T_0$ is large near the source and falls off with increasing distance from the source. $T - T_0$ is about proportional to the flux ratio $\Phi_{\text{epi}}/\Phi_{th}$, which was determined in the same set of experiments by the cadmium difference method. If one eliminates the diffusion cooling effect with the help of Eq. (10.2.20) and fits the data to Eq. (15.3.1), one finds that $A_{\text{eff}}=40\pm10$, a result that agrees with the work of COATES within experimental error.

15.3.3. Other Moderators

JOHANSSON, LAMPA, and SJÖSTRAND have made very careful chopper measurements of the spectrum from the central canal of a D_2O-natural-uranium reactor. The measured spectrum was resolved into a thermal and an epithermal part according to Eq. (10.2.10d). Fig. 15.3.10 shows the joining function so obtained; it resembles that found for graphite systems by COATES.

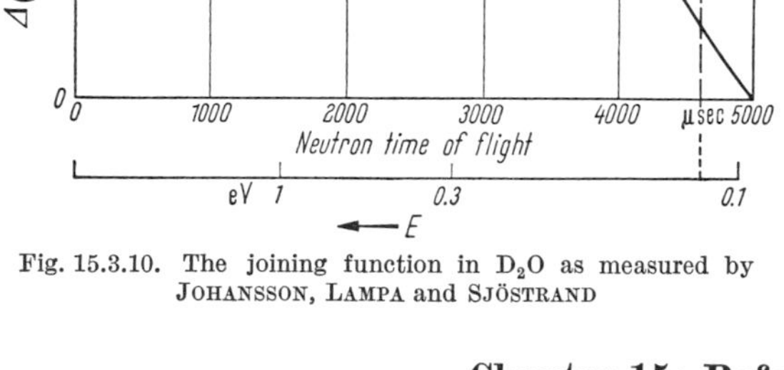

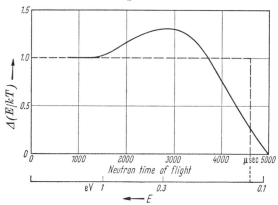

Fig. 15.3.10. The joining function in D_2O as measured by JOHANSSON, LAMPA and SJÖSTRAND

There are additional individual results on various other moderators, mainly hydrogenous substances, such as zirconium hydride and polyethylene, but also D_2O and beryllium oxide (BEYSTER).

Chapter 15: References

General

POOLE, M. J., M. S. NELKIN, and R. S. STONE: The Measurement and Theory of Reactor Spectra, Progress Nucl. Energy Ser. I, Vol. 1, p. 91 (1958).
BEYSTER, J. R., *et al.*: Integral Neutron Thermalization, GA-2544 (1961).
BEYSTER, J. R.: Status of Neutron Spectra Measurements, BNL-719, p. RC 1 (1962).

[1] Here $\Sigma_a(kT)$ is the absorption cross section averaged over the lattice.

Special[1]

BECKURTS, K. H., and W. REICHARDT: in Neutron Time-of-Flight Methods, p. 239; Brussels: Euratom 1961.
BEYSTER, J. R., et al.: Nucl. Sci. Eng. **9**, 168 (1961).
MARSEGUERRA, M., and G. PAULI: Nucl. Instrum. Methods **4**, 140 (1959).
MOSTOVOI, V. I., M. I. PEVZNER, and A. P. TSITOVICH: Geneva 1955 P/640, Vol. 4, p. 12.
POOLE, M. J.: Neutron Time-of-Flight Methods, p. 221; Brussels: Euratom 1961
STONE, R. S., and R. E. SLOVACEK: KAPL-1499 (1956).
STONE, R. S., and R. E. SLOVACEK: Nucl. Sci. Eng. **6**, 466 (1959).

General Facts about the Use of the Time-of-Flight Method to Determine Neutron Spectra.

STURM, W. J.: Phys. Rev. **71**, 757 (1947).
TAYLOR, B. T.: AERE-N/R 1005 (1952).
ZINN, W. H.: Phys. Rev. **71**, 752 (1947).

Measurement of the Neutron Spectrum with a Crystal Spectrometer.

FERMI, E., J. MARSHALL, and L. MARSHALL: Phys. Rev. **72**, 193 (1947).
HUGHES, D. J., J. R. WALLACE, and R. H. HOLTZMANN: Phys. Rev. **73**, 1277 (1948).

Temperature Measurement on Neutron Beams by the Transmission Method.

BRANCH, G.: MDDC-747 (1946).
KÜCHLE, M.: Nucl. Sci. Eng. **2**, 87 (1957).

Temperature Measurement by the Transmission Method with Foils.

LAPORTE, O.: Phys. Rev. **52**, 72 (1937).
ZAHN, C. T.: Phys. Rev. **52**, 67 (1937).

Calculation of Thermal Neutron Transmission through Absorbers.

BURKART, K., and W. REICHARDT: BNL-719, 318 (1962).
BURKART, K.: Diploma Thesis, Karlsruhe 1962.
CHIDLEY, B. G., R. B. TURNER, and C. B. BIGHAM: Nucl. Sci. Eng. **16**, 39 (1963).
SCHMID, L. C., and W. P. STINSON: Nucl. Sci. Eng. **7**, 477 (1960); HW-66319 (1960); HW-69475 (1961); HW-64866 (1960).

Temperature Measurement with Lutetium Foils.

BIGHAM, C. B., R. B. TURNER, and B. G. CHIDLEY: Nucl. Sci. Eng. **16**, 85 (1963).
CAMPBELL, C. G., R. G. FREEMANTLE, and M. J. POOLE: Geneva 1958 P/10 Vol. 22, p. 233.
STINSON, W. P., L. C. SCHMID, and R. E. HEINEMAN: Nucl. Sci. Eng. **7**, 435 (1960).

Temperature Measurement by Fission Rate Measurement.

KORPIUM, P., K. RENZ, and T. SPRINGER: BNL-719, 401 (1962) (Temperature Measurement with Europium Foils).

ANDERSON, H. L., et al.: Phys. Rev. **72**, 16 (1947).
GAVIN, G. B.: Nucl. Sci. Eng. **2**, 1 (1957).

Neutron Temperature Measurement by Comparing Reactivity Effects of Various Absorbers.

BEYSTER, J. R., et al.: Nucl. Sci. Eng. **9**, 168 (1961).
BURKART, E., and W. REICHARDT: BNL-719, 318 (1962).
POOLE, M. J.: J. Nucl. Energy **5**, 325 (1957).
STONE, R. S., and R. E. SLOVACEK: Nucl. Sci. Eng. **6**, 466 (1959).

Studies in H_2O and Aqueous Solutions of Absorbers.

CAMPBELL, C. G., R. G. FREEMANTLE, and M. J. POOLE: Geneva 1958 P/10, Vol. 22, p. 233.
COATES, M. S.: in Neutron Time-of-Flight Methods, p. 233; Brussels: Euratom 1961.
KÜCHLE, M.: Nucl. Sci. Eng. **2**, 87 (1957).
PARKS, D. E., J. R. BEYSTER, and N. F. WIKNER: Nucl. Sci. Eng. **13**, 306 (1962).

Studies in Graphite.

JOHANSSON, E., E. LAMPA, and N. G. SJÖSTRAND: Arkiv Fysik **18**, 513 (1960).
JOHANSSON, E., and E. JONSSON: Nucl. Sci. Eng. **13**, 264 (1962).

Studies in D_2O.

[1] Cf. footnote on p. 53.

Part IV

The Determination of Neutron Transport Parameters

16. Slowing-Down Parameters

In this chapter, we consider stationary and non-stationary experiments for the study of neutron moderation and diffusion at energies above the thermal region. By far the most important of these experiments are the stationary ones for determining the mean squared slowing-down distance (Sec. 16.1) since this quantity enters directly into reactor calculations. In addition, comparison of measured and calculated mean squared slowing-down distances allows us to draw conclusions about the validity of our slowing-down theory and the correctness of our nuclear data. In Sec. 16.2, we shall familiarize ourselves with some measurements of the neutron distribution at very large distances from sources; these measurements are particularly important for shielding problems. Finally, in Sec. 16.3, we shall consider some experiments on the time dependence of the slowing-down process.

16.1. Determination of the Mean Squared Slowing-Down Distance

16.1.1. Basic Facts about the Techniques of Measurement

The mean squared slowing-down distance can be determined from foil measurements of the neutron distribution around a point source in a moderating medium. The medium should be so large that the neutron loss by leakage is unimportant. For practical purposes, the most interesting datum is the mean squared slowing-down distance to thermal energy $\overline{r_{th}^2}=6\tau_{th}$. However, according to Sec. 8.4.1, a measurement of the neutron distribution with thermal probes does not give $\overline{r_{th}^2}$ but rather the total mean squared distance $6M^2=6(\tau_{th}+L^2)$. In other words, the diffusion of thermal neutrons that occurs subsequent to the slowing-down process affects the experimental results. In order to determine $\overline{r_{th}^2}=6\tau_{th}$, we need to measure the neutron distribution with resonance probes at a discrete energy immediately above the thermal range. It has become customary to take as a basis for the determination of the mean squared slowing-down distance the neutron distribution at 1.46 ev, which can be measured with indium foils. As we have seen in Sec. 12.1, in a thin indium foil (<10 mg cm^{-2}) under cadmium, 95% of the 54-min activity arises from neutron capture in the 1.46-ev resonance. Thus we can write for the activation of a thin, cadmium-covered indium foil

$$C^{CD}(r) \sim \Phi(r, 1.46 \text{ ev}) \tag{16.1.1}$$

from which it follows that

$$\frac{\int\limits_0^\infty r^2 C^{CD}(r)\,4\pi r^2\,dr}{\int\limits_0^\infty C^{CD}(r)\,4\pi r^2\,dr} = \overline{r^{*2}_{1.46\,\text{ev}}} = 6\,\tau^*_{1.46\,\text{ev}}. \tag{16.1.2}$$

Note that the measurement gives not the age but the flux age at 1.46 ev since the flux at 1.46 ev and not the slowing-down density is determined. The age is smaller by an amount $\Delta\tau^* = \tau^*_{1.46\,\text{ev}} - \tau_{1.46\,\text{ev}}$. This difference can only be taken into account by calculation. According to GOLDSTEIN *et al.*, $\Delta\tau^* = 0.43\ \text{cm}^2$ in H_2O, 3.7 cm² in D_2O, and 1.7 cm² in graphite. As we shall see later, the experimental errors in the measured values of the age are of the same order of magnitude as these corrections.

The integration in Eq. (16.1.2) extends to $r=\infty$, whereas in the practical case the flux can only be measured over a finite domain. Fortunately, in most cases flux measurement over a limited range of distances suffices because we can determine the flux variation for larger distances accurately enough by extrapolation. We explain this by means of Fig. 16.1.1, which shows $r^2 C^{CD}(r)$ for indium foils in the neighborhood of a (Ra—Be) source in water. At large source distances, we clearly have

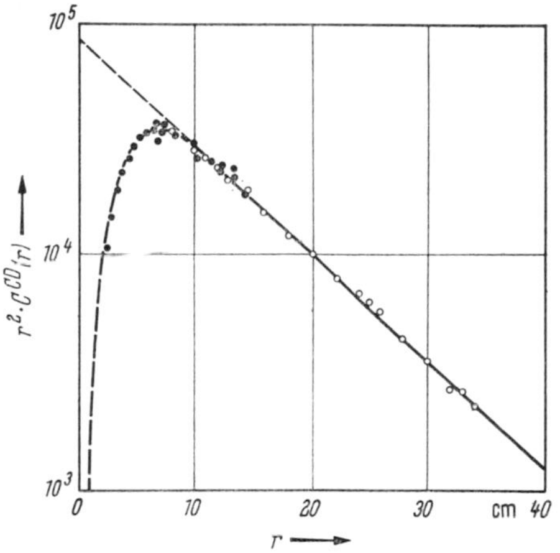

Fig. 16.1.1. $r^2 \cdot C^{CD}(r)$ for cadmium-covered indium foils as a function of distance from a (Ra—Be) source in water

$$r^2 C^{CD}(r) \sim e^{-\Sigma r} \tag{16.1.3}$$

or $C^{CD}(r) \sim e^{-\Sigma r}/r^2$, and we need only extend the measurements far enough to determine the decay constant Σ. For larger source distances, we extrapolate according to Eq. (16.1.3). The explanation of this behavior lies in the first-collision nature of the slowing-down process at large source distances that was repeatedly mentioned in Chapter 8. We shall return to this point in Sec. 16.2.

We now possess a complete "recipe" for the determination of $\tau^*_{1.46\,\text{ev}}$: First we determine $C^{CD}(r)$ for thin indium foils in the neighborhood of a point source. By plotting $r^2 C^{CD}(r)$ on semilogarithmic paper and extrapolating it linearly, we obtain $C^{CD}(r)$ for all r. Then we plot $r^4 C^{CD}(r)$ and $r^2 C^{CD}(r)$ on linear paper and graphically integrate them from $r=0$ to $r=\infty$. The quotient of these two numbers gives $\overline{r^{*2}_{1.46\,\text{ev}}}$.

In the practical use of this procedure, we must take into account a number of sources of error. In the first place, for reasons of intensity one cannot ordinarily use extremely thin indium foils, and surface densities of 100 mg cm⁻² are usual. According to Fig. 12.1.4, for such thick foils about 15% of the activation comes from neutron capture in the higher resonances because the 1.46-ev resonance is already quite strongly self-shielded[1]. This activation by neutrons of higher energy

[1] Fig. 12.1.4 holds for a $1/E$-flux spectrum. Near the source, the spectrum will contain more high-energy neutrons than a $1/E$-spectrum, and the contribution of the higher resonances to the activation will be even larger.

must either be eliminated by calculation or by using the sandwich method of measuring fluxes discussed in Sec. 12.1.3. Additional errors can arise from the finite size of the neutron source: neutrons can be absorbed or inelastically scattered in the source. Finally, deviations from the ideal spherical geometry can cause errors.

In media with large slowing-down lengths, such as D_2O or graphite, it is occasionally difficult to construct a system large enough for age measurements. As Rose has shown — cf. also Weinberg and Noderer — the leakage out of a finite system that is being used for an age measurement can be taken into account by calculation. For this purpose, the material to be investigated is made into a parallelepiped of length a and b along the sides. The height of the parallelepiped must be very large compared to the slowing-down length. A neutron source is placed at the middle of the parallelepiped, and the distribution of the resonance flux on the axis of the parallelepiped is measured as a function of the source distance z. Then we form the moments $\overline{z^{2\nu}}$ of the axial flux distribution according to the formula

$$\overline{z^{2\nu}} = \frac{\int\limits_0^\infty CCD(z)\, z^{2\nu}\, dz}{\int\limits_0^\infty CCD(z)\, dz}.$$

The age may be determined from these moments with the relation

$$\tau^* = \frac{\sum\limits_{\nu=0}^{\infty} \frac{\nu+1}{(2\nu+2)!}\left(\frac{\pi^2}{a^2} + \frac{\pi^2}{b^2}\right)^{\nu} \overline{z^{2\nu+2}}}{\sum\limits_{\nu=0}^{\infty} \frac{1}{(2\nu)!}\left(\frac{\pi^2}{a^2} + \frac{\pi^2}{b^2}\right)^{\nu} \overline{z^{2\nu}}}$$

which has been derived by Weinberg and Noderer.

16.1.2. Measurement of the Mean Squared Slowing-Down Distance of Fission Neutrons in Water

As an example of the method developed in Sec. 16.1.1, let us discuss the measurement of $\tau^*_{1.46\,\mathrm{ev}}$ for fission neutrons in water. This measurement has recently been carried out repeatedly because of a systematic discrepancy between very carefully calculated values ($\approx 26\ \mathrm{cm}^2$) and earlier experimental values ($\approx 30\ \mathrm{cm}^2$). Recently, especially careful measurements have been carried out by Doerner *et al.* and by Lombard and Blanchard, among others.

Fig. 16.1.2 shows schematically the apparatus used by Lombard and Blanchard. Thermal neutrons from the thermal column of a swimming pool reactor irradiate the fission source, a nearly square (*ca.* 8×8 cm) 1-cm-thick slab of uranium-zirconium alloy containing about 1.4×10^{21} U^{235} atoms per cm^3. Indium foils ($100\ \mathrm{mg\ cm}^{-2}$) covered by 1-mm-thick cadmium are fixed in a device hung from a bridge that permits precise and reproducible positioning. The entire apparatus is immersed in the reactor pool. Fig. 16.1.3 shows the epicadmium activity of the indium foils as a function of their distance from the source plate. This distribution was measured with a distance of 1.7 cm between the end of the thermal column and the source plate. An increase of this distance does not affect the relative distribution of the neutrons, from which one can conclude that slowing-down in the graphite of the thermal column does not noticeably affect the experimental results. The values shown were obtained from a difference measurement in which C^{CD} was measured once with and once without the source plate. In this way that fraction of the foil activation due to the direct flux from the thermal column was eliminated. The activation by photoneutrons that are produced by the reaction of energetic γ-rays from the reactor with the deuterium present in the water also disappears from the difference. In contrast, energetic γ-rays from fission in the source plate can produce photoneutrons in the water and falsify the neutron distribution. However, measurements in which the γ-rays leaving the source plate were shielded by a 1.25-cm-thick bismuth slab showed no such effect.

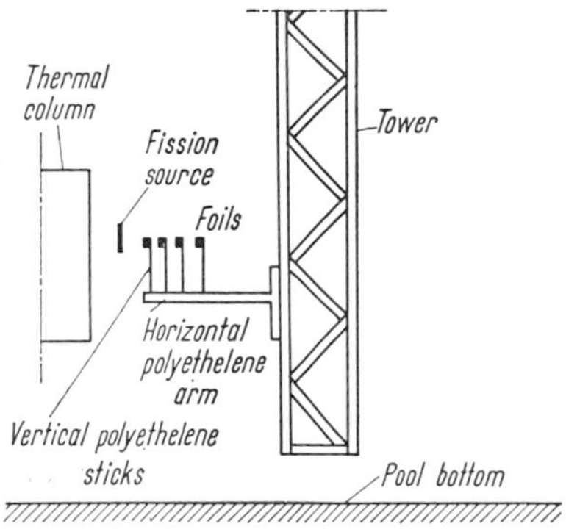

Fig. 16.1.2. The arrangement used by LOMBARD and BLANCHARD for the measurement of the fission neutron age in water

In order to estimate the effect of foil activation due to neutron capture in the higher indium resonances, measurements were carried out in which the indium foils were covered on both sides by cadmium *and* indium covers. It turned out that even under indium covers the foils showed a considerable activation; thus neutrons of energy other than 1.46 ev contribute heavily to the activation. However, the spatial variation of the activity obtained with the indium covers is the same as that obtained without them, i.e., Fig. 16.1.3 holds rigorously for 1.46-ev neutrons.

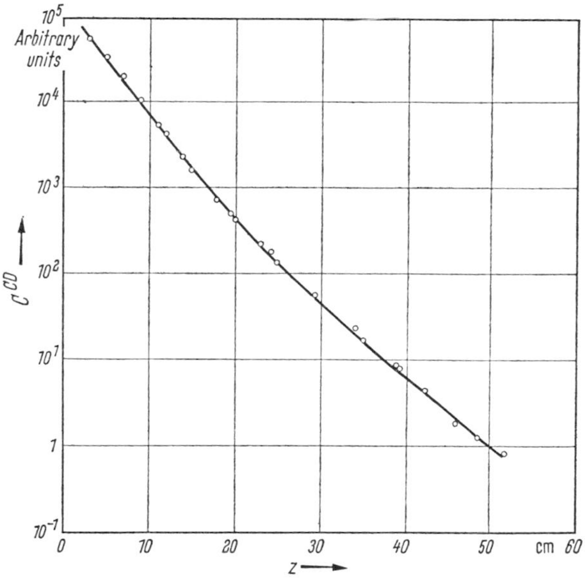

Fig. 16.1.3. The epicadmium activity of Indium foils as a function of distance from the fission source plate

In order to see what effect the finite thickness of the source plate has on the neutron distribution, the measurements near the source were repeated with source plate thicknesses of 2 and 3 cm. It turned out that the flux (normalized to its value at a source distance of 10 cm) decreased with increasing thickness of the source plate for source distances < 7 cm. This decrease is due to the absorption of epithermal neutrons in the plate. This perturbing effect was eliminated by extrapolating to zero source plate thickness.

In calculating the age from the flux distribution we must note that according to its definition the age is determined by the neutron distribution $\Phi(\varrho)$ around a point source, whereas what has been measured is the distribution $F(z)$ in the neighborhood of a source plate. Between these two quantities, the following relation exists:

$$F(\boldsymbol{r}=z)=\int S(\boldsymbol{r}')\,\Phi\big(\varrho=|\boldsymbol{r}-\boldsymbol{r}'|\big)\,dA\,.$$

Here $S(\boldsymbol{r}')$ is the source strength per unit area of the plate; the integration is over the entire area of the fission plate. $S(\boldsymbol{r}')$ was determined by measurement with thermal foils at various places on the source plate. In order to obtain $\Phi(\varrho)$ from $F(z)$, the above integral was carried out with SIMPSON's rule. From the values of $\Phi(\varrho)$ so determined, LOMBARD and BLANCHARD obtained $\tau^*_{1.46\,\mathrm{ev}}=26.8\pm0.9\ \mathrm{cm}^2$ from Eq. (16.1.2). This value is an average of the ages obtained by β-counting the activity of the front and back sides of the foils (relative to their orientation to the source plate during the irradiation). Averaging is necessary when we count a β-activity because otherwise — as we saw in Sec. 11.2.5 — current effects influence the result of a foil measurement. Even after this averaging, the activation of a thick indium foil is not strictly proportional to the flux at 1.46 ev. As was shown in Sec. 12.2.1, all the even coefficients in the expansion of the vector flux in Legendre polynomials contribute to the activation of a foil. Since the $F_l(\boldsymbol{r})$ are generally small for $l\geq 2$ and since the corresponding functions $\varphi_l(\mu_a\delta)$ have small values for thin foils, the contributions of the higher terms in the expansion can almost always be neglected. For 100-mg/cm² indium foils, however, $\mu_a\delta\gg 1$ in the resonance, and thus $\varphi_0(\mu_a\delta)=1$ and $\varphi_2(\mu_a\delta)=\frac{1}{4}$. In this case, $C\sim F_0(\boldsymbol{r})\left(1+\frac{5}{4}\,\frac{F_2(\boldsymbol{r})}{F_0(\boldsymbol{r})}\right)$, and even a small P_2-part in the vector flux would have an effect on the foil activation. It turns out that because of this P_2-fraction a correction of 0.5 cm² must be applied to the flux age. The result of LOMBARD and BLANCHARD is then increased to $\tau^*_{1.46\,\mathrm{ev}}=27.3\pm0.9\ \mathrm{cm}^2$.

Table 16.1.1. *Measured Values of the Flux Age of Fission Neutrons in Water*

Year	$\tau^*_{1.46\,\mathrm{ev}}$ [cm²]	Author
1944	32.3	ANDERSON, FERMI, and NAGLE
1948	30.8	HILL, ROBERTS, and FITCH
1950	30.03	HOOVER and ARNETTE
1956	31.0	WADE
1959	26.7	BLOSSER and TRUBEY
1960	27.3±0.9	LOMBARD and BLANCHARD
1960	27.0±0.9	PETTUS
1961	27.86±0.1	DOERNER *et al.*

In Table 16.1.1, various measured values of $\tau^*_{1.46\,\mathrm{ev}}$ are summarized. The earlier investigations gave rather high values owing in part to insufficient accounting for the effect of source size on the flux distribution. In contrast, the results obtained in the last few years agree with one another rather well. The theoretical value is 26.0 cm². Information on the higher moments of 1.46-ev neutrons in water ($\overline{r^4}$, $\overline{r^6}$, etc.) is given by DOERNER *et al.* (cf. also HILL, ROBERTS, and FITCH).

16.1.3. Results for $\overline{r^{*2}_{1.46\,\mathrm{ev}}}=6\,\tau^*_{1.46\,\mathrm{ev}}$ for Various Media and Sources[1].

Water. Table 16.1.2 shows measured values of $\tau^*_{1.46\,\mathrm{ev}}$ for various neutron sources in H_2O. The values for the radioactive (α, n) sources are not very well

[1] MCARTHY *et al.* and PERSIANI *et al.* both give more complete tables of age measurements.

defined since the spectra of such sources depend on their composition, which may vary according to the process of manufacture. In addition, there has been a series of investigations of the age in homogeneous mixtures of water with aluminum, zirconium, iron, bismuth, and lead. Results are given by MUNN and PONTECORVO, by REIER, OBENSHAIN and HELLENS, as well as by HILL, ROBERTS, and FITCH.

Table 16.1.2. *The Flux Age to Indium Resonance in H_2O*

Source	$\bar{E}$	$\tau^*_{1.46\,\mathrm{ev}}$ [cm²]	Author	Comments
(Sb—Be)	25 kev	5.48±0.15	BARKOV, MAKARIN, and MUKHIN	These measurements still contain uncorrected errors . For example, the finite size of the source is not taken into account
(Na—Be)	970 kev	13.9±0.2	FOSTER	
$H^2(d, n)He^3$ Thick Target $E_d = 250$ kev	2.6 Mev	34.6±2.2	SPIEGEL, OLIVER, and CASWELL	Average value over the neutron distribution measured in different directions from the source
(Ra—Be)	5 Mev	54.4	DUGGAL *et al.*	
(Po—Be)	5 Mev	57.3±2.0	VALENTE and SULLIVAN	
$H^3(d, n)He^4$ Thick Target $E_d = 150$ kev	14.1 Mev	150±6	CASWELL *et al.*	

In Fig. 16.1.4, $\tau_{1.46\,\mathrm{ev}}$ for pure H_2O is plotted as a function of the average energy of the neutron source. In the figure, we compare approximate calculations done with the Selengut-Goertzel method (Sec. 8.3.1), exact[1] calculations, and the measured values. Whereas the exact calculations accurately reproduce the

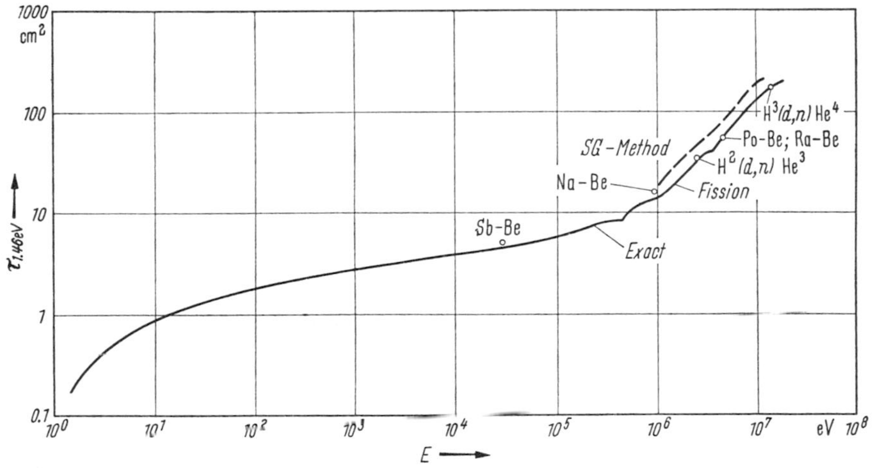

Fig. 16.1.4. $\tau_{1.46\,\mathrm{ev}}$ for pure H_2O as a function of source neutron energy. ——— Exact calculation by GOLDSTEIN *et al.*; — — — approximate calculation by the Selengut-Goertzel method; ∘ Experimental values for various sources

[1] Actually, the calculations were performed by the Monte Carlo method taking into account anisotropic scattering in oxygen.

measured values, the Selengut-Goertzel method always gives too large a value for the age.

SPRINGER has suggested an interesting application of the fact that the age in H_2O increases sharply with the energy of the source. Obviously, we can determine the average energy of the source neutrons quite accurately by measuring the age. If we surround the source with a concentric spherical annulus of a heavy substance, the average energy of the neutrons reaching the moderator will be smaller since energy losses due to inelastic scattering occur. One can determine the decrease in the average energy very accurately by an age measurement and thereby obtain information about the energy loss in an inelastic scattering in the test substance. Such measurements have been done with (Ra—Be), $H^2(d, n)He^3$, and $H^3(d, n)He^4$ neutrons.

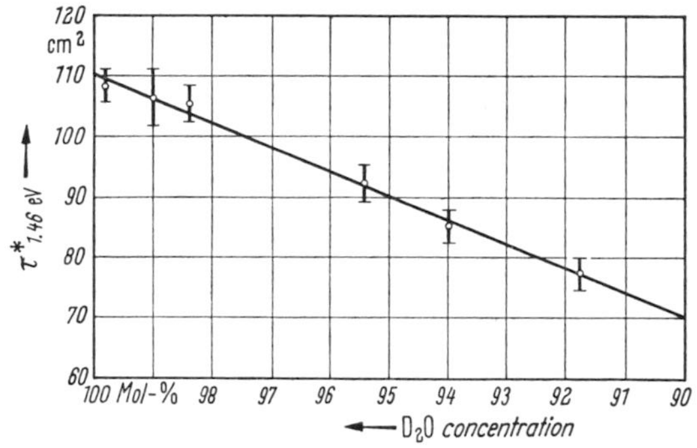

Fig. 16.1.5. The flux age of fission neutrons to indium resonance as a function of the purity of heavy water

Heavy Water. Table 16.1.3 shows some measured flux ages in D_2O. The measured values apply to a D_2O concentration of 99.8%. WADE has investigated the flux age of fission neutrons as a function of the D_2O concentration; Fig. 16.1.5 shows his results. In the concentration range studied, the age decreases about 4 cm² per 1% increase in the light water concentration.

Table 16.1.3. *Flux Age to Indium Resonance in D_2O* (99.8%)

Source	$\bar{E}$	$\tau^*_{1.46\,ev}$ [cm²]	Author	Comments
Uranium Fission	2 Mev	109±3	WADE	Thermal fission of U^{235}
$H^2(d, n)He^3$ Thick Target E_d = 250 kev	2.6 Mev	119.1±1.5	SPIEGEL and RICHARDSON	Average value over the neutron distributions measured in various directions from the source

A Monte-Carlo calculation of the flux age $\tau^*_{1.46\,ev}$ of fission neutrons in 99.8% D_2O gives 112.2 cm² (GOLDSTEIN *et al.*), and an approximate calculation by the Greuling-Goertzel method 108 cm² (LEVINE *et al.*).

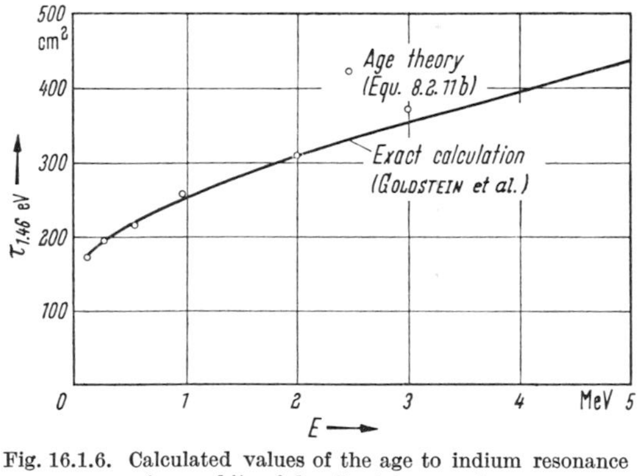

Fig. 16.1.6. Calculated values of the age to indium resonance in graphite of density 1.6 g · cm⁻³

Graphite. Table 16.1.4 contains measured values for the flux age to indium resonance of various sources in graphite of density 1.6 g/cm³. Fig. 16.1.6 shows calculated values of $\tau^*_{1.46\,ev}$ as a function of the source energy. In the figure, a precise calculation (GOLDSTEIN *et al.*) is compared with the simple approximation of age theory [Eq. (8.2.11b)]. The calculated value of $\tau^*_{1.46\,\mathrm{ev}}$ for fission neutrons is 304 cm².

Table 16.1.4. $\tau^*_{1.46\,\mathrm{ev}}$ *in Graphite of Density 1.6 g/cm³*

Source	$\bar{E}$	$\tau^*_{1.46\,\mathrm{ev}}$ [cm²]	Author
(Sb—Be)	25 kev	≈ 142	HILL, ROBERTS, and McCAMMON
Uranium Fission	2 Mev	312.5 ± 0.5	HENDRIE *et al.*
(Ra—Be)	5 Mev	≈ 380	FERMI
(Pu—Be)	5 Mev	≈ 416	STEICHEN

Since age theory holds very accurately in graphite, one can represent the slowing-down density in the neighborhood of a source analytically. For example, the flux distribution at 1.46 ev in the neighborhood of a monoenergetic point source in a very large medium is given by

$$\Phi(1.46\text{ ev}, r) \sim e^{-\frac{r^2}{4\tau^*_{1.46\,\mathrm{ev}}}} \qquad (16.1.4)$$

where $\tau^*_{1.46\,\mathrm{ev}}$ is the flux age determined from the mean squared slowing-down distance. Eq. (16.1.4) no longer holds for a source with a broad energy spectrum. We show this in Fig. 16.1.7, where the flux distribution measured near a (Ra—Be) source in graphite is compared with a calculation based on Eq. (16.1.4) with $\tau = 380$ cm². Better agreement is obtained when the spectrum of the source is approximated by several age groups; then

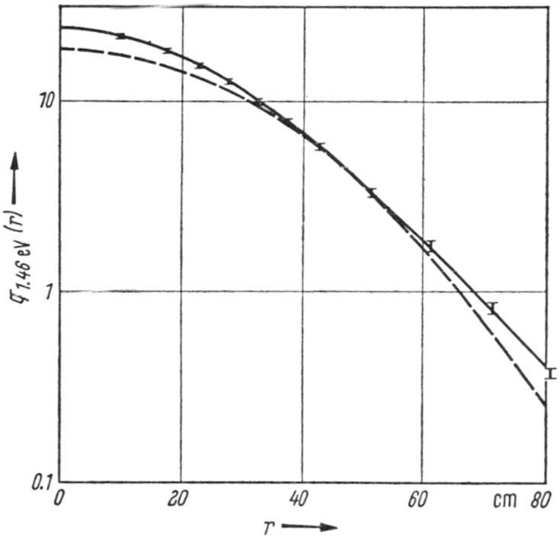

Fig. 16.1.7. The slowing-down density versus distance from a (Ra—Be) source in graphite. I Measured values; — — — calculated for one age group; —— calculated for three age groups

$$\Phi(1.46\text{ ev}, r) = \sum_i g_i\, e^{-\frac{r^2}{4\tau^*_{i,\,1.46\,\mathrm{ev}}}}. \qquad (16.1.5)$$

According to FERMI, (Ra—Be) sources can be represented in graphite by three age groups:

$$\begin{array}{llll} g_i = & 0.15 & 0.69 & 0.16 \\ \tau^*_{i,\,1.46\,\mathrm{ev}} = & 130\text{ cm}^2 & 340\text{ cm}^2 & 815\text{ cm}^2. \end{array}$$

Slightly different values result from measurements of the authors:

$$\begin{array}{llll} g_i = & 0.11 & 0.68 & 0.21 \\ \tau^*_{i,\,1.46\,\mathrm{ev}} = & 175\text{ cm}^2 & 340\text{ cm}^2 & 860\text{ cm}^2. \end{array}$$

The flux distribution corresponding to these values is also given in Fig. 16.1.7, where it is seen to reproduce the measured values quite well.

Other Moderators. Table 16.1.5 contains values of $\tau^*_{1.46\,\mathrm{ev}}$ for an additional series of moderators and neutrons sources. The values stem in part from older measurements and are not always reliable. In some places, calculated values are also given. There are still not many reliable τ^*-measurements in the metal hydrides and in organic moderators.

Table 16.1.5. $\tau^*_{1.46\,\mathrm{ev}}$ *for Various Moderators and Sources*

Substance	Density	Source	$\overline{E}$	$\tau^*_{1.46\,\mathrm{ev}}$ [cm²]	Author	Remarks
Be	1.85 g/cm³	Uranium Fission	2 Mev	80.2±0.2	NOBLES and WALLACE	
Be	1.78 g/cm³	(Ra—Be)	5 Mev	120±23	GERASEVA *et al.*	
BeO	2.96 g/cm³	Uranium Fission	2 Mev	93.4±4.7	GOODJOHN and YOUNG	
Diphenyl	C: 4.595× H: 3.829× 10^{22} atoms/cm³	Uranium Fission	2 Mev	44.8	CAMPBELL and PASCHALL	
ZrH	4.8 g/cm³	Uranium Fission	2 Mev	49.12	LUNDY and GROSS	Calculated Values
Zr_2H	5.0 g/cm³			81.45		
ZrH_2	4.5 g/cm³			25.85		
Zr_2H_3	4.6 g/cm³			34.96		

16.1.4. The Age for Slowing Down from 1.46 ev to Thermal Energy

Let us define the quantity

$$\Delta\tau = \tau_{th} - \tau^*_{1.46\,\mathrm{ev}}. \tag{16.1.6}$$

If $\Delta\tau$ is known, we can calculate the age for moderation to thermal energy from the experimentally determined flux age to indium resonance energy. According to the elementary theory,

$$\Delta\tau \approx \frac{1}{3\xi\left(1-\frac{2}{3A}\right)} \int_{E_{ET}}^{1.46\,\mathrm{ev}} \lambda_s^2(E)\,\frac{dE}{E}. \tag{16.1.7}$$

Here $E_{ET} \approx 3.5\,kT$ is the thermal cut-off energy discussed in Sec. 12.2.2. Table 16.1.6 contains some values of $\Delta\tau$ determined in this way. This kind of calculation is not particularly accurate since the effect of chemical binding on the slowing-down process is not taken into account. However, we can easily determine $\Delta\tau$ experimentally. To do this we note that the following expression holds for the migration area, which can be determined from the mean squared distance to absorption of thermal neutrons:

$$M^2 = L^2 + \tau_{th} = L^2 + \tau^*_{1.46\,\mathrm{ev}} + \Delta\tau. \tag{16.1.8}$$

Therefore

$$\Delta\tau = M^2 - L^2 - \tau^*_{1.46\,\mathrm{ev}} \tag{16.1.9}$$

and we can determine $\Delta\tau$ if we know M^2, L^2, and $\tau^*_{1.46\,\mathrm{ev}}$. We can measure M^2 and $\tau^*_{1.46\,\mathrm{ev}}$ with the methods of this section; in the next chapter, we shall become familiar with methods for determining L^2. A difficulty in such measurements is the fact that as a rule $\Delta\tau$ is small compared to M^2 and L^2. Thus we must form the small difference of large numbers and consequently obtain a less exact result. For this reason, it is better to use an (Sb—Be) source for such measurements (in order to make $\tau^*_{1.46\,ev}$ small) and to poison the moderator with an absorber (in order to make L^2 small). Some measured values of $\Delta\tau$ obtained in this way are given in Table 16.1.6.

Table 16.1.6. $\Delta\tau=\tau_{th}-\tau^*_{1.46\,ev}$ *at* $T=20\,°C$

Substance	$\Delta\tau$ [cm²] According to Eq. (16.1.7)	$\Delta\tau$ [cm²] According to Measurement	Author
H_2O	0.8	1.0 ± 0.5	Barkov, Makarin, and Mukhin
D_2O (99.8%)	14		
Be (1.85 g/cm³) . . .	10		
Graphite (1.6 g/cm³)	40	56 ± 6.5	Duggal and Martelly

16.2. The Behavior of the Neutron Flux at Large Distances from a Point Source of Fast Neutrons

A large number of very careful studies of the spreading out of thermal and 1.46-ev neutrons at large distances from a point source in *water* have been carried out. Figs. 16.2.1, 2, and 3 show some of the results found (cf. also Fig. 16.1.1). Either the quantity $r^2\,\Phi_{1.46\,ev}(r)$ or the quantity $r^2\,\Phi_{th}(r)$ has been plotted on semilog paper. In first approximation, all the curves show a linear decrease at large source distances. In other words $\Phi(r)\sim e^{-\Sigma r}/r^2$. The thermal and epithermal fluxes behave in very similar manners. The explanation of this behavior is very simple if we note that the scattering cross section of hydrogen falls off very sharply with energy (Sec. 1.4.2): If a neutron emitted from the source once collides and experiences a large energy loss, the cross section for additional collisions is large. Thus the probability of the neutron's being moderated and finally absorbed near the site of its first collision is large. The thermal and epithermal fluxes should then behave like the flux of primary neutrons. The latter is proportional to $e^{-\Sigma r}/r^2$, where Σ is the total cross section at the source energy. If the source emits neutrons with a broad energy spread, a hardening effect occurs. Since the cross section for the first collision increases with decreasing energy, the more energetic neutrons in the spectrum of primaries are attenuated less. With increasing distance from the source, these energetic neutrons become more and more important in determining the spreading out of the neutrons. These facts explain the slight flattening of the decay curve for fission neutrons in Fig. 16.2.1.

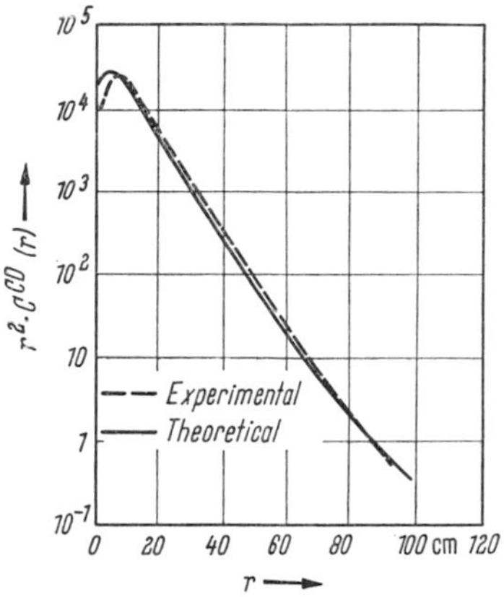

Fig. 16.2.1. $r^2 \cdot C^{CD}(r)$ for cadmium-covered indium foils as a function of distance from a point fission source in water (from Goldstein). The theoretical curve is based on moments method calculations at 2.03 ev. The experimental curve is from Hill, Roberts, and Fitch

More careful scrutiny of the decay curves determined with monoenergetic sources (cf. Figs. 16.2.2 and 3) likewise shows deviations from straight-line behavior. In these cases, the decay becomes somewhat steeper with increasing distance from the source. This originates in the fact that the picture of the spreading-out process developed above, namely, that a secondary radiation component (those neutrons which have had at least one collision) is in equilibrium with a primary component (the uncollided source neutrons), is not quite right. We must note that neutron scattering by hydrogen is strongly anisotropic in the forward direction in the laboratory system. Therefore, a neutron can with an appreciable

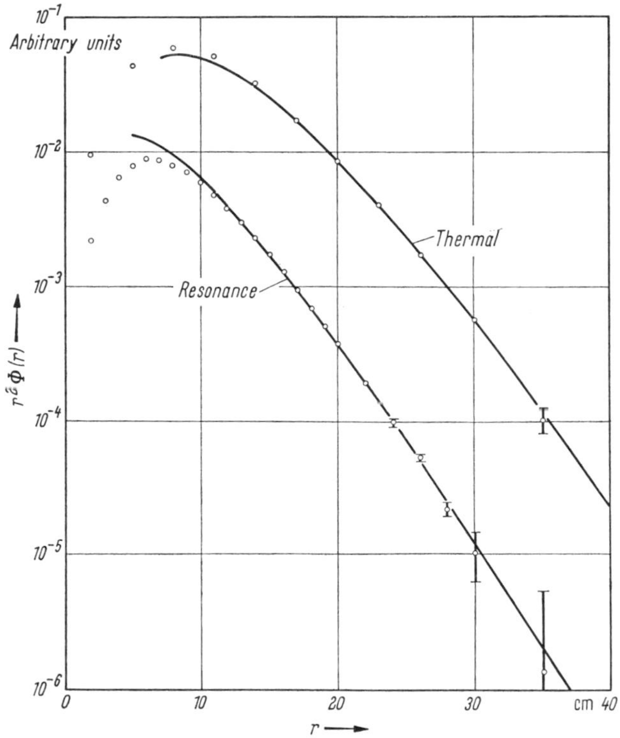

Fig. 16.2.2. $r^2\,\Phi(r)$ for thermal and indium resonance neutrons as functions of distance from a (Na—Be) source in water (as given by FOSTER).

probability suffer only a small change in angle and energy in its first collision. Then after the collision it still belongs to the "primaries" and the distinction between the primary and secondary radiation components is less sharply defined. With increasing distance from the source, the primary component will contain more and more neutrons that have made some small-angle collisions and its average energy will sink somewhat. This explains the steeper decay of the neutron flux at large distances. If the source neutrons have a broad energy spectrum, the last effect described can partly offset the hardening effect described above.

The theoretical treatment of the spreading-out process at large distances is difficult. A complete description of the neutron distribution can be obtained with the moments method (cf. Sec. 8.1.3) or by Monte Carlo calculations; such calculations involve considerable numerical labor. WICK and HOLTE have given some analytic approximation methods.

In spite of the deviations from simple behavior just discussed, it is customary to approximate the measured neutron distribution in the neighborhood of a

Table 16.2.1. *Relaxation Lengths $1/\Sigma$ for Various Neutron Sources in Water*

Source	$\bar{E}$	$1/\Sigma$ [cm]	Range	Author
(Na—Be)	0.97 Mev	3.3 (1.46 ev)	15— 30 cm	FOSTER
(Ra—Be)	5 Mev	9.29 (1.46 ev) 9.46 (thermal)	12— 40 cm	RUSH
Uranium Fission	2 Mev	8.79 (1.46 ev)	10— 40 cm	HILL, ROBERTS, and FITCH
(Po—Be)	5 Mev	10.3±0.7 (thermal)	10— 40 cm	BAER
$H^3(d, n)He^4$ Thick Target E_d=150 kev	14.1 Mev	14 (thermal)	40—120 cm	CASWELL *et al.*

point source by a law of the form $e^{-\Sigma r}/r^2$. In water this is, as we have seen in Figs. 16.2.1—3, reasonably accurate, at least over limited intervals. Table 16.2.1 shows experimental values of $1/\Sigma$ for various neutron sources in water. Also specified there is the range over which the measured flux distribution was fitted with an expression of the form $e^{-\Sigma r}/r^2$. The measurements were made partly with indium resonance foils and partly with thermal foils. Any differences between measurements made with the two kinds of foils are probably due to experimental errors.

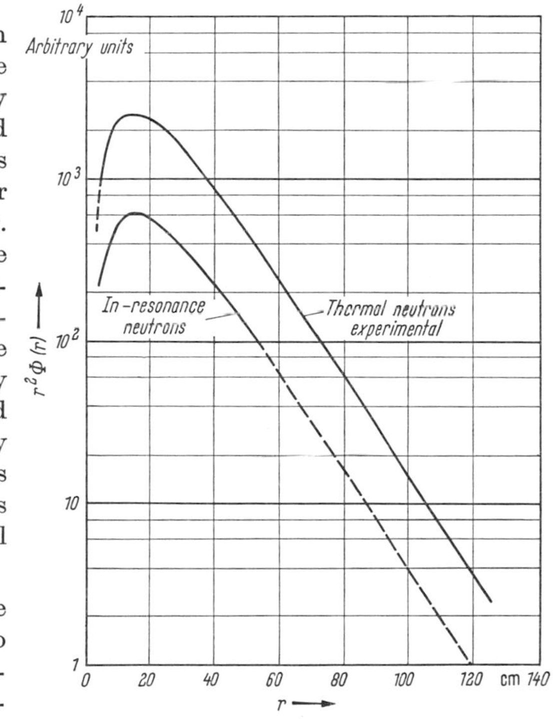

Fig. 16.2.3. $r^2\,\Phi(r)$ for thermal and indium resonance neutrons as a function of distance from a 14.1 Mev-neutron source in water (according to CASWELL *et al.*)

In moderators in which the scattering cross section shows no clear decrease with increasing neutron energy, the concept of radiation equilibrium is even less appropriate than in water. Nevertheless, here also the neutron distribution is occasionally approximated by an expression of the form $e^{-\Sigma r}/r^2$. For example, WADE found $1/\Sigma$=8.7 cm in 99.8% D_2O in the range from 30 to 60 cm from a fission neutron source using indium resonance detectors.

It should be noted that in some cases, e.g., in pure D_2O or in graphite, the primaries do *not* determine the spreading out of the neutrons at large distances from the source. Instead, in such weakly absorbing media there is a pure thermal neutron field that decays like $e^{-r/L}/r$ at large distances from the source.

Detailed experimental and numerical data on the spreading out and slowing down of fast neutrons at large source distances can be found in GOLDSTEIN's book.

16.3. The Time Dependence of the Slowing-Down Process

16.3.1. Slowing Down to Indium and Cadmium Resonance in Hydrogenous Moderators

We can study the time dependence of the slowing-down process by injecting short pulses of neutrons from a pulsed source into a medium and then measuring the time dependence of the capture rate in resonance detectors. In this way, we obtain the time- and space-dependent flux $\Phi(\boldsymbol{r}, E_R, t)$ or, if the medium is large

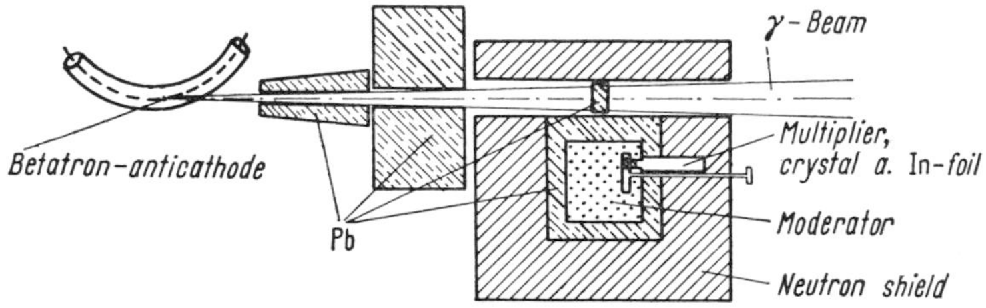

Fig. 16.3.1. The arrangement used by ENGELMANN for measurement of slowing down to indium resonance energy in lead acetate

enough and a spatial integration of the capture rate is performed, the time-dependent flux $\Phi(E_R, t)$, which can be compared with the calculations of Chapter 9. Reliable measurements of this kind have as yet only been carried out in hydrogenous moderators.

Fig. 16.3.1 shows the apparatus ENGELMANN used for studying moderation to indium resonance energy in hydrogen. The actual moderator was lead acetate with a proton density of 1.01×10^{22} cm^{-3}. This proton density, seven times lower than that in water, was chosen in order to make the mean free path and therefore the time scale of the slowing-down process larger. The slowing-down time is thus large compared to the duration of the primary neutron pulse. A 31-Mev betatron served as the neutron source; it produced 2-μsec-long pulses of 2-Mev neutrons

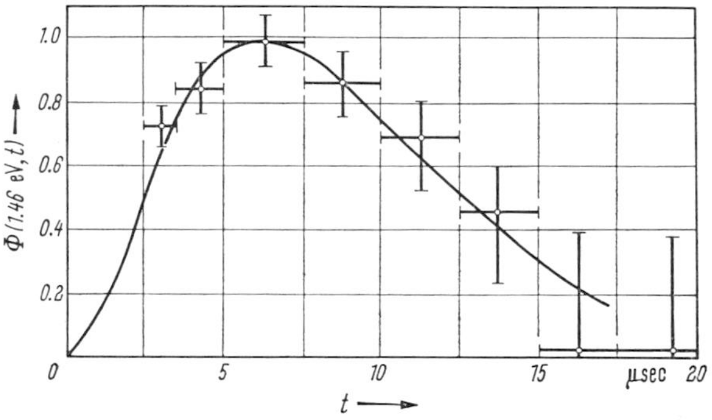

Fig. 16.3.2. The time distribution of neutron captures at 1.46 ev. Measured values; —— calculated with Eq. (9.1.6)

by means of the (γ, n) reaction in lead. The neutrons were detected by means of the capture γ-radiation of an indium foil, which was covered on all sides by an 0.15-g/cm^2-thick boron layer to suppress capture of thermal neutrons. This indium foil, together with an organic scintillator crystal that detected the capture γ-radiation, was housed in the moderator volume. The crystal is also sensitive to γ-rays arising from the capture of thermal neutrons by the surrounding protons and to bremsstrahlung from the betatron. Thus there is a strong background present, and the measurements must be carried out as indium difference measurements, i.e., measurements done once with and once without the indium foils. Fig. 16.3.2 shows the behavior of the indium resonance capture rate obtained in such a difference measurement. A correction that takes into account the leakage of neutrons during the slowing-down process has been applied to the measured values. For this reason, the measurements are comparable with the infinite-medium distributions calculated in Sec. 9.1. The curve in the figure has been calculated with Eq. (9.1.6) for free protons of the density occurring here. The calculated distribution agrees with the measured points within experimental error, from which we can conclude that binding effects play no role in moderation above 1.46 ev[1]. This conclusion is to be expected because of the large number of different modes of proton motion above 1.46 ev to which the neutron can transfer energy.

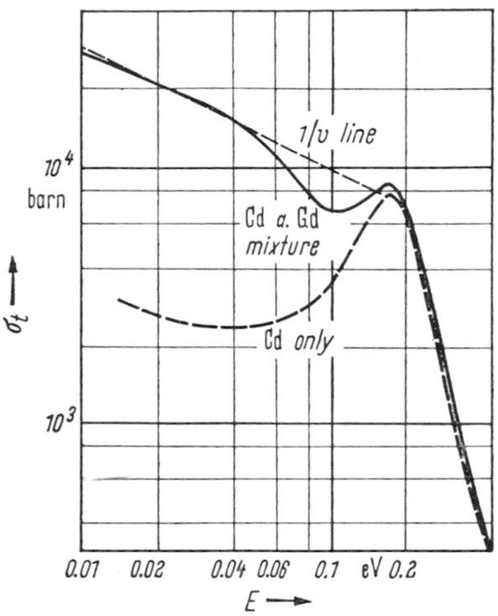

Fig. 16.3.3. The total neutron cross section of a 1:0.365 mixture of cadmium and gadolinium

MÖLLER and SJÖSTRAND have studied the time behavior of moderation to the cadmium cut-off energy in water. For this purpose they shot neutron bursts from a van de Graaff generator into the middle of a 1-m³ tank of water and followed the time dependence of the capture γ-radiation from a small volume filled with aqueous solutions of either cadmium sulfate or gadolinium nitrate. The energy dependence of the capture cross section of cadmium and gadolinium is such that a suitable mixture of both absorbers has a cross section like the one shown in Fig. 16.3.3, viz., it is zero above 0.2 ev and behaves like $1/v$ below 0.2 ev. Therefore, by properly mixing the time-dependent reaction rates measured with cadmium and gadolinium solutions, we obtain a quantity proportional to the neutron density below 0.2 ev. This quantity attains a constant value after a sufficiently long time (if we disregard the fact that the neutron density decays with time due to absorption processes). By dividing through by this constant value, we can obtain $f(t)$, the fraction of the neutron density below 0.2 ev at time t. Fig. 16.3.4 shows $f(t)$ measured in this way by MÖLLER and SJÖSTRAND. These authors defined a slowing-down time to the cadmium

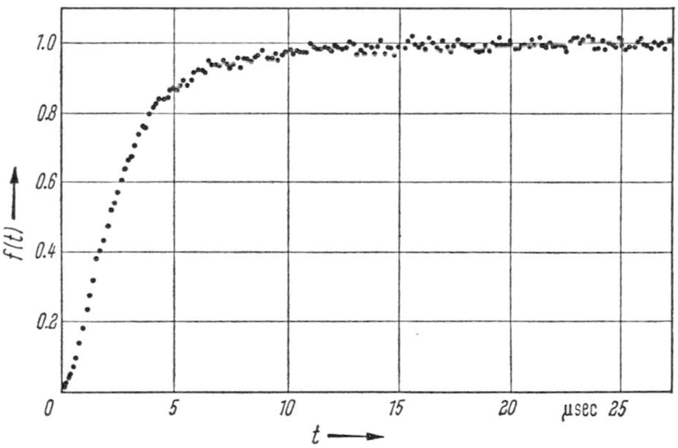

Fig. 16.3.4. The fraction of the neutron density below 0.2 ev as a function of time after burst injection

[1] We thank Dr. ENGELMANN for pointing out an error in his original paper, where the existence of a strong binding effect was concluded on the basis of an incorrectly calculated distribution.

cut-off energy by the relation

$$t_s = \int_0^\infty \left(1 - f(t)\right) dt. \tag{16.3.1}$$

Using the $f(t)$ in Fig. 16.3.4, and with some minor corrections which we shall not discuss here, we find $t_s = 2.5 \pm 0.5\ \mu\text{sec}$. In order to compare t_s with the theory, we cannot resort to Eq. (9.1.7b) since this gives the slowing-down time for a sharply defined energy, whereas we want the average time spent above a particular energy. Obviously,

$$1 - f(t) = \frac{\int_{v_{CD}}^{\infty} \Phi(v, t) \frac{dv}{v}}{\int_0^{\infty} \Phi(v, t) \frac{dv}{v}} \tag{16.3.2}$$

holds for the fraction of the neutron density above the cadmium cut-off energy; if we insert $\Phi(v, t)$ for hydrogen from Eq. (9.1.6), we find $1 - f(t) = (1 + \Sigma_s v_{CD} t)\, e^{-\Sigma v_{CD} t}$. Thus

$$t_s = \frac{2}{v_{CD} \Sigma_s} \tag{16.3.3}$$

according to Eq. (16.3.1). Introducing v_{CD} corresponding to $E_{CD} = 0.2$ ev and the scattering cross section corresponding to the proton density in water, we find that $t_s = 2.42\ \mu\text{sec}$. The experiment and the calculation thus agree within the limits of experimental error, and we can conclude that the chemical binding does not influence the time scale of slowing down to 0.2 ev in water.

De Juren and later Grosshög have shown that the average time which a neutron spends above the cadmium cut-off energy may be obtained from simple stationary measurements. In an infinite medium containing a homogeneously distributed source of strength S [$\text{cm}^{-3}\ \text{sec}^{-1}$] and a $1/v$-absorber, the total neutron density n [cm^{-3}] equals $S \cdot l_0$, where $l_0 = 1/v \cdot \Sigma_a(v)$ is the average lifetime against capture. The counting rate of a $1/v$-detector, e.g., a boron counter, is given by $Z \sim n = S l_0$. If we furthermore measure the counting rate Z^{CD} of the $1/v$-detector under a cadmium cover, we then take into account only that part of the density n_{epi} that comes from neutrons above the cadmium cut-off energy. Now $n_{\text{epi}} = S \cdot t_s$ and therefore

$$\frac{Z^{CD}}{Z} = \frac{t_s}{l_0}. \tag{16.3.4}$$

Thus we can obtain t_s from a measurement of the cadmium ratio if the absorption cross section of the medium is known. One can obtain the values of Z and Z^{CD} that characterize the infinite medium with homogeneously distributed sources by integrating over all space the values $Z(r)$ and $Z^{CD}(r)$ determined in the neighborhood of a point source in a fairly extensive medium. Using this method, de Juren obtained $t_s = 1.54 \pm 0.13\ \mu\text{sec}$ for water; this result was later corroborated by Grosshög, who found $t_s = 1.65 \pm 0.10\ \mu\text{sec}$. On first view, these values seem to be in disagreement with the result of Möller and Sjöstrand. However, while the cadmium cut-off energy was 0.2 ev in their experiment (cf. Fig. 16.3.3), it is usually much higher for detectors enclosed in thick cadmium covers (cf.

Sec. 12.2.2); for the experiments of DE JUREN and of GROSSHÖG, we can assume $E_{CD} \approx 0.5$ ev; for this energy Eq. (16.3.3) yields $t_s = 1.58$ μsec in good agreement with the experimental values.

There are thus four independent experimental results that show chemical binding only slightly affects the time behavior of moderation in water above cadmium resonance energy and that therefore corroborate the simple theory of Sec. 9.1. In contrast, CROUCH finds experimentally a strong influence of chemical binding on the time scale of moderation to cadmium cut-off energy ($t_s \approx 5$ μsec). The reason for this discrepancy is unknown.

16.3.2. Slowing Down in Heavy Moderators; The Slowing-Down-Time Spectrometer

BERGMAN *et al.* have studied the time dependence of neutron moderation over a wide range of energies in *lead*. For this purpose, they injected pulses of 14-Mev neutrons [from the $H^3(d, n)He^4$ reaction] about 1 μsec long into a lead cube about 2 meters on a side (i.e., about 90 tons of lead) and then measured the capture rate in various resonance absorbers as a function of the time. They determined the capture rates by means of the capture γ-radiation, which they detected with a proportional counter. Fig. 16.3.5 shows the counting rate obtained from a silver foil as a function of time. The curve has a series of maxima that occur when the neutrons pass the energies of the silver resonances during moderation. Since the resonance energies are well known from chopper measurements, we can get a relation between the slowing-down time and the neutron energy from such measurements. Fig. 16.3.6 shows such a relation; values measured with Cu, Mn, and Zn are plotted. The smooth curve represents the theoretical relation[1]

$$\bar{E}\,[\text{kev}] = \frac{183}{t^2} \quad (t \text{ in } \mu\text{sec}) \tag{16.3.5}$$

which we obtain from Eq. (9.1.19a) with $A = 207$ and $\Sigma_s = 0.346\ \text{cm}^{-1}$.

A special feature of the slowing-down process in a heavy moderator is that the average energy is sharply defined at all times. According to Eq. (9.1.19b), $\sqrt{\overline{\Delta E^2}}/\bar{E} = \sqrt{8/3A} = 11.4\%$ for lead. In practice, we must take into account two other effects that smear out the energy distribution. Neutrons with energies above 0.57 Mev can be inelastically scattered in lead and can thereby lose large amounts of energy. Owing to this inelastic scattering, the energy distribution of the fast neutrons is broadened considerably during their first few collisions. However, this initial broadening becomes smaller as moderation continues because the slower neutrons make fewer collisions per unit time than the fast neutrons. According to BERGMAN *et al.*, the energy definition is still about 35% at 10 kev and only reaches the value of 11.4% at about 1 kev. At very small energies, the thermal motion of the lead atoms makes itself felt and leads to an additional broadening of the energy distribution. In spite of these effects, the resolution is sufficient in many cases to permit the determination of reaction cross sections as functions of energy with this "slowing-down-time spectrometer". To do this one introduces the substance to be studied into the lead block and measures the

[1] A more accurate relation is $\bar{E}\,[\text{kev}] = \frac{183}{(t+0.3)^2}$ (t in μsec); for $E \lesssim 0.5$ kev, the difference between this and Eq. (16.3.5) is unimportant.

rate of neutron capture as a function of time. Since one can always make a flux measurement with the help of a BF_3-counter, one can determine the energy variation of the cross section directly. Additional normalization measurements are necessary for an absolute determination. In this way one obtains cross sections which because of the limited resolving power are averaged over an energy band whose relative width is about 12—30%. For many nuclei, this resolving power is sufficient to separate the resonances; the resonance parameters can be determined by methods similar to those described in Sec. 4.1.3. BERGMAN *et al.*, KASHUKEEV,

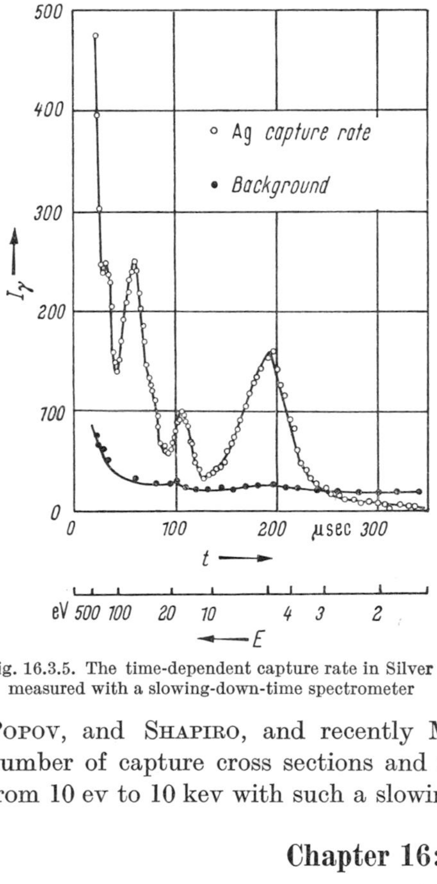

Fig. 16.3.5. The time-dependent capture rate in Silver measured with a slowing-down-time spectrometer

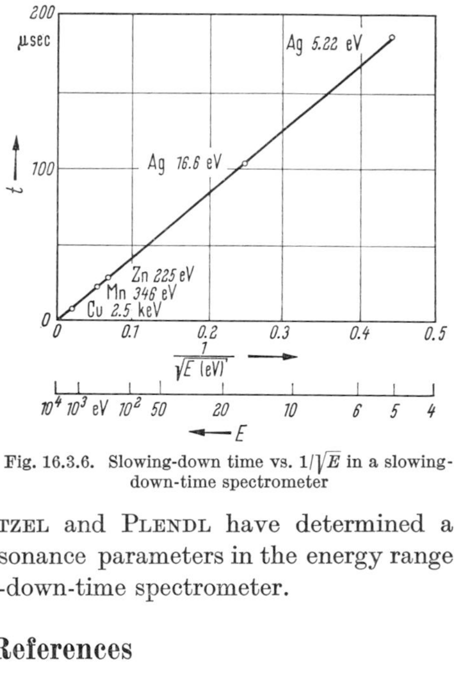

Fig. 16.3.6. Slowing-down time vs. $1/\sqrt{E}$ in a slowing-down-time spectrometer

POPOV, and SHAPIRO, and recently MITZEL and PLENDL have determined a number of capture cross sections and resonance parameters in the energy range from 10 ev to 10 kev with such a slowing-down-time spectrometer.

Chapter 16: References

General

AMALDI, E.: loc. cit., especially section IV, p. 306ff.

GOLDSTEIN, H., J. G. SULLIVAN jr., R. R. COVEYOU, W. E. KINNEY, and R. R. BATE: Calculations of Neutron Age in H_2O and Other Materials, ORNL-2639 (1961).

TEMPLIN, L. J. (ed.): Reactor Physics Constants, ANL-5800, Second Edition (1963); especially Section 3.4: Slowing-Down Parameters in Well-Moderated Media.

McARTHEY, A. E., P. J. PERSIANI, B. I. SPINRAD, and L. J. TEMPLIN: Neutron Resonance Integral and Age Data, Argonne Reactor Physics Constants Center Newsletter **1** (1961).

PERSIANI, P. J., J. J. KAGANOVE, and A. E. McARTHEY: Neutron Resonance Integral and Age Data, Argonne Reactor Physics Constants Center Newsletter **10** (1963).

Special[1]

ROSE, M. E., and A. M. WEINBERG: Mon-P 297. } Age Measurements in

WEINBERG, A. M., and L. C. NODERER: AECD-3471, III-54 (1951). } Finite Geometry.

[1] Cf. footnote on p. 53.

ANDERSON, H. L., E. FERMI, and D. NAGLE: CP-1531 (1944).
BLOSSER, T. V., and D. K. TRUBEY: ORNL-2842, 109 (1959).
DOERNER, R. C., *et al.*: Nucl. Sci. Eng. **9**, 221 (1961).
HILL, J. E., L. D. ROBERTS, and T.E. FITCH: ORNL-181 (1948), J. Appl. Phys. **26**, 1013 (1955).
HOOVER, J. I., and T. I. ARNETTE: ORNL-641 (1950).
LOMBARD, D. B., and C. H. BLANCHARD: Nucl. Sci. Eng. **7**, 448 (1960).
PETTUS, W. G.: BAW-146 (1960).

Measurement of the Age of Fission Neutrons in H_2O.

GOLDSTEIN, H., P. F. ZWEIFEL, and D. G. FOSTER: Geneva 1958 P/2375, Vol. 16, p. 379.
WILKINS, J. E., R. L. HELLENS, and P. F. ZWEIFEL: Geneva 1955 P/597, Vol. 5, p. 62.

Calculation of the Age of Fission Neutrons in Water.

BARKOV, L.M., V.K.MAKARIN, and K.N.MUKHIN: J. Nucl. Energy **4**, 94 (1957).
CASWELL, R. S., *et al.*: Nucl. Sci. Eng. **2**, 143 (1957).
DUGGAL, V. P., S. M. PURI, and K. S. RAM: Geneva 1958 P/1640. Vol. 13, p. 85
FOSTER, D. G.: Nucl. Sci. Eng. **8**, 148 (1960).
HEUSCH, C. A., and T. SPRINGER: Nucl. Sci. Eng. **10**, 151 (1961).
MUNN, A. M., and B. PONTECORVO: Can. J. Res. A **25**, 157 (1947).
REIER, M., F. OBENSHAIN, and R. L. HELLENS: Nucl. Sci. Eng. **4**, 1 (1958).
SPIEGEL, V., D.W. OLIVER, and R. S. CASWELL: Nucl. Sci. Eng. **4**, 546 (1958).
SPRINGER, T.: Nukleonik **1**, 41 (1958).
VALENTE, F. A., and R. E. SULLIVAN: Nucl. Sci. Eng. **6**, 162 (1959).

Age Measurements for Various Sources in H_2O.

LEVINE, M. M., *et al.*: Nucl. Sci. Eng. **7**, 14 (1960).
SPIEGEL, V., and A. C. B. RICHARDSON: Nucl. Sci. Eng. **10**, 11 (1961).
WADE, J.: DP-163 (1956).

The Age for Various Sources in D_2O.

DAVEY, W. G., *et al.*: AERE R/R 2501 (1958).
DUGGAL, V. P., and J. MARTELLY: Geneva 1955 P/358, Vol. 5, p. 28.
FERMI, E. (Ed. G. BECKERLEY): AECD-2664 (1951).
HENDRIE, J. M., *et al.*: Geneva 1958 P/601, Vol. 12, p. 695.
HILL, J. E., L. D. ROBERTS, and G. MCCAMMON: ORNL-187 (1949).
STEICHEN, C. U.: TID-15614 (1960).

The Age for Various Sources in Graphite.

CAMPBELL, R. W., and R. K. PASCHALL: Trans. Am. Nucl. Soc. **4/2**, 280 (1961).
GERASEVA, L. A., *et al.*: Geneva 1955 *P*/662, Vol. 5, p. 13.
GOODJOHN, A. J., and J. C. YOUNG: Trans. Am. Nucl. Soc. **3/2**, 488 (1960).
LUNDY, T. S., and E. E. GROSS: ORNL-2891 (1960).
NOBLES, R., and J. WALLACE: ANL-4076, 10 (1947).

Age Measurements in Various Moderators.

GOLDSTEIN, H.: "Fundamental Aspects of Reactor Shielding", Reading: Addison-Wesley Publishing Company, 1959.
HOLTE, G.: Arkiv Fysik **2**, 523 (1950); **3**, 209 (1951).
VERDE, M., and G. C. WICK: Phys. Rev. **71**, 852 (1947).
WICK, G. C.: Phys. Rev. **75**, 738 (1949).

The Spreading Out of Neutrons at Large Distances.

BAER, W.: J. Appl. Phys. **26**, 1235 (1955).
CASWELL, R. S., *et al.*: loc. cit.
FOSTER, D. G.: HW-64866 (1960).
RUSH, J. H.: Phys. Rev. **73**, 271 (1948).
WADE, J. W.: loc. cit.

Measurement of the Relaxation Length of Fast Neutrons.

CROUCH, M. F.: Nucl. Sci. Eng. **2**, 631 (1957).
DE JUREN, J. A.: Nucl. Sci. Eng. **9**, 408 (1961).
ENGELMANN, P.: Nukleonik **1**, 125 (1958).
GROSSHÖG, G.: Chalmers Teksniska Högskola, Göteborg, Report CTH-RF-4 (1963).
MÖLLER, E., and N. G. SJÖSTRAND: BNL-719, 966 (1962); AB Atomenergi Report RF X 248 (1963).

The Time Dependence of Moderation in Water.

Bergman, A. A., *et al.*: Geneva 1955 P/642, Vol. 4, p. 135.
Jsakov, A. I., Yu. P. Popov, and F. L. Shapiro: Soviet Physics J.E.T.P. **11**, 712 (1960).
Kashukeev, N. T., Yu. P. Popov, and F. L. Shapiro: J. Nucl. Energy A **14**, 76 (1961).
Mitzel, F., and H. S. Plendl: Unpublished Karlsruhe Report, 1964.
Popov, Yu. P., and F. L. Shapiro: Soviet Physics J.E.T.P. **13**, 1132 (1961); **15**, 683 (1962).
} The Slowing-Down-Time Spectrometer.

17. Investigation of the Diffusion of Thermal Neutrons by Stationary Methods

In this chapter, we shall become familiar with stationary methods of determining the diffusion parameters of thermal neutrons. First we shall consider in Sec. 17.1 the classical methods of measuring the diffusion length. Sec. 17.2 treats the so-called poisoning method, which makes it possible to determine simultaneously the diffusion length and the transport mean free path of a given medium. If the absolute value of the absorption cross section of one substance is known, it is possible to find the absorption cross sections of other substances by comparison measurements; some methods for making these measurements are discussed in Sec. 17.3.

17.1. Measurement of the Diffusion Length

17.1.1. Point Source in an Infinite Medium

All methods of directly measuring the diffusion length are based on observing the variation of the thermal neutron flux in the source-free part of a medium. The linear dimensions of the medium must be large, of the order of magnitude of several diffusion lengths at least.

The situation is simplest in an effectively infinite medium, i.e., a medium whose dimensions are so large that practically all the neutrons emitted from the source are absorbed. This happens when the diameter of the medium is about 30 diffusion lengths; in water this diameter is 80 cm, in graphite 15 m (corresponding to a cube of graphite weighing about 5000 tons!). Thus we can only realize an infinite medium of water or some other hydrogenous moderator with a short diffusion length.

When we use a point thermal source, it suffices to measure the flux variation along a radius with foils. According to Sec. 6.1.3 we have in this case

$$\Phi(r)=\frac{Q}{4\pi D}\cdot\frac{e^{-r/L}}{r} \qquad (6.1.13)$$

and we obtain L by plotting $\log(r\cdot\Phi(r))$ against r. However, since most neutron sources emit fast neutrons, the case just discussed cannot be realized immediately. Nevertheless, one can create a negative thermal source through cadmium difference measurements. One measures the neutron flux near the source once with and once without the source closely surrounded by a shell of material, preferably cadmium, that captures thermal neutrons strongly. The difference of the two measured values is caused by the thermal neutrons absorbed in the shell; it should satisfy the formula given above rigorously. Beckurts and Klüber, among others, have measured the diffusion length in light water by this method. Only moderate accuracy is attainable since the flux values underlying the evaluation, being the differences of two large numbers, contain sizeable uncertainties.

The measurement of the flux distribution in the neighborhood of a (Sb—Be) source yields considerably more accurate results. Because of the low neutron energy ($\bar{E} \approx 25$ kev), the neutrons become thermal relatively close to the source, and the flux variation shows only slight deviations from Eq. (6.1.13), which can easily be corrected theoretically. Thus, no cadmium difference measurements are required. Reier and de Juren, among others, have carried out such studies. Fig. 17.1.1 shows the thermal neutron flux variation observed by them in water at 23 °C at distances of from 10 to 24 cm from an (Sb—Be) source. $\log\left(r \cdot A_{th}(r)\right)$ has been plotted against r, where $A_{th}(r)$ is the activity of the indium foils after elimination of the epithermal activation (cf. Sec. 12.2). We see clearly that for $r < 20$ cm the decrease of $r \cdot A_{th}(r)$ is not rigorously exponential: in this range, source neutrons still play an appreciable role. We must either apply calculated corrections or limit ourselves in the determination of the diffusion length to evaluating the flux variation in the range $r > 20$ cm. The latter alternative is hardly feasible since for reasons of intensity precise flux measurements for $r > 20$ cm are very difficult. According to de Juren, the necessary corrections can be calculated as follows. The source term appearing in the diffusion equation

$$\bar{D}\nabla^2\Phi(r) - \bar{\Sigma}_a\Phi(r) - q_{th}(r) = 0 \tag{17.1.1}$$

is given empirically by the expression

$$q_{th}(r) = \frac{K e^{-\Sigma r}}{r^2}$$

for distances from an (Sb—Be) source greater than 12 cm in water. Here $1/\Sigma = 1.58 \pm 0.02$ cm. The solution of Eq. (17.1.1) with this source term is

$$\Phi(r) = \frac{K L}{2\bar{D}} \frac{e^{-r/L}}{r}\left[C - E_1\left(\left[\Sigma - \frac{1}{L}\right] r\right) + e^{2r/L} E_1\left(\left[\Sigma + \frac{1}{L}\right] r\right)\right]. \tag{17.1.2}$$

Here C is a constant[1]. A correction $F(r)$ factor follows immediately from Eq. (17.1.2):

$$F(r) = \frac{1}{C - E_1\left(\left[\Sigma - \frac{1}{L}\right] r\right) + e^{2r/L} E_1\left(\left[\Sigma + \frac{1}{L}\right] r\right)}. \tag{17.1.3}$$

Now $\Phi(r) \cdot F(r) \sim e^{-r/L}/r$; thus by multiplication of the measured values of $A_{th}(r)$ by $F(r)$ the influence of the source neutrons is eliminated. In order to calculate $F(r)$, C must be experimentally determined. Obviously,

$$\frac{\Phi(r)}{q_{th}(r)} = \frac{rL}{2\bar{D}} e^{\left(\Sigma - \frac{1}{L}\right) r}\left[C - E_1\left(\left[\Sigma - \frac{1}{L}\right] r\right) + e^{2r/L} E_1\left(\left[\Sigma + \frac{1}{L}\right] r\right)\right]. \tag{17.1.4}$$

If for constant r we determine Φ/q_{th} (from the cadmium ratio — cf. Sec. 12.2.3) and if we know Σ, $\bar{D}$, and L, we can calculate C using Eq. (17.1.4). However, since L and $\bar{D}$ are not known initially, we must proceed iteratively. First we make

[1] To calculate C, we must know $q_{th}(r)$ all the way to $r = 0$.

an estimate of L and $\bar{D}$, then we determine C and $F(r)$, and then we correct the measured data and determine an improved value for L, etc. REIER and DE JUREN find that $C=0.0606$ for water at 23 °C, and obtain the values shown in Fig. 17.1.2 for $F(r)$. The value $L=2.776\pm0.009$ cm follows from their corrected data. Later in Sec. 17.1.4, we shall learn of additional experimental results for water and other hydrogenous moderators, including some at higher temperatures.

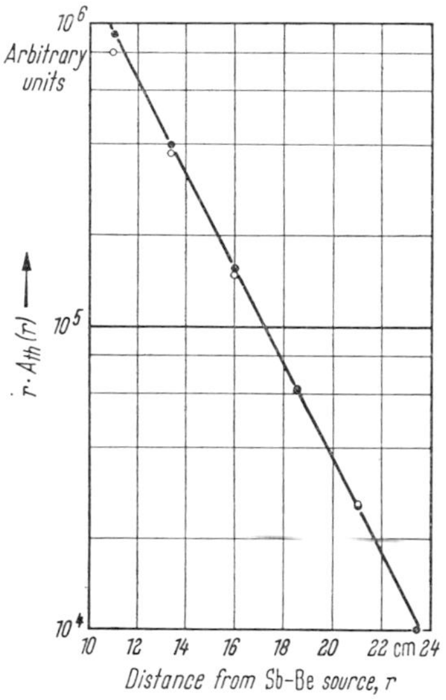

Fig. 17.1.1. The thermal neutron flux versus distance from a (Sb−Be) source. ○ Directly measured data; • corrected for slowing-down sources

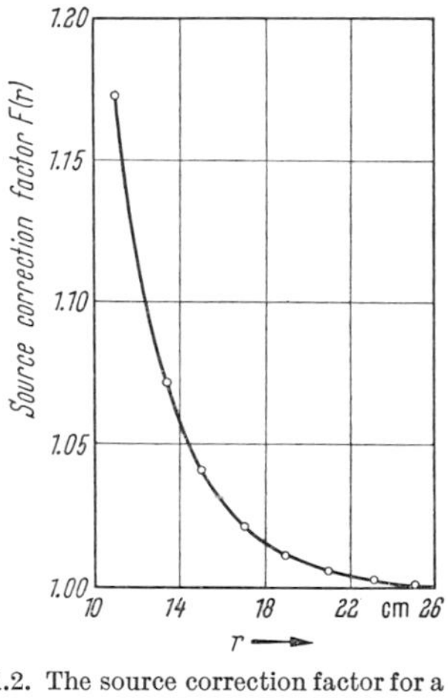

Fig. 17.1.2. The source correction factor for a (Sb−Be) source in H_2O at 23 °C

17.1.2. Finite Media; the Sigma Pile

The method of measuring diffusion lengths just discussed is not applicable to substances with large diffusion lengths, such as D_2O, graphite, and beryllium. In this case we must build an assembly that is finite in comparison with the diffusion length and take the neutron leakage through the surface into account. The standard arrangement for the measurement of diffusion lengths in this case is the so-called sigma pile. A sigma pile is a column of the material being investigated with a cylindrical or square cross section that is fed through one end with neutrons. The diffusion length follows from an analysis of the flux distribution in the pile. Diffusion length measurements in a sigma pile have been carried out on graphite by HEREWARD *et al.*, by CARLBLOM, and by HENDRIE *et al.*; on beryllium by GERASEVA *et al.* and by HUGHES; on beryllium oxide by KOECHLIN *et al.*; and on D_2O by SARGENT *et al.* and by MEIER and LUTZ. In addition, a series of investigations in water and other hydrogenous moderators have been carried out in this geometry.

Fig. 17.1.3 shows a typical apparatus for measurements on graphite. The cross sectional dimensions a and b (generally $a=b$) must be at least 2 to 3, and preferably 3 to 4, diffusion lengths. The same is true for the length c of the pile. One uses as a neutron source either a radioactive source or an accelerator target, which is located on the central axis of the pile close to its end. These sources emit fast neutrons, which again leads to complications in the evaluation of the measured flux distribution, and we must provide equipment for cadmium difference measurements[1]. For this purpose we may profitably use a cadmium plate that covers the entire cross section of the pile and that can easily be inserted and removed. One can avoid the difficulties caused by fast neutrons by feeding well-thermalized neutrons from the thermal column of a nuclear reactor into the sigma pile. The lateral and upper surfaces of the sigma pile should be carefully covered with cadmium (or another neutron absorber) in order to provide a clean boundary condition for the thermal neutrons. When a fast neutron source is being used, neutrons with energies above the cadmium cut-off energy can leave the pile unhindered, scatter on the walls of the room or on some other reflector, and again enter the pile; this leads to a distortion of the flux distribution near the surface. One must therefore avoid placing any strong reflectors near a sigma pile. The flux distribution is measured with foils along the central axis of the pile and usually along the midlines of several cross sections at various distances from the source. Devices are necessary to insert the foils in precisely reproducible positions. Since the diffusion parameters and thus the flux distribution depend on the moderator temperature, the room temperature should be kept reasonably constant ($\pm 2\,^\circ$C). Special heating devices are necessary for measurements at higher temperatures.

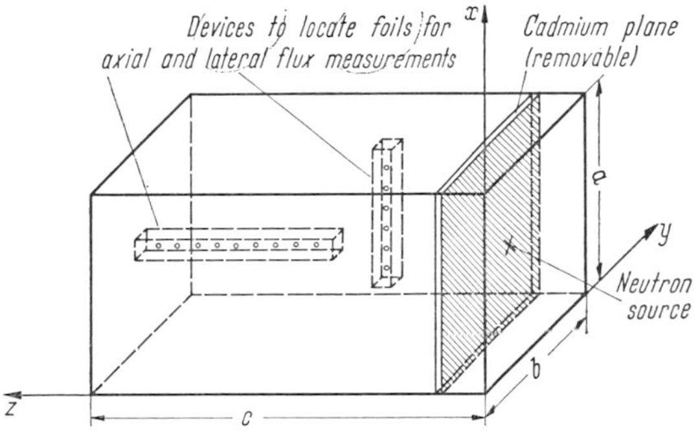

Fig. 17.1.3. A sigma pile for diffusion length measurements in graphite

According to Sec. 6.2.4, if only thermal neutrons are present the flux distribution is given by

$$\Phi(x, y, z) = \sum_{l, m} A_{lm} \sinh\left(\frac{c-z}{L_{lm}}\right) \sin\left(\frac{l\pi x}{a}\right) \sin\left(\frac{m\pi y}{b}\right). \qquad (17.1.5)$$

Here the A_{lm} are source-dependent constants that are unimportant for our purpose; the relaxation lengths L_{lm} are given by

$$\frac{1}{L_{lm}^2} = \frac{1}{L^2} + \pi^2\left(\frac{l^2}{a^2} + \frac{m^2}{b^2}\right). \qquad (17.1.6)$$

a, b, and c are the effective edge lengths, i.e., the actual edge lengths augmented by twice the extrapolation length. Sufficiently far from the source, the contribution of the higher Fourier components of the flux will be small, and we shall have

$$\Phi(x, y, z) \sim \sinh\left(\frac{z-c}{L_{11}}\right) \qquad (17.1.7)$$

on the central axis of the pile ($x = \frac{1}{2}a$, $y = \frac{1}{2}b$). Using Eq. (17.1.7), one can immediately obtain L from the decrease of the flux along the central axis in this region, and then by means of Eq. (17.1.6) one can determine the diffusion length L. In practice, one proceeds in the following way. First one determines at what source distance the contribution of the higher Fourier components may be neglected. Flux measurements along the midlines of the various cross sections serve for this purpose. Fig. 17.1.4 shows such a distribution measured in the pile shown in Fig. 17.1.3; we see that it can very accurately be fit by a sine function, from which we may conclude the absence of any higher Fourier components.

[1] The photoneutrons that are ejected from Be or D by energetic γ-rays from the source are another perturbation that is eliminated in cadmium difference measurements.

In addition, the lateral measurements also yield information on the extrapolated endpoint. If we are certain that a single sine function $\sin(\pi x/a)$ completely describes the flux distribution, we can determine a by the method of least squares and thus determine the extrapolated endpoint d from the relation a = actual edge length $+2d$. It turns out that the relation $d=0.71\,\lambda_{tr}$ is not always exactly fulfilled; some possible causes for the deviations were discussed in Sec. 10.3.4. Table 17.1.1 contains some directly determined values of d. Since a is usually $\gg d$, the deviations from the simple $0.71\,\lambda_{tr}$-law cause no difficulties in the calculation of the diffusion length with Eq. (17.1.5). In other words, it is usually sufficient to determine the quantity a from the simple expression a = actual edge length $+2\cdot(0.71\,\lambda_{tr})$.

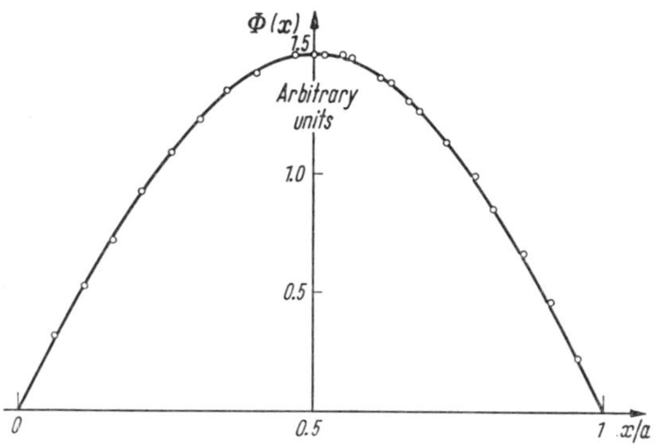

Fig. 17.1.4. The lateral neutron flux distribution in a sigma pile. ○ Measured flux; ——— $\Phi(x) \sim \sin\frac{\pi x}{a}$

Next, the flux decrease along the central axis is determined in that region where the cross sectional measurements show that there are no higher Fourier components (cf. Fig. 17.1.5). Except

Table 17.1.1. *Experimental Values of the Extrapolated Endpoint*

Substance	d [cm]	Author	$d/0.71$ [cm]	λ_{tr} from Poisoning and Pulsed-Neutron Experiments [cm]
Graphite (1.6 g/cm³)	1.97 ± 0.05	HENDRIE *et al.*	2.77	2.6
D_2O [1] (99.4%) . . .	1.64 ± 0.06	AUGER, MUNN, and PONTECORVO	2.31	2.5
H_2O	0.32	SISK	0.45	0.432

[1] A new evaluation of this experiment by KASH and WOODS gave $d=1.81$ cm, $d/0.71=2.55$ cm.

for the region close to $z=c$, the flux decays exponentially, and it is customary to introduce the so-called "endpoint correction" in order to eliminate the influence of the surface, whose presence leads to deviations from the simple exponential law. Therefore let us note that Eq. (17.1.7) can also be written in the form

$$\Phi(z) \sim e^{-z/L_{11}}\left(1 - e^{-2\frac{c-z}{L_{11}}}\right). \tag{17.1.8}$$

If we divide the measured values by $1 - e^{-2\frac{c-z}{L_{11}}}$ — which we evaluate using a guess for L_{11} that will be improved iteratively later — the corrected values should follow the simple exponential law. On a semilog plot we then get a straight line whose slope gives L_{11} (cf. Fig. 17.1.5, curve II). Naturally, this entire procedure can be programmed for a computing machine.

It will be instructive to estimate the quantity of material necessary for an accurate measurement of the diffusion length in a sigma pile. According to Eq. (17.1.6), we have

$$L = L_{11}(1-\beta^2 L_{11}^2)^{-\frac{1}{2}} \tag{17.1.9}$$

where we have set $\beta^2 = \pi^2\left(\frac{1}{a^2} + \frac{1}{b^2}\right)$ by way of abbreviation. If we assume that the error in the determination of β^2 is negligible, then the relative error $\Delta L/L$ is given by

$$\frac{\Delta L}{L} = \frac{\Delta L_{11}}{L_{11}}(1+\beta^2 L^2). \tag{17.1.10}$$

When the cross sectional dimensions are small, an error in L_{11} affects L very strongly: when $a = b = L$, $1+\beta^2 L^2 \approx 20$. Therefore, as mentioned above, the cross sectional dimensions must be at least 2 to 3, and even better 3 to 4, diffusion lengths. The length of the pile follows from the requirement that there must be a region of at least four relaxation lengths in which the decay along the axis is purely exponential. As a rule, we can achieve this by choosing the height to be about 1.5 times the cross sectional dimensions. Finally, let us mention again that these considerations hold only for the flux produced by thermal sources outside the region of measurement. Thus we must either use a thermal column as the neutron source or eliminate the sources lying inside the region of measurement by the cadmium difference method. In media in which age theory holds (graphite, Be, BeO), we can also eliminate the effect of these latter sources theoretically (cf. HEREWARD *et al.*).

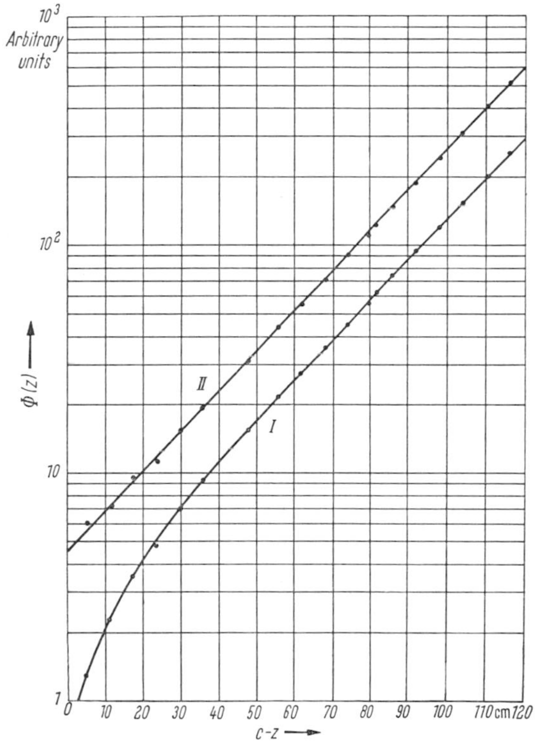

Fig. 17.1.5. The axial neutron flux distribution in a sigma pile (cadmium difference measurement). Curve I: without; Curve II: with endpoint correction

17.1.3. Determination of the Diffusion Length from the Flux Distribution in a Medium Surrounded by a Surface Source

One needs smaller quantities of material than required for a sigma pile when a method suggested by HEISENBERG is used. The substance to be investigated is placed in the interior of a sphere that is surrounded by a reflector of paraffin or water. In the center of the sphere there is a source of fast neutrons, e.g., a (Ra—Be) source. One first measures the flux distribution $\Phi'(r)$ along a radius vector; then one surrounds the sphere with a cadmium shell but leaves the reflector unchanged and again measures the flux $\Phi''(r)$. $\Phi(r)=\Phi'(r)-\Phi''(r)$ then is the flux due to a thermal surface source. In the interior of the sphere Φ is the spherically symmetric solution of elementary diffusion theory

$$\Phi(r)=\frac{\Phi_0}{r/L}\sin(r/L). \tag{17.1.11}$$

There is a condition on the practical applicability of this method, namely, that the radius of the sphere be large enough to allow $\Phi(r)$ to change sufficiently between the center and the surface. This method was used during the 1940's for the determination of the diffusion lengths of heavy water (HEISENBERG and DÖPEL), beryllium (BOTHE and FÜNFER), and graphite (JENSEN and BOTHE). It has the disadvantage of being limited to spheres, and in the case of graphite, for example, requires considerable machining of the individual graphite slugs. This suggests that it may be worthwhile to generalize the method to cubes.

Source on the Surface of a Cube. The substance to be studied has the form of a cube surrounded by a reflector of paraffin or water (cf. Fig. 17.1.6). The (Ra—Be) source is located at the center of the cube. The flux distribution is measured once with and once without a cadmium covering around the cube. The difference represents the effect of a surface source on the cube.

The diffusion equation in the interior of the cube must be solved by approximation. The symmetry of the system around the center suggests that we take as our solution a form containing only even functions of the coordinates (the origin is at the center of the cube):

$$\left.\begin{aligned}\Phi(x,y,z)=\Phi_0 S_0+\Phi_2 S_2+\Phi_4 S_4+\Phi_{22}S_{22}+\Phi_6 S_6+\Phi_{42}S_{42}+\\ +\Phi_{222}S_{222}+\Phi_8 S_8+\Phi_{62}S_{62}+\Phi_{44}S_{44}+\Phi_{422}S_{422}.\end{aligned}\right\} \tag{17.1.12}$$

(It can be shown that it is not necessary to take into account terms of order higher than the eighth.)

Here the S are the elementary symmetric functions

$$\left.\begin{aligned}&S_0=1;\quad S_2=x^2+y^2+z^2;\quad S_4=x^4+y^4+z^4;\\ &S_6=x^6+y^6+z^6;\quad S_8=x^8+y^8+z^8;\quad S_{22}=x^2y^2+y^2z^2+z^2x^2;\\ &S_{42}=x^4y^2+x^4z^2+y^4z^2+y^4x^2+z^4x^2+z^4y^2;\quad S_{222}=x^2y^2z^2;\\ &S_{62}=x^6y^2+x^6z^2+y^6x^2+y^6z^2+z^6x^2+z^6y^2;\\ &S_{44}=x^4y^4+x^4z^4+y^4z^4;\quad S_{422}=x^4y^2z^2+y^4x^2z^2+z^4x^2y^2\end{aligned}\right\} \tag{17.1.13}$$

and the Φ are constants (free, at first). This general form must satisfy the diffusion equation. Application of the Laplacian operator to the various S-functions gives the following relations:

$$\left.\begin{aligned}&\nabla^2 S_0=0\\ &\nabla^2 S_2=6\\ &\nabla^2 S_4=12S_2 \qquad \nabla^2 S_{22}=4S_2\\ &\nabla^2 S_6=30S_4 \qquad \nabla^2 S_{42}=24S_{22}+4S_4 \qquad \nabla^2 S_{222}=2S_{22}\\ &\nabla^2 S_8=56S_6 \qquad \nabla^2 S_{62}=30S_{42}+4S_6\\ &\qquad\qquad\qquad \nabla^2 S_{44}=12S_{42} \qquad \nabla^2 S_{422}=36S_{222}+2S_{42}.\end{aligned}\right\} \tag{17.1.14}$$

The introduction of the form 17.1.12 into the diffusion equation $\nabla^2\Phi - \Phi/L^2 = 0$ and use of the Eq. (17.1.14) gives a series of relations among the coefficients Φ_{ijk}; thus in place of Eq. (17.1.12) we have the somewhat simplified result

$$\left.\begin{aligned}\Phi = \Phi_2\cdot A(x, y, z, L) + \Phi_4\cdot B(x, y, z, L) + \\ + \Phi_6\cdot C(x, y, z, L) + \Phi_8\cdot D(x, y, z, L)\,.\end{aligned}\right\} \qquad (17.1.15)$$

Here

$$\left.\begin{aligned}
&A(x, y, z, L) \\
&\quad = \left[6 + \frac{1}{L^2}\left[S_2 + \frac{1}{4L^2}\left[S_{22} + \frac{1}{2L^2}\left[S_{222} + \frac{1}{36L^2}\left[S_{422} - \frac{S_{44}}{6}\right]\right]\right]\right]\right] \\
&B(x, y, z, L) = S_4 - 3\,S_{22} + \frac{1}{L^2}\left[\frac{S_{42}}{4} - 4.5\,S_{222} + \frac{1}{8L^2}\left[-S_{422} + \frac{S_{44}}{3}\right]\right] \\
&C(x, y, z, L) = S_6 - 7.5\,S_{42} + 90\,S_{22} + \frac{1}{3L^2}\left[\frac{3}{4}\,S_{62} + 7.5\,S_{422} - 5\,S_{44}\right] \\
&D(x, y, z, L) = S_8 - 14\,S_{62} + 35\,S_{44}\,.
\end{aligned}\right\} \qquad (17.1.16)$$

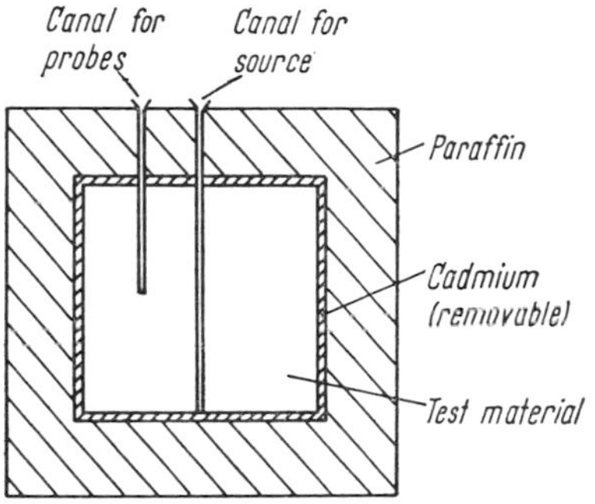

Fig. 17.1.6. An assembly for diffusion length measurements with plane sources on the surface of a cube

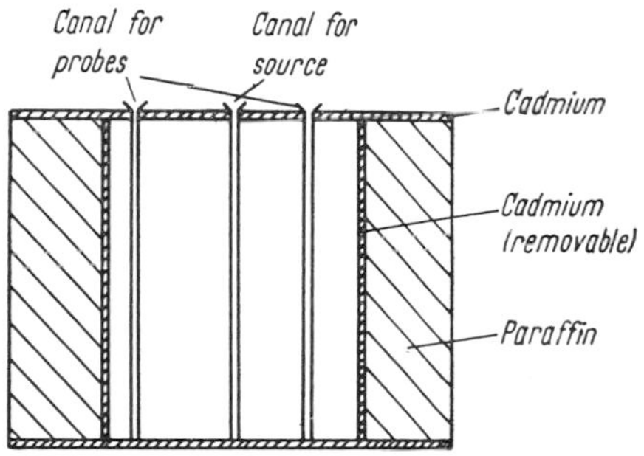

Fig. 17.1.7. An assembly for diffusion length measurements with cylindrical surface sources

We obtain the diffusion length in the following way:

The values Φ^i of the flux determined at a series of points $(xyz)_i$ are fit to Eq. (17.1.15) with the method of least squares. Thus we obtain Φ_2, Φ_4, Φ_6, and Φ_8. These calculations are carried out for a series of L-values that lie near the expected value of the diffusion length. Next we form the sum of the squares of the residuals:

$$Q = \sum_i \{\Phi^i - \Phi((xyz)_i, L)\}^2\,. \qquad (17.1.17)$$

It is obviously a function of L and has a minimum when L is equal to the diffusion length. The method can be simplified by linearizing the L-dependence of the quantities A, B, and C. This method has been applied by FITÉ and by SCHLÜTER to graphite. The detailed numerical work was done on an electronic computing machine.

Cylindrical Surface Source. We can combine the principle, described in Sec. 17.1.2, of the sigma pile whose surface is black to thermal neutrons with that of surface sources. Fig. 17.1.7 shows an arrangement for such a combination. The substance to be studied has the form of a cylinder and is always covered with cadmium on top and bottom. The (Ra—Be) source is located on the axis of the cylinder halfway up. The neutron distribution along various lines parallel to the axis is measured with and without a cadmium sleeve surrounding the cylinder. The difference distribution is due to a thermal surface source on the curved surface of the cylinder. In view of the boundary conditions at $z=0$ and $z=a$, we have

$$\Phi=\sum_n A_n I_0(\lambda_n \cdot r)\sin\frac{n\pi z}{a}, \tag{17.1.18}$$

$$\lambda_n^2=\frac{1}{L^2}+\frac{n^2\pi^2}{a^2}. \tag{17.1.19}$$

Here $I_0(x)$ is the zero-order modified Bessel function of the first kind and the A_n are constants.

The flux distribution along a line parallel to the axis and separated from it by a distance r_0 is given by

$$\Phi(z)=\sum_n A_n I_0(\lambda_n \cdot r_0)\sin\frac{n\pi z}{a}=\sum_n B_n(r_0)\sin\frac{n\pi z}{a}.$$

The quantities

$$B_n(r_0)=A_n I_0(\lambda_n \cdot r_0)$$

can be determined by Fourier inversion of the flux measured along this line. Because $I_0(0)=1$, we have on the axis of the cylinder

$$\Phi=\sum_n A_n \sin\frac{n\pi z}{a}.$$

It follows from Fourier inversion of the flux distribution measured on the axis that

$$B_n(0)=A_n.$$

Thus

$$I_0(\lambda_n \cdot r_0)=\frac{B_n(r_0)}{B_n(0)} \tag{17.1.20}$$

and L can be determined from this transcendental equation. The method can also be used on substances in the form of a prism or a cube. It was used by FITÉ to determine the diffusion length in graphite powder.

17.1.4. Results of Various Diffusion Length Measurements

Ordinary Water. Some recent experimental results at room temperature are collected in Table 17.1.2. The values were referred to 22 °C with the temperature

Table 17.1.2. *The Diffusion Length of Water at 22 °C*

Author	Method	L [cm]	Comments
BECKURTS and KLÜBER (1958)	Cadmium difference in an infinite medium	2.74 ±0.03	
STARR and KOPPEL (1961)	Thermal column and a sigma pile	2.76 ±0.008	
DE JUREN and REIER (1961)	(Sb—Be) source in an infinite medium	2.775±0.006	Corrected for distributed sources according to Sec. 17.1.1
ROHR (1962)	(Sb—Be) source in an infinite medium	2.778±0.011	Corrected for distributed sources according to Sec. 17.1.1
ROCKEY and SKOLNIK (1961)	(Sb—Be) source in an infinite medium	2.835±0.018	No correction for distributed sources

coefficient given in Eq. (17.1.21). The cleanest measurements to date are probably those of DE JUREN and REIER and of STARR and KOPPEL; the average of their values is

$$L = 2.767 \pm 0.008 \text{ cm}$$

which we shall take as the best value available. Fig. 17.1.8 shows some measured values of the diffusion length in water as a function of the temperature up to $T = 250$ °C. Measurements above 100 °C must be carried out in a pressure tank, and the introduction of the neutron source causes difficulties. "P"-metal

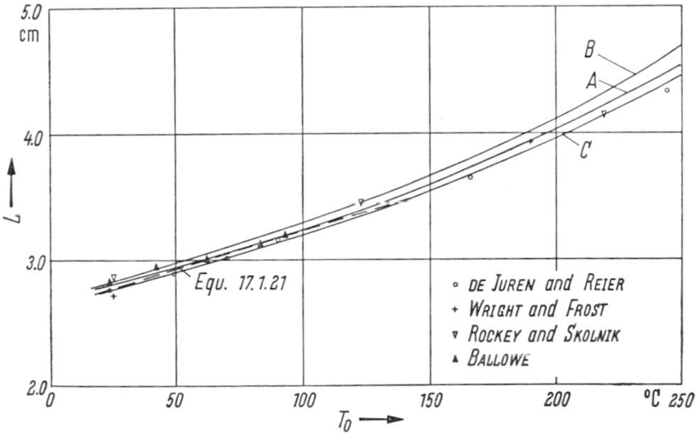

Fig. 17.1.8. The diffusion length of thermal neutrons in H_2O as a function of temperature. ○ + ▿ ▴ Measurements; — — — Eq. (17.2.21); ——— A: calculated using RADKOWSKY's prescription; B: calculated with $\sigma_{tr}(v) \sim 1/v$; C: calculated with $\sigma_{tr}(E)$ from the Nelkin model

(71% Mn, 18% Cu, 10% Ni) and mixed ceramics composed of DyO_2 and Al_2O_3 have proven themselves useful as temperature-resistant foil materials. The temperature dependent measured values agree with one another fairly well. In the vicinity of room temperature, a good approximation is

$$L = 2.77 + 0.006\,[T - 22] \qquad (L \text{ in cm, } T \text{ in °C}). \qquad (17.1.21)$$

The temperature dependence of the diffusion length in H_2O can be interpreted in the following way. To begin with

$$L^2(T) = \frac{\overline{D}}{\overline{\Sigma}_a} = \frac{1}{3N^2} \cdot \frac{\int\limits_0^\infty \frac{1}{\sigma_{tr}(E)} \frac{E}{kT} e^{-E/kT} \frac{dE}{kT}}{\int\limits_0^\infty \sigma_a(E) \frac{E}{kT} e^{-E/kT} \frac{dE}{kT}}. \qquad (17.1.22)$$

The temperature dependence of the atomic density N is known from density measurements. The temperature dependence of the absorption term in the denominator of Eq. (17.1.22) follows simply from the $1/v$-law for σ_a. In order to calculate the temperature dependence of the transport term in the numerator of Eq. (17.1.22) we must make some assumption about the energy dependence of $\sigma_{tr}(E)$. The following cases have been treated:

a) RADKOWSKY's Prescription. $\sigma_{tr}(E)$ is given by $\sigma_s(E)\,\big(1 - \overline{\cos\vartheta}(E)\big)$. $\overline{\cos\vartheta}$ is equal to $2/3A$ for scattering on free nuclei. RADKOWSKY has suggested that this relation be generalized to the protons bound in water by introducing a suitably defined energy-dependent effective mass. This mass is defined in the following way. The scattering cross section of a free proton is 20 barn. According to Eq. (1.4.3), the scattering cross section of a proton bound in a molecule of mass A_{eff} is $\sigma_s = 20\left(\frac{2A_{\text{eff}}}{A_{\text{eff}}+1}\right)^2$ barn. We can therefore derive an effective mass

$$A_{\text{eff}}(E) = \frac{1}{\sqrt{\frac{80\text{ barn}}{\sigma_s(E)}} - 1} \tag{17.1.23}$$

from the measured scattering cross section of water (cf. Fig. 1.4.8) and then calculate $\sigma_{tr}(E)$ from $\sigma_s(E)$ and $A_{\text{eff}}(E)$. Drozdov *et al.* have improved this prescription somewhat by taking the thermal motion of the molecules (Doppler effect, cf. Sec. 10.1.1) into account. The result of this calculation is shown in Fig. 17.1.8 as curve A (in this connection cf. also Esch).

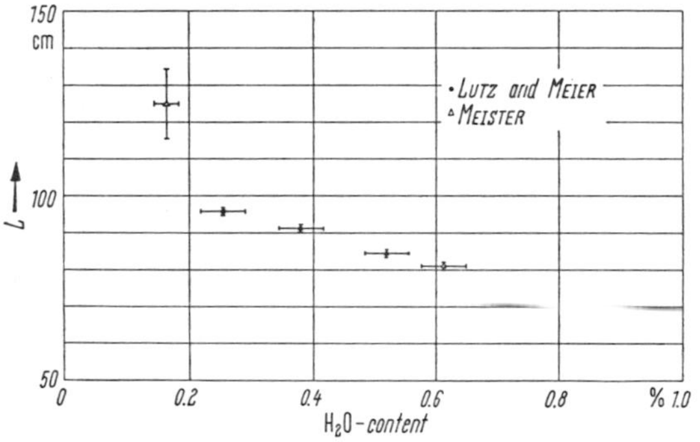

Fig. 17.1.9. The diffusion length in heavy water versus H_2O content

b) According to Deutsch we obtain a good approximation to the temperature dependence of the diffusion length if we simply take $\sigma_{tr}(E) \sim \frac{1}{\sqrt{E}}$; this corresponds to curve B in Fig. 17.1.8.

c) Curve C in Fig. 17.1.8 has been calculated on the basis of the Nelkin model for water (Sec. 10.1.3). $\overline{\cos\vartheta}(E)$ was taken from Fig. 10.1.9 and $\sigma_s(E)$ from Fig. 10.1.8. (Actually, we ought to calculate $\overline{\cos\vartheta}(E)$ and $\sigma_s(E)$ for each water temperature since the state of thermal excitation and thus the scattering properties change with increasing temperature.)

Other Hydrogenous Moderators. Table 17.1.3 contains some measured values of the diffusion length in various hydrogenous moderators; the measurements refer in part to room temperature and in part to higher temperatures.

Table 17.1.3. *The Diffusion Length in Various Hydrogenous Moderators*

Substance	Temperature [°C]	L [cm]	Author
Dowtherm A (26.81% diphenyl, 73.19% diphenyl oxide)	36	4.214 ± 0.006	Bata, Kiss, and Pál
	92	4.941 ± 0.002	
	146	4.934 ± 0.005	
	192	5.245 ± 0.004	
Diphenyl	85	4.82 ± 0.07	Brown *et al.*
Lucite ($C_5H_8O_2$, 1.18 g/cm³)	20	3.14 ± 0.03	Heintzé
Paraffin	61	2.751 ± 0.053	Esch
	70	2.807 ± 0.055	
	81	2.833 ± 0.052	
	95	2.942 ± 0.050	
	123	3.120 ± 0.054	

Heavy Water. SARGENT *et al.*, LUTZ and MEIER, and MEISTER have performed sigma-pile measurements of the diffusion length of heavy water at room temperature. Fig. 17.1.9 shows the various results at 20 °C plotted as a function of the H_2O content of the D_2O. We can calculate the diffusion length of D_2O with the formula

$$L^2 = \frac{\bar{D}}{P_{H_2O}\,\bar{\Sigma}_{a_{H_2O}} + P_{D_2O}\,\bar{\Sigma}_{a_{D_2O}}} = \frac{L^2(P_{H_2O}=0)}{P_{H_2O}\left(\dfrac{\bar{\Sigma}_{a_{H_2O}}}{\bar{\Sigma}_{a_{D_2O}}}\right) + P_{D_2O}} \qquad (17.1.22)$$

where P_{H_2O} and P_{D_2O} are the respective molar concentrations of the light and heavy water: $P_{H_2O} + P_{D_2O} = 1$[1]. With σ_a(2200 m/sec) equal to 327 mbarn for H and 0.5 mbarn for D, one can use Eq. (17.1.22) to extrapolate the measured diffusion length to 100% pure D_2O. In this way, MEIER and LUTZ found 136±7 cm, MEISTER found 168±18 cm, and SARGENT obtained 171±20 cm for $L(P_{H_2O}=0)$. The cause of the strong deviation of LUTZ and MEIER's value from the others is unknown. We should note, however, that besides the errors in the measurement of the diffusion length errors also occur in the determination of the D_2O concentration. Furthermore, the O^{17} content of different heavy water samples can fluctuate considerably and thereby appreciably affect the absorption and thus the diffusion length.

Table 17.1.4. *The Diffusion Length in Be, BeO, and Graphite at Room Temperature (Sigma-pile measurements)*

Substance	Density [g/cm³]	Author	L [cm]
Be	1.85	NOBLES and WALLACE	20.8±0.5
Be	1.78	GERASEVA *et al.*	22.1±1 (21.3±1 at a density of 1.85 g/cm³)
BeO	2.92	KOECHLIN *et al.*	32.7±0.5
C	1.65	CARLBLOM	45.4±0.2

Other Moderators. Table 17.1.4 contains the results of diffusion length measurements on beryllium, beryllium oxide, and graphite. There are many more measurements on graphite than are reported in the table (cf., e.g., HEREWARD *et al.*, WINZELER, BECKURTS, SCHLÜTER, ERTAND *et al.*, REICHARDT, KLOSE, HENDRIE *et al.*, and RICHEY and BLOCK). Such measurements have very frequently been undertaken in order to account for the effect on the absorption cross section of impurities that are difficult to determine chemically. MILLS, as well as LLOYD, CLAYTON, and RICHEY, has studied the temperature dependence of the diffusion length in graphite. Fig. 17.1.10 shows $L(T)$ in the range

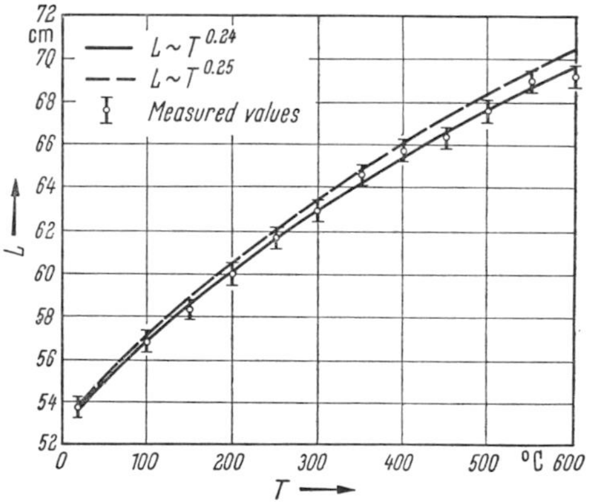

Fig. 17.1.10. The diffusion length in graphite versus temperature (LLOYD, CLAYTON and RICHEY)

[1] The dependence of the diffusion coefficient on the hydrogen concentration has been neglected in Eq. (17.1.22), but in the region of high deuterium concentration that we are considering here this is permissible.

20 to 600 °C. The measured points follow a $T^{0.25}$-law quite accurately, from which we may conclude that the diffusion coefficient in graphite does not depend appreciably on the temperature.

17.2. Measurement of the Transport Mean Free Path in Poisoning Experiments

17.2.1. Principle of the Method

If the diffusion length and the absorption cross section of a medium are known, one can calculate the diffusion coefficient from the relation $L^2=\overline{D}/\overline{\Sigma}_a$ and thus immediately obtain the transport mean free path $\lambda_{tr}=3\overline{D}$. Now it is easy to measure the diffusion length by the methods of Sec. 17.1, but it is not possible to determine absolutely the absorption cross section by stationary methods[1]. The poisoning method offers a way out of this difficulty.

In the pure moderator we have $1/L^2=\overline{\Sigma}_a/\overline{D}$. If we now increase the absorption cross section by homogeneously mixing an absorber of known absorption cross section with the scatterer, we have

$$\frac{1}{L'^2}=\frac{\overline{\Sigma}_a}{\overline{D}}+\frac{\overline{\Sigma}'_a}{\overline{D}}. \tag{17.2.1}$$

It is assumed here that the scattering properties of the moderator and thus $\overline{D}$ do not change upon addition of the absorber. This condition is surely fulfilled if we use a strong absorber like boron since then very small amounts, which hardly affect the average scattering cross section, drastically increase the absorption cross section. If we now measure L' at various absorber concentrations and plot $1/L'^2$ versus $\overline{\Sigma}'_a$, we obtain a straight line from whose slope we can determine $1/\overline{D}$ and from whose intercept we can determine $\overline{\Sigma}_a$ (cf. Fig. 17.2.1).

Application of this method presupposes that we can determine the poisoning very precisely. The first prerequisite is that the added material be a strong $1/v$-absorber with a very precisely known cross section. When liquid moderators are used, it is preferable to add natural boron, for example in the form of boric acid; the boron content is then determined pycnometrically or by titration. The recommended value of the absorption cross section of natural boron (19.81% B^{10}) is 760.8±1.9 barn at 2200 m/sec according to Prosdocimi and Deruytter. Homogeneous poisoning is not possible in solid moderators like graphite. There we must poison heterogeneously with wires or foils that are usually made of copper [σ_a(2200 m/sec)$=3.81\pm0.03$ barn]. The copper thickness should be so small that no self-shielding occurs. In order to approach homogeneous poisoning as closely as possible, the mutual distance of the wires or foils must be small; it should not exceed a transport mean free path.

Eq. (17.2.1) holds under the assumption that the spectrum in the moderator is not affected by the addition of the absorber and is always a Maxwell distribution with the temperature of the moderator. In the absence of sources — which we always assume here — and in pure or only slightly poisoned moderators, this

[1] Cf., however, the non-stationary methods described in Chapter 18. Indirect absolute absorption cross section measurements can be done by stationary comparision methods, cf. Sec. 17.3.

assumption is always justified. In the case of strong absorption, however, the diffusion heating effect discussed in Sec. 10.3 should occur. Then instead of Eq. (17.2.1), the more general relation

$$\frac{1}{L'^2} = \left(\frac{\overline{\Sigma}_a}{\overline{D}} + \frac{\overline{\Sigma}'_a}{\overline{D}}\right)\left(1 - \frac{\sqrt{\pi}}{2v_T} \cdot \frac{\overline{\Sigma}_a + \overline{\Sigma}'_a}{\overline{D}^2} \cdot C\right) \tag{17.2.2}$$

holds [cf. Eq. (10.3.19e)]. Here C is the diffusion cooling constant[1]. When the absorption is strong, a downward curvature appears in the plot of $1/L'^2$ against $\overline{\Sigma}'_a$. In principle, it should therefore be possible to determine $\overline{D}$, $\overline{\Sigma}_a$, *and* C from a poisoning experiment. With the exception of the experiments of STARR and KOPPEL (Sec. 17.2.2), the determination of C in this way has hitherto been impossible; all other authors have striven to keep the absorber concentration so small that there was no deviation from the Maxwell spectrum.

17.2.2. Some Experiments on D_2O, H_2O, and Graphite

The first poisoning experiments were carried out in 1953 by KASH and WOODS on *heavy water*. Using a cylindrical sigma pile, these authors determined the diffusion length in pure D_2O and in boric acid solutions with concentrations up to 146.8 mg B_2O_3/liter. Fig. 17.2.1 shows $1/L'^2$ as a function of $\overline{\Sigma}'_a$. The value $\lambda_{tr} = 3\overline{D} = 2.49 \pm 0.04$ cm follows from the slope of the line[2]. The water temperature was 23 °C, the D_2O concentration 99.4%. We see from the straight-line behavior of the measured points in Fig. 17.2.1 that spectral effects can play no role; we can also conclude the same from Eq. (17.2.2) (with $C = 5.25 \times 10^5\ \mathrm{cm^4\,sec^{-1}}$; cf. Sec. 18.1.3). Extrapolation to 100% D_2O gives $\lambda_{tr} = 2.52 \pm 0.04$ cm.

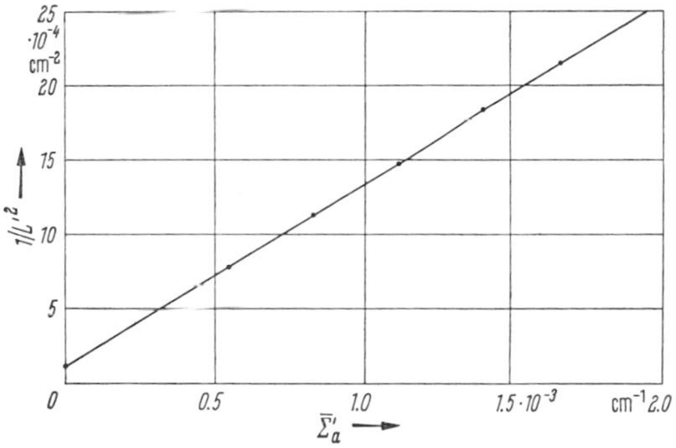

Fig. 17.2.1 $1/L'^2$ vs. $\overline{\Sigma}'_a$ for boric acid solutions in D_2O (after KASH and WOODS)

BROWN and HENNELLY have studied the temperature dependence of the diffusion coefficient of D_2O by the poisoning method. In this case, the poisoning was with copper wires. Fig. 17.2.2 shows $\overline{D} = \frac{\lambda_{tr}}{3}$ in the temperature range from 20 to 250 °C. The smooth curve was calculated according to the Radkowsky prescription (cf. Sec. 17.1.4) and reproduces the measured values surprisingly well.

HENDRIE *et al.* have studied the transport mean free path in *graphite* by heterogeneous poisoning with copper foils. The diffusion length was measured in sigma piles, which were constructed by alternating 25.4-mm-thick layers of

[1] According to Sec. 10.3.3, C contains a contribution due to transport-theoretic effects; as a rule, however, this part is negligible compared to the contribution due to the spectral shift.

[2] The value $\overline{\Sigma}_a(D_2O) = (1.64 \pm 0.1) \times 10^{-4}\ \mathrm{cm^{-1}}$ follows from the intercept. However, the measurement of the H_2O concentration in the D_2O was not sufficiently accurate to permit any conclusion about the absorption cross section of deuterium to be drawn.

graphite with copper foils. The copper thickness increased from 0 to 0.008″. The density of the graphite was 1.674 g/cm³. The result of the measurements was $\lambda_{tr} = 2.62 \pm 0.03$ cm at room temperature. This value is much higher than all the values obtained from non-stationary measurements (Chapter 18).

Several authors (Beckurts and Klüber, Reier, Ballowe, Miller) have studied ordinary *water*, Starr and Koppel having carried out a particularly careful experiment. Boron poisoning was used throughout, on occasions in such high concentrations that spectral effects were noticeable. Fig. 17.2.3 shows Starr and Koppel's values of $1/L'^2$ as a function of the cross section of the added boron. The diffusion length measurements were carried out in a cylindrical water tank (165 cm in diameter, 150 cm high) into which thermal neutrons from a reactor were introduced from below. The downward curvature caused by diffusion heating is clearly recognizable. Least-squares evaluation by means of Eq. (17.2.2) gives $\lambda_{tr} = 0.434 \pm 0.001$ cm, $C = 2900 \pm 350$ cm⁴ sec⁻¹ (at 21 °C), and σ_a(2200 m/sec) $= 326.9 \pm 1.6$ mbarn per proton.

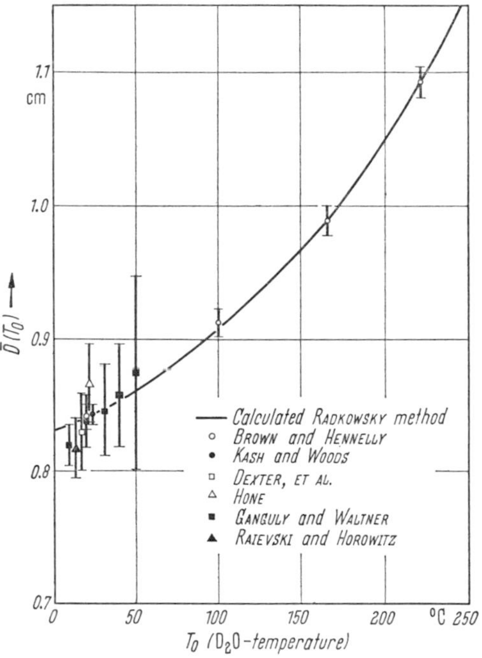

Fig. 17.2.2. $\bar{D}$ *vs.* T_0 for heavy water. The values of Ganguly and Waltner and of Raievski and Horowitz stem from non-stationary measurements; the references are given after chapter 18

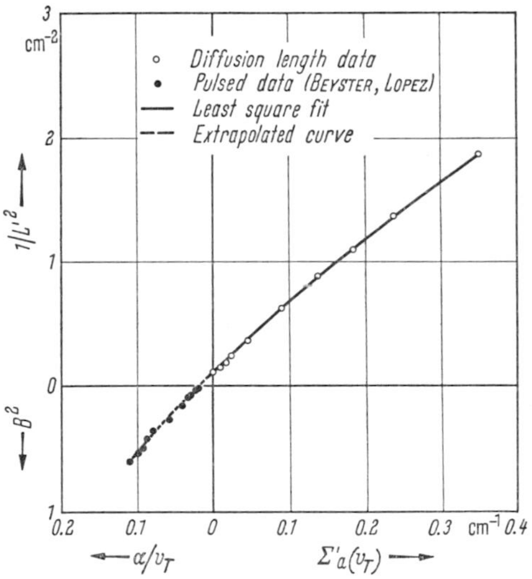

Fig. 17.2.3. $1/L'^2$ *vs.* $\Sigma'_a(v_T)$ in H_2O. ○ Measured by Starr and Koppel, poisoning method. Some values from pulsed source experiments in H_2O (cf. Sec. 18.1) have been included in order to demonstrate the close analogy between these two methods

17.3. Determination of the Absorption Cross Section by Integral Comparison Methods

We can calculate the absolute value of the absorption cross section of a medium from the diffusion length and the transport mean free path. We can determine it independently by the method of pulsed neutron sources (Sec. 18.1). In this section we shall become familiar with some procedures which make it possible to relate to one another the absorption cross sections of various substances (not necessarily only moderator substances). By using standards we can then also obtain the absolute values of the absorption cross sections. Such methods have many advantages compared to measurements with a sigma pile since we need far less material and can frequently carry the measurements out much more simply and quickly. Furthermore, comparison methods, particularly the pile oscillator method, offer nearly the only possibility[1] of measuring the absorption cross sections of substances for which the conditions for an absorption measurement via the diffusion length ($\sigma_a \ll \sigma_s$, good moderation properties) or via a transmission experiment ($\sigma_a \gg \sigma_s$) are not fulfilled.

17.3.1. The Method of Integrated Neutron Flux

Let a source that emits Q neutrons per second be located in a medium that is so large that for practical purposes no neutrons escape. Then since all the neutrons are absorbed in the medium

$$Q = \Sigma_a \int \Phi \, dV \tag{17.3.1}$$

where for simplicity we shall at first ignore spectral effects. A corresponding relation holds when the same source is located in another medium with the absorption cross section Σ_a'. Then

$$\frac{\Sigma_a'}{\Sigma_a} = \frac{\int \Phi \, dV}{\int \Phi' \, dV'}. \tag{17.3.2}$$

Thus we can relate the absorption cross sections of various substances to one another by comparing the flux integrals around the same source. The flux can be measured with foils, and relative measurements obviously suffice. Under some circumstances, it is necessary to take into account the fact that the foil correction is different in different media. The classical application of this method is the measurement of the ratio of the absorption of boron to that of hydrogen (cf. e.g., WHITEHOUSE and GRAHAM, HAMMERMESH, RINGO and WEXLER, BAKER and WILKINSON): The flux integral is measured around the source once in pure water and a second time in a boric acid solution. Then[2] $\Sigma_a = N_H \sigma_H$, $\Sigma_a' = N_H' \sigma_H + N_B \sigma_B$ and thus

$$\frac{N_H'}{N_H} + \frac{N_B \sigma_B}{N_H \sigma_H} = \frac{\int \Phi \, dV}{\int \Phi' \, dV}. \tag{17.3.3}$$

Since N_H, N_H', and N_B are known accurately, σ_B/σ_H follows immediately from comparison of the flux integrals. The accuracy of the method can be increased by making the measurements at various absorber concentrations. In principle we can determine the absorption cross sections of many substances this way, but in comparison with the pulsed neutron method (cf. Sec. 18.1.5) this method is rather complicated and is therefore hardly used any more.

One can also use the method of integrated neutron flux to determine the absorption cross section of an extremely weakly absorbing substance like graphite or beryllium. To do so one must modify it slightly since the requirement of negligibly small leakage from the test body would lead to absurdly large amounts

[1] Other methods, which are, however, limited in their applicability, are activation (cf. Sec. 4.3 and 14.1) and pulsed neutron measurements on absorber solutions (cf. Sec. 18.1.5).

[2] The absorption of oxygen can be neglected.

of material. Fig. 17.3.1 shows a possible arrangement. The test body (Σ'_a) is surrounded by a reflector of the comparison substance (Σ_a); a "strongly" absorbing substance like paraffin serves as the comparison substance and can easily be made thick enough to preclude any appreciable neutron leakage. The comparison measurement is carried out in a sufficiently large sample of the comparison substance; Eq. (17.3.1) again holds for it. On the other hand, for the measurement on the test body and reflector we have

$$Q = \Sigma'_a \int_{\text{test body}} \Phi\, dV + \Sigma_a \int_{\text{reflector}} \Phi\, dV \tag{17.3.4}$$

and we easily obtain

$$\Sigma'_a = \Sigma_a \frac{\int_{\substack{\text{comparison}\\ \text{body}}} \Phi\, dV - \int_{\text{reflector}} \Phi\, dV}{\int_{\text{test body}} \Phi\, dV}. \tag{17.3.5}$$

BÖCKHOFF has done measurements on graphite in this way, and BROSE has done measurements on aluminum. It turns out that one can achieve adequate precision when the test body has linear dimensions of about two diffusion lengths. Paraffin was used as the comparison substance. The flux integration in the test-body-reflector assembly is tedious if the test substance is not in the form of a sphere. One must determine the flux at many points and successively integrate over x, y, and z[1]. On the other hand, in a single, sufficiently large medium, flux measurement along one radius vector suffices; thereafter $4\pi \times \int \Phi(r)\, r^2\, dr$ is calculated.

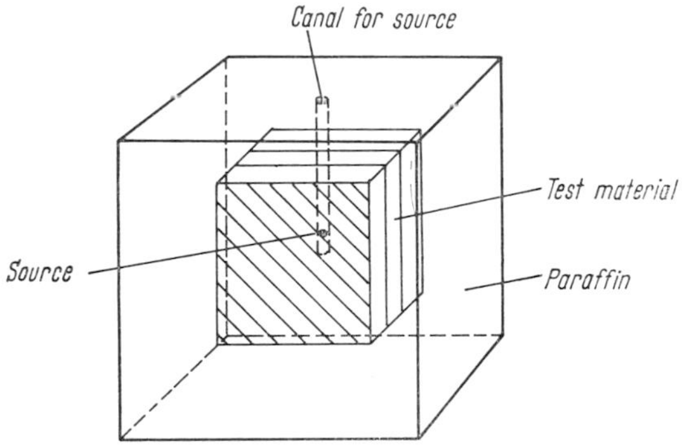

Fig. 17.3.1. An assembly for the measurement of the absorption cross section by comparing integrated neutron fluxes

Spectral effects play only a small role in the method of integrated neutron flux. If the cross sections of the comparison and test substances both have $1/v$-behavior and if the activation cross section of the detector substance follows the $1/v$-law, the neutron spectrum does not enter at all; thus we obtain Σ'_a(2200 m/sec) if we start with Σ_a(2200 m/sec). If there are resonances in the epithermal or fast neutron range, the integrals of just the thermal fluxes are determined by cadmium difference measurements and absorption above the cadmium cut-off energy is taken into account by introducing a resonance escape probability p (cf. Sec. 7.2.3).

17.3.2. The Mireille Pile Method

The Mireille method was first used by RAIEVSKI and later refined by REICHARDT. It allows rapid measurement of the absorption cross section of small quantities of weakly absorbing substances like graphite and beryllium and is particularly well suited to routine measurements (industrial purity testing). Fig. 17.3.2 shows

[1] Some methods for such integrations were discussed in the first edition of this book, p. 207ff.

a typical assembly for graphite measurements. Near the two endfaces of a $160 \times 160 \times 280$-cm prism of comparison graphite (diffusion length L) are located a (Ra—Be) source and a BF_3 counter. The comparison graphite can be replaced by the test graphite being studied (diffusion length L') in a volume V. If Z is the counting rate when comparison graphite is in V, then according to perturbation theory, Z', the counting rate after insertion of the test graphite, is given by

$$\frac{Z'-Z}{Z} = -A\left(\frac{L^2}{L'^2} - 1\right). \tag{17.3.6a}$$

The constant A can be determined by calculation or by normalization measurements on different kinds of graphite. Eq. (17.3.6a) then makes the determination of L^2/L'^2 and thus Σ_a'/Σ_a possible.

We shall content ourselves here with an elementary derivation of Eq. (17.3.6a) under simplified conditions; a more accurate calculation can be found in REICHARDT. Let us assume that a point source of neutrons of strength Q is located at $\boldsymbol{r}_0$ and emits purely thermal neutrons. Then with comparison graphite in V, the thermal flux obeys the equation

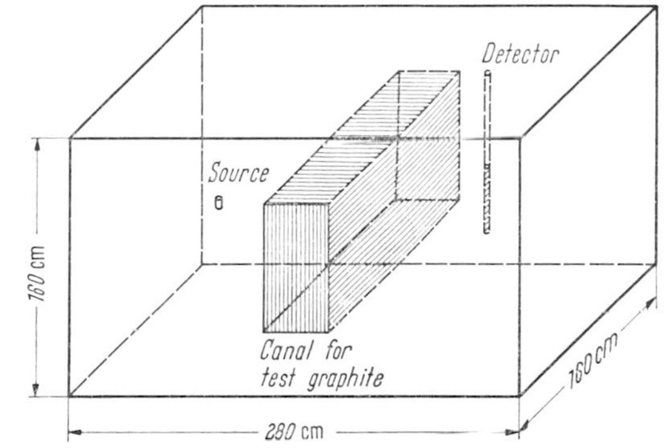

Fig. 17.3.2. A Mireille pile for graphite purity tests

$$\nabla^2 \Phi(\boldsymbol{r}) - \frac{1}{L^2}\Phi(\boldsymbol{r}) = -\frac{Q}{D}\,\delta(\boldsymbol{r}-\boldsymbol{r}_0). \tag{17.3.6b}$$

If we denote the diffusion kernel for a point source in a finite pile by $G(\boldsymbol{r}_0, \boldsymbol{r})$, then

$$\Phi(\boldsymbol{r}) = Q\cdot G(\boldsymbol{r}_0, \boldsymbol{r}) \tag{17.3.6c}$$

is the solution of Eq. (17.3.6b). In particular, the flux $\Phi(\boldsymbol{r}_1)$ at the point $\boldsymbol{r}_1$ at which the counter is located is given by $QG(\boldsymbol{r}_0, \boldsymbol{r}_1)$. Z is proportional to this quantity. If the test substance is in the volume V, then we have

$$\nabla^2 \Phi'(\boldsymbol{r}) - \frac{1}{L^2}\Phi'(\boldsymbol{r}) = -\frac{Q}{D}\,\delta(\boldsymbol{r}-\boldsymbol{r}_0) \quad \text{outside } V, \tag{17.3.7a}$$

$$\nabla^2 \Phi'(\boldsymbol{r}) - \frac{1}{L'^2}\Phi'(\boldsymbol{r}) = 0 \quad \text{inside } V. \tag{17.3.7b}$$

Setting $\Phi'(\boldsymbol{r}) = \Phi(\boldsymbol{r}) + \delta\Phi(\boldsymbol{r})$, we obtain from Eqs. (17.3.6b and c) and Eqs. (17.3.7a and b)

$$\nabla^2 \delta\Phi(\boldsymbol{r}) - \frac{1}{L^2}\delta\Phi(\boldsymbol{r}) = 0 \quad \text{outside } V, \tag{17.3.7c}$$

$$\nabla^2 \delta\Phi(\boldsymbol{r}) - \frac{1}{L^2}\delta\Phi(\boldsymbol{r}) = -\left(\frac{1}{L^2} - \frac{1}{L'^2}\right)\left(\Phi(\boldsymbol{r}) + \delta\Phi(\boldsymbol{r})\right) \quad \text{inside } V. \tag{17.3.7d}$$

On the right-hand side of Eq. (17.3.7d) stand the characteristic sources of the perturbation $\delta\Phi(\boldsymbol{r})$, which exist because of the difference in L and L'. When L and L' only differ little, $\delta\Phi(\boldsymbol{r}) \ll \Phi(\boldsymbol{r})$ and we can neglect $\delta\Phi$ compared to Φ

on the right-hand side of Eq. (17.3.7d). Furthermore, we can combine Eqs. (17.3.7c and d) into

$$\nabla^2 \delta\Phi(\boldsymbol{r}) - \frac{1}{L^2}\,\delta\Phi(\boldsymbol{r}) = -\left(\frac{1}{L^2} - \frac{1}{L'^2}\right)\Phi(\boldsymbol{r})\,\delta(V) \qquad (17.3.7\text{e})$$

where $\delta(V)=1$ inside V and 0 otherwise. The solution of Eq. (17.3.7e) is

$$\delta\Phi(\boldsymbol{r}) = D\left(\frac{1}{L^2} - \frac{1}{L'^2}\right)\int\limits_V \Phi(\boldsymbol{r}')\,G(\boldsymbol{r}',\boldsymbol{r})\,d\boldsymbol{r}'. \qquad (17.3.8\text{a})$$

$G(\boldsymbol{r}',\boldsymbol{r})$ is again the diffusion kernel for the pile; the source point $\boldsymbol{r}'$ now lies inside the volume V. It follows from Eq. (17.3.6c) that

$$\frac{Z'-Z}{Z} = \frac{\delta\Phi(\boldsymbol{r}_1)}{\Phi(\boldsymbol{r}_1)} = -\left(\frac{L^2}{L'^2} - 1\right)\cdot\frac{D}{L^2}\int\limits_V \frac{G(\boldsymbol{r}_0,\boldsymbol{r}')\,G(\boldsymbol{r}',\boldsymbol{r}_1)}{G(\boldsymbol{r}_0,\boldsymbol{r}_1)}\,d\boldsymbol{r}'. \qquad (17.3.8\text{b})$$

Thus Eq. (17.3.6) has been derived and the constant A determined. In order to carry out the integration the diffusion kernel of the pile must be known; it can be obtained from Eqs. (6.2.21) and (6.2.25). In an exact calculation we must take into account the fact that the source emits non-thermal neutrons. However, if the distance between the source and the test volume is sufficiently large compared to the slowing-down length, we obtain the same result as for a thermal source. Fig. 17.3.3 shows A for volumes V of various sizes in the pile shown in Fig. 17.3.2. We see that even for small test volumes good sensitivity can be achieved (e.g., $A=0.1$ for $V\approx 130$ liter). Since we can determine Z and Z' with a precision of about 0.1%, $A=0.1$ means that differences between L and L' of about 0.5% can be detected. However, additional errors are introduced by material inhomogeneities (density fluctuations, anisotropy effects), and in practice L/L' is rarely determined to better than 2%. Fluctuations in the graphite temperature are an important source of error, but they can be eliminated by putting the pile in a temperature-controlled room. It is also worthwhile to cover the pile surface with cadmium in order that changes in the backscattering conditions not affect the counting rate.

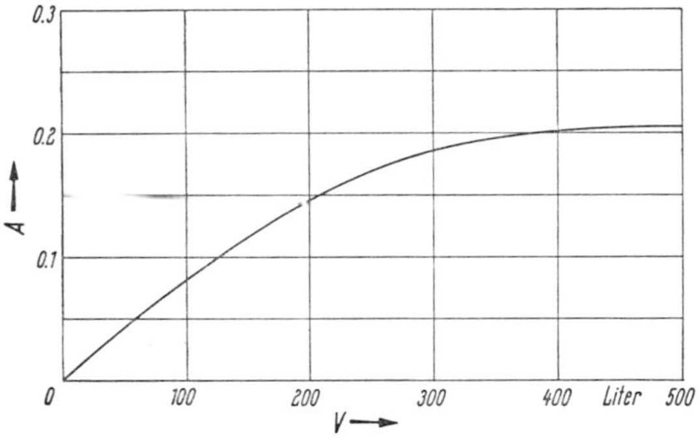

Fig. 17.3.3. The sensitivity of a graphite Mireille pile arrangement as a function of the test graphite volume

The procedure in the form developed here is only suitable for the comparison of substances with similar scattering properties (i.e., for the comparison of different samples of graphite in a graphite pile or different samples of beryllium in a beryllium pile, etc.). When there is a large difference between L and L', simple first-order perturbation theory, which leads to Eq. (17.3.6a), is no longer applicable, and we must then introduce terms of higher order. The evaluation is then more difficult and less accurate.

17.3.3. The Pile Oscillator

There are two kinds of pile oscillator experiments, the *local* kind and the *integral* kind. The local kind is based on observing the flux depression near an absorbing sample in a non-multiplying medium. A nuclear reactor generally serves to provide the neutron field, although this is not necessarily a prerequisite of the method. On the other hand, the integral method is based on the effect of an absorber on the reactivity of a reactor; for its precise understanding a detailed knowledge of reactor theory is necessary, and we shall limit ourselves here to a discussion of the fundamentals.

Fig. 17.3.4 shows schematically a *local* pile oscillator. In the graphite reflector of a reactor is located an annular ionization chamber which has been made sensitive to neutrons by boron coating. Using a suitable mechanical device, we can make a small sample of the test substance move back and forth through the interior cavity of the chamber. The frequency of this motion is about 1 cycle/sec. Owing to the scattering and absorption of neutrons by the test substance, a periodic signal is produced which is superimposed on the steady-state current of the chamber. This signal can be separated and amplified by a sensitive amplifier and is ultimately recorded. Fig. 17.3.5 shows the typical time behavior of such a signal over a full period of oscillation. The signal in case *a* is produced by a cadmium sample (pure absorption) while that in case *b* is produced by a graphite sample (pure scattering). The zero time-point corresponds to the inner turning point of the oscillatory motion. There are two signals each time since the chamber is traversed twice during a period. The absorption signal is negative. On the other hand, the scattering signal is positive; it also has a somewhat different form than the absorption signal and is shifted somewhat in time. The scattering signal comes mainly from those neutrons which stream through the canal and in the absence of the test sample would pass right through the hole in the ionization chamber. The signal forms in Fig. 17.3.5 are idealized, and Fig. 17.3.6 shows signal forms as they are actually observed. The distortions are due to the frequency characteristics of the amplifier, which is given the smallest possible bandwidth in order not to amplify the background noise, particularly that due to statistical fluctuations in the chamber current. As Fig. 17.3.6 shows, these distortions have the convenient effect that a time interval Δt can be found during which the average scattering signal vanishes. We therefore connect the output of the amplifier to a current integrator that is only sensitive during the time interval Δt. In this way, the scattering signal can be largely eliminated. In practice, a sensitivity ratio of 500 has been achieved by suitable choice of the integration range Δt. That is to say, the oscillation of a scatterer with a given "scattering surface"

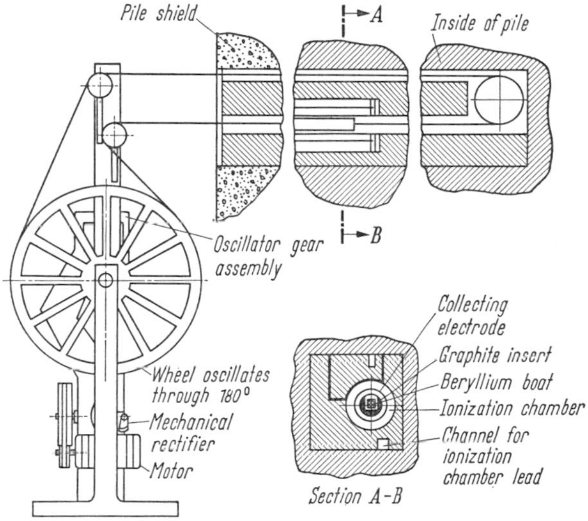

Fig. 17.3.4. A local pile oscillator (HOOVER *et al.*)

(defined as the product of the total number of atoms and the atomic cross section $NV\sigma_s = V\Sigma_s$) gives about a 500 times smaller effect than the oscillation of an absorber with an equal absorption surface $V\Sigma_a$.

The absorption signal is proportional to the absorption surface $V\Sigma_a$ (unless the sample is so large that self-shielding occurs). We measure the absorption cross section by comparing its absorption surface with that of a standard substance (gold, boron). The procedure is very sensitive; the detection limit in a good pile oscillator is about 0.1 mm² in $V\Sigma_a$. (Smaller signals cannot be distinguished from the fluctuations in the chamber current caused by fluctuations in the reactor power.) Thus, for example, we only need about 20 g of aluminum ($\sigma_a = 0.24$ barn) in order to determine the absorption cross section to an accuracy of 1%. With more strongly absorbing substances ($\sigma_a > 1$ barn) the effect of scattering can be completely neglected because of the small scattering sensitivity of the method. When the absorption is weaker, on the other hand, we must apply a correction that can be obtained with the help of the scattering sensitivity determined with a standard scatterer (graphite, D_2O) and the known scattering cross section. We obviously cannot determine the absorption of extremely weakly absorbing substances like graphite, D_2O, and beryllium very precisely with a local pile oscillator since even with a sensitivity ratio of 500 the contribution of the scattering to the total signal dominates that of the absorption. Pomerance, Hoover *et al.*, Small and Spurway, and Fuketa, among others, have described experiments with local pile oscillators. Fuketa lowered the detection limit of the method considerably (to $\Sigma_a V = 10^{-3}$ mm²) by largely eliminating the effect of fluctuations in the reactor power. This was achieved by using two ionization chambers and recording the difference of their outputs.

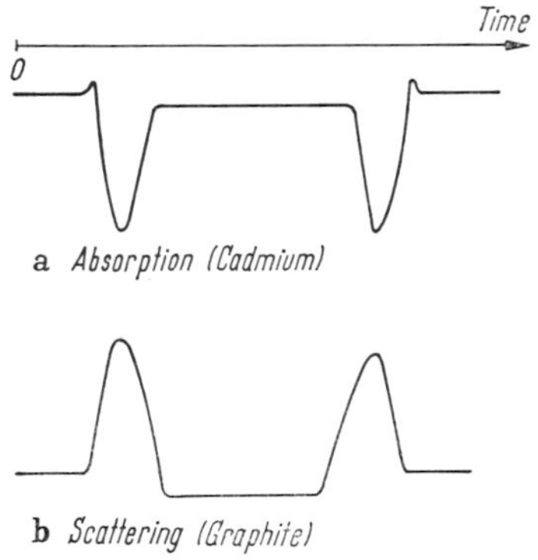

Fig. 17.3.5. Idealized signal forms in a local pile oscillator

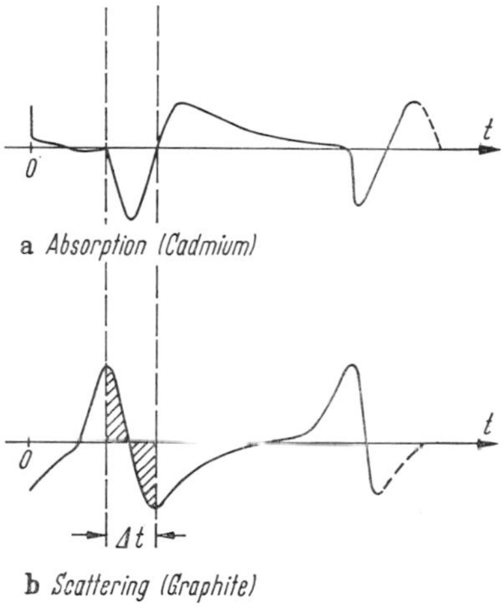

Fig. 17.3.6. Actual signal forms in a local pile oscillator

In an *integral* pile oscillator, the test sample is periodically moved in and out of the core of a reactor. The reactor is run at a constant average power; this average power is kept very low in order to avoid thermal effects. The reactor power then exhibits characteristic oscillations around its average value with the frequency of the test sample's motion. The amplitude of these oscillations is proportional to the change in the multiplication factor caused by the sample. The sample affects the multiplication factor in various ways: Thermal and epithermal neutrons are absorbed. Neutrons are scattered and the leakage is thereby affected. Test substances like graphite and beryllium contribute to the moderation. A complete analysis and separation of the individual effects requires detailed theoretical and experimental work. For example, we can determine thermal and epithermal absorption separately by making measurements with various reactor spectra or by using the cadmium difference method. Neutron streaming effects can be kept small by oscillating the test body between points with vanishing flux gradients (e.g., from outside the reactor, where the flux vanishes, to the middle of the core, where it has a maximum). We shall not go into the details of this method here; more information on the method of the integral pile oscillator can be found in Weinberg and Schweinler, in Langsdorf, in Breton, and in Rose, Cooper, and Tattersall.

The limit of detection in the measurement of thermal absorption cross sections is similar to that of the local pile oscillator. However, the effect of scattering can be more effectively separated, so that measurements on graphite and beryllium are possible. Resonance absorption integrals and η-values (of fissionable substances) can also be determined (cf. Rose, Cooper, and Tattersall). There are a number of reactors which were expressly built for pile oscillator measurements and have special facilities for them (e.g., Dimple, Gleep, Minerve).

With the integral method, oscillation of the sample is in principle not necessary; the change in the multiplication constant due to the introduction of the sample can be measured otherwise, e.g., by compensation with a calibrated shim rod. This static method, the so-called danger coefficient method, was formerly used very often. However, it is much less sensitive than the oscillation method on account of long-term drifts in the reactor power. These drifts are caused by temperature and air-pressure effects that largely cancel out in the oscillator measurements but that limit the accuracy of static measurements of the multiplication constant.

Chapter 17: References

General

Templin, L. J. (ed.): Reactor Physics Constants, ANL-5800, Second Edition (1963); especially Section 3.3: Thermal-Group Diffusion Parameters.

Corngold, N. (ed.): Proceedings of the Brookhaven Conference on Neutron Thermalization, BNL-719 (1962); especially Volume III: Experimental Aspects of Transient and Asymptotic Phenomena.

Special[1]

Ballowe, W. C.: BNL-719, 799 (1962).
Barkov, L., V. K. Makarin, and K. N. Mukhin: J. Nucl. Energy 4, 94 (1957).
Beckurts, K. H., and O. Klüber: Z. Naturforschung 13a, 822 (1958).
Heintzé, L. R.: Nucleonics 14, No. 5, 108 (1956).
Juren, J. A. de, and M. Rosenwasser: J. Res. Nat. Bur. Stand. 51, 203 (1951).
Reier, M., and J. A. de Juren: J. Nucl. Energy A 14, 18 (1961).
Rockey, K. S., and W. Skolnick: Nucl. Sci. Eng. 8, 66 (1960).
Rohr, G.: Unpublished Karlsruhe report (1962).
Sisk, F. J.: ORNL-933 (1951).
Starr, E., and J. U. Koppel: BNL-719, 1012 (1962).
Wilson, V. C., E. W. Bragdon, and H. Kanner: CP-2306 (1944).
Wright, W. B., and R. T. Frost: KAPL-M-WBW 2 (1956).

Measurement of the Diffusion Length in H_2O.

[1] Cf. footnote on p. 53.

Temperature Dependence of the Transport Mean Free Path in Hydrogenous Moderators.

Deutsch, R. W.: Nucl. Sci. Eng. **1**, 252 (1956).
Drozdov, S. I., *et al.*: Progr. Nucl. Energy Ser. I, Vol. 2, p. 207 (1959).
Esch, L. J.: Nucl. Sci. Eng. **16**, 196 (1963).
Petrie, C. D., M. L. Storm, and P. F. Zweifel: Nucl. Sci. Eng. **2**, 728 (1957).
Radkowsky, A.: ANL-4476, 89 (1950).

Diffusion Length Measurements on Graphite with the Sigma-Pile Method.

Carlbom, L.: Arktiv Fysik **6**, 335 (1952).
Ertaud, A., *et al.*: Rapport CEA-3 (1948).
Hendrie, J. M., *et al.*: Progr. Nucl. Energy Ser. I, Vol. 3, p. 270 (1959). Cf. also the diploma thesis of H. Winzeler, Göttingen (1952), K. H. Beckurts, Göttingen (1954), W. Reichardt, Karlsruhe (1960), H. Klose, Karlsruhe (1962).
Hereward, H. G., *et al.*: Canad. J. Res. A **25**, 15 (1947).
Laubenstein, R. A.: Geneva 1958, P/594, Vol. 12, p. 689.
Lee, M. T.: HW-51175 (1957).
Richey, C. R., and E. Z. Block: HW-45035 (1956).

Temperature Dependence of the Diffusion Length in Graphite.

Lloyd, R. C., E. D. Clayton, and C. R. Richey: Nucl. Sci. Eng. **4**, 690 (1958).
Mills, J. E. C.: AERE-RP/R-1618 (1955).

Diffusion Length Measurements in Various Hydrogenous Moderators.

Bata, L., I. Kiss, and L. I. Pál: Geneva 1958, P/1730, Vol. 12, p. 509.
Brown, W. W., *et al.*: Geneva 1958, P/595, Vol. 12, p. 514.
Esch, L. J.: Nucl. Sci. Eng. **16**, 196 (1963).
Heintzé, L. R.: Nucleonics **14**, No. 5, p. 108 (1956).
Tittle, C. W.: Phys. Rev. **80**, 756 (1950).

Sigma-Pile Measurement of the Diffusion Length in D_2O.

Hereward, H. G., *et al.*: Can. J. Res. A **25**, 26 (1947).
Lutz, H. R., and R. W. Meier: Nukleonik **4**, 108 (1962).
Meister, H.: Unpublished Karlsruhe report (1960).
Sargent, B. W., *et al.*: Can. J. Res. A **25**, 134 (1947).

Sigma-Pile Measurement of the Diffusion Length in Be and BeO.

Geraseva, L. A., *et al.*: Geneva 1955 P/662, Vol. 5, p. 13.
Koechlin, J. C., J. Martelly, and V. P. Duggal: Geneva 1955, P/359, Vol. 5, p. 20.
Nobles, R., and J. Wallace: ANL-4076, 10 (1947).

Measurement of the Extrapolated Endpoint of Thermal Neutrons.

Auger, P., A. M. Munn, and B. Pontecorvo: Can. J. Res. A **25**, 143 (1947).
Hendrie, J. M., *et al.*: Progr. Nucl. Energy Ser I, Vol. 3, p. 270 (1959).
Hone, D. W.: J. Nucl. Energy A **11**, 34 (1959).
Sisk, F. J.: ORNL-933 (1951).

Diffusion Length Measurements by Means of Surface Sources.

Bothe, W., and E. Fünfer: Z. Phys. **122**, 769 (1944).
Bothe, W., and P. Jensen: Z. Phys. **122**, 749 (1944).
Fité, J. G.: Thesis, Göttingen 1954.
Schlüter, H.: Diploma Thesis, Göttingen 1956.

Poisoning Experiments.

Ballowe, W. C., and W. R. Morgan: Trans. Am. Nucl. Soc. **4**, 281 (1961).
Beckurts, K. H., and O. Klüber: Z. Naturforschung **13a**, 822 (1958).
Brown, H. D., and E. J. Hennelly: BNL-719, 879 (1962).
Dexter, A. H.: ANL-4746, 14 (1951).
Hendrie, J. M., *et al.*: Progr. Nucl. Energy Ser. I, Vol. 3, p. 270 (1959).
Kash, S. W., and D. C. Woods: Phys. Rev. **90**, 564 (1953).
Miller, J.: Trans. Am. Nucl. Soc. **4**, 282 (1961).
Reier, M.: J. Nucl. Energy A **14**, 186 (1961).
Starr, E., and J. U. Koppel: BNL-719, 1012 (1962).

Prosdocimi, A., and A. I. Deruytter: J. Nucl. Energy A & B **17**, 83 (1963) (The Absorption Cross Section of Boron).

Baker, A. R., and D. H. Wilkinson: Phil. Mag. Ser. VII **3**, 647 (1958).
Böckhoff, K. H.: Diploma Thesis, Göttingen 1956.
Brose, M.: Unpublished Karlsruhe Report, 1960.
Brose, M., and K. H. Beckurts: Nukleonik **2**, 139 (1960).
Hamermesh, B., G. R. Ringo, and S. Wexler: Phys. Rev. **90**, 603 (1953).
Wirtz, K., and K. H. Beckurts: Elementare Neutronenphysik. Berlin-Göttingen-Heidelberg: Springer 1958.
Whitehouse, W. J., and G. A. Graham: Can. J. Res. A **25**, 261 (1947).
} Method of Integrated Neutron Flux.

Raievski, V., *et al.*: Geneva 1958 P/1204, Vol. 13, p. 65.
Reichardt, W.: Diploma Thesis, Karlsruhe 1960.
} Mireille Pile.

Fuketa, T.: Nucl. Instr. Methods **13**, 35 (1961).
Hoover, J. I., *et al.*: Phys. Rev. **74**, 864 (1948).
Pomerance, H.: Phys. Rev. **83**, 641 (1951).
Small, V. G., and A. H. Spurway: AERE RP/R 1439 (1956).
} Local Pile Oscillator.

Breton, D.: Geneva 1955 P/356, Vol. 4, p. 127.
Langsdorf, A.: AECD 3194 (1951).
Rose, H., W. A. Cooper, and R. B. Tattersall: Geneva 1958 P/14. Vol. 16, p. 34.
Weinberg, A. M., and H. C. Schweinler: Phys. Rev. **74**, 851 (1948).
} Integral Pile Oscillator.

18. Investigation of the Diffusion, Absorption, and Thermalization of Neutrons by Non-Stationary Methods

During the last ten years non-stationary methods have gained increasing significance in neutron physics. This is because they frequently permit cleaner and more accurate measurements than stationary methods and because they also give much deeper insight into the physical nature of the problems. This conclusion is especially true for the method of pulsed sources, whose great potentialities were first recognized by von Dardel and which has since become a standard method. We shall become more familiar with it in Secs. 18.1 and 18.3. Closely related to the pulsed-source method but less usual is the method of modulated sources introduced by Raievski and Horowitz; it is discussed in Sec. 18.2.

We once again limit ourselves to the description of experiments on non-multiplying media. Both methods mentioned above, particularly the method of pulsed sources, have a wide range of application in the study of subcritical reactors.

18.1. Measurements in Thermal Neutron Fields by the Pulsed-Source Method

Fig. 18.1.1 shows schematically the apparatus for a pulsed-source experiment. To begin with, a pulsed source emits a short (compared to the time duration of the phenomenon being studied) burst of neutrons into the substance being studied. The time behavior of the neutron flux is followed with detectors in the medium or on its surface and recorded by a multichannel analyzer. When the intensity has decayed sufficiently, another neutron pulse is injected and the entire procedure repeated. After each pulse, the neutrons are first slowed down (some pulsed slowing-down experiments have already been discussed in Sec. 16.3) and then thermalized; ultimately we are left with a thermalized neutron field that decays by leakage and absorption according to the laws of diffusion theory. The theory of this decay has already been discussed in detail in Secs. 9.2. and 10.3; here we only need the single result that after the decay of all higher spatial and energetic modes, the field decays according to the law $\Phi(t) \sim e^{-\alpha t}$, where

$$\alpha(B^2) = \alpha_0 + D_0 B^2 - C B^4 + \text{higher terms}. \tag{18.1.1}$$

The goal of a pulsed neutron experiment in a thermalized field is the determination of the $\alpha(B^2)$-curve and its analysis to determine the diffusion parameters of the medium.

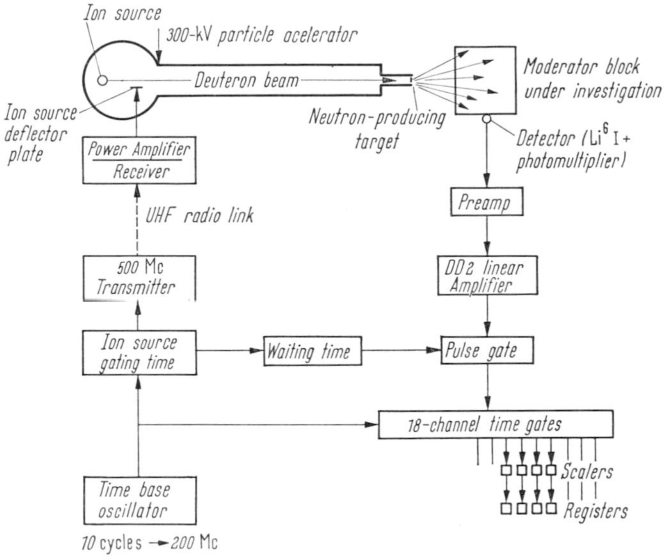

Fig. 18.1.1. The arrangement used by DE SAUSSURE and SILVER for pulsed neutron experiments

18.1.1. Instrumentation of a Pulsed Neutron Experiment

We restrict the discussion here to the most necessary instrumentation; more detailed information on the apparatus used in pulsed neutron experiments can be found in VON DARDEL and SJÖSTRAND and in BECKURTS.

Today, the most frequently used *neutron sources* are small, flexible deuteron accelerators with voltages of from 150 to 400 kev that are operated with tritium targets. The pulse length must be variable in the range from 10 μsec to 1 msec for use in thermal systems; the duty ratio is in the range from 1:20 to 1:100. With a pulse current of about 1 mamp, i.e., an average current of from 10 to 50 μamps, quite adequate intensity can be obtained from a fresh tritium target. An extremely high signal-to-background ratio is important. Thus the current during the pulse should be very large compared to the current in the intervals between pulses (at least 10^5 times larger). A number of suitable sources are presently commercially available.

Neutron detectors offer no particular problem in experiments on non-multiplying, thermal system. Most authors use BF_3-counters, but experiments have been done with fission chambers, Li^6(Eu)I-crystals, and Li-glass scintillators. BF_3-counters and fission chambers that work reliably up to temperatures of

300 °C are commercially available; at higher temperatures, cooling apparatus must be provided. It is important that the detector and its associated circuitry have a small dead time that is also accurately known. The amplifier must be able to handle reliably the high counting rate that occurs immediately after the neutron pulse.

The *multichannel time analyzer* should have about 50 to 100 channels, whose widths are variable in the range from 1 μsec to 1 msec. The analyzers developed for time-of-flight measurements are usually not very suitable because they have too large a dead time. There are a number of circuits that have been built expressly for pulsed neutron experiments (von Dardel, Glass, Gatti). Very good instruments are also commercially available.

In pulsed neutron experiments, neutron backscattering can produce a very disturbing background. It is therefore necessary to take care that as little backscattering as possible occurs. For this reason, a pulsed neutron facility should not be housed in too small a room. The room temperature must be carefully monitored and perhaps even held constant.

18.1.2. Measurement of the $\alpha(B^2)$-Curve

In order to determine the $\alpha(B^2)$-curve, we must measure the decay constant α for assemblies of various sizes and then calculate the geometrical buckling B^2 for these assemblies. Some formulas for the calculation of B^2 were given in Sec. 9.2.1. We usually use the value $d=0.71\,\lambda_{tr}$ for the extrapolated endpoint. However, in water and other hydrogenous moderators, d depends on the size of the assembly because of the strong energy dependence of the scattering cross section. In this case, it is better to use the results of Gelbard (Sec. 10.3.4).

α is then determined from the decay curve. This determination is complicated by the fact that the decay curve is purely exponential only over a limited time interval (Fig. 18.1.2). For short times the effect of the higher spatial harmonics and energy transients is appreciable, while for long times the background contributes significantly to the flux.

The influence of the *energy* transients can be eliminated by waiting long enough after the beginning of the experiment for the neutron field to become completely thermalized. We must thus allow a certain delay time t_w to elapse between the source pulse and the time from which the decay curve is measured. We can determine t_w by observing the decay of the field with two detectors with extremely different energy dependences, e.g., a thin ("$1/v$") and a thick ("black") boron counter. As long as the spectrum still changes with time, the ratio of the counting rates of the two detectors continues to change; it first becomes constant for times longer than t_w (Starr and de Villiers). We can also determine t_w by determining α from the decay curve, beginning the evaluation successively with later and later times t_0. α decreases at first and only becomes constant for $t_0 > t_w$. In this way, we find, for example, that for graphite $t_w \approx 2$ msec. In larger assemblies (linear dimensions greater than 2 to 3 diffusion lengths), spatial harmonics rather than thermalization effects are mainly responsible for the delay time. In extremely small assemblies, there are circumstances under which there is no stationary energy spectrum even after an arbitrarily long delay time. This is clearly evident in the example of a very cold beryllium block (Fig. 18.1.3; Silver), where the value of α decreases steadily with increasing t_0. This phenomenon is connected with the fact, discussed in Sec. 10.3.3, that under certain extreme conditions no asymptotic spectrum of thermalized neutrons exists.

The effect of *spatial* harmonics can be very greatly reduced by a skillful arrangement of the neutron source and detector. In a parallelepipedal assembly the source is placed at the midpoint of one of the side surfaces so that in the x- and y-directions only odd ($l, m=1, 3, 5, \ldots$) harmonics are excited. If the detector is now placed at the midpoint of an adjacent side surface, then of the harmonics excited in the z-direction only the odd ($n=1, 3, 5 \ldots$) ones will be detected. Then it follows that the next higher harmonics after the fundamental mode 111 to be detected are the 311-, the 131-, and the 113-modes, which usually decay considerably faster. If we use two detectors connected in parallel and place them in the moderator as shown in Fig. 18.1.4, we can also eliminate all harmonics with l, m, or $n=3$. Thus the next modes after the fundamental mode are the 511-, 151-, and 115-modes. Similar considerations are possible for cylindrical geometry. As a general rule, the elimination of the higher harmonics is more difficult, the larger the system in comparison to its diffusion length. Under some circumstances, we must convince ourselves by evaluating the decay curve with a variable delay time (see above) that the harmonics have decayed sufficiently.

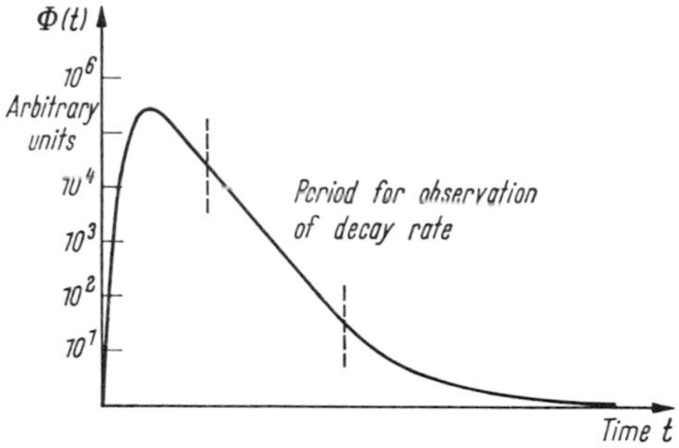

Fig. 18.1.2. The time behaviour of the neutron flux after injecting a fast neutron burstinto a moderator assembly

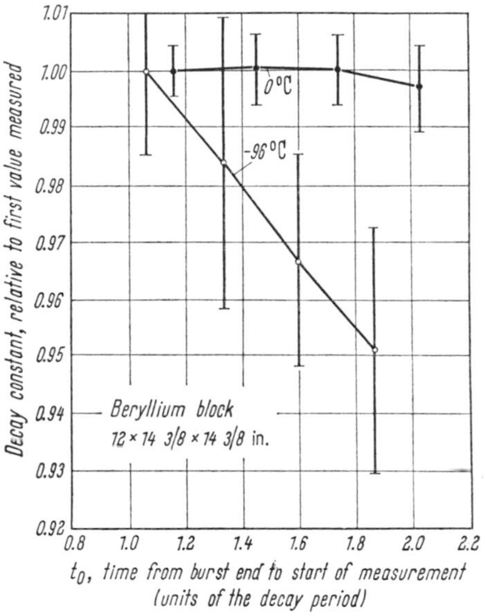

Fig. 18.1.3. The decay constant measured in a small beryllium block as a function of waiting time

In an extended geometry, one can also proceed by the method of Fourier analysis. A small counter is moved through the system and the decay curve is measured at a large number of points $\boldsymbol{r}_\nu$. Now

$$\Phi(\boldsymbol{r}, t)=\sum_n A_n R_n(\boldsymbol{r})\, \Phi_n(t). \tag{18.1.2a}$$

Here the $R_n(\boldsymbol{r})$ is the spatial eigenfunction of the *n-th* harmonic and $\Phi_n(t)$ is the associated decay curve. Obviously,

$$\Phi_n(t) \sim \int \Phi(\boldsymbol{r}, t)\, R_n(\boldsymbol{r})\, d\boldsymbol{r}. \tag{18.1.2b}$$

Thus we can isolate the decay curves of the individual harmonics by integrating the measured space-time distribution as indicated in Eq. (18.1.3). Lopez and

BEYSTER have done this for H_2O systems, and MEISTER has done it for D_2O systems.

The *background* is due to neutron emission from the source during the intervals between pulses and to backscattering of neutrons. We can substantially reduce the effect of backscattering by covering the surface of the moderator block with cadmium or even better with thick boron sheets. Nevertheless, there almost always remains a slowly varying background caused by fast neutrons that we must carefully determine and subtract before the evaluation of α from the measured decay curve.

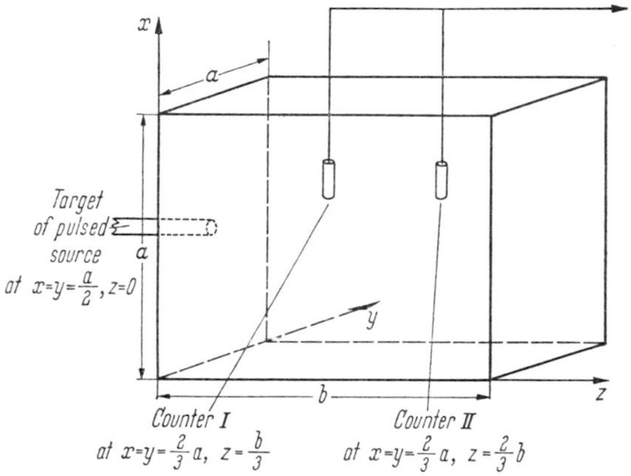

Fig. 18.1.4. An arrangement of neutron source and two neutron detectors in a parallelepipedal assembly that will suppress all even modes and all odd modes with l, m or $n=3$

After the background has been subtracted and the delay time determined, we can determine α from the decay curve. In order to obtain good accuracy, the useable part of the decay curve should stretch over several (2 to 3) decades with good statistical accuracy. The calculation of α is usually done with either PEIERLS's or CORNELL's method. In favorable cases, one can achieve an accuracy of 0.2% in α.

18.1.3. Results of Pulsed Neutron Studies on Various Moderators

The pulsed neutron method primarily yields the $\alpha(B^2)$-curve for a moderator substance. However, we are usually not content with this curve, but using Eq. (18.1.1) try to determine from it the diffusion parameters $\alpha_0=\overline{v\Sigma_a(v)}$, D_0, and C. For this purpose, the measured curve is approximated by a parabola whose coefficients are determined by the method of least squares[1]. This method gives α_0 and D_0 with satisfactory accuracy. In contrast, the achievable accuracy in C is only moderate since the deviation of the $\alpha(B^2)$-curve from a straight line is small and since in small assemblies still higher terms $(FB^6+GB^8+\cdots)$ contribute to α. Either we must limit ourselves to a range of B^2 in which these contributions are not important, or we must apply calculated corrections for the higher terms. However, theoretical estimation of these terms, which depend sensitively on the thermalization and transport properties of the moderator, is difficult.

Light Water. Table 18.1.1 contains the diffusion parameters determined by various authors from three-parameter fits to the $\alpha(B^2)$-curve. In the region $B^2<0.9\ \text{cm}^{-2}$, the role of the B^6-term is small, and there is no strong dependence of the parameters on the evaluation scheme. The values have been corrected to a temperature of 22 °C. The results of ANTONOV *et al.*, of BRACCI and COCEVA, of DIO, and of KÜCHLE agree within experimental error and all agree well with the

[1] One usually performs the fit in such a way that the sum of the squares of the relative (not the absolute) deviations of the measured points from the curve is a minimum.

poisoning experiments of STARR and KOPPEL (Sec. 17.2.2). The results of LOPEZ and BEYSTER are somewhat higher. In Fig. 18.1.5, we compare the $\alpha(B^2)$-curve measured by these latter authors with that observed by KÜCHLE (which almost coincides with those of the other authors above). The values obtained by LOPEZ and BEYSTER are larger throughout than those of KÜCHLE, especially at large bucklings. A possible explanation for this discrepancy is that LOPEZ and BEYSTER used cubic geometry for their assemblies while KÜCHLE (and the other

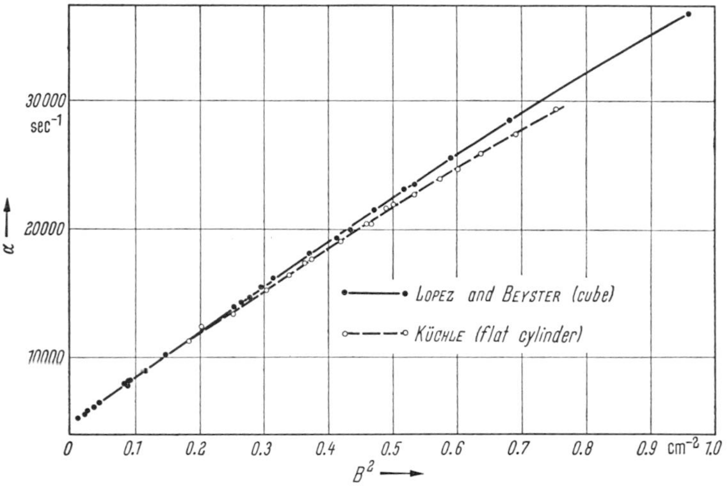

Fig. 18.1.5. The decay constant α *vs.* geometrical buckling B^2 for water at 26.7 °C. KÜCHLE'S results were taken at 22 °C and corrected to 26.7 °C, the temperature at which LOPEZ and BEYSTER'S results were obtained

authors above) used flat cylinders. We might conclude from these facts that there is a strong dependence of the extrapolated endpoint on the geometric shape of the assembly[1] (cf. HALL, SCOTT, and WALKER in this context).

Table 18.1.1. *Diffusion Parameters in Water (22 °C) According to the Pulsed Neutron Method*

Author	Year	α_0 (sec^{-1})	D_0 (cm^2 sec^{-1})	C (cm^4 sec^{-1}) *
v. DARDEL and SJÖSTRAND	1954	4892	36,340 ± 750	7,300 ± 1,500
ANTONOV *et al.*	1955	4831	35,000 ± 1,000	4,000 ± 1,000
BRACCI and COCEVA	1956	4950	34,850 ± 1,100	3,000 ± 1,000
DIO	1959	4808	35,450 ± 600	3,700 ± 700
KÜCHLE	1960	4785	35,400 ± 700	4,200 ± 800
LOPEZ and BEYSTER	1961	4768	36,892 ± 400	5,116 ± 776

* The values of C have not been corrected to 22 °C.

ANTONOV *et al.* have investigated the neutron diffusion in H_2O up to temperatures of 280 °C by the pulsed source method. Their experimental results for the diffusion constant can be approximated by the following formula:

$$\left.\begin{aligned}\frac{D_0(T_0)}{D_0(21\,°\mathrm{C})} &= (0.934 \pm 0.028) + (0.289 \pm 0.009) \times 10^{-2}\, T_0 + \\ &\quad + (0.106 \pm 0.03) \times 10^{-4}\, T_0^2\end{aligned}\right\} \quad (18.1.3\,\mathrm{a})$$

[1] In the curves of Fig. 18.1.5, B^2 is calculated with the extrapolated endpoint d of GELBARD (cf. Sec. 10.3.4); had we simply set $d = 0.71\,\lambda_{tr}$, we would have obtained essentially the same result.

Here T_0 is the water temperature in °C. The measured values of C can be represented by

$$\left.\begin{aligned}\frac{C(T_0)}{C(21^\circ\,\mathrm{C})} = (0.987\pm0.098) + (0.619\pm0.031)\times10^{-3}\,T_0 + \\ + (0.348\pm0.104)\times10^{-4}\,T_0^2.\end{aligned}\right\} \qquad (18.1.3\,\mathrm{b})$$

HONECK has calculated the diffusion parameters in H_2O at room temperature by numerically solving the transport equation for the Nelkin model (cf. Table 10.3.4). He obtained $D_0 = 37{,}460$ cm² sec⁻¹ and $C = 2{,}878$ cm⁴ sec⁻¹. We can also estimate C by the elementary method of Sec. 10.3.1. If we assume that in H_2O molecular rotation is strongly hindered but molecular translation is not hindered at all, we can treat water in first approximation as a gas with $A = 18$ and $\sigma_s = 80$ barn/H-atom. Using the measured value of D_0 and assuming $1/v$-behavior for σ_{tr}, we obtain $C = 4250$ cm⁴ sec⁻¹ (cf. Table 10.3.1).

Heavy Water. Fig. 18.1.6 shows the $\alpha(B^2)$-curve in heavy water (99.82% D_2O, 22 °C). KUSSMAUL and MEISTER obtained their data in the range $B^2 < 5\times10^{-3}$ cm⁻² by Fourier analysis of the space-time distribution in a large cylindrical tank. They obtained the following values from a three-parameter fit of the data:

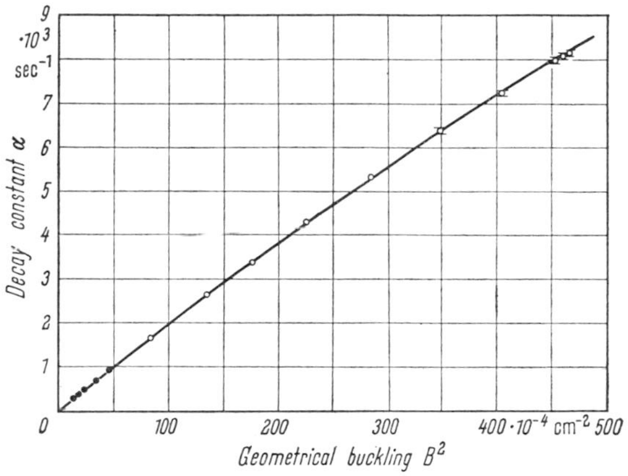

Fig. 18.1.6. The decay constant α *vs.* geometrical buckling B^2 for D_2O (99.82%; 22 °C) measured by KUSSMAUL and MEISTER. • • Measured by Fourier analysis of the space-time neutron distribution in a large D_2O tank. ○ ○ Measured by observing the fundamental mode decay in smaller vessels

$$D_0 = (2.00\pm0.01)\times10^5\ \mathrm{cm^2/sec}$$
$$\alpha_0 = 19.0\pm2.5\ \mathrm{sec^{-1}}$$
$$C = (5.25\pm0.25)\times10^5\ \mathrm{cm^4/sec}.$$

KUSSMAUL and MEISTER were able to show that in the region of B^2 covered by their measurements the B^6-term was negligible. HONECK obtained theoretically (i.e., by numerical solution of the transport equation for the Nelkin model adapted to D_2O) $D_0 = 2.069\times10^5$ cm² sec⁻¹ and $C = 4.852\times10^5$ cm⁴ sec⁻¹, cf. Table 10.3.4.

GANGULY and WALTNER have investigated the temperature dependence of the diffusion coefficient in D_2O in the temperature range from 20 to 50 °C by the pulsed neutron method; their results have already been shown in Fig. 17.2.2.

Graphite. ANTONOV *et al.*, BECKURTS, LALANDE, STARR and PRICE, and KLOSE, KÜCHLE, and REICHARDT have done pulsed neutron experiments on graphite. Fig. 18.1.7 shows the $\alpha(B^2)$-curve according to two recent experiments[1]. The two

[1] The experiments were carried out with graphite samples of different purities and densities. In order to eliminate the differences thereby produced, the results were corrected to a (hypothetical) non-absorbing graphite of density 1.6 g/cm³, i.e., $\frac{1.6}{\varrho}(\alpha-\alpha_0)$ was plotted against $\left(\frac{1.6}{\varrho}\right)^2 B^2$.

series of measurements are mutually consistent; some of the earlier measurements showed a weaker downward curvature at large B^2 which is presumably traceable to erroneous α-measurements resulting from an insufficient delay time. The evaluation of the diffusion parameters from the $\alpha(B^2)$-curve in graphite has still not been fully clarified. We obtain different values of D_0 and C according to the length of the B^2-interval we use in the analysis. This is demonstrated in Fig. 18.1.8 for a three- and a four-parameter evaluation of the data of KLOSE, KÜCHLE and REICHARDT. In the three-parameter evaluation, obviously it is only sensible to use the curve out to values of B^2 of 6×10^{-3} cm^{-2}; we then obtain[1] $\alpha_0 = 88.3 \pm 1.2$ sec^{-1}, $D_0 = (2.13 \pm 0.02) \times 10^5$ cm^2/sec, and $C = (26 \pm 5) \times 10^5$ cm^4/sec (density $= 1.6$ g/cm^3, $T_0 = 20$ °C). A four-parameter fit is obviously possible over the entire B^2-interval and gives

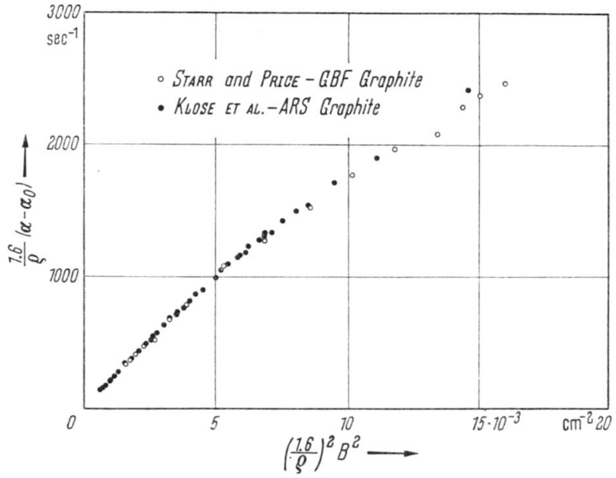

Fig. 18.1.7. The decay constant α *vs.* geometrical buckling B^2 in non-absorbing graphite of density 1.6 g · cm^{-3} at room temperature

$$\alpha_0 = 88.6 \pm 1.6 \text{ sec}^{-1} \qquad D_0 = (2.11 \pm 0.02) \times 10^5 \text{ cm}^2/\text{sec}$$
$$C = (16 \pm 5) \times 10^5 \text{ cm}^4/\text{sec} \qquad F = -(20 \pm 10) \times 10^7 \text{ cm}^6/\text{sec}.$$

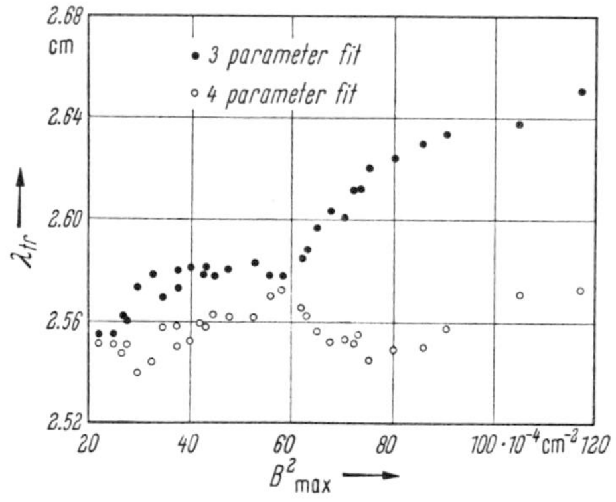

Fig. 18.1.8a. The transport mean free path for thermal neutrons in graphite of density 1.6 g · cm^{-3} as a function of the largest value of B^2 used in evaluating the α *vs.* B^2 curve

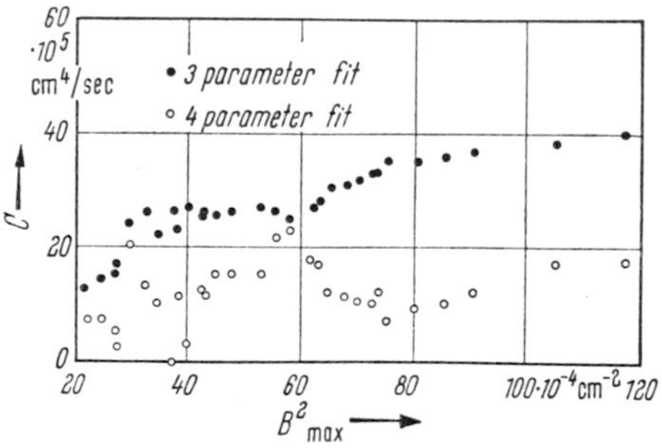

Fig. 18.1.8b. The diffusion cooling constant in graphite of density 1.6 g · cm^{-3} as a function of the largest value of B^2 used in evaluating the α *vs.* B^2 curve

Table 18.1.2 contains some additional data which were obtained by making three-parameter fits to the measured values of $\alpha(B^2)$. The reason for the low values of C reported by ANTONOV *et al.* and by STARR and PRICE (1959) is that, as just

[1] In the evaluation of the diffusion parameters, the pulsed neutron data was augmented by a measured value of the diffusion length obtained in a sigma pile.

mentioned, there were probably errors in the measurements on small assemblies; in contrast, the low C-value reported by BECKURTS can conceivably be explained by his use of too small a B^2-interval.

Table 18.1.2. *The Diffusion Parameters of Graphite (Density: 1.6 g/cm³) as Obtained from Three-Parameter Fits of the $\alpha(B^2)$-Curve*[1]

Author	Year	α_0 [sec⁻¹]	D_0 [cm² sec⁻¹]	C [cm⁴ sec⁻¹]	B^2-Interval [cm⁻²]
ANTONOV *et al.*	1955	82 ± 3	$(2.07 \pm 0.03) \times 10^5$	$(12.5 \pm 2.0) \times 10^5$	$(1—40) \times 10^{-3}$
BECKURTS .	1956	127 ± 1	$(2.13 \pm 0.02) \times 10^5$	$(16.3 \pm 2.5) \times 10^5$	$(0.7— 5.5) \times 10^{-3}$
STARR and PRICE . .	1959	71.2 ± 0.9	$(2.06 \pm 0.02) \times 10^5$	$(12.4 \pm 2.2) \times 10^5$	$(1.6—27.5) \times 10^{-3}$
STARR and PRICE (GBF graphite) .	1962	75.0 ± 0.6	$(2.14 \pm 0.01) \times 10^5$	$(39 \pm 3) \times 10^5$	$(1.76—18.9) \times 10^{-3}$

[1] $\alpha_0(\varrho_1) = \frac{\varrho_1}{\varrho_2}\,\alpha_0(\varrho_2)$; $D_0(\varrho_1) = \frac{\varrho_2}{\varrho_1}\,D_0(\varrho_2)$; $C(\varrho_1) = \left(\frac{\varrho_2}{\varrho_1}\right)^3 C(\varrho_2)$.

For graphite of density 1.6 g/cm³, HONECK obtains theoretically $D_0 = 2.178 \times 10^5$ cm² sec⁻¹, $C = 24.6 \times 10^5$ cm⁴ sec⁻¹, and $F = -8.3 \times 10^7$ cm⁶ sec⁻¹ using PARKS' model, cf. Table 10.3.4. TAKAHASHI's method (cf. Sec. 10.3.2) yields $C = 2.48 \times 10.8 \times 10^5$ cm⁴ sec⁻¹ $= 26.8 \times 10^5$ cm⁴ sec⁻¹, while the elementary theory for a free gas of mass 12 would yield $C = 4 \times 10^5$ cm⁴ sec⁻¹ (cf. Table 10.3.1).

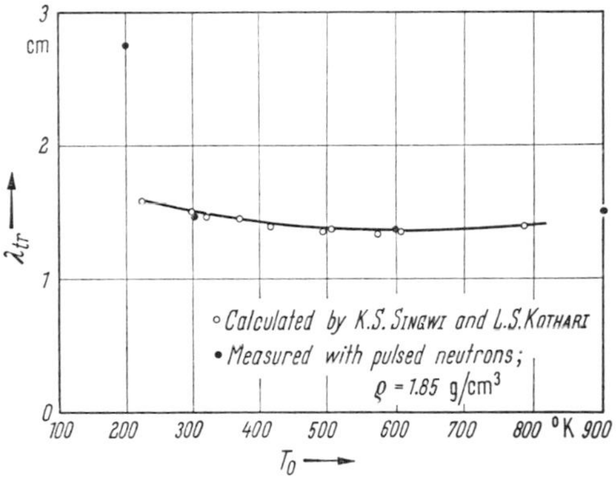

Fig. 18.1.9. The transport mean free path of thermal neutrons in beryllium as a function of the moderator temperature as measured by DE SAUSSURE and SILVER

Beryllium. Table 18.1.3 contains some data on the diffusion parameters of beryllium metal at room temperature. There are marked discrepancies in the diffusion cooling constants, which are probably due to inaccurate measurements at large B^2.

Table 18.1.3. *Diffusion Parameters of Beryllium Metal (Density: 1.85 g/cm³) as Obtained from Three-Parameter Fits of the $\alpha(B^2)$-Curve*

Author	Year	α_0 [sec⁻¹]	D_0 [cm² sec⁻¹]	C [cm⁴ sec⁻¹]	B^2-Interval [cm⁻²]
ANTONOV *et al.* . .	1955		$(1.17 \pm 0.05) \times 10^5$	$(2.93 \pm 1.0) \times 10^5$	0.008—0.062
CAMPBELL and STELSON	1956	150	1.25×10^5	0	0.005—0.036
KLOVERSTROM . .	1958	270 ± 19	$(1.24 \pm 0.04) \times 10^5$	$(3.90 \pm 0.8) \times 10^5$	0.003—0.041
DE SAUSSURE and SILVER	1959	288 ± 60	$(1.25 \pm 0.06) \times 10^5$	$(1.40 \pm 1.0) \times 10^5$	0.008—0.072
ANDREWS	1960	285 ± 8	$(1.235 \pm 0.013) \times 10^5$	$(2.80 \pm 0.3) \times 10^5$	0.003—0.075

Using the method of TAKAHASHI (Sec. 10.3.2) we find $C = 3.36 \times 9 \times 10^4$ cm⁴ sec⁻¹ $\approx 3 \times 10^5$ cm⁴ sec⁻¹, while for a gas of mass 9 (Table 10.3.1) we find $C = 5.6 \times 10^4$ cm⁴ sec⁻¹. DE SAUSSURE and SILVER and also ANDREWS have studied the temperature dependence of diffusion in beryllium by the pulsed neutron method.

Fig. 18.1.9 shows the observed temperature dependence of λ_{tr} compared to a calculation of SINGWI and KOTHARI.

Other Moderators. A variety of results of pulsed neutron experiments on additional moderators are collected in Table 18.1.4. The temperature dependence was also measured for Dowtherm *A* and BeO up to 200 °C.

Table 18.1.4. *The Diffusion Parameters of Various Moderators at Room Temperature Obtained by the Pulsed Neutron Method*

Substance	Density [g/cm³]	α_0 [sec⁻¹]	D_0 [cm² sec⁻¹]	C [cm⁴ sec⁻¹]	B^2-Interval [cm⁻²]	Author
Dowtherm *A*	1.062	2,870±40	49,200±600	11,900±2100	0.025—0.25	KÜCHLE
Plexiglass	1.18	4,300	34,000	—	0.04 —0.7	SEEMANN
BeO	2.96	131.5	(1.18±0.02) $\times 10^5$	(3.85±0.08) $\times 10^5$	0.02 —0.04	IYENGAR
Zirconium Hydride	3.49	3,765±89	57,900±320	(1.58±0.27) $\times 10^5$	0.03 —0.11	MEADOWS and WHALEN
Paraffin	0.87	5,620±150	27,000±900	2,000±1,200	0.071—0.672	KÜCHLE
Polyethylene	0.918	5,900±90	26,600±500	2,600±800	0.098—0.143	SJÖSTRAND *et al.*

18.1.4. Determination of the Geometrical Buckling by the Pulsed Neutron Method

The normal procedure in pulsed neutron experiments is to measure the decay constant α of a geometrically simple assembly for which B^2 can be easily calculated. If we have once determined the $\alpha(B^2)$-curve for a particular moderator in this way, we can then invert the procedure. We can build assemblies with an arbitrarily complicated geometry, measure their decay constants, and by means of the $\alpha(B^2)$-curve determine their geometric buckling. Such experimental determinations are of interest for assemblies in which a solution of the equation $\nabla^2\Phi + B^2\Phi = 0$ is either impossible or possible only at the expense of a great deal of computational labor. Among such assemblies are homogeneous bodies with complicated surfaces (e.g., cylinders with rounded ends) or assemblies in which localized strong absorbers are found (e.g., rods of cadmium that are supposed to simulate the effect of control rods in reactors). Such experiments have been carried out by BECKURTS and by SJÖSTRAND *et al.*, among others. We shall not go into the results, which apply only to special configurations, here.

18.1.5. Absorption Measurements with the Pulsed Neutron Method

a) Large Assemblies. We can determine the absorption cross section of a substance by means of the decay of the neutron field in an "infinite medium" of the substance (practically speaking, such "infinite media" can only be realized in hydrogenous substances). If we integrate the diffusion equation $\frac{1}{v}\frac{\partial\Phi}{\partial t} = D\nabla^2\Phi - \Sigma_a\Phi$ over all space, the diffusion term vanishes and we obtain the following equation for the integrated flux:

$$\frac{1}{\bar{v}}\frac{\partial\bar{\Phi}}{\partial t} = -\bar{\Sigma}_a\bar{\Phi} \tag{18.1.4a}$$

i.e.,

$$\bar{\Phi}(t) = e^{-\overline{v\Sigma_a}t}. \tag{18.1.4b}$$

Thus we must determine the flux $\Phi(\mathbf{r}, t)$ in the vicinity of a pulsed source, integrate it over all space, and plot the logarithm of the integrated flux against t; the result is a straight line with the slope $\overline{v\Sigma_a}$. VON DARDEL and WALTNER, MEADS *et al.*, and RAMANNA *et al.* have carried out such experiments on water. If boron counters are used for the flux measurements, under some circumstances a correction is necessary to account for the effect of the additional absorption in the counter. We can circumvent this correction (MEADS) by using as a detector a liquid scintillator that is mounted on a plexiglas light pipe. This detector does not detect neutrons but only the 2.2-Mev γ-rays produced by neutron capture in hydrogen. Since these γ-rays have a mean free path in water of *ca.* 20 cm, an integration over a certain limited region is already automatically performed. Using this method, MEADS *et al.* found that in water $1/\overline{v\Sigma_a} = 203.3 \pm 2.6$ μsec.

b) Measurements on Dissolved Absorber Substances. The pulsed neutron method permits direct absolute determination of the absorption cross section of any absorber substance that can be dissolved in water. First we make a measurement in the absence of the absorber, i.e., we observe the decay of the neutron field in a vessel filled with pure water (naturally, other solvents are possible) and determine the decay constant α^0. Then the measurement is repeated in the same geometry with an absorber solution. If we are dealing with a $1/v$-absorber, so that no spectral effects occur, then the decay constant α^1 of the absorber solution is given by

$$\alpha^1 = \alpha^0 + N\,\overline{v\sigma_a} + \Delta N_H \overline{v\sigma_H} + \Delta D_0 B^2. \qquad (18.1.5)$$

Table 18.1.5. *Absorption Cross Sections of 1/v-Absorbers Obtained from Pulsed Neutron Measurements* (MEADOWS and WHALEN)

Element	σ_a (2200 m/sec) [barn]
Li	70.4 ± 0.4
B	758 ± 4
B^{10}	3843 ± 17
Na	0.47 ± 0.06
Cl	33.6 ± 0.3
Mn	13.2 ± 0.1
Co	36.3 ± 0.6
Se	11.7 ± 0.1
Br	6.82 ± 0.06
I	6.22 ± 0.15
Nd	$49.9^{+0.3}_{-2.2}$
Dy	936 ± 20

Here σ_a is the absorption cross section of the dissolved substance, N is the number of absorber atoms per cm³, ΔN_H is the change in the number of H-atoms per cm³ due to the addition of the absorber substance, and ΔD_0 is the change in the diffusion constant. N and ΔN_H can be determined by chemical analysis and density measurement, and ΔD_0 can be calculated from the scattering cross section. [For strong absorbers, even a small absorber concentration has a large effect on the decay constant, and the last two terms in Eq. (18.1.5) can be neglected compared to $N\overline{v\sigma_a}$.] Thus, having measured α^0 and α^1, we can determine $\overline{v\sigma_a}$ from Eq. (18.1.5) and thus obtain[1] $\sigma_a(2200\text{ m/sec}) = \frac{\overline{v\sigma_a}}{2200\text{ m/sec}}$. VON DARDEL and SJÖSTRAND determined the absorption cross section of boron in this way (cf. also SCOTT, THOMSON, and WRIGHT). Recently, MEADOWS and WHALEN carried out very careful measurements on a number of substances. Some of their results are summarized in Table 18.1.5.

Additional corrections must be applied to the data for substances whose absorption cross section shows deviations from the $1/v$-law. To begin with, at

[1] In principle, separate determination of α^1 and α^0 is not necessary. It suffices to plot the ratio of the two decay curves point by point; the curve so obtained corresponds to the decay constant $\alpha^1 - \alpha^0$. This procedure presumes that higher spatial harmonics do not contribute strongly to the measurement.

high concentrations of a strong non-$1/v$-absorber (like Cd, Gd, or Sm) there is a clear effect on the neutron spectrum. We must therefore carry out measurements of the absorption cross section at various absorber concentrations and then extrapolate to zero absorber concentration (cf. Fig. 18.1.10). However, the absorption cross section obtained in this way is not an average over a Maxwell spectrum, but rather over the asymptotic neutron spectrum in the experimental vessel, which according to the size of the vessel is shifted by the diffusion cooling effect more or less strongly in the direction of lower energies. Fortunately, it is quite simple to apply a correction for the diffusion cooling effect (cf. MEADOWS and WHALEN). We then obtain $\overline{v\sigma_a} = v_0 \cdot g(T)\ \sigma_a(2200\ \text{m/sec})$. Here $g(T)$ is WESTCOTT's factor, and $v_0 = 2200$ m/sec. Table 18.1.6 contains some results of measurements by MEADOWS and WHALEN.

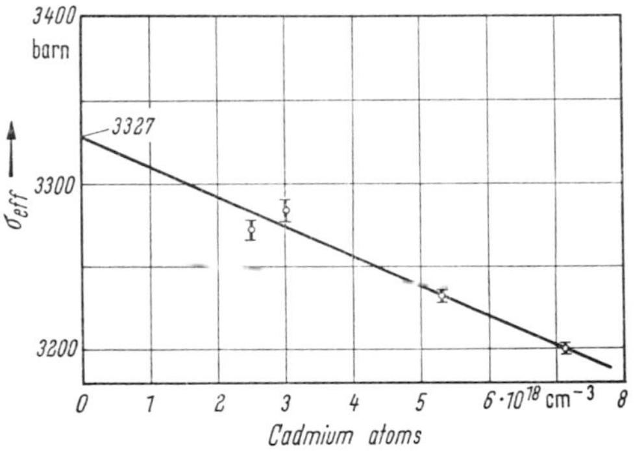

Fig. 18.1.10. The effective thermal neutron absorption cross section of cadmium *vs.* cadmium concentration in water (MEADOWS and WHALEN)

Table 18.1.6. *Absorption Cross Sections of Non-1/v-Absorbers Obtained from Pulsed Neutron Measurements* (MEADOWS and WHALEN)

Substance	$g(T)$ at $T = 20$ °C	σ_a (2200 m/sec) [barns]
Ag	1.004	64.8 ± 0.4
Cd	1.338	2537 ± 9
In	1.020	194 ± 2
Sm	1.638	5828 ± 30
Eu	0.999	4406 ± 30
Gd	0.888	46617 ± 100
Hf	1.020	101.4 ± 0.5
Au	1.005	98.2 ± 0.5
Hg	0.998	374 ± 5

18.2. Measurement of the Diffusion Constant by the Method of Modulated Sources

The study of diffusion by the propagation of neutron waves through scattering media was suggested by WIGNER (cf. also WEINBERG and NODERER) and first carried out by RAIEVSKI and HOROWITZ. The principle of the method follows immediately from the treatment of neutron waves in Sec. 9.4. At a distance r from a sinusoidally modulated source $\left(Q(t) = Q_0 + \delta Q e^{i\omega t}\right)$ in an infinite medium the response is given by

$$\frac{\delta\Phi(r,t)}{\Phi_0(r)} = \frac{\delta Q}{Q_0} e^{r/L} \exp\left[-\frac{\left(1+\frac{\overline{v\Sigma_a}}{2\omega}\right)}{\sqrt{2D_0/\omega}} r + i\left(\omega t - \frac{\left(1-\frac{\overline{v\Sigma_a}}{2\omega}\right)}{\sqrt{2D_0/\omega}} r\right)\right]. \tag{18.2.1}$$

Eq. (18.2.1) holds under the condition that the modulation frequency ω is large compared to the absorption rate $\overline{v\Sigma_a}$ but small compared with the collision rate $v\Sigma_s$. It was derived in Sec. 9.4 for monoenergetic neutrons; in the following we shall assume that it also holds for modulated thermal neutron fields. For the *absolute value* of the response we clearly have

$$\ln\left|\frac{\delta\Phi}{\Phi_0}\right| = \ln\frac{\delta Q}{Q_0} + \frac{r}{L} - \frac{r}{\sqrt{2D_0}}\sqrt{\omega}\left(1+\frac{\overline{v\Sigma_a}}{2\omega}\right) \tag{18.2.2}$$

while for the phase angle we have

$$\varphi = \operatorname{arc}\frac{\delta\Phi}{\Phi_0} = -\frac{r}{\sqrt{2D_0}}\sqrt{\omega}\left(1-\frac{v\Sigma_a}{2\omega}\right). \tag{18.2.3}$$

If we measure $\delta\Phi/\Phi_0$ at a fixed point r for various frequencies ω and either plot $\ln\left|\frac{\delta\Phi}{\Phi_0}\right|$ as a function of $x=\sqrt{\omega}\left(1+\frac{v\Sigma_a}{2\omega}\right)$ or φ as a function of $y=\sqrt{\omega}\left(1-\frac{v\Sigma_a}{2\omega}\right)$, the result will be a straight line with slope $-\frac{r}{\sqrt{2D_0}}$. If we repeat the measurements at various points r and plot the slopes so obtained against r, we again obtain a straight line, but this time with the slope $-\frac{1}{\sqrt{2D_0}}$. In order to calculate x and y, we must know the absorption cross section of the moderator, but since $v\Sigma_a \ll 2\omega$, an approximate value is enough.

18.2.1. Production and Measurement of Modulated Neutron Fields

Fig. 18.2.1 shows the apparatus used by HOROWITZ and RAIEVSKI for experiments on graphite. By periodically narrowing a thermal neutron beam, a

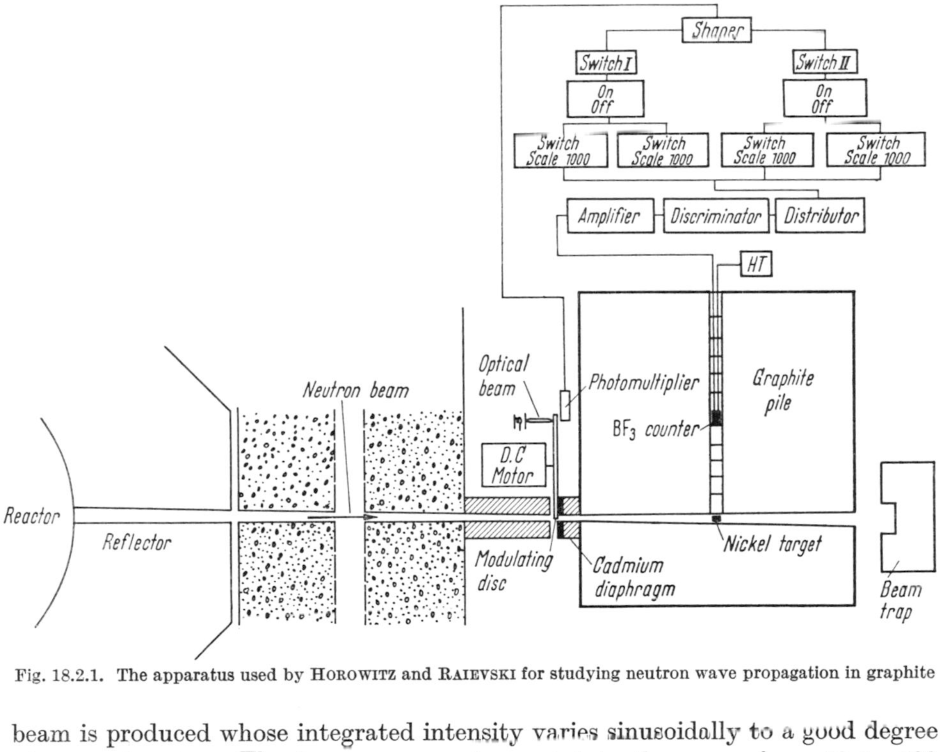

Fig. 18.2.1. The apparatus used by HOROWITZ and RAIEVSKI for studying neutron wave propagation in graphite

beam is produced whose integrated intensity varies sinusoidally to a good degree of approximation. The frequency can be varied in the range from 50 to 500 cycles/sec. The neutron beam is incident on a scatterer in the middle of the moderator being studied. In first approximation, the scatterer can be considered as a sinusoidally modulated point source. Nickel (and later polyethylene) was used as the scatterer.

We can also modulate a radioactive (γ, n) source. To do so, we can, for example, make the source out of an antimony core and an outer mantle consisting of several segments of beryllium. Between the core and mantle we place a rotating absorber (Fig. 18.2.2) which periodically attenuates the γ-intensity falling on the beryllium. Such a device has been used for experiments in heavy water.

Also shown in Fig. 18.2.1 is the electronic apparatus for recording and analyzing the time-dependent neutron flux. The neutron flux is measured with a BF_3-counter whose pulses are led into a bank of four scalers after amplification. These scalers are turned on and off by an electronic system photoelectrically synchronized with the modulator in such a way that they respectively record Z_1+Z_2, Z_3+Z_4, Z_2+Z_3, and Z_4+Z_1. Here Z_i is the counting rate in the i-*th* quarter of the modulation cycle. Since

$$Z_i \sim \int_{(i-1)\frac{\pi}{2\omega}}^{i\frac{\pi}{2\omega}} \Phi(t)\, dt \tag{18.2.4}$$

we have

$$\varphi = \arctan \frac{(Z_2+Z_3)-(Z_4+Z_1)}{(Z_1+Z_2)-(Z_3+Z_4)} \tag{18.2.5a}$$

and

$$\left|\frac{\delta\Phi}{\Phi_0}\right|^2 = \frac{[(Z_1+Z_2)-(Z_3+Z_4)]^2+[(Z_2+Z_3)-(Z_4+Z_1)]^2}{[(Z_1+Z_2)+(Z_3+Z_4)]^2} \cdot \frac{\pi^2}{4}. \tag{18.2.5b}$$

In this method of integration, all even harmonics (which can be excited by deviations of the source modulation from strict sinusoidal form) are automatically eliminated.

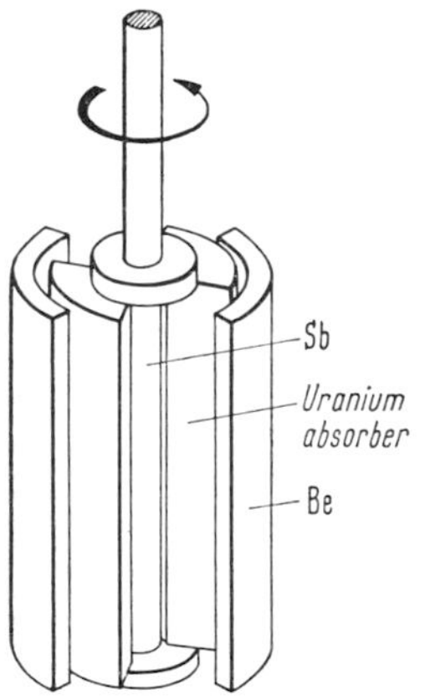

Fig. 18.2.2. A periodically modulated (Sb—Be) source (RAIEVSKI)

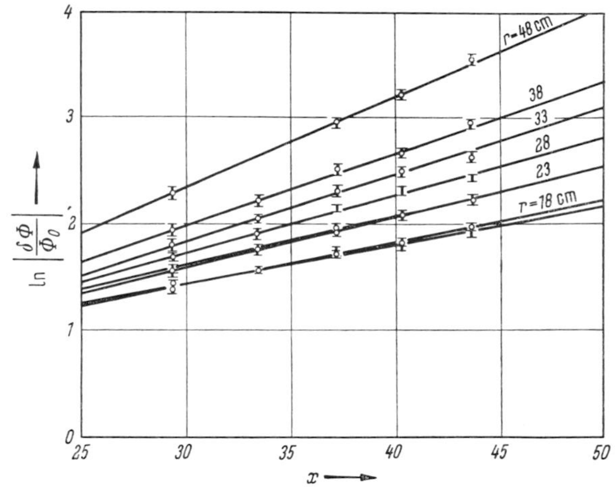

Fig. 18.2.3. $\ln\left|\frac{\delta\Phi}{\Phi_0}\right|$ *vs.* x at various source distances in graphite

18.2.2. Modulation Experiments on Graphite and D_2O

Graphite was first studied by RAIEVSKI and HOROWITZ and later by DROULERS, LACOUR, and RAIEVSKI. The experimental apparatus has already been shown in Fig. 18.2.1. We assume that the graphite pile is sufficiently large that it may justifiably be considered an infinite medium. Fig. 18.2.3 shows the measured values of $\ln\left|\frac{\delta\Phi}{\Phi_0}\right|$ plotted as a function of $x=\sqrt{\omega}\left(1+\frac{v\Sigma_a}{2\omega}\right)$ for various distances r. The plots are quite straight lines, whose slopes are plotted as a function of r in

Fig. 18.2.4. From the slope of the line in Fig. 18.2.4, we obtain $D_0 = (2.09 \pm 0.03) \times 10^5$ cm^2/sec for graphite of density 1.6 g/cm^3 at room temperature. We can also determine D_0 from the phase angle φ. To do so, we plot the measured values of φ against r for various values of $y = \sqrt{\omega}\left(1 - \frac{v\Sigma_a}{2\omega}\right)$ (Fig. 18.2.5), determine the slope of the resulting lines, and plot these slopes as a function of y (Fig. 18.2.6). From the slope of the resulting line follows the value $D_0 = (2.09 \pm 0.03) \times 10^5$ cm^2 sec^{-1}, in good agreement with the value derived from the absolute value of the response. The reason that we proceed differently in evaluating the phase measurements than we do in evaluating the response measurements is the following. The measurement of the phase angle is obviously done relative to the phase of the mechanical modulator. However, Eq. (18.2.3) does not apply to this phase angle; owing to the neutron flight time between the modulator and the scatterer, an additional, frequency-dependent phase shift enters. This latter term is eliminated, however, when we plot against r first.

In a similar way, Raievski and Horowitz found $D_0 = (2.00 \pm 0.05) \times 10^5$ cm^2 sec^{-1} for *heavy water* (100%) at 13 °C.

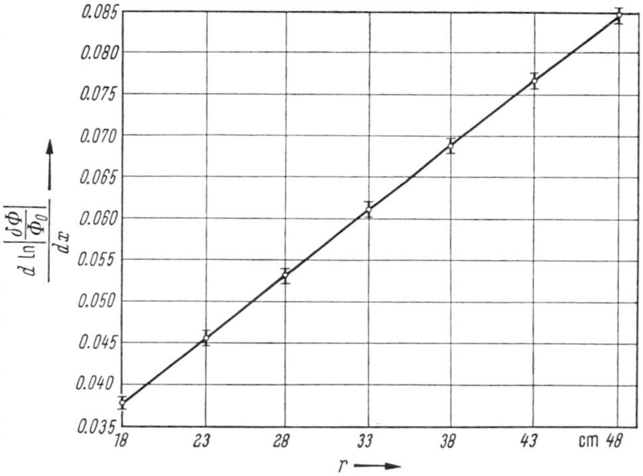

Fig. 18.2.4. $\frac{d \ln \left|\frac{\delta\Phi}{\Phi_0}\right|}{dx}$ *vs.* source distance r in graphite

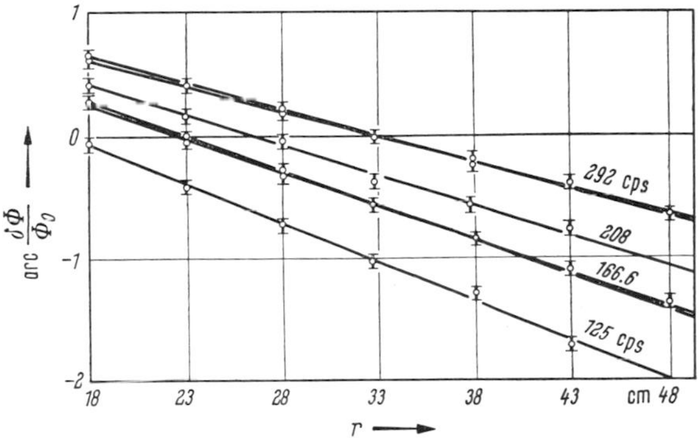

Fig. 18.2.5. The phase angle arc $\frac{\delta\Phi}{\Phi_0}$ *vs.* source distance r at various modulation frequencies in graphite

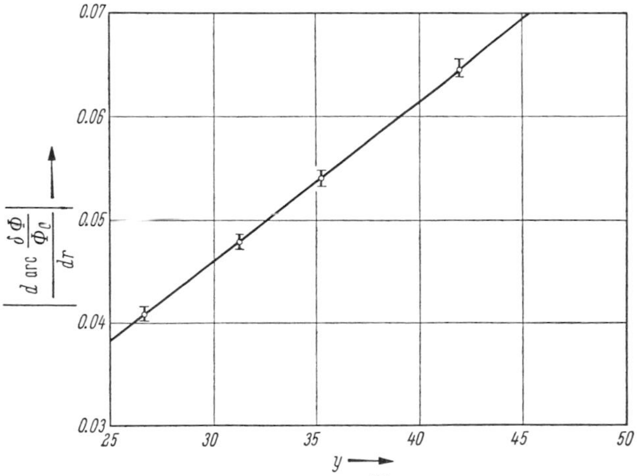

Fig. 18.2.6. $\left|\frac{d \operatorname{arc} \frac{\delta\Phi}{\Phi_0}}{dr}\right|$ *vs.* y

The measured value of the diffusion constant in graphite is somewhat lower than the value obtained in most pulsed neutron measurements. This is conceivably connected with the fact that spectral effects also occur in modulation experiments so that the simple theory developed here no longer holds and additional corrections are necessary. A thorough theoretical study of such effects has only recently been done (cf. PEREZ and UHRIG).

18.3. The Study of Neutron Thermalization by the Pulsed Source Method

Fig. 18.3.1 shows schematically the "life history" of a neutron in a moderating medium. We differentiate the epithermal or slowing-down region, the much longer-lasting thermalization region, and the asymptotic region, in which the average neutron energy no longer changes with time. The pulsed neutron method is eminently well suited for the study of this life history since it allows us to determine the spectrum, or at least certain averages over the spectrum, as a function of the chronological "age" of the neutrons. We have already dealt with some such experiments in the epithermal region (time dependence of the slowing down to indium and cadmium resonance) in Sec. 16.3.1. These studies serve principally to corroborate the elementary theory of slowing down by free, stationary atomic nuclei. In contrast, the experiments in the thermalization and asymptotic ranges to be discussed here have a deeper physical significance in that they demonstrate clearly the role played by chemical binding and, within certain limits, allow us to check the various theories of thermalization.

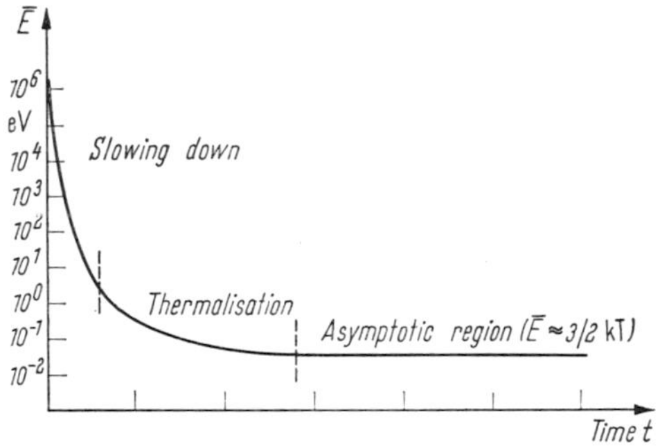

Fig. 18.3.1. The life history of a neutron in a moderating medium

18.3.1. Integral Investigation of the Time-Dependent Spectrum

The classical method of studying the time dependence of the thermalization process is to observe the time-dependent transmission of the neutrons. To this end, the time variation of the intensity of the neutrons leaving a medium is observed once with (Z^+) and once without (Z^0) an absorber foil between the medium and the detector and the ratio $Z^+(t)/Z^0(t)$ formed. The absorber of choice is a $1/v$-absorber such as boron-containing glass. As an example of the results of such an experiment, Fig. 18.3.2 shows transmission curves of water and ice at various temperatures (VON DARDEL). We can see clearly the establishment of the stationary spectrum at the end of the thermalization process, which with decreasing temperature requires more and more time. Similar studies have been carried out on beryllium by ANTONOV *et al.*, on graphite by ANTONOV *et al.* and by BECKURTS, on beryllium oxide by IYENGAR *et al.*, and on zirconium hydride by MEADOWS and WHALEN.

While transmission experiments give us useful qualitative information, it is difficult to obtain reliable quantitative data from them. Most authors have taken a Maxwell energy distribution for granted and calculated neutron temperatures from the measured transmission (cf. Sec. 15.2.1). This procedure yields curves of the neutron temperature T as a function of the time. As an example, $T(t)$ in graphite is shown in Fig. 18.3.3. Using the elementary theory of Sec. 10.4.1, we can approximate this curve by

$$T(t) - T_0 \sim e^{-t/t_T} \quad (10.2.15\text{f})$$

and thus obtain the *temperature relaxation time*, t_T. Table 18.3.1 shows values of t_T obtained in this way.

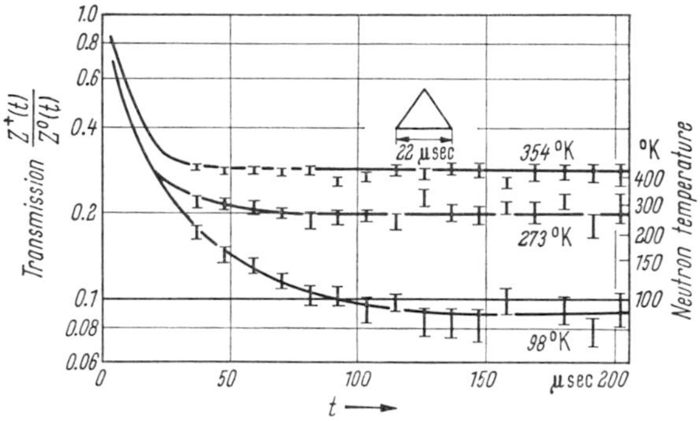

Fig. 18.3.2. The time-dependent transmission of neutrons from ice at 98 and 273 °K and from water at 354 °K through an absorber containing 24.4 mg/cm² boron (von Dardel)

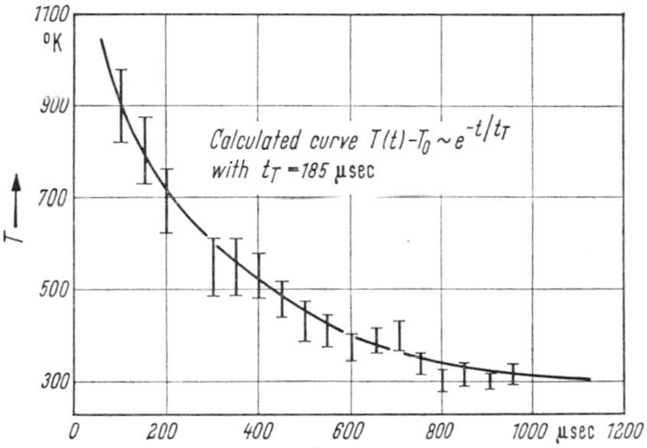

Fig. 18.3.3. $T(t)$ in graphite as calculated from the time-dependent transmission through silver absorbers

There is a whole series of rather serious objections to this procedure. The assumption of a Maxwell spectrum is certainly only permissible in the immediate neighborhood of the equilibrium state, and then only in the interior rather than at the surface of the medium, where most experiments have been carried out. Also the angular distribution of the neutrons leaving the surface, about which we know little, affects many transmission experiments. Spatial harmonics and time-of-flight effects influence the measurement in a way that is hard to account for. Finally, the concept of the temperature relaxation time is itself very crude. One should not therefore attach too much significance to the results in Table 18.3.1, with which other measurements (see below) are in contradiction and which furthermore do not agree at all well with the theoretically calculated values (cf. Table 10.4.1).

Table 18.3.1. *The Temperature Relaxation Time as Obtained from Transmission Experiments*

Substance	t_T (µsec)	Author
H_2O	≈ 7	von Dardel
Be (1.85 g/cm³) . .	121 ± 50	Antonov *et al.*
Graphite (1.6 g/cm³)	183 ± 38	Antonov *et al.*
	185 ± 54	Beckurts
BeO (2.96 g/cm³) .	165 ± 10	Iyengar *et al.*

In contrast to the temperature relaxation time t_T, the *thermalization time* t_{th} is a well-defined quantity. According to Sec. 10.4.2, it is $1/\alpha_1$, where α_1 is the first eigenvalue of the thermalization operator for the time-dependent, absorption-free problem. t_{th} can be cleanly measured. To see how this is done, let us consider an infinite medium with $1/v$-absorption. The time dependence of the spectrum is then given by[1]

$$\left.\begin{aligned}\Phi(E,t) &= M(E)\,e^{-\overline{v\Sigma_a}t} + \Phi_1(E)\,e^{-\left(\overline{v\Sigma_a}+\frac{1}{t_{th}}\right)t} + \\ &\quad + \text{higher terms which decay rapidly.}\end{aligned}\right\} \tag{18.3.1}$$

If we observe the flux with a detector whose energy-dependent sensitivity is $\Sigma(E)$, then the time-dependent counting rate is given by

$$\left.\begin{aligned}Z(t) &= Z_0\,e^{-\overline{v\Sigma_a}t} + Z_1\,e^{-\left(\overline{v\Sigma_a}+\frac{1}{t_{th}}\right)t} + \\ &\quad + \text{higher terms which decay rapidly}\end{aligned}\right\} \tag{18.3.2}$$

with

$$Z_0 = \int \Sigma(E)\,M(E)\,dE; \qquad Z_1 = \int \Sigma(E)\,\Phi_1(E)\,dE. \tag{18.3.3}$$

We now obtain t_{th} by determining the decay constant of the first energy mode in addition to that of the fundamental mode. The accuracy of this determination can be greatly increased by making measurements with a second detector with another sensitivity function $\Sigma'(E)$. Then

$$Z'(t) = Z_0'\,e^{-\overline{v\Sigma_a}t} + Z_1'\,e^{-\left(\overline{v\Sigma_a}+\frac{1}{t_{th}}\right)t}.$$

Since $Z_1'/Z_0' \neq Z_1/Z_0$ we can find a constant a such that

$$\left.\begin{aligned}Z'(t) - a\,Z(t) &\sim e^{-\left(\overline{v\Sigma_a}+\frac{1}{t_{th}}\right)t} + \\ &\quad + \text{higher terms which decay rapidly}\end{aligned}\right\} \tag{18.3.4}$$

i.e., we can eliminate the fundamental mode.

Möller and Sjöstrand have determined the thermalization time in a large water assembly by this method. The neutrons were detected by means of the capture γ-radiations of cadmium and gadolinium; these substances have absorption cross sections with extremely different energy dependences. They were present as solutions in a 250-cm³ vessel which could be moved freely inside the water tank. Fig. 18.3.4 shows the observed capture rates. They have already been normalized to yield the same fundamental mode decay for times $> 20\ \mu$sec. Thus if the difference of both curves is formed, the fundamental mode cancels; by plotting the difference on semilog paper, $t_{th} \approx 4\ \mu$sec was found by Möller and Sjöstrand. The value calculated from the Nelkin model (in the $k=2$ approximation, cf. Sec. 10.4.2 and Table 10.4.1) is also $4\ \mu$sec, in good agreement with the experiment.

Schweikert has done similar studies on graphite. Since an "infinite" medium could not be realized in this case, the relaxation constant of the first energy mode was determined in assemblies of various size; Fig. 18.3.5 shows the results as a function of B^2. One must be very careful in this experiment to eliminate the spatial harmonics. Schweikert used the arrangement shown in Fig. 18.1.4 with two counters. The neutron flux was measured once with bare BF_3 counters

[1] The case of a non-absorbing medium was considered in Sec. 10.4.2. We can easily convince ourselves that the introduction of a $1/v$-absorption does not change the eigenfunctions and merely increases the eigenvalues by $\overline{v\Sigma_a}$.

and once with BF_3 counters covered by silver absorbers in order to change their sensitivity function. Linear extrapolation of the data shown in Fig. 18.3.5 to $B^2=0$ gives $t_{th}\approx 550$ μsec (for density 1.6 g/cm³). The calculation by the method of TAKAHASHI yields $t_{th}=3\times 140$ μsec $=420$ μsec (cf. Sec. 10.4.2 and Table 10.4.1).

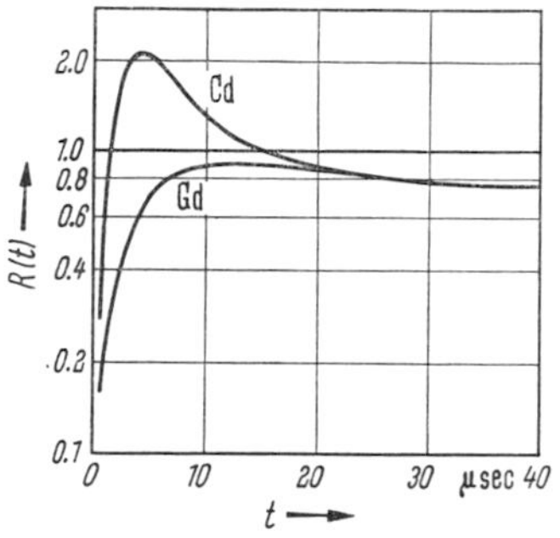

Fig. 18.3.4. The time-dependent reaction rate of cadmium and gadolinium absorbers dissolved in water (MÖLLER and SJÖSTRAND)

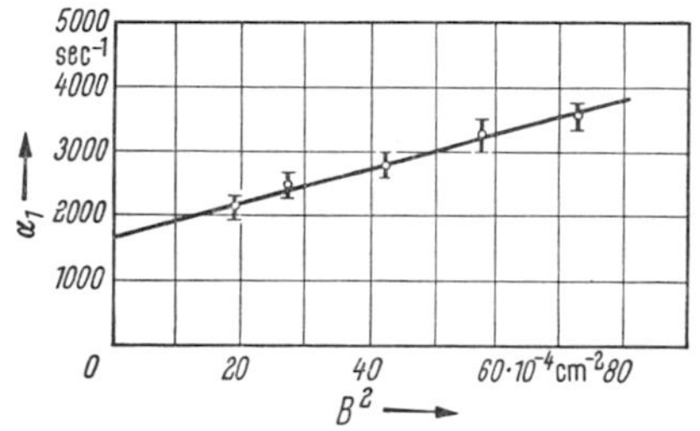

Fig. 18.3.5. The first higher eigenvalue α_1 in graphite as a function of geometrical buckling. Note that the two largest values of α_1 are above the "critical limit" (cf. Table 10.3.3)

In view of the rough extrapolation procedure and of the great inaccuracies of the measured data, the measured and calculated values of t_{th} are compatible.

MEADOWS and WHALEN have done similar studies on zirconium hydride; they found $t_{th}=194\pm 32$ μsec at a density of 3.48 g/cm³.

18.3.2. Direct Observations of the Time-Dependent Spectrum

BARNARD *et al.* have carried out detailed investigations of the time-dependent neutron spectrum in graphite. Their apparatus is shown in Fig. 18.3.6. The pulsed source — in this case a linear accelerator — is synchronized with a chopper in such a way that between each pulse injection and chopper opening a (variable)

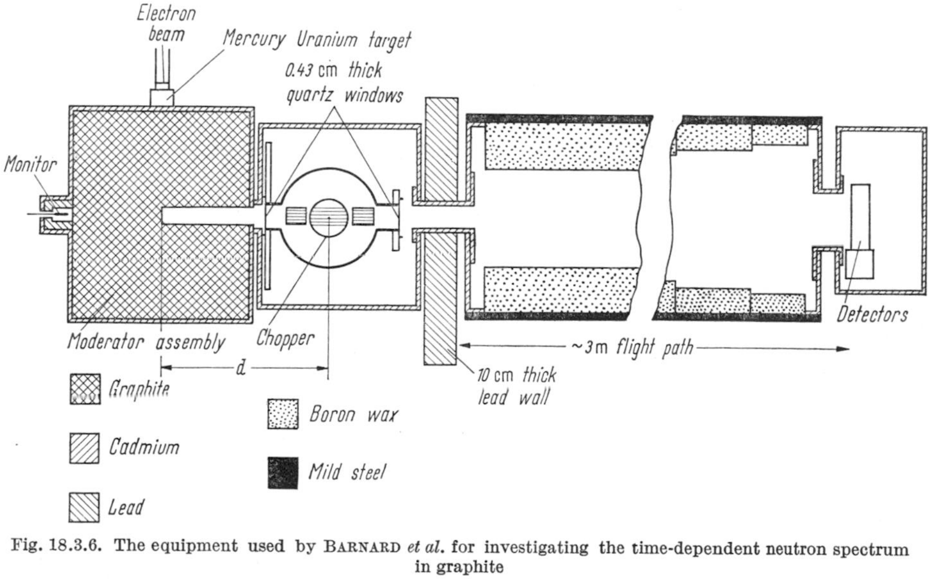

Fig. 18.3.6. The equipment used by BARNARD *et al.* for investigating the time-dependent neutron spectrum in graphite

delay time t elapses. The neutrons are taken out of the middle of a graphite block ($60 \times 62 \times 71$ cm^3) where there is no flux gradient, assuring that the spectrum of the emerging neutron beam corresponds to that of the flux. The spectrum is determined by the time-of-flight distribution between the chopper and the detector bank, as in the stationary chopper experiments described in Chapter 15. Such a measurement does not directly yield $\Phi(E, t)$ but instead gives the spectrum

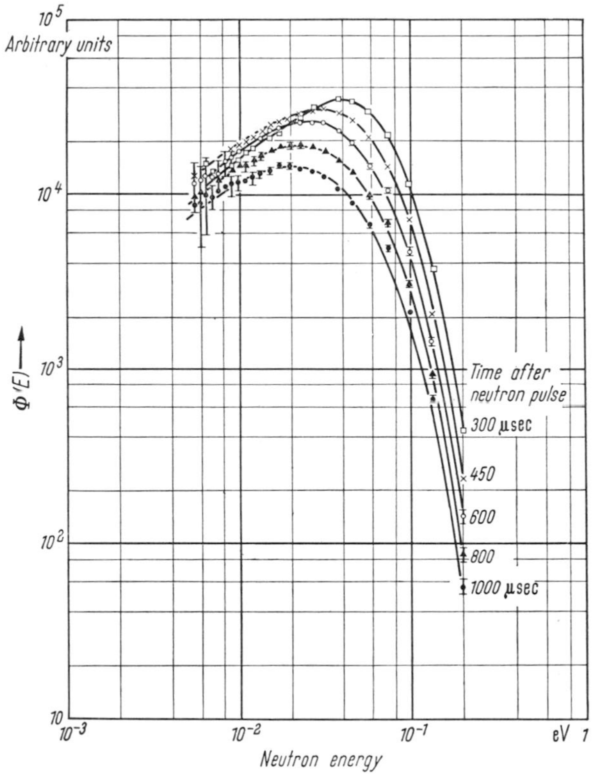

Fig. 18.3.7. The neutron spectrum in graphite at different times after the neutron burst injection

distorted by the neutron flight time between the middle of the block and the chopper, i.e., $\Phi(E, t - d/v)$. However, it is possible to get $\Phi(E, t)$ from the spectra observed at various delay times t. Fig. 18.3.7 shows measured values of $\Phi(E, t)$ in the range from $t = 300$ μsec to $t = 1000$ μsec. We can see clearly the increasing shift of the spectra to smaller energies; even for $t = 1000$ μsec the thermalization process is not completely finished. For reasons of intensity, measurements at longer times were not possible. In Figs. 18.3.8a—c the measured spectra are compared with calculations that were done by a multigroup method using experimental values of the inelastic scattering cross section of graphite; except in the case $t = 1000$ μsec the agreement is good.

If we wish to know the *asymptotic* spectrum at the end of the thermalization process, we can proceed as we did above, save that we must wait a sufficiently long time. In media with short neutron lifetimes like water, however, another procedure is possible. Instead of a chopper, a simple rotating shutter is used that prevents neutron leakage out of the medium during the slowing-down and thermalization processes. When the asymptotic spectrum is established, the shutter opens. Since the field decays in a time that is short compared to the flight time of the neutrons, it is not necessary to close the shutter again. BECKURTS has studied the asymptotic spectrum in *water* in this way. The experimental arrangement was otherwise similar to that of Fig. 18.3.6. The length of the flight path was 335 cm; a high-current deuteron accelerator with a tritium target served as the pulsed neutron source. Flight-time delays also occur in this experiment, but they can easily be eliminated by calculation. Fig. 18.3.9 shows the spectrum of the neutron flux in a "large" vessel of water 170 μsec after injection of a neutron pulse. Also plotted is a Maxwell distribution with the temperature of the moderator. The measured points fit this curve very well; since no diffusion cooling occurs because of the large size of the vessel, this

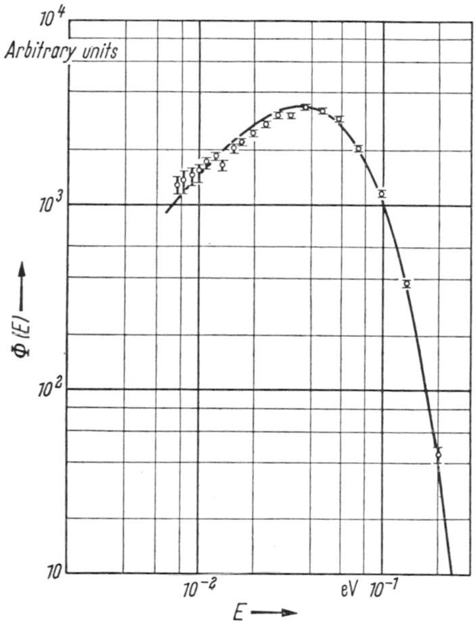

a

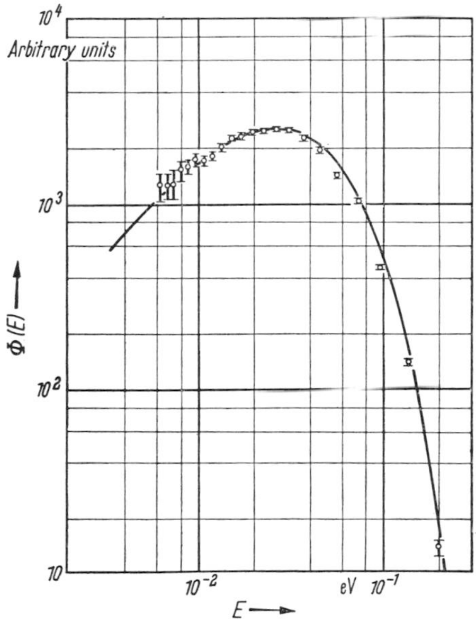

b

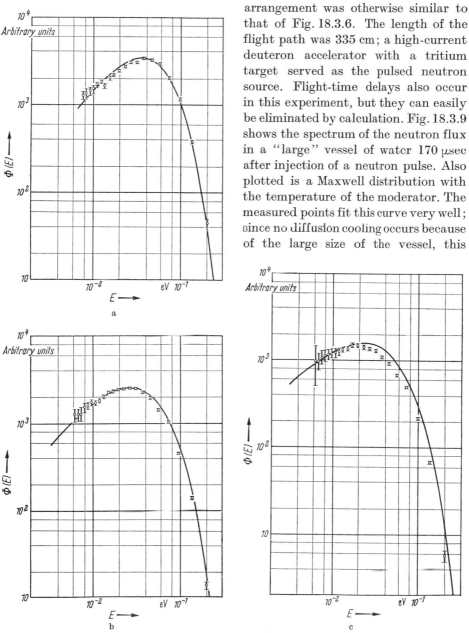

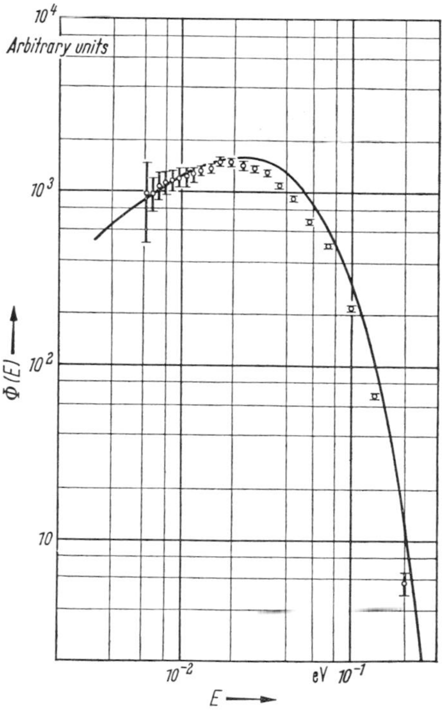

c

Fig. 18.3.8a–c. Comparison of measured and calculated neutron spectra in graphite at different times after the neutron burst injection. ☿ measurements; —— calculation. a: Delay time 300 μsec; b: delay time 600 μsec; c: delay time 1000 μsec

agreement is to be expected. In contrast, the spectrum in a small assembly (Fig. 18.3.10) shows very clearly the effect of diffusion cooling. Fig. 18.3.11 shows the *leakage spectrum* from the surface of a large water tank. Because of

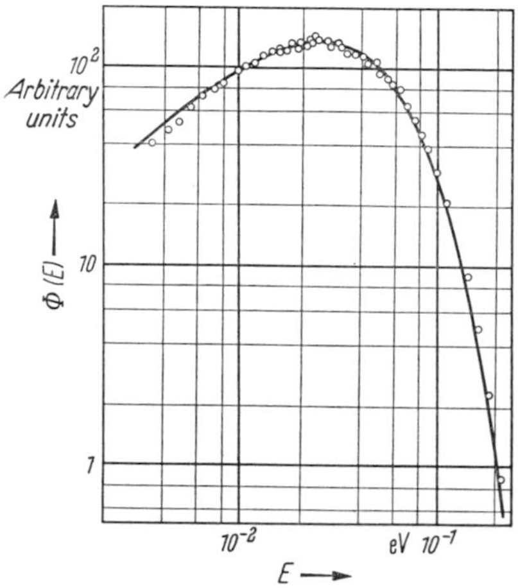

Fig. 18.3.9. The neutron spectrum in a "large" (15 × 15 × 15 cm³) water vessel 170 μsec after the burst injection. ∘∘ Measured; —— Maxwell distribution with the moderator temperature

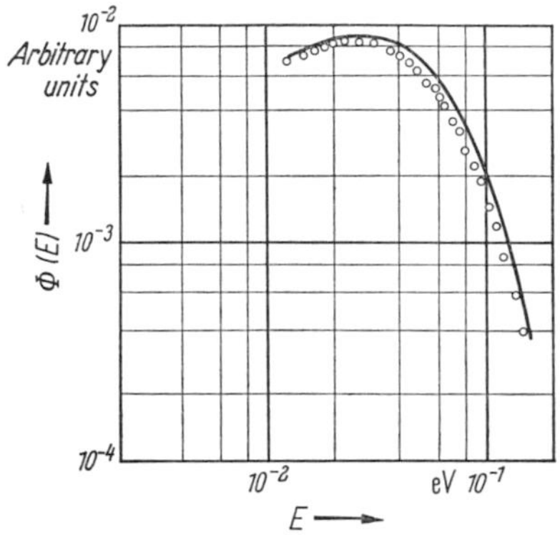

Fig. 18.3.10. The asymptotic neutron spectrum in a "small" (5 × 5 × 5 cm³) water vessel. ∘∘ Measured; —— Maxwell distribution with the moderator temperature

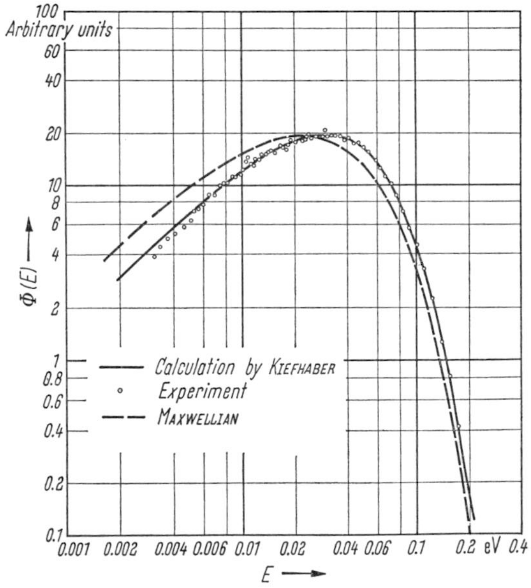

Fig. 18.3.11. The leakage spectrum from the surface of a "large" water vessel (15 × 15 × 15 cm³)

the strong energy dependence of the transport mean free path, this spectrum is much hotter than a Maxwell spectrum. The leakage spectrum is reproduced well by a calculation of KIEFHABER in which the Nelkin model for scattering on H_2O was used.

Chapter 18: References

General

AMALDI, E.: loc. cit., especially § 130—132.

BECKURTS, K. H.: Reactor Physics Research with Pulsed Neutron Sources, Nucl. Instrum. Methods **11**, 144 (1961).

DARDEL, G. F. VON: The Interactions of Neutrons with Matter Studied with a Pulsed Neutron Source, Trans. Roy. Inst. Technol. Stockholm **75** (1954).

DARDEL, G. F. VON, and N. G. SJÖSTRAND: Diffusion Measurements with Pulsed Neutron Sources, Progr. Nucl. Energy Ser. I, Vol. 2, p. 183 (1958).

Proceedings of the Brookhaven Conference on Neutron Thermalization, BNL-719 (1962) especially Vol. III, Experimental Aspects of Transient and Asymptotic Phenomena.

RAIEVSKI, V.: Physique des Piles Atomiques. Paris: Presses Universitaires de France 1960.

Special[1]

DARDEL, G. F. VON: Appl. Sci. Res. B **3**, 35 (1953).
GATTI, E.: Rev. Sci. Instr. **24**, 345 (1953).
GLASS, F. M.: ORNL 2480, 22 (1957).
} Instrumentation of Pulsed Neutron Experiments.

BECKURTS, K. H.: BNL-719, RE 1 (1962).
CORNELL, R. G.: ORNL-2120 (1956).
PEIERLS, R.: Proc. Roy. Soc. (London) A **149**, 467 (1935).
SILVER, E. G.: BNL-719, 981 (1962).
STARR, E., and J. W. L. DE VILLIERS: BNL-719, 997 (1962).
} Measurement of the $\alpha(B^2)$-Curve.

ANTONOV, A. V., *et al.*: Geneva 1955 P/661, Vol. 5, p. 3.
ANTONOV, A. V., *et al.*: Atomnaya Energiya **12**, 22 (1962).
BRACCI, A., and C. COCEVA: Nuovo Cimento **4**, 59 (1956).
DARDEL, G. F. VON, and N. G. SJÖSTRAND: Phys. Rev. **96**, 1245 (1954).
DIO, W. H.: Nukleonik **1**, 13 (1958).
HALL, R. S., S. A. SCOTT, and J. WALKER: Proc. Phys. Soc. (London) **79**, 257 (1962).
KÜCHLE, M.: Nukleonik **2**, 131 (1960).
LOPEZ, W. M., and J. R. BEYSTER: Nucl. Sci. Eng. **12**, 190 (1962).
} The Study of Neutron Diffusion in H_2O.

GANGULY, N. K., and A. W. WALTNER: Trans. Am. Nucl. Soc. **4** No. 2, 282 (1961).
KUSSMAUL, G., and H. MEISTER: J. Nuclear Energy A & B **17**, 411 (1963).
} The Study of Neutron Diffusion in D_2O.

ANTONOV, A. V., *et al.*: Geneva 1955 P/661, Vol. 5, p. 3.
BECKURTS, K. H.: Nucl. Sci. Eng. **2**, 516 (1957).
KLOSE, H.: Diploma Thesis, Karlsruhe (1962).
KLOSE, H., M. KÜCHLE, and W. REICHARDT: BNL-719, 935 (1962).
LALANDE, I.: Industries Atomiques **5**, No. 5/6, 71 (1961).
STARR, E., and G. PRICE: Trans. Am. Nucl. Soc. **2** No. 2, 125 (1959).
STARR, E., and G. PRICE: BNL-719, 1034 (1962).
} The Study of Neutron Diffusion in Graphite.

ANDREWS, W. M.: UCRL-6083 (1960).
ANTONOV, A. V., *et al.*: Geneva 1955 P/661, Vol. 5, p. 3.
CAMPBELL, E. C., and P. H. STELSON: ORNL-2076, 32 (1956).
KOMOTO, T. T., and F. KLOVERSTROM: Trans. Am. Nucl. Soc. **1**, No. 1, 18 (1958).
SAUSSURE, G. DE, and E. G. SILVER: ORNL-2641 (1959); ORNL-2842, 115 (1959).
SINGWI, K. S., and L. S. KOTHARI: Geneva 1958 P/1638, Vol. 16, p. 325.
} The Study of Neutron Diffusion in Beryllium.

---

[1] Cf. footnote on p. 53.

Iyengar, S. B. D., *et al.*: Proc. Indian Acad. Sci. **45**, 215, 224 (1957).
Küchle, M.: Nukleonik **2**, 131 (1960).
Meadows, J. W., and J. F. Whalen: Nucl. Sci. Eng. **13**, 230 (1962).
Seemann, K. W.: Trans. Am. Nucl. Soc. **2** No. 1, 69 (1959).
Sjöstrand, N. G., J. Mednis, and T. Nilsson: Arkiv Fysik **15**, 471 (1959).
} The Study of Neutron Diffusion in Other Moderators.

Beckurts, K. H.: Z. Naturforsch. **11**a, 881 (1956).
Sjöstrand, N. G., J. Mednis, and T. Nilsson: Arkiv Fysik **15**, 471 (1959).
} B^2-Measurement by the Pulsed Neutron Method.

Collie, C. H., R. E. Meads, and E. E. Lockett: Proc. Phys. Soc. (London) A **69**, 464 (1956).
Dardel, G. F. von, and A. W. Waltner: Phys. Rev. **91**, 1284 (1953).
Meads, R. E., *et al.*: Proc. Phys. Soc. (London) A **69**, 469 (1956).
Ramanna, A., *et al.*: J. Nucl. Energy **2**, 145 (1956).
} Absorption Measurements in Large Assemblies.

Dardel, G. F. von, and N. G. Sjöstrand: Phys. Rev. **96**, 1566 (1954).
Meadows, J. W., and J. F. Whalen: Nucl. Sci. Eng. **9**, 132 (1961).
Scott, F. R., D. B. Thomson, and W. Wright: Phys. Rev. **95**, 582 (1954).
} Absorption Measurements on Dissolved Substances.

Droulers, Y., J. Lacour, and V. Raievski: J. Nucl. Energy **7**, 210 (1958).
Perez, R. B., and R. E. Uhrig: Nucl. Sci. Eng. **17**, 90 (1963).
Raievski, V., and J. Horowitz: Compt. Rend. **238**, 1993 (1954); Geneva 1955 P/360, Vol. 5, p. 42.
Weinberg, A. M., and L. C. Noderer: AECD-3471 (1951), p. 1—88.
} The Method of Modulated Neutron Sources.

Antonov *et al.*: Geneva 1955 P/661, Vol. 5, p. 3.
Beckurts, K. H.: Nucl. Sci. Eng. **2**, 516 (1957).
Dardel, G. F. von, and N. G. Sjöstrand: Phys. Rev. **96**, 1245 (1954).
Iyengar, S. B. D., *et al.*: Proc. Indian Acad. Sci. **45**, 215 (1957).
Meadows, J. W., and J. F. Whalen: Nucl. Sci. Eng. **13**, 230 (1962).
} Transmission Experiments on Time-Dependent Neutron Spectra.

Möller, E., and N. G. Sjöstrand: BNL-719, 966 (1962).
Schweickert, E.: Diploma Thesis, Karlsruhe 1964.
} Measurement of the Thermalization Time.

Barnard, E., *et al.*: BNL-719, 805 (1962).
Beckurts, K. H.: Z. Naturforsch. **16**a, 611 (1961).
Kiefhaber, E.: Nucl. Sci. Eng. **18**, 404 (1964).
} Observation of Time-Dependent Neutron Spectra.

Appendix I

Table of Thermal Neutron Cross Sections of the Isotopes

The following table contains measured values of the thermal neutron absorption, reaction, activation, and scattering cross sections of the isotopes[1]. The data were taken from BNL-325, second edition 1958 and supplement 1960; only in a few places were more recent values used. The unit of cross section, unless otherwise noted, is 1 barn $=10^{-24}$ cm^2.

The first column of the table lists the elements to which the isotopes listed in the second column belong. After each isotope, its abundance in the naturally occurring element is listed. For radioactive isotopes, the half-life is given.

The third column contains values of the absorption cross section σ_a at the neutron velocity $v_0=2200$ m/sec. σ_a includes all reactions in which a neutron disappears (n, γ; n, p; n, d; n, α; fission). The listed values are the results of direct absorption measurements, i.e., pile oscillator measurements, pulsed and other integral experiments, or transmission experiments (the latter with a suitable correction for neutron scattering).

The next two columns contain data on reaction cross sections. If neutron absorption leads to radioactivity, column four lists its half-life. Usually, this activity is caused by radiative capture of neutrons; in those few cases where other reactions lead to radioactivity, the symbol of the resulting nucleus is given after the half-life. Sometimes, radiative neutron capture leads to several activities owing to the formation of isomeric states. In this table the upper (metastable) state is listed above the ground state (where the order is known) and the cross sections refer to the direct formation of each state. Where the decay of the metastable state does not go completely to the ground state, the branching ratio is given.

In some cases, reaction cross sections have been determined directly, i.e., not by measurement of the neutron-induced activity. This is the case for fission, for some (n, ϱ) and (n, α) reactions in which the protons and α-particles emitted in the reactions were directly observed, and for some reactions leading to stable or long-lived nuclides in which a mass-spectroscopic analysis of the reaction products was performed. In all these cases, column four lists the reaction type.

In column five, the respective activation or reaction cross sections are given. They refer to a neutron velocity $v_0=2200$ m/sec with the exception of those, marked by an asterisk, which were determined in a broad "pile spectrum" of more or less well-thermalized neutrons.

Finally, the last column lists the scattering cross sections. Most of these values are averages over a thermal neutron spectrum. Where this value was not available — denoted by a double asterisk — the bound atom scattering cross section was listed, i.e., $\left(\frac{A+1}{A}\right)^2$ times the free atom cross section observed in the ev region [cf. Eq. (1.4.7)].

[1] Scattering and absorption cross sections for the elements in their naturally occurring isotopic mixture are given in Table 1.4.1, p. 20.

Element	Isotope (%, $T_{\frac{1}{2}}$)	$\sigma_a(v_0)$ [barn]	$T_{\frac{1}{2}}$	$\sigma_{act}(v_0)$ [barn]	$\overline{\sigma_s}$ [barn]
$_1$H	H^1 ($\approx$ 100)	$(327 \pm 2) \cdot 10^{-3}$			
	H^2 (0.015)	$(0.46 \pm 0.10) \cdot 10^{-3}$	12. 4 y	$(0.57 \pm 0.01) \cdot 10^{-3}$	7 ± 1
$_2$He	He^3 (0.00013)	5327 ± 10	n, p	5400 ± 200	1.0 ± 0.7
	He^4 ($\approx$ 100)	0		0	
$_3$Li	Li^6 (7.52)	936 ± 6	n, γ	0.028 ± 0.008	
	Li^7 (92.48)		0.890 sec	0.033 ± 0.005	1.4 ± 0.2**
$_4$Be	Be^7 (54 d)		n, p	$54{,}000 \pm 8{,}000$	
			n, α	< 1	
	Be^9 (100)	10 mbarn	$2.7 \cdot 10^6$ y	0.009 ± 0.003	7 ± 1
$_5$B	B^{10} (19.8)	$3{,}840 \pm 11$	n, γ	0.5 ± 0.2	4.0 ± 0.5**
			n, p	< 0.2	
	B^{11} (80.2)		0.03 sec	< 0.050	4.4 ± 0.3**
$_6$C	C^{12} (98.89)		n, γ	0.0038 ± 0.0002	
	C^{13} (1.11)	$(0.5 \pm 0.2) \cdot 10^{-3}$	5,570 y	$(0.9 \pm 0.3) \cdot 10^{-3}$	5.5 ± 1.0**
	C^{14} (5,570 y)	< 200	2.3 sec	$< 10^{-6}$	
$_7$N	N^{14} (99.63)		n, p	1.75 ± 0.05	
			n, γ	0.08 ± 0.02	
	N^{15} (0.37)		7.4 sec	24 ± 8 μbarn	
$_8$O	O^{16} (99.76)				
	O^{17} (0.037)		5,570y (C^{14})	0.4 ± 0.1	
	O^{18} (0.204)		29 sec	$(0.21 \pm 0.04) \cdot 10^{-3}$	
$_9$F	F^{19} (100)	< 10 mbarn	11.2 sec	0.009 ± 0.002	3.9 ± 0.2
$_{10}$Ne	Ne^{20} (90.92)				
	Ne^{21} (0.26)				
	Ne^{22} (8.82)		38 sec	0.034 ± 0.010	
$_{11}$Na	Na^{23} (100)	0.528 ± 0.009	14.97 h	0.531 ± 0.008	4.0 ± 0.5
$_{12}$Mg	Mg^{24} (78.60)	0.034 ± 0.010			
	Mg^{25} (10.11)	0.280 ± 0.090			
	Mg^{26} (11.29)	0.060 ± 0.060	9.39 min	0.027 ± 0.005	
	Mg^{27} (9.39 min)		21.3 h	< 0.030*	
$_{13}$Al	Al^{27} (100)		2.30 min	0.21 ± 0.02	1.4 ± 0.1
$_{14}$Si	Si^{28} (92.27)	0.080 ± 0.030			
	Si^{29} (4.68)	0.28 ± 0.09			
	Si^{30} (3.05)	0.4 ± 0.4	2.62 h	0.110 ± 0.010	
$_{15}$P	P^{31} (100)	0.20 ± 0.02	14.3 d	0.19 ± 0.01	5 ± 1
$_{16}$S	S^{32} (95.018)		n, α	0.0018 ± 0.0010	
	S^{33} (0.750)		25.1 d (P^{33})	0.015 ± 0.010	
			n, α	$< 8 \cdot 10^{-3}$	
	S^{34} (4.215)		87 d	0.26 ± 0.05	
	S^{36} (0.017)		5.1 min	0.14 ± 0.04	
$_{17}$Cl	Cl^{35} (75.4)		$3.08 \cdot 10^5$ y	30 ± 20	
			87 d (S^{35})	0.19 ± 0.05	
			14.3 d (P^{32})	$< 0.05 \cdot 10^{-3}$	
	Cl^{36} ($3.08 \cdot 10^5$ y)			90 ± 30	
	Cl^{37} (24.6)		1.0 sec	0.005 ± 0.003*	
			37.5 min	0.56 ± 0.12	
$_{18}$A	A^{36} (0.34)		35 d	6 ± 2	
	A^{38} (0.063)		265 y	0.8 ± 0.2	
	A^{40} (99.600)		109 min	0.53 ± 0.02	
	A^{41} (109 min)		> 3.5 y	> 0.060	
$_{19}$K	K^{39} (93.08)	1.94 ± 0.15	$1.3 \cdot 10^9$ y	3 ± 2*	
	K^{40} (0.012)	70 ± 20	n, p	< 1	
	K^{41} (6.91)	1.24 ± 0.10	12.46 h	1.20 ± 0.15	

Element	Isotope (%, $T_{\frac{1}{2}}$)	$\sigma_a(v_0)$ [barn]	$T_{\frac{1}{2}}$	$\sigma_{act}(v_0)$ [barn]	$\overline{\sigma_s}$ [barn]
$_{20}$Ca	Ca^{40} (96.97)	0.22 ± 0.04			3.1 ± 0.3**
	Ca^{42} (0.64)	42 ± 3			
	Ca^{43} (0.145)				
	Ca^{44} (2.06)		164 d	0.70 ± 0.08	
	Ca^{46} (0.0033)		4.8 d	0.25 ± 0.10	
	Ca^{48} (0.185)		8.5 min	1.1 ± 0.1	
$_{21}$Sc	Sc^{45} (100)	24.0 ± 1.0	20 sec	10 ± 4	24 ± 2
			85 d	12 ± 6	
			20 sec + 85 d	22 ± 2	
$_{22}$Ti	Ti^{46} (7.95)	0.6 ± 0.2			2 ± 2
	Ti^{47} (7.75)	1.7 ± 0.3			4 ± 1
	Ti^{48} (73.45)	8.3 ± 0.6			4 ± 2
	Ti^{49} (5.51)	1.9 ± 0.5			1 ± 1
	Ti^{50} (5.34)	< 0.2	5.8 min	0.14 ± 0.03	3 ± 1
$_{23}$V	V^{50} (0.24)		n, γ	250 ± 200	
	V^{51} (99.76)		3.76 min	4.5 ± 0.9	
$_{24}$Cr	Cr^{50} (4.31)	17.0 ± 1.4	27.7 d	14.6 ± 1.5	
	Cr^{52} (83.76)	0.76 ± 0.06			
	Cr^{53} (9.55)	18.2 ± 1.5			
	Cr^{54} (2.38)	< 0.3	3.6 min	0.38 ± 0.04	
$_{25}$Mn	Mn^{55} (100)	13.2 ± 0.1	2.587 h	13.2 ± 0.1	2.3 ± 0.3
$_{26}$Fe	Fe^{54} (5.84)	2.3 ± 0.2	2.96 y	2.7 ± 0.4	2.5 ± 0.3**
	Fe^{56} (91.68)	2.7 ± 0.2			12.8 ± 0.2**
	Fe^{57} (2.17)	2.5 ± 0.2			2.0 ± 0.5**
	Fe^{58} (0.31)	2.5 ± 2.0	44.3 d	1.00 ± 0.1	
			3.5 min (Cr^{55})	< 0.0015	
$_{27}$Co	Co^{59} (100)	37.1 ± 1.0	10.4 min	16.9 ± 1.5	7 ± 1
			5.28 y	20.2 ± 1.9	
			10.4 min + 5.28 y	36.3 ± 1.5	
			(99.7% of 10.4 min → 5.28 y)		
	Co^{60m} (10.4 min)		1.75 h	100 ± 50	
	Co^{60} (5.28 y)		1.75 h	6 ± 2	
$_{28}$N	Ni^{58} (67.76)	4.4 ± 0.3			24.4 ± 0.5**
	Ni^{60} (26.16)	2.6 ± 0.2			1.0 ± 0.1**
	Ni^{61} (1.25)	2.0 ± 1.0			
	Ni^{62} (3.36)	15 ± 2			9 ± 1**
	Ni^{64} (1.16)		2.56 h	1.52 ± 0.14	
	Ni^{65} (2.56 h)		56 h	20 ± 2	
$_{29}$Cu	Cu^{63} (69.1)	4.5 ± 0.1	12.87 h	4.51 ± 0.23	
	Cu^{65} (30.9)	2.2 ± 0.2	5.15 min	1.8 ± 0.4	
	Cu^{66} (5.15 min)		59 h	130 ± 40*	
$_{30}$Zn	Zn^{64} (48.89)		246.4 d	0.47 ± 0.05	
			12.8 h (Cu^{64})	< 10 μbarn	
			n, α	15 ± 10 μbarn	
	Zn^{66} (27.81)		80 y (Ni^{63})	< 20 μbarn	
	Zn^{67} (4.11)		n, α	6 ± 4 μbarn	
	Zn^{68} (18.56)		13.8 h	0.090 ± 0.010	
			52 min	1.10 ± 0.15	
			n, α	< 20 μbarn	
	Zn^{70} (0.62)		2.2 min	0.100 ± 0.020	
			4.1 h	0.009	
$_{31}$Ga	Ga^{69} (60.2)	2.1 ± 0.2	20.2 min	1.4 ± 0.3	
	Ga^{71} (39.8)	5.1 ± 0.4	14.2 h	4.3 ± 0.3	

Element	Isotope (%, $T_{\frac{1}{2}}$)	$\sigma_a(v_0)$ [barn]	$T_{\frac{1}{2}}$	$\sigma_{act}(v_0)$ [barn]	$\overline{\sigma_s}$ [barn]
$_{32}$Ge	Ge^{70} (20.55)	3.4 ± 0.3	12 d	3.42 ± 0.35	
	Ge^{72} (27.37)	0.98 ± 0.09			
	Ge^{73} (7.67)	14 ± 1			
	Ge^{74} (36.74)	0.62 ± 0.06	48 sec	0.040 ± 0.008	
			82 min	0.21 ± 0.08	
	Ge^{76} (7.67)	0.36 ± 0.07	57 sec	0.080 ± 0.020	
			12 h	0.080 ± 0.020	
			(≈ 50% of 57 sec → 12 h)		
$_{33}$As	As^{75} (100)	4.3 ± 0.2	27 h	5.4 ± 1.0	6 ± 1
$_{34}$Se	Se^{74} (0.87)	50 ± 7	123 d	26 ± 6	
	Se^{76} (9.02)	85 ± 7	18 sec	7 ± 3	
	Se^{77} (7.58)	42 ± 4			
	Se^{78} (23.52)	0.4 ± 0.4			
	Se^{80} (49.82)	0.61 ± 0.06	57 min	0.030 ± 0.010	
			18 min	0.5 ± 0.1	
	Se^{82} (9.19)	2.1 ± 1.5	67 sec	0.050 ± 0.025	
			25 min	0.004 ± 0.002	
			(None of 67 sec → 25 min)		
$_{35}$Br	Br^{79} (50.52)		4.6 h	2.9 ± 0.5	
			18 min	8.5 ± 1.4	
			4.6 h + 18 min	10.4 ± 1.0	
	Br^{81} (49.48)		35.9 h	3.2 ± 0.4	
$_{36}$Kr	Kr^{78} (0.35)		34.5 h	2.0 ± 0.5	
	Kr^{80} (2.27)		13 sec + 2×10^5 y	95 ± 15	
	Kr^{82} (11.56)		n, γ	45 ± 15	
	Kr^{83} (11.55)		n, γ	210 ± 30	
	Kr^{84} (56.90)		4.4 h	0.10 ± 0.03	
			9.4 y	0.060 ± 0.020	
			(23% of 4.4 h → 9.4 y)		
	Kr^{85} (9.4 y)		n, γ	< 15	
	Kr^{86} (17.37)		77 min	0.060 ± 0.020	
	Kr^{87} (77 min)		2.8 h	< 600	
$_{37}$Rb	Rb^{85} (72.15)		18.7 d	0.85 ± 0.08	
	Rb^{87} (27.85)		17.8 min	0.12 ± 0.03	
	Rb^{88} (17.8 min)		15.4 min	1.0 ± 0.2	
$_{38}$Sr	Sr^{84} (0.56)	< 3	70 min	< 1	
			65 d	1.1 ± 0.3	
			(80% of 70 min → 65 d)		
	Sr^{86} (9.86)		2.80 h	1.5 ± 0.2	
	Sr^{87} (7.02)				
	Sr^{88} (82.56)		53 d	0.005 ± 0.001	
	Sr^{89} (53 d)		19.9 y	0.5 ± 0.1	
	Sr^{90} (19.9 y)		9.7 h	1.0 ± 0.6	
$_{39}$Y	Y^{89} (100)	1.31 ± 0.08	64.8 h	1.26 ± 0.08	3 ± 2
	Y^{90} (64.8 h)		61 d	< 7	
$_{40}$Zr	Zr^{90} (51.46)	0.10 ± 0.07			
	Zr^{91} (11.23)	1.58 ± 0.12			
	Zr^{92} (17.11)	0.25 ± 0.12			
	Zr^{93} (1.1×10^6 y)	< 4			
	Zr^{94} (17.40)	0.08 ± 0.06	65 d	0.09 ± 0.03	
	Zr^{96} (2.80)	1.1 ± 0.1	17.0 h	0.10 ± 0.05	
$_{41}$Nb	Nb^{93} (100)	1.16 ± 0.02	6.6 min	1.0 ± 0.5	5 ± 1
	Nb^{94} (2.2×10^4 y)		36 d	15 ± 4	

Element	Isotope (%, $T_{\frac{1}{2}}$)	$\sigma_a(v_0)$ [barn]	$T_{\frac{1}{2}}$	$\sigma_{act}(v_0)$ [barn]	$\overline{\sigma_s}$ [barn]
$_{42}$Mo	Mo^{92} (15.86)	<0.3	6.9 h	<0.006	
	Mo^{94} (9.12)				
	Mo^{95} (15.70)	13.9 ± 1.4			
	Mo^{96} (16.50)	1.2 ± 0.6			
	Mo^{97} (9.45)	2.2 ± 0.7			
	Mo^{98} (12.75)	0.4 ± 0.4	67 h	0.50 ± 0.06	
	Mo^{100} (9.62)	0.5 ± 0.5	14.3 min	0.20 ± 0.05	
$_{43}$Tc	Tc^{99} (2.1×10^5 y)	22 ± 3			5 ± 1**
$_{44}$Ru	Ru^{96} (5.7)		2.8 d	0.21 ± 0.02	
	Ru^{98} (2.2)				
	Ru^{99} (12.8)				
	Ru^{100} (12.7)				
	Ru^{101} (17.0)				
	Ru^{102} (31.3)		41 d	1.44 ± 0.16	
	Ru^{104} (18.3)		4.5 h	0.7 ± 0.2	
$_{45}$Rh	Rh^{103} (100)	149 ± 4	4.4 min	12 ± 2	
			42 sec	140 ± 30	5 ± 1
			(99.9% of 4.4 min → 42 sec)		
$_{46}$Pd	Pd^{102} (0.8)		17.0 d	4.8 ± 1.5	
	Pd^{104} (9.3)				
	Pd^{105} (22.6)				
	Pd^{106} (27.1)				
	Pd^{108} (26.7)		4.8 min	0.26 ± 0.04	
			13.6 h	10 ± 1	
	Pd^{110} (13.5)		23.6 min	0.25 ± 0.05	
			5.5 h	<0.05	
$_{47}$Ag	Ag^{107} (51.35)	31 ± 2	2.3 min	45 ± 4	10 ± 2**
	Ag^{109} (48.65)	87 ± 7	253 d	3.2 ± 0.4	6 ± 1**
			24.2 sec	110 ± 10	
			(5% of 253 d → 24.2 sec)		
$_{48}$Cd	Cd^{106} (1.22)		6.7 h	1.0 ± 0.5	
	Cd^{108} (0.87)		1.3 y	1.41 ± 0.35	
	Cd^{110} (12.39)		49 min	0.15 ± 0.05	
	Cd^{111} (12.75)				
	Cd^{112} (24.07)		5.1 y	0.030 ± 0.015	
	Cd^{113} (12.26)	20,000 ± 300			
	Cd^{114} (28.86)		43 d	0.14 ± 0.03	
			53 h	1.1 ± 0.3	
			(None of 43 d → 53 h)		
	Cd^{116} (7.58)		2.9 h	1.5 ± 0.3	
			50 min	<0.008	
$_{49}$In	In^{113} (4.23)		49 d	56 ± 12	
			72 sec	3.0 ± 1.0	
			(96.5% of 49 d → 72 sec)		
	In^{115} (95.77)		54.12 min	160 ± 2	
			14.1 sec	42 ± 1	
			(None of 54.12 min → 14.1 sec)		
$_{50}$Sn	Sn^{112} (0.95)		112 d	1.3 ± 0.3	
	Sn^{114} (0.65)				
	Sn^{115} (0.34)				
	Sn^{116} (14.24)		14.5 d	0.006 ± 0.002	
	Sn^{117} (7.57)				
	Sn^{118} (24.01)		250 d	0.010 ± 0.006	
	Sn^{119} (8.58)				
	Sn^{120} (32.97)		400 d	1 ± 1 mbarn	
			27.5 h	0.14 ± 0.03	

Element	Isotope (%, $T_{\frac{1}{2}}$)	$\sigma_a(v_0)$ [barn]	$T_{\frac{1}{2}}$	$\sigma_{act}(v_0)$ [barn]	$\overline{\sigma_s}$ [barn]
	Sn^{122} (4.71)		130 d	1.0 ± 0.5 mbarn	
			40 min	0.16 ± 0.04	
	Sn^{124} (5.98)		10 min	0.2 ± 0.1	
			10 d	4 ± 2 mbarn	
			(None of 10 min → 10 d)		
$_{51}Sb$	Sb^{121} (57.25)	5.9 ± 0.5	2.8 d	6.8 ± 1.5	
			3.3 min	0.19 ± 0.03	
	Sb^{123} (42.75)	4.1 ± 0.3	21 min	0.030 ± 0.015	
			1.3 min	0.030 ± 0.015	
			60 d	2.5 ± 0.5	
$_{52}Te$	Te^{120} (0.089)	70 ± 70			
	Te^{122} (2.46)	2.8 ± 0.9	110 d	1.1 ± 0.5	
	Te^{123} (0.87)	410 ± 30			
	Te^{124} (4.61)	6.8 ± 1.3	58 d	5 ± 3	
	Te^{125} (6.99)	1.56 ± 0.16			
	Te^{126} (18.71)	0.8 ± 0.2	110 d	0.090 ± 0.020	
			9.3 h	0.90 ± 0.15	
			(98% of 110 d → 9.3 h)		
	Te^{128} (31.79)	0.3 ± 0.3	33 d	0.015 ± 0.005	
			72 min	0.15 ± 0.04	
	Te^{130} (34.49)	0.5 ± 0.3	30 h	< 0.008	
			25 min	0.18 ± 0.04	
			(22% of 30 h → 25 min)		
$_{53}I$	I^{127} (100)	6.22 ± 0.25	25.0 min	5.6 ± 0.3	3.6 ± 0.5
	I^{129} (1.7 × 10^7 y)		12.6 h	20 ± 5	
	I^{131} (8.1 d)		2.4 h	50 ± 40	
$_{54}Xe$	Xe^{124} (0.096)	74 ± 1			
	Xe^{126} (0.090)				
	Xe^{128} (1.92)		n, γ	< 5	
	Xe^{129} (26.44)		n, γ	45 ± 15	
	Xe^{130} (4.08)		n, γ	< 5	
	Xe^{131} (21.18)		n, γ	120 ± 15	
	Xe^{132} (26.89)		5.3 d	0.2 ± 0.1	
	Xe^{133} (5.3 d)		n, γ	190 ± 90	
	Xe^{134} (10.44)		9.13 h	0.2 ± 0.1	
	Xe^{135} (9.13 h)	(2.72 ± 0.11) · 10^6			
	Xe^{136} (8.87)		3.9 min	0.15 ± 0.8	
$_{55}Cs$	Cs^{133} (100)	29.0 ± 1.0	2.9 h	0.017 ± 0.004	7 ± 1**
			2.3 y	30 ± 1	
			(≈ 99% of 2.9 h → 2.3 y)		
	Cs^{134} (2.3 y)		2.6 × 10^6 y	134 ± 12*	
	Cs^{135} (2.6 × 10^6 y)		13.7	13 ± 4*	
	Cs^{137} (26.6 y)		33 min	< 2	
$_{56}Ba$	Ba^{130} (0.101)		12.0 d	10 ± 1	
	Ba^{132} (0.097)		7.2 y	7 ± 2	
	Ba^{134} (2.42)	2 ± 2			
	Ba^{135} (6.59)	5.8 ± 0.9			
	Ba^{136} (7.81)	0.4 ± 0.4			
	Ba^{137} (11.32)	5.1 ± 0.4			
	Ba^{138} (71.66)	0.7 ± 0.1	85 min	0.5 ± 0.1	
	Ba^{139} (85 min)		12.8 d	4 ± 1	
	Ba^{140} (12.8 d)		18 min	12 ± 4*	
$_{57}La$	La^{138} (0.089)				
	La^{139} (99.911)		40 h	8.2 ± 0.8	
	La^{140} (40 h)		3.7 h	3.1 ± 1.0	

Element	Isotope (%, $T_{\frac{1}{2}}$)	$\sigma_a(v_0)$ [barn]	$T_{\frac{1}{2}}$	$\sigma_{act}(v_0)$ [barn]	$\overline{\sigma_s}$ [barn]
$_{58}$Ce	Ce^{136} (0.19)	25 ± 25	34.5 h	0.6 ± 0.2	
			8.7 h	6.3 ± 1.5	
	Ce^{138} (0.26)	9 ± 6	55 sec	0.007 ± 0.005	
			140 d	0.6 ± 0.3	
	Ce^{140} (88.48)	0.66 ± 0.06	32 d	0.31 ± 0.10	2.8 ± 0.5**
	Ce^{142} (11.07)	1.0 ± 0.2	32 h	0.94 ± 0.05	2.6 ± 0.5**
	Ce^{143} (32 h)		290 d	6.0 ± 0.7*	
$_{59}$Pr	Pr^{141} (100)	11.3 ± 0.2	19.3 h	10.8 ± 1.0	4.0 ± 0.4**
	Pr^{142} (19.2 h)		13.7 d	18 ± 3*	
$_{60}$Nd	Nd^{142} (27.13)	18 ± 2			
	Nd^{143} (12.32)	324 ± 10	n, γ	240 ± 50	76 ± 7**
	Nd^{144} (23.87)	5.0 ± 0.6			
	Nd^{145} (8.29)	60 ± 6			12 ± 2**
	Nd^{146} (17.1)	10 ± 1	11.3 d	1.8 ± 0.6	
	Nd^{148} (5.72)	3.4 ± 1.0	1.8 h	3.7 ± 1.2	
	Nd^{150} (5.60)	3.0 ± 1.5	15 min	1.5 ± 0.2	
$_{61}$Pm	Pm^{147} (2.5 y)		5.3 d	60 ± 20	
$_{62}$Sm	Sm^{144} (3.16)		400 d	< 2	
	Sm^{147} (15.07)	87 ± 60			
	Sm^{148} (11.27)				
	Sm^{149} (13.84)	40,800 ± 900			
	Sm^{150} (7.47)				
	Sm^{151} (73 y)	10,000 ± 2,000 *			
	Sm^{152} (26.63)	216 ± 6	47 h	140 ± 40	
	Sm^{154} (22.53)		24 min	5.5 ± 1.1	
$_{63}$Eu	Eu^{151} (47.77)	7,700 ± 80	9.2 h	1,400 ± 300*	
	Eu^{152} (13 y)		n, γ	5,500 ± 1,500*	
	Eu^{153} (52.23)	450 ± 20	16 y	420 ± 100*	
	Eu^{154} (16 y)		1.7 y	1 ,500 ± 400*	
	Eu^{155} (1.7 y)		15.4 d	14,000 ± 4,000*	
$_{64}$Gd	Gd^{152} (0.20)		230 d	< 125	
	Gd^{154} (2.15)				
	Gd^{155} (14.73)	56,200 ± 1,000	n, γ	70,000 ± 20,000*	
	Gd^{156} (20.47)				
	Gd^{157} (15.68)	242,000 ± 4,000	n, γ	160,000 ± 60,000*	
	Gd^{158} (24.87)		18.0 h	3.9 ± 0.4	
	Gd^{160} (21.90)		3.6 min	0.8 ± 0.3	
$_{65}$Tb	Tb^{159} (100)	46 ± 3	73d	> 22	
	Tb^{160} (73 d)		7.0 d	525 ± 100*	
$_{66}$Dy	Dy^{156} (0.052)				
	Dy^{158} (0.090)				
	Dy^{160} (2.298)				
	Dy^{161} (18.88)				
	Dy^{162} (25.53)				
	Dy^{163} (24.97)				
	Dy^{164} (28.18)		1.3 min	2,000 ± 200	
			139 min	800 ± 100	
	Dy^{165} (139 min)		82 h	5,000 ± 2,000*	
$_{67}$Ho	Ho^{165}	65 ± 3	27.3 h	60 ± 12	
$_{68}$Er	Er^{162} (0.136)		75 min	2.03 ± 0.20	
	Er^{164} (1.56)		10 h	1.65 ± 0.17	
	Er^{166} (33.4)				
	Er^{167} (22.9)				
	Er^{168} (27.1)		9.4 d	2.0 ± 0.4	
	Er^{170} (14.9)		2.5 sec + 7.5 h	9 ± 2	

Element	Isotope (%, $T_{\frac{1}{2}}$)	$\sigma_a(v_0)$ [barn]	$T_{\frac{1}{2}}$	$\sigma_{act}(v_0)$ [barn]	$\overline{\sigma_s}$ [barn]
$_{69}$Tm	Tm^{169} (100)	127 ± 4	129 d	130 ± 30	7 ± 3
	Tm^{170} (129 d)		1.9 y	150 ± 20*	
$_{70}$Yb	Yb^{168} (0.140)		32 d	11,000 ± 3,000*	
	Yb^{170} (3.03)				
	Yb^{171} (14.31)				
	Yb^{172} (21.82)				
	Yb^{173} (16.13)				
	Yb^{174} (31.84)		101 h	60 ± 40	
	Yb^{176} (12.73)		1.8 d	5.5 ± 1.0	
$_{71}$Lu	Lu^{175} (97.40)		3.7 h	35 ± 15	
	Lu^{176} (2.60)		6.8 d	4,000 ± 800	
$_{72}$Hf	Hf^{174} (0.18)	1,500 ± 1,000			
	Hf^{176} (5.15)	15 ± 15			
	Hf^{177} (18.39)	380 ± 30			
	Hf^{178} (27.08)	75 ± 10			
	Hf^{179} (13.78)	65 ± 15			
	Hf^{180} (35.44)	14 ± 5	46 d	10 ± 3	
$_{73}$Ta	Ta^{181} (99.988)	21.0 ± 0.7	16.4 min	0.030 ± 0.01	5 ± 1
			111 d	19 ± 7	
			(≈ 95% of	16.4 min → 111 d)	
	Ta^{182} (111 d)		5.5 d	17,000 ± 2,000*	
$_{74}$W	W^{180} (0.14)	60 ± 60	140 d	10 ± 10	
	W^{182} (26.4)	20 ± 2			
	W^{183} (14.4)	11 ± 1			
	W^{184} (30.6)	2.0 ± 0.3	73 d	2.20 ± 0.25	
			185 min	0.022 ± 0.004	
	W^{186} (28.4)	35 ± 3	24 h	34 ± 7	
	W^{187} (24 h)		65 d	90 ± 40	
$_{75}$Re	Re^{185} (37.07)	104 ± 8	91 h	110 ± 10	
	Re^{187} (62.93)	66 ± 5	17 h	70 ± 7	
	Re^{188} (17 h)		150 d	< 2	
$_{76}$Os	Os^{184} (0.018)		97 d	< 200	
	Os^{186} (1.59)				
	Os^{187} (1.64)				
	Os^{188} (13.3)				
	Os^{189} (16.1)		10 min	0.008 ± 0.002	
	Os^{190} (26.4)		16.0 d	8 ± 3	
	Os^{192} (41.0)		31 h	1.6 ± 0.4	
	Os^{193} (31 h)		700 d	600 ± 200	
$_{77}$Ir	Ir^{191} (38.5)		1.4 min	280 ± 50	
			74 d	650 ± 100	
	Ir^{192} (74 d)		n, γ	700 ± 200	
	Ir^{193} (61.5)		19.0 h	130 ± 30	
$_{78}$Pt	Pt^{190} (0.012)	150 ± 150	18 h	0.76 ± 0.10	
	Pt^{192} (0.78)	8 ± 8	4.3 d	90 ± 40	
	Pt^{194} (32.8)	1.2 ± 0.9			
	Pt^{195} (33.7)	27 ± 2			
	Pt^{196} (25.4)	0.7 ± 0.7	18 h	0.87 ± 0.09	
			1.3 min	0.069 ± 0.014	
	Pt^{198} (7.2)	4.0 ± 0.5	14.1 sec	0.028 ± 0.005	
			31 min	3.9 ± 0.8	
	Pt^{199} (31 min)		11.5 h	15 ± 10	
$_{78}$Au	Au^{197} (100)	98.6 ± 0.3	2.7 d	96 ± 10	9.3 ± 1.0
	Au^{198} (2.7 d)		3.15 d	26,000 ± 1,200	
	Au^{199} (3.15 d)		48 min	30 ± 15*	

Element	Isotope (%, $T_{\frac{1}{2}}$)	$\sigma_a(v_0)$ [barn]	$T_{\frac{1}{2}}$	$\sigma_{act}(v_0)$ [barn]	$\overline{\sigma_s}$ [barn]
$_{80}$Hg	Hg^{196} (0.146)		n, γ	3,100 ± 1,000*	
	Hg^{198} (10.02)				
	Hg^{199} (16.84)		n, γ	2,500 ± 800*	
	Hg^{200} (23.13)		n, γ	< 60*	
	Hg^{201} (13.22)		n, γ	< 60*	
	Hg^{202} (29.80)		47 d	3.8 ± 0.8	
	Hg^{204} (6.85)		5.5 min	0.43 ± 0.10	
$_{81}$Tl	Tl^{203} (29.50)	11.4 ± 0.9	2.7 y	8 ± 3	
	Tl^{205} (70.50)	0.80 ± 0.08	4.2 min	0.10 ± 0.03	
$_{82}$Pb	Pb^{204} (1.48)	0.8 ± 0.6	5 × 10^7 y	0.7 ± 0.2*	
	Pb^{206} (23.6)	0.025 ± 0.005			
	Pb^{207} (22.6)	0.70 ± 0.03			
	Pb^{208} (52.3)	< 0.030	3.2 h	0.0006 ± 0.0002	
$_{83}$Bi	Bi^{209} (100)	0.034 ± 0.002	5.0 d	0.019 ± 0.002	9 ± 1
$_{86}$Rn	Rn^{220} (54 sec)		25 min	< 0.2*	
	Rn^{222} (3.83 d)		11.7 d	0.72 ± 0.07*	
			(Ra^{223})		

Element	Isotope (%, $T_{\frac{1}{2}}$)	$\sigma_a(v_0)$ [barn]	$T_{\frac{1}{2}}$	$\sigma_{act}(v_0)$ [barn]	$\sigma_{n,f}$ [barn]	$\overline{\sigma_s}$ [barn]
$_{88}$Ra	Ra^{223} (11.2 d)		3.64 d	130 ± 20*	< 100	
	Ra^{224} (3.64 d)		14.8 d	12.0 ± 0.5*		
	Ra^{226} (1,620 y)		41.2 min	20 ± 3*	< 0.0001	
	Ra^{228} (6.7 y)		< 10 min	36 ± 5*	< 2	
$_{89}$Ac	Ac^{227} (22 y)	510 ± 40	6.13 h	520 ± 50	< 2	
$_{90}$Th	Th^{227} (18.6 d)				1,500 ± 1,000	
	Th^{228} (1.90 y)		7.3 × 10^3 y	123 ± 15*	≦ 0.3	
	Th^{229}				45 ± 11	
	(7.3 × 10^3 y)					
	Th^{230}	27 ± 2	25.6 h	35 ± 10*	≦ 0.001	
	(8.0 × 10^4 y)					
	Th^{232} (100)	7.56 ± 0.11	23.3 min	7.33 ± 0.12	< 0.0002	12.6 ± 0.2
	(1.45 × 10^{10} y)					
	Th^{233} (23.3 min)		24.1 d	1,400 ± 200*	15 ± 2*	
	Th^{234} (24.1 d)		≦ 10 min	1.8 ± 0.5*	< 0.010	
$_{91}$Pa	Pa^{230} (17.3 d)				1,500 ± 250	
	Pa^{231}		1.31 d	200 ± 15	0.010 ± 0.005	
	(3.4 × 10^4 y)					
	Pa^{232} (1.31 d)		27.4 d	760 ± 100*	700 ± 100	
	Pa^{233} (27.4 d)		1.18 min	43 ± 5	< 0.1	
			6.7 h	25 ± 4		
	Pa^{234} (UX2)				≦ 500	
	(1.18 min)					
	(UZ) (6.7 h)				≦ 5,000	
$_{92}$U	U^{230} (20.8 d)				25 ± 10	
	U^{231} (4.3 d)				400 ± 300	
	U^{232} (73 y)		1.62 × 10^5 y	300 ± 200*	80 ± 20	
	U^{233}	73 ± 5	2.52 × 10^5 y	52 ± 2	524 ± 4	
	(1.61 × 10^5 y)					
	U^{234} (0.0057)	105 ± 4	7.1 × 10^8 y	90 ± 30	≦ 0.65	
	(2.52 × 10^5 y)					
	U^{235} (0.714)	690 ± 8	2.40 × 10^7 y	107 ± 5	586 ± 7	10 ± 2
	(7.1 × 10^8 y)					
	U^{236}	7 ± 2	6.7 d	6 ± 1		
	(2.40 × 10^7 y)					

Element	Isotope (%, $T_{\frac{1}{2}}$)	$\sigma_a(v_0)$ [barn]	$T_{\frac{1}{2}}$	$\sigma_{act}(v_0)$ [barn]	$\sigma_{n,f}$ [barn]	$\overline{\sigma_s}$ [barn]
	U^{238} (99.3)	2.71 ± 0.02	23.5 min	2.74 ± 0.06	< 0.0005	
	(4.50×10^9 y)					
	U^{239} (23.5 min)		17 h	22 ± 5*	14 ± 3*	
$_{93}$Np	Np^{234} (4.4 d)				900 ± 300	
	Np^{236} (7,500 y)				$2{,}800 \pm 800$	
	Np^{237}	170 ± 5	2.10 d	169 ± 6	0.019 ± 0.003	
	(2.2×10^6 y)					
	Np^{238} (2.10 d)				$1{,}600 \pm 100$	
	Np^{239} (2.3 d)		7.3 min	35 ± 10*	< 3	
			60 min	25 ± 15*		
			($< 5\%$ of 7.3 min $\rightarrow$ 60 min)			
$_{94}$Pu	Pu^{236} (2.7 y)				170 ± 35	
	Pu^{237} (46 d)				$2{,}500 \pm 500$	
	Pu^{238} (89.6 y)		2.44×10^4 y	403 ± 10	16.8 ± 0.3	
	Pu^{239}	$1{,}030.1 \pm 8$	6.6×10^3 y	315 ± 16	748.2 ± 4.9	9.6 ± 0.5
	(2.44×10^4 y)					
	Pu^{240}	286 ± 7	13.2 y	250 ± 40	0.030 ± 0.045	
	(6.6×10^3 y)					
	Pu^{241} (13.2 y)	$1{,}400 \pm 80$	3.7×10^5 y	390 ± 50*	$1{,}010 \pm 13$	
	Pu^{242}	30 ± 2	4.98 h	19 ± 1	< 0.2	
	(3.7×10^5 y)					
	Pu^{243} (4.98 h)		7.5×10^7 y	170 ± 90*		
	Pu^{244}		10 h	1.8 ± 0.3*		
	(7.5×10^7 y)					
	Pu^{245} (10 h)		11.2 d	260 ± 150*		
$_{95}$Am	Am^{241} (461 y)	630 ± 35	16.0 h	750 ± 80*	3.1 ± 0.2	
			100 y	50 ± 40*		
			($< 6\%$ of 16.0 h $\rightarrow$ 100 y)			
	Am^{242m} (16 h)				$2{,}500 \pm 1{,}000$	
	Am^{242} (100 y)	$8{,}000 \pm 1{,}000$*			$6{,}400 \pm 500$*	
	Am^{243}		26 min	74 ± 4	< 0.075	
	(7.6×10^3 y)					
$_{96}$Cm	Cm^{242} (162.5 d)		35 y	20 ± 10*	< 5*	
	Cm^{243} (35 y)	500 ± 300*	18 y	250 ± 50	700 ± 50	
	Cm^{244} (18 y)		2×10^4 y	15 ± 10*		
	Cm^{245} (2×10^4 y)		6.6×10^3 y	200 ± 100*	$1{,}900 \pm 200$	
	Cm^{246}		$< 10^6$ y	15 ± 10*		
	(6.6×10^3 y)					
	Cm^{248}		65 min	6 ± 4*		
	(4.2×10^5 y)					
$_{97}$Bk	Bk^{249} (290 d)		3.1 h	500 ± 200*		
$_{98}$Cf	Cf^{249} (470 y)	900 ± 400*	10 y	270 ± 100*	600 ± 400*	
	Cf^{250} (10 y)		≈ 700 y	$1{,}500 \pm 1{,}000$*		
	Cf^{251} (≈ 700 y)		2.2 y	$3{,}000 \pm 2{,}000$*		
	Cf^{252} (2.2 y)		18 d	28 ± 7		
	Cf^{254} (55 d)			< 2*		
$_{99}$E	E^{253} (20 d)		38 h	300 ± 150*		
	E^{254} (480 d)	$2{,}700 \pm 600$*	24 d	< 40*		

Appendix II

Table of Resonance Integrals for Infinitely Dilute Absorbers (from McArthy, Persiani et al.[1])

The following table, which is reprinted with kind permission of the Argonne National Laboratory, contains measured and calculated values of resonance absorption, activation, and fission integrals for infinitely dilute absorbers. The tabulated quantity is

$$I_x = \int_{E_c}^{E_{\max}} \sigma_x(E) \frac{dE}{E}$$

and thus includes the $1/v$-part. The values of $E_{\max}$ and E_c depend on the experimental conditions. No attempt has been made to reduce the experimental I_x to the value $E_{\max} = 2$ Mev, which was proposed in Sec. 12.3.1; however, this neglect is permissible since only a very small contribution to the resonance integral stems from the high energy region. For the calculations, $E_{\max} = \infty$ was assumed in most cases. A value of E_c was estimated for each experiment and is listed in the table. The values of E_c employed in the calculations are also listed; $E_c = 0.44$ ev was used in most cases.

Measured values are listed according to the method by which the measurement was made, i.e., according to whether the activation or absorption method was used. Brief comments, pertinent to the results, are indicated by literal superscripts, and they are presented at the end of the table. References are listed at the end of the table.

Resonance Integrals for Dilute Solutions (in barn).
Notes: (c) = capture integral, (f) = fission integral, (c + f) = capture plus fission integrals.

Element	Nat. Abund., %	Measured		Calculated	Lower Energy, (ev)	Ref.
		Activation	Absorption			
Li			28		0.4	1
B			280 ± 40		0.49	2
N			4.8 ± 2.4		0.49	2
F^{19}	100		2.3 ± 0.5		0.49	2
Na^{23}	100	≈ 0.24	0.27		0.4	1
Na^{23}	100	0.30 ± 0.01			0.5	5
Mg			0.9		0.4	1
Mg			0.072[b]			3
Al^{27}	100		< 0.18[b]			3
Al^{27}	100	≈ 0.16	0.18		0.4	1
Si			0.5		0.4	1
P^{31}	100		< 2		0.4	1
S			0.6		0.4	1
Cl			12		0.4	1
Cl			12.8 ± 1.7		0.49	2
K			3.5 ± 1.1		0.49	2
Ca			2		0.4	1

[1] McArthy, A. E., P. J. Persiani, B. I. Spinrad and L. J. Templin: Argonne National Laboratory, Reactor Physics Constants Center Newsletter No. 1 (1961); Persiani, P. J., J. J. Kaganove, and A. E. McArthy: Ibid. No. 10 (1963).

Element	Nat. Abund., %	Measured		Calculated	Lower Energy, (ev)	Ref.
		Activation	Absorption			
Sc^{45}	100	≈ 10.7			0.4	1
Ti			3		0.4	1
Ti			3.8 ± 0.9		0.49	2
V			4.1[b]			3
V			3.3 ± 0.8		0.49	2
V^{51}	99.76	≈ 2.2			0.4	1
Cr			2.6 ± 1.1		0.49	2
Mn^{55}	100	15.7 ± 0.6[l]			0.55	14
Mn^{55}	100		15.4[b, c]			3
Mn^{55}	100	15.6 ± 0.6[r]			0.55	4
Mn^{55}	100	14.2 ± 0.6			0.5	5
Mn^{55}	100	14.0 ± 0.3				6
Mn^{55}	100			17[n]	0.4	
Fe			2.1		0.4	1
Fe			1.8[b]			3
Fe			2.3 ± 0.25		0.5	7
Fe			2.3 ± 0.4		0.49	2
Co^{59}	100	81 ± 4[r]				8
Co^{59}	100	75 ± 5			0.5	9
Co^{59}	100	72.3 ± 5			0.5	5
Ni			4		0.4	1
Ni			3.1[b]			3
Ni			3.2 ± 0.5		0.49	2
Cu			3.3 ± 0.3		0.5	7
Cu		3.7	4		0.4	1
Cu		3.13			0.64	10
Cu			3.7 ± 0.8		0.49	2
Cu		3.12 ± 0.07			0.64	31
Cu^{63}	69.1	3.71 ± 0.13			0.64	31
Cu^{63}	69.1	4.4			0.4	1
Cu^{63}	69.1	5.1 ± 0.15			0.5	5
Cu^{65}	30.9	1.82 ± 0.21			0.64	31
Cu^{65}	30.9	2.2			0.4	1
Cu^{65}	30.9	2.3 ± 0.23			0.5	5
Zn			2		0.4	1
Zn			3.4 ± 0.8		0.49	2
Ga			11.7 ± 2.7		0.49	2
Ga^{69}	60.2	9.2			0.4	1
Ga^{71}	39.8	15			0.4	1
Ge			3.5 ± 2.9		0.49	2
As^{75}	100	36.8	33	170[a]	0.4	1
Se			9.6 ± 1.2		0.49	2
Se^{77}	7.58			33[s]	0.4	12
Se^{80}	49.82			0.9[s]	0.4	12
Br			118 ± 14		0.49	2
Br^{79}	50.52	147			0.4	1
Br^{81}	49.48			61.6[s]	0.4	12
Kr^{82}	11.56			200[s]	0.4	12
Kr^{83}	11.55			240[s]	0.4	12
Kr^{84}	56.90			3.4[s]	0.4	12
Rb			9.0 ± 2.8		0.49	2
Rb^{85}	72.15			0.55[s]	0.4	12
Rb^{87}	27.85			0.17[s]	0.4	12
Sr			16		0.4	1
Sr			10.0 ± 2.6		0.49	2
Y^{89}	100	0.91			0.4	1
Y^{89}	100			0.73[s]	0.4	12
Zr			3		0.4	1
Zr			0.69 ± 0.09	0.63[b, f]		3
Zr			3.7 ± 0.5		0.49	2
Zr			0.94 ± 0.15			32

Element	Nat. Abund., %	Measured		Calculated	Lower Energy, (ev)	Ref.
		Activation	Absorption			
Zr^{91}	11.23		5.4 ± 1.6^{r}		0.55	13
Zr^{91}	11.23			5.6^{s}	0.4	12
Nb^{93}	100		14^{b}	$16^{b,g}$		3
Nb^{93}	100		14.5 ± 2.3		0.55	14
Nb^{93}	100		8.62 ± 0.65		0.60	33
Nb^{93}	100	8.4 ± 1.3^{q}				34
Nb^{93}	100			10.9^{d}	0.44	
Nb^{94}		500 ± 200^{q}				34
Mo			32.1 ± 3.1			37
Mo				26.75^{d}	0.44	
Mo			20^{b}	$22^{b,g}$		3
Mo^{95}	15.70		$107^{b,h}$	$107^{b,g}$		3
Mo^{95}	15.70			102.9^{d}	0.44	
Mo^{96}	16.5			43.4^{d}	0.44	
Mo^{97}	9.45			12.5^{d}	0.44	
Mo^{98}	23.75			5.43^{d}	0.44	
Mo^{98}	23.75	10.8 ± 2.5			0.5	5
Mo^{100}	9.62	3.73 ± 0.20			0.5	15
Mo^{100}	9.62			10.0^{d}	0.44	
Tc^{99}			72^{b}			3
Rh^{103}	100	656	575	$1{,}146^{a}$	0.4	1
Rh^{103}	100			$1{,}095^{s}$	0.4	12
Pd			23		0.4	1
Ag			> 650		0.4	1
Ag			841^{b}	$761^{b,g}$		3
Ag^{107}	51.35	74			0.4	1
Ag^{109}	48.65	1,160			0.4	1
Ag^{109}	48.65		$1{,}910^{b,h}$	$1{,}440^{b,g}$		3
In			$3{,}700^{b}$	$3{,}100^{b,g}$		3
In		$2{,}615 \pm 125$			0.6	38
In			$3{,}300 \pm 850$	3,181	0.40	39
In				$3{,}150^{d}$	0.44	
In^{113}	4.23			877.4^{d}	0.44	
In^{115}	95.77			$3{,}250^{d}$	0.44	
In^{115}	95.77			3,337		40
In^{115}	95.77	$2{,}540^{i}$			0.4	1
In^{115}	95.77	$2{,}630 \pm 133$		—	—	6
Sn			8.8^{b}	$5.3^{b,g}$		3
Sn			5.7 ± 0.7		0.49	2
Sn^{120}	32.97			0.82^{s}	0.4	12
Sn^{124}	5.98			12^{s}	0.4	12
Sb			106 ± 13		0.49	2
Sb			115 ± 12		0.5	7
Sb^{121}	57.25	162			0.4	1
Sb^{121}	57.25			203^{s}	0.4	12
Sb^{123}	42.75			162^{s}	0.4	12
Sb^{123}	42.75	≈ 138			0.4	1
Te			$\approx 74^{b,j}$	$> 49^{b,g}$		3
Te			50.0 ± 6.0		0.49	2
Te^{122}	2.46			56^{s}	0.4	12
Te^{123}	0.87			$6{,}765^{s}$	0.4	12
Te^{124}	4.61			5.5^{s}	0.4	12
Te^{125}	6.99			24^{s}	0.4	12
Te^{126}	18.71			9.0^{s}	0.4	12
Te^{128}	31.79			10^{s}	0.4	12
I^{127}	100		130 ± 18		0.5	7
I^{127}	100		183^{b}	$158^{b,g}$		3
I^{127}	100	140			0.4	1
I^{127}	100			173°		
I^{129}				42^{s}	0.4	12
I^{129}		36 ± 4			0.6	16

Element	Nat. Abund., %	Measured		Calculated	Lower Energy, (ev)	Ref.
		Activation	Absorption			
Xe^{129}	26.44			260^{s}	0.4	12
Xe^{131}	21.18			810^{s}	0.4	12
Cs^{133}	100	400 ± 25^{r}			0.55	17
Cs^{133}	100		504^{b}	$394^{b,g}$		3
Cs^{135}		62 ± 2			0.6	16
Ba			12.6 ± 1.7		0.49	2
Ba^{136}	7.81			13^{s}	0.4	12
La			11		0.4	1
La^{139}	99.911			13^{s}	0.4	12
Ce			3.7 ± 1.7		0.49	2
Ce^{144}		2.60 ± 0.26			0.5	41
Pr^{141}	100			25^{s}	0.4	12
Pr^{141}	100	15.5 ± 3^{r}			0.55	17
Nd			40 ± 6		0.5	7
Nd^{143}	12.32		$< 218^{b}$	$> 181^{b,f}$		3
Nd^{145}	8.29		$156^{b,h}$	$> 273^{b,f}$		3
Pm^{147}				$2{,}190^{s}$	0.4	12
Pm^{147}		$1{,}700 \pm 250$ (to Pm^{148}) $1{,}520 \pm 230$ (to Pm^{148m})			0.6	42
Sm				$1{,}524.7^{d}$	0.44	
Sm			$1{,}790 \pm 270$		0.49	2
Sm^{147}	15.07		$< 1{,}385^{b}$			3
Sm^{147}	15.07			572.6^{d}	0.44	
Sm^{149}	13.84			$3{,}249^{d}$	0.44	
Sm^{150}	7.47		255 ± 25		0.54	43
Sm^{150}	7.47			407.9^{d}	0.44	
Sm^{151}				$2{,}373^{d}$	0.44	
Sm^{152}	26.63			$3{,}598^{d}$	0.44	
Sm^{152}	26.63	$2{,}740 \pm 150^{r}$			0.55	17
Sm^{152}	26.63			$2{,}210^{s}$	0.4	12
Sm^{152}	26.63		$2{,}960^{b}$			3
Eu^{151}	47.77		$< 6{,}900^{b}$	$7{,}320^{b}$		3
Eu^{153}	52.23		$1{,}500^{b}$	$1{,}760^{b,g}$		3
Gd			67 ± 8.0		0.49	2
Tb^{159}	100			393^{s}	0.13	12
Dy			$1{,}390 \pm 220$	1,240	0.40	39
Dy				$1{,}264^{d}$	0.44	
Dy^{161}	18.88			947.8^{d}	0.44	
Dy^{162}	25.53			$2{,}610^{d}$	0.44	
Dy^{163}	24.97			$1{,}244.8^{d}$	0.44	
Dy^{164}	28.18			382.1^{d}	0.44	
Dy^{164}	28.18	420 ± 50			0.4	11
Dy^{164}	28.18	482 ± 33				6
Lu			720 ± 70		0.5	7
Lu^{175}	97.40	463 ± 15				6
Lu^{176}	2.60	887 ± 65				6
Hf			$2{,}800 \pm 600$		0.5	7
Hf			$2{,}900^{b}$			3
Hf			$2{,}150 \pm 480$	1,906	0.40	39
Hf			$2{,}000 \pm 350$		0.55	14
Hf				$2{,}258^{d}$	0.44	
Hf^{174}	0.18			195.0^{d}	0.44	
Hf^{177}	18.39			$7{,}476^{d}$	0.44	
Hf^{178}	27.08			$2{,}999^{d}$	0.44	
Hf^{179}	13.78			429.7^{d}	0.44	
Hf^{180}	35.44			33.3^{d}	0.44	
Hf^{180}	35.44	21.8			0.4	1
Ta			$1{,}110^{b}$	$750^{b,g}$		3
Ta			474 ± 62		0.49	2

Element	Nat. Abund., %	Measured: Activation	Measured: Absorption	Calculated	Lower Energy, (ev)	Ref.
Ta^{181}	99.988	590		555[e]	0.4	1
W			290 ± 35		0.49	2
W			340[b]			3
W			367 ± 33		0.56	37
W			370 ± 60	320	0.40	39
W				333.5[d]	0.44	
W^{182}	26.4			531.0[d]	0.44	
W^{183}	14.4			389.3[d]	0.44	
W^{184}	30.6		7.5 ± 2.0	6.3	0.40	39
W^{184}	30.6			8.83[d]	0.44	
W^{186}	28.4		420 ± 100	460	0.40	39
W^{186}	28.4			473.5[d]	0.44	
W^{186}	28.4	355			0.4	1
W^{186}	28.4	396 ± 59				6
Re		694 ± 28			0.6	38
Re^{185}	37.07	1,160		1,630[a]	0.4	1
Re^{187}	62.93	305		310[e]	0.4	1
Os			180 ± 20		0.49	2
Ir			2,000 ± 490		0.49	2
Ir^{191}	38.5	3,500[i]		7,680[a]	0.4	1
Ir^{193}	61.5	1,370[i]		976[a]	0.4	1
Pt			69		0.4	1
Pt			≈ 189[b, j]	114[b, j]		3
Au^{197}	100		1,558[b]	1,591[b, g]		3
Au^{197}	100	1,533 ± 40			0.5	18
Au^{197}	100	1,558			0.4	1
Au^{197}	100	1,553			0.64	10
Au^{197}	100			1,584[o]	0.4	
Hg			73 ± 5		0.5	7
Hg			72.4 ± 8		0.49	2
Tl^{203}		129			0.4	1
Tl^{205}		0.5			0.4	1
Pb			0.1		0.4	1
Pb			0.13[b]			3
Bi^{209}	100		0.5		0.4	1
Bi^{209}	100		0.07[b]			3
Th^{232}	100		110[b, h]			3
Th^{232}	100	85			0.5	9
Th^{232}	100			94.0[d]	0.44	
Th^{233}		500 ± 150[k]			0.4	20
Th^{233}		400 ± 100			0.5	9
Pa^{231}				1,560 ± 55	0.1	26
Pa^{233}		700 ± 200[k]			0.4	20
Pa^{233}		770 ± 90			0.6	16
U			224 ± 40		0.49	2
U^{233}			100 ± 4 (c)		0.4	20
U^{233}			1,000 ± 200 (c + f)		0.4	20
U^{233}				813 (f)[p]		21
U^{233}				812 (f)[p]	0.4	22
U^{233}		870 (f)				23
U^{233}				129.1 (c)[d]	0.44	
U^{233}				686 (f)[d]	0.44	
U^{234}	0.0057			661.4[e]	0.44	
U^{235}	0.714			188.3 (c)[d]	0.44	
U^{235}	0.714			264.4 (f)[d]	0.44	
U^{235}	0.714	271 (f)			0.4	1
U^{235}	0.714	274 ± 11 (f)			0.5	25
U^{235}	0.714			340 (f)[p]		21
U^{236}		415 ± 40[r]			0.55	17
U^{236}			400 ± 50[k]		0.4	20
U^{236}		350 ± 90		340 ± 40[m]	0.4	24

Element	Nat. Abund., %	Measured		Calculated	Lower Energy, (ev)	Ref.
		Activation	Absorption			
U^{238}	99.3	280 ± 10[k]			0.4	20
U^{238}	99.3		287[b]			3
U^{238}	99.3	281 ± 20			0.4	19
U^{238}	99.3			279 ± 7[m]	0.4	24
Np^{237}			955[b, c]			3
Pu^{238}		$3{,}260 \pm 280$			0.5	35
Pu^{238}				111.2[d]	0.44	
Pu^{239}				181.5 (c)[d]	0.44	
Pu^{239}				280.2 (f)[d]	0.44	
Pu^{239}				1,137 (c)[d]	0.15	
Pu^{239}				1,843 (f)[d]	0.15	
Pu^{239}				$3{,}600 \pm 1{,}000$ (c + f)[k]	0.15	20
Pu^{239}		327 ± 22 (f)			0.5	25
Pu^{240}				8,269[d]	0.44	
Pu^{240}				8,467[d]	0.133	
Pu^{240}			$8{,}400 \pm 1{,}100$[t]		0.133	36
Pu^{240}			11,450[b, c]			3
Pu^{240}		$8{,}700 \pm 800$			0.5	27
Pu^{240}		$9{,}000 \pm 3{,}000$			0.4	28
Pu^{240}			$10{,}000 \pm 2{,}800$		0.2	29
Pu^{241}		557 ± 33 (f)			0.5	25
Pu^{242}		$1{,}275 \pm 30$			0.5	30
Pu^{242}				1,034[d]	0.44	
Am^{241}				1,614 (c)[e]	0.44	
Am^{241}				4.07 (f)[d]	0.44	
Am^{243}				1,374[d]	0.44	
Am^{243}		$2{,}290 \pm 50$			0.5	30

Comments

a) Estimated from the parameters of the first large resonance and the thermal cross section; does not include a correction for unresolved levels.

b) The resonance integral given is not significantly dependent on what cut-off is used since the material has a cross section dependence that is closely $1/v$ in the cut-off region.

c) Values were deduced from measurements in the Dimple Maxwellian spectrum and with the Gleep oscillator.

d) Values computed using IBM 704 code ANL-RE-266, including negative energy levels.

e) Estimated from level parameters; does not include a correction for unresolved levels.

f) Calculated from the parameters of the first resonance only; includes unresolved resonance contributions.

g) Calculated from parameters given in BNL-325; includes unresolved level contributions.

h) Measurements on single solutions only; corrected for screening.

i) A resonance near thermal leads to considerable dependence on the details of the cadmium absorber that was used.

j) Only one sample of these materials was available and the estimated screening was large. Values listed must be treated with caution.

k) Values preferred by the authors after analysis of the available data.

l) Value for a 1-mil foil; the value for a 2-mil foil is 15.5 ± 0.6 barn.

m) Calculated from level parameters and the thermal cross section; includes an unresolved level correction.

n) Calculated by P. Persiani of Argonne from level parameters given in Supplement 1, Second Edition, BNL-325 (1960); includes unresolved level corrections. The entry for Dy^{164} includes the bound level. The parameters for this level were obtained from R. Sher (private communication).

o) Calculated by P. Persiani of Argonne from level parameters given in Ref. 28; includes an unresolved level correction.

p) Calculated by numerical integration of the fission cross section.

q) Preliminary estimate of value and error.

r) A gold resonance integral (including the $1/v$-part) of 1534 barns was used as standard.

s) Calculated from level parameters in the second edition of BNL-325 or its supplement (1960). The number of levels listed is the number of resonance levels for which separate calculations were carried out. If the number of resolved levels is three or more, the resonance integral listed includes a contribution from the unresolved resonances calculated using average resonance parameters.

t) An average value involving extrapolation and screening considerations.

References

1. Macklin, R. L., and H. S. Pomerance: Resonance Capture Integrals. Proc. 1st Intern. Conf. Peaceful Uses Atomic Energy, Geneva, P/833, **5**, 96 (1955).
2. Klimentov, V. B., and V. M. Griazev: Some Neutron Resonance Absorption Integrals, J. Nuclear Energy **9**, 20—27 (1959).
3. Tattersall, R. B., *et al.*: Pile Oscillator Measurements of Resonance Absorption Integrals AERE-R-2887 (Aug. 1959).
4. Feiner, F.: KAPL, personal communication (1961).
5. Dahlberg, R., *et al.*: Measurements of Some Resonance Activation Integrals. J. Nuclear Energy, 14 (No. 1), 53 (April 1961)
6. Crandall, J. L.: Savannah River, personal communication (1960) (Work by G. M. Jacks).
7. Spivak, P. E., *et al.*: Measurements of the Resonance Absorption Integrals for Various Materials and ν_{eff} — the Multiplication Coefficient of Resonance Neutrons for Fissionable Isotopes. Proc. 1st Intern. Conf. Peaceful Uses Atomic Energy, Geneva, P/659, **5**, 91 (1955).
8. Feiner, F., and L. J. Esch: Cobalt Resonance Integral. KAPL-2000-12 (Dec. 1960).
9. Johnston, F. J., *et al.*: The Thermal Neutron Cross Section of Th^{233} and the Resonance Integrals of Th^{232}, Th^{233} and Co^{59}. J. Nuclear Energy A **11**, 95—100 (1960).
10. Bennett, R. A.: Effective Resonance Integrals of Cu and Au. HW-63576 (1960).
11. Sher, R.: BNL, personal communication (1960).
12. Walker, W. H.: Yields and Effective Cross Sections of Fission Products and Pseudo Fission Products, CRRP-913 (March 1960).
13. Feiner, F.: A Measurement of the Resonance Integral of Zr^{91}. KAPL-2000-8 (Dec. 1959).
14. Feiner, F.: Resonance Integrals of Manganese, Hafnium, and Niobium. KAPL-2000-16 (Dec. 1961).
15. Cabell, M. J.: The Thermal Neutron Capture Cross Section and the Resonance Capture Integral of Mo^{100}. J. Nuclear Energy A **12**, 172—176 (1960).
16. Eastwood, T. A., *et al.*: Radiochemical Methods Applied to the Determination of Cross Sections of Reactor Interest. Proc. 2nd Intern. Conf. Peaceful Uses Atomic Energy, Geneva, P/203, **16**, 54 (1958).
17. Eiland, H. M., *et al.*: Determination of Thermal Cross Sections and Resonance Integrals of Some Fission Products. KAPL 2000-11 (Sept. 1960).
18. Jirlow, K., and E. Johansson: The Resonance Integral of Gold. J. Nuclear Energy A **11**, 101—107 (1960).

19. MACKLIN, R. L., and H. S. POMERANCE: Resonance Activation Integrals of U^{238} and Th^{232}. J. Nuclear Energy **2**, 243—246 (1956).
20. HALPERIN, J., and R. W. STOUGHTON: Some Cross Sections of Heavy Nuclides Important to Reactor Operation. Proc. 2nd Intern. Conf. Peaceful Uses Atomic Energy, Geneva, P/1072, **16**, 64 (1958).
21. TERASAWA, S.: The Effect of Epithermal Fission on Aqueous Homogeneous Reactors. ORNL 2553 (Aug. 1958).
22. FLUHARTY, R. G., *et al.*: Total and Fission Cross Sections of U^{233} below 1 kev. Proc. 2nd Intern. Conf. Peaceful Uses of Atomic Energy, Geneva, P/645, **15**, 111 (1958).
23. HALPERIN, J., *et al.*: The Resonance Integral for Neutron Fission of U^{233}. ORNL-2983 (Sept. 1960).
24. HARVEY, J. A., and R. B. SCHWARTZ: Thermal and Resonance Cross Sections of Heavy Nuclei, in Progress in Nuclear Energy, Pergamon Press Ltd., London. Series I: Physics and Mathematics, II, p. 51 (1958).
25. HARDY jr., J., *et al.*: The Resonance Fission Integrals of U^{235}, Pu^{239}, and Pu^{241}. Nuclear Sci. Eng. **9**, 341—345 (March 1961).
26. SIMPSON, F. B., *et al.*: Total Neutron Cross Section of Pa^{231}. Nuclear Sci. and Eng. **12**, 243—249 (February 1962).
27. CORNISH, F. W., and M. LOUNSBURY: Cross Sections of Pu^{239} and Pu^{240} in the Thermal and Epicadmium Regions. CRC-633 (March 1956).
28. EROZOLIMSKY, B. G., *et al.*: Measurement of the Average η_{eff} of Neutrons Emitted per Single Capture Event for Samples of Pu^{239} with Impurities of the Isotope Pu^{240}. Soviet J. Atomic Energy, No. 3, 311—315 (1957): Also J. Nuclear Energy **4**, 86 (1957).
29. KRUPCHINSKY, P. A.: The Thermal Neutron Absorption Cross Section and Resonance Absorption Integral of Pu^{240}. J. Nuclear Energy 6, 155—162 (1957).
30. BUTLER, J. P., *et al.*: The Neutron Capture Cross Sections of Pu^{238}, Pu^{242}, and Am^{243}. Can. J. Phys. **35**, 147 (Feb. 1957).
31. BENNETT, R. A.: Note on the Resonance Integrals of Natural Copper, Cu^{63} and Cu^{65}. HW-68389 (Jan. 1961).
32. HELLSTRAND, E., *et al.*: A Study of the Resonance Integral of Zirconium. EANDC (OR) 15 "U" (June 1961).
33. HELLSTRAND, E., *et al.*: Studies of the Effective Total and Resonance Absorption Cross Sections for Zircaloy 2 and Zirconium. Arkiv Fysik **20**, 41 (1962).
33. HELLSTRAND, E., and G. LUNDGREN: The Resonance Integral of Niobium. A.E.-81 (Aug. 1962).
34. SCHUMAN, R. P.: Activation Cross Sections of Nb-93 and Nb-94. IDO-16760, MTR-ETR Technical Branches Quarterly Report, Oct. 1—Dec. 1, 1961.
35. BUTLER, J. P., *et al.*: The Neutron Capture Cross Sections of Pu^{238}, Pu^{242}, and Am^{243}. Can. J. Phys. **35**, 147 (Feb. 1957).
36. TATTERSALL, R. B.: Thermal Cross-Section and Resonance Absorption Integral of Pu^{240}. AEEW-R 115 (March 1962).
37. LONG, R. L.: Resonance Integral Measurements. ANL-6580, Reactor Development Program Progress Report (June 1962).
38. KARAM, R. A., and T. F. PARKINSON: The Nuclear Properties of Rhenium. Eight Quarterly Technical Progress Report, University of Florida, Department of Nuclear Engineering NP-11239 (December 1961).
39. SCOVILLE, J. J., E. FAST, and D. W. KNIGHT: Resonance Absorption Integrals of Dysprosium and Tungsten. Trans. Am. Nucl. Soc. **5** No 2, 377 (1962).
40. KLOPP, D. A., and W. E. ZAGOTTA: Effective Resonance Integral of Indium Foils. Trans. American Nuclear Soc. **5**, No. 2, 377.
41. LANTZ, P. M.: Thermal Neutron Capture Cross Section and Resonance Capture Integral of Ce^{144}. Nuclear Sci. and Eng. **13**, 289—294 (July 1962).
42. SCHUMAN, R. P., and J. R. BERRETH: Neutron Activation Cross Sections of Pm^{147}, Pm^{148}, and Pm^{148m}. Nuclear Sci. and Eng. **12**, 519—522 (April 1962).
43. HALPERIN, J., *et al.*: Thermal Neutron Cross Section and Resonance Integral of Sm-150. Trans. Am. Nucl. Soc. **5**, No. 2, 376—377 (Nov. 1962).

Appendix III

Table of the Functions $E_n(x)$

The table contains values of

$$E_n(x) = \int_1^\infty \frac{e^{-ux}}{u^n}\, du = \int_0^1 t^{n-2} e^{-\frac{x}{t}}\, dt$$

(cf. Sec. 11.2) for $n = 0, 1, 2$, and 3. A general discussion of the properties of the $E_n(x)$ functions and values for higher n can be found in the Canadian report NRC-1547 by G. Placzek.

x	$E_0(x)$	$E_1(x)$	$E_2(x)$	$E_3(x)$
.00			1.000000	.500000
.01	99.004983	4.037930	0.949670	.490277
.02	49.009934	3.354708	.913104	.480968
.03	32.348184	2.959119	.881672	.471998
.04	24.019736	2.681264	.853539	.463324
.05	19.024588	2.467898	.827834	.454919
.06	15.696076	2.295307	.804046	.446761
.07	13.319912	2.150838	.781835	.438833
.08	11.538954	2.026941	.760961	.431120
.09	10.154791	1.918745	.741244	.423610
.10	9.048374	1.822924	.722545	.416292
.11	8.143947	1.737107	.704752	.409156
.12	7.391004	1.659542	.687775	.402194
.13	6.754580	1.588899	.671538	.395398
.14	6.209702	1.524146	.655978	.388761
.15	5.738053	1.464462	.641039	.382276
.16	5.325899	1.409187	.626674	.375938
.17	4.962734	1.357781	.612842	.369741
.18	4.640390	1.309796	.599507	.363680
.19	4.352416	1.264858	.586636	.357749
.20	4.093654	1.222650	.574201	.351945
.21	3.859925	1.182902	.562175	.346264
.22	3.647813	1.145380	.550535	.340700
.23	3.454494	1.109883	.539260	.335252
.24	3.277616	1.076235	.528331	.329914
.25	3.115203	1.044283	.517730	.324684
.26	2.965583	1.013889	.507440	.319558
.27	2.827332	0.984933	.497448	.314534
.28	2.699228	.957308	.487737	.309609
.29	2.580219	.930918	.478297	.304779
.30	2.469394	.905677	.469115	.300042
.31	2.365958	.881506	.460180	.295396
.32	2.269216	.858335	.451482	.290837
.33	2.178557	.836101	.443010	.286365
.34	2.093442	.814746	.434757	.281976
.35	2.013394	.794215	.426713	.277669
.36	1.937990	.774462	.418870	.273442
.37	1.866850	.755441	.411221	.269291
.38	1.799635	.737112	.403759	.265216
.39	1.736043	.719437	.396477	.261126

x	$E_0(x)$	$E_1(x)$	$E_2(x)$	$E_3(x)$
.40	1.675800	.702380	.389368	.257286
.41	1.618659	.685910	.382427	.253428
.42	1.564397	.669997	.375648	.249637
.43	1.512812	.654613	.369025	.245914
.44	1.463719	.639733	.362554	.242256
.45	1.416951	.625331	.356229	.238662
.46	1.372356	.611386	.350046	.235131
.47	1.329792	.597877	.344000	.231661
.48	1.289132	.584784	.338087	.228251
.49	1.250258	.572089	.332303	.224899
.50	1.213061	.559774	.326644	.221604
.51	1.177442	.547822	.321106	.218366
.52	1.143309	.536220	.315686	.215182
.53	1.110575	.524952	.310381	.212052
.54	1.079163	.514004	.305186	.208974
.55	1.049000	.503364	.300100	.205948
.56	1.020016	.493020	.295118	.202972
.57	0.992150	.482960	.290238	.200045
.58	.965342	.473173	.285458	.197166
.59	.939538	.463650	.280774	.194335
.60	.914686	.454380	.276184	.191551
.61	.890739	.445353	.271686	.188811
.62	.867652	.436562	.207276	.186117
.63	.845384	.427997	.262954	.183466
.64	.823894	.419652	.258715	.180857
.65	.803147	.411517	.254560	.178291
.66	.783108	.403586	.250484	.175766
.67	.763744	.395853	.246487	.173281
.68	.745025	.388309	.242567	.170836
.69	.726922	.380950	.238721	.168429
.70	.709408	.373769	.234947	.166061
.71	.692457	.366760	.231244	.163730
.72	.676045	.359918	.227611	.161436
.73	.660149	.353237	.224046	.159178
.74	.644748	.346713	.220546	.156955
.75	.629822	.340341	.217111	.154767
.76	.615351	.334115	.213739	.152612
.77	.601316	.328032	.210428	.150492
.78	.587700	.322088	.207178	.148404
.79	.574487	.316277	.203986	.146348
.80	.561661	.310597	.200852	.144324
.81	.549208	.305042	.197774	.142331
.82	.537112	.299611	.194750	.140368
.83	.525361	.294299	.191781	.138436
.84	.513941	.289103	.188864	.136532
.85	.502841	.284019	.185999	.134658
.86	.492049	.279045	.183183	.132812
.87	.481554	.274177	.180417	.130994
.88	.471344	.269413	.177699	.129204
.89	.461411	.264750	.175029	.127440
.90	.451744	.260184	.172404	.125703
.91	.442334	.255714	.169825	.123992
.92	.433173	.251336	.167290	.122306
.93	.424251	.247050	.164798	.120646
.94	.415562	.242851	.162348	.119010

x	$E_0(x)$	$E_1(x)$	$E_2(x)$	$E_3(x)$
.95	.407096	.238738	.159940	.117399
.96	.398847	.234708	.157573	.115811
.97	.390807	.230760	.155246	.114247
.98	.382970	.226891	.152958	.112706
.99	.375330	.223100	.150708	.111188
1.00	.367879	.219384	.148496	.109692
1.01	.360613	.215742	.146320	.108218
1.02	.353524	.212171	.144180	.106765
1.03	.346609	.208671	.142076	.105334
1.04	.339860	.205238	.140007	.103924
1.05	.333274	.201873	.137971	.102534
1.06	.326845	.198572	.135969	.101164
1.07	.320569	.195335	.134000	.099814
1.08	.314440	.192160	.132062	.098484
1.09	.308456	.189046	.130156	.097173
1.10	.302610	.185991	.128281	.095881
1.11	.296900	.182994	.126436	.094607
1.12	.291321	.180052	.124621	.093352
1.13	.285870	.177167	.122835	.092115
1.14	.280543	.174335	.121078	.090895
1.15	.275336	.171555	.119348	.089693
1.16	.270247	.168828	.117646	.088508
1.17	.265271	.166150	.115971	.087340
1.18	.260406	.163522	.114323	.086189
1.19	.255648	.160942	.112701	.085054
1.20	.250995	.158408	.111104	.083935
1.21	.246444	.155921	.109532	.082832
1.22	.241992	.153479	.107986	.081744
1.23	.237636	.151081	.106463	.080672
1.24	.233374	.148726	.104964	.079615
1.25	.229204	.146413	.103488	.078572
1.26	.225122	.144142	.102035	.077545
1.27	.221127	.141911	.100605	.076532
1.28	.217217	.139719	.099197	.075533
1.29	.213388	.137566	.097811	.074548
1.30	.209640	.135451	.096446	.073576
1.31	.205970	.133373	.095102	.072619
1.32	.202375	.131331	.093778	.071674
1.33	.198855	.129325	.092475	.070743
1.34	.195407	.127354	.091191	.069825
1.35	.192030	.125417	.089928	.068919
1.36	.188721	.123513	.088683	.068026
1.37	.185480	.121642	.087457	.067145
1.38	.182303	.119803	.086250	.066277
1.39	.179191	.117996	.085061	.065420
1.40	.176141	.116219	.083890	.064576
1.41	.173151	114473	.082736	.063742
1.42	.170221	.112756	.081600	.062921
1.43	.167349	.111068	.080481	.026110
1.44	.164533	.109409	.079379	.061311
1.45	.161773	.107777	.078293	.060523
1.46	.159066	.106174	.077223	.059745
1.47	.156412	.104596	.076169	.058978
1.48	.153809	.103045	.075131	.058222
1.49	.151257	.101520	.074108	.057476

x	$E_0(x)$	$E_1(x)$	$E_2(x)$	$E_3(x)$
1.50	.148753	.100020	.073101	.056740
1.51	.146298	.098544	.072108	.056014
1.52	.143889	.097094	.071130	.055297
1.53	.141527	.095666	.070166	.054591
1.54	.139208	.094263	.069216	.053894
1.55	.136934	.092882	.068281	.053206
1.56	.134703	.091524	.067359	.052528
1.57	.132513	.090188	.066450	.051859
1.58	.130364	.088874	.065555	.051199
1.59	.128255	.087580	.064673	.050548
1.60	.126185	.086308	.063803	.049906
1.61	.124154	.085057	.062946	.049272
1.62	.122160	.083825	.062102	.048647
1.63	.120202	.082613	.061270	.048030
1.64	.118280	.081421	.060450	.047421
1.65	.116394	.080248	.059641	.046821
1.66	.114542	.079093	.058845	.046228
1.67	.112723	.077957	.058059	.045644
1.68	.110937	.076838	.057285	.045067
1.69	.109183	.075738	.056523	.044498
1.70	.107461	.074655	.055771	.043937
1.71	.105770	.073588	.055029	.043383
1.72	.104108	.072539	.054299	.042836
1.73	.102478	.071506	.053579	.042297
1.74	.100874	.070490	.052869	.041764
1.75	.099299	.069489	.052169	.041239
1.76	.097753	.068503	.051479	.040721
1.77	.096233	.067534	.050799	.040210
1.78	.094740	.066579	.050128	.039705
1.79	.093274	.065639	.049467	.039207
1.80	.091833	.064713	.048815	.038716
1.81	.090417	.063802	.048173	.038231
1.82	.089025	.062905	.047539	.037752
1.83	.087658	.062021	.046915	.037280
1.84	.086314	.061152	.046299	.036814
1.85	.084993	.060295	.045692	.036354
1.86	.083695	.059452	.045093	.035900
1.87	.082419	.058621	.044502	.035452
1.88	.081165	.057803	.043920	.035010
1.89	.079932	.056998	.043346	.034576
1.90	.078720	.056204	.042780	.034143
1.91	.077529	.055423	.042222	.033718
1.92	.076358	.054654	.041672	.033299
1.93	.075206	.053896	.041129	.032885
1.94	.074074	.053150	.040594	.032476
1.95	.072961	.052414	.040066	.032073
1.96	.071866	.051690	.039546	.031675
1.97	.070790	.050977	.039032	.031282
1.98	.069732	.050274	.038526	.030894
1.99	.068691	.049582	.038027	.030511
2.0	6.76676 (−2)	4.89005 (−2)	3.75343 (−2)	3.01334 (−2)
2.1	5.83126	4.26143	3.29663	2.66136
2.2	5.03651	3.71911	2.89827	2.35207
2.3	4.35908	3.25023	2.55036	2.08002
2.4	3.77991	2.84403	2.24613	1.84054

x	$E_0(x)$	$E_1(x)$	$E_2(x)$	$E_3(x)$
2.5	3.28340 (—2)	2.49149 (—2)	1.97977 (—2)	1.62954 (—2)
2.6	2.85668	2.18502	1.74630	1.44349
2.7	2.48909	1.91819	1.54145	1.27939
2.8	2.17179	1.68553	1.36152	1.13437
2.9	1.89735	1.48240	1.20336	1.00629
3.0	1.65957	1.30484	1.06419	0.89306
3.1	1.45320	1.14944	0.94165	0.79290
3.2	1.27382	1.01330	0.83366	0.70425
3.3	11.17672 (—3)	8.93904 (—3)	7.38433 (—3)	6.25744 (—3)
3.4	9.81567	7.89097	6.54396	5.56190
3.5	8.62782	6.97014	5.80189	4.94538
3.6	7.58992	6.16041	5.14623	4.39865
3.7	6.68203	5.44782	4.56658	3.91860
3.8	5.88705	4.82025	4.05383	3.48310
3.9	5.19023	4.26715	3.60004	3.10087
4.0	4.57891	3.77935	3.19823	2.76136
4.1	4.04212	3.34888	2.84226	2.45969
4.2	3.57038	2.96876	2.52678	2.19156
4.3	3.15548	2.63291	2.24704	1.95315
4.4	2.79030	2.33601	1.99890	1.74110
4.5	2.46867	2.07340	1.77869	1.55244
4.6	2.18518	1.84101	1.58321	1.38454
4.7	1.93517	1.63525	1.40960	1.23507
4.8	1.71453	1.45299	1.25538	1.10197
4.9	1.51971	1.29148	1.11831	0.98342
5.0	1.34759	1.14830	0.99647	0.87780
5.1	1.19544	1.02130	0.88812	0.78368
5.2	10.6088 (—4)	9.0862 (—4)	7.9173 (—4)	6.9978 (—4)
5.3	9.4181	8.0861	7.0591	6.2498
5.4	8.3640	7.1980	6.2964	5.5827
5.5	7.4305	6.4093	5.6168	4.9877
5.6	6.6033	5.7084	5.0116	4.4569
5.7	5.8701	5.0855	4.4725	3.9832
5.8	5.2199	4.5316	3.9922	3.5604
5.9	4.6431	4.0390	3.5641	3.1830
6.0	4.1313	3.6008	3.1826	2.8460
6.1	3.6768	3.2109	2.8424	2.5451
6.2	3.2733	2.8638	2.5390	2.2763
6.3	2.9148	2.5547	2.2683	2.0362
6.4	2.5962	2.2795	2.0269	1.8217
6.5	2.3130	2.0343	1.8115	1.6300
6.6	2.0612	1.8158	1.6192	1.4586
6.7	1.8372	1.6211	1.4475	1.3055
6.8	1.6379	1.4476	1.2942	1.1685
6.9	1.4606	1.2928	1.1573	1.0461
7.0	1.3027	1.1548	1.0351	0.9366
7.1	1.1621	1.0317	0.9259	0.8386
7.2	10.3692 (—5)	9.2188 (—5)	8.2831 (—5)	7.5100 (—5)
7.3	9.2540	8.2387	7.4112	6.7261
7.4	8.2602	7.3640	6.6319	6.0247
7.5	7.3745	6.5831	5.9353	5.3970
7.6	6.5849	5.8859	5.3125	4.8352
7.7	5.8809	5.2633	4.7556	4.3323
7.8	5.2530	4.7072	4.2576	3.8821
7.9	4.6930	4.2104	3.8122	3.4790

x	$E_0(x)$	$E_1(x)$	$E_2(x)$	$E_3(x)$
8.0	4.1933 (−5)	3.7666 (−5)	3.4138 (−5)	3.1181 (−5)
8.1	3.7474	3.3700	3.0573	2.7949
8.2	3.3494	3.0155	2.7384	2.5054
8.3	2.9942	2.6986	2.4530	2.2461
8.4	2.6770	2.4154	2.1975	2.0138
8.5	2.3937	2.1621	1.9689	1.8057
8.6	2.1408	1.9356	1.7642	1.6192
8.7	1.9148	1.7331	1.5810	1.4521
8.8	1.7129	1.5519	1.4169	1.3024
8.9	1.5325	1.3898	1.2700	1.1682
9.0	1.3712	1.2447	1.1384	1.0479
9.1	1.2271	1.1150	1.0205	0.9400
9.2	10.9825 (−6)	9.9881 (−6)	9.1492 (−6)	8.4335 (−6)
9.3	9.8306	8.9485	8.2033	7.5668
9.4	8.8004	8.0179	7.3558	6.7896
9.5	7.8791	7.1848	6.5965	6.0927
9.6	7.0551	6.4388	5.9160	5.4677
9.7	6.3179	5.7709	5.3061	4.9071
9.8	5.6583	5.1727	4.7595	4.4044
9.9	5.0681	4.6369	4.2695	3.9533
10.0	4.5400	4.1570	3.8302	3.5488

Appendix IV

Tables of the Functions $\varphi_0(\nu, \beta)$ and $\varphi_1(\nu, \beta)$

The tables contain values of $\varphi_0(\nu, \beta)$, $\varphi_1(\nu, \beta)$ and φ_1/φ_0. These functions were introduced in Sec. 11.2 [cf. Eq. (11.2.21 a) and (11.2.21 b)]. $\varphi_0(\nu, \beta=0)$ is identical with $\varphi_0(\nu)=1-2E_3(\nu)$.

IV. 1. Values of φ_0 as a Function of ν and β

β \ ν	0.2	0.4	0.6	0.8	1.0	1.2	1.4	1.6	1.8	2.0
0.0	0.2961	0.4854	0.6169	0.7114	0.7806	0.8321	0.8709	0.9002	0.9226	0.9397
0.2	0.2684	0.4399	0.5591	0.6447	0.7076	0.7545	0.7896	0.8162	0.8366	0.8522
0.4	0.2441	0.4002	0.5087	0.5867	0.6440	0.6867	0.7188	0.7432	0.7620	0.7762
0.6	0.2227	0.3653	0.4644	0.5358	0.5883	0.6275	0.6571	0.6795	0.6969	0.7102
0.8	0.2039	0.3346	0.4254	0.4911	0.5394	0.5755	0.6030	0.6238	0.6405	0.6522
1.0	0.1873	0.3075	0.3911	0.4516	0.4965	0.5298	0.5555	0.5754	0.5902	0.6023
1.2	0.1726	0.2835	0.3608	0.4170	0.4584	0.4897	0.5137	0.5321	0.5469	0.5579
1.4	0.1596	0.2622	0.3339	0.3861	0.4249	0.4542	0.4768	0.4944	0.5082	0.5194
1.6	0.1480	0.2433	0.3101	0.3588	0.3951	0.4228	0.4442	0.4609	0.4744	0.4851
1.8	0.1378	0.2264	0.2888	0.3344	0.3686	0.3948	0.4151	0.4312	0.4443	0.4546
2.0	0.1284	0.2114	0.2698	0.3127	0.3450	0.3698	0.3893	0.4048	0.4173	0.4275
2.2	0.1201	0.1979	0.2528	0.2933	0.3239	0.3476	0.3662	0.3812	0.3934	0.4035
2.4	0.1127	0.1857	0.2375	0.2758	0.3049	0.3276	0.3455	0.3601	0.3720	0.3820
2.6	0.1060	0.1748	0.2238	0.2600	0.2878	0.3095	0.3269	0.3410	0.3528	0.3627
2.8	0.0999	0.1649	0.2113	0.2459	0.2724	0.2933	0.3101	0.3239	0.3355	0.3452
3.0	0.0943	0.1560	0.2001	0.2330	0.2585	0.2786	0.2949	0.3084	0.3198	0.3296

IV. 2. Values of φ_1 as a Function of ν and β

β \ ν	0.2	0.4	0.6	0.8	1.0	1.2	1.4	1.6	1.8	2.0
0.2	0.0009	0.0030	0.0056	0.0085	0.0114	0.0132	0.0164	0.0189	0.0214	0.0235
0.4	0.0016	0.0052	0.0098	0.0147	0.0198	0.0249	0.0297	0.0342	0.0389	0.0431
0.6	0.0021	0.0071	0.0133	0.0201	0.0271	0.0339	0.0405	0.0468	0.0527	0.0582
0.8	0.0026	0.0086	0.0162	0.0243	0.0330	0.0411	0.0491	0.0569	0.0640	0.0708
1.0	0.0030	0.0098	0.0185	0.0278	0.0375	0.0471	0.0561	0.0644	0.0734	0.0808
1.2	0.0033	0.0108	0.0202	0.0305	0.0413	0.0516	0.0616	0.0714	0.0800	0.0891
1.4	0.0036	0.0115	0.0217	0.0327	0.0441	0.0552	0.0660	0.0762	0.0859	0.0948
1.6	0.0037	0.0121	0.0227	0.0343	0.0463	0.0579	0.0692	0.0801	0.0900	0.0995
1.8	0.0038	0.0125	0.0235	0.0355	0.0480	0.0600	0.0718	0.0829	0.0932	0.1033
2.0	0.0040	0.0128	0.0241	0.0364	0.0491	0.0615	0.0735	0.0850	0.0958	0.1060
2.2	0.0040	0.0130	0.0245	0.0370	0.0499	0.0625	0.0747	0.0864	0.0973	0.1077
2.4	0.0041	0.0132	0.0247	0.0374	0.0504	0.0631	0.0755	0.0872	0.0983	0.1087
2.6	0.0041	0.0132	0.0248	0.0375	0.0506	0.0634	0.0758	0.0877	0.0988	0.1093
2.8	0.0041	0.0132	0.0248	0.0375	0.0506	0.0634	0.0759	0.0877	0.0989	0.1096
3.0	0.0040	0.0132	0.0247	0.0374	0.0504	0.0632	0.0757	0.0875	0.0987	0.1091

IV. 3. Values of φ_1/φ_0 as a Function of ν and β

β \ ν	0.2	0.4	0.6	0.8	1.0	1.2	1.4	1.6	1.8	2.0
0.2	0.003	0.007	0.010	0.013	0.016	0.018	0.021	0.023	0.026	0.028
0.4	0.007	0.013	0.019	0.025	0.031	0.036	0.041	0.046	0.051	0.056
0.6	0.010	0.019	0.029	0.038	0.046	0.054	0.062	0.069	0.076	0.082
0.8	0.013	0.026	0.038	0.049	0.061	0.071	0.081	0.092	0.100	0.109
1.0	0.016	0.032	0.047	0.062	0.075	0.089	0.101	0.112	0.124	0.134
1.2	0.019	0.038	0.056	0.073	0.090	0.105	0.120	0.134	0.146	0.154
1.4	0.022	0.044	0.065	0.085	0.104	0.122	0.138	0.154	0.169	0.183
1.6	0.025	0.050	0.073	0.096	0.117	0.137	0.156	0.174	0.190	0.205
1.8	0.028	0.055	0.081	0.106	0.130	0.152	0.173	0.192	0.210	0.237
2.0	0.031	0.061	0.089	0.116	0.142	0.166	0.189	0.210	0.230	0.248
2.2	0.033	0.066	0.097	0.126	0.154	0.180	0.204	0.227	0.248	0.267
2.4	0.036	0.071	0.104	0.136	0.165	0.193	0.218	0.242	0.264	0.284
2.6	0.039	0.075	0.111	0.145	0.176	0.204	0.232	0.257	0.280	0.302
2.8	0.041	0.080	0.117	0.153	0.186	0.216	0.244	0.271	0.295	0.318
3.0	0.043	0.085	0.123	0.160	0.195	0.227	0.256	0.284	0.309	0.331

Appendix V

Table of Time Factors

The following table lists λt, $e^{\lambda t}$, $e^{-\lambda t}$ and $1-e^{-\lambda t}$ where t is measured in units of the half-life $T_{\frac{1}{2}}$; $\lambda = \frac{\ln 2}{T_{\frac{1}{2}}}$. These data are helpful in the evaluation of foil measurements (cf. Sec. 11.1.1).

$t/T_{\frac{1}{2}}$	$e^{\lambda t}$	$e^{-\lambda t}$	$1-e^{-\lambda t}$	λt	$t/T_{\frac{1}{2}}$	$e^{\lambda t}$	$e^{-\lambda t}$	$1-e^{-\lambda t}$	λt
0.01	1.0069	0.9931	0.0069	0.00693	0.51	1.4241	0.7022	0.2978	0.35351
0.02	1.0140	0.9862	0.0138	0.01386	0.52	1.4340	0.6974	0.3026	0.36044
0.03	1.0210	0.9794	0.0206	0.02079	0.53	1.4439	0.6926	0.3074	0.36737
0.04	1.0281	0.9726	0.0274	0.02773	0.54	1.4540	0.6878	0.3122	0.37430
0.05	1.0353	0.9659	0.0341	0.03466	0.55	1.4647	0.6830	0.3170	0.38123
0.06	1.0425	0.9593	0.0407	0.04159	0.56	1.4743	0.6783	0.3217	0.38816
0.07	1.0497	0.9526	0.0474	0.04852	0.57	1.4845	0.6736	0.3264	0.39509
0.08	1.0570	0.9461	0.0539	0.05545	0.58	1.4949	0.6690	0.3310	0.40203
0.09	1.0644	0.9395	0.0605	0.06238	0.59	1.5053	0.6643	0.3357	0.40896
0.10	1.0718	0.9330	0.0670	0.06931	0.60	1.5157	0.6597	0.3403	0.41589
0.11	1.0792	0.9266	0.0734	0.07625	0.61	1.5263	0.6552	0.3448	0.42282
0.12	1.0867	0.9202	0.0798	0.08318	0.62	1.5369	0.6507	0.3493	0.42975
0.13	1.0943	0.9138	0.0862	0.09011	0.63	1.5476	0.6462	0.3538	0.43668
0.14	1.1019	0.9075	0.0925	0.09704	0.64	1.5583	0.6417	0.3583	0.44361
0.15	1.1096	0.9013	0.0987	0.10397	0.65	1.5692	0.6373	0.3627	0.45055
0.16	1.1173	0.8950	0.1050	0.11090	0.66	1.5801	0.6329	0.3671	0.45748
0.17	1.1251	0.8888	0.1112	0.11784	0.67	1.5911	0.6285	0.3715	0.46441
0.18	1.1329	0.8827	0.1173	0.12477	0.68	1.6021	0.6242	0.3758	0.47134
0.19	1.1408	0.8766	0.1234	0.13170	0.69	1.6133	0.6199	0.3801	0.47827
0.20	1.1487	0.8705	0.1295	0.13863	0.70	1.6245	0.6156	0.3844	0.48520
0.21	1.1567	0.8645	0.1355	0.14556	0.71	1.6358	0.6113	0.3887	0.49213
0.22	1.1647	0.8586	0.1414	0.15249	0.72	1.6472	0.6071	0.3929	0.49907
0.23	1.1728	0.8526	0.1474	0.15942	0.73	1.6586	0.6029	0.3971	0.50600
0.24	1.1810	0.8467	0.1533	0.16636	0.74	1.6702	0.5987	0.4013	0.51293
0.25	1.1892	0.8409	0.1591	0.17329	0.75	1.6818	0.5946	0.4054	0.51986
0.26	1.1975	0.8351	0.1649	0.18022	0.76	1.6935	0.5905	0.4095	0.52679
0.27	1.2058	0.8293	0.1707	0.18715	0.77	1.7053	0.5864	0.4136	0.53372
0.28	1.2142	0.8236	0.1764	0.19408	0.78	1.7171	0.5824	0.4176	0.54065
0.29	1.2226	0.8179	0.1821	0.20101	0.79	1.7291	0.5783	0.4217	0.54759
0.30	1.2311	0.8122	0.1878	0.20794	0.80	1.7411	0.5744	0.4256	0.55452
0.31	1.2397	0.8066	0.1934	0.21488	0.81	1.7532	0.5704	0.4296	0.56145
0.32	1.2483	0.8011	0.1989	0.22181	0.82	1.7654	0.5664	0.4336	0.56838
0.33	1.2570	0.7955	0.2045	0.22874	0.83	1.7777	0.5625	0.4375	0.57531
0.34	1.2658	0.7900	0.2100	0.23567	0.84	1.7901	0.5586	0.4414	0.58224
0.35	1.2746	0.7846	0.2154	0.24260	0.85	1.8025	0.5548	0.4452	0.58918
0.36	1.2834	0.7792	0.2208	0.24953	0.86	1.8151	0.5509	0.4491	0.59611
0.37	1.2923	0.7738	0.2262	0.25646	0.87	1.8277	0.5471	0.4529	0.60304
0.38	1.3013	0.7684	0.2316	0.26340	0.88	1.8404	0.5434	0.4566	0.60997
0.39	1.3104	0.7631	0.2369	0.27033	0.89	1.8532	0.5388	0.4612	0.61690
0.40	1.3195	0.7579	0.2421	0.27726	0.90	1.8661	0.5359	0.4641	0.62383
0.41	1.3287	0.7526	0.2474	0.28419	0.91	1.8791	0.5322	0.4678	0.63076
0.42	1.3379	0.7474	0.2526	0.29112	0.92	1.8921	0.5285	0.4715	0.63770
0.43	1.3471	0.7423	0.2577	0.29805	0.93	1.9053	0.5249	0.4751	0.64463
0.44	1.3566	0.7371	0.2629	0.30498	0.94	1.9185	0.5212	0.4788	0.65156
0.45	1.3660	0.7320	0.2680	0.31192	0.95	1.9319	0.5176	0.4824	0.65849
0.46	1.3755	0.7270	0.2730	0.31885	0.96	1.9453	0.5141	0.4859	0.66542
0.47	1.3851	0.7220	0.2780	0.32578	0.97	1.9588	0.5105	0.4895	0.67235
0.48	1.3947	0.7170	0.2830	0.33271	0.98	1.9725	0.5070	0.4930	0.67928
0.49	1.4044	0.7120	0.2880	0.33964	0.99	1.9862	0.5035	0.4965	0.68622
0.50	1.4142	0.7071	0.2929	0.34657	1.00	2.0000	0.5000	0.5000	0.69315

$t/T_{\frac{1}{2}}$	$e^{\lambda t}$	$e^{-\lambda t}$	$1-e^{-\lambda t}$	λt
1.02	2.0279	0.4931	0.5069	0.70701
1.04	2.0562	0.4863	0.5137	0.72087
1.05	2.0705	0.4830	0.5170	0.72780
1.06	2.0849	0.4796	0.5204	0.73474
1.08	2.1140	0.4730	0.5270	0.74860
1.10	2.1435	0.4665	0.5335	0.76246
1.12	2.1735	0.4601	0.5399	0.77632
1.14	2.2038	0.4538	0.5462	0.79019
1.15	2.2191	0.4506	0.5494	0.79712
1.16	2.2345	0.4475	0.5525	0.80405
1.18	2.2658	0.4413	0.5587	0.81791
1.20	2.2974	0.4353	0.5647	0.83178
1.22	2.3295	0.4293	0.5707	0.84564
1.24	2.3620	0.4234	0.5766	0.85950
1.25	2.3784	0.4204	0.5796	0.86643
1.26	2.3950	0.4175	0.5825	0.87337
1.28	2.4284	0.4118	0.5882	0.88723
1.30	2.4623	0.4061	0.5939	0.90109
1.32	2.4967	0.4005	0.5995	0.91495
1.34	2.5315	0.3950	0.6050	0.92882
1.35	2.5491	0.3923	0.6077	0.93575
1.36	2.5668	0.3896	0.6104	0.94268
1.38	2.6027	0.3842	0.6158	0.95654
1.40	2.6390	0.3789	0.6211	0.97041
1.42	2.6758	0.3737	0.6263	0.98427
1.44	2.7132	0.3685	0.6315	0.99813
1.45	2.7321	0.3660	0.6340	1.00506
1.46	2.7511	0.3635	0.6365	1.01199
1.48	2.7895	0.3585	0.6415	1.02586
1.50	2.8285	0.3536	0.6464	1.03972
1.52	2.8679	0.3487	0.6513	1.05358
1.54	2.9080	0.3439	0.6561	1.06745
1.55	2.9282	0.3415	0.6585	1.07438
1.56	2.9486	0.3391	0.6609	1.08131
1.58	2.9897	0.3345	0.6655	1.09517
1.60	3.0314	0.3299	0.6701	1.10904
1.62	3.0738	0.3253	0.6747	1.12290
1.64	3.1167	0.3209	0.6791	1.13676
1.65	3.1383	0.3186	0.6814	1.14369
1.66	3.1602	0.3164	0.6836	1.15062
1.68	3.2043	0.3121	0.6879	1.16449
1.70	3.2490	0.3078	0.6922	1.17835
1.72	3.2944	0.3035	0.6965	1.19221
1.74	3.3403	0.2994	0.7006	1.20608
1.75	3.3636	0.2973	0.7027	1.21301
1.76	3.3870	0.2953	0.7047	1.21994
1.78	3.4342	0.2912	0.7088	1.23380
1.80	3.4822	0.2872	0.7128	1.24766
1.82	3.5308	0.2832	0.7168	1.26153
1.84	3.5801	0.2793	0.7207	1.27539
1.85	3.6050	0.2774	0.7226	1.28232
1.86	3.6301	0.2755	0.7245	1.28925
1.88	3.6808	0.2717	0.7283	1.30312
1.90	3.7322	0.2679	0.7321	1.31698
1.92	3.7842	0.2643	0.7357	1.33084
1.94	3.8371	0.2606	0.7394	1.34471
1.95	3.8638	0.2588	0.7412	1.35164
1.96	3.8906	0.2570	0.7430	1.35857

$t/T_{\frac{1}{2}}$	$e^{\lambda t}$	$e^{-\lambda t}$	$1-e^{-\lambda t}$	λt
1.98	3.9449	0.2535	0.7465	1.37243
2.00	4.0000	0.2500	0.7500	1.38629
2.02	4.0558	0.2466	0.7534	1.40016
2.04	4.1124	0.2432	0.7568	1.41402
2.05	4.1410	0.2415	0.7585	1.42095
2.06	4.1698	0.2398	0.7602	1.42788
2.08	4.2280	0.2365	0.7635	1.44175
2.10	4.2871	0.2333	0.7667	1.45561
2.12	4.3469	0.2300	0.7700	1.46947
2.14	4.4076	0.2269	0.7731	1.48334
2.15	4.4382	0.2253	0.7747	1.49027
2.16	4.4691	0.2238	0.7762	1.49720
2.18	4.5316	0.2207	0.7793	1.51106
2.20	4.5948	0.2176	0.7824	1.52492
2.22	4.6590	0.2146	0.7854	1.53879
2.24	4.7240	0.2117	0.7883	1.55265
2.25	4.7569	0.2102	0.7898	1.55958
2.26	4.7899	0.2088	0.7912	1.56651
2.28	4.8568	0.2059	0.7941	1.58038
2.30	4.9246	0.2031	0.7969	1.59424
2.32	4.9933	0.2003	0.7997	1.60810
2.34	5.0630	0.1975	0.8025	1.62196
2.35	5.0982	0.1961	0.8039	1.62890
2.36	5.1337	0.1948	0.8052	1.63583
2.38	5.2054	0.1921	0.8079	1.64969
2.40	5.2780	0.1895	0.8105	1.66355
2.42	5.3517	0.1869	0.8131	1.67742
2.44	5.4264	0.1843	0.8157	1.69128
2.45	5.4642	0.1830	0.8170	1.69821
2.46	5.5021	0.1817	0.8183	1.70514
2.48	5.5789	0.1792	0.8208	1.71901
2.50	5.6569	0.1768	0.8232	1.73287
2.52	5.7359	0.1744	0.8256	1.74673
2.54	5.8159	0.1719	0.8281	1.76059
2.55	5.8564	0.1708	0.8292	1.76753
2.56	5.8971	0.1696	0.8304	1.77446
2.58	5.9794	0.1673	0.8327	1.78832
2.60	6.0629	0.1649	0.8351	1.80218
2.62	6.1475	0.1627	0.8373	1.81605
2.64	6.2333	0.1604	0.8396	1.82991
2.65	6.2767	0.1593	0.8407	1.83684
2.66	6.3203	0.1582	0.8418	1.84377
2.68	6.4086	0.1560	0.8440	1.85763
2.70	6.4980	0.1539	0.8461	1.87150
2.72	6.5887	0.1518	0.8482	1.88536
2.74	6.6807	0.1497	0.8503	1.89922
2.75	6.7271	0.1487	0.8513	1.90615
2.76	6.7739	0.1476	0.8524	1.91309
2.78	6.8685	0.1456	0.8544	1.92695
2.80	6.9643	0.1436	0.8564	1.94081
2.82	7.0615	0.1416	0.8584	1.95468
2.84	7.1603	0.1397	0.8603	1.96854
2.85	7.2101	0.1387	0.8613	1.97547
2.86	7.2602	0.1377	0.8623	1.98240
2.88	7.3616	0.1358	0.8642	1.99626
2.90	7.4643	0.1340	0.8660	2.01013
2.92	7.5685	0.1321	0.8679	2.02399

$t/T_{\frac{1}{2}}$	$e^{\lambda t}$	$e^{-\lambda t}$	$1-e^{-\lambda t}$	λt	$t/T_{\frac{1}{2}}$	$e^{\lambda t}$	$e^{-\lambda t}$	$1-e^{-\lambda t}$	λt
2.94	7.6741	0.1303	0.8697	2.03785	5.10	34.297	0.0292	0.9708	3.53505
2.95	7.7275	0.1294	0.8706	2.04478	5.20	36.759	0.0272	0.9728	3.60437
2.96	7.7813	0.1285	0.8715	2.05172	5.30	39.397	0.0254	0.9746	3.67368
2.98	7.8899	0.1267	0.8733	2.06558	5.40	42.224	0.0237	0.9763	3.74299
3.00	8.0000	0.1250	0.8750	2.07944	5.50	45.255	0.0221	0.9779	3.81231
					5.60	48.503	0.0206	0.9794	3.88162
3.05	8.2821	0.1207	0.8793	2.11410	5.70	51.984	0.0192	0.9808	3.95094
3.10	8.5741	0.1166	0.8834	2.14876	5.80	55.715	0.0179	0.9821	4.02025
3.15	8.8765	0.1127	0.8873	2.18341	5.90	59.715	0.0167	0.9833	4.08957
3.20	9.1897	0.1088	0.8912	2.21807	6.00	64.000	0.0156	0.9844	4.15888
3.25	9.5137	0.1051	0.8949	2.25273					
3.30	9.8492	0.1015	0.8985	2.28739	6.20	73.517	0.0136	0.9864	4.29751
3.35	10.196	0.0981	0.9019	2.32204	6.40	84.448	0.0118	0.9882	4.43614
3.40	10.556	0.0948	0.9052	2.35670	6.60	97.007	0.0103	0.9897	4.57477
3.45	10.928	0.0915	0.9085	2.39136	6.80	111.43	0.0090	0.9910	4.71340
3.50	11.314	0.0884	0.9116	2.42602	7.00	128.00	0.0078	0.9922	4.85203
					7.20	147.03	0.0068	0.9932	4.99066
3.55	11.713	0.0854	0.9146	2.46067	7.40	168.90	0.0059	0.9941	5.12929
3.60	12.126	0.0825	0.9175	2.49533	7.60	194.01	0.0052	0.9948	5.26792
3.65	12.553	0.0797	0.9203	2.52999	7.80	222.86	0.0045	0.9955	5.40655
3.70	12.996	0.0770	0.9230	2.56464	8.00	256.00	0.0039	0.9961	5.54518
3.75	13.454	0.0743	0.9257	2.59930					
3.80	13.929	0.0718	0.9282	2.63396	8.20	294.07	0.0034	0.9966	5.68381
3.85	14.420	0.0693	0.9307	2.66862	8.40	337.79	0.0030	0.9970	5.82244
3.90	14.929	0.0670	0.9330	2.70327	8.60	388.03	0.0026	0.9974	5.96107
3.95	15.455	0.0647	0.9353	2.73793	8.80	445.72	0.0022	0.9978	6.09970
4.00	16.000	0.0625	0.9375	2.77259	9.00	512.00	0.0020	0.9980	6.23832
					9.20	588.14	0.0017	0.9983	6.37695
4.10	17.148	0.0583	0.9417	2.84190	9.40	675.59	0.0015	0.9985	6.51558
4.20	18.379	0.0544	0.9456	2.91122	9.60	776.05	0.0013	0.9987	6.65421
4.30	19.698	0.0508	0.9492	2.98053	9.80	891.44	0.0011	0.9989	6.79284
4.40	21.112	0.0474	0.9526	3.04985	10.00	1024.0	0.0010	0.9990	6.93147
4.50	22.628	0.0442	0.9558	3.11916					
4.60	24.252	0.0412	0.9588	3.18848	10.50	1448.2	0.0007	0.9993	7.27805
4.70	25.992	0.0385	0.9615	3.25779	11.00	2048.0	0.0005	0.9995	7.62462
4.80	27.857	0.0359	0.9641	3.32711	11.50	2896.3	0.0004	0.9996	7.97119
4.90	29.857	0.0335	0.9665	3.39642	12.00	4096.0	0.0002	0.9998	8.31777
5.00	32.000	0.0312	0.9688	3.46574	13.00	8192.0	0.0001	0.9999	9.01091

Appendix VI

List of Symbols

A	mass number
A [cm^{-2} sec^{-1}]	activity
B^2 [cm^{-2}]	geometric buckling
C [cm^{-2} sec^{-1}]	activation
C^{CD} [cm^{-2} sec^{-1}]	activation of a cadmium-covered foil
C_0 [cm^{-2} sec^{-1}]	unperturbed activation
C_{th} [cm^{-2} sec^{-1}]	thermal neutron activation
C_{epi} [cm^{-2} sec^{-1}]	epithermal neutron activation
C [cm^4 sec^{-1}]	diffusion cooling constant
$c=\frac{C}{D_0}$ [cm^2]	reduced diffusion cooling constant
c [cm sec^{-1}]	velocity of light
$D, D(E)$ [cm]	diffusion coefficient
$\bar{D}$ [cm]	average diffusion coefficient for thermal neutrons
D_0 [cm^2 sec^{-1}]	diffusion constant
d [cm]	extrapolated endpoint
d [cm]	thickness of absorber slab
E, E_n [ev] or [Mev]	neutron energy
$E_T=\frac{kT}{2}$ [ev]	$\frac{1}{2}$ times the most probable energy of thermal neutrons
E_{ET} [ev]	thermal cut-off energy
E_C [ev]	lower cut-off energy in resonance integrals
E_{CD} [ev]	cadmium cut-off energy
E_R [ev]	resonance energy
E_Q [Mev]	source neutron energy
E_s^{eff} [Mev]	effective threshold energy
$\frac{dE}{dx}$ [ev cm^{-1}]	slowing-down power of target substance for charged particles
$E_n(x)$	exponential integral
$F(\boldsymbol{r}, \boldsymbol{\Omega}, E)$ [cm^{-2} sec^{-1} steradian^{-1} ev^{-1}], $F(\boldsymbol{r}, \boldsymbol{\Omega}, u)$ [cm^{-2} sec^{-1} steradian^{-1}]	differential neutron flux
$F(\boldsymbol{r}, \boldsymbol{\Omega})$ [cm^{-2} sec^{-1} steradian^{-1}]	vector flux
$F_l(\boldsymbol{r}, u)$ [cm^{-2} sec^{-1}], $F_l(\boldsymbol{r})$ [cm^{-2} sec^{-1}]	coefficients of the expansion of $F(\boldsymbol{r}, \boldsymbol{\Omega}, u)$, $F(\boldsymbol{r}, \boldsymbol{\Omega})$ in Legendre polynomials
F_{CD}	cadmium correction factor
$F_1(E)$ [ev^{-1}]	Horowitz-Tretiakoff spectral function
$f(\omega, \tau)$, $f(\omega, E)$	Fourier transform of the slowing-down density around a point source in an infinite medium
g	statistical factor arising in the Breit-Wigner formula
$g, g(T)$	WESTCOTT correction factor for non-$1/v$-absorbers

$g(E' \to E)$ [ev^{-1}]	probability that neutron energy changes from E' to E in a collision[1]
$\hbar$ [ev sec]	PLANCK's constant divided by 2π
I_∞ [barn]	resonance integral at infinite dilution
I_{eff} [barn]	effective resonance integral
I_x [barn]	resonance integral
I'_x [barn]	excess resonance integral
J, $\boldsymbol{J}$ [cm^{-2} sec^{-1}]	neutron current density
$K = \frac{2\pi}{\lambda}$ [cm^{-1}]	wave number
k [ev (°K)$^{-1}$]	BOLTZMANN's constant
L [cm]	diffusion length
L [cm^{-1}]	thermalization operator
L_s [cm]	slowing-down length
M [g]	mass of a scattering atom
$M(E) = \frac{E}{(kT_0)^2} e^{-E/kT_0}$ [ev^{-1}]	equilibrium thermal neutron spectrum in a moderator with temperature T_0
M_2 [barn]	mean squared energy transfer in an equilibrium Maxwellian spectrum
m, m_n [g]	neutron mass
N [cm^{-3}]	number of atoms per cubic centimeter
$N(E)$ [Mev^{-1}]	fission neutron spectrum
$n(\boldsymbol{r}, \boldsymbol{\Omega}, E)$ [cm^{-3} steradian^{-1} ev^{-1}]	differential neutron density
$n(\boldsymbol{r}, \boldsymbol{\Omega})$ [cm^{-3} steradian^{-1}]	vector density
$n(\boldsymbol{r})$, n [cm^{-3}]	neutron density
$n(E)$ [cm^{-3} ev^{-1}]	neutron density per unit energy
$\bar{n}$	average occupation number of an oscillator
p, $p(E)$, $p(u)$	resonance escape probability
Q [sec^{-1}]	source strength
Q [Mev]	reaction energy
$q(E)$, $q(u)$ [cm^{-3} sec^{-1}]	slowing-down density at energy E (lethargy u)
$q_{th}(\boldsymbol{r})$ [cm^{-3} sec^{-1}]	slowing-down density just above the thermal energy range
R [cm]	nuclear radius
R_{CD}	cadmium ratio
r	spectral index arising in the Westcott convention
$\overline{r_E^2}$ [cm^2]	mean squared slowing-down distance
$S(\boldsymbol{r}, \boldsymbol{\Omega}, E)$ [cm^{-3} sec^{-1} steradian^{-1} ev^{-1}]	differential source density
S, $S(\boldsymbol{r})$ [cm^{-3} sec^{-1}]	source density
$S(E)$ [cm^{-3} sec^{-1} ev^{-1}]	source density per unit energy
S_β	effective β-self-absorption factor
s	reduced excess resonance integral
$T = \frac{Z^+}{Z^0}$	neutron transmission
T	time factor

[1] More precisely, that the energy after the collision will be in an unit energy interval around E.

T [°K]	neutron temperature
T_0 [°K]	moderator temperature
$T_{\frac{1}{2}}$ [sec, min, h, d, y]	half-life
$\overline{t_E}$ [sec]	slowing-down time to energy E
t_T [sec]	temperature relaxation time
t_{th} [sec]	thermalization time
u	lethargy
V [cm sec^{-1}]	scattering atom velocity
v [cm sec^{-1}]	neutron velocity
v_0 [cm sec^{-1}]	most probable velocity in a thermal neutron Maxwell spectrum with $T = 293.6$ °K
v_T [cm sec^{-1}]	most probable velocity in a thermal neutron Maxwell spectrum with temperature T
$\bar{v} = \frac{2}{\sqrt{\pi}} v_T$ [cm sec^{-1}]	average velocity in a thermal neutron Maxwell spectrum with temperature T
α, $\alpha(B^2)$ [sec^{-1}]	time decay constant of a pulsed neutron field
$\alpha = \frac{\sigma_{n,\gamma}}{\sigma_{n,f}}$	capture-to-fission ratio
$\alpha = \left(\frac{A-1}{A+1}\right)^2$	maximum relative energy change in an elastic collision
α [cm^2 g^{-1}]	β-ray absorption coefficient
$\alpha_0 = \overline{v\Sigma_a}$ [sec^{-1}]	absorption probability
β	albedo
Γ [ev]	total width of a neutron resonance
Γ_n [ev]	neutron width of a neutron resonance
Γ_γ [ev]	radiation width of a neutron resonance
$\frac{\Delta t}{l}$ [μsec m^{-1}]	resolving power of a time-of-flight spectrometer
$\Delta(E/kT)$	joining function
$\delta = \varrho \cdot d$ [g cm^{-2}]	surface mass loading
ε	correction for neutrons incident on the edges of a foil
η	neutron yield per neutron absorbed in a fissionable material
ϑ_0	scattering angle in the laboratory system
$\varkappa$ [cm^{-1}]	asymptotic relaxation constant of a neutron field
$\boldsymbol{\varkappa}$, $\varkappa$ [cm^{-1}]	change of the neutron wave number in a collision
$\varkappa_c$	activation correction
$\varkappa_\Phi(\boldsymbol{r})$	flux correction
λ [sec^{-1}]	radioactive decay constant
λ [cm]	mean free path
λ_s [cm]	scattering mean free path
λ_a [cm]	absorption mean free path
λ_{tr} [cm]	transport mean free path
λ_i	fraction of absorptions in the *i-th* resonance of a resonance detector that leads to the desired activity
$\lambda\!\!\!^{-}$ [cm]	reduced de Broglie wavelength of a neutron

μ_t, μ_a, μ_{act}, μ_s $[\mathrm{cm^2\,g^{-1}}]$	total collision, absorption, activation, scattering coefficient
μ_{a_0} $[\mathrm{cm^2\,g^{-1}}]$	absorption coefficient at the peak of a resonance
$\bar{\nu}$	average number of secondary neutrons in fission
ξ	average logarithmic energy loss
$\xi\,\Sigma_s$ $[\mathrm{cm^{-1}}]$	slowing-down power
ϱ $[\mathrm{g\cdot cm^{-3}}]$	density
Σ $[\mathrm{cm^{-1}}]$	macroscopic cross section (t=total, a=absorption, act=activation, s=scattering, tr=transport)
$\bar{\Sigma}_a$ $[\mathrm{cm^{-1}}]$	average macroscopic absorption cross section for thermal neutrons
$\frac{\xi\Sigma_s}{\bar{\Sigma}_a}$	moderating ratio
$\Sigma_s(\mathbf{\Omega}'\to\mathbf{\Omega}, E'\to E)$ $[\mathrm{cm^{-1}\,steradian^{-1}\,ev^{-1}}]$	(macroscopic) cross section for scattering processes in which the neutron direction changes from $\mathbf{\Omega}'$ to $\mathbf{\Omega}$ and the neutron energy changes from E' to E [1]
$\Sigma_s(\mathbf{\Omega}'\to\mathbf{\Omega}, u'\to u)$ $[\mathrm{cm^{-1}\,steradian^{-1}}]$	(macroscopic) cross section for scattering processes in which the neutron direction changes from $\mathbf{\Omega}'$ to $\mathbf{\Omega}$ and the lethargy changes from u' to u [1]
$\Sigma_s(\mathbf{\Omega}'\to\mathbf{\Omega})$ $[\mathrm{cm^{-1}\,steradian^{-1}}]$	(macroscopic) cross section for scattering process in which the neutron direction changes from $\mathbf{\Omega}'$ to $\mathbf{\Omega}$ [2]
$\Sigma_{sl}(u'\to u)$ $[\mathrm{cm^{-1}}]$, Σ_{sl} $[\mathrm{cm^{-1}}]$	coefficients of the expansion of $\Sigma_s(\mathbf{\Omega}'\to\mathbf{\Omega}, u'\to u)$, $\Sigma_s(\mathbf{\Omega}'\to\mathbf{\Omega})$ in Legendre polynomials
$\Sigma_s(E'\to E)$ $[\mathrm{cm^{-1}\,ev^{-1}}]$	(macroscopic) cross section for scattering processes in which the neutron energy changes from E' to E [2]
σ [barn]	microscopic cross section (t=total, s=scattering, a=absorption, act=activation, ne=non-elastic, tr=transport)
σ_{a_0} [barn]	peak absorption cross section in a resonance
σ_0 [barn]	peak total cross section in a resonance
σ_{sf} [barn]	free atom scattering cross section
σ_{sb} [barn]	bound atom scattering cross section
$\bar{\sigma}$ [barn]	average cross section in a fission spectrum
$\hat{\sigma}$ [barn]	effective cross section in WESTCOTT's convention
σ_{coh} [barn]	cross section for coherent scattering
σ_0 [barn]	plateau value of threshold detector cross section
$\sigma_{n,x}$ [barn]	reaction cross section (α=(n,α) process; γ=radiative capture; n=elastic scattering; n'=inelastic scattering p=(n,p) process; $2n$=$(n,2n)$ process; f=fission)

[1] More precisely, the direction after the collision should be in the unit solid angle around $\mathbf{\Omega}$ and the energy in the unit energy range around E (or the lethargy in the unit lethargy range around u).

[2] Cf. preceding footnote.

$\sigma_{n,n}(\cos\vartheta_0)$ [barn]	cross section for an elastic scattering process with the scattering angle ϑ_0 [1]
$\sigma_{n,n'}(E, E_j)$ [barn]	inelastic excitation cross section
$\sigma_{n,n'}(E'\to E)$ [barn Mev^{-1}]	cross section for an inelastic scattering process in which the neutron energy changes from E' to E [2]
$\sigma_s(E'\to E, \cos\vartheta_0)$ [barn ev^{-1}]	cross section for a scattering process with the scattering angle ϑ_0 and an energy change from E' to E [1-3]
$\sigma_s(E'\to E)$ [barn ev^{-1}]	cross section for a scattering process with an energy change from E' to E [2,3]
$\overline{\sigma_s\cdot(\Delta E)^\nu}$ [barn $\cdot$ (ev)$^\nu$]	energy transfer moments of the scattering cross section[3]
τ_E [cm^2]	age
$\tau(E)$, τ [cm^2]	Fermi age
τ_E^* [cm^2]	flux age
$\Phi(r)$, Φ [cm^{-2} sec^{-1}]	neutron flux
$\Phi(E)$ [cm^{-2} sec^{-1} ev^{-1}]	neutron flux per unit energy interval
Φ_{th} [cm^{-2} sec^{-1}]	thermal neutron flux
Φ_{epi} [cm^{-2} sec^{-1}]	epithermal neutron flux per logarithmic energy unit
Φ_f [cm^{-2} sec^{-1}]	fast neutron flux
$\varphi_0(\mu_a\delta)$	absorption probability of a foil in an isotropic neutron flux
$\left.\begin{array}{l}\varphi_0(\nu,\beta)\\ \varphi_1(\nu,\beta)\end{array}\right\}$	functions describing the β-activities of foils
$\chi(\varkappa, t)$	intermediate scattering function
$\chi_0(\Sigma_a R)$	absorption probability of a cylinder in an isotropic neutron flux
ψ [cm^{-3} sec^{-1}]	collision density, i.e., number of collisions per cm^3 and second
ψ	scattering angle in the center-of-mass system
$\psi(E)$ [cm^{-3} sec^{-1} (ev)$^{-1}$]	collision density per unit energy interval
$\psi_0(\Sigma_a R)$	absorption probability of a sphere in an isotropic neutron flux

[1] More precisely, the cosine of the scattering angle must be in the unit interval around $\cos\vartheta_0$.

[2] More precisely, the energy after the scattering process must be in the unit energy interval around E.

[3] Only used in connection with the molecular inelastic scattering of slow neutrons, cf. section 10.1.

Subject Index